CONTROL TECHNIQUES FOR COMPLEX NETWORKS

Power grids, flexible manufacturing, cellular communications: interconnectedness has consequences. This remarkable book gives the tools and philosophy you need to build network models detailed enough to capture essential dynamics but simple enough to expose the structure of effective control solutions and to clarify analysis.

Core chapters assume only prior exposure to stochastic processes and linear algebra at the undergraduate level; later chapters are for advanced graduate students and researchers/practitioners. This gradual development bridges classical theory with the state of the art. The workload model that is the basis of traditional analysis of the single queue becomes a foundation for workload relaxations used in the treatment of complex networks. Lyapunov functions and dynamic programming equations lead to the celebrated MaxWeight policy along with many generalizations. Other topics include methods for synthesizing hedging and safety stocks, stability theory for networks, and techniques for accelerated simulation.

Examples and figures throughout make ideas concrete. Solutions to end-of-chapter exercises are available on a companion Web site.

Sean Meyn is a professor in the Department of Electrical and Computer Engineering and director of the Division and Control Laboratory of the Coordinated Science Laboratory at the University of Illinois. He has served on the editorial boards of several journals in areas of systems and control and applied probability, and he is coauthor with Richard Tweedie of *Markov Chains and Stochastic Stability*, which won the 1994 ORSA/TIMS Best Publication in Applied Probability Award.

CONTROL TECHNIQUES FOR COMPLEX NETWORKS

SEAN MEYN
University of Illinois, Urbana-Champaign

CAMBRIDGE
UNIVERSITY PRESS

CAMBRIDGE UNIVERSITY PRESS
Cambridge, New York, Melbourne, Madrid, Cape Town, Singapore, São Paulo, Delhi

Cambridge University Press
32 Avenue of the Americas, New York, NY 10013-2473, USA

www.cambridge.org
Information on this title: www.cambridge.org/9780521884419

First published 2008

Printed in the United States of America

A catalog record for this publication is available from the British Library.

Library of Congress Cataloging in Publication Data

Meyn, S. P. (Sean P.)
Control techniques for complex networks / Sean Meyn.
p. cm.
Includes bibliographical references and index.
ISBN 978-0-521-88441-9 (hardback)
1. Computer networks. 2. Control theory. I. Title.
TK5105.5.M49 2008
004.6–dc22 2007035250

ISBN 978-0-521-88441-9 hardback

Contents

List of Illustrations

Preface

A representative of a major publishing house is on her way home from a conference in Singapore, excited about the possibility of a new book series. On the flight home to New York she opens her blackberry organizer, adding names of new contacts, and is disappointed to realize she may have caught the bug that was bothering her friend Alex at the café near the conference hotel. When she returns home she will send Alex an email to see how she's doing and to make sure this isn't a case of some new dangerous flu.

Of course, the publisher is aware that she is part of an interconnected network of other business men and women and their clients: Her value as an employee depends on these connections. She depends on the transportation network of taxis and airplanes to get her job done and is grateful for the most famous network today that allows her to contact her friend effortlessly even when separated by thousands of miles. Other networks of even greater importance escape her consciousness, even though consciousness itself depends on a highly interconnected fabric of neurons and vascular tissue. Communication networks are critical to support the air traffic controllers who manage the airspace around her. A supply chain of manufacturers makes her book business possible, as well as the existence of the airplane on which she is flying.

Complex networks are everywhere. Interconnectedness is as important to business men and women as it is to the viruses who travel along with them.

Much of the current interest in networks within physics and the biological sciences is phenomenological. For example, given a certain degree of connectivity between individuals, what is the likelihood that a virus will spread to the extinction of the planet? Degree and mode of connectivity in passive agents can combine to form images resembling crystals or snowflakes [463].

The main focus within our own bodies is far more utilitarian. Endocrine, immune, and vascular systems adjust chemical reactions to maintain equilibria in the face of ongoing attacks from disease and diet. In biology this is called *homeostasis*. In this book, the regulation of a network is called *control*.

It is not our goal to take on biology, computer science, communications, and operations research in a single volume. Rather, the intended purpose of this book is an introduction to a rapidly evolving engineering discipline. The examples come from applications in which complexity is real, but less daunting than that found in the human brain. We describe methods to model networks in order to capture essential structure, dynamics, and uncertainty. Based on these models we explore ways to visualize network behavior so that effective control techniques can be synthesized and evaluated.

Modeling and control. The operator of an electric power grid hopes to find a network model that will help form predictions of supply and demand to maintain stability of the power network. This requires the expertise of statisticians, economists, and power engineers. The resulting model may provide useful simulations for forecasting, but will fail entirely for our purposes. This book is about control, and for this it is necessary to restrict to models that capture essential behavior, but no more.

Modeling for the purposes of control and the development of control techniques for truly complex networks has become a major research activity over the past two decades. Breakthroughs obtained in the stochastic networks community provide important tools that have had real impact in some application areas, such as the implementation of MaxWeight scheduling for routing and scheduling in communications. Other breakthroughs have had less impact due in part to the highly technical and mathematical language in which the theory has developed. The goal of this book is to expose these ideas in the simplest possible setting.

Most of the ideas in this book revolve around a few concepts.

(i) The *fluid model* is an idealized deterministic model. In a communication network a unit of "fluid" corresponds to some quantities of packets; in a power network this might correspond to a certain number of megawatts of electricity.

A fluid model is often a starting point to understand the impact of topology, processing rates, and external arrivals on network behavior. Based on the fluid model we can expose the inherent conflict between short-sighted control objectives, longer-range issues such as recovery from a singular external disruption, and truly long-range planning such as the *design* of appropriate network topology.

(ii) Refinements of the fluid model are developed to capture variability in supply, demand, or processing rates. The *controlled random walk* model favored in this book is again a highly stylized model of any real network, but contains enough structure to give a great deal of insight and is simple enough to be tractable for developing control techniques.

For example, this model provides a vehicle for constructing and evaluating *hedging* mechanisms to limit exposure to high costs, and to ensure that valuable resources can operate when needed.

(iii) The concept of *workload* is developed for the deterministic and stochastic models. Perhaps the most important concept in this book is the *workload relaxation* that provides approximations of a highly complex network by a far simpler one. The approximation may be crude in some cases, but its value in attaining intuition can be outstanding.

(iv) Methods from the stability theory of Markov models form a foundation in the treatment of stochastic network models. Lyapunov functions are a basis of dynamic programming equations for optimization, for stability and analysis, and even for developing algorithms based on simulation.

What's in here? The book is divided into three parts. The first part, entitled Modeling and Control, contains numerous examples to illustrate some of the basic concepts developed in the book, especially those topics listed in (i) and (ii) concerning the fluid and CRW models. Lyapunov functions and the dynamic programming equations are introduced; based on these concepts we arrive at the MaxWeight policy along with many generalizations.

Workload relaxations are introduced in Part II. In these three chapters we show how a cost function defined for the network can be "projected" to define the *effective cost* for the relaxation. Applications to control involve first constructing a policy for the low-dimensional relaxation, and then translating this to the original physical system of interest. This translation step involves the introduction of hedging to guard against variability.

Most of the control techniques are contained in the first two parts of the book. Part III, entitled Stability and Performance, contains an in-depth treatment of Lyapunov stability theory and optimization. It contains approximation techniques to explain the apparent solidarity between control solutions for stochastic and deterministic network models. Moreover, this part of the book develops several approaches to performance evaluation for stochastic network models.

Who's it for? The book was created for several audiences. The gradual development of network concepts in Parts I and II was written with the first-year graduate student in mind. This reader may have had little exposure to operations research concepts, but some prior exposure to stochastic processes and linear algebra at the undergraduate level.

Many of the topics in the latter chapters are at the frontier of stochastic networks, optimization, simulation, and learning. This material is intended for the more advanced graduate student, as well as researchers and practitioners in any of these areas.

Acknowledgments

This book has been in the making for 5 years and over this time has drawn inspiration and feedback from many. Some of the ideas were developed in conjunction with students, including Mike Chen, Richard Dubrawski, Charuhas Pandit, Rong-Rong Chen, and David Eng. In particular, the numerics in Section 7.2 are largely taken from Dubrawski's thesis [151], and the *diffusion heuristic* for hedging is based on a paper with Chen and Pandit [103]. Section 9.6 is based in part on research conducted with Chen [105].

My collaborators are a source of inspiration and friendship. Many of the ideas in this book revolve around stochastic Lyapunov theory for Markov processes, which is summarized in the appendix. This appendix is essentially an abridged version of my book coauthored with Richard Tweedie [368]. Vivek Borkar's research on Markov decision theory (as summarized in [70]) has had a significant influence on my own view of optimization. My interest in networks was sparked by a lecture presented by P. R. Kumar when he was visiting the Australian National University in 1988 while I resided there as a postdoctoral fellow. He became a mentor and a coauthor when I joined the University of Illinois the following year. I learned of the beauty of simulation theory from the work of Peter Glynn and his former student Shane Henderson. More recent collaborators are Profs. In-Koo Cho, David Gamarnik, Ioannis Kontoyiannis, and Eric Moulines, who have provided inspiration on a broad range of topics. I am grateful to Devavrat Shah and Damon Wischik for sharing insights on the "input-queued switch," and for allowing me to adapt a figure from their paper [435] that is used to illustrate workload relaxations in Section 6.7.1. Pierre L'Ecuyer shared his notes on simulation from his course at the University of Montréal, and Bruce Hajek at the University of Illinois shared his lecture notes on communication networks.

Profs. Cho, Kontoyiannis, Henderson, and Shah have all suggested improvements on exposition, or warned of typos. Sumit Bhardwaj, Jinjing Jiang, Shie Mannor, Eric Moulines, and Michael Veatch each spent significant hours pouring through selected chapters of draft text and provided valuable feedback. Early input by Veatch moved the book toward its present organization, with engineering techniques introduced first, and harder mathematics postponed to later chapters.

Any remaining errors or awkward prose are, of course, my own.

It would be impossible to write a book like this without financial support for graduate students and release-time for research. I am sincerely grateful to the National Science Foundation, in particular, the Division of Electrical, Communications & Cyber Systems, for ongoing support during the writing of this book. The DARPA ITMANET initiative, the Laboratory for Information and Decision Systems at MIT, and United Technologies Research Center provided support in the final stages of this project during the 2006–2007 academic year.

Equally important has been support from my family, especially during the last months, when I have been living away from home. Thank you Belinda! Thank you Sydney and Sophie! And thanks also to all the poodles at South Harding Drive.

Dedication

It was a sad day on June 7, 2001, when Richard Tweedie died at the peak of his career. A brief survey of his contributions to applied probability and statistics can be found in [154].

In memory of his friendship and collaboration, and in honor of his many contributions to our scientific communities, this book is dedicated to Richard.

CONTROL TECHNIQUES FOR COMPLEX NETWORKS

1

Introduction

Network models are used to describe power grids, cellular telecommunications systems, large-scale manufacturing processes, computer systems, and even systems of elevators in large office buildings. Although the applications are diverse, there are many common goals:

(i) In any of these applications one is interested in controlling delay, inventory, and loss. The crudest issue is *stability*: do delays remain bounded, perhaps in the mean, for all time?

(ii) Estimating performance, or comparing the performance of one policy over another. Performance is of course context-dependent, but common metrics are average delay, loss probabilities, or backlog.

(iii) Prescriptive approaches to policy synthesis are required. A policy should have reasonable complexity; it should be flexible and robust. *Robustness* means that the policy will be effective even under significant modeling error. *Flexibility* requires that the system respond appropriately to changes in network topology, or other gross structural changes.

In this chapter we begin in Section 1.1 with a survey of a few network applications, and the issues to be explored within each application. This is far from comprehensive. In addition to the network examples described in the Preface, we could fill several books with applications to computer networks, road traffic, air traffic, or occupancy evolution in a large building.[1]

Although complexity of the physical system is both intimidating and unavoidable in typical networks, for the purposes of control design it is frequently possible to construct models of reduced complexity that lead to effective control solutions for the physical system of interest. These idealized models also serve to enhance intuition regarding network behavior.

Section 1.2 contains an outline of the modeling techniques used in this book for control design and performance evaluation. Section 1.3 reviews some of the mathematical prerequisites required to read the remainder of this book.

---

[1] *Egress* from a building is in fact a topic naturally addressed using the techniques described in Chapter 7. See the 1981 paper by Smith and Towsley [452], and the collection of papers [424].

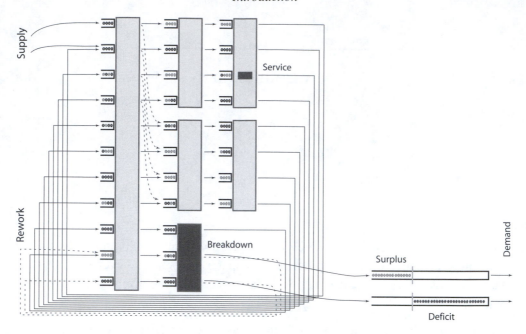

Figure 1.1. Control issues in a production system.

1.1 Networks in practice

1.1.1 Flexible manufacturing

Within the manufacturing domain, complexity is most evident in the manufacture of semiconductors.

A factory where semiconductors are produced is known as a wafer fabrication facility, or *wafer fab* [267, 410]. A schematic of a typical wafer fab is shown in Fig. 1.1. A large wafer fab will produce thousands of wafers each month, and a single wafer can hold thousands of individual semiconductor chips, depending on the size of the chips.

Control of a wafer fab or any other complex manufacturing facility involves many issues, including

 (i) *Resource allocation:* Scheduling to minimize inventory, and satisfy constraints such as deadlines, finite buffers, and maximum processing rates. A key constraint in manufacturing applications is that one machine can only process one set of products at a time. This is significant in semiconductor manufacturing where one product (e.g., a wafer) may visit a single station repeatedly in the course of manufacture, and must complete with other products with similar requirements.

 (ii) *Complexity management:* In the manufacture of semiconductors there may be hundreds of processing steps, and many different products. The control solution should have reasonable complexity in spite of the complexity of the system.

(iii) *Visualization of control solutions:* It is not enough for a solution to "spit out a sequence of numbers" representing service allocations at the stations in the manufacturing facility. Solutions should be tunable and provide some intuition to the user.

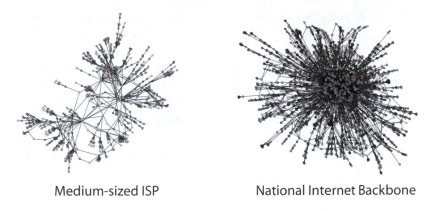

Medium-sized ISP   National Internet Backbone

Figure 1.2. The Internet is one of the most complex man-made networks.

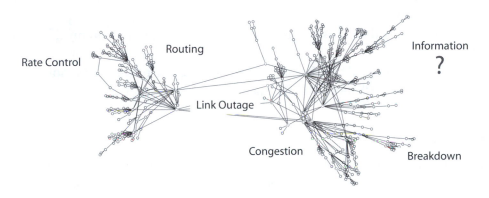

Rate Control   Routing   Link Outage   Congestion   Information ?   Breakdown

Figure 1.3. Control issues in the Internet.

(iv) *Recovery from crisis:* When machine failures occur, or preventative maintenance is called for, the control solution should be modified automatically.
 (v) *Priorities:* The lifetime of a typical semiconductor wafer in a wafer fab may be more than 1 month. These typical wafers may sometimes compete with special *hot lots* that are given high priority due to customer demand or testing.

The International Semiconductor Roadmap for Semiconductors (ITRS) maintains a website describing the current challenges facing the semiconductor industry [279].

Scheduling policies are developed with each of these goals in mind in Chapters 4–7. Sometimes we are so fortunate that we can formulate policies that are nearly *optimal* when the network is highly loaded (see Chapter 9). Methods for approximating performance indicators such as mean delay are developed in Chapter 8.

1.1.2 The Internet

Figure 1.2 shows two subsets of the global Internet. Even a network representing a small Internet service provider (ISP) can show fantastic complexity.

As illustrated in Fig. 1.3, many issues arising in control of the Internet or more general communication networks are similar to those seen in production systems. In particular, decision making involves scheduling and routing of packets from node to node across a network consisting of links and buffers. Key differences are:

(i) Individual nodes do not have access to global information regarding buffer levels and congestion throughout the network. Routing or scheduling decisions must then be determined using only that information which can be made available.

(ii) Design is thus constrained by limited information. It is also constrained by protocols such as TCP/IP that are an inherent component of the existing Internet.

(iii) The future Internet will carry voice, audio, and data traffic. How can network resources be distributed fairly to a heterogenous customer population?

Burstiness and periodicity have been observed in Internet communications traffic [274, 504]. Part of the reason for these difficulties lies in the complex dynamics resulting from a large number of interconnected computers that are controlled based on limited local information.

In the future it will be possible to obtain much greater relevant information at each node in the network through *explicit congestion notification algorithms* [274, 182]. The system designer must devise algorithms to make use of this global information regarding varying congestion levels and network topology.

1.1.3 Wireless networks

It is evident today that wireless networks are only beginning to impact communications and computer networking. In a wireless network there are scheduling and routing decisions that are nearly identical to those faced in management of the Internet. The *resources* in a multiple-access wireless network include transmission power and bandwidth, as well as multiple paths between users and stations.

Wireless networks are subject to significant variability due to fading and path losses. Consequently, maximal transmission rates can be difficult to quantify, especially in a multiuser setting.

One significant difference between manufacturing and communication applications is that achievable transmission rates in a communication system depend upon the specific coding scheme employed. High transmission rates require long block-lengths for coding, which corresponds to long delays.

A second difference is that errors resulting from mutual interference from different users need not result in disaster, as would be the case in, say, transportation. Errors arising through collisions can be repaired through the miracle of coding, up to a point. These features make it difficult to quantify the capacity region in a communication networks, and wireless networks in particular.

1.1.4 Power distribution

Shown in Fig. 1.4 is a map of the California transmission network, which is of course embedded within the highly complex North American power grid.

OREGON

Figure 1.4. California power grid.

Regulation of power networks is further complicated by deregulation. Private power generators now provide a significant portion of electricity in the United States, whose owners seek to extract the maximal profit from the utilities who serve as their clients. However, the transmission network remains regulated by independent system operators (ISOs) who attempt to distribute transmission access fairly, and maintain system reliability.

Among the stated goals of deregulation are increased innovation, efficiency of power procurement, and reliability of power delivery. The results are often disappointing:

(i) During the period of 2000–2001, utilities in California saw historic price fluctuations and rolling blackouts. Suspicion of price manipulation was confirmed following the release of phone conversations in which ENRON employees discuss shutting down power plants in order to drive up prices [450, 91].

(ii) Reliability of the power grid is also dependent on the reliability of the electric transmission network. We are reminded of its importance by the major blackout of August 2003 that swept the north-eastern United States and parts of Canada [179, 168]. Similarly, wildfires in California in 2001 resulted in damaged

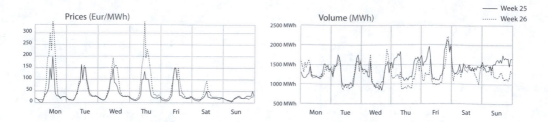

Figure 1.5. Market prices and power demand in continental Europe during the 25th and 26th weeks of 2003 (taken from the APX website [15]). Demand is periodic and shows high variability.

transmission lines that subsequently drove up power prices. Private conversations between ENRON employees reveal that they predicted these natural events would lead to increased profits [91].

(iii) A contributing factor to high power prices in California was the unusually hot and dry summer in 2000 [87]. This led to higher demand for power, and lower hydropower reserves. In a decentralized setting it is difficult to ensure that alternate sources of power reserves will be made available in the face of unexpected environmental conditions.

Even under average conditions, price and demand for power are periodic, and both exhibit significant variability. This is illustrated in Fig. 1.5 where demand and prices are plotted based on data collected in continental Europe during 2 weeks in 2003 [15]. The high volatility shown in these plots is *typical* behavior that has persisted for many years.

A power grid differs from many other network systems in that capacity must meet demand at every instant of time. If not, the transmission system may become unstable and collapse, with severe economic consequences to follow. For instance, according to the U.S. Department of Energy, the overall cost of the blackout of August 2003 was over 4 billion dollars [179, 168]. To ensure reliable operation it is necessary to schedule power generation capacity beyond the expected demand, called power reserves. Hence operation of the power grid is based on algorithms for forecasting demand, along with rules to determine appropriate power reserves.

The operational aspects of scheduling generation capacity in most power markets can be delineated into two general stages.

STAGE 1 The hour-by-hour demand for power can be predicted reasonably accurately over the upcoming 24 hour period. The high-capacity generators are scheduled in the first stage, on a day-ahead basis, based on these predictions. Advance planning is necessary since the high-capacity generators require time to ramp up or ramp down power production.

STAGE 2 The predicted demand is inevitably subject to error. To ensure system reliability, smaller generators are called upon in the second stage to provide a margin of *excess generation capacity*. These generators can ramp up or ramp down power production on a relatively short time scale, and hence can be scheduled on an hour-ahead basis.

A typical transmission network such as the California network may have hundreds of nodes, so a detailed model is far too complex to provide any insight into planning or design. On the other hand, the deterministic *DC power flow model* that is favored in many recent economic studies ignores important dynamic issues such as limited ramp-up rates, delayed information, and variability. The DC model can be viewed as a fluid equilibrium model, of the form used to define network load (see e.g. Chapter 4).

One of the goals of this book is to formulate compromise models that are simple enough for control design, and for performance approximation to compare control solutions.

1.2 Mathematical models

In each of these applications one is faced with a control problem: *what is the best way to sequence processing steps, or routing and scheduling decisions to obtain the best performance?*

An essential aspect of control theory is its flexible approach to modeling. For the purposes of design one typically considers a finite-dimensional, linear, deterministic system, even if the physical system is obviously nonlinear, with significant uncertainty with respect to modeling and disturbances. The idea is that the control system should be robust to uncertainty, so one should consider the simplest model that captures essential features of the system to be controlled.

1.2.1 A range of probabilistic models

The networks envisioned here consist of a finite set of stations, each containing a finite set of buffers. A *customer* residing at one of the buffers may represent a wafer, a packet, or a unit of power reserve. One or more *servers* process customers at a given station, after which a customer either leaves the network, or visits another station. Customers arrive from outside the network to various buffers in the network. The interarrival and service times all exhibit some degree of irregularity.

Consider a network with ℓ buffers, and ℓ_u different *activities* that may include scheduling, routing, or release of raw material into the system. Some of these buffers may be *virtual*. For example, in a manufacturing model, they may correspond to back-log or excess inventory. In a power distribution system, a buffer level is interpreted as the difference between the capacity and demand for power.

A general stochastic model can be described as follows: the vector-valued queue-length process Q evolves on $\mathbb{R}_+^\ell$, and the vector-valued cumulative allocation process Z evolves on $\mathbb{R}_+^{\ell_u}$. The ith component of $Z(t)$, denoted $Z_i(t)$, is equal to the cumulative time that the activity i has run up to time t. The evolution of the queue-length process is described by the vector equation

$$Q(t) = x + B(Z(t)) + A(t), \qquad t \geq 0, \quad Q(0) = x, \tag{1.1}$$

where the process A may denote a combination of exogenous arrivals to the network, and exogenous demands for materials *from* the network. The function $B(\cdot)$ represents the effects of (possibly random) routing and service rates.

The cumulative allocation process and queue-length process are subject to several hard constraints:

(i) The queue-length process is subject to the state space constraint

$$Q(t) \in \mathsf{X}, \qquad t \geq 0, \tag{1.2}$$

where $\mathsf{X} \subset \mathbb{R}_+^\ell$ is used to model finite buffers.

(ii) The control rates are subject to linear constraints

$$C(Z(t_1) - Z(t_0)) \leq \mathbf{1}(t_1 - t_0), \quad Z(t_1) - Z(t_0) \geq \mathbf{0}, \qquad 0 \leq t_0 \leq t_1, \tag{1.3}$$

where the *constituency matrix* C is an $\ell_m \times \ell_u$ matrix. The rows of C correspond to *resources* $r = 1, \ldots, \ell_m$, and the constraint (1.3) expresses the assumption that resources are shared among activities, and they are limited.

Stochastic models such as (1.1) have been by far the most popular in queueing theory. An idealization is the *linear fluid model*, described by the purely deterministic equation

$$q(t; x) = x + Bz(t) + \alpha t, \qquad t \geq 0, \ x \in \mathbb{R}_+^\ell, \tag{1.4}$$

where the state q evolves in the state space $\mathsf{X} \subset \mathbb{R}_+^\ell$, and the (cumulative) allocation process z evolves in $\mathbb{R}_+^{\ell_u}$. We again assume that $z(0) = \mathbf{0}$, and for each $0 \leq t_0 \leq t_1 < \infty$,

$$C[z(t_1) - z(t_0)] \leq (t_1 - t_0)\mathbf{1}, \quad \text{and} \quad z(t_1) - z(t_0) \geq \mathbf{0}. \tag{1.5}$$

The fluid model can also be described by the differential equation

$$\tfrac{d^+}{dt} q = B\zeta + \alpha, \tag{1.6}$$

where $\zeta = \zeta(t)$ denotes the right derivative, $\zeta = \frac{d^+}{dt} z$.

Two different symbols are used to denote the state processes for the stochastic and fluid models since much of the development to follow is based on the relationship between the two models. In particular, the fluid model can be interpreted as the mean flow of the stochastic model (1.1) on writing

$$Q(t) = x + A(t) - B(Z(t)) = x - BZ(t) + \alpha t + N(t), \qquad t \geq 0, \tag{1.7}$$

where α and B are interpreted as average values of (A, B), and

$$N(t) := [A(t) - \alpha t] - [B(Z(t)) - BZ(t)].$$

Typical assumptions on the stochastic model (1.1) imply that the mean of the process $\{N(t)\}$ is bounded as a function of time, and its variance grows linearly with t. Under these conditions (1.1) can be loosely interpreted as a fluid model subject to the additive disturbance N.

1.2.2 What is a good model?

It is impossible to construct a model that provides an entirely accurate picture of network behavior. Statistical models are almost always based on idealized assumptions, such as independent and identically distributed (i.i.d.) interarrival times, and it is often difficult to capture features such as machine breakdowns, disconnected links, scheduled repairs, or uncertainty in processing rates.

In the context of economic modeling, Milton Friedman writes ... *theory is to be judged by its predictive power for the class of phenomena which it is intended to "explain."* The choice of an appropriate network model is also determined by its intended use. For long-range prediction the linear fluid model has little value, and for prediction alone a detailed model may be entirely suitable. Conversely, a model that gives an accurate representation of network behavior is likely to be far too complex to be useful for control design. Fortunately, it is frequently possible to create policies that are insensitive to modeling error, so that a design based on a relatively naive model will be effective in practice.

The controlled differential equation (1.6) can be viewed as a *state space model*, as frequently used in control applications, with control ζ, state q, and state space X. It is instructive to consider how control is typically conceptualized for linear systems without state-space constraints. Typically, a deterministic "fluid" model similar to (1.6) is taken as a starting point. If a successful design is obtained, then refinements are constructed based on a more detailed model that includes prior knowledge regarding uncertainty and noise. Once these issues are understood, the next step is to consider response to major structural uncertainty, such as component failure.

If the control engineers at NASA had not understood this point we never would have made it to the moon! In virtually *every* application of control, from flight control to cruise control, design is based on a fluid model described by an ordinary differential equation. This design is then refined to account for variability and other unmodeled quantities.

Throughout much of this book we adopt this control-theoretic viewpoint. We find that understanding a simple network model leads to practical solutions to many network control problems.

(i) Stability of the model of interest, in the sense of ergodicity, is essentially equivalent to a finite draining time for a fluid model. These connections are explored in Chapter 10.

(ii) Optimality of one is closely related to optimality of the other, with appropriate notions of "cost" for either model. In particular, the value function for the fluid control problem approximates the relative value function (the solution to Poisson's equation) for the discrete model (see Chapters 8 and 9).

In the control of linear state space models, Poisson's equation is known as the *Lyapunov equation*, and the solution is known to be a quadratic function of the state process when the cost is quadratic (see e.g. [329]). Remarkably, the solution is completely independent of variability, and moreover it coincides with the value function for an associated "fluid model."

(iii) In the case of network models, the value function for the deterministic fluid model is known as the *fluid value function*. This is a piecewise quadratic function when the cost function is linear. The solution to Poisson's equation for a stochastic network does not coincide with the fluid value function in general, but the two functions are approximately equal for large state values. This motivates the development of algorithms to construct quadratic or piecewise quadratic *approximations* to Poisson's equation for stochastic networks to bound steady-state performance. Deterministic algorithms are described in Chapter 8.

Approximate solutions to Poisson's equation such as a carefully chosen quadratic function, or the fluid value function, are used to construct fast simulation algorithms to estimate performance in Chapter 11.

(iv) A convenient approach to the analysis of buffer overflow or any similar disaster is through the analysis of a fluid model (see Section 3.5).

Again, when it comes to control design (i.e., policy synthesis), a solution obtained from an idealized model (deterministic or probabilistic) must be refined to account for unmodeled behavior. This refinement step for networks is developed over Chapters 4–11.

1.3 What do you need to know to read this book?

This book makes use of several different sets of tools from probability theory, control theory, and optimization.

1.3.1 Linear programs

In the theory of linear programming the standard *primal problem* is defined as the optimization problem

$$\textbf{max} \quad c^\mathsf{T} x \tag{1.8}$$

$$\textbf{s.t.} \quad \begin{aligned} \sum_j a_{ij} x_j &\leq b_i, & \text{for } i = 1, \dots, m; \\ x_j &\geq 0, & \text{for } j = 1, \dots, n. \end{aligned}$$

Its dual is the linear program

$$\textbf{min} \quad b^\mathsf{T} w \tag{1.9}$$

$$\textbf{s.t.} \quad \begin{aligned} \sum_j a_{ji} w_j &\geq c_i, & \text{for } i = 1, \dots, n; \\ w_j &\geq 0, & \text{for } j = 1, \dots, m. \end{aligned}$$

The primal is usually written in matrix notation, $\max c^\mathsf{T} x$ subject to $Ax \leq b$, $x \geq 0$; and the dual as $\min b^\mathsf{T} w$ subject to $A^\mathsf{T} w \geq c$, $w \geq 0$.

Any linear programming problem can be placed in the standard form (1.8). For example, a minimization problem can be reformulated as a maximization problem by changing the sign of the objective function. An equality constraint $y = b$ can be represented as two inequality constraints, $y \leq b$ and $-y \leq -b$. In the resulting dual one finds that the two corresponding variables can be replaced by one variable that is unrestricted in sign.

Lemma 1.3.1. *Let x and w be feasible solutions to (1.8) and (1.9), respectively. Then, $c^{\mathsf{T}}x \leq b^{\mathsf{T}}w$.*

Proof. Feasibility of x implies that $Ax \leq b$ and $x \geq 0$, while feasibility of w implies that $c \leq A^{\mathsf{T}}w$ and $w \geq 0$. Putting these four inequalities together gives the desired bound:

$$x^{\mathsf{T}}c \leq x^{\mathsf{T}}A^{\mathsf{T}}w = (Ax)^{\mathsf{T}}w \leq b^{\mathsf{T}}w. \qquad \square$$

The fundamental theorem of linear program states that the bound obtained in Lemma 1.3.1 is tight:

Theorem 1.3.2 (Duality theorem of linear programming). *If either of the linear programs (1.8) or (1.9) has a finite optimal solution, so does the other, and the corresponding values are equal. If either linear program has an unbounded optimal value, then the other linear program has no feasible solution.*

Luenberger is an excellent introduction to these topics [342].

1.3.2 Some probability theory

Until Part III this book requires little knowledge of advanced topics in probability. It is useful to outline some of this advanced material here since, for example, the Law of Large Numbers and the Central Limit Theorem for martingales and renewal processes serves as motivation for the idealized network models developed in Parts I and II.

The starting point of probability theory is the *probability space*, defined as the triple $(\Omega, \mathcal{F}, \mathsf{P})$ with Ω an abstract set of points, $\mathcal{F}$ a σ-field of subsets of Ω, and P a probability measure on $\mathcal{F}$. A mapping $X : \Omega \to \mathsf{X}$ is called a *random variable* if

$$X^{-1}\{B\} := \{\omega : X(\omega) \in B\} \in \mathcal{F}$$

for all sets $B \in \mathcal{B}(\mathsf{X})$: that is, if X is a measurable mapping from Ω to X.

Given a random variable X on the probability space $(\Omega, \mathcal{F}, \mathsf{P})$, we define the σ-field *generated by* X, denoted $\sigma\{X\} \subseteq \mathcal{F}$, to be the smallest σ-field on which X is measurable.

If X is a random variable from a probability space $(\Omega, \mathcal{F}, \mathsf{P})$ to a general measurable space $(\mathsf{X}, \mathcal{B}(\mathsf{X}))$, and h is a real-valued measurable mapping from $(\mathsf{X}, \mathcal{B}(\mathsf{X}))$ to the real line $(\mathbb{R}, \mathcal{B}(\mathbb{R}))$ then the composite function $h(X)$ is a real-valued random variable on $(\Omega, \mathcal{F}, \mathsf{P})$: note that some authors reserve the term "random variable" for such real-valued mappings. For such functions, we define the *expectation* as

$$\mathsf{E}[h(X)] = \int_{\Omega} h(X(\omega))\mathsf{P}(dw).$$

The set of real-valued random variables Y for which the expectation is well defined and finite is denoted $L^1(\Omega, \mathcal{F}, \mathsf{P})$. Similarly, we use $L^\infty(\Omega, \mathcal{F}, \mathsf{P})$ to denote the collection of essentially bounded real-valued random variables Y; that is, those for which there is a bound M and a set $A_M \subset \mathcal{F}$ with $\mathsf{P}(A_M) = 0$ such that $\{\omega : |Y(\omega)| > M\} \subseteq A_M$.

Suppose that $Y \in L^1(\Omega, \mathcal{F}, \mathsf{P})$ and $\mathcal{G} \subset \mathcal{F}$ is a sub-σ-field of $\mathcal{F}$. If $\hat{Y} \in L^1(\Omega, \mathcal{G}, \mathsf{P})$ and satisfies

$$\mathsf{E}[YZ] = \mathsf{E}[\hat{Y}Z] \quad \text{for all } Z \in L_\infty(\Omega, \mathcal{G}, \mathsf{P})$$

then $\hat{Y}$ is called the *conditional expectation* of Y given $\mathcal{G}$, and denoted $\mathsf{E}[Y \mid \mathcal{G}]$. The conditional expectation defined in this way exists and is unique (modulo P-null sets) for any $Y \in L^1(\Omega, \mathcal{F}, \mathsf{P})$ and any sub-σ-field $\mathcal{G}$.

Suppose now that we have another σ-field $\mathcal{H} \subset \mathcal{G} \subset \mathcal{F}$. Then

$$\mathsf{E}[Y \mid \mathcal{H}] = \mathsf{E}[\mathsf{E}[Y \mid \mathcal{G}] \mid \mathcal{H}]. \tag{1.10}$$

The identity (1.10) is often called "the smoothing property of conditional expectations."

1.3.3 Random walks

Random walks are used in this book to model the cumulative arrival process to a network, as well as cumulative service at a buffer. The reflected random walk is a model for storage and queueing systems.

Both are defined by taking successive sums of independent and identically distributed (i.i.d.) random variables.

Definition 1.3.3 (Random walks). Suppose that $X = \{X(k); k \in \mathbb{Z}_+\}$ is a sequence of random variables defined by

$$X(k + 1) = X(k) + \mathcal{E}(k + 1), \qquad k \in \mathbb{Z}_+,$$

where $X(0) \in \mathbb{R}$ is independent of $\mathcal{E}$, and the sequence $\mathcal{E}$ is i.i.d., taking values in $\mathbb{R}$. Then X is called a *random walk* on $\mathbb{R}$.

Suppose that the stochastic process Q is defined by the recursion

$$Q(k + 1) = [Q(k) + \mathcal{E}(k + 1)]_+ := \max(0, Q(k) + \mathcal{E}(k + 1)), \qquad k \in \mathbb{Z}_+,$$

where again $Q(0) \in \mathbb{R}$, and $\mathcal{E}$ is an i.i.d. sequence of random variables taking values in $\mathbb{R}$. Then Q is called the *reflected random walk*. ∎

Consider the following two models for comparison: For a fixed constant $a > 0$, let L^u denote the uniform distribution on the interval $[0, a]$, and L^d the discrete distribution supported on the two points $\{a/3, a\}$ with

$$L^d\{a/3\} = 1 - L^d\{a\} = 3/4.$$

Then L^u and L^d have common first and second moments given by $a/2$ and $a^2/3$, respectively.

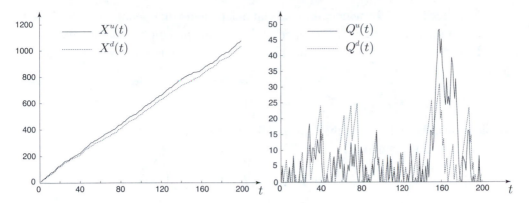

Figure 1.6. Simulation of all four random walks with increment distributions supported on $[0, 10]$. The plot on the left shows two unreflected random walks, and the plot on the right shows the two reflected processes.

Let $\mathcal{E}^u$ denote an i.i.d. sequence with marginal distribution L^u. We then construct an i.i.d. sequence $\mathcal{E}^d$ with marginal distribution L^d consistently as follows:

$$\mathcal{E}^d(k) = \begin{cases} a/3 & \text{if } \mathcal{E}^u(k) \leq 3/4 \\ a & \text{otherwise.} \end{cases}$$

We then simulate the four random walk models

$$X^u(t+1) = X^u(t) + \mathcal{E}^u(t+1),$$
$$Q^u(t+1) = [Q^u(t) + \mathcal{E}^u(t+1) - 1 - a/2]_+,$$

$$X^d(t+1) = X^d(t) + \mathcal{E}^d(t+1),$$
$$Q^d(t+1) = [Q^d(t) + \mathcal{E}^d(t+1) - 1 - a/2]_+, \qquad t \geq 0,$$

with common initial condition $X^u(0) = Q^u(0) = X^d(0) = Q^d(0) = 0$. The introduction of the constant term in the reflected processes $\boldsymbol{Q}^u$ and $\boldsymbol{Q}^d$ is used to enforce the negative drift condition

$$\mathsf{E}[Q^u(t+1) - Q^u(t)|Q^u(t) = x]$$
$$= \mathsf{E}[Q^d(t+1) - Q^d(t)|Q^d(t) = x]$$
$$= \mathsf{E}[\mathcal{E}^u(t+1) - 1 - a/2] = -1, \qquad t \geq 0, \ x \geq 1 + a/2.$$

The plots shown in Fig. 1.6 illustrate results from a simulation of all four processes in the special case $a = 10$, based on a single sample path of the i.i.d. process $\mathcal{E}^u$. The two unreflected random walks show nearly identical behavior, and each grows linearly with slope given by the mean value $\frac{1}{2}a = 5$. The sample paths of the two reflected random walks also show very similar qualitative behavior.

For a stochastic process in continuous time there is a natural analog of the random walk: A continuous time, real-valued stochastic process $\boldsymbol{X}$ has *independent increments* if for each choice of time points $0 \leq t_0 < t_1 < \cdots < t_n$, the random variables $\{X(t_{k+1}) - X(t_k) : k = 0, \ldots, n-1\}$ are independent.

A one-dimensional *Brownian motion* is an independent increment process with continuous sample paths. The independence of the increments implies that the variance of $X(t)$ is increasing with t. In the standard Brownian motion the mean of $X(t)$ is taken to be zero, and we have

$$\mathsf{Var}[X(t_1) - X(t_0)] = \mathsf{E}[(X(t_1) - X(t_0))^2] = t_1 - t_0, \qquad 0 \leq t_0 \leq t_1 < \infty.$$

We also consider in this book the *reflected Brownian motion* (RBM) with drift $-\delta \in \mathbb{R}$ defined through the equation

$$W(t) = w - \delta t + I(t) + \sigma X(t), \qquad t \geq 0, \quad W(0) = w \in \mathbb{R}_+, \qquad (1.11)$$

where X is a standard Brownian motion, $\sigma > 0$ is a constant, and I is an increasing process that defines the reflection. The following two properties characterize this reflection process:

$$(i) \quad W(t) \geq 0 \text{ for all } t \geq 0, \qquad (ii) \quad \int_0^\infty W(t)\, dI(t) = 0.$$

The second requirement means that $I(t)$ cannot increase when $W(t) > 0$.

For background on Brownian motion and RBMs see [233, 72, 283, 445]. Although models based on Brownian motion often have explicit solutions, which is a great asset in analysis, the theory of Brownian motion is not really a prerequisite to understanding the concepts in this book. Chapters in this book are organized so that sections that treat models based on Brownian motion can be skipped without breaking continuity.

1.3.4 Renewal processes

Renewal processes are used to model service-processes as well as arrivals to a network in standard books on queueing theory [114, 19]. The general renewal process is defined as follows.

Definition 1.3.4 (Renewal process). Let $\{\mathcal{E}(1), \mathcal{E}(2), \ldots\}$ be a sequence of independent and identical random variables with distribution function Γ on $\mathbb{R}_+$, and let T denote the associated random walk defined by $T(n) = \mathcal{E}(1) + \cdots + \mathcal{E}(n), n \geq 1$, with $T(0) = 0$. Then the (undelayed) renewal process is the continuous-time stochastic process, taking values in $\mathbb{Z}_+$, defined by

$$R(t) = \max\{n : T(n) \leq t\}.$$

The sample paths of a renewal process are piecewise constant, with jumps at the *renewal times* $\{T(n) : n \geq 1\}$. ∎

A renewal process R takes on integer values and is nondecreasing, so that the quantity $R(t_1) - R(t_0)$, $t_0, t_1 \in \mathbb{R}_+$, can be used to model the number of arrivals during the time interval $(t_0, t_1]$, or the number of service completions for a server that is busy during this time-interval.

The most important example of a renewal process is the standard *Poisson process*, in which the process $\mathcal{E}$ has an exponential marginal distribution. The Poisson process is also another example of a stochastic process with independent increments, whose distribution is expressed as follows: For each $k \geq 0$ and $0 \leq t_0 \leq t_1 < \infty$,

$$P\{R(t_1) - R(t_0) = k\} = \frac{\left(\mu(t_1 - t_0)\right)^k}{k!} e^{\mu(t_1 - t_0)}.$$

Proposition 1.3.5 summarizes some basic results. More structure is described in Asmussen [19].

Proposition 1.3.5. *Let R be a renewal process, and suppose that the increments $\mathcal{E}$ have finite mean $\mu^{-1} = E[\mathcal{E}(t)]$, and finite variance $\sigma^2 = E[\mathcal{E}(t)^2] - \mu^{-2}$. Then, as $t \to \infty$,*

  (i) *$t^{-1}R(t) \to \mu$ with probability one;*
  (ii) *$t^{-1}E[R(t)] \to \mu$;*
  (iii) *$t^{-1}\mathsf{Var}[R(t)] \to \sigma^2 \mu^3$.*

1.3.5 Martingales

A sequence of integrable random variables $\{M(t) : t \in \mathbb{Z}_+\}$ is called *adapted* to an increasing family of σ-fields $\{\mathcal{F}_t : t \in \mathbb{Z}_+\}$ if $M(t)$ is $\mathcal{F}_t$-measurable for each t. The sequence is called a *martingale* if $E[M(t+1) \mid \mathcal{F}_t] = M(t)$ for all $t \in \mathbb{Z}_+$, and a *supermartingale* if $E[M(t+1) \mid \mathcal{F}_t] \leq M(t)$ for $t \in \mathbb{Z}_+$.

A *martingale difference sequence* $\{Z(t) : t \in \mathbb{Z}_+\}$ is an adapted sequence of random variables such that the sequence $M(t) = \sum_{i=0}^{t} Z(i), t \geq 0$, is a martingale.

The following result is basic.

Theorem 1.3.6 (Martingale convergence theorem). *Let M be a supermartingale, and suppose that*

$$\sup_t E[|M(t)|] < \infty.$$

Then $\{M(t)\}$ converges to a finite limit with probability one.

If $\{M(t)\}$ is a positive, real-valued supermartingale then by the smoothing property of conditional expectations (1.10),

$$E[|M(t)|] = E[M(t)] \leq E[M(0)] < \infty, \qquad t \in \mathbb{Z}_+.$$

Hence we have as a direct corollary to the Martingale Convergence Theorem.

Theorem 1.3.7. *A positive supermartingale converges to a finite limit with probability one.*

Since a positive supermartingale is convergent, it follows that its sample paths are bounded with probability one. The following result gives an upper bound on the magnitude of variation of the sample paths of both positive supermartingales and general martingales.

Theorem 1.3.8 (Kolmogorov's inequality for martingales).

(i) *If M is a martingale then for each $c > 0$, $n, p \geq 1$,*

$$\mathsf{P}\{\max_{0 \leq t \leq n} |M(t)| \geq c\} \leq \frac{1}{c^p} \mathsf{E}[|M(n)|^p].$$

(ii) *If M is a positive supermartingale then for each $c > 0$*

$$\mathsf{P}\{\sup_{0 \leq t \leq \infty} M(t) \geq c\} \leq \frac{1}{c} \mathsf{E}[M(0)].$$

Closely related is the Strong Law of Large Numbers for martingales.

Theorem 1.3.9 (i) follows from the Martingale Convergence Theorem 1.3.6 applied to the martingale,

$$M_b(t) = \sum_{i=1}^{t} \frac{M(i) - M(i-1)}{i}, \qquad t \geq 1.$$

The conditions of Theorem 1.3.6 hold, so that M_b is convergent with probability one. The result then follows from Kronecker's Lemma, or summation by parts,

$$\frac{1}{T}(M(T) - M(0)) = \frac{1}{T} \sum_{t=1}^{T} \{M(t) - M(t-1)\} = M_b(T) - \frac{1}{T} \sum_{t=0}^{T-1} M_b(t).$$

The right-hand side vanishes whenever M_b is convergent. Theorem 1.3.9 (ii) is Azuma's Inequality [28, 215].

These results, and related concepts, can be found in Billingsley [61], Chung [109], Hall and Heyde [225], and of course Doob [143].

Theorem 1.3.9 (Laws of large numbers for martingales). *Suppose that M is a martingale. Then:*

(i) *If M satisfies the L_2 bound*

$$\sup_t \mathsf{E}[|M(t) - M(t-1)|^2] < \infty,$$

then the Strong Law of Large Numbers holds,

$$\lim_{T \to \infty} \frac{1}{T} M(T) = 0 \qquad a.s.$$

(ii) *Suppose moreover that the increments of the martingale are* uniformly bounded: *For some constant $b_M < \infty$ and each $t \geq 1$,*

$$|M(t) - M(t-1)| \leq b_M \quad a.s.$$

Then, for each $\varepsilon \in (0, 1]$, $T > 0$,

$$\mathsf{P}\{M(T) \geq T\varepsilon\} \leq \exp\left(-\frac{1}{2} \frac{\varepsilon^2}{b_M^2} T\right).$$

We now turn to approximation of a martingale with a Brownian motion. Consider the sequence of continuous functions on $[0, 1]$

$$X^n(t) := M(\lfloor nt \rfloor) + (nt - \lfloor nt \rfloor)\big[M(\lfloor nt \rfloor + 1) - M(\lfloor nt \rfloor)\big], \quad 0 \leq t \leq 1. \quad (1.12)$$

The function $X^n(t)$ is piecewise linear, and is equal to $M(i)$ when $t = i/n$ for $0 \leq t \leq 1$. In Theorem 1.3.10 below we give conditions under which the normalized sequence $\{n^{-1/2} X^n(t) : n \in \mathbb{Z}_+\}$ converges to a standard Brownian motion on $[0, 1]$. This result requires some care in the definition of convergence for a sequence of stochastic processes.

Let $C[0, 1]$ denote the normed space of all continuous functions $\phi \colon [0, 1] \to \mathbb{R}$ under the uniform norm, which is defined as

$$\|\phi\|_\infty = \sup_{0 \leq t \leq 1} |\phi(t)|.$$

The vector space $C[0, 1]$ is a complete, separable metric space, and hence the theory of weak convergence may be applied to analyze measures on $C[0, 1]$.

The stochastic process $\mathbf{X}^n$ possesses a distribution μ^n, which is a probability measure on $C[0, 1]$. We say that $\mathbf{X}^n$ *converges in distribution* to a stochastic process $\mathbf{X}$ as $n \to \infty$, which is denoted $\mathbf{X}^n \xrightarrow{\text{w}} \mathbf{X}$, if the sequence of measures μ^n converge weakly to the distribution μ of $\mathbf{X}$. That is, for any bounded continuous functional h on $C[0, 1]$,

$$\mathsf{E}[h(\mathbf{X}^n)] \to \mathsf{E}[h(\mathbf{X})] \qquad \text{as } n \to \infty. \quad (1.13)$$

To prove convergence we use the following key result which is a consequence of Theorem 4.1 of [225]. Assumption (i) is the existence of a finite limiting variance for the increments of the martingale, and (ii) provides a bit more control on the tails.

Theorem 1.3.10 (Functional central limit theorem). *Let $(M(t), \mathcal{F}_t)$ be a square integrable martingale, so that for all $n \in \mathbb{Z}_+$*

$$\mathsf{E}[M(n)^2] = \mathsf{E}[M(0)^2] + \sum_{k=1}^n \mathsf{E}[(M(k) - M(k-1))^2] < \infty,$$

and suppose that the following conditions hold with probability one:

(i) *For some constant $0 < \gamma^2 < \infty$,*

$$\lim_{n \to \infty} \frac{1}{n} \sum_{k=1}^n \mathsf{E}[(M(k) - M(k-1))^2 | \mathcal{F}_{k-1}] = \gamma^2 \quad (1.14)$$

(ii) *For each $\varepsilon > 0$,*

$$\lim_{n \to \infty} \frac{1}{n} \sum_{k=1}^n \mathsf{E}[(M(k) - M(k-1))^2 \mathbf{1}_{\{(M(k)-M(k-1))^2 \geq \varepsilon n\}} | \mathcal{F}_{k-1}] = 0. \quad (1.15)$$

Then $(\gamma^2 n)^{-1/2} \mathbf{X}^n \xrightarrow{\text{w}} \mathbf{X}$, where $\mathbf{X}$ is a standard Brownian motion on $[0, 1]$.

1.3.6 Markov models

The Markov chains that we consider evolve on a countable state space, denoted X. The chain itself is denoted $X = \{X(t) : t \in \mathbb{Z}_+\}$, with transition law defined by the *transition matrix P*:

$$\mathsf{P}\{X(t+1) = y \mid X(0), \ldots, X(t)\} = P(x, y), \qquad x, y \in \mathsf{X}. \tag{1.16}$$

Examples of Markov chains include both the reflected and unreflected random walks defined in Section 1.3.3. The independence of the $\mathcal{E}$ guarantees the Markovian property (1.16).

The transition matrix is viewed as a (possibly infinite-dimensional) matrix. Likewise, a function $c\colon \mathsf{X} \to \mathbb{R}$ can be viewed as a column vector, and we can express conditional expectations as a matrix–vector product,

$$\mathsf{E}[c(X(t+1)) \mid X(t) = x] = Pc\,(x) := \sum_{y \in \mathsf{X}} P(x, y)c(y), \qquad x \in \mathsf{X}.$$

More generally, the matrix product is defined inductively by $P^0(x, y) = \mathbf{1}_{\{x=y\}}$ and for $n \geq 1$,

$$P^n(x, y) = \sum P(x, z) P^{n-1}(z, y), \qquad x, y \in \mathsf{X}. \tag{1.17}$$

Based on this we obtain the representation

$$\mathsf{E}[c(X(t+n)) \mid X(t) = x] = P^n c\,(x), \qquad x \in \mathsf{X},\ t \geq 0,\ n \geq 1. \tag{1.18}$$

Central to the theory of Markov chains is the following generalization, known as the *strong Markov property*. Recall that a random time τ is called a *stopping time* if there exists a sequence of functions $f_n\colon \mathsf{X}^{n+1} \to \{0, 1\}$, $n \geq 0$, such that the event $\{\tau = n\}$ can be expressed as a function of the first n samples of X,

$$\{\tau = n\} = f_n(X(0), \ldots, X(n)), \qquad n \geq 0.$$

We write this as $\{\tau = n\} \in \mathcal{F}_n$, where $\{\mathcal{F}_k : k \geq 0\}$ is the filtration generated by X. We let $\mathcal{F}_\tau$ denote the σ-field generated by the events "before τ": that is,

$$\mathcal{F}_\tau := \{A : A \cap \{\tau \leq n\} \in \mathcal{F}_n, n \geq 0\}.$$

Proposition 1.3.11. *Let τ be any stopping time and $f\colon \mathsf{X} \to \mathbb{R}$ any bounded function. Then, for each initial condition*

$$\mathsf{E}[f(X(t+\tau)) \mid \mathcal{F}_\tau] = \mathsf{E}_{X(\tau)}[f(X(t))].$$

In particular, for the simple stopping time $\tau \equiv 1$ we have

$$\mathsf{E}[f(X(t+1)) \mid X(0), X(1)] = \mathsf{E}_{X(1)}[f(X(t))] = P^t f\,(X(1)). \tag{1.19}$$

A self-contained treatment of stability theory for Markov chains is contained in Appendix. Much of the discussion is devoted to the following two linear equations:

(i) *Invariance equation*:

$$\pi P = \pi, \tag{1.20}$$

where π is seen as a row vector, and P as a matrix. Hence (1.20) may be expressed

$$\sum_{x \in \mathsf{X}} \pi(x) P(x,y) = \pi(y), \quad y \in \mathsf{X}.$$

We seek a solution that is positive, $\pi(x) \ge 0$ for $x \in \mathsf{X}$, with $\sum_x \pi(x) = 1$ so that π defines a probability distribution on X. Given a π-integrable function $f : \mathsf{X} \to \mathbb{R}$ we denote the *steady-state mean* by $\pi(f) := \sum_{x \in \mathsf{X}} \pi(x) f(x)$.

(ii) *Poisson's equation*: Given a π-integrable function $f : \mathsf{X} \to \mathbb{R}$, called the *forcing function*, we let $\eta := \pi(f)$ and seek a function $h \colon \mathsf{X} \to \mathbb{R}$ satisfying

$$Ph = h - f + \eta. \tag{1.21}$$

A solution h to Poisson's equation is also called a *relative value function*.
In (1.21) the relative value function is seen as a column vector, so that (1.21) is equivalently expressed

$$\sum_y P(x,y) h(y) = h(x) - f(x) + \eta \qquad x \in \mathsf{X}.$$

Letting $\mathcal{D} = P - I$ denote the difference operator, the two equations can be written

$$\pi \mathcal{D} = 0 \quad \text{and} \quad \mathcal{D} h = -f + \eta. \tag{1.22}$$

Solutions to (1.22) are used to address issues surrounding stability and performance in a Markov model. The existence of π leads to ergodic theorems such as

$$\begin{aligned}
n^{-1} \sum_{t=0}^{n-1} f(X(t)) &\to \eta, & n \to \infty \\
\mathsf{E}[f(X(t))] &\to \eta, & t \to \infty.
\end{aligned} \tag{1.23}$$

The existence of h leads to finer results:

(i) The relative value function is central to optimal control where f is a one-step cost function (see Chapter 9).

(ii) Approximate solutions to Poisson's equation lead to direct performance bounds (e.g., estimates of η). Results of this kind are developed in Section 8.6 and Section 8.7.2, based on the general theory of Markov chains surveyed in Appendix.

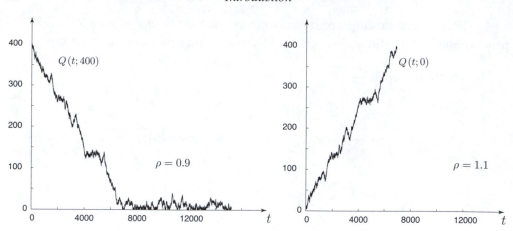

Figure 1.7. The M/M/1 queue: In the stable case on the left we see that the process $Q(t)$ appears piecewise linear, with a relatively small high-frequency "disturbance." The process explodes linearly in the unstable case shown on the right.

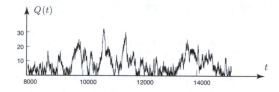

Figure 1.8. A closeup of the trajectory shown on the left-hand side of Fig. 1.7 with load $\rho = 0.9 < 1$. After a transient period, the queue length oscillates around its steady-state mean of 9.

(iii) The solution to Poisson's equation allows us to construct the martingale

$$M(t) = \left(\sum_{i=0}^{t-1} f(X(i)) \right) + h(X(t)) - \eta t, \qquad t \geq 1.$$

Under suitable stability conditions on X, this leads to the *Central Limit Theorem*,

$$\frac{1}{\sqrt{n}} \sum_{0}^{n-1} (f(X(t)) - \eta) \xrightarrow{\text{w}} N(0, \gamma^2), \qquad n \to \infty,$$

where the *asymptotic variance* has several representations. In terms of the relative value function it can be expressed

$$\gamma^2 = \pi(2h(f - \eta) - \pi((f - \eta)^2).$$

Hence Poisson's equation provides tools for addressing performance of simulators. These ideas are developed in Chapter 11.

A standard queueing model is one of the simplest examples of a Markov chain, and is also an example of the reflected random walk described in Definition 1.3.3.

Example 1.3.12 (The M/M/1 queue). The transition function for the *M/M/1 queue* is defined as

$$P(Q(t+1) = y \mid Q(t) = x) = P(x,y) = \begin{cases} \alpha & \text{if } y = x+1 \\ \mu & \text{if } y = (x-1)_+, \end{cases} \tag{1.24}$$

where α denotes the arrival rate to the queue, μ is the service rate, and these parameters are normalized so that $\alpha + \mu = 1$.

The invariant measure π exists if and only if the *load condition* holds, $\rho := \alpha/\mu < 1$, and in this case $\pi(k) = (1 - \rho)\rho^k$ for $k \geq 0$. The existence of an invariant measure π is interpreted as a form of stability for the queueing model, so that the sample-path behavior looks like that shown in the left-hand side of Fig. 1.7 and in Fig. 1.8. ∎

1.4 Notes

Prediction and control in networks has been the subject of intense research over the past 10 years, but network modeling is far from a mature research area. Each subdiscipline tends to focus on a particular class of models. In particular, equilibrium fluid models are favored in analysis of routing algorithms, and in economic studies of electric power reserves. Research on scheduling has focused primarily on stochastic models of the form (1.1), or specialized versions of this model for the purposes of performance approximation.

The formulation of control solutions for stochastic network models such as (1.1) based on a simple fluid model (1.4) has a significant history – see for Vandergraft [480], Chen and Mandelbaum [94], Perkins and Kumar [396, 397, 395], and their references. Optimization of the fluid model is treated in a series of papers since Anderson and Nash's 1987 book on infinite-dimensional linear programming [13]. Some history is included at the end of Chapter 4.

Extensions of the linear fluid model (1.4) that include bursty arrival rates and bursty service rates are developed for the purposes of network control and design in Newell's monograph [385] and Kleinrock's monograph [304]. Similar models are considered in the work of Cruz et. al. [121, 2] in applications to computer and communications systems.

Theory to support this approach for policy synthesis is relatively new. An important milestone is Dai [124], following Rybko and Stolyar in [420]. The foundation of this work is the *multistep drift criterion* for stability of Markov chains of Malyšhev and Men'šikov [347, 368]. The *fluid limit criterion* for stability that emerges from this work is one focus of Chapter 10, based on Lyapunov theory developed in Chapter 8, along with stability theory for Markov chains surveyed in the appendix.

The RBM model (1.11) is one foundation of the *heavy-traffic* approach to stochastic networks. The concept as well as the term "heavy traffic" goes back to Kingman [301, 302], in which the waiting time in the single server queue is approximated by an exponential distribution through a central limit scaling. In his review of [302][2], Newell writes that *Queueing theory, which has endured a long period in which people treated one example after another, is finally breaking out of its confinement to independent arrivals, service times, etc.* Newell's book [385], which emphasizes fluid and diffusion models for the purposes of design, was influenced in part by Kingman's contributions.

Diffusion models have grown in importance in a variety of related fields. In the operations research community the application of RBM models began to expand following the 1985 monograph of Harrison [231, 233]. Similar in flavor is the multiplicative Brownian motion introduced by Black and Scholes to solve problems in options pricing [62]. Here again, this class of models may not have great predictive power, but the insight gained through this framework has generated wealth among some, and a Nobel prize for Merton and Scholes in 1997.

---

[2] AMS Mathematical Reviews MR0148146.

More history on Brownian models is contained in the Notes for Chapter 5.

This book provides foundations for resource allocation and performance evaluation, but cannot go too deeply into specific issues in each possible application. A notable example is the area of Internet congestion control where there are many constraints due to the reliance on architecture and algorithms designed in the 1970s. Srikant's monograph [459] treats this problem in-depth using a range of techniques, including variants of methods described in this book.

Although much of this book concerns the construction and analysis of algorithms to construct feedback laws for control, to bound performance, or to improve simulation, this book does not contain any theory of algorithms. In particular, we do not touch upon complexity theory for algorithms as described in [390, 391, 392, 115, 42, 194], although this theory is the most important motivation for the approximation techniques developed in the book.

The optimal control problems posed in this book are primarily *centralized* in the sense that there is a centralized decision maker that possesses complete information. A *decentralized* control solution is one that can be implemented based on local information, such as nearby congested links.

For a physical network such as the Internet, or the North American power grid, a centralized control framework is absurd. For example, in a power distribution system generators may be owned by different companies, who supply power to various utilities, using power lines managed by different system operators. Methods from game theory can be applied to study the consequences of potential outcomes in a decentralized noncooperative setting [31, 412]. We do not address any of these game-theoretic issues. However, the centralized optimal policy can be used as a benchmark against which the performance of a decentralized system is evaluated.

Moreover, we do consider classes of policies that can be implemented using only local information. One example is the class of MaxWeight policies introduced in Section 4.8. These are a subset of *myopic policies*. In some cases it can be shown that a myopic policy is approximately optimal if the network is congested, or the network load is high (see Chapter 9 and Theorem 10.0.2).

Part I

Modeling and Control

2

Examples

In this chapter we introduce many of the modeling and control concepts to be developed in this book through several examples. The examples in this chapter are extremely simple, but are intended to convey key concepts that can be generalized to more complex networks. We will return to each of these examples over the course of the book to illustrate various techniques.

A natural starting point is the single server queue.

2.1 Modeling the single server queue

The single server queue illustrated in Fig. 2.1 is a useful model for a range of very different physical systems. The most familiar example is the single-line queue at a bank: Customers arrive to the bank in a random manner, wait until they reach the head of the line, are served according to their needs and the abilities of the teller, and then exit the system. In the single server queue we assume that there is a single line, and only one bank teller. To understand how delays develop in this system we must look at average time requirements of customers and the rate of arrivals to the bank. Also, variability of service times or interarrivals times of customers has a detrimental effect on average delays.

Even in this very simple system there are control and design issues to consider. Is it in the best interest of the bank to reserve a teller to take care of customers with short time requirements? Or, perhaps it is more profitable to invest in customers who have large accounts to provide high service to those customers that are most valuable to the bank. In a very different application the queue models an unreliable machine, such as one used for lithography in a wafer fab, and it is necessary to schedule routine maintenance based on environment and the history of prior maintenance.

These control issues will be revisited later in the book. Here we concentrate on modeling the single server queue in a variable but stationary environment.

Typical mathematical models assume that the interarrival times are i.i.d., service times are i.i.d., and that service and arrival processes are mutually independent. This is known as the *GI/G/1 queue*.

Figure 2.1. Single server queue.

Definition 2.1.1 (The GI/G/1 queue). Suppose the following assumptions hold:

(i) Customers arrive to the queue at time points $T^a(0) = 0$, $T^a(1) = \mathcal{E}^a(1)$, $T^a(2) = \mathcal{E}^a(1) + \mathcal{E}^a(2), \ldots$ where the interarrival times $\{\mathcal{E}^a(i) : i \geq 1\}$, are independent and identically distributed random variables, distributed as a random variable $\mathcal{E}^a$ with distribution defined by $G^a(-\infty, r] = \mathsf{P}(\mathcal{E}^a \leq r)$, $r \geq 0$.

(ii) The nth customer brings a job requiring service $\mathcal{E}^s(n)$ where the service times are independent of each other and of the interarrival times, and are distributed as a variable $\mathcal{E}^s$ with distribution $G^s(-\infty, r] = \mathsf{P}(\mathcal{E}^s \leq r)$, $r \geq 0$.

(iii) The arrival and service rates are denoted α and μ, respectively, so that a typical service time $\mathcal{E}^s$ has mean μ^{-1}, and a typical interarrival time $\mathcal{E}^a$ has mean α^{-1}.

(iv) Customers are served in order of arrival. This is the *first-in first-out* (FIFO) service discipline.

Let $\{R^a(t), R^s(t)\}$ denote the renewal processes associated with the i.i.d. sequences $\{\mathcal{E}^a(i), \mathcal{E}^s(i) : i \geq 1\}$, defined in Definition 1.3.4. Then, the queue length at time t in the GI/G/1 queue is expressed

$$\mathcal{Q}(t) = x - R^s(Z(t)) + R^a(t), \qquad t \geq 0, \tag{2.1}$$

where $\mathcal{Q}(0) = x$ is the initial condition, the *cumulative allocation process* (or cumulative busy time) $\boldsymbol{Z}$ is continuous with $Z(0) = 0$, and

$$0 \leq Z(t_1) - Z(t_0) \leq t_1 - t_0, \qquad 0 \leq t_0 \leq t_1 < \infty.$$

In the GI/G/1 queue it is assumed that $\boldsymbol{Z}$ is defined by the *nonidling policy*

$$\frac{d^+}{dt} Z(t) = \begin{cases} 1 & \text{if } \mathcal{Q}(t) \geq 1. \\ 0 & \text{otherwise.} \end{cases} \qquad \blacksquare$$

The notation was introduced by Kendall [293, 294]: GI for general independent input, G for general service time distributions, and 1 for a single server system.

A theme in this book is that a detailed model may be too complex to synthesize and analyze control solutions. Conversely, it is often possible to construct tractable models that capture essential dynamics, and in particular carry enough structure to generate and evaluate control solutions.

A basic example is the *Controlled Random Walk* (CRW) model. The CRW model for the single server queue is defined by the one-dimensional recursion

$$Q(t+1) = Q(t) - S(t+1)U(t) + A(t+1), \qquad t \in \mathbb{Z}_+, \tag{2.2}$$

where the sequence $(\boldsymbol{A}, \boldsymbol{S})$ is i.i.d., and we typically assume that each of these processes takes integer values. The *allocation sequence* $\boldsymbol{U}$ satisfies $0 \leq U(t) \leq 1$. The nonidling policy is defined for $t \geq 0$ by $U(t) = \mathbf{1}(Q(t) \geq 1)$. Multidimensional versions of the CRW model are developed in subsequent chapters.

If the arrival process for the general continuous time model (2.1) is *Poisson* then the queue is called an *M/G/1 queue*. We will find that in this case a CRW model can be obtained via sampling: A special case is considered in the following subsection, and generalizations are discussed in Sections 4.1 and 5.2. In Section 2.1.2, it is argued that the CRW recursion is a reasonable *approximation* for many queueing models.

In Sections 2.1.3 and 2.1.4, we will see how the GI/G/1 queue can be approximated by a simpler continuous-time model under "heavy traffic" conditions, or for a large initial condition.

2.1.1 Sampling

The *M/M/1 queue* is the special case in which the renewal processes $\boldsymbol{R}^a$ and $\boldsymbol{R}^s$ are independent Poisson processes, so that

$$G^a(x) = 1 - e^{-\alpha x}, \quad G^s(x) = 1 - e^{-\mu x}, \quad x \geq 0.$$

A sampling technique known as *uniformization* is described as follows: Let $\boldsymbol{R}$ be a Poisson process with rate $\alpha + \mu$, and renewal times denoted

$$T(n) = \mathcal{E}(1) + \cdots + \mathcal{E}(n), \quad n \geq 1.$$

Based on $\boldsymbol{T}$ we construct two Poisson processes $\boldsymbol{R}^a$ and $\boldsymbol{R}^s$ with parameter α and μ, respectively, so that $\boldsymbol{R}$ is expressed as the pointwise sum

$$R(t) = R^a(t) + R^s(t), \quad t \geq 0.$$

The two processes are initialized by $R^a(0) = R^s(0) = 0$, and remain constant on the time interval $[T(k), T(k+1))$ for $k \geq 0$. At each time $T(k)$ with $k \geq 1$, a (biased) coin is flipped, independent of $\boldsymbol{R}$, with probability of heads given by $\alpha/(\mu + \alpha)$. If a "heads" appears, then the time $T(k)$ is interpreted as an arrival, so that $R^a(T(k)) = R^a(T(k)^-) + 1$ and $R^s(T(k)) = R^s(T(k)^-)$. Otherwise, the time is interpreted as a (potential) service completion, and $R^s(T(k)) = R^s(T(k)^-) + 1$. The resulting processes $\boldsymbol{R}^a$ and $\boldsymbol{R}^s$ are Poisson and have the desired properties. We then have the following:

Proposition 2.1.2. *Consider the M/M/1 queue (under the nonidling policy) defined by the renewal processes $\boldsymbol{R}^a$ and $\boldsymbol{R}^s$. The sampled process defined by $Q(k) := \mathcal{Q}(T(k))$, $k \geq 0$, can be described by the CRW model (2.2), where the statistics of the i.i.d. sequence $(\boldsymbol{A}, \boldsymbol{S})$ are described as follows: $\boldsymbol{A}$ is i.i.d. with Bernoulli marginal and $S(k) = 1 - A(k)$. The queue-length process $\boldsymbol{Q}$ is a Markov chain, with transition matrix*

$$P(x, y) = \begin{cases} \frac{\alpha}{\alpha+\mu} & y = x + 1 \\ \frac{\mu}{\alpha+\mu} & y = [x - 1]_+. \end{cases}$$

2.1.2 Approximation

A renewal process can be approximated by a random walk, and in this way the GI/G/1 queue is approximated by a CRW recursion.

Consider a general renewal process R whose increment process possesses a finite second moment. By Proposition 1.3.5, we know that

$$t^{-1}\mathsf{E}[R(t)] \to \mu, \quad \text{and} \quad t^{-1}\mathsf{Var}[R(t)] \to \sigma^2\mu^3, \quad t \to \infty,$$

where μ^{-1} and σ^2 denote the common mean and variance of the increment variables $\{\mathcal{E}(i) : i \geq 1\}$.

We wish to construct a stochastic process $X = \{X(k) : k \geq 0\}$ that shares these asymptotic properties. Let X denote the random walk defined by $X(0) = 0$, and $X(k) = \sum_{i=1}^{k} \mathcal{E}^x(i)$ for $k \geq 1$, where $\mathcal{E}^x = (\mathcal{E}^x(i) : i \geq 1)$ consists of i.i.d. non-negative random variables. Given a sampling increment $T_s > 0$ we define the sampling times $t_k = kT_s$ for $k \geq 0$, and choose the distribution of $\mathcal{E}^x$ so that the random variables $R(kT_s)$ and $X(k)$ have approximately the same distribution for $k \geq 1$. Consideration of large k suggests the two restrictions

$$\mathsf{E}[\mathcal{E}^x(1)] = \mu T_s \quad \text{and} \quad \mathsf{Var}[\mathcal{E}^x(1)] = \sigma^2\mu^3 T_s. \tag{2.3}$$

Note that the random walk X is not an exact representation of R, but shares some of its asymptotic properties.

This building block can be used to construct a controlled random-walk model to approximate virtually any queueing system, even when the service times are deterministic (see Exercise 3.3).

2.1.3 Heavy traffic and Brownian models

The *load* for a single server queue is defined to be the ratio of the arrival and service rates, $\rho := \alpha/\mu$. If $\rho > 1$ then the mean interarrival time is less than the mean service time, and the queue cannot possibly keep up with the flow of customers. Consequently, the queue length $Q(t)$ diverges as $t \to \infty$.

Shown in Fig. 2.2 is a sample path of the sampled M/M/1 queue started at zero with load $\rho = 0.9$. The queue length decreases linearly, and then "oscillates" near the boundary of the state space $\mathbb{Z}_+$. The trajectory for the fluid model shown on the right is similar.

The approximation of a renewal process R by a random walk X may be highly accurate in *heavy traffic*, which means that $\rho < 1$ is very nearly unity. To make this precise we demonstrate that the GI/G/1 queue can be approximated by a *reflected Brownian motion* in which the driving noise is Gaussian. This approximation is illustrated in Fig. 1.8 in the stable regime with $\rho = 0.9$, where the sample path behavior does look similar to that of the RBM (1.11) with $\delta > 0$.

Consider a GI/G/1 queue in which the service times and interarrival times that define R^s and R^a have finite first and second moments. It is assumed that $\rho = 1$, so that $\alpha = \mu$, and we fix a constant $\delta > 0$. From this single model, we construct a parameterized family of models with queue-length process denoted $\{\widetilde{\boldsymbol{Q}}^{\kappa} : \kappa \geq 1\}$, and load equal to $\rho^{\kappa} = 1 - \kappa^{-1}\mu^{-1}\delta < 1$.

For each $\kappa \geq 1$, the service process is defined by the renewal process $\boldsymbol{R}^s$, and the arrival process is scaled as follows:

$$R^{a\kappa}(t) = R^a(t(1 - \kappa^{-1}\mu^{-1}\delta)), \qquad t \geq 0.$$

The arrival rate is thus $\alpha^{\kappa} = \mu - \kappa^{-1}\delta$, which gives the desired form for the load ρ^{κ}.

Define for $\kappa \geq 1$,

$$\begin{aligned}
Q^{\kappa}(t; x) &= \kappa^{-1}\widetilde{Q}^{\kappa}(\kappa^2 t; \kappa x), \\
I^{\kappa}(t; x) &= \kappa^{-1}[R^s(\kappa^2 t) - R^s(Z(\kappa^2 t; \kappa x))], \qquad t \geq 0,\ x \in \mathbb{R}_+,
\end{aligned} \tag{2.4}$$

where in this definition the initial condition for $\widetilde{\boldsymbol{Q}}^q$ is taken to be κx. Consequently, we must restrict to x satisfying $\kappa x \in \mathbb{Z}_+$. The process $\boldsymbol{I}^{\kappa}$ is interpreted as the idleness process for the scaled queue-length process $\boldsymbol{Q}^{\kappa}$.

From the representation (2.1) for $\widetilde{\boldsymbol{Q}}$ we obtain the following representation for $\boldsymbol{Q}^{\kappa}$:

$$\begin{aligned}
Q^{\kappa}(t; x) = x + I^{\kappa}(t; x) &- \kappa^{-1}\Big(R^s(\kappa^2 t) - \kappa^2 t\mu\Big) \\
&+ \kappa^{-1}\Big(R^a((\kappa^2 - \kappa\mu^{-1}\delta)t) - (\kappa^2 - \kappa\mu^{-1}\delta)\mu t\Big) \\
&- \kappa^{-1}\Big(\kappa^2\mu - (\kappa^2 - \kappa\mu^{-1}\delta)\mu\Big)t, \qquad t \geq 0.
\end{aligned}$$

The centering of $\boldsymbol{R}^a$ and $\boldsymbol{R}^s$ is done so that the scaled process can be expressed as

$$Q^{\kappa}(t; x) = x + I^{\kappa}(t; x) - \delta t + N^{\kappa}(t),$$

where $\boldsymbol{N}^{\kappa}$ is the sum of the scaled and centered renewal processes

$$\begin{aligned}
N^{\kappa}(t) &:= N^{a\kappa}(t) - N^{s\kappa}(t) \\
&:= \kappa^{-1}\Big(R^a((\kappa^2 - \kappa\mu^{-1}\delta)t) - (\kappa^2 - \kappa\mu^{-1}\delta)\mu t\Big) - \kappa^{-1}\Big(R^s(\kappa^2 t) - \kappa^2 t\mu\Big).
\end{aligned}$$

The Central Limit Theorem implies that the distribution of $N^{\kappa}(t)$ is approximately Gaussian for large κ, with first and second moment approximated by

$$\mathsf{E}[N^{\kappa}(t)] \approx 0, \quad \mathsf{Var}[N^{\kappa}(t)] \approx \sigma_N^2 t,$$

$\sigma_N^2 = (\sigma_a^2 + \sigma_s^2)\mu^3$, and $\{\sigma_a^2, \sigma_s^2\}$ the respective variances of interarrivals and service.

These arguments are used in Section 3.2.2 to show that $(\boldsymbol{Q}^{\kappa}, \boldsymbol{I}^{\kappa}, \boldsymbol{N}^{\kappa})$ converge in distribution as $\kappa \to \infty$ to a triple $(\boldsymbol{Q}, \boldsymbol{I}, \boldsymbol{N})$, where $\boldsymbol{N}$ is Brownian motion with instantaneous variance σ_N^2, and $\boldsymbol{Q}$ is the RBM defined by

$$Q(t) = x - \delta t + I(t) + N(t), \qquad t \geq 0, \quad Q(0) = x \in \mathbb{R}_+.$$

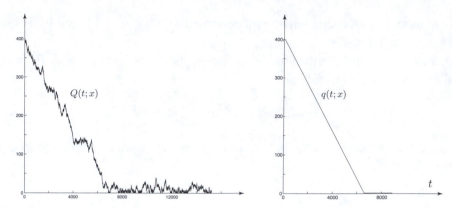

Figure 2.2. On the left is a sample path $Q(t)$ of the M/M/1 queue with $\rho = \alpha/\mu = 0.9$ and $Q(0) = 400$. On the right is a solution to the fluid model equation $\frac{d}{dt}q = (-\mu + \alpha)\zeta$, starting from the same initial condition.

2.1.4 Transient behavior and fluid models

The simplest approximation of the single server queue is the *fluid model*, defined for an initial condition $q(0) = x \in \mathbb{R}_+$ by the linear equation

$$q(t) = x - \mu z(t) + \alpha t, \qquad t \geq 0, \tag{2.5}$$

where z is the *cumulative allocation process*. It is subject to the linear constraints

$$0 \leq z(t_1) - z(t_0) \leq t_1 - t_0, \qquad 0 \leq t_0 \leq t_1 < \infty,$$

with $z(0) = 0$. Letting $\zeta(t)$ denote the right derivative, $\zeta(t) := \frac{d^+}{dt}z(t)$, the fluid-model equation (2.8) is expressed as an ODE model,

$$\frac{d^+}{dt}q(t) = -\mu\zeta(t) + \alpha, \qquad t \geq 0. \tag{2.6}$$

Typically, it assumed that q is controlled using the nonidling policy so that $\zeta(t) = 1$ when $q(t) > 0$, and hence $q(t; x) = [x - (\mu - \alpha)t]_+$ for all $t \geq 0$.

The fluid model can be obtained by scaling the GI/G/1 model initialized at a large initial condition. This is illustrated in Fig. 2.2 where a sample path of the fluid model is compared with a sample path of the GI/G/1 queue. The behavior of the stochastic and deterministic processes look similar when viewed on this large spatial/temporal scale.

Consider again the GI/G/1 queue $\widetilde{Q}$, and define for $\kappa \geq 1$

$$\begin{aligned} q^\kappa(t; x) &= \kappa^{-1}\widetilde{Q}(\kappa t; \kappa x), \\ z^\kappa(t; x) &= \kappa^{-1}Z(\kappa t; \kappa x), \qquad t \geq 0, \; x \in \mathbb{R}_+. \end{aligned} \tag{2.7}$$

On letting $\kappa \to \infty$ we obtain convergence of the scaled processes to a pair (q, z) by Proposition 1.3.5. This limiting process is deterministic and evolves according to the continuous-time equation

$$q(t) = x - \mu z(t) + \alpha t, \qquad t \geq 0. \tag{2.8}$$

Since $\boldsymbol{Z}$ was defined using the nonidling policy it follows that the same holds for $\boldsymbol{z}$, so that $\zeta(t) = 1$ when $q(t) > 0$.

While (2.8) can be obtained as a limit of the scaled processes $(\boldsymbol{q}^\kappa, \boldsymbol{z}^\kappa)$, it can also be taken as a simplified model of the single server queue for capturing transient behavior.

In conclusion, whether or not the fluid model, a CRW model, the GI/G/1 model, or a complex refinement is appropriate depends upon the particular application and the purpose of the model. Our primary motivation for consideration of the CRW model (2.2) is its flexibility in modeling network behavior, combined with its tractability in analysis.

2.2 Klimov model

We now consider a redesign of the single teller bank in which we explicitly model the fact that there are several different classes of customers arriving to the bank. The customers have different service requirements, and different requirements with respect to delay. Figure 2.3 shows a special case of the *Klimov model* in which there is a single station with a single server and ℓ buffers in which customers await service. Each buffer is fed by one of ℓ arrival processes.

This single-station model can describe many very different physical systems. As another important application, consider the following model of a queue in which service requirements are modeled explicitly. At each time $t = 0, 1, 2, \ldots$ at most one customer comes to the queue with a job that requires service. We let $A(t)$ denote the total number of service completions required to complete this job. Since this information is available at time t, the server can discriminate based on job length. Suppose that for each $m \geq 1$ a buffer is created to hold customers whose job length is m. We thus arrive at the Klimov model in which μ_i does not depend upon i, and $A_m(t) = m\mathbf{1}\{A(t) = m\}$, $m \geq 1$, $t \geq 1$. This special case is the subject of Section 5.1 and Exercise 5.4.

In the general Klimov model there are ℓ different customer classes, and we are faced with the choice of policy at the station: *At which buffer should the server work at a given time, based on observations of the ℓ buffer levels?*

It is simplest to begin with consideration of a fluid model. Consider the special case shown in Fig. 2.3 with $\ell = 4$, and let $\boldsymbol{q}$ denote the four-dimensional queue-length process. For a given initial condition $x \in \mathbb{R}_+^4$ this has continuous paths, with $q(0) = x$, and is determined by the cumulative allocation process $\boldsymbol{z}$ which also evolves on $\mathbb{R}_+^4$. The quantity $z_i(t)$ represents the total time that the station has worked on buffer i up

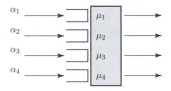

Figure 2.3. Single-station scheduling problem.

to time t, for any $t \geq 0$ and $i \in \{1, 2, 3, 4\}$. Consequently, the components of z are nondecreasing, and since the server must split its effort across the four buffers we have the additional linear constraints

$$0 \leq \sum_{i=1}^{4} \Big(z_i(t_1) - z_i(t_0) \Big) \leq t_1 - t_0, \qquad 0 \leq t_0 \leq t_1 < \infty. \tag{2.9}$$

We again let $\zeta(t) = \frac{d^+}{dt} z(t)$, $t \geq 0$, denote the four-dimensional process of allocation rates. The cumulative allocation process is called *nonidling* if $\sum \zeta_i(t) = 1$ whenever $\sum q_i(t) > 0$.

Let $\alpha \in \mathbb{R}_+^4$ denote the vector of long-term arrival rates to the station, and let B denote the diagonal matrix, $B := -\mathrm{diag}\,(\mu_1, \ldots, \mu_4)$. For a given initial condition $x \in \mathbb{R}_+^4$, the queue-length process evolves according to

$$q(t) = x + Bz(t) + \alpha t, \qquad t \geq 0,$$

which is equivalently expressed as the ODE model

$$\frac{d^+}{dt} q(t) = B\zeta(t) + \alpha, \qquad t \geq 0.$$

Because this can be interpreted as a state space model, for a given allocation process z, the resulting queue-length process q is sometimes called the state trajectory.

Suppose that a linear cost function $c\colon \mathbb{R}_+^4 \to \mathbb{R}_+$ is given. The time derivative of the cost is given by

$$\frac{d^+}{dt} c(q(t)) = c^{\mathsf{T}}(B\zeta + \alpha) = \sum_i c_i(\alpha_i - \mu_i \zeta_i(t)).$$

A *myopic* (or *greedy*) policy is defined to be any allocation process that minimizes the right-hand side for each $t \geq 0$. In the Klimov model, the myopic policy is known as the c–μ *rule*.

Definition 2.2.1 (The c–μ rule). Consider the Klimov model consisting of a single station and ℓ buffers. The *Klimov indices* $\{i_1, \ldots, i_\ell\}$ are defined so that $c_{i_k}\mu_{i_k} \geq c_{i_l}\mu_{i_l}$ whenever $k \leq l$. The c–μ rule for the fluid model is the policy defined so that, for each $1 \leq n \leq \ell$,

$$\sum_{j=1}^{n} \zeta_{i_j} = 1 \quad \text{whenever} \quad \sum_{j=1}^{n} q_{i_j} > 0. \tag{2.10}$$

$\blacksquare$

In this instance, the myopic policy is *path-wise optimal* for this single-station model. That is, for a given initial condition, if q^* denotes the state trajectory under the c–μ rule, and q denotes any other state trajectory from the same initial condition, obtained with a different allocation process z, then

$$c(q^*(t)) \leq c(q(t)), \qquad 0 \leq t < \infty.$$

Consider now a four-dimensional CRW model of the form analogous to (2.2). This is defined in discrete time as before, and we let $Q(t)$ denote the four-dimensional vector of buffer lengths, restricted to the integer lattice $\mathbb{Z}_+^4$. Let $(\boldsymbol{A}, \boldsymbol{B})$ denote an i.i.d. process whose marginal distribution is defined as follows. For each t the distribution of $A(t) = (A_1(t), \ldots, A_4(t))^{\mathsf{T}}$ is supported on $\mathbb{Z}_+^4$, and $B(t) = -\mathrm{diag}\,(M_1(t), \ldots, M_4(t))$ is diagonal. The distribution of $M_i(t)$ has a Bernoulli distribution for each i, with $\mu_i = \mathsf{E}[M_i(t)] = \mathsf{P}\{M_i(t) = 1\}$, and the arrival rate to buffer i is given by $\alpha_i = \mathsf{E}[A_i(t)]$ for each i and t. The assumption that the random service variables $M_i(t)$ take on binary values implies that $\mu_i \le 1$. This may be assumed without loss of generality by slowing the natural time scale.

Given these primitive stochastic processes, the CRW model is defined as in (2.2),

$$Q(t+1) = Q(t) + B(t+1)U(t) + A(t+1), \qquad t \in \mathbb{Z}_+,$$

with initial condition $Q(0) \in \mathbb{Z}_+^4$. The allocation process $\boldsymbol{U}$ is defined as $U_i(t) = 1$ if and only if server i is busy. The physical description of the model leads to linear constraints as in the fluid model:

$$U_i(t) \ge 0 \text{ for each } i, \text{ and } U_1(t) + \cdots + U_4(t) \le 1, \qquad t \ge 0.$$

The constraint that $\boldsymbol{Q}$ evolves on the integer lattice implies additional constraints on $\boldsymbol{U}$: The components of the allocation process must take on binary values, $U_i(t) \in \{0, 1\}$, for each $i \in \{1, \ldots, 4\}$ and $t \ge 0$. Let $\mathsf{U}_\diamond \subset \{0, 1\}^4$ denote the set of all allowable allocations, and $\mathsf{U}_\diamond(x) := \{u \in \mathsf{U}_\diamond : u_i = 0 \text{ whenever } x_i = 0\}$.

The myopic policy for the CRW model is defined in analogy with Definition 2.2.1,

$$U(t) \in \arg\min \mathsf{E}[c(Q(t+1)) \mid Q(0), \ldots, Q(t)], \qquad t \ge 0, \ Q(t) = x, \qquad (2.11)$$

where the minimum is over $U(t) \in \mathsf{U}_\diamond(x)$. Since c is assumed linear, the conditional expectation can be expressed

$$\mathsf{E}[c(Q(t+1)) \mid Q(0), \ldots, Q(t)] = c\big(\mathsf{E}[Q(t+1) \mid Q(t) = x]\big) = \langle c, x + BU(t) + \alpha \rangle$$

and hence the allocation at time t is given by

$$U(t) \in \arg\min_{u \in \mathsf{U}_\diamond(x)} \langle c, x + Bu + \alpha \rangle, \qquad t \ge 0, \ Q(t) = x.$$

A solution to this minimization is expressed in terms of the Klimov indices,

$$U_{i_j}(t) = 1 \text{ if and only if } Q_{i_j}(t) \ge 1 \text{ and } Q_{i_l}(t) = 0 \text{ for } l < j, \qquad (2.12)$$

and this solution is unique if the Klimov indices are uniquely specified (i.e., $c_{i_k}\mu_{i_k} > c_{i_l}\mu_{i_l}$ whenever $k < l$). The policy (2.12) is known as the c–μ rule for the CRW model.

The Klimov model is a rare case in which the myopic policy is optimal for the CRW model – this optimality is explained in Section 5.5.1. Myopic policies are a theme in this book because they are easily computable, and have attractive properties for an appropriate choice of c. For example, in many cases it is possible to choose the cost function so that the resulting policy is in some sense "universally stabilizing", and can be implemented based on local information. An example is the class of MaxWeight policies introduced in Section 4.8.

2.3 Capacity and queueing in communication systems

The single server queue can describe the flow of customers through a bank. A very different application is a communication model in which packets are sent through a communication channel to a receiver. The *service rate* is equal to the rate at which packets can be sent through the channel. This is limited due to noise, as well as interference from other users. Factors that contribute to delay again include arrival rates, service rates, and variability.

Consider the queueing model in which data arrives for transmission in the form of fixed-length packets $A = (A(1), A(2), \dots)$. The arrival process is i.i.d., with finite mean $\alpha = \mathsf{E}[A(t)]$. Once a packet arrives for transmission, the data in that packet is queued until it can be coded and sent via the input sequence X.

The basic additive white Gaussian noise (AWGN) channel with real-valued input sequence X and output sequence Y is described by

$$Y(t) = X(t) + N(t), \qquad t \geq 0,$$

where N is independent of X. It is assumed that the noise N is i.i.d. and Gaussian with zero mean and variance σ^2, and that the input sequence is subject to the average power constraint $\mathsf{E}[X(t)^2] \leq \sigma_X^2$ for all t, where σ_X^2 is a finite constant. The maximum rate at which data can be transmitted is given by Shannon's formula,

$$C_{\sigma_N^2}(\sigma_X^2) = \tfrac{1}{2} \log_2 \left(1 + \frac{\sigma_X^2}{\sigma^2} \right) \quad \text{bits per time slot.} \tag{2.13}$$

This model describes a version of the single server queue in which the arrival rate is α and the maximal service rate is $\mu = C_{\sigma_N^2}(\sigma_X^2)$. Any achievable transmission rate $R < C_{\sigma_N^2}(\sigma_X^2)$ can be interpreted as $R = \zeta \mu$ where $\zeta \in [0, 1]$ is the allocation rate. However, there is one important difference between the communication model and standard queueing models. To implement the allocation rule $\zeta \approx 1$, so that data is sent at rate $R \approx C_{\sigma_N^2}(\sigma_X^2)$, it is necessary to use a coding scheme consisting of very long block lengths. In a sense then, the variability of the service process increases with the mean allocation rate ζ. Nevertheless, as long as a strict bound on ζ is enforced, this system can be described reasonably accurately using a GI/G/1 or CRW model.

2.4 Multiple-access communication

Consider the multiple-access system illustrated in Fig. 2.4, where two users transmit to a single receiver. The two users share a single channel which is corrupted by AWGN. The output of the system seen at the receiver is given by

$$Y(t) = X_1(t) + X_2(t) + N(t), \tag{2.14}$$

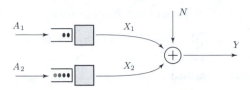

Figure 2.4. Multiple-access communication system with two users and a single receiver.

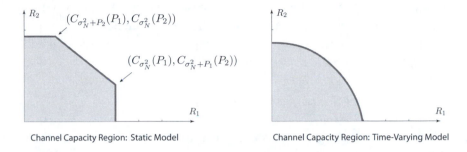

Channel Capacity Region: Static Model   Channel Capacity Region: Time-Varying Model

Figure 2.5. Achievable rates in two multidimensional communication models.

where N is again i.i.d. Gaussian, and independent of the two inputs $\{X_1, X_2\}$. It is assumed that user i is subject to the average power constraint $E[X_i(t)^2] \leq \sigma_{X_i}^2$ for all t. This is known as the *ALOHA model* when the two users send data independently [3, 335, 43]. Stability of ALOHA systems based on queueing models and channel coding theory is addressed in [10, 406, 354].

The queueing system associated with the ALOHA model has two buffers that receive arriving packets of data modeled as the i.i.d. sequences $A_i = (A_i(1), A_i(2), \dots)$, with finite mean $\alpha_i = E[A_i(t)]$, $i = 1, 2$. Data at queue i is stored in its respective queue until it is coded and sent to the receiver using the respective input sequence X_i.

The set of all possible data rates is given by the Cover-Wyner region U illustrated on the left in Fig. 2.5. Any pair $(R_1, R_2) \in \mathsf{U}$ within this region can be achieved through independent coding schemes at the two buffers. Additional information, such as knowledge of the state of the other users' queue, or joint-coding of data, does not improve the achievable rate region (see [354]).

The capacity region U depends critically on the channel model statistics. If, for example, the noise N is not Gaussian then there is in general no closed-form expression for U. In wireless communication systems there is typically random fading, so that the model (2.14) is refined through the introduction of fading processes $\{G_i : i = 1, 2\}$,

$$Y(t) = G_1(t)X_1(t) + G_1(t)X_2(t) + N(t). \qquad (2.15)$$

Again, the capacity region is not known except in very special circumstances. Shown on the right in Fig. 2.5 is the form of U that might be expected in a fading environment.

2.5 Processor sharing model

The *processor sharing* model shown in Fig. 2.6 represents a pair of stations. Arriving to the system are two separate streams of customers. It is assumed that the two classes of customers have different service requirements, as in the Klimov model considered in Section 2.2. However, in this network there are two servers, and the second server has extra flexibility so that it is able to assist either station.

The myopic policy is again easily computed, and we again find that this policy is path-wise optimal for the fluid model when the cost is linear. This is most easily seen by reducing the control problem to a version of the Klimov model.

Consider the special case $\alpha_a \geq \mu_a$, and suppose that the policy is nonidling at Station 1. Under these conditions we define

$$\alpha_1 = \alpha_a - \mu_a, \quad \alpha_2 = \alpha_b, \quad \mu_1 = \mu_b, \quad \mu_2 = \mu_c,$$

so that the processor sharing model is equivalent to the model shown on the right in Fig. 2.6, described by the ODE model

$$\tfrac{d^+}{dt} q_i(t) = -\mu_i \zeta_i(t) + \alpha_i, \qquad i = 1, 2. \tag{2.16}$$

If $c \colon \mathbb{R}_+^2 \to \mathbb{R}_+$ is a linear cost function it then follows that $c_1 > 0$ and hence the myopic policy is nonidling at Station 1 as previously assumed. The optimal policy is a priority policy, the c–μ rule, in which Station 2 gives strict priority to buffer 1 if and only if $c_1\mu_1 > c_2\mu_2$. This priority policy is again pathwise optimal since the fluid model is identical to the Klimov model analyzed previously.

In general, if the processor sharing model is described using a stochastic model, such as the CRW model, then there is no exact correspondence with a Klimov model. However, as described in detail in Example 4.2.7, the fluid-model analysis reveals structure of the optimal policy for the stochastic model.

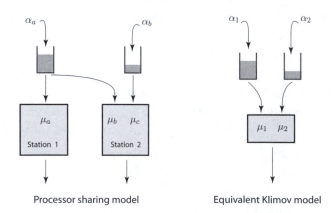

Processor sharing model   Equivalent Klimov model

Figure 2.6. On the left is a fluid model for the processor-sharing network. On the right is an equivalent Klimov model.

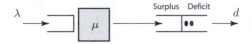

Figure 2.7. Simple make-to-stock system.

2.6 Inventory model

Figure 2.7 shows a *make-to-stock* production facility in which a single product is produced. The recurrent demand is not entirely predictable, but it is assumed that some statistics such as the mean and variance are known. In this simple example the only control decisions are whether to produce or to remain idle at the station, and whether to order new raw material.

If the supplier does not maintain adequate inventory of finished work, then it is likely that she will be forced to pay high backorder costs, or cost from lost sales. Conversely, an excessively large inventory is costly due to holding costs or perishability of the product produced.

A standard policy for a system of this form is based on a *base stock*, or *hedging point* $\bar{x}$. When there is deficit, or surplus inventory falls below $\bar{x}$, then production proceeds at the fastest rate possible, modeled as μ in Fig. 2.7. Production ceases when surplus inventory is above the hedging-point.

If production and demand are perfectly predictable then a fluid model is justifiable, and an optimal hedging-point is $\bar{x} = 0$. In practice, the production system is subject to variability due to breakdown or preventative maintenance, as well as uncertainty with respect to demand. When taking this randomness into account one finds that the best hedging-point value may be large.

After understanding how to control this simple model we will find that it serves as a vital building block for understanding truly complex networks. Techniques for computing optimal hedging-point values in this simple model and in network models are developed in Chapters 7–9.

The inventory model is also a first step toward understanding resource allocation in a power grid.

2.7 Power transmission network

Figure 1.4 shows the California power grid. Demand and supply of electric power are subject to sudden and unexpected changes due to local weather conditions, equipment failure, or generator outages. It is common practice for the system operator to maintain a substantial amount of reserve production capacity to meet seasonal and hourly fluctuations in demand, and in order to respond to unexpected changes in the market environment.

Secondary sources of power generation used to ensure adequate reserves are known as *ancillary services*. In the simple example shown in Fig. 2.8 there is a single customer (a large city) that is served by a single primary generator, together with ancillary

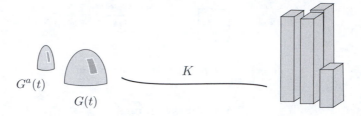

Figure 2.8. Power model with two suppliers providing power to a single utility.

service. The marginal cost of power production is higher for ancillary service, but its ramping rate is significantly higher than that of the primary generator. Both the primary and the ancillary generators are located away from the city, and connected by a transmission line with finite capacity.

Reliability and efficiency of the power system is intimately related to power reserves. Too little reserve increases the chance of very costly blackouts, while too much reserve may waste valuable resources. Hence, in spite of the fact that electric power is not stored, choosing appropriate reserves is a decision problem very similar to the choice of hedging-points in an inventory model.

In fact, by viewing reserve as a "queue level" we can construct a network model that is very similar to an inventory model. Given current generation capacities $G^p(t)$ and $G^a(t)$ from primary and ancillary services, and given the demand for power $D(t)$, the reserve at time t is denoted

$$Q(t) = G^p(t) + G^a(t) - D(t). \tag{2.17}$$

A two-dimensional CRW model can be used to describe the behavior of this system with state process $X(t) = (Q(t), G^a(t))^\mathsf{T}$, subject to conditions on demand and generation. The two-dimensional allocation sequence is defined as

$$U(t) = (U^p(t), U^a(t))^\mathsf{T}, \qquad t \geq 0,$$

where $U^p(t) = G^p(t+1) - G^p(t)$, and $U^a(t) = G^a(t+1) - G^a(t)$. Demand is modeled as a random walk of the form

$$D(t+1) = D(t) + \mathcal{E}(t+1), \qquad t \geq 0, \quad D(0) = 0,$$

in which the increment process $\mathcal{E}$ is a bounded, i.i.d. sequence. On defining the two-dimensional increment process

$$A(t) := -(\mathcal{E}(t), 0)^\mathsf{T}, \qquad t \geq 0,$$

and the 2×2 matrix

$$B = \begin{bmatrix} 1 & 1 \\ 0 & 1 \end{bmatrix}, \tag{2.18}$$

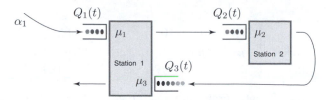

Figure 2.9. Simple re-entrant line.

the state process X is expressed for $t \geq 0$ in the recursive form

$$X(t+1) = X(t) + BU(t) + A(t+1), \qquad t \geq 0. \tag{2.19}$$

The initial condition $X(0) = x \in \mathsf{X} := \mathbb{R} \times \mathbb{R}_+$ is interpreted as $x_1 = Q(0) = G^p(0) + G^a(0)$, and $x_2 = G^a(0)$.

In the CRW model (2.19) the allocation sequence U is adapted to X, and is subject to the constraint $U(t) \in \mathsf{U}(X(t))$ for all $t \in \mathbb{Z}_+$, where

$$\mathsf{U} := \{u = (u^p, u^a)^\mathsf{T} \in \mathbb{R}^2 : -\zeta^{p-} \leq u^p \leq \zeta^{p+} \,, \ -\zeta^{a-} \leq u^a \leq \zeta^{a+}\},$$
$$\mathsf{U}(x) := \{u \in \mathsf{U} : x + Bu \in \mathsf{X}\}, \qquad x \in \mathsf{X}. \tag{2.20}$$

An effective policy for this network can be constructed based on hedging points $\{\overline{x}^p, \overline{x}^a\}$ as follows: Power from the primary source is ramped up whenever reserves fall below a threshold $\overline{x}^p$, and power from the ancillary source is ramped up if reserves fall below a threshold $\overline{x}^a$ that is strictly less than $\overline{x}^p$.

We revisit this example in Example 7.5.2 where we introduce a cost function on X designed to take into account the costs of power generation, excess power reserves, and the high cost of not meeting demand. We find that the optimal thresholds scale roughly linearly with variability in demand.

Also, $\overline{x}^a$ scales inversely with ζ^{a+}, and $\overline{x}^p$ scales inversely with the sum of the rates $\zeta^{a+} + \zeta^{p+}$. In this way we quantify the value of responsive power generation.

2.8 Optimization in a simple re-entrant line

A *re-entrant line* is a network model in which there is a single arrival stream and routing is deterministic. The buffers are numbered so that an arriving customer visits the first queue, then the second, and so on until it is processed by the last queue and leaves the system. Customers may revisit a station several times before exiting.

A simple example is illustrated in Fig. 2.9. We begin with consideration of the fluid model q evolving on $\mathbb{R}_+^3$. This deterministic process is described by the ODE model

$$\tfrac{d^+}{dt}q_1(t) = \alpha_1 - \mu_1\zeta_1(t) \,, \ \tfrac{d^+}{dt}q_2(t) = \mu_1\zeta_1(t) - \mu_2\zeta_2(t) \,, \ \tfrac{d^+}{dt}q_3(t) = \mu_2\zeta_2(t) - \mu_3\zeta_3(t) \,, \tag{2.21}$$

or in matrix form,

$$\tfrac{d^+}{dt}q(t) = B\zeta(t) + \alpha, \qquad t \geq 0,$$

with

$$B = \begin{bmatrix} -\mu_1 & 0 & 0 \\ \mu_1 & -\mu_2 & 0 \\ 0 & \mu_2 & -\mu_3 \end{bmatrix}, \qquad \alpha = \begin{bmatrix} \alpha_1 \\ 0 \\ 0 \end{bmatrix}. \tag{2.22}$$

The constraints on the allocation rates are given by $\zeta(t) \in U$, $t \geq 0$, where

$$U := \{u \in \mathbb{R}_+^3 : \zeta_1 + \zeta_3 \leq 1,\ \zeta_2 \leq 1\}.$$

A policy for (2.8) is called *nonidling* if

$$\zeta_1(t) + \zeta_3(t) = 1 \qquad \text{whenever } q_1(t) + q_3(t) > 0;$$
$$\zeta_2(t) = 1 \qquad \text{whenever } q_2(t) > 0.$$

Control of networks is more subtle in situations where there is re-entry. A short survey of control approaches is given here.

Buffer priority policies In a buffer priority policy the buffers at a station are ordered in preference for service, exactly as in the Klimov model. Priority policies are always assumed nonidling.

Two priority policies defined for any re-entrant line are:

(i) The *Last-Buffer First-Served* (LBFS) policy in which buffers nearest the exit buffer receive priority. In this example, this means that $\zeta_3(t) = 1$ whenever $q_3(t) > 0$, and $\zeta_2(t) = 1$ whenever $q_2(t) > 0$. Once the third buffer empties for some $t > 0$ it stays empty, so that $\frac{d}{dt} q_3(t) = 0 = \mu_2 \zeta_2(t) - \mu_3 \zeta_3(t)$.

(ii) The *First-Buffer First-Served* (FBFS) policy gives priority to buffers in the opposite order. Hence, in this example, the LBFS policy gives priority to buffer 1 over buffer 3 at Station 1 until $q_1(t) = 0$, and henceforth $\zeta_1(t) = \alpha_1/\mu_1$.

Myopic policies A *myopic policy* is defined exactly as in the Klimov model. If $X = \mathbb{R}_+^3$ (i.e., there are no buffer constraints), then for a given cost function $c\colon X \to \mathbb{R}_+^3$, the allocation rate at time t for the myopic policy is defined by

$$\zeta(t) \in \arg\min \tfrac{d^+}{dt} c(q(t)), \tag{2.23}$$

where the minimum is over all allocation rates in U, subject to the constraint that $\frac{d^+}{dt} q_i(t) \geq 0$ whenever $q_i(t) = 0$.

A special case is the ℓ_1-*norm*, $c(x) = x_1 + x_2 + x_3$, $x \in \mathbb{R}_+^3$, so that $c(q(t))$ represents the total customer population at time t. For this linear cost function the derivative is expressed

$$\tfrac{d^+}{dt} c(q(t)) = (\alpha_1 - \mu_1 \zeta_1(t)) + (\mu_1 \zeta_1(t) - \mu_2 \zeta_2(t)) + (\mu_2 \zeta_2(t) - \mu_3 \zeta_3(t))$$
$$= \alpha_1 - \mu_3 \zeta_3(t), \qquad t \geq 0.$$

We conclude that the LBFS priority policy is precisely the nonidling myopic policy when c is the ℓ_1 norm.

Time optimality Consider the two *workload vectors* defined by

$$\xi^1 = (\mu_1^{-1} + \mu_3^{-1}, \mu_3^{-1}, \mu_3^{-1})^{\mathsf{T}}$$
$$\xi^2 = (\mu_2^{-1}, \mu_2^{-1}, 0)^{\mathsf{T}},$$

and associated load parameters

$$\rho_1 = \langle \xi^1, \alpha \rangle = \alpha_1(\mu_1^{-1} + \mu_3^{-1})$$
$$\rho_2 = \langle \xi^2, \alpha \rangle = \alpha_1 \mu_2^{-1}.$$

It is shown in Proposition 4.2.5 below that the minimum time that q can reach the origin is given by

$$T^*(x) = \max_{s=1,2} \frac{w_s(0)}{1 - \rho_s} = \max_{s=1,2} \frac{\langle \xi^s, x \rangle}{1 - \rho_s}, \qquad \text{when } q(0) = x.$$

A policy is called *time optimal* if this minimum time is achieved for each initial condition.

Station 2 is called the *bottleneck* if $\rho_1 < \rho_2 < 1$. This model is most interesting in this case, since then a tradeoff must be made between draining the system, and avoiding starvation at the bottleneck.

The two-dimensional *workload process* (in units of time) is defined by $w(t) = (\langle \xi^1, q(t) \rangle, \langle \xi^2, q(t) \rangle)^{\mathsf{T}}, t \geq 0$. Under any admissible allocation process z,

$$\frac{d^+}{dt} w_s(t) \geq -1 + \rho_s, \qquad s = 1, 2, \ t \geq 0,$$

and this lower bound is attained for $w_s(t)$ if and only if Station i is working at maximal rate. For example, $\frac{d^+}{dt} w_2(t) = -1 + \rho_2$ if and only if $\zeta_2(t) = 1$. Consequently, the minimum time that $w_s(\cdot)$ can reach zero is given by $w_s(0)/(1 - \rho_s)$. Based on these observations, it is not difficult to show that a policy is time optimal if and only if the following two properties hold whenever $q(t) \neq 0$,

$$\frac{d^+}{dt} w_1(t) = -1 + \rho_1 \qquad \textit{whenever} \qquad \frac{w_1(t)}{1 - \rho_1} \geq \frac{w_2(t)}{1 - \rho_2} \tag{2.24}$$
$$\frac{d^+}{dt} w_2(t) = -1 + \rho_2 \qquad \textit{whenever} \qquad \frac{w_2(t)}{1 - \rho_2} \geq \frac{w_1(t)}{1 - \rho_1}.$$

A proof for general networks is presented in Proposition 4.3.4.

The FBFS policy satisfies these conditions and is hence time optimal in this example.

The LBFS policy is not necessarily time optimal if $\rho_1 < \rho_2 < 1$. Suppose that $\mu_1 = \mu_3$ so that $2\mu_1^{-1}\alpha_1 = \rho_1 < \rho_2 = \mu_2^{-1}\alpha_1$, or $\mu_1 > 2\mu_2$. Consider an initial condition satisfying $q_1(0) = q_2(0) = 0$, and $q_3(0) = x_3 > 0$. The minimal draining time is then given by

$$T^*(x) = \max_{s=1,2} \frac{\langle \xi^s, x \rangle}{1 - \rho_s} = \frac{\langle \xi^1, x \rangle}{1 - \rho_1} = \frac{x_3}{\mu_1 - 2\alpha_1}.$$

Under the LBFS policy, the first buffer grows in size until the first time that $q_3(t)$ reaches zero, given by $T_1 = x_3/\mu_1$. Hence, during this time interval the work destined

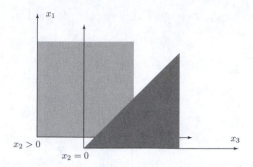

Figure 2.10. Optimal fluid policy for the three buffer re-entrant line with parameters defined in (2.26). In this illustration, the grey regions indicate those states for which buffer three is given exclusive service.

for Station 2 grows, yet Station 2 remains idle. When work commences at buffer 1 at time T_1 we have $q_1(T_1) = \alpha_1 T_1$ and $q_2(T_1) = 0$. Consequently, under the LBFS policy, the first time at which $\boldsymbol{q}$ reaches the origin, is given by

$$T^\circ(x) = T_1 + \frac{\alpha_1 T_1}{\mu_2 - \alpha_1} = \frac{\mu_2}{\mu_2 - \alpha_1} \frac{x_3}{\mu_1}. \tag{2.25}$$

This shows that LBFS is not time optimal since $T^\circ(x) > T^*(x)$ from this initial condition.

Infinite-horizon optimal control Another approach to policy synthesis is through consideration of the *total cost*, defined by

$$J(x) = \int_0^\infty c(q(t))\, dt, \qquad q(0) = x.$$

A policy is called infinite-horizon optimal for the fluid model if it minimizes the total cost $J(x)$ for each initial condition. We let $J^*(x)$ denote the minimal value.

An infinite-horizon optimal policy for the simple re-entrant line is described as follows, where m_x^* is a positive constant determined by the parameters of the network:

(i)   Serve $q_3(t)$ exclusively ($\zeta_3(t) = 1$) whenever $q_2(t) > 0$ and $q_3(t) > 0$;
(ii) Serve $q_3(t)$ exclusively whenever $q_2(t) = 0$, and $q_3(t)/q_1(t) > m_x^*$;
(iii) Give $q_1(t)$ partial service with $\zeta_1(t) = \mu_2/\mu_1$ whenever $q_2(t) = 0$, and

$$q_3(t) < m_x^* q_1(t).$$

For example, taking the service rates and arrival rates as follows:

$$\mu_1 = \mu_3 = 22, \ \mu_2 = 10, \ \text{and} \ \alpha^{\mathsf{T}} = (9, 0, 0), \tag{2.26}$$

the constant m_x^* is equal to one, and hence the optimal policy is of the form illustrated in Fig. 2.10.

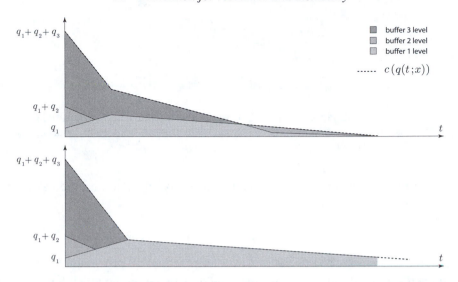

Figure 2.11. Trajectory of buffer levels and evolution of the cost $c(q(t)) = |q(t)|$ for the simple re-entrant line. The first figure illustrates the optimal policy. The second shows the myopic policy, which is precisely the last-buffer first served priority policy. The optimal policy holds off on draining the last buffer to avoid starvation at the bottleneck.

Shown in Fig. 2.11 are buffer levels as a function of time for this network under the optimal policy, and under the myopic policy.

Consideration of a stochastic model brings further subtleties. In particular, it is difficult to ensure that Station 2 does not idle while providing partial service to buffer 3. These issues are addressed in Chapters 4, 8, 9, and 10, among others.

2.9 Contention for resources and instability

The model shown in Fig. 2.12 is a multiclass network with multiple arrivals. It is similar to the simple re-entrant line considered previously, and we can once again construct effective policies for a fluid model with little effort. However, for certain values of the service rates, a nonidling policy similar to LBFS is *destabilizing* for this model even though there exist many stabilizing policies.

Assume that the service rates at the four buffers satisfy

$$\mu_1 > \mu_2 \text{ and } \mu_3 > \mu_4. \tag{2.27}$$

The myopic policy is again defined as in (2.23). If the cost function is defined as the total inventory, $c(x) = |x| = \sum x_i$, then at Station 1 the myopic policy will set $\zeta_4 = 1$ provided $q_4(t; x) > 0$, and $q_2(t; x) > 0$. As soon as $q_2(t; x) = 0$, the draining rate of inventory from the system is compromised unless the second buffer is fed via buffer one. If $q_1(t; x) > 0$ and $q_2(t; x) = 0$, then the myopic policy sets $\zeta_1 = \mu_2/\mu_1$, so that the second buffer can continue to drain at maximum rate. The myopic policy is stabilizing whenever a stabilizing policy exists.

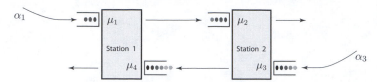

Figure 2.12. Kumar–Seidman–Rybko–Stolyar (KSRS) network. This is a multiclass network multiple arrivals. At each station the scheduling problem amounts to determine processing rates of the two materials waiting in queue.

Consider now the adversarial case: the two stations schedule in a mutually myopic fashion. Station 1 schedules to minimize

$$\frac{d^+}{dt}\, q_4(t; x),$$

and Station 2 chooses the analogous rule to determine its own processing rates. Assuming that each station is nonidling, this gives the priority policy in which buffers 2 and 4 have strict priority at their respective stations,

$$\zeta_4(t; x) = 1 \text{ when } q_4(t; x) > 0; \qquad \zeta_2(t; x) = 1 \text{ when } q_2(t; x) > 0. \qquad (2.28)$$

How does this policy perform?

From the initial condition $x = (1, 0, 0, 0)^{\mathsf{T}}$ the state trajectory can be computed for small t,

$$q(t; x) = x + t(\alpha_1 - \mu_1, \mu_1 - \mu_2, \alpha_3, 0)^{\mathsf{T}}.$$

At time $T_1 = (\mu_1 - \alpha_1)^{-1}$ the first buffer empties, and we then have, for $t > T_1$, $t \approx T_1$,

$$q(t; x) = q(T_1; x) + t(0, \alpha_1 - \mu_2, \alpha_3, 0)^{\mathsf{T}}.$$

At time $T_2 = T_1 + q(T_1; x)/(\mu_2 - \alpha_1)$ the second buffer will drain, and all of the work will be at buffer 3.

The main point is, during the entire time interval $[0, T_2]$ the exit buffer at Station 1 is starved of work. Starting from time T_2, an analogous situation arises, where now the exit buffer at Station 2 is temporarily starved. We can conclude that either $\zeta_4(t) = 0$ or $\zeta_2(t) = 0$ for all $t \geq 0$. This implies that

$$\zeta_2 + \zeta_4 \leq 1, \qquad (2.29)$$

so that buffers 2 and 4 behave as if they are located at a single station. The inequality (2.29) resulting from this policy is known as the *virtual station* constraint. The "virtual load" is defined by

$$\rho_v = \frac{\alpha_1}{\mu_2} + \frac{\alpha_3}{\mu_4}. \qquad (2.30)$$

The virtual load has nothing in common with the usual notion of network load (see e.g. Section 2.8.) The definition of ρ_v is dependent upon the specific policy under consideration.

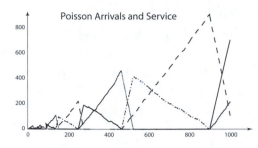
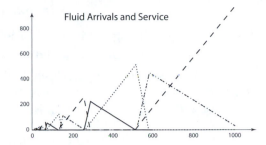

Figure 2.13. Sample paths of the KSRS model. Each plot shows a simulation of the four buffer levels in the KSRS model for identical initial conditions. Shown on the left are results in which the service times have exponential distributions, and the two arrival processes are Poisson processes. In the figure shown on the right the fluid model was simulated in which there is no variability.

To understand the dynamics of q we consider an associated *virtual workload* process,

$$w_v(t) = \frac{q_1(t) + q_2(t)}{\mu_2} + \frac{q_3(t) + q_4(t)}{\mu_4}, \qquad t \geq 0. \tag{2.31}$$

We can compute

$$\frac{d^+}{dt} w_v(t) = \frac{\alpha_1 - \mu_2 \zeta_2}{\mu_2} + \frac{\alpha_3 - \mu_4 \zeta_4}{\mu_4} = \frac{\alpha_1}{\mu_2} + \frac{\alpha_3}{\mu_4} - (\zeta_2 + \zeta_4).$$

If $\rho_v > 1$ then $\frac{d^+}{dt} w_v(t) \geq -(1 - \rho_v) > 0$ for all t, so that $|q(t; x)| \to \infty$ as $t \to \infty$.

A specific example is given by

$$\mu_1 = \mu_3 = 10; \qquad \mu_2 = \mu_4 = 3; \qquad \alpha_1 = \alpha_3 = 2,$$

in which the virtual load is given by $\rho_v = 2(1/3 + 1/3) > 1$. We conclude that, for these parameters, the controlled system is unstable in the strongest possible sense. The network load, defined in Chapter 4, is given by $\rho_\bullet = 2(1/3 + 1/10) < 1$, and we will see that this implies that there are many stabilizing policies.

Figure 2.13 shows two simulation of the KSRS model based on these parameters with common initial condition. In the simulation shown on the left the service times have exponential distributions, and the two arrival processes are Poisson processes. In the figure shown on the right the fluid model was simulated in which there is no variability. In each case the total inventory diverges to infinity linearly with time.

This example also illustrates the potentially disastrous impact of boundary constraints. Suppose that a CRW model of this network is constructed, and the allocation sequence U is defined using a myopic policy as in the Klimov model (2.11), with $c(x) = |x|$. This is precisely the adversarial priority policy given in (2.28). The myopic policies for the fluid and stochastic models differ because a server cannot work on an empty buffer in a discrete state-space model.

An important lesion from this section is that a myopic policy can be destabilizing in the CRW model.

2.10 Routing model

The model shown in Fig. 2.14 represents a network with controlled routing. There are three buffers – one at the router and two at the down-stream nodes. However, it is assumed that the service rate μ_r at the router is relatively fast so that we may set this buffer level equal to zero in considering the model's dynamic behavior.

The two-dimensional fluid model is described by the pair of equations

$$\tfrac{d^+}{dt} q_i(t) = -\mu_i \zeta_i(t) + \mu_r \zeta_i^r(t), \qquad t \geq 0,\ i = 1, 2.$$

Under the assumption that the buffer $q_r(t)$ at the router remains empty, we must have $\tfrac{d}{dt} q_r(t) = 0$ for each t, and hence $\zeta_1^r(t) + \zeta_2^r(t) \equiv \alpha_r / \mu_r$.

Consider first how a time-optimal policy is constructed. First, observe that on maintaining the assumption that the buffer at the router remains empty, we have

$$\tfrac{d^+}{dt} |q(t)| = -\mu_1 \zeta_1(t) - \mu_2 \zeta_2(t) + \mu_r (\zeta_1^r(t) + \zeta_2^r(t))$$
$$= -\mu_1 \zeta_1(t) - \mu_2 \zeta_2(t) + \alpha_r, \qquad t \geq 0.$$

Suppose that $\alpha_r > \min_i \mu_i$. Under this assumption the router can maintain nonzero inventory at each downstream buffer up until $|q(t)| = 0$, so that $\tfrac{d}{dt} |q(t)| = -\mu_1 - \mu_2 + \alpha_r$ when $q(t) \neq 0$ and the two queues are operating under the nonidling policy. On integrating this identity from 0 to T we conclude that, whenever $q(t) \neq 0$ on $[0, T]$,

$$0 < |q(T)| = |q(0)| + \int_0^T \tfrac{d}{dt} |q(t)|\, dt = x - (\mu_1 + \mu_2 - \alpha_r) T.$$

For a given initial condition $q(0) = x \in \mathbb{R}_+^2$, it follows that the minimal draining time is given by

$$T^*(x) = \frac{x}{\mu_1 + \mu_2 - \alpha_r}, \qquad x \in \mathbb{R}_+^2. \tag{2.32}$$

Moreover, a policy for this network is time optimal if and only if $\zeta_1(t) = \zeta_2(t) = 1$ for all $t \in [0, T^*(x))$.

Consider now a myopic policy based on a linear cost function $c(x) = c^\mathsf{T} x$ with $c \in \mathbb{R}_+^2$. The rate of decrease of cost is expressed

$$\tfrac{d^+}{dt} c(q(t)) = [-\mu_1 \zeta_1(t) + \mu_r \zeta_1^r(t)] c_1 + [-\mu_2 \zeta_2(t) + \mu_r \zeta_2^r(t)] c_2, \qquad t \geq 0. \tag{2.33}$$

Based on this expression we conclude that the myopic policy is nonidling at the two buffers, so that $\zeta_i = 1$ whenever $q_i > 0$.

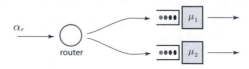

Figure 2.14. Simple routing model.

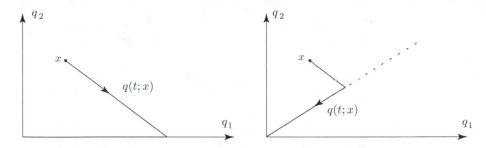

Figure 2.15. Optimization of the simple routing model. For a given linear cost function on the two down-stream buffer levels, both trajectories are pathwise optimal when $c_1 = c_2$. When $c_1 < c_2$ then the trajectory shown on the left is the unique pathwise optimal solution.

The allocation rates at the router depend upon the specific values of the cost parameters. We consider two special cases:

CASE 1 If $c_1 < c_2$ then the myopic policy that minimizes (2.33) over $\zeta \in \mathsf{U}$ sends all arrivals to buffer 1, up until the first time that $q_2(t) = 0$. From this time up until the emptying time for the network, the policy maintains $q_2(t) = 0$, but sends material to this buffer at rate μ_2, so that $\zeta_1 + \zeta_2 = 1$ whenever $q(t) \neq 0$.

CASE 2 When the cost parameters are equal then (2.33) may be simplified

$$\tfrac{d^+}{dt} c(q(t)) = [-\mu_1 \zeta_1(t) - \mu_2 \zeta_2(t) + \alpha_r] c_1, \qquad t \geq 0.$$

In this case the router has extra flexibility: A myopic policy is any rule that dictates $\zeta_1(t) = \zeta_2(t) = 1$ whenever $q(t) \neq 0$.

In each case we conclude that the myopic policy is path-wise optimal, and hence also time optimal. A typical optimal state trajectory in Case 1 is illustrated on the left in Fig. 2.15.

The priority policy described in Case 1 is also optimal in Case 2. However, another policy is perhaps more easily justified, or more "fair." Suppose that each increment of fluid that enters the network attempts to minimize its own delay, where the delay through the ith downstream queue is equal to

$$D_i = \frac{q_i}{\mu_i}, \qquad i = 1, 2.$$

Fluid is routed to buffer 1 whenever $D_1 < D_2$, and to buffer 2 when $D_1 > D_2$. Hence, whenever $q \neq 0$,

$$\zeta^{\mathsf{T}} = \begin{cases} (1, 1, 1, 0) & \mu_1^{-1} q_1 < \mu_2^{-1} q_2 \\ (1, 1, 0, 1) & \mu_1^{-1} q_1 > \mu_2^{-1} q_2 \\ \left(1, 1, \frac{\mu_1}{\mu_1 + \mu_2} \frac{\alpha_r}{\mu_r}, \frac{\mu_2}{\mu_1 + \mu_2} \frac{\alpha_r}{\mu_r}\right) & \mu_1^{-1} q_1 = \mu_2^{-1} q_2. \end{cases}$$

This is known as the *shortest expected delay* policy.

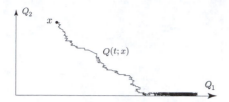 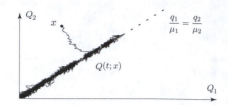

Figure 2.16. A naive translation of the myopic policy to the stochastic model may not be stabilizing. In the simulation shown on the left, the queue length process Q explodes along the Q_1-axis.

A state trajectory resulting from this policy is shown on the right in Fig. 2.15. This policy appears more reasonable from a "fairness" point of view. However, other considerations might lead to a different switching curve in $\mathbb{R}_+^2$.

Consider now the following two-dimensional CRW network model,

$$Q(t+1) = Q(t) - S_1(t+1)U_1(t)\mathbf{1}^1 - S_2(t+1)U_1(t)\mathbf{1}^2$$
$$+ A(t+1)U_1^r(t)\mathbf{1}^1 + A(t+1)U_2^r(t)\mathbf{1}^2, \qquad t \geq 0, \tag{2.34}$$

where $\mathbf{1}^j$ denotes the jth basis element in $\mathbb{R}^2$ (e.g., $\mathbf{1}^1 = [1,0]^\mathsf{T}$). It is assumed that A and S are independent processes. The marginal distribution of the arrival process A has finite support on $\mathbb{Z}_+$, and the marginal distribution of the service process S_i is Bernoulli for each $i = 1, 2$.

Consider Case 1 in which $c_1 < c_2$. In the myopic policy constructed for the fluid model, the router sends all customers to buffer 1 whenever buffer 2 is nonempty. A direct translation of this policy is expressed

$$U_1^r(t) = 1 \text{ whenever } Q_2(t) \geq 1. \text{ When } Q_2(t) = 0, \text{ then } U_2^r(t) = 1. \tag{2.35}$$

How does this perform in the face of variability?

Consider the model obtained via uniformization of a network with Poisson arrivals and exponential service. The resulting CRW model is of the form (2.34) with

$$\mathsf{P}\{(S_1(t), S_2(t), A_3(t))^\mathsf{T} = \mathbf{1}^i\} = \mu_i, \ i = 1, 2, \quad \text{and} \quad \alpha_r \text{ for } i = 3. \tag{2.36}$$

As illustrated in the simulation on the left in Fig. 2.16, the policy (2.35) does not perform well at all. The problem is, when $Q_2(t) = 0$ there is a delay before new work can be routed to the second queue. This *starvation* of a critical resource results in instability in this example, just as in the KSRS network model.

The policy considered in Case 2 above is defined by the interior switching curve

$$U_1^r(t) = 1 \text{ whenever } Q_2(t)/\mu_2 \geq Q_1(t)/\mu_1.$$

As shown on the right in Fig. 2.16, this policy avoids idleness of the two down-stream stations, and the overall network is stable.

2.11 Braess' paradox

The myopic policy is path-wise optimal for the fluid model of the simple routing network if $\alpha_r \geq \min_i \mu_i$. This optimality holds for both the global myopic policy, and the individual-myopic policy in which traffic attempts to minimize its own delay.

We have also seen that these conclusions fail drastically when variability is introduced in the network model. Similar conclusions hold with the introduction of delay.

Consider a fluid model of the form (1.6) for the network shown in Fig. 2.17: The proportion of traffic sent to the upper line is denoted ζ_1^r, and the proportion directed to the lower line is ζ_2^r. Given the current vector of queue lengths $q(t) \in \mathbb{R}_+^2$, the delay along the upper path is $D_1(t) = \frac{q_1(t)}{\mu_1} + d$. The delay along the lower path is

$$D_2(t) = d + \frac{1}{\mu_2} q_2(t+d) = d + \frac{1}{\mu_2}\big[q_2(t) + (z_2^r(t) - z_2^r(t-d))\alpha_r - (z_2(t+d) - z_2(t))\mu_2\big].$$

The more complicated expression is a consequence of the memory resulting from delay.

Note that the value of $q_2(t+d)$ can be computed at time t based on the history of routing decisions. The individual-myopic policy is essentially unchanged:

(i) when $D_1 < D_2$ then $\zeta_1^r = 1$,
(ii) when $D_2 > D_1$ then $\zeta_2^r = 1$, and
(iii) when $D_1 = D_2$ then ζ_1^r, ζ_2^r are both positive, and are computed according to the formula $\frac{d}{dt}D_1 = \frac{d}{dt}D_2$, or

$$\frac{1}{\mu_1}\frac{d}{dt}q_1(t) = \frac{1}{\mu_2}\frac{d}{dt}q_2(t) + \frac{1}{\mu_2}\Big((\zeta_2^r(t) - \zeta_2^r(t-d))\alpha_r - (\zeta_2(t+d) - \zeta_2(t))\mu_2\Big).$$

This gives a nonlinear trajectory as shown on the left in Fig. 2.19. Note that eventually $q(t) = 0$, so that the delay experienced by arriving traffic is exactly d.

Suppose now that we present the traffic with an additional choice. On exiting buffer 1, traffic can proceed to the delay line as before, or it may move to buffer 2. This new model is illustrated in Fig. 2.18. Given this additional choice, if traffic behaves

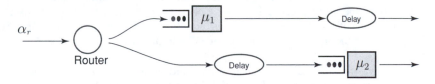

Figure 2.17. A routing model with delay.

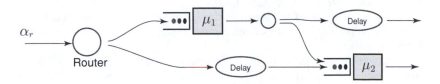

Figure 2.18. Routing model with delay, and an additional choice for one traffic class: more opportunities for optimization.

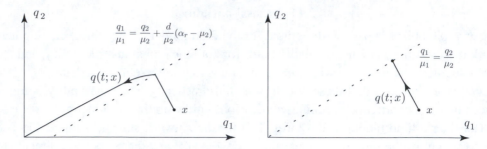

Figure 2.19. Trajectory for the routing model with delay. Steady-state delay is doubled when additional choices are presented.

in an individually-myopic fashion, then on exiting buffer 1 traffic will flow to buffer 2 whenever $\frac{q_2(t)}{\mu_2} < d$ since it will then experience decreased delay.

Consider the case where $\alpha_r > \mu_i$, $i = 1, 2$, but $\alpha_r < \mu_1 + \mu_2$. The latter condition is required to ensure that the model may be stabilized. As shown on the right in Fig. 2.19, this individual-myopic policy will result in an equilibrium (following a transient period) satisfying

$$\frac{q_1}{\mu_1} = \frac{q_2}{\mu_2} = d.$$

Consequently, at equilibrium the traffic will experience a delay of $2d$. This is precisely double the delay found at equilibrium for the model shown in Fig. 2.17 in which communication traffic is presented with fewer choices.

2.12 Notes

It would take a volume as large as this book to give a complete history of all the concepts surveyed in this chapter. More background on optimization, dynamics in networks, and other general concepts introduced here is provided in the Notes sections of subsequent chapters.

The treatment in Section 2.1 barely scratches the surface. Modeling for the single server queue is treated in the encyclopedic book of Cohen [114]. Whitt [497] contains a survey on diffusion approximations for the single server queue, as well as generalizations. See also Glynn [211], and the monographs Kleinrock [303], Asmussen [20], Chen and Yao [96], and Whittle [502].

The c–μ rule was introduced by Smith in [453], and optimality was established for a single station model in the monograph of Cox and Smith [119]. In the half century since its introduction, the c–μ rule has been generalized to many different models and cost citeria (e.g. [18, 305, 306, 230, 337, 86, 138, 92, 436, 477, 488, 298, 264, 96, 323].) Priority policies arise in the theory of *bandit processes* of Nash [380], Gittens [207], and Whittle [499, 502]. See Tsitsiklis [473] for a streamlined treatment of these problems, and more recent extensions of Niño-Mora [387]. Altman [6] contains a survey on applications to queues, and an extensive history on priority policies.

Versions of the processor sharing model have been considered in [235, 239, 35, 324].

More on models for multiple access communication can be found in Shamai and Wyner [439], Gallager [193], and Hanly and Tse [470, 226]. Cover and Thomas is an excellent reference [118], as well as Cover's survey on broadcast channels [117]. For more on resource allocation in wireless models see the Notes section in Chapter 6.

Bertsekas and Gallager's book [43] tells much more about Braess' paradox, and routing in communication networks.

Inventory models based on the simple example contained in Section 7.4 can be found in thousands of articles since the seminal work of Clark and Scarf [18, 422, 110, 423, 203]. The discussion on power transmission networks is adapted from [101, 107].

The network described in Section 2.9 is among several examples introduced in the 1990s to show that a network can be unstable under seemingly innocuous conditions. Instability of the KSRS fluid model under a particular nonidling policy was established by Kumar and Seidman in [318]. The corresponding CRW model under a priority policy was treated in Rybko and Stolyar in [420] (see also [341]). The *virtual station condition* introduced in [130] helped to clarify the dynamics of instability in these examples.

Instability under a global first-in first-out (FIFO) policy was established in particular examples constructed by Bramson and by Seidman [80, 79, 428]. Some positive results are contained in [97].

In certain network/policy combinations it is possible that stability depends upon detailed statistics of the network. This is evident in policies based upon safety stocks where the appropriate safety stock level depends on higher order statistics (beyond the steady-state means of arrival and service processes). More exotic behavior is reported in [186, 127].

3

The Single Server Queue

In this chapter we consider the CRW model for the single server queue introduced in Section 2.1. Recall that this is defined by the recursion

$$Q(t+1) = Q(t) - S(t+1)U(t) + A(t+1), \qquad t \in \mathbb{Z}_+, \qquad (3.1)$$

with given initial condition $Q(0) = x \in \mathbb{Z}_+$.

It is assumed that the joint arrival-service process $\{A(t), S(t) : t \geq 1\}$ is i.i.d. with a finite second moment; the common marginal distribution of $S = \{S(t) : t \geq 1\}$ is Bernoulli; and the marginal distribution of A is supported on $\mathbb{Z}_+$. Unless stated otherwise, the allocation sequence U is defined as the nonidling policy (i.e., $U(t) = 1$ if $Q(t) \geq 1$).

Our first task in this chapter is to obtain a sample path representation of Q using the *Skorokhod map*. This representation has many consequences, such as an explicit expression for the process in steady state, and approximations by simpler continuous time processes, such as the fluid model introduced in Section 2.1.4.

The single server queue is a Markov chain on $\mathbb{Z}_+$ whose transition probability is denoted P, and expectation operator E_x when $Q(0) = x \in \mathbb{Z}_+$. The transition probability has the explicit representation, for any $x, y \in \mathbb{Z}_+$,

$$P(x,y) = \mathsf{P}\{Q(t+1) = y \mid Q(t) = x\} = \mathsf{P}\{[x - S(t+1)]_+ + A(t+1) = y\}. \quad (3.2)$$

We then ask, when is the chain stable? Is the invariant measure π computable? What are the steady-state statistics, such as the mean queue length and average delay?

We find that stability of the stochastic model is equivalent to stability of the deterministic fluid model, expressed under the nonidling policy by

$$q(t;x) = [x - (\mu - \alpha)t]_+, \qquad t \geq 0,$$

where $x \in \mathbb{R}_+$ is given as the initial condition. For either model, stability is characterized in terms of the system *load* $\rho := \alpha/\mu$. If $\rho < 1$ then $\mu - \alpha > 0$, so that $q(t;x) = 0$ for all sufficiently large t. In this case we find that several Lyapunov functions for the fluid model also serve as Lyapunov functions for the stochastic model Q.

We recall here the basic Lyapunov drift condition (V3) for any Markov chain on a denumerable state space X: for a nonnegative-valued function V on X, a finite set

$S \subset X$, $b < \infty$, and a function $f \colon X \to [1, \infty)$,

$$\mathcal{D}V(x) \leq -f(x) + b\mathbf{1}_S(x), \qquad x \in X. \tag{V3}$$

The generator for the Markov chain Q is defined as the "discrete derivative": For each $x \in \mathbb{Z}_+$,

$$\mathcal{D}V(x) := \mathsf{E}[V(Q(t+1)) - V(Q(t)) \mid Q(t) = x] = \sum_{y=0}^{\infty} P(x, y)[V(y) - V(x)]. \tag{3.3}$$

When $f \equiv 1$ then (V3) is known as *Foster's criterion*.

For the fluid model, a natural "generator" is defined for differentiable functions V by

$$\mathcal{D}_0 V(x) := \frac{d^+}{dt} V(q(t; x)) \Big|_{t=0}, \tag{3.4}$$

and the deterministic, continuous time version of (V3) is expressed

$$\mathcal{D}_0 V(x) \leq -f(x) \qquad x \neq 0,$$

where $V \colon \mathbb{R}_+ \to \mathbb{R}_+$ is differentiable, and $f(x) > 0$ for $x > 0$. Two canonical Lyapunov functions for the fluid model are value functions: the *minimal draining time* and the *total cost*,

$$T^*(x) = \min\{t : q(t; x) = 0\} = \frac{x}{\mu - \alpha} \tag{3.5}$$

$$J^*(x) = \int_0^{\infty} c(q(t; x)) \, dt = \tfrac{1}{2} \frac{x^2}{\mu - \alpha}, \qquad x \in \mathbb{R}_+, \tag{3.6}$$

where in the single queue we take $c(x) \equiv x$. The "$*$" indicates that the nonidling policy is optimal for q, in that it minimizes the time to drain, as well as the total cost, over all admissible allocation processes z. Each of these value functions serves as a Lyapunov function for the fluid model, in the sense that a continuous time version of (V3) holds:

$$\mathcal{D}_0 T^*(x) = -\mathbf{1}\{x > 0\}, \quad \text{and} \quad \mathcal{D}_0 J^*(x) = -x, \tag{3.7}$$

assuming of course that $\rho < 1$ so that T^* and J^* are finite valued.

The identities (3.7) are applied in Section 3.3.1, where it is also shown that either of these value functions serves as a Lyapunov function for the CRW model (3.1). Foster's criterion holds with $V = T^*$, and Condition (V3) is satisfied with $f(x) \equiv 1 + \tfrac{1}{2}x$ for the function $V = J^*$. These bounds combined with Theorem A.2.3 imply that Q is ergodic with finite steady-state mean when $\rho < 1$.

The *Poisson equation* for the CRW queue with c the identity function is expressed

$$\mathcal{D}h(x) = -x + \eta, \qquad x \in \mathbb{Z}_+, \tag{3.8}$$

where η denotes the steady-state mean of $Q(t)$. Poisson's equation arises in the ergodic theory of Markovian models, and a generalization is used to define the dynamic programming equations for average-cost optimal control in Section 4.1.3 and Chapter 9.

Theorem 3.0.1 establishes formulae for the steady-state mean as well as the associated solution to Poisson's equation. The formula (3.10) for the steady-state mean may be viewed as an analog of the celebrated *Pollaczek–Khintchine formula* for the M/G/1 queue. The Pollaczek–Khintchine formula expresses the average queue length in terms of the "mean drift" $\mu - \alpha$, and the variance of the joint service/arrival process.

The identity $\mathcal{D}_0 J^*(x) = -x$ can be interpreted as a version of Poisson's equation for the fluid model. The solution to Poisson's equation for the CRW model given in (3.11) is a perturbation of the total cost J^*.

Theorem 3.0.1 (Stability of the single server queue). *Consider the CRW queueing model (3.1) satisfying $\rho = \alpha/\mu < 1$, and define*

$$m^2 = \mathsf{E}[(S(1) - A(1))^2], \quad m_A^2 = \mathsf{E}[A(1)^2], \quad \sigma^2 = \rho m^2 + (1 - \rho)m_A^2. \quad (3.9)$$

Then,

(i) Q *is positive recurrent: there is a unique invariant measure π on $\mathbb{Z}_+$, with steady-state mean*

$$\eta := \mathsf{E}_\pi[Q(0)] = \tfrac{1}{2}\frac{\sigma^2}{\mu - \alpha} \quad (3.10)$$

(ii) *The function $h^*\colon \mathbb{Z}_+ \to \mathbb{R}_+$ defined in (3.11) solves Poisson's equation (3.8),*

$$h^*(x) = J^*(x) + \tfrac{1}{2}\mu^{-1}\left(\frac{m^2 - m_A^2}{\mu - \alpha}\right)x, \qquad x \in \mathbb{Z}_+. \quad (3.11)$$

Proof. The existence of π follows from Proposition 3.4.3. Proposition 3.4.8 establishes the formula for h^* and the expression for η. Proposition A.3.12 implies that h^* is the unique solution to Poisson's equation that is bounded from below with $h^*(0) = 0$.   □

The higher-order statistics are summarized in the *log moment generating function* (log-MGF) defined as the steady-state expectation,

$$\Lambda_Q(\vartheta) := \log\left(\mathsf{E}_\pi\left[e^{\vartheta Q(0)}\right]\right), \quad \vartheta \in \mathbb{R}. \quad (3.12)$$

The log-MGF for Q can be computed in terms of the two log-MGFs,

$$\Lambda(\vartheta) := \log\left(\mathsf{E}\left[e^{\vartheta(A(1) - S(1))}\right]\right) \quad \Lambda_A(\vartheta) := \log\left(\mathsf{E}\left[e^{\vartheta A(1)}\right]\right), \qquad \vartheta \in \mathbb{R}. \quad (3.13)$$

To avoid complications we assume that the arrival process is bounded.

Theorem 3.0.2 (Large deviations in the single server queue). *Consider the CRW queueing model (3.1) satisfying $\rho = \alpha/\mu < 1$. Suppose that A is a bounded sequence, and that $\mathsf{P}\{A(t) > S(t)\} > 0$. Then*

$$\lim_{n \to \infty} n^{-1} \log\left(\mathsf{P}_\pi\{Q(t) \geq n\}\right) = -\vartheta_0, \quad (3.14)$$

where $\vartheta_0 > 0$ is the second zero of Λ,

$$\vartheta_0 := \max\{\vartheta : \Lambda(\vartheta) \leq 0\}. \tag{3.15}$$

Proof. This follows from Proposition 3.4.4 (i) where the log-MGF Λ_Q is computed. It is shown in particular that $\Lambda_Q(\vartheta)$ is finite for $\vartheta < \vartheta_0$, and infinite for $\vartheta > \vartheta_0$, which implies the desired conclusion. □

In Section 3.5 we find that the parameter ϑ_0 also appears in the analysis of the queue length process in the transient phase prior to a "buffer overflow."

This chapter introduces an approach to performance evaluation to be developed over the course of the book. To establish formulae for the mean queue length or the MGF we do not compute π. Instead, to compute the steady-state mean of a function f on X, we search for solutions to invariance equations similar to Poisson's equation,

$$\mathcal{D}h = -f + g, \tag{3.16}$$

where g is a function on X whose mean is known. Subject to a growth condition on the function h, (3.16) implies that $\pi(f) = \pi(g)$. This result is a refinement of the *Comparison Theorem* given in the appendix. This is recalled in Theorem 3.3.2 since it is the basis of the proofs of many of the main results in this chapter.

Whenever possible, results for the CRW model are stated for the RBM model. These results are collected together in subsections, typically located at the end of each section. This advanced material can be skipped without loss of continuity.

We begin with various sample path representations for Q.

3.1 Representations

The recursion (3.1) can be represented in different forms for different purposes. An immediate representation is via the reflected random walk.

3.1.1 Lindley recursion

The single server queue can be reduced to the reflected random walk to obtain the *Lindley recursion*.

Proposition 3.1.1. *The process $X := Q - A$ is a version of the reflected random walk, with initial condition $X(0) = Q(0)$ and increments $\mathcal{E}(t) = A(t-1) - S(t)$ for $t \geq 1$, where $A(0) := 0$.*

Proof. We have from the definitions,

$$X(t+1) := Q(t+1) - A(t+1) = [Q(t) - S(t+1)]_+ = [X(t) + A(t) - S(t+1)]_+. \tag{3.17}$$

Hence $X(t+1) = [X(t) + \mathcal{E}(t+1)]_+$ as claimed. □

One warning is required here: although it is true that X is a reflected random walk, the process $\mathcal{E}$ is not stationary since $\mathcal{E}(1) = -S(1)$, while $\mathcal{E}(i) = A(i-1) - S(i)$ for $i \geq 2$. Moreoever, although $\{\mathcal{E}(t) : t \geq 2\}$ is stationary, it is not i.i.d. unless A and S are independent. The recursion (3.17) remains a useful representation since we can translate standard results for a reflected random walk to the CRW queue.

The reader will likely ask, *why not take the reflected random walk as the model for Q?* The answer is that the recursion (3.1) is most easily generalized to multidimensional networks.

3.1.2 Skorokhod map

Iteration of the projection $[\,\cdot\,]_+$ that defines X in (3.17) leads to a second useful representation for the single server queue.

Consider an arbitrary reflected random walk X with increment process $\mathcal{E}$ and initial condition $X(0) = x$. Let F denote the unreflected *free process*, defined via $F(0) = x$ and

$$F(t) = x + \sum_{i=1}^{t} \mathcal{E}(i), \qquad t \geq 1. \tag{3.18}$$

The *Skorokhod map* is then defined as the mapping

$$[\boldsymbol{F}]_{\mathrm{s}}(t) := \max_{0 \leq i \leq t} \big(\max[F(t), F(t) - F(i)]\big), \qquad t \geq 0. \tag{3.19}$$

Proposition 3.1.2 (i) asserts that this is indeed a representation of X. This will be generalized to continuous time models, and is applied to construct fluid and diffusion approximations for the queue process. For approximation it is useful that the map is also continuous.

Proposition 3.1.2. *The Skorokhod map (3.19) has the following properties:*

(i) Consistency*: The solution X is precisely the reflected random walk, $X(t+1) = [X(t) + \mathcal{E}(t+1)]_+, t \geq 0$, with $X(0) = x$ given.*
(ii) Continuity*: Suppose that F, F' are two different processes giving rise to two reflected processes $X = [F]_{\mathrm{s}}$ and $X' = [F']_{\mathrm{s}}$. We then have, for each $t \geq 1$,*

$$|X(t) - X'(t)| \leq \max_{0 \leq i \leq t} \big(\max\big[|F(t) - F'(t)|, |(F(t) - F'(t)) - (F(i) - F'(i))|\big]\big).$$

(iii) Monotonicity*: Suppose that F, F' are two different processes giving rise to two reflected processes X and X' in which $F'(t_1) - F'(t_0) \geq F(t_1) - F(t_0)$ for all $t_1 \geq t_0 \geq 0$, and $F'(0) \geq F(0)$. Then $X'(t) \geq X(t)$ for all t.*

Proof. We first establish the bound $[F]_{\mathrm{s}}(t) \leq X(t)$ for any t. We begin with two obvious properties,

$$F(t) \leq X(t) \quad \text{and} \quad X(t+1) - X(t) \geq \mathcal{E}(t+1) = F(t+1) - F(t), \qquad t \geq 0.$$

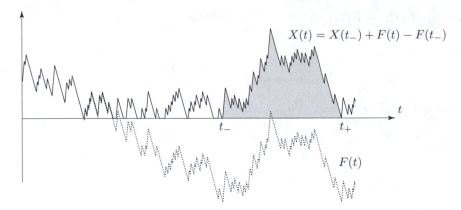

Figure 3.1. Skorokhod representation for the single server queue. $X(t) = F(t) - F(t_-)$ for time points $t = t_-, \ldots, t_+$, where t_- is the largest time not exceeding t satisfying $X(t_-) = 0$, and t_+ is the next time that $X(t_+) = 0$.

The second bound can be iterated to conclude that all increments of $\boldsymbol{F}$ are bounded above by the increments of $\boldsymbol{X}$. Since $\boldsymbol{X}$ is nonnegative-valued, we obtain for each $0 \le i \le t$,

$$X(t) = X(i) + X(t) - X(i) \ge X(i) + F(t) - F(i) \ge F(t) - F(i).$$

Maximizing over i gives $[\boldsymbol{F}]_s(t) \le X(t)$.

The fact that this lower bound is attained is illustrated in Fig. 3.1. Let t_- denote the largest integer $i \le t$ such that $X(i) = 0$. If no such i exists then $X(t) = F(t)$, so that $[\boldsymbol{F}]_s(t) = X(t)$ as claimed. Otherwise, it is clear from the figure that

$$X(t) = X(t_-) + F(t) - F(t_-) = F(t) - F(t_-).$$

This shows that the maximum in (3.19) is indeed attained at t_-.

The continuity result is then immediate from the bound

$$\left| \left(\max_{0 \le i \le k} a_i \right) - \left(\max_{0 \le j \le k} b_j \right) \right| \le \max_{0 \le i \le k} |a_i - b_i|$$

which holds for any collection of real numbers $\{a_i, b_j\}$.

Monotonicity follows directly from the definition. $\square$

3.1.3 Loynes construction

Our first application of the Skorokhod map (3.19) is an explicit representation for the invariant measure of a reflected random walk $\boldsymbol{X}$. The *Loynes construction* of a stationary version of the queue length process is based on initializing the process at a time in the distant past, denoted $-T$, and then letting $T \to \infty$.

Suppose that the i.i.d. process $\boldsymbol{\mathcal{E}}$ is defined on the two-sided time interval $\mathbb{Z}$, and let $X_T(t)$ denote the random walk on the time set $\{-T, -T+1, \ldots\}$, initialized with $X_T(-T) = 0$. The Skorokhod map provides a representation of $X_T(0)$ in terms of the free process $F_T(t) := \sum_{i=-T+1}^{t} \mathcal{E}(i)$,

$$X_T(0) = \max_{-T \le i \le 0} \left(\max[F_T(0), F_T(0) - F_T(i)] \right).$$

On writing $F_T(0) - F_T(i) = \sum_{j=i+1}^{0} \mathcal{E}(j)$ for $-T \leq i \leq -1$ we obtain

$$X_T(0) = \max_{-T \leq i \leq -1} \left[\sum_{j=i+1}^{0} \mathcal{E}(j) \right]_+. \tag{3.20}$$

Proposition 3.1.3. *The random walk satisfies, for any $y \in \mathbb{Z}_+$,*

$$\mathsf{P}\{X(t) \geq y \mid X(0) = 0\} \uparrow \mathsf{P}\{X_\infty^+ \geq y\}, \qquad t \uparrow \infty,$$

where

$$X_\infty^+ := \sup_{0 \leq i < \infty} \left[\sum_{j=0}^{i} \mathcal{E}(j) \right]_+. \tag{3.21}$$

Proof. The proof is based on the representation

$$\mathsf{P}\{X(t) \geq y \mid X(0) = 0\} = \mathsf{P}\{X_T(0) \geq y\}. \tag{3.22}$$

Applying (3.20), we find that $X_T(0)$ is equal in distribution to the random variable

$$X_T^+ := \sup_{0 \leq i < T-1} \left[\sum_{j=0}^{i+1} \mathcal{E}(j) \right]_+.$$

We obviously have $X_T^+ \uparrow X_\infty^+$ as $T \to \infty$ (where the limit may take on infinite values). This combined with (3.22) shows that the desired convergence holds when $X(0) = 0$. $\qquad\qquad\square$

As an immediate consequence of Proposition 3.1.3 (using $Q(t) = X(t) + A(t)$) we obtain an ergodic theorem for the CRW queue: For any $y \in \mathbb{Z}_+$,

$$\lim_{t \to \infty} \mathsf{P}\{Q(t) \geq y \mid Q(0) = 0\} = \mathsf{P}\{X_\infty^+ + A_\infty \geq y\},$$

where A_∞ is independent of X_∞^+, and distributed as a typical arrival increment. When $\rho < 1$ we will find that $X_\infty^+ < \infty$ a.s., and the ergodic theorem can be generalized to arbitrary initial conditions through *coupling* (see also Section A.5).

3.2 Approximations

In this section we provide several applications of the Skorokhod representation for the CRW queue, combined with an extension of the continuity result Proposition 3.1.2, to obtain approximations of the queue by a deterministic fluid model, or the reflected Brownian motion model described informally in the introduction.

The definition of the Skorokhod map is extended as follows: Suppose that $F \colon \mathbb{R}_+ \to \mathbb{R}$ is any function of continuous time, and define the Skorokhod map as in (3.19) by

$$[\boldsymbol{F}]_s(t) := \sup_{0 \leq s \leq t} \left(\max[F(t), F(t) - F(s)] \right), \qquad t \geq 0. \tag{3.23}$$

The proof of the following result is similar to the proof of Proposition 3.1.2.

Proposition 3.2.1. *The Skorokhod map (3.23) is continuous: Suppose that $\mathbf{F}, \mathbf{F}'$ are two different processes giving rise to two reflected processes $\mathbf{X} = [\mathbf{F}]_s$ and $\mathbf{X}' = [\mathbf{F}']_s$. We then have, for each $t \geq 0$,*

$$|X(t) - X'(t)| \leq \sup_{0 \leq s \leq t} \left(\max\left[|F(t) - F'(t)|, |(F(t) - F'(t)) - (F(s) - F'(s))|\right] \right).$$

3.2.1 Fluid models

Recall that the *load* for the queue is defined as $\rho = \alpha/\mu$. The following result is immediate:

Proposition 3.2.2. *The following hold for the single server queue:*

(i) *$\rho < 1$ if and only if the queue is stabilizable. That is, for each $x \geq 0$ there exists z such that $q(t) = 0$ for all t sufficiently large. In this case the minimal draining time $T^*(x)$ is finite for each initial condition, and is given by (3.5).*

(ii) *$\rho \leq 1$ if and only if the origin is a potential equilibrium. That is, z can be chosen so that $q(t) = 0$ for all t when $q(0) = 0$*

   In Section 2.1.4 we described how the fluid model approximates the transient behavior of the queue, viewed from a large initial condition. This observation can be refined through a limiting argument based on the Strong law of large numbers. We extend the definition of the domain of Q to arbitrary initial conditions in $\mathbb{R}_+$ and arbitrary time points $t \in \mathbb{R}_+$ as follows: Let $Q(t; x)$ denote the continuous, piecewise linear process that coincides with the CRW model at the integer time points $t = \{0, 1, 2, \dots\}$, with initial condition $\lfloor x \rfloor$ (the integer part of x). The domain of the random walk X is defined similarly. The free process $F(t; x)$ is the continuous, piecewise linear function with initial condition $F(0; x) = Q(0; x)$ satisfying

$$F(t; x) = \lfloor x \rfloor + \sum_{i=1}^{t} \mathcal{E}(i), \qquad t = 1, 2, \dots. \tag{3.24}$$

Recall that we define $\mathcal{E}(i) = -S(i) + A(i - 1)$ in the CRW model (see Proposition 3.1.1).

   Let $\kappa \geq 1$ denote a scaling parameter and define the scaled processes

$$q^\kappa(t; x) = \kappa^{-1} Q(\kappa t; \kappa x),$$
$$x^\kappa(t; x) = \kappa^{-1} X(\kappa t; \kappa x),$$
$$f^\kappa(t; x) = \kappa^{-1} F(\kappa t; \kappa x) \qquad t \in \mathbb{R}_+, \ x \in \mathbb{R}_+.$$

The process X can be represented using the continuous time version of the Skorokhod map,

$$X(t; x) = [\mathbf{F}]_s(t; x) := \max_{0 \leq s \leq t} \left(\max[F(t; x), F(t; x) - F(s; x)] \right), \qquad t \geq 0.$$

It immediately follows that the scaled process $\boldsymbol{x}^\kappa$ has an analogous representation,

$$x^\kappa(t;x) = [\boldsymbol{f}^\kappa]_s(t;x) := \max_{0 \le s \le t} \left(\max[f^\kappa(t;x), f^\kappa(t;x) - f^\kappa(s;x)] \right), \qquad t \ge 0, \quad (3.25)$$

which is very similar to the representation of $\boldsymbol{q}$,

$$q(t;x) = [\boldsymbol{f}]_s(t;x) := \max_{0 \le s \le t} \left(\max[f(t;x), f(t;x) - f(s;x)] \right), \qquad t \ge 0, \quad (3.26)$$

where $f(t;x) := x - (\mu - \alpha)t$.

These representations combined with continuity of the Skorokhod map lead to the following conclusions.

Proposition 3.2.3. *For each $x \in \mathbb{R}_+$ we have with probability one,*

$$\lim_{\kappa \to \infty} q^\kappa(t;x) = \lim_{\kappa \to \infty} x^\kappa(t;x) = q(t;x),$$

where $\boldsymbol{q}$ is the deterministic, continuous time solution to the fluid model equations (2.5) under the nonidling policy for $\boldsymbol{z}$.

Proof. Applying the Strong law of large numbers theorem A.5.8, we have with probability one,

$$f^\kappa(t;x) \to f(t) := x - (\mu - \alpha)t, \qquad \kappa \to \infty, \ t \ge 0.$$

The Skorokhod map (3.25) is continuous, so we must have $x^\kappa(t;x) \to q(t;x)$. In fact, on applying Proposition 3.2.1 and the triangle inequality we obtain the explicit bound

$$|x^\kappa(t;x) - q(t;x)| \le 2|f^\kappa(t;x) - f(t;x)| + \sup_{0 \le s \le t} |f^\kappa(s;x) - f(s;x)|. \quad (3.27)$$

The fact that $q^\kappa(t;x) \to q(t;x)$ then follows from the representation $\boldsymbol{X} := \boldsymbol{Q} - \boldsymbol{A}$. $\square$

3.2.2 Heavy traffic and Brownian models

In perhaps the most common application of the Skorokhod map, the free process is given directly as Brownian motion on $\mathbb{R}$, with initial condition x. The resulting process $\boldsymbol{X}$ is precisely the *reflected Brownian motion* (RBM) described in Section 1.3.3. Here we provide a formal definition.

Definition 3.2.4 (Reflected Brownian motion). Let $F(t) = x - \delta t + N(t), t \ge 0$, where $x \in \mathbb{R}_+$, $\delta \in \mathbb{R}$, and $\boldsymbol{N}$ is a driftless Brownian motion. The associated *reflected Brownian motion* is defined as $\boldsymbol{X} := [\boldsymbol{F}]_s$, or equivalently

$$X(t) = \max_{0 \le s \le t} \left(\max[x - \delta t + N(t), N(t) - N(s) - (t-s)\delta] \right), \qquad t \ge 0. \quad (3.28)$$

$\blacksquare$

In Proposition 3.2.3 we saw that the fluid model can be obtained through a "Law of large numbers scaling" by appealing to continuity of the Skorokhod map. Alternatively, the free process (3.24) for the CRW model can be scaled to obtain a Brownian motion, and the resulting queue length process converges to an RBM.

This limiting argument is most natural in *heavy traffic*, which means that $\rho < 1$ is very nearly unity. The approximation is illustrated in Fig. 1.8 in the stable regime with $\rho = 0.9$, where the sample path behavior does look similar to that of the RBM (3.28).

To illustrate the construction of a limiting RBM model consider first a queue satisfying $\mathsf{E}[A(t)] = \mathsf{E}[S(t)] = \frac{1}{2}$ so that $\rho = 1$. As always, we assume that $A(t)$ has finite variance. Given a constant $\delta > 0$ we define a "thinning" of $\boldsymbol{A}$ as follows: At each time t a weighted coin is flipped, independent of the past, with a probability of "heads" given by $1 - 2\kappa^{-1}\delta$. With the occurrence of a heads we take $A^\kappa(t) = A(t)$, and otherwise $A^\kappa(t) = 0$. That is, for each t and each $\kappa \geq 2\delta$,

$$A^\kappa(t) = \begin{cases} A(t) & \text{with prob. } 1 - 2\kappa^{-1}\delta \\ 0 & \text{with prob. } 2\kappa^{-1}\delta. \end{cases}$$

We then have $\alpha^\kappa := \mathsf{E}[A^\kappa(t)] = (1 - 2\kappa^{-1}\delta)\mathsf{E}[A(t)] = \frac{1}{2} - \kappa^{-1}\delta$, and $\mu - \alpha = \kappa^{-1}\delta > 0$.

The scaled queue-length process and free process are then defined by

$$Q^\kappa(t; x) = \kappa^{-1} Q^\kappa(\kappa^2 t; \kappa x),$$

$$F^\kappa(t; x) = \kappa^{-1}\left(\lfloor \kappa x \rfloor + \sum_{i=1}^{\kappa^2 t} \mathcal{E}^\kappa(i)\right), \qquad t = \kappa^{-2}, 2\kappa^{-2}, \ldots, \tag{3.29}$$

where $\mathcal{E}^\kappa(i) = A^\kappa(i-1) - S(i)$. For arbitrary $t \in \mathbb{R}_+$, each of these processes is defined by linear interpolation to form a continuous function.

The process $\boldsymbol{F}^\kappa$ converges to a Brownian motion with drift $-\delta$. To see this, decompose the free process into three parts: $F^\kappa(t; x) = x^\kappa - \delta^\kappa(t) + N^\kappa(t)$, where $x^\kappa = \kappa^{-1}\lfloor \kappa x \rfloor$,

$$\delta^\kappa(t) = -\kappa^{-1}\sum_{i=1}^{\kappa^2 t}(A^\kappa(i-1) - A(i-1)), \quad \text{and} \quad N^\kappa(t) = \kappa^{-1}\sum_{i=1}^{\kappa^2 t}(A(i-1) - S(i)).$$

We obviously have $x^\kappa \to x$ as $\kappa \to \infty$. The Central Limit Theorem asserts that the distribution of $N^\kappa(t)$ is approximately Gaussian for large κ. An application of the functional form Theorem 1.3.10 gives

$$\boldsymbol{N}^\kappa \to \boldsymbol{N}, \qquad \kappa \to \infty,$$

where $\boldsymbol{N}$ is a driftless Brownian motion with instantaneous covariance $\sigma^2 = \mathsf{E}[(A(t) - S(t))^2]$, and the convergence is in distribution.

By construction we have $\mathsf{E}[\delta^\kappa(t)] = \delta t$ when $\kappa^2 t$ is an integer, and

$$\mathsf{Var}(\delta^\kappa(t)) = \kappa^{-2}\sum_{i=1}^{\kappa^2 t}\mathsf{Var}[A^\kappa(i-1) - A(i-1)] \leq t(2\delta\kappa^{-1})\mathsf{E}[A(1)^2].$$

The last bound uses $\mathsf{Var}[A^\kappa(i) - A(i)] \leq \mathsf{E}[(A^\kappa(i) - A(i))^2] = \mathsf{P}\{A^\kappa(i) \neq A(i)\} \times \mathsf{E}[A(i)^2]$, $i \geq 1$. Hence $\delta^\kappa(t) \to \delta t$ in the L_2 sense since $\mathsf{Var}(\delta^\kappa(t)) \to 0$ as $\kappa \to \infty$.

We conclude that $F^\kappa(t;x) \to x - \delta t + N(t)$ in distribution. To apply the same technique used in the proof of the simpler fluid limit Proposition 3.2.3 requires some magic. A technique known as the *Skorokhod embedding* permits the construction of an entirely new probability space for which the convergence of $F^\kappa(t;x)$ to $x - \delta t + N(t)$ is with probability one. This can be deduced from Billingsley [61, Theorem 37.7]. In this way we obtain,

Proposition 3.2.5. *For each $x \in \mathbb{R}_+$, the scaled process $Q^\kappa(\cdot\,;x)$ converges in distribution to the RBM on $\mathbb{R}_+$ with drift δ and instantaneous covariance $\sigma^2 = \mathsf{E}[(A(1) - S(1))^2]$.*

3.3 Stability

The Loynes construction summarized in Proposition 3.1.3 implies a general ergodic theorem for the CRW queue. Though stated for i.i.d. service and arrivals, it is easily extended to stationary, possibly dependent (S, A). Of course, to obtain a meaningful limit one requires $X_\infty^+ < \infty$ a.s., which requires $\rho < 1$.

3.3.1 Fluid and stochastic Lyapunov functions

When $\rho < 1$ the queue length process is stable, in the sense that it admits a stationary version, with marginal distribution π. It is also stable in the sense that the process drifts downward whenever it is large. To quantify this observation we construct a Lyapunov function $V \colon \mathsf{X} \to \mathbb{R}_+$ satisfying (V3).

We first revisit the generator for the fluid model (3.4) applied to the value functions T^* and J^*. The following identity follows from the definition of T^*, and the fact that $\frac{d}{dt}q(t;x) = -(\mu - \alpha)$ when $q(t;x) > 0$:

$$\frac{d}{dt}T^*(q(t;x))\Big|_{t=0} = -1, \qquad x > 0. \tag{3.30}$$

This can also be interpreted as an instance of the *fundamental theorem of calculus*. We have

$$T^*(q(t;x)) = \int_t^\infty \mathbf{1}\{q(s;x) > 0\}\,ds,$$

so that $\frac{d}{dt}T^*(q(t;x)) = -\mathbf{1}\{q(t;x) > 0\}$. Similarly, for the total cost

$$J^*(q(t;x)) = \int_t^\infty q(s;x)\,ds, \qquad x \in \mathbb{R}_+,\ t \geq 0,$$

so that by the fundamental theorem of calculus,

$$\frac{d}{dt}J^*(q(t;x)) = -q(t;x), \qquad x \in \mathbb{R}_+,\ t \geq 0. \tag{3.31}$$

The "drift conditions" (3.30) and (3.31) for the fluid model suggest that we apply the generator for Q to these value functions. We have

$$\mathsf{E}[Q(t+1) \mid Q(t) = x] = x - (\mu - \alpha), \qquad x \neq 0, \tag{3.32}$$

and hence the function T^* satisfies (V3) in the form of Foster's criterion with $S = \{0\}$:

$$\mathcal{D}T^*(x) := \mathsf{E}[T^*(Q(t+1)) - T^*(Q(t)) \mid Q(t) = x] = -1, \qquad x \neq 0. \qquad (3.33)$$

That is, $T^*(Q(t))$ decreases on average by 1 at each time increment, up until the first time that the queue empties. This behavior for $\boldsymbol{Q}$ is entirely analogous to the behavior of the fluid model, as captured by the identity (3.30).

We obtain similar conclusions for the total cost J^*. First observe that for nonzero x,

$$\mathsf{E}[(Q(t+1))^2 \mid Q(t) = x] = \mathsf{E}[(x - S(t+1) + A(t+1))^2]$$

$$= x^2 - 2(\mu - \alpha)x + \mathsf{E}[(-S(t+1) + A(t+1))^2]. \qquad (3.34)$$

We thus obtain a version of (V3),

$$\mathcal{D}J^*(x) = -x + \tfrac{1}{2}\frac{m^2}{\mu - \alpha}, \qquad x \neq 0, \qquad (3.35)$$

where $m^2 = \mathsf{E}[(-S(t) + A(t))^2]$.

Equation (3.35) forms the first step in constructing a solution to Poisson's equation in Section 3.4. We first consider the consequences of the bound obtained for T^*.

3.3.2 Consequences of Foster's criterion

Based on (3.33), Proposition 3.3.1 establishes solidarity between the mean hitting time to the origin for the CRW model,

$$\tau_0 := \min\{t \geq 1 : Q(t) = 0\},$$

and the minimal draining time for the fluid model.

The *coefficient of variation* of a random variable X is defined to be the ratio of the variance and the first moment squared: $\mathsf{CV}(X) := \mathsf{Var}[X]/(\mathsf{E}[X])^2$. If the coefficient of variation is small, then the variability of X is small compared to its mean, and we might say that X is "nearly deterministic." For the random time τ_0, we obtain from Proposition 3.3.1 the following expression for the coefficient of variation:

$$\mathsf{CV}_x(\tau_0) := \frac{\mathsf{Var}_x[\tau_0]}{(\mathsf{E}_x[\tau_0])^2} = \frac{\sigma_1^2}{\mu - \alpha}\frac{1}{x}, \qquad x \geq 1,$$

where $\sigma_1^2 = \mathsf{Var}[S(1) - A(1)]$. It is evident that the coefficient of variation $\mathsf{CV}_x(\tau_0)$ is small, and hence the approximation $\tau_0 \approx T^*(x)$ is relatively accurate, provided x is much larger than $\sigma_1^2(\mu - \alpha)^{-1}$. The Pollaczek–Khintchine formula (3.10) implies that $\sigma_1^2(\mu - \alpha)^{-1}$ is approximately twice the steady-state mean of $Q(t)$ when $\rho \approx 1$.

The ideas used in the proof of Proposition 3.3.1 are applied elsewhere in the book to obtain performance bounds for network models. In particular, this is the focus of Section 8.2.

Proposition 3.3.1. *Consider the CRW queueing model (3.1) satisfying $\rho = \alpha/\mu < 1$. The following hold for any nonzero initial condition $x \in \mathbb{Z}_+$, with τ_0 equal to the first hitting time to the origin,*

$$\mathsf{E}_x[\tau_0] = T^*(x), \qquad \mathsf{Var}_x[\tau_0] = \frac{\sigma_1^2}{(\mu - \alpha)^2} T^*(x), \qquad x \geq 1,$$

where $\sigma_1^2 = \mathsf{Var}[A(1) - S(1)]$.

The proof is based on the Comparison Theorem from the appendix which we specialize here for the single server queue:

Theorem 3.3.2 (Comparison theorem for the single server queue). *Consider the single server queue with generator defined in (3.3). Suppose that nonnegative functions V, f, g on $\mathbb{Z}_+$ satisfy the bound*

$$\mathcal{D}V \leq -f + g. \qquad x \in \mathbb{Z}_+. \tag{3.36}$$

Then for each $x \in \mathbb{Z}_+$ and any stopping time τ we have

$$\mathsf{E}_x\left[V(Q(\tau)) + \sum_{t=0}^{\tau-1} f(Q(t))\right] \leq V(x) + \mathsf{E}_x\left[\sum_{t=0}^{\tau-1} g(Q(t))\right].$$

Proof of *Proposition 3.3.1.* The identity (3.33) can be equivalently expressed,

$$\mathsf{E}[T^*(Q(t+1)) \mid \mathcal{F}_t] = T^*(Q(t)) - 1 \quad \text{if } \tau_0 > t, \tag{3.37}$$

where the history (filtration) is defined by $\mathcal{F}_t := \sigma\{Q(0), \ldots, Q(t)\}$. This implies that $M(t) := t \wedge \tau_0 + T^*(Q(t \wedge \tau_0))$, $t \geq 0$, is a martingale: If $Q(0) \geq 1$, then

$$\mathsf{E}[M(t+1) \mid \mathcal{F}_t] = M(t) \qquad t \geq 0.$$

If the martingale M is uniformly integrable, this implies the formula for the mean hitting time to the origin via

$$\mathsf{E}[M(\tau_0)] = M(0) = T^*(x),$$

combined with the identity $M(\tau_0) = \tau_0$.

To see that M is uniformly integrable we obtain a uniform bound as follows:

$$|M(t) - M(0)| = \left|\sum_{i=0}^{t-1} (M(i+1) - M(i))\right| \leq \tau_0 + \frac{1}{\mu - \alpha} \sum_{i=0}^{\tau_0 - 1} |Q(i+1) - Q(i)|.$$

$$\tag{3.38}$$

We have the bound $\mathsf{E}_x[\tau_0] \leq T^*(x) < \infty$ from (3.37) and the Comparison Theorem 3.3.2. Moreover,

$$\mathsf{E}_x\left[\sum_{i=0}^{\tau_0-1} |Q(i+1) - Q(i)|\right] = \sum_{i=0}^{\infty} \mathsf{E}_x\left[\mathbf{1}\{\tau_0 > i\}|A(i+1) - S(i+1)|\right]$$

$$= \sum_{i=0}^{\infty} \mathsf{E}_x\left[\mathbf{1}\{\tau_0 > i\}\right] \mathsf{E}\left[|A(i+1) - S(i+1)|\right],$$

where the last equality follows from the fact that (A, S) is i.i.d., and $\{\tau_0 > i\}$ is $\mathcal{F}_i$ measurable. An equivalent expression follows:

$$\mathsf{E}_x\left[\sum_{i=0}^{\tau_0-1} |Q(i+1) - Q(i)|\right] = \mathsf{E}\big[|A(1) - S(1)|\big]\mathsf{E}_x[\tau_0],$$

and the right-hand side is finite by Foster's criterion and Theorem A.4.1. It follows that $\mathsf{E}_x[\max_{t\geq 0} |M(t)|] < \infty$, which implies uniform integrability.

For the variance representation we first write

$$\mathsf{Var}_x[\tau_0] = \mathsf{Var}_x\big[M(\tau_0)\big] = \mathsf{E}_x[(M(\tau_0) - M(0))^2]$$

Exactly as in (3.38) we have the telescoping series representation

$$M(\tau_0) - M(0) = \sum_{i=0}^{\tau_0-1}(M(i+1) - M(i)) = \sum_{i=0}^{\infty}(M(i+1) - M(i))\mathbf{1}(\tau_0 > i).$$

The random variables $\{(M(i+1) - M(i))\mathbf{1}(\tau_0 > i) : i \geq 0\}$ are uncorrelated, with

$$\mathsf{E}_x\big[((M(i+1) - M(i))\mathbf{1}(\tau_0 > i))^2 \mid \mathcal{F}_i\big]$$
$$= \mathsf{E}_x\big[(-1 + (\mu - \alpha)^{-1}(A(i+1) - S(i+1)))^2 \mid \mathcal{F}_i\big]\mathbf{1}(\tau_0 > i)$$
$$= (\mu - \alpha)^{-2}\mathsf{Var}[A(1) - S(1)]\mathbf{1}(\tau_0 > i), \qquad i \geq 0.$$

Consequently, using $\sigma_1^2 = \mathsf{Var}[A(1) - S(1)]$,

$$\mathsf{Var}_x[\tau_0] = \sum_{i=0}^{\infty}\mathsf{E}_x\Big[((M(i+1) - M(i))\mathbf{1}(\tau_0 > i))^2\Big]$$
$$= (\mu - \alpha)^{-2}\sigma_1^2\mathsf{E}_x\left[\sum_{i=0}^{\tau_0-1}\mathbf{1}(\tau_0 > i)\right]$$
$$= (\mu - \alpha)^{-2}\sigma_1^2\mathsf{E}_x[\tau_0].$$

The formula $\mathsf{E}_x[\tau_0] = T^*(x)$ established in (i) completes the proof of (ii). □

3.3.3 Stability of the RBM model

To extend the Lyapunov function criteria for stability to the RBM model (3.28) we require a generator. This is described in detail in Section 8.7.1 for a more general Brownian network model. The definition of the generator is based on *Itô's formula*.

In the RBM model considered here, Itô's formula provides a representation of $h(X(t))$ for any twice continuously differentiable (C^2) function $h\colon \mathbb{R} \to \mathbb{R}$,

$$h(X(t)) = h(x) + \int_0^t [-\delta h'(X(s)) + \tfrac{1}{2}\sigma^2 h''(X(s))]\,ds$$

$$+ \int_0^t h'(X(s))\,dN(s) + \int_0^t h'(0)\,dI(s). \tag{3.39}$$

Borrowing from McKean [289, 353], this can be interpreted as a second-order Taylor series expansion,

$$\text{``}h(X(t+dt)) = h(X(t)) + h'(X(t))dX(t) + \tfrac{1}{2}h''(X(t))(dX(t))^2\text{'',}$$

with the convention that $(dX(t))^2 = \sigma^2 dt$. We have used the fact that the idleness process can only increase when X is zero, giving

$$\int_0^t h'(X(s))\, dI(s) = \int_0^t h'(0)\, dI(s) = h'(0)I(t).$$

The representation (3.39) leads to the following formulation of the generator: For a C^2 function h,

$$\mathcal{D}h = -\delta h' + \tfrac{1}{2}\sigma^2 h''. \tag{3.40}$$

The point is, Itô's formula implies the representation

$$M(t) := h(X(t)) - h(x) - \int_0^t [\mathcal{D}h\,(X(s))]\, ds - h'(0)I(t) = \int_0^t h'(X(t))\, dN(t).$$

It follows that M is a martingale whenever h' has polynomial growth.

A version of the Comparison Theorem holds for this model: If the bound (3.36) holds, $\mathcal{D}V \le -f + g$ where V, f, g are nonnegative-valued and C^2 (not necessarily with polynomial growth), then for any stopping time τ,

$$\mathsf{E}_x\left[V(X(\tau)) + \int_0^\tau f(X(t))\, dt\right] \le V(x) + \mathsf{E}_x\left[\int_0^\tau g(X(t))\, dt + V'(0)I(\tau)\right].$$

Condition (V3) for the RBM model is expressed exactly as in the CRW model: $\mathcal{D}V \le -f + b\mathbf{1}_S$ where $f \ge 1$, $b < \infty$ is constant, and $S = [0, x_0]$ for some $x_0 > 0$. This is again a necessary and sufficient condition for positive recurrence of X [369, 367]. A natural candidate is the fluid value function

$$J^*(x) := \tfrac{1}{2}\delta^{-1}x^2, \qquad x \ge 0. \tag{3.41}$$

Applying the generator (3.40) we obtain

$$\mathcal{D}J^*(x) = -\delta[\delta^{-1}x] + \tfrac{1}{2}\sigma^2[\delta^{-1}]. \tag{3.42}$$

Hence J^* solves (V3) provided $\delta > 0$.

3.4 Invariance equations

We now turn to properties of the invariant measure π for a stable queue, and return to the Pollaczek–Khintchine formula.

3.4.1 The M/M/1 queue

For the M/M/1 queue the invariant measure is geometric:

Proposition 3.4.1. *When $\rho = \alpha/\mu < 1$, the invariant measure for the M/M/1 queue is given by*

$$\pi(x) = (1 - \rho)\rho^x, \qquad x = 0, 1, 2, \ldots. \tag{3.43}$$

Moreover, the detailed balance equations hold

$$\pi(x)P(x,y) = \pi(y)P(y,x), \qquad x, y = 0, 1, 2, \ldots. \tag{3.44}$$

Proof. We begin with the detailed balance equations: For $x \geq 1$ we have $P(x, x-1) = \mu$, $P(x, x+1) = \alpha$, and hence

$$\pi(x)P(x, x-1) = (1-\rho)\rho^x \mu = (1-\rho)\rho^{x-1}\alpha = \pi(x-1)P(x-1, x),$$

where we have used $\rho := \alpha/\mu$. Similarly, for any $x \in \mathbb{Z}_+$,

$$\pi(x)P(x, x+1) = (1-\rho)\rho^x \alpha = (1-\rho)\rho^{x+1}\mu = \pi(x+1)P(x+1, x).$$

This establishes (3.44).

To see that π is invariant we sum (3.44) over all $x \in \mathbb{Z}_+$:

$$\sum_{x=0}^{\infty} \pi(x)P(x,y) = \sum_{x=0}^{\infty} \pi(y)P(y,x) = \pi(y),$$

where the final equation follows from the fact that P is a transition matrix, satisfying $P(y, \mathbb{Z}_+) = 1$ for $y \in \mathbb{Z}_+$. $\qquad\square$

Since the invariant measure for the M/M/1 queue is geometric, the formula for the steady-state mean queue length can be computed directly:

Proposition 3.4.2. *The steady-state mean and solution to Poisson's equation (3.8) for the M/M/1 queue are, respectively,*

$$\eta := \mathsf{E}_\pi[Q(t)] = \frac{\rho}{1-\rho}, \qquad h^*(x) = \frac{1}{2}\frac{x^2 + x}{\mu - \alpha}, \qquad x \in \mathbb{R}_+. \tag{3.45}$$

Proof. The steady-state mean is by definition $\eta = \sum_{n=0}^{\infty} n\pi(n) = \frac{\rho}{1-\rho}$. The fact that the function h^* given in (3.45) solves Poisson's equation follows from routine calculations that are worked out in the proof of the more general Theorem 3.0.1. $\qquad\square$

To see that Proposition 3.4.2 is consistent with Theorem 3.0.1 observe that the variance constant m^2 given in (3.9) is $m^2 = \mathsf{E}[(S(1)-A(1))^2] = 1$ since $A(1) = 1 - S(1)$ with probability one. Also, since $A(1)$ has a Bernoulli distribution we have $m_A^2 = \mathsf{E}[A(1)^2] = \alpha$, which gives

$$\sigma^2 = \rho m^2 + (1-\rho)m_A^2 = \rho(1-\alpha) + \alpha.$$

Using $1 - \alpha = \mu$ and $\rho = \alpha/\mu$ we obtain $\sigma^2 = 2\alpha$, and Theorem 3.0.1 gives

$$\eta := \mathsf{E}_\pi[Q(t)] = \tfrac{1}{2}\frac{\sigma^2}{\mu - \alpha} = \frac{\alpha}{\mu - \alpha} = \frac{\rho}{1 - \rho},$$

$$h^*(x) = J^*(x) + \tfrac{1}{2}\mu^{-1}\left(\frac{m^2 - m_A^2}{\mu - \alpha}\right)x = \tfrac{1}{2}\frac{x^2 + x}{\mu - \alpha}, \qquad x \in \mathbb{R}_+.$$

3.4.2 Existence of π and its MGF

We have seen in Proposition 3.1.3 the form of an invariant measure, should one exist. Foster's criterion is satisfied in the form (3.33) with $V = T^*$, so that existence follows from Kac's Theorem A.2.2, which states that the invariant measure can be expressed

$$\pi(x) = \left(\mathsf{E}_x[\tau_x]\right)^{-1}, \qquad x \in \mathbb{Z}_+.$$

Proposition 3.4.3. *Consider the CRW queueing model (3.1) satisfying $\rho = \alpha/\mu < 1$. Then, there is a unique invariant measure π on $\mathbb{Z}_+$.*

It is in principle possible to calculate the invariant measure for the CRW model provided $\rho < 1$ and the marginal distribution of A has finite support. However, this calculation is typically complex, so that in practice then we seek partial statistics of π rather than a complete characterization. In particular, to obtain the formula for the steady-state mean described in the Pollaczek–Khintchine formula, we find that it is more convenient to obtain steady-state statistics using the Comparison Theorem.

The most direct application of the Comparison Theorem involves computation of the log-MGF Λ_Q for Q. The formula obtained in Proposition 3.4.4 is based on the two log-MGFs defined in (3.13). Recall that $\vartheta_0 > 0$ is defined in (3.15) as the second zero, $\Lambda(\vartheta_0) = 0$. The MGFs (rather than logarithmic MGFs) are given by $\lambda = e^\Lambda$, $\lambda_Q = e^{\Lambda_Q}$, etc.

Proposition 3.4.4. *Under the assumptions of Theorem 3.0.2 the MGF for Q is given by*

$$\lambda_Q(\vartheta) := \mathsf{E}_\pi[e^{\vartheta Q(0)}] = \begin{cases} (1 - \rho)\dfrac{\lambda_A(\vartheta) - \lambda(\vartheta)}{1 - \lambda(\vartheta)}, & \vartheta < \vartheta_0 \\ \infty, & \vartheta \geq \vartheta_0 \end{cases}$$

We first establish two identities with respect to the generator.

Lemma 3.4.5. *Suppose that the conditions of Theorem 3.0.2 hold. We then have the following identities,*

(i) *On setting $h(x) = e^{\vartheta x}$, $f(x) = (1 - \lambda(\vartheta))h(x)$, and $g(x) = \mathbf{1}\{x = 0\}[\lambda_A(\vartheta) - \lambda(\vartheta)]$ we have $\mathcal{D}h = -f + g$ whenever $\lambda_A(\vartheta) < \infty$.*

(ii) *$\mathcal{D}h^0 = -f^0 + g^0$ with $h^0(x) = \mu^{-1}x$, $f^0(x) = \mathbf{1}(x = 0)$, and $g^0(x) = 1 - \rho$.*

Proof. For any $\vartheta < \vartheta_0$ and $x \in \mathbb{Z}_+$ we have, on setting $u = \mathbf{1}\{x \geq 1\}$,

$$\mathsf{E}[e^{\vartheta Q(t+1)} \mid Q(t) = x] = \mathsf{E}[e^{\vartheta(x - S(1)u + A(1))}]$$
$$= [u\lambda(\vartheta) + (1 - u)\lambda_A(\vartheta)]e^{\vartheta x}.$$

Using the nonidling assumption we have $(1 - u)e^{\vartheta x} = 1 - u$, and thus for any x,

$$\mathsf{E}[e^{\vartheta Q(t+1)} \mid Q(t) = x] = \lambda(\vartheta)e^{\vartheta x} + (1 - u)[\lambda_A(\vartheta) - \lambda(\vartheta)], \qquad (3.46)$$

which gives (i).

Part (ii) is essentially given in (3.32). This can be refined to give $\mathcal{D}h^0 := \mu^{-1}\mathsf{E}[Q(t+1) - Q(t) \mid Q(t) = x] = \mu^{-1}(-\mu\mathbf{1}(x \neq 0) + \alpha)$, which implies (ii). $\qquad\square$

Proof of Proposition 3.4.4. We first explain why the second zero ϑ_0 exists.

The function Λ is convex, finite-valued, and satisfies $\Lambda(0) = 0$. Its derivative at zero coincides with the mean of $A(1) - S(1)$,

$$\Lambda'(0) = \lambda'(0) = \mathsf{E}\big[(A(1) - S(1))e^{\vartheta(A(1) - S(1))}\big]\Big|_{\vartheta = 0} = -(\mu - \alpha),$$

which is negative since $\rho < 1$. Finally, $\Lambda(\vartheta) \to \infty$ as $\vartheta \to \infty$ under the assumption that $\mathsf{P}\{A(1) - S(1) \geq 1\} > 0$ using the bound

$$\Lambda(\vartheta) \geq \log\big(\mathsf{P}\{A(1) - S(1) \geq 1\}e^{\vartheta}\big), \qquad \vartheta \geq 0.$$

It follows that the second zero $\vartheta_0 > 0$ exists as claimed.

The remainder of the proof is based on Lemma 3.4.5. The function f given in Lemma 3.4.5 (i) is nonnegative-valued when $\vartheta < \vartheta_0$. Hence we can apply the Comparison Theorem 3.3.2 to obtain the *bound* $\pi(f) \leq \pi(g)$, and hence $\pi(h) = \mathsf{E}_\pi[e^{\vartheta Q(0)}] < \infty$. This bound justifies the conclusion $\pi(\mathcal{D}h') = \pi(Ph' - h') = 0$ for any function h' that is bounded by a constant times h. Parts (i) and (ii) of Lemma 3.4.5 then give, respectively,

$$0 = \pi(\mathcal{D}h) = \pi(-f + g) = -(1 - \lambda(\vartheta))\mathsf{E}_\pi[e^{\vartheta Q(0)}] + \pi(0)[\lambda_A(\vartheta) - \lambda(\vartheta)]$$
$$0 = \pi(\mathcal{D}h^0) = \pi(-f^0 + g^0) = -\pi(0) + 1 - \rho.$$

Substitution gives $\lambda_Q(\vartheta) = \mathsf{E}_\pi[e^{\vartheta Q(0)}] = (1 - \rho)(1 - \lambda(\vartheta))^{-1}[\lambda_A(\vartheta) - \lambda(\vartheta)]$, which is the desired formula for $\lambda_Q(\vartheta)$ when $\vartheta < \vartheta_0$.

It is clear that $\lambda_Q(\vartheta) \to \infty$ as $\vartheta \to \vartheta_0$, and hence $\lambda_Q(\vartheta) = \infty$ for all $\vartheta \geq \vartheta_0$. $\qquad\square$

3.4.3 Refined comparison theorems

Provided $\Lambda(\vartheta) < \infty$ for some $\vartheta > 0$, the mean queue length η can be computed by differentiating the log-MGF,

$$\eta = \Lambda'_Q(0),$$

with Λ_Q given in Proposition 3.4.4. This does give the formula for the mean queue length (3.10) after lengthy calculations.

The same conclusion can be obtained more efficiently and under milder assumptions using the following refinement of the Comparison Theorem. This result is generalized to network models in Proposition 8.5.4, where a proof is also provided.

Proposition 3.4.6. *Consider the CRW model satisfying $\rho < 1$. Suppose that h, f, g are functions on $\mathbb{Z}_+$ satisfying the identity (3.16). Suppose moreover g is bounded, and that h has at most quadratic growth:*

$$\limsup_{x \to \infty} \frac{|h(x)|}{x^2} < \infty.$$

Then $\pi(f) = \pi(g)$.

A direct application of Proposition 3.4.6 gives an expression for the probability of an empty queue:

Proposition 3.4.7. *If $\rho < 1$ then $\pi(0) = 1 - \rho$.*

Proof. Lemma 3.4.5 (ii) is valid, giving $\mathcal{D}h^0 = -f^0 + g^0$ with $h^0(x) = \mu^{-1}x$, $f^0(x) = \mathbf{1}(x = 0)$, and $g^0(x) = 1 - \rho$, so that $\pi(0) = \pi(f^0) = \pi(g^0) = 1 - \rho$ by Proposition 3.4.6. $\qquad\square$

We now verify that the function h^* given in Theorem 3.0.1 solves the invariance equation (3.8). Proposition 3.4.6 then implies that η given in (3.10) is the steady-state mean queue length.

Proposition 3.4.8. *The function h^* defined in (3.11) solves (3.8) with $\eta = \frac{1}{2}\frac{\sigma^2}{\mu-\alpha}$.*

Proof. The following is a refinement of (3.32) and (3.34): For each $x \in \mathbb{Z}_+$, on setting $u = \mathbf{1}\{x \geq 1\}$,

$$\mathsf{E}[Q(t + 1) \mid Q(t) = x] = x - \mu u + \alpha$$
$$\mathsf{E}[Q(t + 1)^2 \mid Q(t) = x] = x^2 - 2(\mu - \alpha)x + (m^2 - m_A^2)u + m_A^2. \tag{3.47}$$

We have noted that the second identity implies a version of the drift inequality (V3). The function h^* defined in (3.10) is a quadratic function of the form $h^*(x) = ax^2 + bx$, with

$$a = \frac{1}{2}\frac{1}{\mu - \alpha}, \qquad b = \frac{1}{2}\frac{1}{\mu - \alpha}\left(\frac{m^2 - m_A^2}{\mu}\right).$$

Consequently,

$$\mathcal{D}h^*(x) := \mathsf{E}_x[h^*(Q(1)) - h^*(Q(0))]$$
$$= a[-2(\mu u - \alpha)x + um^2 + (1 - u)m_A^2] + b[-\mu u + \alpha] \tag{3.48}$$
$$= -2a(\mu u - \alpha)x + b[-\mu + \alpha] + am^2,$$

based on the fact that the precise values of a and b result in $a[m^2 - m_A^2] = b\mu$.

We have $xu = x$ under the nonidling policy, so that by (3.48)

$$\mathcal{D}h^*(x) = -x + \frac{1}{2}\frac{\mu^{-1}(\alpha - \mu)(m^2 - m_A^2) + m^2}{\mu - \alpha}, \qquad x \in \mathbb{Z}_+,$$

which implies the desired conclusion. $\qquad\qquad\qquad\qquad\qquad\qquad\qquad\qquad$ □

3.4.4 Discounted cost

In many situations it is the transients that are most important, such as the mean clearing time $\mathsf{E}_x[\tau_0]$ computed in Proposition 3.3.1. The most common performance criterion designed to capture transient behavior is the *discounted cost*, defined here as the sum

$$h_\gamma^*(x) := \sum_{t=0}^{\infty}(1 + \gamma)^{-t-1}\mathsf{E}_x[Q(t)], \qquad Q(0) = x \in \mathbb{Z}_+, \qquad (3.49)$$

where $\gamma > 0$ is the discount rate.

The expression (3.49) is not standard in the operations research literature. Usually we are given a *discount factor* $\beta \in (0, 1)$ and define the value function by

$$V_\beta^*(x) := \sum_{t=0}^{\infty}\beta^t\mathsf{E}_x[Q(t)], \qquad Q(0) = x \in \mathbb{Z}_+. \qquad (3.50)$$

Setting $\beta = (1 + \gamma)^{-1}$ we obtain $V_\beta^* = \beta h_\gamma^*$, so the definitions are essentially equivalent. The form (3.49) is preferred in part because it helps bridge continuous time and discrete-time theory.

In particular, Proposition 3.4.9 establishes the dynamic programming equation

$$\mathcal{D}h_\gamma^*(x) = -x + \gamma h_\gamma^*(x), \qquad x \in \mathbb{Z}_+. \qquad (3.51)$$

This can be expressed as $[\gamma I - \mathcal{D}]h_\gamma^* = c$ with $c(x) \equiv x$, or equivalently

$$h_\gamma^* = [\gamma I - \mathcal{D}]^{-1}c. \qquad (3.52)$$

The formula (3.52) will look familiar to readers familiar with continuous time Markov models; the inverse $R_\gamma := [\gamma I - \mathcal{D}]^{-1}$ is called the *resolvent* (see Section A.2).

The dynamic programming equation (3.51) also holds for the fluid model – see (3.53). This is the basis of the explicit expression for h_γ^* obtained in Section 3.4.5.

We first establish the dynamic programming equation and uniqueness of its solution.

Proposition 3.4.9. *The discounted cost satisfies the dynamic programming equation (3.51). Conversely, if h is any function satisfying (3.51) with linear growth,*

$$\limsup_{x \to \infty} \frac{|h(x)|}{x} < \infty,$$

then $h = h_\gamma^*$ everywhere.

Proof. Applying the transition matrix to h_γ^* gives

$$Ph_\gamma^*(x) = \sum_{k=0}^{\infty} (1+\gamma)^{-k-1} P^{k+1} c(x),$$

where $c(x) \equiv x$. Making the change of variables $j = k+1$ gives

$$Ph_\gamma^*(x) = \sum_{j=1}^{\infty} (1+\gamma)^{-j} P^j c(x) = -c(x) + (1+\gamma) \sum_{j=0}^{\infty} (1+\gamma)^{-j-1} P^j c(x).$$

This combined with the definition (3.49) gives $Ph_\gamma^*(x) = -c(x) + (1+\gamma)h_\gamma^*(x)$, which is (3.51).

To see the converse, assume that $\mathcal{D}h = -c + \gamma h$, which can be written $Ph = -c + (1+\gamma)h$, or

$$(1+\gamma)^{-1} Ph = -(1+\gamma)^{-1} c + h.$$

Applying $(1+\gamma)^{-1} P$ to both sides of this displayed equation gives

$$(1+\gamma)^{-2} P^2 h = -(1+\gamma)^{-2} Pc + (1+\gamma)^{-1} Ph$$
$$= -\left[(1+\gamma)^{-1} c + (1+\gamma)^{-2} Pc \right] + h,$$

and by induction

$$(1+\gamma)^{-(n+1)} P^{n+1} h = h - \sum_{k=0}^{n} (1+\gamma)^{-k-1} P^k c.$$

Now we use the assumption that h has linear growth: for some constant $b_0 < \infty$ we have $|h(y)| \leq b_0(1+y)$ for $y \geq 0$, and consequently

$$P^n |h|(x) = \mathsf{E}_x[|h(Q(n))|] \leq b_0 \mathsf{E}\left[1 + x + \sum_{1}^{n} A(i) \right] = b_0(1+x+n\alpha), \qquad x \in \mathsf{X}.$$

It follows that $(1+\gamma)^{-n} P^n h(x) \to 0$ as $n \to \infty$ for each x, and hence from the foregoing,

$$0 = h(x) - \sum_{k=0}^{\infty} (1+\gamma)^{-k-1} P^k c(x) = h(x) - h_\gamma^*(x). \qquad \square$$

3.4.5 Computation of the discounted cost

Motivated by our success in the analysis of Poisson's equation, to compute the discounted cost for the CRW model we begin with the fluid model.

The discounted value function for the fluid model is given by

$$J_\gamma^*(x) = \int_0^\infty e^{-\gamma s} c(q(s; x)) \, ds, \qquad x \in \mathsf{X},$$

where $c(x) \equiv x$. We have for any time $t > 0$, $x \in \mathsf{X}$,

$$J_\gamma^*(q(t;x)) = \int_0^\infty e^{-\gamma s} c(q(s+t;x))\, ds = e^{\gamma t} \int_t^\infty e^{-\gamma s} c(q(s;x))\, ds.$$

We thus obtain the dynamic programming equation

$$\mathcal{D}_0 J_\gamma^* (x) := \frac{d^+}{dt} J_\gamma^*(q(t;x)) \Big|_{t=0} = -x + \gamma J_\gamma^*(x). \tag{3.53}$$

This is very similar to the dynamic programming equation (3.51) for the CRW model.

To solve (3.53) we postpone integration. First consider the simple linear function

$$J_\gamma^0(x) = \gamma^{-1}x - \gamma^{-2}(\mu - \alpha), \qquad x \in \mathbb{R}_+. \tag{3.54}$$

Applying the generator for the fluid model gives

$$\mathcal{D}_0 J_\gamma^0 (x) = -(\mu - \alpha)\left(\frac{d}{dx} J_\gamma^0 (x) \right) = -\gamma^{-1}(\mu - \alpha)$$

Substituting $-\gamma^{-1}(\mu-\alpha) = \gamma J_\gamma^0(x) - x$ we obtain the dynamic programming equation (3.53).

However, there are many other solutions. The discounted value function for the fluid model can be computed by direct calculation, giving

$$J_\gamma^*(x) := \int_0^\infty e^{-\gamma t} q(t;x)\, dt = J_\gamma^0(x) + \gamma^{-2}(\mu - \alpha)e^{-\gamma T^*(x)}, \qquad x \in \mathbb{R}_+.$$

The functions J_γ^0 and J_γ^* solve the same dynamic programming equation (3.53), since the function $g(x) = e^{-\gamma T^*(x)}$ solves $\mathcal{D}_0 g = \gamma g$.

We now turn to the CRW model. To construct a solution to (3.51) we modify the fluid value function J_γ^* by replacing the decaying term $e^{-\gamma T^*(x)}$ with another,

$$h_\gamma^*(x) = J_\gamma^0(x) + b_\gamma e^{-\vartheta_\gamma x}, \qquad x \in \mathbb{Z}_+, \tag{3.55}$$

where $b_\gamma > 0$ is a constant, and ϑ_γ is the unique positive solution to

$$\Lambda(-\vartheta_\gamma) = \log(1 + \gamma). \tag{3.56}$$

Recall that the log-MGFs Λ and Λ_A are defined in (3.13).

Proposition 3.4.10. *The function h_γ^* defined in (3.55) solves the dynamic programming equation (3.51) with*

$$b_\gamma = \mu\gamma^{-1}\big(e^{\Lambda(-\vartheta_\gamma)} - e^{\Lambda_A(-\vartheta_\gamma)}\big)^{-1}.$$

We first consider the geometrically decaying term in (3.55):

Lemma 3.4.11. *The exponential function $g_\gamma(x) = e^{-\vartheta_\gamma x}$ solves the discounted dynamic programming equation*

$$\mathcal{D} g_\gamma (x) = -b_{\gamma 0}\mathbf{1}\{x = 0\} + \gamma g_\gamma(x), \qquad x \in \mathbb{Z}_+,$$

with $b_{\gamma 0} := e^{\Lambda(-\vartheta_\gamma)} - e^{\Lambda_A(-\vartheta_\gamma)}$.

Proof. We have for $x \geq 1$,

$$Pg_\gamma(x) = \mathsf{E}[e^{-\vartheta_\gamma(x-S(1)+A(1))}] = e^{\Lambda(-\vartheta_\gamma)} g_\gamma(x).$$

We also have $e^{\Lambda(-\vartheta_\gamma)} = 1 + \gamma$ by (3.56), giving $\mathcal{D}g_\gamma(x) = \gamma g_\gamma(x)$.
  For $x = 0$,

$$Pg_\gamma(0) = \mathsf{E}[e^{-\vartheta_\gamma A(1)}] = e^{\Lambda_A(-\vartheta_\gamma)},$$

which gives

$$\mathcal{D}g_\gamma(0) = Pg_\gamma(0) - g_\gamma(0) = e^{\Lambda_A(-\vartheta_\gamma)} - 1 = -b_{\gamma 0} + \gamma,$$

where the last equation uses (3.56) and the definition of $b_{\gamma 0}$. The desired result for $x = 0$ follows from $g_\gamma(0) = 1$. $\qquad\square$

Proof of Proposition 3.4.10. If f is any affine function, $f(x) = ax + b$ for constants a, b, then the two generators agree,

$$\mathcal{D}f(x) = \mathcal{D}_0 f(x) = -(\mu - \alpha)a, \qquad x > 0.$$

It thus follows from (3.53) that (3.51) does hold when $x \neq 0$.
  However, when $x = 0$ the identity fails since $\mathcal{D}_0 J_\gamma^0(0) = 0$, yet $\mathcal{D}J_\gamma^0(0) = \gamma^{-1}\alpha$. Hence, from the definition (3.54) of J_γ^0, we have for all x,

$$\mathcal{D}J_\gamma^0(x) = -x + \gamma J_\gamma^0(x) + \gamma^{-1}\mu\mathbf{1}\{x = 0\}.$$

Thanks to Lemma 3.4.11, when computing $\mathcal{D}h_\gamma^*$ the geometrically decaying term in (3.55) will annihilate the indicator function above. $\qquad\square$

3.4.6 Invariance equations for the RBM model

A complete analysis of the RBM model is a bit out of reach. We sketch the main ideas since the conclusions as well as the proofs are so similar to those obtained for the CRW model.

Proposition 3.4.12. *Consider the RBM satisfying $\delta > 0$.*

(i) *There is a unique invariant measure on $\mathbb{R}_+$ such that $X(t) \sim \pi$ for all $t \geq 0$ provided $X(0) \sim \pi$. It is exponential,*

$$\mathsf{P}_\pi\{X(t) \geq r\} = \exp(-r/\eta), \qquad r \geq 0, \tag{3.57}$$

where η is the steady-state mean,

$$\eta = \mathsf{E}_\pi[X(t)] = \int_0^\infty x\,\pi(dx) = \tfrac{1}{2}\frac{\sigma^2}{\delta}.$$

(ii) *Poisson's equation for the RBM model is expressed*

$$\mathcal{D}h^* = -c + \eta.$$

A solution for $c(x) \equiv x$ is the fluid value function defined in (3.41).

(iii) *Letting τ_0 denote the first hitting time to the origin, we have*

$$\mathsf{E}_x[\tau_0] = T^*(x),$$

where $T^*(x) = \delta^{-1}x$, $x \in \mathbb{R}_+$.

Proof. Part (ii) follows from (3.42).

The proof of (i) is the same as the proof of Proposition 3.4.4: To compute the MGF $\lambda_X(\vartheta) := \mathsf{E}_\pi[e^{\vartheta X(0)}]$ we consider the family of "test functions,"

$$V_\vartheta(x) = e^{\vartheta x} - \vartheta x, \qquad x \geq 0.$$

We have $V_\vartheta : \mathbb{R}_+ \to \mathbb{R}_+$, $\frac{d}{dx}V_\vartheta(0) = 0$, and

$$\mathcal{D}V_\vartheta = -\delta\vartheta(e^{\vartheta x} - 1) + \tfrac{1}{2}\sigma^2\vartheta^2 e^{\vartheta x}. \tag{3.58}$$

As in the proof of Proposition 3.4.4 we can argue that $\pi(\mathcal{D}V_\vartheta) = 0$ whenever $\pi(V_\vartheta) < \infty$. The latter holds by the Comparison Theorem for X provided $-\delta\vartheta e^{\vartheta x} + \tfrac{1}{2}\sigma^2\vartheta^2 < 0$, since in this case (3.58) implies a version of (V3). In this case,

$$0 = \int_0^\infty \left(-\delta\vartheta(e^{\vartheta x} - 1) + \tfrac{1}{2}\sigma^2\vartheta^2 e^{\vartheta x}\right)\pi(dx)$$

so that on rearranging terms the MGF is expressed

$$\lambda_X(\vartheta) = \int_0^\infty e^{\vartheta x}\pi(dx) = \frac{1}{1 - \vartheta\eta}.$$

This is the MGF for an exponential random variable with mean η, which is (i).

We have $\mathcal{D}T^*(x) = -\delta\frac{d}{dx}T^*(x) \equiv -1$, and hence Itô's formula gives

$$T^*(X(t)) = T^*(x) + \int_0^t [-1]\,ds + \int_0^t [\delta^{-1}]\,dI(s) + \int_0^t [\delta^{-1}]\,dN(s),$$

or $T^*(X(t)) = T^*(x) - t + \delta^{-1}(I(t) + N(t))$ for $t \geq 0$. Letting τ_0 denote the first hitting time to the origin, we obtain $0 = T^*(X(\tau_0)) = T^*(X(0)) - \tau_0 + \delta^{-1}N(\tau_0)$. Taking expectations proves (iii). □

The discounted value function for the RBM model is defined by

$$h_\gamma^*(x) = \mathsf{E}\left[\int_0^\infty e^{-\gamma t}X(t;x)\,dt\right]. \tag{3.59}$$

It is the solution to the same dynamic programming equation analyzed for the fluid and CRW models, based on the differential generator (3.40).

Proposition 3.4.13. *The discounted cost for the RBM model can be expressed*

$$h_\gamma^*(x) = J_\gamma^0(x) + b_\gamma e^{-\vartheta_\gamma x}, \qquad x \in \mathsf{X},$$

where ϑ_γ is the positive solution to

$$\gamma - \delta\vartheta - \tfrac{1}{2}\sigma^2\vartheta^2 = 0,$$

$b_\gamma = \gamma^{-1}\vartheta_\gamma^{-1}$, *and J_γ^0 is defined in (3.54).*

Proof. Let $g(x) = e^{-\vartheta_\gamma x}$ for $x \geq 0$. Applying the differential generator gives

$$\mathcal{D}g(x) = -\delta g'(x) + \tfrac{1}{2}\sigma^2 g''(x) = \left(\delta\vartheta_\gamma + \tfrac{1}{2}\sigma^2(\vartheta_\gamma)^2\right)e^{-\vartheta_\gamma x} = \gamma g(x), \qquad (3.60)$$

where the last equation uses the assumed form of ϑ_γ. On applying the generator to J_γ^0 we obtain

$$\mathcal{D}J_\gamma^0 = -\gamma^{-1}(\mu - \alpha) = -x + \gamma J_\gamma^0. \qquad (3.61)$$

Combining these two identities gives

$$\mathcal{D}h_\gamma^*(x) = -x + \gamma h_\gamma^*(x), \qquad x \in \mathbb{R}_+.$$

The constant b_γ is chosen so that $\frac{d}{dx}h_\gamma^*$ vanishes at $x = 0$. Itô's formula can be applied to obtain the following identity for any $T > 0$,

$$h_\gamma^*(x) = \mathsf{E}\left[e^{-\gamma T}h_\gamma^*(X(T)) + \int_0^T e^{-\gamma t}X(t;x)\,dt\right].$$

Letting $T \to \infty$ completes the proof. $\qquad\square$

3.5 Big queues

This section describes finer sample path properties of the queue, focusing on the way that large queue lengths can arise in a stable model. The analysis is based on theory of large deviations for i.i.d. processes, which we only describe superficially. References are given for results beyond the scope of this book.

Our central question is, *given that at time t_0 we observe $Q(t_0) \geq n_0$, where $n_0 \gg 1$ is a large number, what was the most likely behavior of Q prior to time t_0?* There are several approaches to an answer. Consider first the simplest setting.

3.5.1 Time reversal and reversibility

Suppose that Q is the stationary CRW queue defined on the two-sided time interval with $\rho < 1$, and let $\widetilde{Q}$ denote the time-reversed process $\widetilde{Q}(t) = Q(-t)$ for $t \in \mathbb{Z}$. This is a Markov chain whose transition matrix can be computed using Bayes' rule as follows:

$$\begin{aligned}
\widetilde{P}(x,y) &= \mathsf{P}\{\widetilde{Q}(t+1) = y \mid \widetilde{Q}(t) = x\} \\
&= \frac{\mathsf{P}\{\widetilde{Q}(t+1) = y, \widetilde{Q}(t) = x\}}{\mathsf{P}\{\widetilde{Q}(t) = x\}} \\
&= \frac{\mathsf{P}\{Q(-t-1) = y, Q(-t) = x\}}{\mathsf{P}\{Q(-t) = x\}} = \frac{\pi(y)}{\pi(x)}P(y,x).
\end{aligned} \qquad (3.62)$$

In general, the time-reversed process is not easily identified since we do not have a formula for π. However, in the special case of the M/M/1 queue π is computable, and

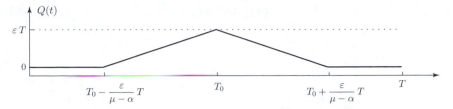

Figure 3.2. Typical trajectory of the queue length process in the M/M/1 queue, conditioned on a very large value at time T_0.

moreover Q is *reversible* in the sense that the process viewed in "rewind mode" with time reversed is statistically identical to the process viewed in forward time. That is,

Proposition 3.5.1. *Suppose that Q is the stationary M/M/1 queue defined on the two-sided time interval, and define the time-reversed process $\widetilde{Q}(t) = Q(-t)$ for $t \in \mathbb{Z}$. Then $\widetilde{Q}$ is identical to Q in law: The one-dimensional marginal distribution of $\widetilde{Q}$ is given by π, and for any $x, y \in \mathbb{Z}_+$,*

$$\mathsf{P}\{\widetilde{Q}(t+1) = y \mid \widetilde{Q}(t) = x\} = \mathsf{P}\{Q(t+1) = y \mid Q(t) = x\}.$$

Proof. It is obvious that $\widetilde{Q}$ has marginal distribution π when this holds for Q. To see that the transition probabilities agree we apply (3.62),

$$\mathsf{P}\{\widetilde{Q}(t+1) = y \mid \widetilde{Q}(t) = x\} = \frac{\pi(y)}{\pi(x)} P(y, x).$$

The right-hand side is precisely $P(x, y)$, by the detailed balance equations (3.44). □

Proposition 3.5.1 provides a precise answer to our question regarding the most likely path to a buffer overflow: If $Q(t_0) \geq n_0$ then equivalently $\widetilde{Q}(-t_0) \geq n_0$, and any questions regarding the past of Q can be translated to questions regarding the future of $\widetilde{Q}$. In particular, it is overwhelmingly likely that a buffer overflow occurs following a linear divergence of Q with slope approximately equal to $\mu - \alpha > 0$.

The typical behavior of Q predicted by Proposition 3.5.2 is illustrated in Fig. 3.2.

Proposition 3.5.2. *Under the assumptions of Proposition 3.5.1, the following limit holds for each $x \in \mathbb{R}_+$, $\varepsilon > 0$, and $T > 0$:*

$$\lim_{\kappa \to \infty} \mathsf{P}_\pi\left\{ \sup_{-T \leq t \leq 0} |q^\kappa(t) - \tilde{q}(t)| \geq \varepsilon \,\Big|\, Q(0) = \lfloor \kappa x \rfloor \right\} = 0,$$

where $\tilde{q}$ denotes the time-reversed fluid model,

$$\tilde{q}(t) = \tilde{q}(t; x) := q(-t) = [x + (\mu - \alpha)t]_+, \qquad t \leq 0.$$

Proof. With $\tilde{q}(t) = \tilde{q}(t; \lfloor \kappa x \rfloor)$ for $t \le 0$, and $q(t) = q(t; \lfloor \kappa x \rfloor)$ for $t \ge 0$ we have

$$P_\pi \Big\{ \sup_{-\kappa T \le t \le 0} |Q(t) - \tilde{q}(t)| \ge \varepsilon \kappa \,\Big|\, Q(0) = \lfloor \kappa x \rfloor \Big\}$$

$$= P_\pi \Big\{ \sup_{0 \le t \le \kappa T} |\widetilde{Q}(t) - \tilde{q}(-t)| \ge \varepsilon \kappa \,\Big|\, \widetilde{Q}(0) = \lfloor \kappa x \rfloor \Big\}$$

$$= P_\pi \Big\{ \sup_{0 \le t \le \kappa T} |Q(t) - q(t)| \ge \varepsilon \kappa \,\Big|\, Q(0) = \lfloor \kappa x \rfloor \Big\},$$

where the first identity follows by the definition $\widetilde{Q}(t) = Q(-t)$, and the last identity follows from reversibility and the definitions of q and $\tilde{q}$.

The proof is completed on noting that the final probability is equivalent to

$$P\Big\{ \sup_{0 \le t \le T} |q^\kappa(t; x) - q(t; x)| \ge \varepsilon \Big\}$$

which is convergent to zero by Proposition 3.2.3. □

3.5.2 Most likely behavior in the general model

Proposition 3.5.2 indicates the most likely behavior prior to a "buffer overflow" in the M/M/1 queue. A very similar conclusion is possible in the CRW model provided the log-MGF $\Lambda(\vartheta)$ defined in (3.13) is finite for some $\vartheta > 0$. One proof is obtained as in the case of the M/M/1 by identifying the time-reversed process. The ideas are contained in Exercise 3.12.

An alternative approach is to translate large deviation limit theory from the free process F to the process Q using properties of the Skorokhod map. This approach is more easily applied than reversibility since we do not have to identify the dynamics of the time-reversed process $\widetilde{Q}$.

Suppose that $\mathcal{E}$ is an i.i.d. sequence with distribution G on $\mathbb{R}$, and let F denote the partial sums

$$F(t) = \sum_{i=1}^{t} \mathcal{E}(i), \qquad t = 1, 2, \dots. \tag{3.63}$$

Define $F(0) = 0$, and extend the definition of F to $t \in \mathbb{R}_+$ by linear interpolation. The scaled process is then defined by

$$f^\kappa(t) = \frac{1}{\kappa} F(\kappa t), \qquad \kappa \ge 1, \ t \in \mathbb{R}_+.$$

Letting $-\delta$ denote the common mean of the $\{\mathcal{E}(i)\}$, the *typical* behavior of the sequence is given by the Strong Law of Large Numbers, $f^\kappa(t) \to f^\infty(t) = -\delta t, t \ge 0$. Chernoff's bound and Cramer's Theorem give bounds on the error based on the *rate function* $I : \mathbb{R} \to \mathbb{R}_+ \cup \{\infty\}$, expressed as the convex dual of Λ,

$$I(r) = \sup_{\vartheta \in \mathbb{R}} (\vartheta r - \Lambda(\vartheta)), \qquad r \in \mathbb{R}. \tag{3.64}$$

Proposition 3.5.3. *Suppose that $\mathcal{E}$ is a bounded i.i.d. sequence with mean $-\delta$ and log-MGF Λ. We then have for each $r > -\delta$,*

Chernoff's bound: *For each $\kappa \geq 1$ and $t = 0, \kappa^{-1}, 2\kappa^{-1}, \ldots$,*

$$\frac{1}{\kappa} \log P\{f^\kappa(t) \geq rt\} \leq -tI(r)$$

Cramer's Theorem: *Chernoff's bound is asymptotically tight,*

$$\lim_{\kappa \to \infty} \frac{1}{\kappa} \log P\{f^\kappa(t) \geq rt\} = -tI(r).$$

These results do not quite explain why I is called a rate function. Consider the following question of *tracking*. Let $\{f(t) : 0 \leq t \leq T\}$ be a function that is continuous as a function of t with $f(0) = 0$. Assume that it is also piecewise smooth, in the sense that there is a finite sequence of numbers $0 = t_0 < t_1 < \cdots < t_n = T$ such that $f'(t)$ is continuous and bounded on (t_i, t_{i+1}) for each $0 \leq i \leq n-1$. We can then establish the following functional large deviations limit under the assumptions of Proposition 3.5.3,

$$\lim_{\varepsilon \downarrow 0} \lim_{\kappa \to \infty} \frac{1}{\kappa} \log \left(P\left\{ \sup_{0 \leq t \leq T} |f^\kappa(t) - f(t)| \leq \varepsilon \right\} \right) = -\int_0^T I(f'(s))\, ds. \qquad (3.65)$$

For a proof see [140]. Hence $I(f'(s))$ measures the cost of the deviation of the slope (or rate) $f'(s)$ from normal behavior.

Using the limit (3.65) together with convexity of the rate function we can obtain answers to our question regarding typical behavior. Convexity implies the bound

$$\frac{1}{T} \int_0^T I(f'(s))\, ds \geq I(\bar{f}')$$

where $\bar{f}'$ denotes the average slope, $\bar{f}' = T^{-1} \int_0^T f'(s)\, ds = T^{-1}(f(T) - f(0))$. Hence the most likely path from $f(0) = 0$ to $f(T)$ is by the straight line $f^*(t) = t\bar{f}'$ for $0 \leq t \leq T$. In this case (3.65) can be applied to obtain the limit

$$\lim_{\kappa \to \infty} \frac{1}{\kappa} \log \left(P\left\{ \inf_{0 \leq t \leq T} (f^\kappa(t) - rt) \geq 0 \right\} \right) = -TI(r), \qquad (3.66)$$

for any $r > -\delta$. The limit is precisely the same limit obtained using Cramer's Theorem for the probability of the simpler event $P\{f^\kappa(T) \geq rT\}$. Applying Bayes' rule gives the following most likely behavior, conditioned on a large value for $f^\kappa(T)$:

$$\lim_{\kappa \to \infty} \frac{1}{\kappa} \log \left(P\left\{ \inf_{0 \leq t \leq T} (f^\kappa(t) - rt) \geq 0 \mid f^\kappa(T) \geq rT \right\} \right) = 0.$$

These arguments and another application of the Skorokhod map are the basis of the following generalization of Proposition 3.5.2. The most likely path to overflow remains linear, but the slope is now given by $\Lambda'(\vartheta_0) > 0$, with $\vartheta_0 > 0$ given in (3.15) the second zero of Λ.

Proposition 3.5.4. *The following limit holds under the assumptions of Theorem 3.0.2: For each $\varepsilon > 0, T > 0$:*

$$\lim_{\kappa \to \infty} \kappa^{-1} \log \mathsf{P}_\pi \Big\{ \sup_{-T \leq t \leq 0} |q^\kappa(t) - \tilde{q}(t)| \geq \varepsilon \,\Big|\, Q(0) = \lfloor \kappa x \rfloor \Big\} = 0,$$

where $\tilde{q}(t) = [x - (\tilde{\mu} - \tilde{\alpha})t]_+$ for $t \leq 0$, with

$$-(\tilde{\mu} - \tilde{\alpha}) := \Lambda'(\vartheta_0) > 0.$$

For the proof we refer the reader to [198]. The specific form for $\tilde{q}$ is explained in Exercises 3.11 and 3.12.

Theorem 3.0.2 implies that the event on which the conditional probability is conditioned admits the asymptotic

$$\lim_{\kappa \to \infty} \kappa^{-1} \log \mathsf{P}\{Q(0) = \lfloor \kappa x \rfloor\} = -\vartheta_0. \tag{3.67}$$

Hence the second zero determines both the exponent for the probability of overflow, and the most likely path to overflow.

Consider for example the CRW model with $(\boldsymbol{S}, \boldsymbol{A})$ satisfying

$$\begin{pmatrix} S(t) \\ A(t) \end{pmatrix} = \begin{cases} \binom{1}{0} & \text{with prob. } p, \\ \binom{0}{a} & \text{with prob. } 1 - p, \end{cases}$$

where $a \geq 1$, and $p \in (0, 1)$ is chosen so that the mean of $\mathcal{E}(t) = A(t) - S(t)$ is independent of a, with

$$\mathsf{E}[\mathcal{E}(t)] = -p + (1 - p)a = -(\mu - \alpha).$$

Solving this equation we obtain $p = (1+a)^{-1}(\mu - \alpha + a)$. When $a = 1$ and $\mu + \alpha = 1$ this gives $p = \mu$.

The MGF and its derivative are expressed

$$\lambda(\vartheta) = pe^{-\vartheta} + (1 - p)e^{a\vartheta}, \qquad \lambda'(\vartheta) = -pe^{-\vartheta} + (1 - p)ae^{a\vartheta}, \qquad \vartheta \in \mathbb{R}.$$

A plot of Λ is shown in Fig. 3.3 for $a = 5, 15, 50$. As expected from (3.67), the value of ϑ_0 vanishes as $a \to \infty$. Note however that the slope of Λ at ϑ_0 remains bounded for all $a \geq 1$. This is the most likely slope preceding $t = 0$ as given in Proposition 3.5.4.

We close this chapter with a few words on constructing a model.

3.6 Model selection

The details of the model (3.1) will be determined based on observations of the physical system. In Section 2.1.2 it is argued that a renewal model can be approximated using the discrete-time model by fitting first and second moments. The results of Section 3.2.2 refine this approximation, since either the GI/G/1 or the CRW model can be approximated by an RBM on $\mathbb{R}_+$ when $\rho \approx 1$, and the parameters in the approximation depend only on the first and second moments of the discrete state-space queueing model.

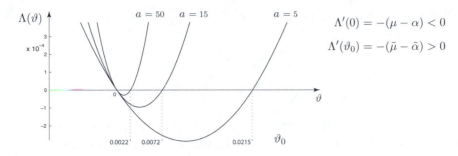

Figure 3.3. Log moment generating function for three distributions with common mean $E[\mathcal{E}] = -(\mu - \alpha)$, supported on $\{-1, a\}$ with $a = 5, 15, 50$.

Rather than use the GI/G/1 queue as an intermediate model, in practice one might fit low-order moments to those observed in the physical queue. This leaves a great deal of freedom in choosing the distribution of (S, A) in the CRW model.

Up until the first time that the buffer is empty, the process Q evolves as the random walk defined in Definition 1.3.3. Figure 1.6 shows a simulation of two random walks X^u and X^d based on two different i.i.d. processes $\{\mathcal{E}^u, \mathcal{E}^d\}$ with common first and second moments. The *typical behavior* shown in these two plots is comparable. Likewise, the Pollaczek–Khintchine formula for the steady-state mean queue length depends critically on both load and the covariance of (A, B), but is insensitive to the precise distribution of these processes.

However, two distributions with common first and second moments may have very different higher-order statistics. Hence the probability of a rare event, such as a buffer overflow, may be substantially different.

Consider the following worst-case construction: We are told that the distribution G of a random variable $\mathcal{E}$ is supported on a bounded interval $[0, a]$, and the first and second moments m_1, m_2 are given. We may then ask, for a given $\vartheta > 0$, what is the worst-case value for the MGF? The answer is independent of the particular $\vartheta > 0$ chosen, and is realized by a distribution G^* with just two points of support. We can also specify just the first moment and arrive at a similar conclusion.

The log-MGFs shown in Fig. 3.3 were obtained using the worst-case marginal distribution (3.68). The process X^d that is illustrated in Fig. 1.6 was constructed using the marginal distribution specified by (3.69).

Proposition 3.6.1. *The MGF is maximized by a binary distribution on $[0, a]$ in each of the two situations:*

(i) *If the first moment $m_1 = \int x\, G(dx)$ is specified, then $\lambda_G(\vartheta)$ is maximized, for each $\vartheta > 0$, by the distribution supported on the extremal points $\{0, a\}$ with consistent mean:*

$$G^* = p^* \delta_0 + (1 - p^*)\, \delta_a\,, \qquad\qquad (3.68)$$

where $p^* = 1 - m_1/a$.

(ii) *If two moments are specified, $m_i = \int x^i G(dx), i = 1, 2,$ then $\lambda_G(\vartheta)$ is maximized uniquely for each $\vartheta > 0$ with*

$$G^* = p^* \delta_{x_0} + (1 - p^*) \delta_a, \qquad (3.69)$$

where $x_0 = [a - m_1]^{-1}(m_1 a - m_2)$ and $p^* = [a^2 + m_2 - 2m_1 a]^{-1}(a - m_1)^2$.

Proposition 3.6.1 (i) is the essential ingredient in Hoeffding's inequality [266], and (ii) is essentially Bennett's Lemma [40]. The significance of Proposition 3.6.1 is that a very simple model can be constructed that captures worst-case behavior in the CRW model. The worst-case model simultaneously minimizes the exponent ϑ_0 defined in Theorem 3.0.2, and maximizes the most likely slope $\tilde{\mu} - \tilde{\alpha} := \Lambda'(\vartheta_0)$ appearing in Proposition 3.5.4.

3.7 Notes

The title of this chapter is borrowed from Cohen's classic monograph [114]. Much more on the single server queue can be found in this book, as well as Asmussen [19] and Bertsekas and Gallager [43].

The embedded regeneration time approach has been enormously significant since its introduction by Kendall in [293, 294]. Similar concepts are used elsewhere in the book. For example, sampling is used to prove that stability of a stochastic network follows from stability of its corresponding fluid model (see Proposition 10.4.1 and discussion in Chapter 10).

Generalization of the Skorokhod map to multidimensional models is considered in Section 5.3 and subsequent chapters. Generalizations are not always obvious. In particular, the representation of a reflected Brownian motion on a bounded interval was resolved only recently [313], and the Skorokhod map in this case is relatively complex when compared to the RBM on $\mathbb{R}_+$.

The variance formula in Proposition 3.3.1 is taken from [362]. The identity $\mathsf{E}_x[\tau_0] = T^*(x)$ is due to Foster [191].

The derivation of the heavy-traffic limit to an RBM in Proposition 3.2.5 is unfortunately terse. For a comprehensive treatment see Kushner [323] or Whitt [498]. Introductory presentations can be found in several other books, including [385, 231, 233, 96].

The title of Section 3.5 is borrowed from Ganesh, O'Connell, and Wischik [198], along with some of its contents. Although an important topic with a significant literature, large deviations asymptotics and blocking probabilities are not central to this book. The reader is referred to [155, 140] for a serious look at large deviations. See also Takagi [466] and Shwartz and Weiss [446] for more on applications.

Exercises

3.1   The M/M/1 queue can be represented as the CRW model by sampling using a specially constructed Poisson process. A similar construction holds for the M/G/1 queue by sampling the number in the queue immediately after the nth service time is completed.

(a) Explain why the sampled process Q is Markovian, and has the recursive form (2.2).

(b) Verify that the marginal distribution of A in a CRW model is given by

$$P\{A(1) = j\} = \int_0^\infty \{e^{-\alpha t}(\alpha t)^j / j!\}\, G^s(dt) \qquad j \geq 0, \qquad \text{(E3.1)}$$

where G^s is the distribution of a typical service time.

3.2   The GI/M/1 queue is the special case in which service is exponentially distributed. Discuss how this continuous-time model might be sampled to obtain a countable state space, discrete-time Markov model.

3.3   Consider the GI/D/1 queue, where the "D" means that service is deterministic: In the notation of Definition 2.1.1 we have $\mathcal{E}^s(t) \equiv \mu^{-1}$. Take $\mu = 1$, and assume that the distribution of $\mathcal{E}^a(t)$ is uniform on $[0, 2\rho]$. Construct a CRW model by applying the procedure described in Section 2.1.2. Simulate the continuous-time model and the discrete-time approximation for $\rho = 0.85$, $0.9, 0.95$, and in each case compare the steady-state mean queue length using the Monte Carlo estimator

$$\eta(n) := \frac{1}{n} \sum_{t=0}^{n-1} Q(t), \quad n = 1, \ldots, 10^6.$$

3.4   Consider a system that is identical to the M/M/1 queue except that when the system empties, service does not begin again until n_0 customers are present in the system ($n_0 \geq 0$ is given). Find a Markov model for this system, compute the steady-state distribution, and compute the mean of $Q(t)$ in steady state.

3.5   Establish the following under the assumption that the distribution of $A(1)$ has finite support: If $\rho < 1$ then the unique invariant measure π for Q satisfies for some $n_A \geq 1$,

$$\pi(n) = k_0 e^{-\vartheta_0 n}, \qquad n \geq n_A,$$

where $\vartheta_0 > 0$ is the second zero defined in (3.15), and $k_0 > 0$ is a normalizing constant. *Hint*: It suffices to verify that the invariance equation holds,

$$\sum_x e^{-\vartheta_0 x} P(x, y) = e^{-\vartheta_0 y}$$

provided $y \geq n_A$.

3.6   For a Markov chain on $\mathbb{Z}_+$, suppose that the transition function P satisfies the detailed balance equations

$$\pi(n) P(n, m) = \pi(m) P(m, n), \qquad n, m \in \mathsf{X} = \{0, 1, 2, 3, \ldots\},$$

where $\{\pi(n)\}$ is a sequence of positive numbers. In this case the chain is called *reversible*.

(a) Verify that π must be an invariant measure.

(b) Let $X(t)$, $-\infty < t < \infty$, be a stationary Markov process. Verify that $Y(t) = X(-t)$ is also a stationary Markov process by computing the conditional probabilities

$$\widetilde{P}(m,n) = P(Y(t+1) = n \mid Y(t = m)), \qquad n \in \mathbb{Z}_+.$$

What is $\widetilde{P}$ under the detailed balance equations?

3.7   The M/M/1 queue with *finite waiting room* is a Markov chain on a finite state space $\{0, 1, \dots, \overline{x}\}$, with $\overline{x} \geq 1$. Its transition matrix is identical to the $M/M/1$ queue for $x < \overline{x}$, with

$$P(\overline{x}, \overline{x} - 1) = 1 - P(\overline{x}, \overline{x}) = \mu.$$

Show that this process is reversible, and that its invariant measure π is geometric. For a given value of ρ, how large must $\overline{x}$ be to ensure that the loss-probability is less than 10^{-3}.

3.8   Consider the CRW model of the single server queue

$$Q(t+1) = Q(t) - S(t+1)U(t) + A(t+1), \qquad t \geq 0,$$

where (A, S) is i.i.d. The common marginal distribution of S is Bernoulli, and the marginal distribution of A is supported on $\mathbb{Z}_+$. The allocation sequence U is defined by the nonidling policy. Obtain conditions on (A, S) so that the detailed balance equations are satisfied.

3.9   The M/M/n queueing system consists of a single queue and n servers. Jobs arrive according to a Poisson process, with rate α. Service times at each server are exponentially distributed, with mean μ. The service discipline is First Come First Served. Construct a CRW model using uniformization. Compute the invariant measure π when $n = 2$.

For general n, compute $\mathcal{D}V$ for $V(x) = x$ and $V(x) = x^2$, and $V(x) = e^{\beta x}$ with fixed β. Obtain a bound on the steady-state mean of $Q(t)$ using the Comparison Theorem.

3.10   Consider the following version of the M/M/∞ queue based on independent and identically distributed Bernoulli random variables $\{S_i(t) : i \geq 1, \ t \geq 1\}$, and an i.i.d. Bernoulli sequence A satisfying

$$A(t)S_i(t) = 0, \qquad i \geq 1, \ t \geq 1.$$

The queue length process evolves according to

$$Q(t+1) = Q(t) - \sum_{i=1}^{Q(t)} S_i(t) + A(t+1), \qquad t \geq 0.$$

This is a queue with input rate $\alpha := \mathsf{E}[A(t)]$, and a server available for each customer in queue.

(a) Compute $\mathcal{D}V$ for $V(x) = x$, $V(x) = x^2$, and $V(x) = e^x$.

(b) A fluid model is given by $\dot{q} = -\mu q + \alpha$, $q(0) = x \in \mathbb{R}_+$. As $t \to \infty$, from each initial state x, the trajectory $q(t; x)$ converges to some $x_\infty \geq 0$. Compute x_∞.

(c) Using intuition gained from (i) and (ii), construct the solution h to Poisson's equation,

$$\mathcal{D}h = -f + \eta,$$

where $f(x) \equiv x$, h is a polynomial function of x, and η is the steady-state queue length.

(d) Verify that there is an exponential moment

$$\mathsf{E}_\pi[\exp(\vartheta Q(t))] < \infty.$$

How large can ϑ be?

(e) Is there a super-exponential tail $\mathsf{E}_\pi[\exp(\vartheta[Q(t)]^2)] < \infty$, for some $\vartheta > 0$?

3.11 Compute the transition matrix $\widetilde{P}(x, y)$ given in (3.62) for $x, y \geq n_A$ under the assumptions of Exercise 3.5. Compute also the mean drift,

$$\tilde{\Delta}(x) = \mathsf{E}[\widetilde{Q}(t+1) - \widetilde{Q}(t) \mid \widetilde{Q}(t) = x].$$

How does the value of $\tilde{\Delta}(x)$ for $x \geq 1$ compare with the drift for the process $\tilde{q}$ defined in Proposition 3.5.4?

3.12 Based on (3.65) it might not be difficult to convince yourself of the following limit for the single server queue under the assumptions of Theorem 3.0.2, where $\mathbf{Q}$ is the stationary process on the two-sided time interval. For any non-negative constant r let $\tilde{q}$ denote the fluid trajectory $\tilde{q}(t) = [x + rt]_+$, $t \in \mathbb{R}$. Choosing $T = r^{-1}x$ we have

$$\lim_{\varepsilon \downarrow 0} \lim_{\kappa \to \infty} \frac{1}{\kappa} \log \left(\mathsf{P}_\pi \left\{ \sup_{-T \leq t \leq 0} |q^\kappa(t) - \tilde{q}(t)| \leq \varepsilon \right\} \right) = -r^{-1}xI(r). \quad \text{(E3.2)}$$

In this exercise you can take this limit for granted.

(a) From the definition of the rate function given in (3.64), verify that $r^{-1}I(r) \geq \vartheta_0$ (replace the supremum by evaluation at ϑ_0). This is a universal bound, for any $r \geq 0$.

(b) Show that this lower bound is attained using $r_0 = \Lambda'(\vartheta_0)$. This is the most likely slope, in that it minimizes the probability (E3.2).

(c) Show using Theorem 3.0.2 that

$$\lim_{\varepsilon \downarrow 0} \lim_{\kappa \to \infty} \frac{1}{\kappa} \log \left(\mathsf{P}_\pi \{ |q^\kappa(0) - \tilde{q}(0)| \leq \varepsilon \} \right) = -r_0^{-1}xI(r_0).$$

(d) Discuss your findings as they compare to Proposition 3.5.4.

3.13 Consider the queue with deterministic service at rate $\mu > 0$,

$$Q(t+1) = [Q(t) - \mu + A(t+1)]_+,$$

where A is i.i.d. with bounded support. The log-MGF is then given by $\Lambda(\vartheta) = \Lambda_A(\vartheta) - \mu$. Assume that the conclusions of Theorem 3.0.2 hold with this more general model (they do).

(a) For a given $r < \infty$, verify that the performance specification holds,

$$\lim_{n \to \infty} n^{-1} \log\left(\mathsf{P}_\pi\{Q(t) \geq n\}\right) \leq -nr,$$

if and only if $r \leq \vartheta_0$.

(b) Show that $r \leq \vartheta_0$ if and only if $E_A(r) := r^{-1}\Lambda_A(r) \leq \mu$.

(c) Verify the following limits,

$$E_A(r) \downarrow \alpha \text{ as } r \downarrow 0, \qquad E_A(r) \uparrow \bar{A} \text{ as } r \uparrow \infty,$$

where $\alpha = \mathsf{E}[A(1)]$, and $\bar{A}$ is the essential supremum,

$$\bar{A} = \sup\{a : \mathsf{P}\{A(t) \geq a\} > 0\}.$$

In communications applications, the function E_A is known as the *effective bandwidth* of the "source" A.

4

Scheduling

In this chapter we restrict to a special class of network models in which routing is deterministic and uncontrolled. Each customer waits in one of ℓ buffers until it receives processing at the respective station. The processing rate at the ith queue is equal to $\mu_i > 0$, provided the station devotes full priority to this customer. In a stochastic model this means that the mean service time for a customer at this buffer is given by μ_i^{-1}. Once processing is completed, the customer is either routed to another buffer for processing or exits the system. Examples include the simple re-entrant line, the Klimov model, and the KSRS model described in Chapter 2.

Both stochastic and deterministic network models are considered in this chapter, though much of the analysis remains focused on deterministic models in which the concepts are most transparent. Below is an incomplete list of topics to be considered:

(i) The general *Controlled Random Walk* (CRW) model is introduced for the scheduling problem.

(ii) A deterministic fluid model is introduced that describes the mean behavior of its stochastic counterpart. This fluid model describes the transient behavior of a network, as illustrated in Fig. 2.2 for the single server queue.

(iii) Stabilizability and load are defined for the scheduling model.

(iv) Besides modeling, the main goal of this chapter is to demonstrate how the preceding concepts can be applied to gain insight in control design.

We fix the following notation throughout this chapter: There are ℓ buffers, and ℓ_m stations, denoted $s \in \mathcal{S} := \{1, \ldots, \ell_m\}$. For each $i \in \{1, \ldots, \ell\}$ the index $s(i) \in \mathcal{S}$ denotes the station at which buffer i is located, and the set of buffers at Station s is denoted $\mathcal{I}_s \subset \{1, \ldots, \ell\}$ for each $s \in \mathcal{S}$. The constituency matrix is the $\ell_m \times \ell$ matrix C whose (s, i) entry is defined by

$$C_{si} = \mathbf{1}\{s(i) = s\}, \qquad s \in \mathcal{S}, \ i \in \{1, \ldots, \ell\}. \tag{4.1}$$

Routing is deterministic: for each $i \in \{1, \ldots, \ell\}$, after processing at buffer i a customer either enters some buffer $i_+ \in \{1, \ldots, \ell\}$, or exits the system. The *routing matrix* R is the $\ell \times \ell$ matrix defined for $i, j \in \{1, \ldots, \ell\}$ as $R_{ij} = \mathbf{1}_{j=i_+}$. Consequently, the routing matrix can be expressed as the concatenation of row vectors,

$R = [1^{1+} \mid \cdots \mid 1^{\ell+}]^{\mathsf{T}}$, where we adopt the convention that $1^{i+} = 0$ if customers exit the system following service at buffer i.

Section 4.1 contains a brief discussion of modeling and control for the general CRW model, along with an introduction to dynamic programming and optimization. Although appealing, a direct approach to optimization is typically futile in all but the simplest networks. Firstly, if buffers are infinite, then solving the dynamic programming equations using value iteration or policy iteration (algorithms introduced in Chapter 9) leads to an infinite sequence of infinite-dimensional optimization problems. Even when buffers are subject to strict upper bounds, the complexity of the dynamic programming equations grows exponentially with the dimension of the state space. This is known as the *curse of dimensionality*. For example, computation of the optimal policy for the simple re-entrant line in Example 4.5.3 required the solution of a sequence of N-dimensional matrix equations with $N = 45^3 = 91,125$.[1] If one is truly forced to face this complexity in such a simple model then this certainly is a curse! In a more realistic model with, say, 20 buffers the situation becomes much worse with $N = 45^{20} > 10^{33}$.

Moreover, typically optimization is based on minimizing a *single performance objective*, while in almost any interesting application there are a range of objectives to be considered simultaneously, such as minimizing delays or loss for different customer classes, while simultaneously minimizing operational costs and the cost of holding inventory.

In this chapter we begin a discussion on how to construct policies with attractive characteristics that take into account a range of performance criteria. One technique is to construct a policy for the physical network of interest based on a policy for the fluid model.

Modeling and control for the fluid model are developed in Sections 4.2 and 4.3. Through examples it is shown that optimization remains complex, but not nearly as complex as seen in the CRW model. A simple suboptimal approach that explicitly makes a tradeoff between minimizing holding-cost and minimizing customer delay is the *Greedy time optimal policy* described in Section 4.3.3.

We have seen in Section 2.10 and elsewhere that effective policies for a discrete-stochastic model do not exactly mirror those obtained for the simpler fluid model. This is particularly evident on the boundaries of the state space since this is where the discrete model is subject to more constraints than the fluid model. How then can one modify a fluid policy to obtain an effective policy for the discrete network?

The following three general techniques are used to construct policies based on a more realistic network model that includes both variability and complex constraints.

Discrete review In a *discrete-review* policy, control decisions are made at discrete time intervals, and the time period between these decision epochs is called a *planning horizon*. It may be desirable to employ *packetization*: A server concentrates work on a group of similar customers to avoid excessive switch-overs.

---

[1] This computation was based on the value iteration algorithm introduced in Chapter 9.

For example, in the simple re-entrant line a policy for the fluid model might require time-sharing at the first station so that each station can work at capacity. That is, if $q_2(t) = 0$ then it may be desirable to take $\zeta_1 = \mu_2/\mu_1$ and $\zeta_3 = 1 - \zeta_1$. To translate this to a discrete model we impose the constraint at the start of the planning horizon that buffers 1 and 3 should receive these fractions of the total service time over the planning horizon – the precise order of service is not necessarily specified.

Hedging In economic language, *hedging* is a process whereby a "player" deliberately takes on new risk to offset existing risks. Decisions on how to hedge are based upon current forecasts of future trends and volatility in the market. In this book, hedging refers to any mechanism introduced to guard against the risk of potentially high cost.

An example is the hedging point used in the inventory model introduced in Section 2.6, where one source of "risk" in this example is the high cost of falling behind demand.

Safety stocks are also a general technique to guard against a very special form of risk: starvation of resources.

The difference between safety stocks and hedging points is not always obvious. The distinction will be clear when we consider hedging-points for workload models in the following chapters, beginning with an example in Section 5.6.

The need for safety stocks is evident in the examples presented in Section 4.1.2 – it is not difficult to construct examples in which a policy is stabilizing or even optimal for a deterministic model, yet the same policy applied to a stochastic model is destabilizing. Conversely, in examples presented in Section 4.5, we find that an *optimal policy* for the CRW model frequently resembles a perturbation of the optimal policy for the fluid model using safety stocks of the form developed in this book. Section 4.6 contains an introduction to safety stocks with numerical results illustrating this observation. Section 4.7 contains more elaborate methods for employing safety stocks based on the discrete review paradigm.

Application of safety stocks can be complex in large networks. Motivated by the dynamic programming equations considered in Section 4.1.3, myopic policies and related MaxWeight policies are introduced in Sections 4.8 and 4.9. These policies often define safety stocks *implicitly* as an outcome of the one-step optimization that determines the policy.

We begin with the discrete-time scheduling model.

4.1 Controlled random-walk model

In this section we introduce the CRW model and present some superficial generalizations of the concepts introduced in Chapter 3 for the single server queue. We introduce a stability criterion similar to (V3) and also similar to the Poisson equation introduced in Section 3.3. Poisson's equation for the CRW model is an element of the dynamic programming equations described in Section 4.1.3.

4.1.1 Basic model

A stochastic network model is defined in discrete time as follows:

Definition 4.1.1 (CRW scheduling model). This is described by the recursion

$$Q(t+1) = Q(t) + B(t+1)U(t) + A(t+1), \qquad Q(0) = x. \qquad (4.2)$$

The following assumptions are imposed on the policy and parameters.

(i) We say that the allocation sequence U is *adapted* (to (Q, A, B)) if $U(t)$ can be expressed as a function of the random variables $\{Q(0), \ldots, Q(t), A(0), \ldots, A(t), B(0), \ldots, B(t)\}$ for each $t \geq 0$. It is assumed that U is adapted and satisfies the linear constraints

$$U(t) \geq 0, \quad CU(t) \leq 1, \qquad t \geq 1,$$

where 1 is the ℓ_m-dimensional vector of ones, and 0 is the ℓ-dimensional vector of zeros. That is, the allocation sequence is restricted to the polyhedron

$$\mathsf{U} := \{u \in \mathbb{R}^{\ell_m} : u \geq 0, \quad Cu \leq 1\}. \qquad (4.3)$$

(ii) The queue length process Q is similarly constrained, $Q(t) \in \mathsf{X}$, $t \geq 0$, where $\mathsf{X} \subset \mathbb{R}_+^\ell$ is a polyhedron representing both positivity constraints, and bounds on buffer levels if present. The assumption that Q evolves on X imposes implicit constraints on U: We let $\mathsf{U}(x) \subset \mathsf{U}$ denote the set of allowable values for $U(t)$ when $Q(t) = x \in \mathsf{X}$.

(iii) It is assumed that $R^\ell = 0_{\ell \times \ell}$ to ensure that each customer receives at most ℓ services during its lifetime in the network.

(iv) B is an i.i.d. sequence of $\ell \times \ell$ matrices; and A is an i.i.d. sequence of ℓ-dimensional vectors. The matrix $B(t)$ is expressed

$$B(t) = -[I - R^{\mathsf{T}}]M(t), \qquad t \geq 1,$$

where M is an i.i.d. sequence of diagonal matrices. Its diagonal elements are non-negative, and are denoted $M_i(t) \geq 0$ for each i and t. Consequently, the recursion (4.2) can be expressed

$$Q(t+1) = Q(t) + \sum_{i=1}^{\ell} M_i(t+1)[-1^i + 1^{i+}]U_i(t) + A(t+1). \qquad (4.4)$$

(v) Station s is said to be *homogeneous* if the random variables $\{M_j(t) : s(j) = s\}$ are all identical. In this case, the common value is denoted $S_s(t)$. If each station is homogeneous, then this is called the homogeneous CRW scheduling model. ∎

It is assumed in this chapter that the random variables $\{M_j(t)\}$ each have a Bernoulli distribution. That is, $M_j(t) = 0$ or 1 for each j and t.

It is not necessary to restrict Q to an integer lattice, although this constraint can be added if desirable. In this case, it is assumed that A and M have integer entries, and the allocation sequence is restricted to the discrete set

$$U_\diamond := \{u \in \{0,1\}^\ell : Cu \le 1\}. \tag{4.5}$$

We let $X_\diamond := X \cap \mathbb{Z}_+^\ell$ denote the restricted state space, and $U_\diamond(x) \subset U_\diamond$ the set of allowable values for $U(t)$ when $Q(t) = x \in X_\diamond$, so that $Q(t+1) \in X_\diamond$ with probability one.

A special case is the re-entrant line:

Definition 4.1.2 (Re-entrant line). Consider a network with deterministic routing, and a single arrival process to the first buffer. A customer of class j requires service at Station $s(j)$, and then proceeds to buffer $j+1$ or, if $j = \ell$, exits the system. A network of this form is called a *re-entrant line*.

The CRW model for a re-entrant line is described as follows:

$$Q(t+1) = Q(t) + \sum_{i=1}^{\ell} M_i(t+1)[1^{i+1} - 1^i]U_i(t) + A(t+1), \qquad t \ge 0, \tag{4.6}$$

where, as above, 1^i denotes the ith basis vector in $\mathbb{R}^\ell$ for $1 \le i \le \ell$, and $1^{\ell+1} := 0$. ■

Recall that uniformization was a sampling approach introduced in Section 2.1.1 to obtain the CRW model (2.2). Uniformization is easily extended to network models in which the increment distributions of the arrival processes, and the distributions of the service times all have exponential distributions. This leads to a CRW model in which the entries of A and B are Bernoulli, and the additional properties

$$\begin{aligned} M_i(t)A_j(t) &= 0, \quad \text{for all } i, j, \\ M_i(t)M_j(t) &= 0, \quad \text{for all } i \ne j, \qquad t \ge 1. \end{aligned} \tag{4.7}$$

That is, only one service or arrival event can occur at each sampling period.

We illustrate this construction using the following special case:

Example 4.1.3 (Simple re-entrant line). Consider the simple re-entrant line shown in Fig. 2.9, modeled in continuous time. The single arrival stream is assumed to be Poisson with rate α_1, and each of the three i.i.d service distributions is exponential with mean μ_i^{-1}, $i = 1, 2, 3$.

The network is controlled using a three-dimensional cumulative allocation process Z that is subject to the constraints

$$Z(t_1) - Z(t_0) \ge 0, \quad C[Z(t_1) - Z(t_0)] \le (t_1 - t_0)1, \qquad 0 \le t_0 \le t_1 < \infty,$$

where the constituency matrix C is the 2×3 matrix

$$C = \begin{bmatrix} 1 & 0 & 1 \\ 0 & 1 & 0 \end{bmatrix}. \tag{4.8}$$

A three-dimensional CRW model can be constructed in analogy with the M/M/1 queue. Let $\{R^{a1}(t), R^{si}(t), i = 1, 2, 3\}$ denote the independent Poisson processes associated with the arrival stream and the three service processes, and for a given initial condition $Q(0) = x \in \mathbb{Z}_+^3$ define

$$Q_1(t) = x_1 - R^{s1}(Z_1(t)) + R^{a1}(t)$$
$$Q_2(t) = x_2 - R^{s2}(Z_2(t)) + R^{s1}(Z_1(t))$$
$$Q_3(t) = x_3 - R^{s3}(Z_3(t)) + R^{s2}(Z_2(t)), \qquad t \geq 0,$$

Consider now the renewal process defined as the sum

$$R(t) = R^a(t) + R^{s1}(t) + R^{s2}(t) + R^{s3}(t), \qquad t \geq 0,$$

and to simplify the notation, assume that time has been scaled so that $\mu_1 + \mu_2 + \mu_3 + \alpha_1 = 1$. We let $\{T(n) : n \geq 1\}$ denote the associated renewal times at which $\boldsymbol{R}$ exhibits successive jumps, and set $T(0) = 0$. To construct a CRW model via uniformization we impose two additional restrictions on the service discipline: $Z(t)$ is constant on the time interval $[T(n), T(n+1))$ for each $n \geq 0$, and $\frac{d^+}{dt} Z_i(t) \in \{0, 1\}$ for each i and t.

The controlled process sampled at the renewal times $\{T(n)\}$ evolves as the CRW model of a re-entrant line with routing matrix given by

$$R^\mathsf{T} = \begin{bmatrix} 0 & 0 & 0 \\ 1 & 0 & 0 \\ 0 & 1 & 0 \end{bmatrix}. \tag{4.9}$$

That is, (4.6) holds for this example with $\ell = 3$, and each of the random variables $\{M_1(t), M_2(t), M_3(t), A_1(t)\}$ is Bernoulli.

If the service times at buffers 1 and 3 are equal then the model can be simplified. In this case we define the service time at Station 1 using the single renewal process $\boldsymbol{R}^{s1}$, and we redefine the renewal process $\boldsymbol{R}$ as follows:

$$R(t) = R^a(t) + R^{s1}(t) + R^{s2}(t), \qquad t \geq 0.$$

Re-scaling time so $\mu_1 + \mu_2 + \alpha_1 = 1$, and letting $\{T(n) : n \geq 1\}$ denote the associated renewal times, we again obtain a CRW model, but the model is now homogeneous, of the form

$$Q(t+1) = Q(t) + S_1(t+1)\left((-\mathbf{1}^1 + \mathbf{1}^2)U_1(t) - \mathbf{1}^3 U_3(t)\right)$$
$$+ S_2(t+1)(-\mathbf{1}^2 + \mathbf{1}^3)U_2(t) + A_1(t+1)\mathbf{1}^1, \qquad t \geq 0, \tag{4.10}$$

where the three-dimensional i.i.d. process $(\boldsymbol{S}_1, \boldsymbol{S}_2, \boldsymbol{A}_1)$ has marginal distribution defined by

$$\mathsf{P}\{(S_1(t), S_2(t), A_1(t))^\mathsf{T} = \mathbf{1}^i\} = p_i, \qquad t \geq 1, \ i = 1, 2, 3, \tag{4.11}$$

with $p_i = \mu_i$ for $i = 1, 2$, and $p_3 = \alpha_1$.

4.1.2 Policies

In constructing and analyzing policies we embed the CRW model within the general framework of Markov Decision Theory, so that $(\boldsymbol{Q}, \boldsymbol{U})$ is viewed as a *Markov Decision Process*, or MDP. In this setting there are several broad classes of allocation processes, based on the assumption that $\boldsymbol{Q}$ is restricted to the discrete state space $\mathsf{X}_\diamond$:

(i) A (deterministic) *policy* ϕ is a sequence of functions $\{\phi^t : t \in \mathbb{Z}_+\}$, from $\mathsf{X}_\diamond^{t+1}$ to $\mathsf{U}_\diamond$ such that $\phi^t(x_0, \ldots, x_{t-1}, x_t) \in \mathsf{U}_\diamond(x_t)$ for each $t \geq 0$, and each sequence $(x_0, \ldots, x_{t-1}, x_t) \in \mathsf{X}_\diamond^{t+1}$. For each t the allocation vector is defined by

$$U(t) = \phi^t(Q(0), \ldots, Q(t-1), Q(t)),$$

so that $\boldsymbol{U}$ is adapted.

(ii) The policy ϕ is called *Markov* if ϕ^t depends only on x_t for each $t \geq 0$.

(iii) If ϕ is Markov, and if there is a fixed function ϕ such that $\phi^t = \phi$ for all t, then the policy is *stationary*. The function $\phi \colon \mathsf{X}_\diamond \to \mathsf{U}_\diamond$ is called a *feedback law*.

(iv) A *randomized stationary policy* is defined by a mapping $\phi \colon \mathsf{X}_\diamond \to \mathcal{P}(\mathsf{U}_\diamond)$, where $\mathcal{P}(\mathsf{U}_\diamond)$ denotes the set of probability measures on $\mathsf{U}_\diamond$. For each $x \in \mathsf{X}_\diamond$, $\phi(x) = \{\phi_u(x) : u \in \mathsf{U}_\diamond\}$ satisfies $\phi_u(x) = 0$ for $u \notin \mathsf{U}_\diamond(x)$. The control sequence is defined by the family of laws,

$$\mathsf{P}\{U(t) = u \mid Q(0), \ldots, Q(t), U(0), \ldots, U(t-1)\} = \phi_u(Q(t)), \qquad t \geq 0.$$

A stationary policy can be regarded as a randomized stationary policy in which the law ϕ is degenerate, in the sense that $\phi(x)$ assigns all mass to a single point in $\mathsf{U}_\diamond(x)$ for each $x \in \mathsf{X}_\diamond$.

Randomized policies are useful in the theory of multiobjective optimization, and in linear-programming approaches to optimal control. These topics are treated in Chapter 9. Throughout much of this book we restrict to deterministic stationary policies.

In the theory of MDPs policies are typically compared through a given cost function $c \colon \mathbb{R}_+^\ell \to \mathbb{R}_+$. Here we assume that c admits an extension to all of $\mathbb{R}^\ell$ to define a norm. Examples are the ℓ_1-norm $c(x) = |x|$, or any piecewise linear function that is strictly positive on $\mathbb{R}_+^\ell \setminus \{0\}$.

The *controlled transition matrix* is defined for $x, y \in \mathsf{X}_\diamond$ and $u \in \mathsf{U}_\diamond(x)$ by

$$P_u(x, y) := \mathsf{P}\{Q(t+1) = y \mid Q(t) = x, U(t) = u\}. \tag{4.12}$$

When controlled using a randomized stationary policy ϕ, the queue length process is a Markov chain with transition matrix,

$$P_\phi(x, y) := \sum_{u \in \mathsf{U}_\diamond(x)} \phi_u(x) P_u(x, y), \qquad x, y \in \mathsf{X}_\diamond.$$

If the stationary policy is deterministic, then the sum reduces to

$$P_\phi(x, y) := P_{\phi(x)}(x, y), \qquad x, y \in \mathsf{X}_\diamond.$$

Although the state space is typically infinite, many well-known concepts for finite state space Markov chains carry over to this more complex setting. The appendix contains a survey of relevant results from this theory, and Chapter 8 contains applications of this theory to the CRW model.

Several approaches to performance evaluation for Markov models are based on solutions to the following *Poisson inequality*: For a constant $\bar{\eta} < \infty$ and function $h \colon \mathsf{X} \to \mathbb{R}_+$,

$$\mathcal{D}h \leq -c + \bar{\eta}, \tag{4.13}$$

where $\mathcal{D}$ denotes the difference operator $\mathcal{D} = P - I$. This is a relaxation of the Poisson equation (3.8) introduced for the M/M/1 queue.

Many consequences of this bound in the general CRW model are described in Chapter 8. The simplest result is contained in the following application of the Comparison Theorem A.4.3.

Proposition 4.1.4. *Suppose that the CRW scheduling model is controlled using a stationary policy, and that there exist nonnegative-valued functions c and h satisfying (4.13). Then, for each initial condition,*

$$\frac{1}{r} \sum_{t=0}^{r-1} \mathsf{E}\big[c(Q(t; x))\big] \leq \frac{1}{r} h(x) + \bar{\eta}, \qquad r \geq 1, \tag{4.14}$$

and $\quad \eta_x := \limsup_{r \to \infty} \frac{1}{r} \sum_{t=0}^{r-1} \mathsf{E}\big[c(Q(t; x))\big] \leq \bar{\eta}.$

Proof. Although the result follows from Theorem A.4.3 we provide the brief proof here. The bound (4.13) can be written $Ph \leq h - c + \bar{\eta}$, and on applying P to both sides,

$$P^2 h \leq Ph - Pc + \bar{\eta} \leq h - (c + Pc) + 2\bar{\eta}.$$

By induction we obtain for each $r \geq 1$,

$$P^r h \leq h - (c + Pc + \cdots + P^{r-1}c) + r\bar{\eta}.$$

This implies the two bounds (4.14) using nonnegativity of h combined with the identity

$$P^r h(x) + (c(x) + Pc(x) + \cdots + P^{r-1}c(x)) = \mathsf{E}\Big[h(Q(r; x))) + \sum_{i=0}^{r-1} c(Q(i; x))\Big].$$

$\square$

The Poisson inequality is a foundation of optimization theory for the CRW model: The solution to (4.13) with $\bar{\eta}$ *minimal* solves the average-cost optimality equation described next.

4.1.3 Optimization

Given a function $h\colon \mathsf{X} \to \mathbb{R}_+$, the *myopic policy* is the stationary policy defined as the constrained minimization

$$\phi(x) \in \arg\min_{u\in\mathsf{U}_\circ(x)} \mathsf{E}[h(Q(t+1)) \mid Q(t) = x,\ U(t) = u]. \qquad (4.15)$$

Note that we write "$\phi(x) \in \arg\min_{u\in\mathsf{U}_\circ(x)}\cdots$" since the minimizer may not be unique. We call this the *h-myopic policy* when we wish to emphasize the function used in the construction of ϕ. The h-myopic policy can be equivalently expressed in terms of the controlled transition matrix (4.12) by the minimization,

$$\phi(x) \in \arg\min_{u\in\mathsf{U}_\circ(x)} \left(\sum_y P_u(x,y)h(y) \right). \qquad (4.16)$$

Myopic policies are of interest because of their simplicity. They are also "optimal" if the function h is carefully chosen.

The *discounted-cost value function* for a given policy is defined as

$$h_\gamma(x) := \sum_{t=0}^{\infty} (1+\gamma)^{-t-1} \mathsf{E}\big[c(Q(t;x))\big]. \qquad (4.17)$$

For each x we let $h_\gamma^*(x)$ denote the minimum of (4.17) over all policies. A policy is *discounted-cost optimal* if $h_\gamma(x) = h_\gamma^*(x)$ for each x.

The *discounted-cost optimality equation* is expressed in (4.18a); the policy defined in (4.18b) is h-myopic with $h = h_\gamma^*$.

$$(1+\gamma)h_\gamma^*(x) = \min_{u\in\mathsf{U}_\circ(x)} [c(x) + P_u h_\gamma^*(x)] \qquad (4.18a)$$

$$\phi^*(x) \in \arg\min_{u\in\mathsf{U}_\circ(x)} P_u h_\gamma^*(x), \qquad x \in \mathsf{X}. \qquad (4.18b)$$

Under mild conditions we find that ϕ^* does define an optimal policy. Theorem 4.1.5 is a special case of Proposition 9.6.2.

Theorem 4.1.5 (Value function for the scheduling model). *Suppose that* $\mathsf{E}[\|A(1)\|] < \infty$, *and that c is a norm on $\mathbb{R}^\ell$. Then:*

(i) *The function $h_\gamma^*(x)$ defined as the minimum of (4.17) over all policies has linear growth*

$$\sup_{x\in\mathsf{X}} \frac{|h_\gamma^*(x)|}{1+\|x\|} < \infty, \qquad (4.19)$$

and solves the discounted-cost optimality equation (4.18a).

(ii) *The h_γ^*-myopic policy ϕ^* in (4.18b) is discounted-cost optimal.*

(iii) *If h_γ° is any other solution with linear growth then $h_\gamma^\circ = h_\gamma^*$ everywhere.*

Recall that a myopic policy was used to motivate the c–μ rule for the Klimov model in Section 2.2 based on $c = h$. However, we shall see in several examples in Section 4.4

that this approach does not always lead to an effective policy. Ideally we would like to choose h based on a given cost function c so that the Poisson inequality (4.13) holds under the h-myopic policy with a small value of $\bar{\eta}$. In the average-cost optimality criterion considered next we take the "optimal" choice in which $\bar{\eta}$ is minimal.

The *average cost* for a given policy appears as the limit supremum in (4.14),

$$\eta_x := \limsup_{r \to \infty} \frac{1}{r} \sum_{t=0}^{r-1} \mathsf{E}_x[c(Q(t))]. \tag{4.20}$$

A policy is called *stabilizing* if the average cost is finite for each initial condition $Q(0) = x$. We will see that typically η_x is independent of x, and can be expressed as the limiting discounted or average cost,

$$\eta_x = \lim_{\gamma \downarrow 0} \gamma h_\gamma(x) = \lim_{t \to \infty} \mathsf{E}_x[c(Q(t))].$$

The optimal cost η_x^* is defined to be the infimum of (4.20) over all policies; a policy is called *average-cost optimal* if it achieves η_x^* for each initial condition. Under general conditions, described in Chapter 9 and specialized to the scheduling model in Theorem 4.1.6, the average-cost optimal policy is again myopic, based on the solution of the following equations in the pair of variables (h^*, η^*):

$$\eta^* + h^*(x) = \min_{u \in \mathsf{U}(x)} [c(x) + P_u h^*(x)] \tag{4.21a}$$

$$\phi^*(x) \in \operatorname*{arg\,min}_{u \in \mathsf{U}(x)} P_u h^*(x), \qquad x \in \mathsf{X}. \tag{4.21b}$$

The equality (4.21a) is known as the *average-cost optimality equation* (ACOE), and the function h^* is called a *relative value function*. The second equation (4.21b) defines a stationary policy ϕ^* that is myopic with respect to h^*.

When c is a norm on X then h^* has quadratic growth, just as seen in the M/M/1 queue.

Theorem 4.1.6 (Relative value function for the scheduling model). *Suppose that* $\mathsf{E}[\|A(1)\|^2] < \infty$, *and that* c *is a norm on* $\mathbb{R}^\ell$. *Suppose moreover that a stabilizing policy exists. Then:*

(i) *There exists a solution* (h^*, η^*) *to the dynamic programming equation (4.21a) with* η^* *constant, and* h^* *nonnegative with quadratic growth,*

$$\sup_{x \in \mathsf{X}} \frac{h^*(x)}{1 + \|x\|^2} < \infty. \tag{4.22}$$

(ii) *The* h^*-*myopic policy is average-cost optimal.*

(iii) *If* h° *is any other solution with quadratic growth then there is a constant* b *such that* $h^* = h^\circ + b$ *everywhere.*

Proof. The details are postponed to Chapter 9. However, we note that the assumptions of Theorem 9.0.5 hold:

(a) The cost function c is "coercive" since it is a norm.
(b) If there is a stabilizing policy, then there exists a policy satisfying a somewhat stronger condition called *regularity*. □

An important question left unanswered is, *when does a stabilizing policy exist?* We first provide an answer for the simpler fluid model.

4.2 Fluid model

Here a formal definition of the fluid model is defined based on the CRW model (4.2), under the assumption that the following expectations are finite for each i and t,

$$\alpha = \mathsf{E}[A(t)], \quad M = \mathsf{E}[M(t)], \quad \text{and} \quad \mu_i = \mathsf{E}[M_i(t)]. \tag{4.23}$$

The fluid model is a controlled, ordinary differential equation (ODE) model that may be viewed as a description of the *mean* behavior of its discrete-stochastic counterpart.

Definition 4.2.1 (Fluid scheduling model). The ℓ-dimensional, continuous-time process q represents the queue-length process in this deterministic setting, and satisfies the equations,

$$q(t) = x + Bz(t) + \alpha t, \qquad t \ge 0, \ x \in \mathbb{R}_+^\ell. \tag{4.24}$$

It is assumed that the fluid model is consistent with Definition 4.1.1, in the sense that,

(i) The cumulative allocation process z is subject to the linear rate constraints

$$\frac{z(t_1) - z(t_0)}{t_1 - t_0} \in \mathsf{U}, \qquad 0 \le t_0 < t_1 < \infty,$$

where the polyhedral control set U is defined in (4.3).
(ii) The queue length process (or *state trajectory*) q is constrained to X. A deterministic process z satisfying (i) and (ii) is called *admissible*.
(iii) The matrix R in (4.25) is equal to the $\ell \times \ell$ routing matrix.
(iv) The vector $\alpha \in \mathbb{R}_+^\ell$ represents exogenous arrival rates for the fluid model, and coincides with the mean defined via (4.23). Similarly, the matrix B is defined as

$$B := -[I - R^\mathsf{T}]M, \tag{4.25}$$

with $M := \operatorname{diag}(\mu_i) = \mathsf{E}[M(1)]$.
(v) The servers at Station s are said to be *homogeneous* if $\mu_{j_1} = \mu_{j_2}$ whenever $s(j_1) = s(j_2) = s$. If each station is homogeneous, then the fluid model is called homogeneous. ∎

The allocation-rate vector is defined as $\zeta(t) = \frac{d^+}{dt}z(t)$, $t \geq 0$. Using this notation, the fluid model is expressed as the controlled ODE

$$\frac{d^+}{dt}q_i(t) = \alpha_i - \mu_i\zeta_i(t) + \sum_{j=1}^{\ell} \mu_j\zeta_j(t)R_{ji}, \qquad 1 \leq i \leq \ell.$$

It is sometimes convenient to envision this ODE model as a *differential inclusion*,

(i) The state q is constrained to evolve in the polyhedron X.
(ii) For each $t \geq 0$, the velocity vector $v(t) = \frac{d^+}{dt}q(t)$ is constrained to lie in the polyhedron

$$\mathsf{V} := \{v = B\zeta + \alpha : \zeta \in \mathsf{U}\}. \tag{4.26}$$

Control of the fluid model amounts to driving the queue-length process to the origin in an efficient manner. This leads to the following definitions:

Definition 4.2.2 (Stabilizability). The following terminology applies to the fluid model (4.24):

(i) The fluid model is called *stabilizable* if, for any $x \in \mathsf{X}$, there exists $\zeta \in \mathsf{U}$ and a time $T \geq 0$ such that

$$x + (B\zeta + \alpha)T = 0. \tag{4.27}$$

(ii) The *minimal draining time*, denoted $T^*(x)$, is defined for $x \in \mathsf{X}$ as the smallest $T \geq 0$ satisfying (4.27) for some $\zeta \in \mathsf{U}$.
(iii) More generally, for two states $x, y \in \mathsf{X}$, the *minimal time to reach y from x* over all policies is denoted $T^*(x, y)$.
(iv) The minimal draining time for the arrival-free model (in which $\alpha = 0$) is denoted $W^*(x)$. ∎

The minimal draining times T^* and W^* tell a great deal about network congestion, and we shall see that they also provide tools for the design of efficient policies that define z as a function of q.

Stabilizability is intimately connected to workload.

Definition 4.2.3 (Workload in units of time)

(i) The *workload matrix* is defined as the $\ell_m \times \ell$ matrix $\Xi = -CB^{-1} = M^{-1}(I - R^{\mathsf{T}})^{-1}$, where the inverse exists as a power series,

$$[I - R]^{-1} = \sum_{k=0}^{\infty} R^k = \sum_{k=0}^{\ell-1} R^k. \tag{4.28}$$

(ii) The *workload vectors* $\{\xi^s : s \in \mathcal{S}\}$ are defined as the rows of Ξ, so that

$$\Xi = [\xi^1 \mid \cdots \mid \xi^{\ell_m}]^{\mathsf{T}}.$$

(iii) The *workload process* (in units of time) is defined by $w(t) = \Xi q(t)$, $t \geq 0$.
(iv) The *vector load* is defined as $\rho = \Xi\alpha$, and the *network load* is the maximum, $\rho_\bullet = \max_{s \in \mathcal{S}} \rho_s$. Station s is called a *bottleneck* if $\rho_\bullet = \rho_s$. ∎

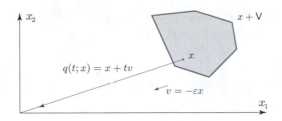

Figure 4.1. A time optimal trajectory in which **q** moves toward the origin in a straight line. The velocity vector v is of the form $v = -\varepsilon x$ for some $\varepsilon > 0$, and also has the form $v = B\zeta^* + \alpha$ with ζ^* given in (4.30).

For example, in the single server queue we have $w(t) = W^*(q(t)) = \mu^{-1}q(t)$, which is precisely the time the server must work to clear all of the inventory *currently in the system at time t*, ignoring future arrivals.

This interpretation can be generalized to the general scheduling model. It is shown in Proposition 4.2.5 below that $w_s(t)$ is the time required at Station s to process all of the customers currently in the system at time t (ignoring future arrivals). The proof is based on the following identity.

Proposition 4.2.4. *The following holds for any* $\zeta \in \mathsf{U}$, $s \in \mathcal{S}$,

$$\langle \xi^s, B\zeta + \alpha \rangle = -(1 - \rho_s) + \iota,$$

where the idleness rate is defined by $\iota = 1 - C\zeta \geq 0$.

Proof. By definition of the workload matrix we have

$$\Xi B\zeta = -CB^{-1}B\zeta = -C\zeta, \qquad \zeta \in \mathsf{U}.$$

The proposition is a restatement of this vector equality, and the definition $\rho_s = \langle \xi^s, \alpha \rangle$. $\qquad \square$

Proposition 4.2.5 contains a characterization of stabilizability, along with justification for the terminology in Definition 4.2.3. A state trajectory under the time optimal allocation (4.30) is illustrated in Fig. 4.1.

Proposition 4.2.5. *The following hold for the scheduling model:*

(i) *The model is stabilizable if, and only if,* $\rho_\bullet < 1$. *If this load condition holds, then the minimal draining time is given by*

$$T^*(x) = \max_{s \in \mathcal{S}} \frac{\langle \xi^s, x \rangle}{1 - \rho_s}, \qquad x \in \mathbb{R}_+^\ell. \tag{4.29}$$

Moreover, an example of a time optimal allocation from the initial condition $x \in \mathsf{X}$ *is given by* $\zeta^*(t) = \zeta^*$ *on* $[0, T^*(x))$, *where*

$$\zeta^* = -B^{-1}\left(\alpha + \frac{1}{T^*(x)}x\right). \tag{4.30}$$

(ii) $W^*(x) = \max_{s \in \mathcal{S}}\langle \xi^s, x \rangle$ *for* $x \in \mathsf{X}$.

Proof. Part (ii) follows from (4.29) since $\rho_s = 0$ for each s in the arrival-free model.

For a given $x \in \mathsf{X}$, let $T^\circ(x) = \max_{s \in \mathcal{S}} (1 - \rho_s)^{-1} \langle \xi^s, x \rangle$. We first demonstrate the bound $T^*(x) \geq T^\circ(x)$.

Consider any admissible cumulative allocation process z, and suppose that for some $T > 0$ we have $q(T) = x + Bz(T) + \alpha T = 0$. Writing $\zeta = T^{-1} z(T)$ we have $\zeta \in \mathsf{U}$, and hence by Proposition 4.2.4 we obtain the vector inequality

$$0 = \Xi q(T) = \Xi\Big(x + (B\zeta + \alpha)T\Big) \geq \Xi x - (1 - \rho)T.$$

That is,

$$T \geq \frac{\langle \xi^s, x \rangle}{1 - \rho_s}, \qquad s \in \mathcal{S},$$

which establishes the desired lower bound $T^*(x) \geq T^\circ(x)$.

To complete the proof we now demonstrate that this lower bound is tight by explicitly constructing a time optimal solution from the initial condition $x \in \mathsf{X}$. This is already given in (4.30): On applying the constant allocation rate $\zeta^*(t) = \zeta^*$ on $[0, T^*]$, the state trajectory is given by

$$q(t) = x + (B\zeta^* + \alpha)t = \frac{T^* - t}{T^*} x, \qquad 0 \leq t \leq T^*,$$

from which it follows that $q(T^*(x)) = 0$.

To ensure admissibility of ζ^*, it is necessary that

$$\text{(a) } \zeta^* \geq 0, \qquad \text{(b) } C\zeta^* \leq 1.$$

The first condition is automatic since B^{-1} has nonnegative entries due to (4.28).

From the definition of T^* we have for each s, $(\Xi x)_s \leq (1 - \rho_s)T^*(x)$ provided $\rho_s < 1$. Consequently, from the definition of ζ^* we obtain the vector bound $C\zeta^* = \rho + (\Xi x)/T^*(x)$. We have $\langle \xi^s, x \rangle \leq (1 - \rho_s)T^*(x)$ when $\rho_\bullet < 1$, so that for each $s \in \mathcal{S}$,

$$(C\zeta^*)_s \leq \rho_s + \frac{1}{T^*(x)} \langle \xi^s, x \rangle \leq \rho_s + (1 - \rho_s) = 1. \qquad \square$$

Example 4.2.6 (Klimov model). The fluid model in the four-buffer model illustrated in Fig. 2.3 is defined through the differential equation (4.24) with $B = -\text{diag}(\mu_1, \ldots, \mu_4)$. The constituency matrix is the row vector $C = [1, 1, 1, 1]$, and hence the single workload vector is given by $\xi^\mathsf{T} = -CB^{-1} = (\mu_1^{-1}, \ldots, \mu_4^{-1})$. The "vector load" is equal to the network load,

$$\rho_\bullet = \langle \xi, \alpha \rangle = \sum_{i=1}^{4} \frac{\alpha_i}{\mu_i},$$

and the minimal draining time is linear,

$$T^*(x) = \frac{\xi^\mathsf{T} x}{1 - \rho_\bullet}, \qquad x \in \mathbb{R}_+^4.$$

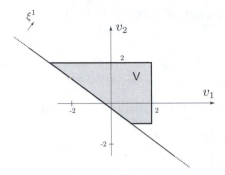

Figure 4.2. Velocity space for the processor sharing model with parameters defined in (4.31).

Any allocation rate process ζ satisfying $\sum_i \zeta_i(t) = 1$ whenever $\sum_i q_i(t) > 0$ is time-optimal for this single-station model. ∎

Example 4.2.7 (**Processor sharing model**). The processor sharing model illustrated in Fig. 2.6 is not a scheduling model since the number of activities is strictly greater than the number of buffers. However, we have seen that the fluid model can be reduced to the two-dimensional version of the Klimov model shown in (2.16). This is expressed as (4.24) with $\ell = 2$, and

$$B = \begin{bmatrix} -\mu_b & 0 \\ 0 & -\mu_c \end{bmatrix}, \quad C = \begin{bmatrix} 1 & 1 \end{bmatrix}, \quad \alpha = \begin{bmatrix} \alpha_a - \mu_a, \\ \alpha_b \end{bmatrix}.$$

Consequently, the vector load is again one-dimensional, and coincides with the network load,

$$\rho_\bullet = -CB^{-1}\alpha = \frac{\alpha_a - \mu_a}{\mu_b} + \frac{\alpha_c}{\mu_c}.$$

The velocity space for the two-dimensional model is defined as $\mathsf{V} = \{v : v = B\zeta + \alpha, \, C\zeta \le 1\}$, or

$$\mathsf{V} = \left\{ \begin{pmatrix} -\mu_a\zeta_a - \mu_b\zeta_b \\ -\mu_c\zeta_c \end{pmatrix} + \begin{pmatrix} \alpha_a - \mu_a, \\ \alpha_b \end{pmatrix} : \zeta \ge (0,0,0)^\mathsf{T}, \, \zeta_a \le 1, \zeta_b + \zeta_c \le 1 \right\}.$$

This is illustrated in Fig. 4.2 for the specific parameters

$$\begin{aligned} \alpha_a &= 2, \, \alpha_b = 2 \\ \mu_a &= 1, \, \mu_b = 4, \, \mu_c = 3. \end{aligned} \tag{4.31}$$

∎

Example 4.2.8 (**Simple re-entrant line**). The fluid model for the simple re-entrant line shown in Fig. 2.9 is expressed as the controlled ODE model in (2.8). Consequently, the differential equation (2.8) can be written as (4.24) with

$$B = -(I - R^\mathsf{T})M = \begin{bmatrix} -\mu_1 & 0 & 0 \\ \mu_1 & -\mu_2 & 0 \\ 0 & \mu_2 & -\mu_3 \end{bmatrix} \quad \alpha = \begin{bmatrix} \alpha_1 \\ 0 \\ 0 \end{bmatrix}.$$

Applying (4.28) we conclude that $B^{-1} = -M^{-1}[I + R^\mathsf{T} + R^{2\mathsf{T}}]$ since

$$[I - R]^{-1} = \sum_{k=0}^{\infty} R^k = I + R + R^2 = \begin{bmatrix} 1 & 0 & 0 \\ 1 & 1 & 0 \\ 1 & 1 & 1 \end{bmatrix}.$$

Based on the definition $\Xi = -CB^{-1}$, the two workload vectors are expressed

$$\begin{bmatrix} \xi^{1\mathsf{T}} \\ \xi^{2\mathsf{T}} \end{bmatrix} = \begin{bmatrix} 1 & 0 & 1 \\ 0 & 1 & 0 \end{bmatrix} \begin{bmatrix} \mu_1^{-1} & 0 & 0 \\ \mu_2^{-1} & \mu_2^{-1} & 0 \\ \mu_3^{-1} & \mu_3^{-1} & \mu_3^{-1} \end{bmatrix} = \begin{bmatrix} \mu_1^{-1} + \mu_3^{-1} & \mu_3^{-1} & \mu_3^{-1} \\ \mu_2^{-1} & \mu_2^{-1} & 0 \end{bmatrix} \tag{4.32}$$

$$\begin{bmatrix} \rho_1 \\ \rho_2 \end{bmatrix} = \Xi\alpha = \begin{bmatrix} \frac{\alpha_1}{\mu_1} + \frac{\alpha_1}{\mu_3} \\ \frac{\alpha_1}{\mu_2} \end{bmatrix}. \tag{4.33}$$

Proposition 4.2.5 implies that this model is stabilizable if, and only if

$$\rho_1 = \frac{\alpha_1}{\mu_1} + \frac{\alpha_1}{\mu_3} < 1 \quad \text{and} \quad \rho_2 = \frac{\alpha_1}{\mu_2} < 1. \qquad \blacksquare$$

Example 4.2.9 (KSRS model). The fluid model for the network shown in Fig. 2.12 is defined by the parameters

$$R^\mathsf{T} = \begin{bmatrix} 0 & 0 & 0 & 0 \\ 1 & 0 & 0 & 0 \\ 0 & 0 & 0 & 0 \\ 0 & 0 & 1 & 0 \end{bmatrix} \qquad C = \begin{bmatrix} 1 & 0 & 0 & 1 \\ 0 & 1 & 1 & 0 \end{bmatrix},$$

and (4.25) then gives

$$B = \begin{bmatrix} -\mu_1 & 0 & 0 & 0 \\ \mu_1 & -\mu_2 & 0 & 0 \\ 0 & 0 & -\mu_3 & 0 \\ 0 & 0 & \mu_3 & -\mu_4 \end{bmatrix}.$$

There are external arrival at buffers 1 and 3 only, so that $\alpha_2 = \alpha_4 = 0$.

We have $[I - R]^{-1} = \sum_{k=0}^{\infty} R^k = I + R$ since $R^k = 0$ for $k \geq 2$. From (4.28) we then have $B^{-1} = -M^{-1}[I + R^\mathsf{T}]$, and hence the workload matrix is given by

$$\Xi = -CB^{-1} = \begin{bmatrix} \mu_1^{-1} & 0 & \mu_4^{-1} & \mu_4^{-1} \\ \mu_2^{-1} & \mu_2^{-1} & \mu_3^{-1} & 0 \end{bmatrix}. \tag{4.34}$$

The vector load is expressed

$$\rho = \Xi\alpha = \begin{bmatrix} \frac{\alpha_1}{\mu_1} + \frac{\alpha_3}{\mu_4} \\ \frac{\alpha_1}{\mu_2} + \frac{\alpha_3}{\mu_3} \end{bmatrix}. \qquad \blacksquare$$

In the remainder of this chapter we survey many of the control techniques to be developed over the course of this book. We begin with the fluid model in Section 4.3. After describing general classes of feedback laws, we then begin what will become an ongoing discussion on how to translate a policy from the fluid model to a more realistic discrete and stochastic network model.

4.3 Control techniques for the fluid model

The fluid model can be viewed as a state space model with state process q, constrained to the polyhedral state space $\mathsf{X} \subset \mathbb{R}_+^\ell$, and controlled by the cumulative allocation process z with rate constraints specified in (6.2). The control approaches described in this section are defined through *state feedback*, $\zeta(t) = \frac{d^+}{dt} z(t; x) = f(q(t))$, where the *feedback law* f is a measurable function from X to the control set U. The feedback law is said to be *stabilizing* if $q(t; x) \to \mathbf{0}$ as $t \to \infty$, from each initial condition x. A stabilizing feedback law can only exist if $\rho_\bullet < 1$.

It is assumed throughout this section that the state space is of the following form,

$$\mathsf{X} = \{x \in \mathbb{R}_+^\ell : x_i \leq b_i, \quad 1 \leq i \leq \ell\}, \tag{4.35}$$

where $0 < b_i \leq \infty$ for each i. When $b_i = \infty$ we interpret $x_i \leq b_i$ as a strict inequality.

Recall the definition of the velocity space V in (4.26). Policies for the fluid model must always respect the following constraints: Given any $x \in \mathsf{X}$, the allocation rate ζ satisfies $\zeta \in \mathsf{U}$, and the corresponding velocity vector $v = B\zeta + \alpha \in \mathsf{V}$ satisfies

$$v_i \begin{cases} \geq 0, & \text{if } x_i = 0; \\ \leq 0, & \text{if } x_i = b_i, \quad i \in \{1, \ldots, \ell\}. \end{cases} \tag{4.36}$$

Many of the policies considered in this book are based on a cost function that reflects our desire to control the dynamic behavior of the network. It is assumed here that the cost function $c \colon \mathsf{X} \to \mathbb{R}_+$ is a linear function of queue lengths, of the form $c(x) = c^\mathsf{T} x$, with $c_i > 0$ for each $i = 1, \ldots, \ell$. A typical choice is the ℓ_1-*norm* or *total inventory* $c(x) = |x| := \sum_i x_i$.

Extensions to piecewise linear or smooth convex cost functions are considered in later chapters. In Section 6.3.2, we discuss control techniques for routing models based on a cost function depending on allocation rates.

We have already considered formulations of optimal control for the fluid model. Here are three criteria that are most useful in the applications considered elsewhere in this book.

Time optimal control For each initial condition $q(0) = x$, find a control that minimizes

$$T(x) = \min\{t : q(t; x) = \mathbf{0}\}.$$

Infinite-horizon optimal control For each initial condition $q(0) = x$, find a control that minimizes the total cost

$$J(x) = \int_0^\infty c(q(t; x)) \, dt. \tag{4.37}$$

Discounted infinite-horizon optimal control Fix a constant $\gamma > 0$. For each initial condition $q(0) = x$, find a control that minimizes the discounted cost

$$J_\gamma(x) = \int_0^\infty e^{-\gamma t} c(q(t; x))\, dt. \tag{4.38}$$

Recall from Proposition 4.2.5 that $T^*(x)$ denotes the minimal draining time for the fluid model from the initial condition $x \in \mathsf{X}$. We let $J^*(x)$ (respectively $J_\gamma^*(x)$) denote the "optimal cost," that is, the infimum over all policies, in the respective infinite-horizon control problems. Each of these three functions on X is known as the (optimal) value function for the associated optimal control problem.

Time optimality is a central theme in this book. A time optimal feedback law is certainly stabilizing, and has the desirable property that congestion is cleared from the system in minimal time. The function $T^* \colon \mathsf{X} \to \mathbb{R}_+$ satisfies the following *dynamic programming equation*:

Proposition 4.3.1. *If q^* is a time optimal state trajectory starting from $x \in \mathsf{X}$, then for each pair of time points satisfying $0 \le t_0 < t_1 \le T^*(x)$:*

$$T^*(q^*(t_1)) = T^*(q^*(t_0)) - (t_1 - t_0). \tag{4.39}$$

Proof. The identity (4.39) simply states that, after $t_1 - t_0$ seconds have passed, the time required to reach the origin is reduced by precisely this amount. $\square$

A lower bound on the cost at a specific time t, given a specific initial condition $x \in \mathsf{X}$, is found by solving the linear program

$$
\begin{aligned}
\underline{c}^*(t; x) := \quad &\textbf{min} \quad \langle c, y \rangle \\
&\textbf{s.t.} \quad y = x + Bz + \alpha t \\
&\qquad\quad Cz \le t1 \\
&\qquad\quad y, z \ge 0.
\end{aligned} \tag{4.40}
$$

A feasible state trajectory q^* starting from x is called *pathwise optimal* if $c(q^*(t; x)) = \underline{c}^*(t; x)$ for every t. A path-wise optimal feedback law is simultaneously time optimal and infinite-horizon optimal. Such a strong form of optimality is rare except in very simple examples, such as the Klimov model introduced in Section 2.2.

Proposition 4.3.2 (iii) introduces the dynamic programming equation for the infinite-horizon control problem. Figure 2.11 illustrates Proposition 4.3.2 (i) and (ii) for the simple re-entrant line shown in Fig. 2.9. The proof of the proposition is left as an exercise – it is similar to the proof of Proposition 4.3.6 that follows.

Proposition 4.3.2. *Let J^* denote the value function for the infinite-horizon optimal control problem. Fix $x \in \mathsf{X}$, and let (z^*, q^*) be any optimal allocation-state trajectory starting from x, in the sense that the minimal infinite-horizon cost $J^*(x)$ is achieved. Then:*

(i) *The cost $c(q^*(t))$ is nonincreasing in t;*

(ii) *The cost $c(q^*(t))$ is convex in t, so that its rate of decrease is maximal when $t \approx 0$.*

(iii) *For each t, whenever the gradient of J^* at $y = q^*(t)$ exists we have*

$$\zeta^*(t) \in \arg\min_{\zeta \in U(y)} \langle \nabla J^*(y), B\zeta + \alpha \rangle \tag{4.41}$$

and $\quad -c(y) = \langle \nabla J^*(y), B\zeta^*(t) + \alpha \rangle.$

We now describe some easily computable policies with desirable properties. We first briefly revisit time optimality.

4.3.1 Time-optimality

We have seen in Proposition 4.2.5 that the minimal draining time T^* is easily computed as a function of the initial condition, and that one optimal solution is given by the constant allocation rate (4.30) on $[0, T^*(x))$, resulting in

$$q^*(t) = \left(\frac{T^*(x) - t}{T^*(x)} \right) x, \qquad 0 \le t \le T^*(x).$$

There are many other time optimal policies. Here we provide a complete characterization of time optimality.

The set of indices for which the maximum in the definition of $T^*(x)$ is attained represent "hot spots" in the network. These hotspots will change as the state q evolves in X.

Definition 4.3.3 (Dynamic bottlenecks for the scheduling model). For each $x \in$ X we denote by $\mathcal{S}^*(x)$ the set of stations $s \in \mathcal{S}$ such that

$$\frac{\langle \xi^s, x \rangle}{1 - \rho_s} = T^*(x).$$

The indices $\mathcal{S}^*(x)$ are called the *dynamic bottlenecks* associated with the state x.   ■

A policy is called *nonidling* if for each station $s \in \mathcal{S}$,

$$\sum_{i:s(i)=s} x_i > 0 \implies \sum_{i:s(i)=s} \zeta_i = 1. \tag{4.42}$$

A time optimal trajectory need not be nonidling. It is however essential that any dynamic bottleneck work at capacity. The conditions introduced in (2.24) for time optimality in the simple re-entrant line can be regarded as an application of Proposition 4.3.4.

Proposition 4.3.4. *Suppose that $\rho_\bullet < 1$. Then, for each initial condition $x^0 \in$ X,*

(i) *An admissible allocation-state pair (z, q) achieves the minimal draining time $T^*(x^0)$ if and only if*

$$\frac{d^+}{dt} \langle \xi^s, q(t; x^0) \rangle = -(1 - \rho_s), \tag{4.43}$$

for $s \in \mathcal{S}^*(q(t; x^0))$ and almost every (a.e.) $0 \le t < T^*(x^0)$.

(ii) *Suppose that* z *is a time optimal allocation. Then, with* q *equal to the resulting state process, the set of dynamic bottlenecks* $\mathcal{S}^*(q(t; x^0))$ *is nondecreasing in* t *for* $t \in [0, T^*(x^0))$.

Proof. Whenever (4.43) holds for each $s \in \mathcal{S}^*(q(t; x^0))$, it immediately follows from the representation (4.29) that $\frac{d^+}{dt} T^*(q(t; x^0)) = -1$. Integrating both sides of this equation from $t = 0$ to $T^*(x^0)$ gives for $0 < T \leq T^*(x^0)$,

$$T^*(q(T; x^0)) - T^*(q(0; x^0)) = \int_0^T \left[\tfrac{d^+}{dt} T^*(q(t; x^0)) \right] dt = -T.$$

On rearranging terms we obtain

$$T^*(q(T; x^0)) = T^* - T,$$

where $T^* = T^*(x^0)$. Letting $T \uparrow T^*$ then gives $T^*(q(T^*; x^0)) = 0$, which implies that $q(T^*; x^0) = 0$, so that q is time optimal from the initial condition x^0.

Conversely, if q is time optimal, then the dynamic programming equation (4.39) holds. The condition (4.43) then follows from (4.39) and the representation (4.29).

To prove (ii) we take any $s \in \mathcal{S}^*(q(t_0; x^0))$, and note that by definition we must have for each $t_1 \in (t_0, T^*(x^0))$,

$$\frac{\langle \xi^s, q(t_1; x^0) \rangle}{1 - \rho_s} \leq T^*(q(t_1))$$
$$= T^*(q(t_0)) - (t_1 - t_0).$$
$$= \frac{\langle \xi^s, q(t_0; x^0) \rangle}{1 - \rho_s} - (t_1 - t_0),$$

where the last equality follows from (i) since $s \in \mathcal{S}^*(q(t_0; x^0))$. Proposition 4.2.4 provides an inequality in the reverse direction: Letting $\zeta = (t_1 - t_0)^{-1}(z(t_1) - z(t_0))$,

$$\frac{\langle \xi^s, q(t_1; x^0) \rangle}{1 - \rho_s} - \frac{\langle \xi^s, q(t_0; x^0) \rangle}{1 - \rho_s} = \frac{\langle \xi^s, B\zeta + \alpha \rangle (t_1 - t_0)}{1 - \rho_s} \geq -(t_1 - t_0).$$

We conclude that all of these inequalities are equalities. In particular,

$$\frac{\langle \xi^s, q(t_1; x^0) \rangle}{1 - \rho_s} = T^*(q(t_1)),$$

which means that $s \in \mathcal{S}^*(q(t_1; x^0))$. $\square$

4.3.2 Myopic policies

While consideration of the infinite-horizon optimal policy is well motivated and often practical, it is not necessary to restrict to optimal policies since there are many simple stabilizing policies with attractive characteristics and low complexity. The *myopic feedback law* is one very simple policy.

Definition 4.3.5 (Myopic policy). For a linear function $c: X \to \mathbb{R}_+$, the c-myopic feedback law $\phi^F: X \to U$ is defined for the fluid model as follows: For each $x \in X$, the allocation rate $\zeta = \phi^F(x)$ is defined to be any $\zeta \in U$ that optimizes the linear program

$$\min \langle c, v \rangle$$

$$\begin{aligned}
\textbf{s.t.} \qquad v_i &\geq 0 && \text{for all } i \text{ such that } x_i = 0 \\
v_i &\leq 0, && \text{for all } i \text{ such that } x_i = b_i \\
v &= B\zeta + \alpha \\
\zeta &\in U.
\end{aligned} \qquad (4.44)$$

More generally, given any smooth function $h: X \to \mathbb{R}_+$ we define the h-myopic policy in the same way with objective function $\langle \nabla h(x), v \rangle$. Proposition 4.3.2 (iii) shows that a J^*-myopic policy is infinite-horizon optimal for the fluid model.

The myopic policy is always stabilizing if the model is stabilizable:

Proposition 4.3.6. *Suppose that $\rho_\bullet < 1$. Then, the network satisfies the following properties under any myopic policy:*

(i) *There exists $k_0 < \infty$ such that*

$$q(t; x) = 0, \qquad t \geq k_0 \|x\|, \ x \in X.$$

(ii) *Suppose that the linear program (4.44) has a unique solution ζ^* for each $x \in X$. If from a given initial condition $x^0 \in X$ there exists a path-wise optimal solution q^* starting from x^0, then $q(t; x^0) = q^*(t; x^0)$ for all $t \geq 0$.*

(iii) *For each $x \in X$, $c(q(t; x))$ is convex and nonincreasing in t.*

Proof. Stabilizability implies that there exists $\varepsilon_0 > 0$ such that, for each $x \in X$, $x \neq 0$, there is $\zeta^x \in U$ such that

$$v^x := B\zeta^x + \alpha = -\varepsilon_0 \frac{x}{|x|},$$

where $|x| = \sum x_i$. For example, we can take the time optimal allocation rate defined in (4.30).

Consider any initial condition $x \in X$, and a solution (ζ, q) to the myopic policy. Let $T^\circ = \min\{t \geq 0 : q(t) = 0\}$, set to ∞ if q never reaches the origin. We have, when $y := q(t) \neq 0$,

$$\tfrac{d^+}{dt} c(q(t)) \leq \langle c, v^y \rangle = -\frac{\varepsilon_0}{|y|} \langle c, y \rangle. \qquad (4.45)$$

The right-hand side is bounded above by $-\varepsilon_1 := -\varepsilon_0 \min_i c_i$. Consequently, on integrating both sides of this inequality we obtain, for all $T < T^\circ$,

$$c(q(T)) - c(q(0)) = \int_0^T \tfrac{d^+}{dt} c(q(t))\, dt \leq -\varepsilon_1 T,$$

or $T \leq \varepsilon_1^{-1}[c(x) - c(q(T))]$. It follows that $T^\circ \leq \varepsilon_1^{-1} c(x)$ for $x \in X$, and this completes the proof of (i).

Part (ii) follows from the fact that the path-wise optimal solution, if it exists, is also a myopic solution.

To prove (iii) we first note that $c(q(t))$ is nonincreasing: This follows from the assumption that the derivative $\frac{d^+}{dt}c(q(t)) = \langle c, B\zeta + \alpha \rangle$ is minimal (see (4.45)).

We now establish convexity. Consider any two time points $0 < t_0 < t_1$, and let $x^0 = q(t_0)$, $x^1 = q(t_1)$. Then, the trajectory q' defined below is also feasible on $[t_0, t_1]$,

$$q'(t) = x^0 + \frac{t - t_0}{t_1 - t_0}(x^1 - x^0), \qquad t \in [t_0, t_1],$$

with $q'(t_0) = x^0$ and $q'(t_1) = x^1$. By minimality of the derivative, on using the myopic policy we obtain

$$\frac{d^+}{dt}c(q(t_0)) \le \frac{d^+}{dt}c(q'(t_0)) = \frac{\langle c, x^1 - x^0 \rangle}{t_1 - t_0} = \frac{c(q(t_1)) - c(q(t_0))}{t_1 - t_0}.$$

Letting $t_1 \downarrow t_0$ we find that this inequality is tight, and we obtain the formula

$$\frac{d^+}{dt}c(q(t))\Big|_{t=t_0} = \inf_{T>t_0} \frac{c(q(T)) - c(q(t_0))}{T - t_0} \le 0, \qquad t_0 \ge 0.$$

We now show that this implies convexity: For each $t_1 > t_0$,

$$
\begin{aligned}
\frac{d^+}{dt}c(q(t))\Big|_{t=t_0} &= \inf_{T>t_0} \frac{c(q(T))-c(q(t_0))}{T-t_0} \\
&\le \inf_{T>t_0} \frac{c(q(T))-c(q(t_1))}{T-t_0} \quad \text{since } c(q(t_1)) \le c(q(t_0)) \\
&\le \inf_{T>t_1} \frac{c(q(T))-c(q(t_1))}{T-t_0} \quad \text{since the infimum is over smaller set} \\
&\le \inf_{T>t_1} \frac{c(q(T))-c(q(t_1))}{T-t_1} \quad \text{since } t_1 > t_0 \\
&= \frac{d^+}{dt}c(q(t))\Big|_{t=t_1}.
\end{aligned}
$$

It follows that $c(q(t))$ is convex as a function of t since the derivative is nondecreasing. $\qquad\square$

To illustrate part (ii), recall that the myopic policy is always path-wise optimal for the single-station fluid model described in Section 2.2.

The simple re-entrant line described in Section 2.8 is the simplest example in which the myopic policy may not be infinite-horizon optimal. When c is equal to the ℓ_1-norm, then the myopic policy is precisely the last-buffer first-served (LBFS) priority policy. The LBFS policy is not necessarily optimal for the fluid model since it may result in excessive starvation at the second station whenever the second station is the bottleneck so that $\rho_2 > \rho_1$ (see Fig. 2.11 and equation (2.25)). The optimal policy is constructed for the fluid model in Example 4.3.10 below.

4.3.3 Balancing long- and short-term costs

The following greedy time optimal (GTO) policy combines desirable properties of the myopic and time optimal feedback laws (*greedy* is a synonym for *myopic* in this book).

The linear program that defines the GTO policy is derived from the myopic policy, with the additional constraint $\langle \xi^s, v \rangle = -(1 - \rho_s)$ for $s \in S^*(x)$. That is, the GTO policy minimizes the derivative $\frac{d}{dt}c(q(t))$, subject to time optimality, and state space constraints.

Definition 4.3.7 (GTO policy). The *GTO policy* defines the feedback law $\phi^\text{F} \colon \mathsf{X} \to \mathsf{U}$ for $x \in \mathsf{X}$ as follows: The allocation $\phi^\text{F}(x)$ is defined to be any ζ that optimizes the linear program

$$\min \langle c, v \rangle$$

$$
\begin{array}{lrll}
\text{s.t.} & v_i & \geq & 0 & \text{for all } i \text{ such that } x_i = 0 \\
& v_i & \leq & 0 & \text{for all } i \text{ such that } x_i = b_i \\
& \langle \xi^s, v \rangle & = & -(1 - \rho_s) & \text{if } s \in S^*(x),\ 1 \leq s \leq \ell_1 \\
& v & = & B\zeta + \alpha & \\
& \zeta & \in & \mathsf{U}. &
\end{array}
\tag{4.46}
$$

∎

The following properties of the GTO policy are analogous to those obtained in Proposition 4.3.6 for the myopic policy.

Proposition 4.3.8. *Suppose that $\rho_\bullet < 1$. Then, the network satisfies the following properties under a GTO policy:*

(i) *The state trajectory q is time optimal. That is,*

$$q(t; x) = 0, \qquad t \geq T^*(x),\ x \in \mathsf{X}.$$

(ii) *Suppose that the linear program (4.46) has a unique solution q for each $x \in \mathsf{X}$. If from a given initial condition $x^0 \in \mathsf{X}$ there exists a path-wise optimal solution q^* starting from x^0, then $q(t; x^0) = q^*(t; x^0)$ for all $t \geq 0$.*

(iii) *$c(q(t; x))$ is convex and nonincreasing in t.*

(iv) *$S^*(q(t; x))$ is nondecreasing in t.*

Proof. Part (i) follows directly from Proposition 4.3.4 and the definition of the GTO policy. Part (ii) follows from the observation that a path-wise optimal solution, if it exists, is time optimal and myopic.

The proof of (iii) is identical to the proof of Proposition 4.3.6 (iii), and (iv) is contained in Proposition 4.3.4 (ii). ☐

Example 4.3.9 (Optimal policies for tandem queues). To illustrate the construction of time optimal, myopic, infinite-horizon optimal, and the GTO policy we consider here the pair of *tandem queues* shown in Fig. 4.3. The fluid model can be written in the standard form (4.24) with

$$B = \begin{bmatrix} -\mu_1 & 0 \\ \mu_1 & -\mu_2 \end{bmatrix}, \qquad \alpha = \begin{bmatrix} \alpha_1 \\ 0 \end{bmatrix}.$$

The two load parameters are $\rho_i = \alpha_1/\mu_i$, $i = 1, 2$.

Figure 4.3. Tandem queues.

The cost function is assumed linear with $c(x) = c_1 x_1 + c_2 x_2$. Except in a few special cases, we find that the myopic, GTO, and optimal policies are each defined by a linear switching curve,

$$\zeta_1(t) = 1 \text{ if and only if } q_2(t) \leq m_x q_1(t) \text{ and } q(t) \neq 0, \qquad (4.47)$$

for some constant $m_x \in [0, \infty]$.

When $c_2 < c_1$ then the myopic policy is nonidling, so that the policy can be described by (4.47) with $m_x = \infty$. The nonidling policy is also myopic when $c_2 = c_1$, though in this case the myopic policy is not unique. When $c_2 > c_1$ then the myopic policy allows Station 1 to idle until buffer 2 is empty. Subsequently, buffer 2 is fed at rate $\min(\mu_1, \mu_2)$, and hence buffer 2 remains empty. This policy is again of the form (4.47) with $m_x = \infty$ provided $\mu_1 \leq \mu_2$.

A time optimal policy is defined by the constraints $\{\zeta_i(t) = 1$ whenever $i \in \mathcal{S}^*(q(t)), i = 1, 2\}$. To compute the function $\mathcal{S}^*$ we must consider the minimal draining time,

$$T^*(x) = \max\left(\frac{x_1}{\mu_1 - \alpha_1}, \frac{x_1 + x_2}{\mu_2 - \alpha_1}\right), \qquad x \in \mathbb{R}_+^2.$$

We have $\mathcal{S}^*(x) = \{1\}$ when $T^*(x) > (\mu_2 - \alpha_1)^{-1}(x_1 + x_2)$, $\mathcal{S}^*(x) = \{2\}$ when $T^*(x) > (\mu_1 - \alpha_1)^{-1} x_1$, and $\mathcal{S}^*(x) = \{1, 2\}$ when x is nonzero and satisfies $T^*(x) = (\mu_2 - \alpha_1)^{-1}(x_1 + x_2) = (\mu_1 - \alpha_1)^{-1} x_1$. Equivalently,

$$\mathcal{S}^*(x) = \{1, 2\} \iff \left\{\frac{x_2}{x_1} = \frac{\mu_2 - \mu_1}{\mu_1 - \alpha_1}\right\}.$$

This is possible only if $\rho_1 \leq \rho_2$ so that Station 1 is a bottleneck. When $\rho_2 > \rho_1$ we have $\mathcal{S}^*(x) = \{2\}$ for all nonzero x.

In the remainder of this example we restrict to the case in which buffer 1 is the unique bottleneck, so that $\mu_1 < \mu_2$. It is also assumed that $c_2 > c_1$. Under these conditions it is necessary to make a compromise between draining the total inventory quickly, and maintaining a low holding cost $c(q(t))$.

The GTO policy coincides with the myopic policy when $\mathcal{S}^*(q(t)) = \{2\}$ so that Station 2 is the unique dynamic bottleneck, and is nonidling when $\mathcal{S}^*(q(t)) = \{1\}$. Consequently, the GTO policy is also of the form (4.47) with

$$m_x^{\text{GTO}} = \frac{\mu_2 - \mu_1}{\mu_1 - \alpha_1}. \qquad (4.48)$$

The parameter defining the infinite-horizon optimal policy in the representation (4.47) satisfies $m_x^* \in [0, m_x^{\text{GTO}}]$. A formula for m_x^* is obtained by applying the dynamic programming equation Proposition 4.3.2 (iii).

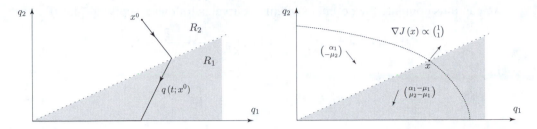

Figure 4.4. Shown on the left is a state trajectory for the policy defined by a switching curve with optimal slope m_x^*. Shown on the right is a level set of the piecewise quadratic value function J^*. The gradient $\nabla J^*(x)$ is orthogonal to the level set of J^* at x, and is parallel to $(1,1)^\mathsf{T}$ for x lying on the optimal switching curve $\{x_2 = m_x^* x_1\}$.

Define the two positive cones determined by the parameter m_x^*:

$$\mathsf{R}_1 := \{x \in \mathbb{R}_+^2 : x_2 \le m_x^* x_1\}, \qquad \mathsf{R}_2 := \{x \in \mathbb{R}_+^2 : x_2 \ge m_x^* x_1\},$$

and let v^i denote the velocity vector under the optimal policy within the interior of each of these regions,

$$v^1 = (\alpha_1 - \mu_1, -\mu_2 + \mu_1)^\mathsf{T}, \quad v^2 = (\alpha_1, -\mu_2)^\mathsf{T}.$$

The value function J^* is piecewise quadratic, and purely quadratic on each of the sets $\{\mathsf{R}_i\}$. Writing $J^*(x) = \frac{1}{2}x^\mathsf{T} D^i x$ and $\nabla J^*(x) = D^i x$ for $x \in \text{interior}(\mathsf{R}_i)$, the dynamic programming equation (4.41) gives

$$-c(x) = (D^1 x)^\mathsf{T} v^1 \le (D^1 x)^\mathsf{T} v^2, \qquad x \in \text{interior}(\mathsf{R}_1)$$
$$-c(x) = (D^2 x)^\mathsf{T} v^2 \le (D^2 x)^\mathsf{T} v^1, \qquad x \in \text{interior}(\mathsf{R}_2).$$

As shown in Fig. 4.4, the value function J^* is continuously differentiable (C^1). Consequently, on considering nonzero $x^* \in \mathsf{R}_1 \cap \mathsf{R}_2$ we have $D^1 x^* = D^2 x^*$, and hence

$$-c(x^*) = (D^1 x^*)^\mathsf{T} v^1 = (D^1 x^*)^\mathsf{T} v^2. \tag{4.49}$$

Using (4.49) we can obtain an expression for the optimal slope m_x^* based on the matrix D^1.

The dynamic programming equation gives $-c(x) = (D^1 x)^\mathsf{T} v^1$ within R_1, which implies that $D^1 v^1 = -c$. Also, from the initial condition $x^0 = (1,0)^\mathsf{T}$ the velocity vector is $v^0 = (\alpha_1 - \mu_1, 0)^\mathsf{T}$ and the dynamic programming equation becomes

$$-c_1 = -c_1 x_1^0 = \langle \nabla J^*(x^0), v^0 \rangle = x^{0\mathsf{T}} D^1 v^0$$

giving $D_{11}^1 v_1^2 = -c_1$. These equations can be solved uniquely to give

$$D^1 = \begin{bmatrix} \frac{c_1}{\mu_1 - \alpha_1} & 0 \\ 0 & \frac{c_2}{\mu_2 - \mu_1} \end{bmatrix}.$$

Similar arguments show that $J(x) = \frac{1}{2}x^\mathsf{T} D^1 x$ for *all* x under the nonidling policy.

We can thus compute the optimal switching curve as follows: Applying (4.49),

$$0 = (D^1 x^*)^\mathsf{T}(v^2 - v^1) = \mu_1 \left(\frac{c_1 x_1^*}{\mu_1 - \alpha_1} - \frac{c_2 x_2^*}{\mu_2 - \mu_1} \right),$$

so that

$$m_x^* = \frac{x_2^*}{x_1^*} = \frac{\mu_2 - \mu_1}{\mu_1 - \alpha_1} \frac{c_1}{c_2} = m_x^{\text{GTO}} \frac{c_1}{c_2}. \tag{4.50}$$

If $c_1 < c_2$ then $m_x^* \in (0, m_x^{\text{GTO}})$, which means that the optimal policy is not time optimal for initial conditions satisfying $x_2 > m_x^* x_1$.

Consider the special case

$$\alpha_1 = 9, \ \mu_1 = 10, \ \mu_2 = 11,$$

giving $\rho_1 = 9/10$ and $\rho_2 = 9/11$. The switching curves defined using (4.48) and (4.50) are

$$m_x^{\text{GTO}} = \frac{\mu_2 - \mu_1}{\mu_1 - \alpha_1} = 1, \qquad m_x^* = m_x^{\text{GTO}} \frac{c_1}{c_2}.$$

We take $c_1 = 1$ and $c_2 = 3$ so that $m_x^* = 1/3$. In this case we can calculate each of the matrices defining J^*,

$$D^1 = \begin{bmatrix} 1 & 0 \\ 0 & 3 \end{bmatrix}, \qquad D^2 = \frac{1}{21} \begin{bmatrix} 16 & 15 \\ 15 & 18 \end{bmatrix} \tag{4.51}$$

In particular, given the initial condition $x = (0, 1)^\mathsf{T}$, under the optimal policy and under the nonidling policy we have, respectively,

$$J^*(x) = \tfrac{1}{2} x^\mathsf{T} D^2 x = \frac{3}{7}, \qquad J(x) = \tfrac{1}{2} x^\mathsf{T} D^1 x = \frac{3}{2}.$$

We conclude that the optimal policy reduces the total cost by more than threefold when compared to the nonidling policy. ∎

Example 4.3.10 (Optimal policies for the simple re-entrant line). We now compute the infinite-horizon optimal policy and several time optimal policies for the simple reentrant line with cost function $c(x) = |x| = x_1 + x_2 + x_3$. It is assumed that the model is homogeneous, and that $\rho_1 < \rho_2 < 1$, where the load parameters are given in (4.33).

Consider first the GTO policy. This gives strict priority to buffer 3 if $q_2(t) > 0$. While $q_2(t) = 0$, then buffer 3 continues to receive strict priority at Station 1 as long $S^*(x) = \{1\}$. From the formula for the two workload vectors given in (4.32) this condition is equivalently expressed

$$\frac{1}{1 - \rho_1} \frac{2x_1 + x_3}{\mu_1} > \frac{1}{1 - \rho_2} \frac{x_1}{\mu_2},$$

or on rearranging terms and applying the definition of $\{\rho_1, \rho_2\}$, this may be written $x_3 > m_x^{\text{GTO}} x_1$, where

$$m_x^{\text{GTO}} := 2 \frac{\rho_1^{-1} - \rho_2^{-1}}{\rho_2^{-1} - 1}.$$

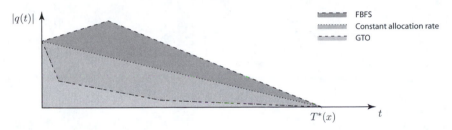

Figure 4.5. Evolution of the cost $c(q(t)) = |q(t)|$ for the simple re-entrant line under three time-optimal policies.

In this special case in which the parameters are given in (2.26) we have $\rho_1^{-1} = 11/9$, $\rho_2^{-1} = 10/9$, so that $m_x^{\text{GTO}} = 2$.

Shown in Fig. 4.5 are the state trajectories obtained from three different time-optimal policies for this model. Under the GTO policy we see that $c(q(t))$ is convex as a function of time, as predicted by Proposition 4.3.8. The FBFS policy is time optimal for this model, and gives rise to the *concave* trajectory shown in the figure. Also, shown is the time optimal solution in which $\zeta(t)$ is the constant value (4.30) on $[0, T^*(x))$.

We now turn to the infinite-horizon optimal policy. The parameter m_x^* that defines the optimal policy can be constructed based on the dynamic programming equation (4.41) exactly as in Example 4.3.9.

To apply Proposition 4.3.2, we first compute the value function. We restrict to initial conditions satisfying $x_2 = 0$ in the definition (4.37), and hence we may assume without loss of generality that $q_2(t) = 0$ for all $t \geq 0$ under the optimal policy. We identify $\mathbb{R}_+^2$ as the resulting restricted state space, where $x \in \mathbb{R}_+^2$ is interpreted as the pair $x = (x_3, x_1)^{\mathsf{T}}$.

Consider the two closed regions defined by

$$\mathsf{R}_1 := \left\{ (x_3, x_1)^{\mathsf{T}} \in \mathbb{R}_+^2 : \frac{x_3}{x_1} \geq m_x^* \right\}, \quad \mathsf{R}_2 := \left\{ (x_3, x_1)^{\mathsf{T}} \in \mathbb{R}_+^2 : \frac{x_3}{x_1} \leq m_x^* \right\},$$

and the feedback policy that is constant within the interior of each of these regions,

$$\zeta_3(t) = 1, \qquad \begin{pmatrix} q_3(t) \\ q_1(t) \end{pmatrix} \in \text{ interior } \mathsf{R}_1,$$

$$\zeta_1(t) = 1 - \zeta_3(t) = \frac{\mu_2}{\mu_1}, \qquad \begin{pmatrix} q_3(t) \\ q_1(t) \end{pmatrix} \in \text{ interior } \mathsf{R}_2.$$

Shown on the left in Fig. 4.6 is an illustration of typical regions $\{\mathsf{R}_i\}$, and a typical state trajectory under this policy.

The value function $J \colon \mathbb{R}_+^2 \to \mathbb{R}_+$ defined in (4.37) on the restricted domain is piecewise quadratic, and purely quadratic on each of the sets $\{\mathsf{R}_i\}$, of the form

$$J(x) = \tfrac{1}{2}(x_3, \ x_1) D^i \begin{pmatrix} x_3 \\ x_1 \end{pmatrix}, \qquad x \in \mathsf{R}_i,$$

where $\{D^i : i = 1, 2\}$ are 2×2 matrices. To compute these matrices consider the drift vectors illustrated on the right in Fig. 4.6, and defined by

$$\frac{d}{dt} \begin{pmatrix} q_3(t) \\ q_1(t) \end{pmatrix} = -\delta^i, \qquad \begin{pmatrix} q_3(t) \\ q_1(t) \end{pmatrix} \in \text{ interior } \mathsf{R}_i.$$

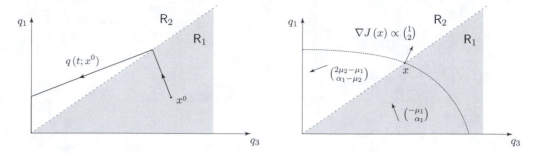

Figure 4.6. On the left is shown a state trajectory within the region $\{x_2 = 0\}$ for a policy defined by a switching curve with optimal slope m_x^*. Shown on the right is a level set of the piecewise quadratic value function. The gradient $\nabla J(x)$ is orthogonal to the level set of J at x.

The matrix D^2 is easily computed by integration, giving

$$D^2 = \begin{bmatrix} (\delta_1^2)^{-1} & 0 \\ 0 & (\delta_2^2)^{-1} \end{bmatrix}.$$

We have $-\delta^1 = (-\mu_1, \alpha_1)^{\mathsf{T}}$ since $\zeta_1(t) = 0$ when $\binom{q_3(t)}{q_1(t)} \in$ interior R_1. To compute δ^2, observe that $\zeta_3(t) = 1 - \zeta_1(t)$ within the interior of R_2, which gives

$$\tfrac{d}{dt} q_3(t) = \mu_1 \zeta_1(t) - \mu_3(1 - \zeta_1(t)) = 2\mu_1 \zeta_1(t) - \mu_1 = 2\mu_2 - \mu_1,$$

where we have applied the homogeneity assumption, and the policy specification $\mu_1 \zeta_1(t) = \mu_2$. This gives an expression for the second drift vector,

$$-\delta^2 = (2\mu_2 - \mu_1, \alpha_1 - \mu_2)^{\mathsf{T}}.$$

We now apply the dynamic programming equation Proposition 4.3.2 (iii) which gives

$$-c(x) = \langle \nabla J^*(x), -\delta^2 \rangle \leq \langle \nabla J^*(x), -\delta^1 \rangle, \qquad x \in \text{interior } R_2.$$

In particular, since J^* is quadratic on R_2,

$$x^{\mathsf{T}} D^2 (\delta^2 - \delta^1) = \langle \nabla J^*(x), \delta^2 - \delta^1 \rangle \geq 0, \qquad x \in \text{interior } R_2, \tag{4.52}$$

where the difference is given by

$$\delta^2 - \delta^1 = -(2\mu_2 - \mu_1, \alpha_1 - \mu_2)^{\mathsf{T}} + (-\mu_1, \alpha_1)^{\mathsf{T}} = \mu_2(-2, 1)^{\mathsf{T}}.$$

As illustrated on the right in Fig. 4.6, the inequality (4.52) becomes an equality as x approaches the right boundary of R_2. The vector $x = (m_x^*, 1)^{\mathsf{T}}$ is on this boundary, and consequently,

$$\begin{aligned}
0 &= (m_x^*, 1) D^2 (\delta^2 - \delta^1) \\
&= \mu_2 (m_x^*, 1) D^2 \binom{-2}{1} \\
&= \mu_2 \left(-2 m_x^* (\mu_1 - 2\mu_2)^{-1} + (\mu_2 - \alpha_1)^{-1} \right).
\end{aligned}$$

Solving for m_x^* gives the final expression for the optimal parameter,

$$m_x^* = \frac{1}{2} \frac{\mu_1 - 2\mu_2}{\mu_2 - \alpha_1} = \frac{\rho_1^{-1} - \rho_2^{-1}}{\rho_2^{-1} - 1}. \tag{4.53}$$

With the specific service rates and arrival rates given in (2.26), the formula (4.53) simplifies to $m_x^* = 1$, which coincides with the policy shown in Fig. 2.10. Hence, under the optimal policy buffer 3 has priority when $x_2 > 0$ or $x_3 > x_1$. Recall that the GTO policy has the same form, but to preserve time-optimality buffer 3 has priority over buffer 1 only if $x_3 > 2x_1$ when $x_2 = 0$. ∎

In the remainder of this chapter we survey policies for the CRW model. We first reconsider the differences between the two models from the standpoint of control.

4.4 Comparing fluid and stochastic models

The simplest relationship between the two models is the following lower bound on the finite time-horizon value functions for the stochastic model.

Proposition 4.4.1. *For each fixed time horizon $T \in \mathbb{Z}_+$, the finite time-horizon value functions for the fluid and stochastic models satisfy*

$$\inf \mathsf{E}\left[\sum_{k=0}^{T} c(Q(k; x)) \right] \geq \inf \int_0^T c(q(t; x)) \, dt, \qquad x \in \mathsf{X},$$

where the infimums are over all admissible allocation sequences U and z for the respective network models.

Proof. Fix the initial condition $x \in \mathsf{X}$, and an admissible allocation U for the CRW model. Define the deterministic allocation process z as the piecewise linear function whose derivative is expressed

$$\tfrac{d^+}{dt} z(t) = \zeta^k := \mathsf{E}[U(k)], \qquad k \leq t < k+1, \ k \geq 0,$$

and let q denote the resulting deterministic process defined through the fluid model equations (4.24). The state trajectory q is piecewise linear, and moreover

$$\begin{aligned}
\mathsf{E}[Q(k+1)] &= \mathsf{E}[Q(k)] + B\mathsf{E}[U(k)] + \alpha \\
&= \mathsf{E}[Q(k)] + B\zeta^k + \alpha \\
&= x + \sum_{i=0}^{k} \left(B\zeta^i + \alpha \right) \\
&= x + Bz(k+1) + (k+1)\alpha = q(k+1), \qquad k \geq 0.
\end{aligned} \tag{4.54}$$

This observation leads to the following bound:

$$\mathsf{E}\left[\int_0^T c(q(t;x))\,dt\right] = \sum_{k=0}^{T-1} \tfrac{1}{2}\Big(c(q(k+1;x)) + c(q(k;x))\Big)$$

$$= \sum_{k=0}^{T-1} \tfrac{1}{2}\mathsf{E}\Big[c(Q(k+1;x)) + c(Q(k;x))\Big]$$

$$= \mathsf{E}\left[\sum_{k=0}^{T} c(Q(k;x))\right] - \tfrac{1}{2}\mathsf{E}\Big[c(Q(0;x)) + c(Q(T;x))\Big].$$

To prove the proposition it suffices to show that z is an admissible allocation, since then

$$\int_0^T c(q^*(t;x))\,dt \leq \mathsf{E}\left[\int_0^T c(q(t;x))\,dt\right] \leq \mathsf{E}\left[\sum_{k=0}^{T} c(Q(k;x))\right],$$

where q^* achieves the minimum over all z.

Admissibility of z follows from the following two observations: (a) Since $U(k) \in \mathsf{U}$ a.s. for each k, and since U is a convex set, it follows that its expectation ζ^k also lies in U. Thus, $\frac{d^+}{dt}z(t) \in \mathsf{U}$ for each $t \geq 0$. (b) Based on the identity (4.54) we can argue that $q(k) \in \mathsf{X}$ for each k since X is convex, and it follows that $q(t) \in \mathsf{X}$ for each $t \in \mathbb{R}_+$. □

Several questions arise. For instance, is the bound obtained in Proposition 4.4.1 tight? That is, can a policy for the stochastic model be constructed that meets this lower bound? This of course depends upon the initial condition, the time horizon T, and the specific network structure. One might hope to reach the lower bound through a direct application of an optimal policy for the fluid model. Unfortunately, this approach is often unsuccessful.

The difference in behavior in the two models based on common policies can be substantial in the simplest settings.

4.4.1 Buffer priority policies

A simple class of policies is defined by a rank ordering of the buffers in the network. Let $\{\theta_1, \ldots, \theta_\ell\}$ denote a permutation of the buffer indices $\{1, \ldots, \ell\}$. For a re-entrant line we always assume that the index $i \in \{1, \ldots, \ell\}$ corresponds to the ith buffer visited by a customer passing through the network.

For the CRW model a buffer priority policy is defined as follows:

Definition 4.4.2 (Buffer priority policy for the CRW model). Let $\{\theta_1, \ldots, \theta_\ell\}$ be a permutation of $\{1, \ldots, \ell\}$. Then:

(i) The buffer priority policy is defined by

$$U_i(t) = 1 \quad \text{if and only if } Q_i(t) \geq 1, \text{ and } Q_j(t) = 0$$
$$\text{for all } j \in \mathcal{I}_{s(i)} \text{ satisfying } \theta_j < \theta_i,$$

where $\mathcal{I}_s \subset \{1, \ldots, \ell\}$ denotes the set of buffers at Station s.

(ii) For a re-entrant line, if $\{\theta_1, \ldots, \theta_\ell\} = \{1, \ldots, \ell\}$ then the buffer priority policy is called *First Buffer-First Served* (FBFS), and if $\{\theta_1, \ldots, \theta_\ell\} = \{\ell, \ldots, 1\}$ it is called *Last Buffer-First Served* (LBFS).

Consider the following equivalent definition: For each $i \in \{1, \ldots, \ell\}$, given $Q(t) = x$, the allocation vector $U(t) = u$ must satisfy

$$\sum\{u_j : j \in \mathcal{I}_{s(i)} \text{ and } \theta_j \leq \theta_i\} = 1$$

whenever $\quad \sum\{x_j : j \in \mathcal{I}_{s(i)} \text{ and } \theta_j \leq \theta_i\} > 0.$

(4.55)

This form is precisely the definition used for the fluid model:

Definition 4.4.3 (Buffer priority policy for the fluid model). For a given permutation $\{\theta_1, \ldots, \theta_\ell\}$ of $\{1, \ldots, \ell\}$, the buffer priority policy for the fluid model is defined for each $t \geq 0$ as follows: Given $q(t) = x$, the allocation rates $u = \zeta(t)$ satisfy the set of constraints (4.55). ■

The following example illustrates the *significant* difference in behavior in the fluid and stochastic models under a buffer priority policy.

Example 4.4.4 (Delay in the simple re-entrant line). Consider the network shown in Fig. 2.9 with $\rho_\bullet < 1$. It is assumed that the model is homogeneous and balanced, so that the service rates are expressed

$$\mu_1 = \mu_3 = 2\frac{\alpha}{\rho_\bullet}, \qquad \mu_2 = \frac{\alpha}{\rho_\bullet}.$$

The fluid model is controlled using the LBFS policy in which buffer 3 has strict priority. This coincides with the myopic policy with respect to the ℓ_1-norm. When $q_1(t) > 0$ and $q_2(t) = q_3(t) = 0$, the myopic policy dictates

$$\zeta_2 = 1, \quad \zeta_1 = \zeta_3 = \tfrac{1}{2}, \tag{4.56}$$

so that buffer 2 is served at maximum rate even though it remains empty.

Consider now the CRW model in which Q is constrained to $\mathbb{Z}_+^3$, and U is constrained to $\mathsf{U}_\diamond$ as defined in (4.5). It is assumed that each of the service processes $\{S_1(t) \equiv M_1(t) = M_3(t), \; S_2(t) \equiv M_2(t)\}$ has a Bernoulli distribution.

The LBFS policy for the CRW model is defined as follows:

$$
\begin{aligned}
U_1(t) &= 1 \quad \text{if and only if } Q_1(t) \geq 1 \text{ and } Q_3(t) = 0; \\
U_2(t) &= 1 \quad \text{if and only if } Q_2(t) \geq 1; \\
U_3(t) &= 1 \quad \text{if and only if } Q_3(t) \geq 1.
\end{aligned}
$$

Suppose that buffers 2 and 3 are empty at time $t = 0$. Then, assuming $Q_1(0) \neq 0$, the policy dictates that $U_1(0) = 1$, yet $U_2(0) = U_3(0) = 0$. Note that this is very different than the fluid allocation (4.56). When service is completed at some time t_0, which will take on average μ_1^{-1} seconds, we have $Q_2(t_0) = 1$ and $Q_3(t_0) = 0$. Now Station 1 must wait on average μ_2^{-1} seconds for this customer to reach buffer 3 so that service at this buffer can proceed.

Suppose that $Q_1(0) \gg 1$, and fix an integer N satisfying $N \gg 1$ and $Q_1(0)\mu_1^{-1} \gg N$. The cumulative allocation on $[0, N)$ can be approximated as follows: The second station is busy for μ_2^{-1} seconds on average during each service, but then must wait $\mu_1^{-1} + \mu_3^{-1}$ seconds on average to receive a new customer, so that

$$\sum_{k=0}^{N-1} U_2(k) \approx \frac{\mu_2^{-1}}{\mu_1^{-1} + \mu_3^{-1} + \mu_2^{-1}} N = \frac{\rho_2}{\rho_1 + \rho_2} N. \tag{4.57}$$

Since the network is assumed balanced with $\rho_1 = \rho_2$, we conclude that *the second station idles about 50% of the time in this overloaded regime.*   ∎

Another example is the KSRS model considered in Section 2.9 using a policy giving priority at the exit buffers, so that $\{\theta_1, \ldots, \theta_4\} = \{3, 1, 4, 2\}$. Based on the simulation results shown in Fig. 2.13 it appears that the controlled network is highly unstable. Yet, in this same example the buffer priority policy for the fluid model is stabilizing – see Example 4.4.5 that follows. Such a large gap in behavior in the two models is possible since a buffer may receive service when it is empty in the fluid model, while this is ruled out in the discrete model.

Among myopic policies we can construct other examples illustrating this fundamental difference.

4.4.2 Myopic policies

Recall from Proposition 4.3.6 that the myopic policy is always stabilizing for the fluid model when c is linear, and this result can be generalized to any convex, monotone cost function. The situation is very different for a stochastic model. The myopic policy (4.16) may or may not be stabilizing, depending upon the particular network and the structure of the cost function.

The following examples show that Proposition 4.3.6 does not have a direct extension to the CRW model.

Example 4.4.5 (KSRS model: Transience of the myopic policy). Consider the KSRS model with $c(x) = |x|$. The myopic policy for the fluid model gives priority to the exit buffers if no machine is starved of work. If for example $q_1(t) > 0$ and $q_4(t) > 0$, yet $q_2(t) = q_3(t) = 0$, then provided (2.27) holds we have

$$\zeta_1(t) = \mu_2 \mu_1^{-1}, \qquad \zeta_4(t) = 1 - \zeta_1(t).$$

The myopic policy for the CRW model is very different: The optimization (4.16) defines $U_4(t) = 1$ if $Q_4(t) \neq 0$, and $U_2(t) = 1$ if $Q_2(t) \neq 0$. This is precisely the policy found to be destabilizing in Section 2.9.   ∎

The myopic policy with linear cost may be entirely irrational:

Example 4.4.6 (Tandem queues: Idling in the myopic policy). Consider the pair of queues in tandem illustrated in Fig. 4.3, whose CRW model is the recursion

$$Q(t+1) = Q(t) + (-1^1 + 1^2)U_1(t)S_1(t) - 1^2 U_2(t)S_2(t) + 1^1 A_1(t+1), \tag{4.58}$$

with $Q(0) = x \in X_\diamond = \mathbb{Z}_+^2$. Suppose that a linear cost function is given $c(x) = c_1 x_1 + c_2 x_2$, with $c_2 > c_1$. The myopic policy for the fluid model described previously in Example 4.3.9 is nonidling at Station 2, while at Station 1,

$$
\zeta_1(t) = \begin{cases} 0 & \text{if } q_2(t) > 0 \\ \min(1, \mu_2 \mu_1^{-1}) & \text{if } q_2(t) = 0, q_1(t) > 0. \end{cases} \tag{4.59}
$$

The myopic policy is pathwise optimal when $\mu_1 \geq \mu_2$.

For the CRW model we have for $x \in \mathbb{Z}_+^2$,

$$
\phi(x) \in \arg \min_{u \in U_\diamond(x)} \mathsf{E}[c(Q(t+1)) \mid Q(t) = x, \, U(t) = u],
$$

and since c is linear this gives $\phi(x) \in \arg \min_{u \in U_\diamond(x)} \big(c_1(\alpha_1 - \mu_1 u_1) + c_2(\mu_1 u_1 - \mu_2 u_2) \big)$. At Station 2 this policy is nonidling, while at Station 1,

$$
\phi_1(x) = \arg \min_{u \in U_\diamond(x)} \big((c_2 - c_1) \mu_1 u_1 \big).
$$

That is, $\phi_1(x) \equiv 0$ under our assumption that $c_2 > c_1$, so that Station 1 is *always idle*. ∎

We now turn to the question, what do *good* policies look like?

4.5 Structure of optimal policies

In the following examples we explore the structure of solutions to the average-cost optimality equation. In each case, the optimal policy resembles a perturbation of the optimal policy for the fluid model.

We begin with the simplest example:

Example 4.5.1 (Single server queue: Solution of the ACOE). With $c(x) \equiv x$, the c-myopic policy for the single server queue is defined by

$$
\begin{aligned}
\phi(x) &= \arg \min_{u \in U_\diamond(x)} \mathsf{E}[Q(t+1) \mid Q(t) = x, \, U(t) = u] \\
&= \arg \min_{u \in U_\diamond(x)} \mathsf{E}[x - S(t+1)u + A(t+1) \mid Q(t) = x, \, U(t) = u] \\
&= \arg \min_{u \in U_\diamond(x)} \{-\mu u\},
\end{aligned}
$$

giving $\phi(x) = \mathbf{1}(x \geq 1)$. The nonidling policy is indeed optimal: on denoting Q^* the resulting queue-length trajectory, and Q the queue-length trajectory using some other policy with the same initial condition, we can couple the two processes using identical sequences (S, A) to obtain $Q^*(t) \leq Q(t)$ for each $t \geq 0$.

Provided the arrival process possesses a second moment, so that a policy with finite average cost exists, the nonidling policy is average-cost optimal, and the solution to the ACOE is the solution to Poisson's equation given in Theorem 3.0.1. It can be expressed $h^* = J^* + L$, with J^* the fluid value function, which is quadratic, and L a linear function of x. ∎

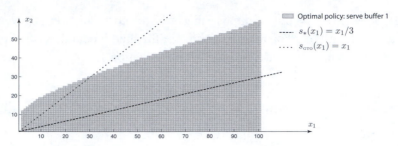

Figure 4.7. Average cost optimal policy for the tandem queues with $c_1 = 1, c_2 = 3, \rho_1 = 9/10$, $\rho_2 = 9/11$.

Example 4.5.2 (**Tandem queues: Structure of optimal policies**). As always, it is assumed that $\Phi(t) := (S_1(t), S_2(t), A_1(t))^{\mathsf{T}}$, $t \geq 1$, is an i.i.d. sequence. Suppose that its marginal distribution is given by

$$\begin{aligned} \mathsf{P}\{\Phi(t) = \mathbf{1}^1\} &= \mu_1, & \mathsf{P}\{\Phi(t) = \mathbf{1}^2\} &= \mu_2, \\ \mathsf{P}\{\Phi(t) = \mathbf{1}^3\} &= \alpha_1, \end{aligned} \tag{4.60}$$

and that $\mu_1 + \mu_2 + \alpha_1 = 1$. We take a linear cost function with $c_1 = 1$, $c_2 = 3$, and assume throughout that $\rho_\bullet = 0.9$.

The higher holding cost at buffer 2 means that the first server will idle unless there is risk that the second server will be starved of work. In each numerical example considered here the optimal policy at the first station can be expressed as

$$\phi_1^*(x) = \mathbf{1}\{x_1 \geq 1, \ x_2 \leq s(x_1)\}, \qquad x \in \mathsf{X}_\diamond, \tag{4.61}$$

where the switching curve $s \colon \mathbb{R}_+ \to \mathbb{R}_+$ depends on relative rates and other parameters.

Consider two separate cases for the service rates:

CASE 1: $\mu_1 < \mu_2$

This is the setting of Example 4.3.9. The infinite-horizon optimal policy for the fluid model is represented by a switching curve of the form (4.47) with $m_x^* > 0$ given in (4.50). For the values $\rho_1 = 9/10$, $\rho_2 = 9/11$ we obtain

$$m_x^* = \tfrac{1}{3},$$

while the time optimal switching curve is defined by $m_x^{\mathrm{GTO}} = 1$. Figure 4.7 shows the average-cost optimal policy for the CRW model. It is defined by a switching curve of the form (4.61) where the slope of s is roughly consistent with the slope $m_x^* = 1/3$.

CASE 2: $\mu_1 \geq \mu_2$

In this case there is a unique path-wise optimal solution for the fluid model defined by the switching curve (4.47) with $m_x = 0$. The average-cost optimal policy for the CRW model is shown in Fig. 4.8 with $\rho_1 = 9/11 < \rho_2$. This policy can be represented by (4.61), where the switching curve s is concave and unbounded in x_1.

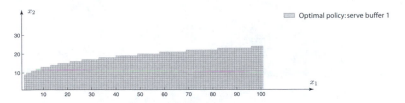

Figure 4.8. Average cost optimal policy for the tandem queues with $c_1 = 1, c_2 = 3, \rho_1 = 9/11$, $\rho_2 = 9/10$.

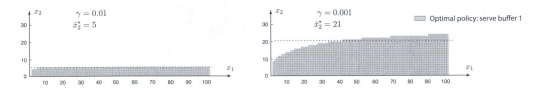

Figure 4.9. Discounted cost optimal policy for the tandem queues in Case 1. The load parameters are $\rho_1 = 9/10$, $\rho_2 = 9/11$, and the linear cost defined by $c_1 = 1$, $c_2 = 3$. On the left $\gamma = 0.001$ and on the right $\gamma = 0.01$.

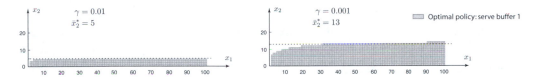

Figure 4.10. Discounted cost optimal policy for the tandem queues in Case 2. The cost parameters are identical to those used in Fig. 4.9, but the load parameters are reversed with $\rho_1 = 9/11$, $\rho_2 = 9/10$.

We conclude with numerical results for the discounted-cost optimal control problem (4.18a, 4.18b). Shown in Figs. 4.9 and 4.10 are optimal policies, in Cases 1 and 2 respectively, with discount rate $\gamma = 0.01$ and $\gamma = 0.001$. In each case the discounted-cost optimal policy is approximated by a static threshold, in the sense that the representation (4.61) holds with $s(x_1)$ quickly approaching a constant value $\bar{x}_2^* > 0$. We return to this model in Example 10.6.3 where we obtain approximations to the threshold $\bar{x}_2^*$ shown in each of these figures. ∎

Example 4.5.3 (Optimization in the simple re-entrant line). Consider the homogeneous CRW model satisfying (4.10) and (4.11). The generator can be expressed

$$\mathcal{D}f(x) = \alpha_1[f(x + 1^1) - f(x)] + \mu_1\phi_1(x)[f(x - 1^1 + 1^2) - f(x)]$$
$$+ \mu_2\phi_2(x)[f(x - 1^2 + 1^3) - f(x)] + \mu_3\phi_3(x)[f(x - 1^3) - f(x)], \tag{4.62}$$

with $\mu_1 = \mu_3$ since the model is homogeneous.

An average-cost optimal policy is shown in Fig. 4.11 for the service rates given in (2.26). The optimal policy was computed for the stochastic model using value iteration,

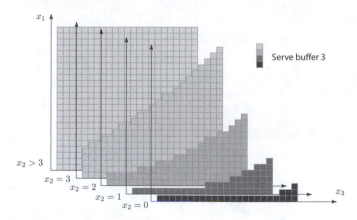

Figure 4.11. Optimal discrete policy for simple re-entrant with arrival and service rates defined in (2.26).

terminated at $n = 7,000$, as described in Chapter 9. The buffer levels were truncated so that $x_i < 45$ for all i. This gives rise to a finite state space Markov Decision Process with $45^3 = 91,125$ states. This is an example of the "curse of dimensionality" described at the start of this chapter.

As in the previous examples, the discrete optimal policy is easily interpreted: It regulates the work waiting at buffer 2 in such a way that Station 2 is rarely starved of work when the network is congested.

The policy shown in Fig. 4.11 is also very similar to the fluid policy shown in Fig. 2.10. Performing some curve fitting, we can approximate this discrete policy as follows: Serve buffer 1 at time t if and only if either buffer 3 is equal to zero, or

$$\phi_1(x) = \mathbf{1}\{x_1 - \overline{x}_1 > x_3 - \overline{x}_3\}\mathbf{1}\{x_2 \le \overline{x}_2\}, \tag{4.63}$$

where the translation $\overline{x}$ positive. The most accurate approximation is obtained when $\overline{x}$ depends upon the current state $Q(t) = x$, say

$$\overline{x}(x) = \overline{x}^0 \log(1 + |x|/\|\overline{x}^0\|), \qquad x \in \mathsf{X}, \tag{4.64}$$

with $\overline{x}^0 \in \mathbb{R}_+^3$ a constant. ∎

4.6 Safety-stocks

The simple routing model introduced in Section 2.10 demonstrates that care that must be taken on translating a policy constructed for the fluid network model to any stochastic counterpart. In this and other examples we have seen that a direct translation of the myopic or optimal policy obtained from the fluid model might be destabilizing. This is plainly seen in the simulation for the routing model shown in Fig. 2.16.

Deterministic and stochastic models behave differently because the two models are subject to different constraints along the boundaries of the state space. In the CRW model, and in a physical network, a station must be idle if each buffer at that station is empty. This is not true in the fluid model. For example, as shown on the left in Fig. 2.15, an optimal policy for the routing model might dictate that $q_2(t)$ remain at zero for all $t \geq \mu_2^{-1}x_2$, and also require that $\zeta_2 \equiv 1$ on $[0, T^*(x))$.

Stability requires that excessive idleness be avoided at each resource in the network. Hence some mechanism must be constructed to predict the onset of starvation, and to respond by sending inventory to any resource that risks starvation. The role of a *safety-stock* is to provide this warning.

The policies considered in Section 4.6.1 involve sending work to a station whenever the total inventory there falls below a specified minimal level, called the safety-stock at this station.

A more elaborate technique is described in Section 4.6.2 in which the safety-stock grows logarithmically with the total customer population. In this way it is possible to construct a policy that is "universally stabilizing" in the sense that the fixed policy stabilizes the network whenever a stabilizing policy exists.

The application of safety stocks can be used to design effective policies in complex network settings. However, it is not straightforward to present a clear roadmap since there are many choices to consider once the "safety-stock signal" is announced that a resource is in need of work. In this section we illustrate the main concepts through examples.

Safety-stocks constitute one component of the discrete-review policies introduced in Section 4.7.

Theory supporting the application of safety stocks is presented in Sections 8.3 and 10.5. These results provide design guidelines that constitute a completely general approach to policy synthesis for virtually any network.

4.6.1 Static safety-stocks

We illustrate the application of static safety stocks using the simple re-entrant line.

Example 4.6.1 (Safety-stocks for the simple re-entrant line). Consider again the CRW model (4.10) for the network shown in Fig. 2.9 satisfying (4.11). It is assumed that the arrival rate and service rates are scaled so that $\mu_1 + \mu_2 + \alpha_1 = 1$. The network parameters are defined in (2.26), so that $\rho_1 = 9/11$ and $\rho_2 = 9/10$.

The infinite-horizon optimal policy for the fluid model is defined by the switching curve illustrated in Fig. 2.10. Consider the following translation of this policy to the CRW model: with $\bar{x}_2 \geq 0$ a given constant,

$$\text{Serve buffer 1 at Station 1 if } x_2 \leq \bar{x}_2 \text{ and } x_1 \geq x_3. \tag{4.65}$$

This policy looks ahead to avoid starvation at the second station: when $Q_2(k) \leq \bar{x}_2$ then Station 1 places emphasis on feeding Station 2, rather than eliminating material from the system through service at buffer 3. The constant $\bar{x}_2$ is a safety-stock for the second station.

Figure 4.12 shows a comparison of the statistics of the controlled network using the following four policies: the average-cost optimal policy, the policy (4.65) with $\bar{x}_2 = 7$,

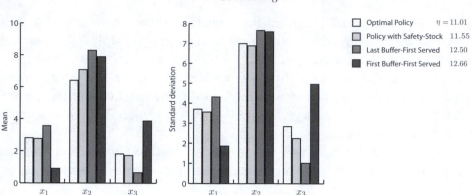

Figure 4.12. A comparison of four policies for the simple re-entrant line with $\rho_1 = 9/11$ and $\rho_2 = 9/10$, with homogeneous servers. The CRW model was simulated in each case, and mean queue-lengths and average cost were obtained through simulation.

and the two buffer priority policies, LBFS and FBFS. The average cost was estimated through simulation for each of these four policies using the standard *Monte Carlo estimator*,

$$\eta(n) := \frac{1}{n} \sum_{k=0}^{n-1} c(Q(k)), \quad n \geq 1. \tag{4.66}$$

The optimal average-cost is approximately $\eta^* = 11.01$, which is similar to the value of $\eta = 11.55$ obtained for the policy (4.65).

The results shown in Fig. 4.12 indicate that the resulting standard deviation at each buffer using the policy (4.65) is *less* than that of the optimal policy. ∎

In the next example we construct safety stocks for the KSRS model.

Example 4.6.2 (KSRS model: Optimizing safety stocks). The KSRS network depicted in Fig. 2.12 can be modeling using the CRW scheduling model with

$$B(t) = \begin{bmatrix} -M_1(t) & 0 & 0 & 0 \\ M_1(t) & -M_2(t) & 0 & 0 \\ 0 & 0 & -M_3(t) & 0 \\ 0 & 0 & M_3(t) & -M_4(t) \end{bmatrix} \qquad A(t) = \begin{bmatrix} A_1(t) \\ 0 \\ A_3(t) \\ 0 \end{bmatrix}. \tag{4.67}$$

The mean $\mu_i := \mathsf{E}[M_i(t)]$ is the service rate and $\alpha_i := \mathsf{E}[A_i(t)]$ is the arrival rate at the ith queue. The constituency matrix for this model given in Example 4.2.9 represents the two constraints $U_1(t) + U_4(t) \leq 1$ and $U_2(t) + U_3(t) \leq 1$.

Consider the specific parameters $\mu = [1, 1/3, 1, 1/3]$ and $\alpha = \rho_\bullet[1/4, 0, 1/4, 0]$, scaled so that the components of μ and α sum to 1. It follows from the discussion in Section 2.9 that the fluid model admits a path-wise optimal solution when c is equal to the ℓ_1-norm.

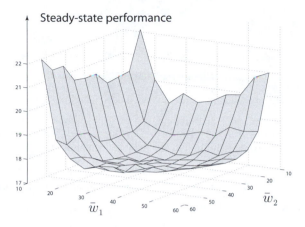

Figure 4.13. Estimates of the steady-state customer population in the KSRS model as a function of 100 different safety-stock levels.

The random variables appearing in (4.67) for the CRW model are Bernoulli with

$$P\{(M_1(t), \dots, M_4(t), A_1(t), A_3(t))^\top = 1^i\} = \begin{cases} \mu_i, & i = 1, \dots, 4, \\ \alpha_1, & i = 5, 6. \end{cases} \quad (4.68)$$

We now construct a policy for this model through the application of safety stocks.

The "time to starvation" at Station 2 is given by $\tau_{02} := \min(t : Q_2(t) + Q_3(t) = 0)$. For a given initial condition $Q(0) = x$, a bound on the mean of τ_{02} is given by

$$E[\tau_{02} \mid Q(0) = x] \geq \mu_2^{-1} x_2 + \mu_3^{-1} x_3. \quad (4.69)$$

The proof is left as Exercise 10.6 in Chapter 10.

The bound (4.69) suggests the following policy using static safety-stock values,

$$\text{Serve buffer 1 if buffer 4 is empty, or } \mu_2^{-1} Q_2 + \mu_3^{-1} Q_3 \leq \overline{w}_2,$$
$$\text{Serve buffer 3 if buffer 2 is empty, or } \mu_1^{-1} Q_1 + \mu_4^{-1} Q_4 \leq \overline{w}_1, \quad (4.70)$$

where $\{\overline{w}_1, \overline{w}_2\}$ are fixed constants, measured in terms of *workload* at each resource.

Figure 4.13 shows estimates of the steady-state customer population for the KSRS model using this policy with static safety-stock values given in (4.70), and with $\rho_\bullet = 0.9$. The plot shows that performance is approximately convex, and it deteriorates rapidly when either component of $\overline{w} \in \mathbb{R}_+^2$ is small. Performance deteriorates for large values, but at a more gentle rate. The optimal value of 17.6 in this plot occurs at $\overline{w} = (35, 30)^\top$. By way of contrast, the safety-stock value $\overline{w} = (60, 60)^\top$ yields a cost of 18.8.

A "safety workload" of 35 at Station 1 is perhaps difficult to interpret. In terms of buffer levels, this corresponds to approximately 11 jobs in buffer 1 (and 0 in buffer 4), or 4 jobs in buffer 4 (and 0 in buffer 1). ∎

These examples suggest that the implementation of static safety stocks may require substantial fine-tuning. For example, in the case of the tandem queues in which Station 2 is a bottleneck, any fixed threshold is destabilizing for sufficiently high load

Figure 4.14. Optimal policy for the processor sharing model. The local buffer at Station 2 receives strict priority when the state lies in the grey region shown.

(see Exercise 4.7). By allowing the safety-stock value to grow with the total customer population we obtain a policy that is stable for a far greater range of network parameters.

4.6.2 Dynamic safety stocks

The approximation (4.64) for the average-cost optimal policy in the simple re-entrant line, and the average-cost optimal policies for the tandem queues described in Example 4.5.2 show that optimal policies frequently resemble a perturbation of the fluid-optimal policy. In these examples and others we find that a switching curve defining an average-cost optimal policy has a concave, logarithmic shape. An optimal policy under the discounted-cost criterion frequently resembles a static safety-stock policy.

The switching curve used to define the policies in the examples that follow is defined as

$$s_\theta(r) := \theta \log(1 + r\theta^{-1}), \qquad r \geq 0. \tag{4.71}$$

For any value of θ the switching curve s_θ is increasing, with $s_\theta(0) = 0$, $s_\theta'(0) = 1$, and $r^{-1} s_\theta(r) \to 0$ as $r \to \infty$.

Example 4.6.3 (Processor sharing model: Optimal policy). In Section 2.5 we introduced the simple processor sharing model. Consider the specific model with ℓ_1 cost, $\alpha = (1, 1)$, and $\mu = (1, 3, 2)$. The CRW model is of the form obtained through uniformization, so that each service and arrival increment is Bernoulli, and no two events occur at the same time.

We saw that the optimal policy for the fluid model is defined by the c–μ rule, with priority given to processor b at Station 2.

The average-cost optimal policy for the CRW model is shown in Fig. 4.14. Within the grey region shown in the figure, the local buffer at Station 2 receives strict priority. The optimal policy is similar to the c–μ priority policy, with the boundary $\{x_2 = 0\}$ shifted to form the convex region shown. It is closely approximated by the logarithmic switching curve (4.71) with $\theta = 5$.   ∎

Example 4.6.4 (Dai–Wang model). The Dai–Wang model shown in Fig. 4.15 consists of two stations and five buffers, numbered in order so that customers arrive to buffer 1 and exit the system following processing at buffer 5.

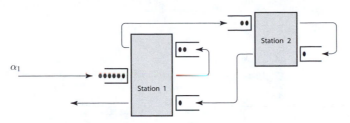

Figure 4.15. The five-buffer model of Dai and Wang.

In the numerical results presented here the second station is a bottleneck, so that $\rho_1 < \rho_2 = \rho_\bullet$, with $\rho_1 = 3\alpha_1/\mu_1$ and $\rho_2 = 2\alpha_1/\mu_2$. The cost function is linear with $c(x) = x_1 + 2(x_2 + x_3 + x_4 + x_5)$, $x \in \mathbb{R}_+^\ell$, so that it is desirable to maintain much of the inventory at buffer 1. However, it is also necessary to feed the bottleneck when starvation is imminent.

We define *two* regions in which starvation avoidance is prioritized:

$$\mathcal{P}_1 = \{x \in \mathbb{Z}_+^5 : x_3 + x_4 \le s_\theta(x_1),\ x_2 \ne 0\},$$
$$\mathcal{P}_2 = \{x \in \mathbb{Z}_+^5 : x_2 + x_3 + x_4 \le s_\theta(x_1)\}. \tag{4.72}$$

The policy defined below is designed to move inventory from Station 1 to Station 2 when $Q(t) \in \mathcal{P}_1$, and from buffer 1 to buffer 2 when $Q(t) \in \mathcal{P}_2$. These goals are captured in the following two drift conditions: for some $\varepsilon_1, \varepsilon_2 > 0$, whenever $x_1 \ne 0$,

$$\mathsf{E}[Q_3(t+1) + Q_4(t+1) \mid Q(t) = x] \ge x_3 + x_4 + \varepsilon_1, \qquad x \in \mathcal{P}_1,$$
$$\mathsf{E}[Q_2(t+1) \mid Q(t) = x] \ge x_2 + \varepsilon_2, \qquad x \in \mathcal{P}_2. \tag{4.73}$$

The following randomized policy with feedback law ϕ is designed so that the bounds (4.73) hold with $\varepsilon_1 = \frac{2}{5}\mu_1 - \frac{1}{2}\mu_2$, and $\varepsilon_2 = \frac{1}{5}\mu_1$.

(i) If $x \in \mathcal{P}_1$ then $\phi_2(x) = 2/5$. Otherwise, $\phi_2(x) = 0$.
(ii) If $x \in \mathcal{P}_2$ and $x_1 \ge 1$ then $\phi_1(x) = 3/5$. Otherwise, $\phi_1(x) = 0$.
(iii) If $x_5 \ne 0$ then $\phi_5(x) = 1 - \phi_1(x) - \phi_2(x)$.
(iv) At Station 2 the policy is nonidling, with priority to buffer 4.

In the following simulation a homogeneous CRW model was used, in which the common distribution of $\Phi(t) := (S_1(t), S_2(t), A_1(t))^\mathsf{T}$ was specified as follows:

$$\mathsf{P}\{\Phi(t) = 1^1\} = \mu_1; \qquad \mathsf{P}\{\Phi(t) = 1^2\} = \mu_2;$$
$$\mathsf{P}\{\Phi(t) = \kappa 1^3\} = \alpha_1 - \mathsf{P}\{\Phi(t) = 0\} = \alpha_1/\kappa. \tag{4.74}$$

The integer $\kappa \ge 1$ models "burstiness" of the arrival process.

Shown in Fig. 4.16 are four plots obtained using $\rho_1 = 0.8$, $\rho_2 = 0.9$, $\kappa = 1, 3, 5, 9$, and $\theta = 1, 3, 5, 7, \ldots, 19$. For each κ, identical sample paths of the service and arrival processes were held fixed in the experiments using these 10 values of θ, and the queue was initially empty, $Q(0) = 0$. The vertical axis shows the average of $c(Q(t))$ for

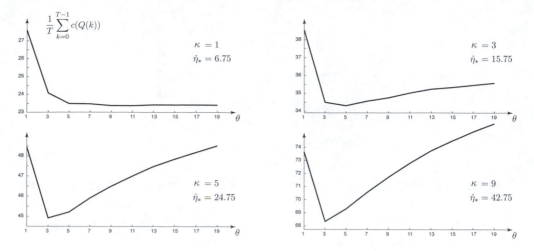

Figure 4.16. Average cost for the model of Dai and Wang with $\rho_1 = 0.8$ and $\rho_2 = 0.9$. Also, shown is the average cost $\widehat{\eta}^*$ for the "one-dimensional relaxation" introduced in Example 5.4.4.

$t \leq T = 10^5$. The optimal value of θ is very insensitive to the parameter κ: In all but the first instance it lies between 3 and 5. However, the sensitivity of steady-state cost with respect to the parameter θ grows with the value of κ.

    The parameter $\widehat{\eta}^*$ indicated in each plot is the average cost for the "one-dimensional relaxation" that will be introduced in Example 5.4.4.   ■

4.7 Discrete review

In practice a stationary policy may not be practical. Consider for example a single resource in a manufacturing system that is required in the production of two separate products. Switching from one type to the other requires some setup time, and it is thus desirable to attempt to minimize the total setup time to avoid idling. It is not clear how to perform this minimization using a stationary policy.

    In this section we describe an alternative setting for policy synthesis in which decisions over a given time horizon are made based on a predetermined solution to the fluid model equations.

Definition 4.7.1 (Discrete-review policy). Assume that the following is given:

(a) A sequence of *review times* $\{0 = T_0 < T_1 < T_2 < T_3 < \cdots\}$. It is assumed that there is a function $\mathcal{T} : \mathsf{X}_\diamond \to \{1, 2, 3, \dots\}$ such that the sequence is defined inductively by

$$T_0 = 0, \quad T_{m+1} = T_m + \mathcal{T}(Q(T_m)), \quad m \geq 0. \tag{4.75}$$

(b) A sequence of policies $\{\phi_m : m \geq 0\}$, such that for each $m \geq 0$ and $t \in [T_m, T_{m+1})$,

$$\phi_m^t(x_0, \dots, x_t) = \phi_m^t(x_{T_m}, \dots, x_t)$$

The *discrete-review* (DR) policy is defined via

$$U(t) = \phi_m^t(Q(T_m), \dots, Q(t)), \qquad T_m \le t \le T_{m+1} - 1, \ m \ge 0. \tag{4.76}$$

■

It is sometimes desirable to allow randomization, in which case (4.76) becomes

$$\begin{aligned}\mathsf{P}\{U(t) = u \mid Q(T_m), \dots, Q(t), U(T_m), \dots, U(t-1)\} \\ = \phi_{m,u}^t(Q(T_m), \dots, Q(t)).\end{aligned} \tag{4.77}$$

A discrete-review policy provides the most natural technique to translate a policy based on the fluid model or some other idealized model for application in a physical network. For example, the GTO policy is defined by the linear program (4.46). The discrete-review *translation* of this policy involves the addition of new constraints to account for variability.

Suppose that a stable solution to the fluid model equations (q, z) is given. Given the review period $T_1 \ge 1$ we then construct an allocation sequence U on $[0, T_1)$ to obtain the approximations

$$T_1^{-1}[Z(T_1)] - z(T_1)] \approx 0 \tag{4.78a}$$

$$T_1^{-1}[Q(T_1; x) - q(T_1; x)] \approx 0, \qquad x \in \mathsf{X}_\diamond, \tag{4.78b}$$

where the cumulative allocation process is defined by

$$Z(t) = \sum_{k=1}^{t} U(k-1), \qquad t \ge 1, \ Z(0) = 0. \tag{4.79}$$

There are many ways to obtain the approximation $\mathsf{E}[Z(T_1)] \approx z(T_1)$. To avoid unnecessary switch-overs a *generalized round-robin* policy can be used: Choose an increasing sequence of times $\{T_1^i : 0 \le i \le m\} \subset \mathbb{Z}_+$, with $T_1^0 = 0$ and $T_1^m = T_1$, along with a set of allocation values $\{u^i : 1 \le i \le m\} \subset \mathsf{U}_\diamond$. On the time interval $[T_1^{i-1}, T_1^i)$ the allocation vectors are defined by $U_j(t) = u_j^i$ when $Q_j(t) \ge 1$ so that this is feasible; If $u_j^i = 1$ and $Q_j(t) = 0$ then some other buffer at Station $s(i)$ can receive service instead. Finally, suppose that the following approximation holds

$$\sum_{i=1}^{m} (T_1^i - T_1^{i-1}) u^i \approx z(T_1).$$

This approximation can be made arbitrarily tight for large values of T_1.

Stability of a network under a DR policy is obtained in Section 8.3 under general conditions, based on the approximations (4.78a), (4.78b). In Section 8.3 these approximations are quantified, and explicit methods to achieve them are spelled-out in greater detail.

The following myopic-DR and GTO-DR policies require the following components:

(a) A convex function $h \colon \mathsf{X} \to \mathbb{R}_+$ satisfying $h(0) = 0$ and $h(x) \to \infty$ as $|x| \to \infty$.
(b) The function $\mathcal{T} \colon \mathsf{X}_\diamond \to \{1, 2, 3, \dots\}$ used to define the review times.

(c) A vector of safety-stock values $\overline{x} \in \mathbb{R}_+^\ell$, possibly also defined as a function on X. For a fixed parameter $\varepsilon_1 \in (0, 1)$ an intermediate safety-stock value is defined by

$$\overline{y} := \min(\overline{x}, x^0 + \varepsilon_1 \overline{x}), \qquad (4.80)$$

where x^0 denotes the initial condition, and the minimum is component-wise.

For an initial condition $x^0 := Q(0)$ we set $T_1 = \mathcal{T}(x^0)$ and construct a constant allocation rate ζ^* so that $z(t) = \zeta^* t$ and $q(t) = x^0 + (B\zeta^* + \alpha)t$ on $[0, T_1)$, with $q(T_1) \geq \overline{y}$. The allocation process U is defined on the discrete-time interval $[0, T_1)$ so that the approximations (4.78a), (4.78b) hold using (q, z).

   The allocation process is constructed on any subsequent review period $[T_m, T_{m+1})$ by "resetting the clock" and defining U exactly as on the first time interval.

   Consider first a generalized myopic policy.

Definition 4.7.2 (Myopic-DR policy). Given the initial condition x^0, the safety-stock vector $\overline{y} = \overline{y}(x^0)$, and the value $T_1 = \mathcal{T}(x^0) \geq 1$, the *myopic-DR policy* defines the allocation process U on $[0, T_1)$ so that the cumulative allocation process (4.79) satisfies $\mathsf{E}[Z(T_1)] \approx \zeta^* T_1$, where ζ^* is an optimizer to the following convex program:

$$\begin{aligned}
\min \quad h(y) \qquad \text{s.t.} \quad y \ &= \ x^0 + (B\zeta + \alpha)T_1, \\
y \ &\in \ \mathsf{X}, \\
y \ &\geq \ \overline{y}, \\
\zeta \ &\in \ \mathsf{U}.
\end{aligned} \qquad (4.81)$$
■

   The constraint set in (4.81) will be feasible through choice of $\mathcal{T}, \overline{x}$, and ε_1.

   The constraint $y \geq \overline{y}$ is imposed so that $Q_i(t) \geq \overline{x}_i$ with high probability for all t and i; The policy attempts to increase the values of buffers for which this lower bound fails. The constraint $y \in \mathsf{X}$ is redundant if $\mathsf{X} = \mathbb{R}_+^\ell$ since $\overline{y} \geq 0$.

   The convex program (4.81) reduces to a linear program when h is linear, or piecewise linear. In general, if the convex program (4.81) is too complex then the algorithm can be modified to obtain a linear program with objective function

$$\langle \nabla h(x^0), B\zeta \rangle. \qquad (4.82)$$

   We now consider a modification of the GTO policy.

   Given the initial condition x^0 and a desired final state x^1, the minimal time to reach x^1 from x^0 is denoted $T^*(x^0, x^1)$. This can be computed using a linear program, exactly as in the construction of the minimal draining time (see Proposition 6.1.4 in Chapter 6.) In the GTO-DR policy we assume that $\mathcal{T}(x) \geq T^*(x, \overline{x})$ for each x.

Definition 4.7.3 (GTO-DR policy). Given the initial condition x^0, the safety-stock vectors $\overline{x}, \overline{y}$, and the value $T_1 = \mathcal{T}(x^0) \geq T^*(x^0, \overline{x})$, the *GTO-DR policy* defines the allocation process U on $[0, T_1)$ so that the cumulative allocation process satisfies

$\mathsf{E}[Z(T_1)] \approx \zeta^* T_1$, where ζ^* is an optimizer to the following convex program:

$$
\begin{aligned}
\textbf{min} \quad h(y) \qquad \textbf{s.t.} \qquad y &= x^0 + (B\zeta + \alpha)T_1, \\
\overline{x} &= y + (B\zeta^2 + \alpha)(T^*(x^0, \overline{x}) - T_1), \\
y &\in \mathsf{X}, \\
y &\geq \overline{y}, \\
\zeta, \zeta^2 &\in \mathsf{U}.
\end{aligned}
\tag{4.83}
$$

Feasibility of (4.83) can again be assured through choice of $\varepsilon_1 \in (0,1)$, $\overline{x}$, and $\mathcal{T}$. And, the objective function (4.82) can be used to obtain a simpler linear program if this is preferable.

Suppose that the outcome of the linear program is applied to the fluid model, so that $\zeta(t) = \zeta^*$ on $[0, T_1)$, and $\zeta(t) = \zeta^{2*}$ on $[T_1, T^*(x^0, \overline{x}))$. Then q reaches $\overline{x}$ in minimal time, subject to the constraint $q(T_1) \geq \overline{y}$. However, for application in the CRW model or a physical network, the allocation rate ζ^{2*} is discarded since it is only an intermediate variable used to represent the time optimality constraint.

The solution ζ^* of the linear program (4.83) coincides with the allocation rate obtained in the GTO policy (4.46) when $\overline{x}$ is set to zero, and the planning horizon T_1 is sufficiently small. The proof of Proposition 4.7.4 follows from the fact that a GTO solution (q, z) can be taken to be piecewise linear as a function of time.

Proposition 4.7.4. *Suppose that $\overline{x} = 0$ in the GTO-DR policy (4.83). Then, for each $x^0 \in \mathsf{X}$, there exists $T_1 > 0$ sufficiently small such that the solution $\zeta^* \in \mathsf{U}$ is a GTO allocation rate on $[0, T_1]$.*

4.8 MaxWeight and MinDrift

The class of policies formulated next do not rely on safety stocks. Instead, a particular function h is constructed so that safety stocks arise automatically in an h-myopic policy.

The policies considered in this section and the next are stationary. However, it will be clear that any of these stationary policies can be modified to define a DR policy.

We have several results and examples to help guide the selection of the function h in an h-myopic policy. Theorem 4.1.6 suggests that we should restrict to functions h that are roughly quadratic if our goal is to approximate the solution to the ACOE.

Suppose that $h(x) = \frac{1}{2} x^{\mathsf{T}} D x$, $x \in \mathbb{R}_+^\ell$, where $D > 0$ is a diagonal matrix. Given any value $U(t) = u \in \mathsf{U}_\circ(x)$ we can compute the conditional expectation in (4.15) to obtain

$$
\begin{aligned}
\mathsf{E}[h(Q(t+1)) - h(Q(t)) \mid Q(t) = x, \, U(t) = u] \\
= \langle Bu + \alpha, Dx \rangle + \mathsf{E}[h(B(t+1)u + A(t+1))].
\end{aligned}
\tag{4.84}
$$

Minimizing the right-hand side of (4.84) defines the myopic policy. However, this is difficult to analyze directly due to the quadratic term $\mathsf{E}[h(B(t+1)u + A(t+1))]$. A simpler policy is obtained if this bounded term is ignored, giving rise to the MaxWeight policy:

Definition 4.8.1 (MaxWeight policy). For a given positive definite diagonal matrix D, the MaxWeight policy is defined as the stationary policy

$$\phi^{\text{MW}}(x) \in \arg\min_{u \in \mathsf{U}_\diamond(x)} \langle Bu + \alpha, Dx \rangle, \qquad x \in \mathsf{X}_\diamond. \tag{4.85}$$

Note that the minimum in (4.85) may not be unique – any minimizer defines a MaxWeight policy.

To obtain a simpler expression for ϕ^{MW}, define for any $j = 1, \dots, \ell$,

$$\Theta_j(x) := \mu_j \sum_i x_i D_{ii}(I_{ji} - R_{ji}) = \mu_j(D_{jj}x_j - D_{j_+j_+}x_{j_+}), \qquad x \in \mathsf{X}_\diamond, \tag{4.86}$$

where $x_{j_+} := 0$ if buffer j is an exit buffer (recall that upon completing service at buffer j a customer enters buffer $j_+ \in \{1, \dots, \ell\}$ or exits the system). For each x and j the coefficient of u_j in (4.85) is precisely $-\Theta_j(x)$. Denote the maximum at a station by $\overline{\Theta}_s(x) := \max_{j \in \mathcal{I}_s} \Theta_j(x)$, $s \in \mathcal{S}$.

The vector α is unimportant in the minimization (4.85), so that the MaxWeight policy can be described as follows:

Proposition 4.8.2. *For each station $s \in \mathcal{S}$, given $Q(t) = x$, the allocation vector $U(t) = \phi^{\text{MW}}(x)$ under the MaxWeight policy satisfies the following:*

(i) *If $\overline{\Theta}_s(x) < 0$ then $U_j(t) = 0$ for each $j \in \mathcal{I}_s(x)$.*
(ii) *If $\Theta_j(x) \geq 0$ for some $j \in \mathcal{I}_s(x)$ with $x_j \geq 1$ then $\overline{\Theta}_s(x) \geq 0$ and,*

$$\sum_{i \in \mathcal{I}_s} \{U_i(t) : \Theta_i(x) = \overline{\Theta}_s, \ x_i \geq 1\} = 1. \tag{4.87}$$

MaxWeight is known as the *back-pressure policy* when $D = I$. In this case $\Theta_j(x) = \mu_j(x_j - x_{j_+})$, and $\overline{\Theta}_s(x)$ is called the *maximal back-pressure* at Station s. Based on (4.87) we conclude that the back-pressure policy is any stationary policy that gives strict priority to buffers achieving the maximal back-pressure: If $x_j - x_{j_+} \geq 0$ and $x_j \geq 1$ for some $j \in \mathcal{I}_s$, then

$$\sum_{i \in \mathcal{I}_s} \{\phi_i^{\text{MW}}(x) : \mu_i(x_j - x_{j_+}) = \overline{\Theta}_s, \ x_i \geq 1\} = 1, \qquad x \in \mathsf{X}, \ s \in \mathcal{S}. \tag{4.88}$$

Proposition 4.8.3 implies that the MaxWeight feedback law ϕ^{MW} also defines the myopic policy for the fluid model since the minimization in (4.85) can be relaxed to a minimization over all of U. It follows that the policy (4.85) minimizes the "drift" for the fluid model, $\frac{d}{dt}h(q(t)) = \langle Bu + \alpha, Dq(t) \rangle$.

Proposition 4.8.3. *For each $x \in \mathsf{X}_\diamond$ any MaxWeight allocation vector $u^* = \phi^{\mathrm{MW}}(x)$ is a solution to the linear program*

$$\mathbf{arg\,max} \quad x^{\mathrm{T}} D(I - R^{\mathrm{T}}) M u$$

$$\mathbf{s.t.} \quad Cu \leq 1, \quad u \geq 0.$$

(4.89)

Proof. Note first that the objective function in (4.85) can be written

$$\arg\min_{u \in \mathsf{U}_\diamond(x)} \langle Bu + \alpha, Dx \rangle = \arg\min_{u \in \mathsf{U}_\diamond(x)} \langle Bu, Dx \rangle$$

$$= \arg\max_{u \in \mathsf{U}_\diamond(x)} \left(\sum_{i,j} x_i D_{ii}(1 - R_{ji}) \mu_j u_j \right)$$

$$= \arg\max_{u \in \mathsf{U}_\diamond(x)} \left(\sum_{j} (D_{jj} x_j - D_{j+j+} x_{j+}) \mu_j u_j \right).$$

Hence maximization of the objective function (4.89) over $\mathsf{U}_\diamond(x)$ gives the MaxWeight policy. To complete the proof we show that this maximization can be relaxed to a maximum over all of U.

Since the matrix D has nonnegative entries we conclude that

$$D_{jj} x_j - D_{j+j+} x_{j+} \leq 0 \text{ whenever } x_j = 0. \tag{4.90}$$

It then follows that the optimizer u^* of (4.89) satisfies without loss of generality $u_i^* = 0$ whenever $x_i = 0$. This shows that $u^* \in \mathsf{U}(x)$ for $x \in \mathbb{Z}_+^\ell$.

To show that u^* can be chosen in $\mathsf{U}_\diamond$ we argue that optimizers of linear programs can be chosen among the extreme points in the constraint region. The extreme points for this linear program are all contained in $\{0, 1\}^\ell$, which proves the proposition. $\square$

Proposition 4.8.3 combined with Proposition 4.3.6 leads to a proof of stability. Recall that by definition of the myopic policy we have

$$P_{\mathrm{myopic}} h \leq P_{\mathrm{MW}} h,$$

so that the Poisson inequality holds for the h-myopic policy if it holds for the MaxWeight policy.

Theorem 4.8.4 (Stability of MaxWeight). *Suppose that $\rho_\bullet < 1$ and $\mathsf{E}[\|A(1)\|^2] < \infty$ in the CRW model, and that the MaxWeight policy is applied for some diagonal matrix $D > 0$. Then, for some sufficiently large $b > 0$, the pair of functions $V(x) = \frac{b}{2} x^{\mathrm{T}} Dx$, $c(x) = |x|$ solve the Poisson inequality (4.13).*

The proof is largely based on the following property of the fluid model:

Lemma 4.8.5. *If $\rho_\bullet < 1$, then there exists $\varepsilon > 0$ such that $v \in \mathsf{V}$ whenever $0 \geq v_i \geq -\varepsilon$ for each i.*

Proof of Theorem 4.8.4. We extend the proof of Proposition 4.3.6 as follows: Applying Lemma 4.8.5, on setting $v_i = -\varepsilon\mathbf{1}(x_i \geq 1)$ for each i, and $u \in \mathsf{U}$ the solution to $(-I + R^{\mathrm{T}})Mu = v$, we obtain under the MaxWeight policy with $V(x) = \frac{b}{2}x^{\mathrm{T}}Dx$,

$$P_{\mathrm{MW}}V(x) := \mathsf{E}_{\mathrm{MW}}[V(Q(t+1)) \mid Q(t) = x] \leq V(x) - \varepsilon b x^T D\mathbf{1} + \bar{\eta},$$

with

$$\bar{\eta} := \frac{b}{2} \max_{u \in \mathsf{U}_\circ(x)} \mathsf{E}[(Q(t+1) - Q(t))^{\mathrm{T}}D(Q(t+1) - Q(t)) \mid Q(t)].$$

Hence the desired conclusion holds with $b^{-1} = \varepsilon(\min_i D_{ii})$. $\qquad\square$

Example 4.8.6 (Simple re-entrant line: MaxWeight policy). The MaxWeight policy for this model is defined through the maximization (4.89), where the term to be maximized can be written

$$x^{\mathrm{T}}D(I - R^{\mathrm{T}})Mu = x_1 D_{11}(\mu_1 u_1) + x_2 D_{22}(-\mu_1 u_1 + \mu_2 u_2) + x_3 D_{33}(-\mu_2 u_2 + \mu_3 u_3)$$

$$= (x_1 D_{11} - x_2 D_{22})\mu_1 u_1 + (x_3 D_{33})\mu_3 u_3 + (x_2 D_{22} - x_3 D_{33})\mu_2 u_2.$$

The MaxWeight policy is thus described as follows:

Station 1: There are two cases to consider, depending upon whether there is inventory at buffer 3: If $x_3 = 0$ then we only need consider the sign of the coefficient of u_1, giving

$$\phi_1(x) = \mathbf{1}\{-x_1 D_{11} + x_2 D_{22} \leq 0\} \quad \text{when } x_3 = 0.$$

Otherwise, a minimization over the coefficients of u_1 and u_3 gives

$$\phi_3(x) = \mathbf{1}\{-x_3 D_{33}\mu_3 \leq -x_1 D_{11}\mu_1 + x_2 D_{22}\mu_1\} \quad \text{when } x_3 \geq 1.$$

Station 2: The second station is nonidling if the coefficient of u_2 is nonnegative:

$$\phi_2(x) = \mathbf{1}\{x_2 D_{22} \geq x_3 D_{33}\} \quad \text{when } x_2 \geq 1.$$

This policy is similar to the policy shown in Fig. 2.10 if D is chosen so that $D_{11} = D_{33}$, and D_{22} is much larger than the other diagonal elements $\qquad\blacksquare$

4.9 Perturbed value function

The key property (4.90) used to establish stability of the MaxWeight policy is a consequence of the derivative condition,

$$\frac{\partial}{\partial x_j}h(x) = 0 \quad \text{when } x_j = 0. \tag{4.91}$$

With h interpreted as a surrogate "value function," the quantity

$$\frac{\partial}{\partial x_j}h(x)$$

represents the "marginal cost" or "marginal dis-utility" of an additional increment of inventory at buffer j. If this marginal cost is zero, then it is reasonable to shift inventory to this buffer when possible.

For any function satisfying (4.91) we define the *h-MaxWeight* policy as the following generalization of (4.85),

$$\phi^{\text{MW}}(x) \in \arg\min_{u \in U_\circ(x)} \langle Bu + \alpha, \nabla h(x) \rangle, \qquad x \in X. \tag{4.92}$$

To obtain a wide class of policies we consider here perturbations of a given function so that (4.91) holds.

For fixed $\theta \geq 1$ denote

$$\tilde{x}_i := x_i + \theta(e^{-x_i/\theta} - 1), \quad \text{for any } i \text{ and } x \in \mathbb{R}^\ell_+, \tag{4.93}$$

and $\tilde{x} = (\tilde{x}_1, \ldots, \tilde{x}_\ell)^{\mathsf{T}} \in \mathbb{R}^\ell_+$. Let h_0 be a smooth (continuously differentiable) function on $\mathbb{R}^\ell$, and consider the perturbation

$$h(x) = h_0(\tilde{x}), \qquad x \in \mathbb{R}^\ell_+. \tag{4.94}$$

We have $\frac{d}{dx_i}\tilde{x}_i\big|_{x_i=0} = 0$ for any i. An application of the chain rule of differentiation shows that h satisfies (4.91):

Proposition 4.9.1. *For any C^1 function h_0, the function h defined in (4.94) satisfies the derivative conditions (4.91). We have the following explicit representations:*

(i) *The first derivative is given by*

$$\nabla h(x) = [I - M_\theta]\nabla h_0(\tilde{x}), \tag{4.95}$$

where

$$M_\theta = \text{diag}\,(e^{-x_i/\theta}), \qquad x \in \mathbb{R}^\ell. \tag{4.96}$$

(ii) *If h_0 is C^2, then the Hessian of h is*

$$\nabla^2 h(x) = [I - M_\theta]\nabla^2 h_0(\tilde{x})[I - M_\theta] + \theta^{-1}\text{diag}\,(M_\theta \nabla h_0(\tilde{x})). \tag{4.97}$$

Hence h is convex provided h_0 is both convex and monotone.

If h_0 is linear then the function h can be expressed

$$h(x) = \sum_{i=1}^{\ell} c_i \tilde{x}_i, \qquad x \in \mathbb{R}^\ell_+, \tag{4.98}$$

The first and second derivatives are given by

$$\begin{aligned}
\nabla h(x) &= (c_1(1 - e^{-x_1/\theta}), \ldots, c_\ell(1 - e^{-x_\ell/\theta}))^{\mathsf{T}}, \\
\nabla^2 h(x) &= \theta^{-1}\text{diag}\,(c_1 e^{-x_1/\theta}, \ldots, c_\ell e^{-x_\ell/\theta}).
\end{aligned} \tag{4.99}$$

It is evident that the derivative condition (4.91) holds, which verifies Proposition 4.9.1 in this special case. Moreover h is monotone, in the sense that $\nabla h(x) \in \mathbb{R}^\ell_+$ for all $x \in \mathbb{R}^\ell_+$. It is also strictly convex since $\nabla^2 h(x) > 0$ for each x.

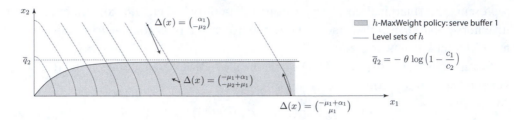

Figure 4.17. The h-MaxWeight policy for the tandem queues is approximated by a static safety-stock at Station 2 when h_0 is linear. In this plot the cost parameters are $(c_1, c_2) = (1, 3)$ and $\theta = 10$. The asymptote (4.101) is $\bar{x}_2 = 10\log(3/2) \approx 4$ in this special case. The contour plots shown are the level sets $\{x : h(x) = r\}$ for $r = 1, 2, \ldots$.

We will see in Proposition 8.4.4 that the myopic policy with respect to h is stabilizing for the CRW scheduling model when h_0 is linear, provided $\theta \geq 1$ is suitably large.

In an attempt to approximate the solution to the ACOE we might take a more sophisticated choice for h_0: In Theorem 8.4.1 it is assumed that for some norm c we have

$$\min_{u \in \mathsf{U}(x)} \langle \nabla h_0(x), Bu + \alpha \rangle = -c(x) \qquad x \in \mathbb{R}_+^\ell. \tag{4.100}$$

The resulting h-MaxWeight policy and the h-myopic policy are again stabilizing, provided $\theta \geq 1$ is suitably large. In some examples the policy is stabilizing for *any* positive θ.

In the tandem queues the h-MaxWeight policy resembles a policy constructed using static safety stocks when h_0 is linear, and the policy resembles a dynamic safety-stock using a logarithmic switching curve when h_0 is a fluid value function.

Example 4.9.2 (Tandem queues: Emergence of a safety-stock policy). We return to the setting of Example 4.4.6 in which the c-myopic policy never allows service at Station 1. Suppose that we replace the linear function c with the convex cost function $h \colon \mathbb{R}_+^2 \to \mathbb{R}_+$ defined in (4.98).

The inner product (4.92) becomes

$$\langle Bu, \nabla h(x) \rangle = -\mu_1 u_1 c_1 (1 - e^{-x_1/\theta}) + (\mu_1 u_1 c_1 - \mu_2 u_2 c_2)(1 - e^{-x_2/\theta}).$$

The h-MaxWeight policy minimizing this expression is nonidling at Station 2, and at Station 1 the policy can be expressed as a switching curve,

$$\phi_1^{\mathrm{MW}}(x) = \mathbf{1}\{-c_1(1 - e^{-x_1/\theta}) + c_2(1 - e^{-x_2/\theta}) \leq 0\}, \qquad x_1 \geq 1.$$

Figure 4.17 illustrates this switching curve when $c_1 = 1$, $c_2 = 3$, and $\theta = 10$.

For small values of x_1 a first-order Taylor series gives the approximation $\phi_1^{\mathrm{MW}}(x) \approx \mathbf{1}\{x_2 \leq (c_1/c_2)x_1\}$. If $x_1 \gg \theta$ then ϕ_1^{MW} can be approximated by a static safety-stock policy, $\phi_1^{\mathrm{MW}}(x) \approx \mathbf{1}\{x_2 \leq \bar{x}_2\}$, where the threshold $\bar{x}_2$ is the solution to the equation $c_2(1 - e^{-\bar{x}_2/\theta}) = c_1$. That is,

$$\bar{x}_2 = \theta \left| \log\left(1 - \frac{c_1}{c_2}\right) \right|. \tag{4.101}$$

We now consider a perturbation of the fluid value function J^* that was computed in Example 4.3.9. The resulting policy is approximated by a dynamic safety stock similar to (4.71), and it is also very similar to the average-cost optimal policies obtained in Example 4.5.2.

With $\rho_1 < \rho_2$ and $c_1 < c_2$ (Case 2 of Example 4.5.2) the fluid value function is purely quadratic,

$$J^*(x) = \tfrac{1}{2}\frac{c_1}{\mu_2 - \alpha_1}(x_1 + x_2)^2 + \tfrac{1}{2}\frac{c_2 - c_1}{\mu_2}x_2^2, \qquad x \in \mathbb{R}_+^2. \qquad (4.102)$$

Letting $h_0 = J^*$, the bound (4.100) is satisfied with equality. The derivative conditions (4.91) fail for h_0, so we do not know if the h_0-MaxWeight policy is stabilizing for the CRW model.

To compute the h-MaxWeight policy we write (4.102) as

$$h_0(x) = J^*(x) = \tfrac{1}{2}d_1(x_1 + x_2)^2 + \tfrac{1}{2}d_2 x_2^2, \qquad x \in \mathbb{R}_+^2,$$

so that the gradient of h can be expressed

$$\nabla h(x) = \begin{pmatrix} d_1(\tilde{x}_1 + \tilde{x}_2)(1 - e^{-x_1/\theta}) \\ (d_1(\tilde{x}_1 + \tilde{x}_2) + d_2\tilde{x}_2)(1 - e^{-x_2/\theta}) \end{pmatrix}.$$

Writing $Bu = (-\mu_1 u_1, \mu_1 u_1 + \mu_2 u_2)^{\mathsf{T}}$, we obtain for any $x \in \mathsf{X}_\diamond, u \in \mathsf{U}(x)$,

$$\langle \nabla h(x), Bu \rangle = \mu_1 u_1\left[d_1(e^{-x_1/\theta} - e^{-x_2/\theta})(\tilde{x}_1 + \tilde{x}_2) + d_2(1 - e^{-x_2/\theta})\tilde{x}_2\right]$$

$$- \mu_2 u_2\left[d_1(1 - e^{-x_2/\theta})(\tilde{x}_1 + \tilde{x}_2) + d_2(1 - e^{-x_2/\theta})\tilde{x}_2\right].$$

Minimizing over u we see that the policy is nonidling at Station 2. At Station 1 we have $u_1 = 1$ if and only if $x_1 \geq 1$ and the coefficient of u_1 is nonpositive. That is, the policy at Station 1 is defined by the switching curve described by the equation

$$d_1(e^{-x_1/\theta} - e^{-x_2/\theta})(\tilde{x}_1 + \tilde{x}_2) + d_2(1 - e^{-x_2/\theta})\tilde{x}_2 = 0. \qquad (4.103)$$

When x_1 is large we obtain the approximation

$$x_2 \approx s(x_1) := \theta \log\left(1 + \frac{d_1}{d_2}x_1\right), \qquad (4.104)$$

where by (4.102),

$$\frac{d_1}{d_2} = \left(\frac{c_2}{c_1} - 1\right)^{-1}\frac{1}{1 - \rho_2}.$$

This is an approximation to (4.103) in the sense that for all sufficiently large x_1 there is a unique x_2 such that (x_1, x_2) solve the equation (4.103), and the ratio $x_2/s(x_1)$ tends to unity as $x_1 \to \infty$. The approximation is also valid for small x_1: if $x_1 = 0$ then $s(x_1) = 0$, and the solution to (4.103) is also $x_2 = 0$. ∎

4.10 Notes

The CRW model is a generalization of the countable state-space model obtained using Lippman's uniformization technique [338]. It is a common model in operations research and in economics. A version of the CRW model, the *binomial option price model*, is developed in Cox, Ross, and Rubinstein [120]. Shreve remarks in his online monograph *Lectures on Stochastic Calculus and Finance* [444] (now published as [445]) that *many models are first developed and understood in continuous time, and then binomial versions are developed for purposes of implementation.*

Fluid approximations for queueing networks are described in the 1982 book of Newell [385]. Interest in this approach grew following the work of Chen and Mandelbaum [94] and Dai [124]. See also [26, 163, 95, 24, 481] and their references.

Optimal control solutions for a stochastic network model are typically defined by switching curves or threshold rules. The prototype example is the class of (S, s) threshold policies introduced by Clark and Scarf approximately 50 years ago [423, 422, 110]. Similar policies are used for elaborations of the single server queue in [372, 144]. Rosberg, Varaiya, and Walrand [414] demonstrate the existence of a switching curve in the tandem queues if the cost is higher in the second queue, as is clearly seen in Figs. 4.9, 4.10 for the discounted cost criterion, and in Figs. 4.7, 4.8 for the average-cost criterion. Hajek in [222] considers a model generalizing the tandem queues and the simple routing model to demonstrate the existence of a switching curve.

Discrete review policies are commonly called *periodic review* in the inventory theory literature [472, 203, 434, 265]. The BIGSTEP approach of Harrison [234] is similar. Also closely related is *model predictive control* and *receding horizon control* for dynamical systems (see Goodwin et al. [218], and the survey by Mayne et al. [351].)

The GTO-DR policy was introduced in [359, 102] for scheduling models, and demand driven models, respectively. Closely related is the work of Dai and Weiss [133] and Bertsimas and Sethuraman [56] where a fluid model is the basis of a policy that approximately minimizes make-span.

Proposition 4.4.1 relating the fluid and stochastic value functions is based on Altman, Jiménez, and Koole [9] which establishes related bounds for multiserver queues in tandem. Related bounds are the basis of [50].

Tassiulas and Ephremides showed in [468] that the MaxWeight policy is universally stabilizing for a class of stochastic networks. This work has been extended in multiple directions over the past 15 years [201, 467, 438, 128], and in particular these policies are known to be approximately optimal in heavy traffic under certain conditions on the network [477, 462, 348].

The h-MaxWeight policy (4.92) was introduced in [365]. It is shown there that a specialized version is approximately average-cost optimal in heavy traffic. The h-MaxWeight policy coincides with the h-myopic policy for the fluid model. The special case in which h is linear is treated by Chen and Yao in [96, Thm. 12.5], whereit is shown that such policies are universally stabilizing for the fluid model. The generalization Proposition 4.3.6 is based on [359, Proposition 11].

Also related to the MaxWeight policy is the sum of squares algorithm for bin-packing [116, 122], and the utility-function based flow control algorithms introduced by Kelly et al. [206, 459].

A logarithmic switching curve similar to (4.71) was proposed in [357, p. 194] to translate a policy from the fluid model to a stochastic model. Further history is contained in the Notes section in Chapter 10.

One motivation for myopic policies comes from considering the dynamic programming equations for the infinite-horizon optimal control problem. For the fluid model, the optimal policy is the solution to (4.44) with c replaced by the value function J^*. The infinite-horizon optimal control problem for q has its origins in the work of Bellman [37]. The general subject of infinite-dimensional linear programming was revitalized after Anderson's PhD thesis [11], and subsequent monographs [12, 13]. The theory has matured recently with the work of Pullan (e.g. [399, 401, 400]), and computational methods have appeared in [224, 25, 344, 395, 492]. See also the references on computational methods for network flow at the end of Chapter 6.

Exercises

4.1 Prove that any nonidling feedback law is time optimal for the single-station fluid model. *Hint: the constituency matrix C is a row vector.*

4.2 Consider the simple re-entrant line shown in Fig. 2.9. Show that the myopic feedback law is path-wise optimal provided the network is balanced, and $c(x) = |x| = x_1 + x_2 + x_3$.

4.3 Compute the fluid value function J^{GTO} for the tandem queues under the conditions of Example 4.3.9. How does it compare to the optimal policy?

4.4 Consider a variation of the single queue in which each customer declares its service requirement in advance. In this case it is not clear that the FIFO service discipline is the best choice. If at time t there are $m = A(t) \geq 1$ arrivals to the queue, they bring with them service requirements denoted $\{G_j(t) : 1 \leq j \leq m\}$. Assume that $\{G_j(t) : j, t \in \mathbb{Z}_+\}$ are i.i.d. with support in the finite set $\{1, \ldots, \ell_g\}$, and independent of A. For each $k \geq 1$ let $\mathcal{L}_k(t)$ denote the number of size-k jobs that arrive at time t,

$$\mathcal{L}_k(t) = \mathbf{1}\{A(t) \geq 1\} \sum_{j=1}^{A(t)} \mathbf{1}\{G_j(t) = k\}$$

Construct a single station network with ℓ_g buffers, deterministic service of rate 1, and arrival process $\{\mathcal{L}(t)\}$ that captures this system. The special case $\ell_g = 4$ is illustrated below:

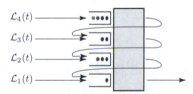

Construct the associated fluid model by computing the mean of $\mathcal{L}_k(t)$. Based on this, what is the workload vector for the resulting ℓ_g-dimensional model? What is the load? Can you find a path-wise optimal policy? Can you find a solution to (V3) for the CRW model under your favorite policy?

We return to this example in Section 5.1

4.5 Write a succinct, complete proof of Proposition 4.3.8 (iii) and (iv).

4.6 The scheduling model shown in Fig. 4.18 is known as the *criss-cross network*. The fluid model is described by the linear differential equation (1.4), where $\alpha = (\alpha_1, 0, \alpha_3)^{\mathsf{T}}$, and the allocation rates $\zeta \in \mathbb{R}^3$ are constrained via

$$\zeta_i \geq 0, \qquad\qquad i = 1, 2, 3;$$
$$\zeta_1 + \zeta_3 \leq 1, \quad \zeta_2 \leq 1. \qquad\qquad \text{(E4.1)}$$

(a) Write down the 3×3 matrix B, and the 3×3 constituency matrix C.
(b) Compute the workload vectors $\{\xi^i\}$, and the network load.
(c) Compute the c-myopic policy with $c(x) = x_1 + 2x_2 + x_3$.
(d) Construct a time optimal policy.

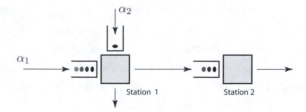

Figure 4.18. Criss-cross network.

4.7   Complete Example 4.3.9: Under the assumption that $\mu_1 < \mu_2$ and $c_2 > c_1$, show that the fluid value function J is purely quadratic under the nonidling policy, and compute $J^*(x)$ for all $x \in \mathsf{R}_2$ under the optimal policy to verify (4.51). Compare $J(x)$ and $J^*(x)$ for $x \in \mathsf{R}_2$ on the boundary $\{x_1 = 0\}$.

4.8   Consider the fluid model for the simple re-entrant line shown in Fig. 2.9. Show that the myopic feedback law is path-wise optimal provided the network is balanced, and $c(x) = |x| = x_1 + x_2 + x_3$.

4.9   This chapter describes several stabilizing policies. With the exception of the MaxWeight policy, the definitions depend on the arrival rate α, which may not be known in advance. For example, if α is not known then it is not possible to determine dynamic bottlenecks that are required in the implementation of the GTO policy.

        For a given $x \in \mathsf{X}$, let $\mathcal{S}_0^*(x)$ denote the set of dynamic bottlenecks when $\alpha = 0$. That is, those stations $s \in \mathcal{S}$ such that

$$\langle \xi^s, x \rangle = W^*(x),$$

where W^* is defined in Definition 4.2.2. Consider any policy that satisfies $(C\zeta(t))_s = 1$ whenever $s \in \mathcal{S}_0^*(q(t))$. Show that this policy is stabilizing, and obtain a bound on the draining time $T^\circ(x)$ under this policy. *Hint*: note that $W^*(q(t)) \leq W^*(Bz(t)) + tW^*(\alpha)$ for $t \geq 0$. Does $W^*(\alpha)$ have any significance? Can you obtain an expression for the derivative of $W^*(Bz(t))$ with respect to time?

4.10   Consider the single server queue

$$Q(t+1) = Q(t) + S(t+1)U(t) - A(t+1), \qquad t \geq 0, \; Q(0) \in \mathbb{Z}_+,$$

where S takes values in $\{0, 1\}$ and A has finite support in $\mathbb{Z}_+$. However, instead of i.i.d., the joint process can be written as a function of a Markov chain,

$$\binom{A(t)}{S(t)} = F(\Phi(t)),$$

where $F \colon \mathbb{Z}_+ \to \mathbb{Z}_+^2$, and Φ is a Markov chain on $\mathbb{Z}_+$. Argue that Q no longer describes the state space for an MDP model. Construct a multidimensional MDP model that does describe this control problem by specifying the controlled transition matrix $P_u(x_0, x_1)$, $x_i \in X_\diamond$, and the state space $X_\diamond$. You will return to this model in Exercises 4.11, 4.3, 4.5, and 4.16 to see how the concepts in Part III can be generalized beyond the CRW model.

4.11   Simulate the KSRS model under the fluid-optimal policy, and under the LBFS policy, for these parameters:

$$\mu_1 = \mu_3 = 10; \qquad \mu_2 = \mu_4 = 3; \qquad \alpha_1 = \alpha_3 = 2.$$

Start the simulation at $Q(0) = 0$, and consider both deterministic and bursty arrival streams.

4.12   Compute the MaxWeight policy for the KSRS model. Observe that the solution gives $u_i = 0$ if $x_i = 0$. Verify that the Poisson inequality (4.13) holds for the quadratic V chosen, provided $\mathsf{E}[\|A(t)\|^2] < \infty$.

4.13   Consider a re-entrant line controlled using the h-MaxWeight policy, with

$$h(x) = \tfrac{1}{2} \sum_{n=1}^{\ell} \mu_n^{-1} \left(\sum_{i=1}^{n} x_i \right)^2.$$

Show that this policy is precisely LBFS. Is the h-MaxWeight policy always stabilizing for this choice of h?

4.14   There are many ways to construct a function satisfying (4.91). For a given $\theta > 0$ consider the following alternate definition of $\tilde{x}$,

$$\tilde{x}_i := (x_i + \theta) \log(x_i + \theta) - (1 + \log(\theta))x_i, \qquad x_i \geq 0, \qquad \text{(E4.2)}$$

and as before $\tilde{x} = (\tilde{x}_1, \ldots, \tilde{x}_\ell)^\mathsf{T} \in \mathbb{R}_+^\ell$.

(a) Obtain a generalization of Proposition 4.9.1. That is, given a C^1 function h_0, the function h defined in (4.94) using the definition (E4.2) is also C^1 and its derivative has a simple form in terms of h_0.

(b) Does (4.91) hold?

(c) Compute the h-MaxWeight policy for the tandem queues with h_0 linear. Sketch the region for which the upstream station is nonidling in the two cases $c_2 = 2c_1$ and $c_2 = c_1/2$.

4.15   Consider the function

$$h(x) := \log \left(\sum_i \exp(V_{D_i}(x)) \right), \qquad x \in \mathbb{R}_+^\ell,$$

where $\{D_i\}$ are $\ell \times \ell$ matrices.

(a) Compute the gradient ∇h, and explain why it is Lipschitz.

(b) Compute the h-MaxWeight policy (4.85).

(c) Show that for any $x \in \mathbb{R}_+^\ell$,

$$\lim_{r \to \infty} r^{-2} V(rx) = \max_i V_{D_i}(x).$$

Part II

Workload

5

Workload and Scheduling

Chapter 4 touches on many of the techniques to be developed in this book for controlling large interconnected networks. The fluid model was highlighted precisely because control is most easily conceptualized when variability is disregarded. The infinite-horizon optimal control problem with objective function defined in (4.37) can be recast as an infinite-dimensional linear program when c is linear. In many examples, such as the simple re-entrant line introduced in Section 2.8, a solution is explicitly computable. The MaxWeight policy and its generalizations are universally stabilizing, in the sense that a single policy is stabilizing for any CRW scheduling model satisfying the load condition $\rho_{\bullet} < 1$ along with the second moment constraint $\mathsf{E}[\|A(1)\|^2] < \infty$.

What is missing at this stage is any intuition regarding the structure of "good policies" for a network with many stations and buffers. In this chapter we introduce one of the most important concepts in this book, the *workload relaxation*. Our main goal is to construct a model of reduced dimension to simplify computation of policies, and to better visualize network behavior.

In the theory of optimization, a relaxation of a given model is simply a new model obtained by removing constraints. In the case of networks there are several classes of constraints that complicate analysis:

(i) The integer constraint on buffer levels.
(ii) Constraints on the allocation sequence determined by the constituency matrix.
(iii) State space constraints, including positivity of buffer levels, as well as strict upper limits on available storage.

The fluid model can be regarded as a relaxation of the integral constraint (i). Beginning in Section 5.3 we introduce relaxations of (ii) to obtain a workload relaxation for the fluid model, and we apply similar techniques to address CRW network models in Section 5.4. Finally, the unconstrained process introduced in Definition 5.6.2 is a relaxation of all state space constraints to construct hedging points for multidimensional workload models.

We have stressed that the CRW network model is far too detailed to be useful in optimization except in the simplest examples. Fortunately, in many cases a lower dimensional CRW workload relaxation may be entirely tractable. It is not possible to obtain an *explicit* expression for the average cost or any of the value functions introduced in Section 4.1.3 except in very special cases. However, for workload models of moderate dimension an optimal policy can be constructed using one of the dynamic programming methods described in Chapter 9.

The most important concepts of this chapter can all be described within the simplest fluid model setting. The *effective cost*, the *effective state*, the concept of the *monotone region* in workload space and its impact on control design are all introduced for the fluid model in Section 5.3, and extended to the CRW model in Section 5.4.

These essential ingredients form the basis for control techniques to be developed for stochastic network models. In Section 5.6, we introduce general techniques to obtain approximations for optimal policies using generalizations of the hedging-point policies introduced in Section 2.6 for the simple inventory model.

We restrict to the scheduling model here. The ideas presented in this chapter will be expanded and refined throughout the remainder of the book.

Before investigating relaxations we first take a closer look at workload for the single server queue, and for more general single-resource models.

5.1 Single server queue

Within the scope of this book we have said almost everything that needs to be said about the single server queue. What is remaining is one more representation of workload, and a consequence known as Little's law.

5.1.1 Workload in units of time

Workload in units of time as defined in Definition 4.2.3 is expressed $w(t) = \mu^{-1} q(t)$ in the single server queue. This is interpreted as the time the server must work to clear all of the inventory *currently in the system at time t*, ignoring future arrivals. Equivalently, an increment of fluid arriving at time t will experience a delay of $w(t)$ time units before exiting the queue. The workload process satisfies the ODE

$$\tfrac{d^+}{dt} w(t) = -(1 - \rho) + \iota(t), \tag{5.1}$$

where $\iota(t) = 1 - \zeta(t)$ is interpreted as the rate of idleness in the fluid model, and the allocation rate is defined as the right derivative, $\zeta(t) := \tfrac{d^+}{dt} z(t)$. When q is defined using the nonidling policy, then $\iota(t) = 0$ when $w(t) > 0$, and $\iota(t) = 1 - \rho$ when $w(t) = 0$.

Consider now the CRW model. The quantity $W(t) = \mu^{-1} x$ represents the mean service time required to process all of the customers residing in the queue at time t, conditioned on $Q(t) = x$. We introduce here the *unconditional workload*, denoted $\mathcal{W}(t)$, which is defined to be the *actual time* required to clear all of the customers in the system at time t, ignoring future arrivals. To obtain a model for $\mathcal{W}$ we require a refined probabilistic representation of the queue.

Suppose that at time t there are $m = A(t) \geq 0$ arrivals to the queue. For $j = 1, \dots, m$ we denote by $\mathcal{G}_j(t)$ the workload requirement of the respective customer. It is assumed that $\{\mathcal{G}_j(t) : j, t \in \mathbb{Z}_+\}$ are i.i.d. with common mean μ^{-1}, and also independent of the process $\boldsymbol{A}$. A refined description of this model is given in Exercise 5.4.

If the model is to be consistent with the CRW model (3.1) then these random variables will have a geometric distribution supported on $\{1, 2, \dots\}$ satisfying

$$P\{\mathcal{G}_j(t) = T + 1 \mid \mathcal{G}_j(t) > T\} = \mu, \qquad t, T \in \mathbb{Z}_+. \tag{5.2}$$

In this case we have for each j, t,

$$\mathsf{E}[\mathcal{G}_j(t)] = \sum_{n=1}^{\infty} \mathsf{P}\{\mathcal{G}_j(t) \geq n\} = \sum_{n=1}^{\infty} (1 - \mu)^{n-1} = \mu^{-1}.$$

The total new workload introduced at time t is denoted

$$\mathcal{L}(t) = \mathbf{1}\{A(t) \geq 1\} \sum_{j=1}^{A(t)} \mathcal{G}_j(t),$$

and the initial condition is similarly defined by

$$\mathcal{W}(0) = \mathbf{1}\{Q(0) \geq 1\} \sum_{j=1}^{Q(0)} \mathcal{G}_j(0).$$

At time $t + 1$ the workload will be increased by $\mathcal{L}(t+1)$, and simultaneously decreased by $U(t)$. Hence the unconditional workload process evolves according to a CRW model of the form (3.1) with $S(t) \equiv 1$,

$$\mathcal{W}(t + 1) = \mathcal{W}(t) - U(t) + \mathcal{L}(t + 1), \qquad t \geq 0. \tag{5.3}$$

Observe that the ODE (5.1) captures the mean behavior of the unconditional workload process since $\mathsf{E}[\mathcal{L}(t)] = \mathsf{E}[A(t)]\mathsf{E}[\mathcal{G}_j(t)] = \rho$.

5.1.2 Workload, delay, and Little's law

What is the delay experienced by a typical customer entering the system? To answer this question consider first the *actual* delay: If a single customer arrives at time T, then the delay experienced before service begins is given by $\mathcal{W}(T)$. Hence the delay is precisely $\mathcal{W}(T)$ plus this customer's service time. However, since the arrival time is random it is not immediately obvious how to use this insight to obtain an expression for the average delay.

Let $\mathcal{D}(t)$ denote the total amount of time spent in the system by the tth customer to arrive to the queue, and let $\bar{D}$ denote the average delay,

$$\bar{D} := \lim_{n \to \infty} n^{-1} \sum_{t=1}^{n} \mathcal{D}(t). \tag{5.4}$$

Calculation of $\bar{D}$ is made possible through a reinterpretation of the Strong Law of Large Numbers (Theorem A.2.3),

$$T^{-1} \sum_{t=0}^{T-1} Q(t) \to \eta, \qquad a.s. \text{ as } T \to \infty. \tag{5.5}$$

The resulting formula is expressed in *Little's law*.

Theorem 5.1.1 (**Little's law**). *The average delay $\bar{D}$ and the mean queue length $\eta = \mathsf{E}_\pi[Q(0)]$ satisfy the linear equation*

$$\eta = \alpha \bar{D}. \tag{5.6}$$

Proof. The result is trivial if $\rho \geq 1$ since both sides of (5.6) are infinite.

In applying the Strong Law of Large Numbers (5.5) with $Q(0) = 0$ and $\rho < 1$ we sample at arrival times of customers to the queue. Let $T(n)$ denote the time of the nth arrival to the queue. It is possible that several arrivals appear at a given time $T \geq 1$, in which case the ordering of the customers arriving at that time is arbitrary.

For each $n \geq 1$, $k \geq 1$, let $\mathcal{D}_n(k)$ denotes the total delay experienced by the kth customer prior to time $T(n+1)$. That is,

$$\mathcal{D}_n(k) := \sum_{t=0}^{T(n+1)-1} \mathbf{1}\{\text{customer } k \text{ is in the queue at time } t\}.$$

A useful relationship between the partial sums of $\{Q(t)\}$ and $\{\mathcal{D}_n(k)\}$ is obtained through a change in order of summation,

$$\sum_{t=0}^{T(n+1)-1} Q(t) = \sum_{t=0}^{T(n+1)-1} \sum_{k=1}^{n} \mathbf{1}\{\text{customer } k \text{ is in the queue at time } t\}$$

$$= \sum_{k=1}^{n} \mathcal{D}_n(k).$$

We have $\mathcal{D}_n(k) \leq \mathcal{D}(k)$, and $\mathcal{D}_n(k) \uparrow \mathcal{D}(k)$ as $n \to \infty$ for each k. It follows that the Cesaro averages also converge,

$$\lim_{n \to \infty} n^{-1} \sum_{k=1}^{n} [\mathcal{D}(k) - \mathcal{D}_n(k)] = 0,$$

and we thus establish the limit

$$\lim_{n \to \infty} n^{-1} \sum_{t=0}^{T(n+1)-1} Q(t) = \lim_{n \to \infty} n^{-1} \sum_{k=1}^{n} \mathcal{D}_n(k) = \bar{D}. \tag{5.7}$$

To complete the proof we must compute the left-hand side of (5.7). At time $T(n)$ at least n customers have arrived to the queue, and at time $T(n) - 1$ the nth customer has not yet arrived. This gives the bound

$$\sum_{i=1}^{T(n)} A(i) \geq n > \sum_{i=1}^{T(n)-1} A(i),$$

and dividing both sides by $T(n)$ and letting $n \to \infty$ we obtain by the Strong Law of Large Numbers for A,

$$\alpha = \lim_{n\to\infty} \frac{1}{T(n)} \sum_{i=1}^{T(n)} A(i) \geq \lim_{n\to\infty} \frac{n}{T(n)} > \lim_{n\to\infty} \frac{1}{T(n)} \sum_{i=1}^{T(n)-1} A(i) = \alpha.$$

This implies that $n^{-1} T(n) \to \alpha^{-1}$ as $n \to \infty$, which together with (5.5) gives

$$\lim_{n\to\infty} n^{-1} \sum_{i=0}^{T(n+1)-1} Q(i) = \lim_{n\to\infty} \left(\frac{T(n+1)}{n} \right) \left(\frac{1}{T(n+1)} \sum_{i=0}^{T(n+1)-1} Q(i) \right) = \alpha^{-1}\eta.$$

This combined with (5.7) completes the proof.　　　$\square$

5.2 Workload for the CRW scheduling model

We now construct workload models for the general CRW scheduling model in analogy with the workload models obtained for the single server queue. We first consider the homogeneous model in which a workload process is defined through a simple matrix inversion. Recall that homogeneity is defined for the scheduling model in Definition 4.1.1.

5.2.1 Workload in units of inventory

Definition 5.2.1 (Workload in units of inventory). Consider the ℓ_m-dimensional process defined by

$$Y(t) = \Xi_Y Q(t), \qquad t \geq 0, \tag{5.8}$$

where $\Xi_Y := C[I - R^{\mathsf{T}}]^{-1}$. For each $s \in \mathcal{S}$, $t \geq 0$, the quantity $Y_s(t)$ is called the *workload (in units of inventory)* at time t, for Station s.　　　■

Applying the formula (4.28) we find that the value $[I - R]_{ij}^{-1}$ is zero or one for each i, j, and it is equal to one if and only if a customer entering the ith queue will receive processing at buffer j prior to exiting the system. It then follows that, for each $t \geq 0$, the quantity

$$[Q(t)^{\mathsf{T}}[I - R]^{-1}]_j = \sum_{i=1}^{\ell} [I - R^{\mathsf{T}}]_{ji}^{-1} Q_i(t)$$

denotes the total number of services that must be completed at buffer j for all of the customers in the system at time t. Finally, the sth entry of $Y(t)$ defined in (5.8) is expressed

$$Y_s(t) := \sum_{j:s(j)=s} [Q(t)^{\mathrm{T}}[I - R]^{-1}]_j, \qquad t \geq 0.$$

Hence, $Y_s(t)$ is equal to the total number of services that must be completed at Station s for all of the customers in the system at time t.

The recursion (4.2) can be applied to express Y through a similar recursion,

$$Y(t + 1) = Y(t) - CM(t + 1)U(t) + L(t + 1), \qquad t \geq 0, \qquad (5.9)$$

where $L(t) := C[I - R]^{-1}A(t)$ for $t \geq 1$. A simplified expression is obtained for the homogeneous model based on the ℓ_m-dimensional *idleness process*,

$$\iota(t) = 1 - CU(t), \qquad t \geq 0. \qquad (5.10)$$

If $U(t) \in \mathsf{U}_\diamond$ then $\iota_s(t)$ is equal to zero or one, and is zero if and only if the server at Station s is active.

Proposition 5.2.2 (Workload and idleness). *Assume that the CRW model is homogeneous. Then, the workload process can be expressed*

$$Y(t + 1) = Y(t) - S(t + 1)\mathbf{1} + S(t + 1)\iota(t) + C[I - R^{\mathrm{T}}]^{-1}A(t + 1), \qquad (5.11)$$

where S is the diagonal matrix sequence $\{S(t) = \operatorname{diag}(S_1(t), \dots, S_{\ell_m}(t)) : t \geq 1\}$ whose entries are defined in Definition 4.1.1 (v).

Proof. Consider any fixed station $s \in \mathcal{S}$, and any time $t \geq 0$. Based on the assumptions of the proposition we obtain

$$[CM(t + 1)U(t)]_s = \sum_j C_{sj}M_j(t + 1)U_j(t)$$

$$= S_s(t + 1) \sum_j C_{sj}U_j(t)$$

$$= (1 - \iota_s(t))S_s(t + 1), \qquad s \in \mathcal{S}, \ t \geq 0,$$

which when combined with (5.9) gives (5.11). $\qquad \square$

Example 5.2.3 (KSRS model). The workload matrix in units of inventory is given by

$$\Xi_Y := C[I - R^{\mathrm{T}}]^{-1} = C[I + R^{\mathrm{T}}] = \begin{bmatrix} 1 & 0 & 1 & 1 \\ 1 & 1 & 1 & 0 \end{bmatrix},$$

where the routing matrix R is also defined in Example 4.2.9. The model is homogeneous if $S_1(t) := M_1(t) = M_4(t)$ and $S_2(t) := M_2(t) = M_3(t)$ for each $t \geq 1$. In this case, the workload process Y is expressed

$$Y(t) := \Xi_Y Q(t) = \begin{bmatrix} Q_1(t) + Q_3(t) + Q_4(t) \\ Q_1(t) + Q_2(t) + Q_3(t) \end{bmatrix},$$

which does evolve according to the recursion (5.11). $\qquad \blacksquare$

5.2.2 Workload in units of time

Before moving to the completely general, possibly nonhomogeneous model we consider two special cases: the Klimov model and the single-station re-entrant line.

5.2.2.1 Klimov model

The fluid model for the Klimov model is the ODE model

$$\tfrac{d^+}{dt} q_i(t) = -\mu_i \zeta_i(t) + \alpha_i, \qquad i = 1, 2, \ldots, \ell, \ t \geq 0. \tag{5.12}$$

The single workload vector is $\xi = (\mu_1^{-1}, \ldots, \mu_\ell^{-1})^\mathsf{T}$, and the workload in units of time is $w(t) = \xi^\mathsf{T} q(t) = \sum \mu_i^{-1} q_i(t), t \geq 0$.

A CRW model is described by the recursion

$$Q(t+1) = Q(t) + A(t+1) - \sum_{i=1}^{\ell} M_i(t+1)U_i(t)\mathbf{1}^i, \qquad t \geq 0, \tag{5.13}$$

where A is i.i.d. with finite mean, and each M_i is Bernoulli with mean μ_i. In the homogeneous model we have $M_i = S$ for each i, and in this case the workload in units of inventory is simply the total customer population $Y(t) = \sum Q_i(t), t \geq 0$.

In the general Klimov model we define workload in units of time as follows. Exactly as in the single server queue it is assumed that customers define their service requirements upon arrival. Let $\mathcal{G}_{ij}(t)$ denote service time required by the jth customer to arrive to buffer i at time t. For each i, the random variables $\{\mathcal{G}_{ij}(t) : j, t \in \mathbb{Z}_+\}$ are assumed i.i.d. with common mean μ_i^{-1}. The unconditional workload process in units of time evolves according to the recursion (5.3) with $U(t) := \sum_{i=1}^{m} U_i(t)$ and

$$\mathcal{L}(t) = \sum_{i=1}^{\ell} \mathbf{1}\{A_i(t) \geq 1\} \sum_{j=1}^{A_i(t)} \mathcal{G}_{ij}(t). \tag{5.14}$$

5.2.2.2 Single-station re-entrant line

The re-entrant line consisting of a single station is defined by the recursion

$$Q(t+1) = Q(t) + A_1(t+1)\mathbf{1}^1 + \sum_{i=1}^{\ell} U_i(t)M_i(t+1)[\mathbf{1}^{i+1} - \mathbf{1}^i], \tag{5.15}$$

where to obtain a compact representation we set $\mathbf{1}^{\ell+1} := 0$.

We list here the three different notions of workload for this model:

(i) Suppose that the model is homogeneous: $M_i(t) = S(t)$ for each i. Then, applying Definition 5.2.1, the workload in units of inventory is expressed

$$Y(t) := \sum_{i=1}^{\ell} (\ell - i + 1)Q_i(t), \qquad t \geq 0.$$

(ii) The (conditional) workload in units of time is defined by

$$W(t) := \sum_{i=1}^{\ell} \xi_i Q_i(t), \qquad t \geq 0,$$

where $\xi_i = \sum_{j=i}^{\ell} \mu_j^{-1}$, $1 \leq i \leq \ell$.

(iii) The unconditional workload in units of time is denoted $\mathcal{W}(t)$. It evolves according to the recursion (5.3), where $U(t) = \sum U_i(t)$, and $\{\mathcal{L}(t) : t \geq 1\}$ is an i.i.d. process on $\mathbb{Z}_+$ with common mean $\mathsf{E}[\mathcal{L}(t)] = \rho_\bullet = \xi_1 \alpha_1$.

To complete the description of $\mathcal{W}(t)$ we now describe the input process $\{\mathcal{L}(t) : t \geq 1\}$. If there are $m = A(t) \geq 1$ arrivals to the queue at time t, we let $\{\mathcal{G}_{ij}(t) : 1 \leq i \leq \ell, 1 \leq j \leq m\}$ denote the total service time required by the jth customer at the ith queue. These random variables are assumed to be mutually independent, with mean consistent with the fluid model. We thus arrive at a formula for the work arriving to the system at time t,

$$\mathcal{L}(t) = \mathbf{1}\{A_1(t) \geq 1\} \sum_{i=1}^{\ell} \sum_{j=1}^{A_1(t)} \mathcal{G}_{ij}(t).$$

The process $\{\mathcal{L}(t)\}$ is i.i.d. with mean

$$\mathsf{E}[\mathcal{L}(1)] = \sum_{i=1}^{\ell} \mathsf{E}[A_1(1)\mathcal{G}_{ij}(1)] = \sum_{i=1}^{\ell} \alpha_1 \mu_i^{-1} = \rho_\bullet.$$

5.2.2.3 *Workload in the general scheduling model*

In the general scheduling model it is possible to formulate an ℓ_m-dimensional workload process $\mathcal{W}$. The specification of the arrival processes $\{\mathcal{L}_s\}$ is complicated in the general model since we must consider up to ℓ separate arrival streams.

At time t suppose that $m = A_r(t) \geq 1$ new customers arrive to buffer r. This collection of customers brings with it a family of service requirements $\{\mathcal{G}_{r,i,j}(t) : 1 \leq i \leq \ell, 1 \leq j \leq m\}$, where for the jth customer to arrive at time t, $\mathcal{G}_{r,i,j}(t)$ denotes the respective service time required at buffer i. These random variables are assumed to be supported on $\{1, 2, \dots\}$ with mean $\mathsf{E}[\mathcal{G}_{r,i,j}(t)] = \mu_i^{-1}$.

The total new workload for Station s at time t is denoted

$$\mathcal{L}_s(t) = \sum_{r=1}^{\ell} \mathbf{1}\{A_r(t) \geq 1\} \sum_{i:s(i)=s} \sum_{j=1}^{A_r(t)} \mathcal{G}_{r,i,j}(t). \tag{5.16}$$

It is assumed that $\mathcal{G}$ and A are independent processes. We then define the unconditional workload process as the ℓ_m-dimensional process whose sth component evolves as

$$\mathcal{W}_s(t+1) = \mathcal{W}_s(t) - 1 + \iota_s(t) + \mathcal{L}_s(t+1), \qquad t \geq 0, \tag{5.17}$$

where ι is the ℓ_m-dimensional idleness process defined in (5.10).

In summary, given any CRW model we obtain an ℓ_m-dimensional description of the evolution of the workload in units of time via (5.17). For consistency with the CRW model (4.2) a geometric distribution is imposed: for each r, i, j, t,

$$\mathsf{P}\{\mathcal{G}_{r,i,j}(t) = T + 1 \mid \mathcal{G}_{r,i,j}(t) > T\} = \mu_i, \qquad T \geq 0, \; i = 1, \ldots, \ell. \qquad (5.18)$$

However, it is not necessary to start with the CRW model (4.2), in which case the distribution of $\{\mathcal{G}_{r,i,j}(t)\}$ may not be geometric. Given a physical network, a workload model of the form (5.17) can be constructed directly based on observed statistics of the physical system.

The evolution equation for workload in the fluid and CRW models is the basis of the relaxations considered next.

5.3 Relaxations for the fluid model

For the purposes of control, typically only a few of the workload vectors impose serious constraints on the dynamic behavior of the network. A much simpler control problem is obtained by relaxing those constraints corresponding to relatively small load.

Definition 5.3.1 (Relaxation for the fluid model). For a given integer $1 \leq n \leq \ell_m$, the *nth relaxation* of the fluid model is composed of a cumulative allocation process $\widehat{z}$, a queue length process $\widehat{q}$, and n-dimensional workload process $\widehat{w}$ defined as follows:

(i) The cumulative allocation process is subject to the linear constraints

$$\widehat{C}[\widehat{z}(t_1) - \widehat{z}(t_0)] \leq (t_1 - t_0)\mathbf{1}, \qquad 0 \leq t_0 \leq t_1 < \infty,$$

where $\widehat{C}$ is the $n \times \ell$ matrix whose ith row coincides with that of C for $1 \leq i \leq n$. The nonnegativity constraints on the increments are relaxed, so that the control set (4.3) is the unbounded polyhedron,

$$\widehat{\mathsf{U}} := \{u \in \mathbb{R}^{\ell_m} : \widehat{C}u \leq \mathbf{1}\}. \qquad (5.19)$$

The allocation rate at time t defined as the right derivative $\widehat{\zeta}(t) := \frac{d^+}{dt}\widehat{z}(t)$ satisfies $\widehat{\zeta}(t) \in \widehat{\mathsf{U}}$ for each t.

(ii) The queue length process is defined exactly as in (4.24) by

$$\widehat{q}(t) = x - B\widehat{z}(t) + \alpha t, \qquad t \geq 0, \; x \in \mathbb{R}_+^\ell, \qquad (5.20)$$

and $\widehat{q}$ is again constrained by $\widehat{q}(t) \in \mathsf{X}$ for $t \geq 0$.

(iii) With $\widehat{\Xi} := -\widehat{C}B^{-1}$, the associated workload process (in units of time) is defined by $\widehat{w}(t) = \widehat{\Xi}\widehat{q}(t)$ for $t \geq 0$.

(iv) The workload process is constrained to the polyhedral *workload space*

$$\mathsf{W} := \{\widehat{\Xi}x : x \in \mathsf{X}\}. \qquad (5.21)$$

(v) If the fluid model is homogeneous, then the workload process in units of inventory is defined by $\widehat{y}(t) = \widehat{C}[I - R^{\mathsf{T}}]^{-1}\widehat{q}(t)$ for $t \geq 0$.
∎

The dynamics of the workload processes are described in Proposition 5.3.2. Part (ii) is an extension of Proposition 4.2.4: The *independence* in (5.23) means that the derivative $\frac{d^+}{dt}\widehat{w}(t)$ can take on any value in $\{x \in \mathbb{R}^n : x \geq -(1-\rho)\}$ depending on the choice of $\zeta(t) \in \widehat{\mathsf{U}}$.

Proposition 5.3.2. *For each $n \leq \ell_m$, the nth relaxation satisfies the following:*

(i) *The velocity space for the relaxation is given by*

$$\widehat{\mathsf{V}} = \{v : \langle \xi^s, v \rangle \geq -(1-\rho_s), 1 \leq s \leq n\}. \tag{5.22}$$

(ii) *The process $\widehat{w}$ is subject to the independent linear constraints*

$$\frac{d^+}{dt}\widehat{w}_s(t) \geq -(1-\rho_s), \qquad 1 \leq s \leq n. \tag{5.23}$$

Equivalently, the workload process can be expressed

$$\widehat{w}(t) = \widehat{w}(0) - \delta t + I(t), \qquad t \geq 0, \tag{5.24}$$

where $\delta_s = 1 - \rho_s$ for $1 \leq s \leq n$; the idleness process I is nonnegative-valued and nondecreasing.

(iii) *Suppose that the fluid model is homogeneous. Then, the process $\widehat{y}$ is subject to the independent linear constraints*

$$\frac{d^+}{dt}\widehat{y}_i(t) \geq -\mu_s + \lambda_s, \qquad 1 \leq s \leq n, \tag{5.25}$$

where $\lambda_s := \mu_s \rho_s$.

Proof. The expression (5.20) for $\widehat{q}$ and the definition of $\widehat{\Xi}$ imply the representation

$$\widehat{w}(t) := \widehat{\Xi}\widehat{q}(t) = \widehat{w}(0) - \widehat{C}\widehat{z}(t) + \widehat{\Xi}\alpha t, \qquad t \geq 0, \ x \in \mathbb{R}^\ell_+.$$

From the definition of ρ given in Definition 4.2.3 we also have

$$\widehat{\Xi}\alpha = (\rho_1, \ldots, \rho_n)^{\mathsf{T}}.$$

This combined with the constraint $\widehat{C}\zeta(t) \leq 1$ completes the proof of (i) and (ii).

Part (iii) follows from (ii) and the representation $\widehat{y}_s = \mu_s \widehat{w}_s$.   □

A one-dimensional relaxation reveals a great deal about network behavior whenever there is a single dominant face in V that is relatively close to the origin. This is illustrated in the processor sharing model:

Example 5.3.3 (Processor sharing model). Consider the model shown in Fig. 2.6 with the parameters specified in Example 4.2.7. Figure 5.1 compares the velocity space V for the fluid model previously shown in Fig. 4.2, with the velocity space $\widehat{\mathsf{V}}$ for a one-dimensional relaxation.

It follows from the geometry illustrated in Fig. 4.2 that T^* is a linear function of $x \in \mathbb{R}^2_+$. In fact, for the numerical values used in Example 4.2.7 the left-hand sloping face of V is given by $\{v : \langle \xi^1, v \rangle = -(1-\rho_1)\} = \{v : 3v_1 + 4v_2 = -1\}$, so that $T^*(x) = 3x_1 + 4x_2$ for $x \in \mathbb{R}^2_+$. The minimal draining time $\widehat{T}^*$ for the one-dimensional relaxation coincides with T^* in this example.   ∎

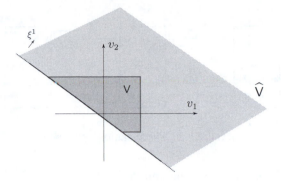

Figure 5.1. Velocity space for the processor sharing model with parameters defined in (4.31) and its relaxation with velocity space $\widehat{V}$.

5.3.1 Minimal process

The most basic control solution for a relaxation is a generalization of the nonidling condition.

Definition 5.3.4 (Minimal solution). Suppose that $R \subset W \subset \mathbb{R}_+^n$ is a closed convex set containing the origin, and $w \in W$ is any given initial condition. An R-*minimal* solution $\widehat{w}^\circ$ starting from w is any feasible solution satisfying the differential constraints (5.23) and the following conditions:

  (i) $\widehat{w}^\circ(t) \in R$ for all $t > 0$.
 (ii) If $\widehat{w}$ is any other solution with identical initial condition w and satisfying $\widehat{w}(t) \in R$ for all $t > 0$, then

$$\widehat{w}^\circ_s(t) \le \widehat{w}_s(t), \qquad t \ge 0, \ s \in \{1, \dots, n\}.$$

When $R = W$ then $\widehat{w}^\circ$ is called *minimal*, or *pointwise minimal*. In this case the process is said to be controlled using the *work-conserving policy*. ∎

An R-minimal solution exists in one or two dimensions since R is a convex subset of the positive orthant. In higher dimensions we show by example that such a strong form of minimality may not be feasible.

Be forewarned that pointwise minimality on W is not necessarily a desirable property. While the work-conserving policy is typically optimal for a one-dimensional relaxation, in Section 5.3.2 we find that optimal solutions for a relaxation are frequently not work-conserving for dimensions two or higher.

In the one-dimensional relaxation there is a minimal solution on $W = \mathbb{R}_+$, defined by the nonidling policy: $\iota(t) = 0$ when $w(t) > 0$. Similarly, an R-minimal solution always exists for the two-dimensional relaxation, regardless of R. For each $w^0 \in \mathbb{R}^2$ define

$$R_{w^0} = \Big\{ w \in W : w_i \ge w_i^0, \quad \text{for all } i \Big\}. \tag{5.26}$$

We denote by $[w]_R$ the projection of w onto the set R_w in the standard ℓ_2 norm.

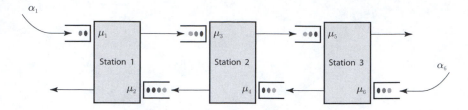

Figure 5.2. Three-station network.

Proposition 5.3.5 follows from Theorem 5.3.7 that follows.

Proposition 5.3.5. *Consider a two-dimensional relaxation of the fluid scheduling model, and suppose that* $\mathsf{R} \subset \mathsf{W}$ *is a closed convex set containing the origin. Then, for each initial condition* $w \in \mathsf{W}$ *there exists an* R-*minimal solution* $\widehat{w}^\circ$ *that can be expressed*

$$\widehat{w}^\circ(t; w) = [[w]_\mathsf{R} - \delta t]_\mathsf{R}, \qquad t > 0, \ w \in \mathsf{W}. \tag{5.27}$$

The existence of a minimal solution is not guaranteed when the dimension is three or greater.

Example 5.3.6 (Minimal solution for a three-station network). Consider the three-station network shown in Fig. 5.2. It is assumed that all service rates are equal to unity, so that the 3×6 workload matrix can be expressed

$$\Xi = \begin{bmatrix} 1 & 1 & 0 & 1 & 0 & 1 \\ 1 & 0 & 1 & 1 & 0 & 1 \\ 1 & 0 & 1 & 0 & 1 & 1 \end{bmatrix}.$$

For simplicity consider the arrival-free model where $\alpha_1 = \alpha_6 = 0$, so that $\rho = 0$.

Consider the pair of vectors

$$w^3 := \widehat{\Xi} 1^3 = [0, \ 1, \ 1]^\mathsf{T}, \quad w^4 := \widehat{\Xi} 1^4 = [1, \ 1, \ 0]^\mathsf{T}.$$

The initial condition $x = 1^3 + 1^4$ has corresponding workload $w = \Xi x = (1, 2, 1)^\mathsf{T}$. From the initial condition x it is possible to reach either 1^3 or 1^4 in exactly 1 sec. Any minimal solution must then satisfy $\widehat{w}^\circ(t; w) \leq w^3 = \Xi 1^3$ and $\widehat{w}^\circ(t; w) \leq w^4 = \Xi 1^4$ at $t = 1$, which implies that $\widehat{w}^\circ(1; x) \leq (0, 1, 0)^\mathsf{T}$.

The only vector in W satisfying this inequality is $w = (0, 0, 0)^\mathsf{T}$. However, the origin is not reachable in 1 sec since the minimal draining time is $W^*(x) = T^*(x) = \max\langle \xi^i, x \rangle = 2$. ∎

We now turn to some positive results. The following result implies Proposition 5.3.5.

Theorem 5.3.7 (Existence of minimal workload). *The following are equivalent for any given closed, convex, polyhedral set* $\mathsf{R} \subset \mathsf{W}$ *containing the origin:*

  (i) *A minimal solution* $\widehat{w}^\circ$ *exists for each initial condition* $w \in \mathsf{W}$.
  (ii) *For each* $w \in \mathsf{W}$ *the set (5.26) contains a pointwise minimal element.*

If either of these equivalent conditions hold, then a pointwise minimal trajectory may be expressed as (5.27), where $[w]_R$, $w \in \mathbb{R}^n$, is the projection of w onto the set R_w in the standard ℓ_2 norm.

The interesting part of the theorem is the representation (5.27) when $[\cdot]_R$ exists. To prove this we first establish some properties for this process.

Proposition 5.3.8. *If the pointwise projection $[\cdot]_R$ exists, then for each $w \in W$ the semigroup property holds for the process $\widehat{w}^\circ$ defined (5.27),*

$$\widehat{w}^\circ(t_1 + t_2; w) = [\widehat{w}^\circ(t_1; w) - \delta t_2]_R, \qquad t_1, t_2 > 0, \ w \in W. \tag{5.28}$$

Moreover, for each $1 \leq i \leq n$ the function $\widehat{w}_i^\circ(t; w)$ is convex as a function of t for $t \in (0, \infty)$.

Proof. We first establish convexity. For any positive t_1, t_2 consider the trajectory

$$\widehat{w}(t) = \widehat{w}_1^\circ + \frac{t - t_1}{t_2}(\widehat{w}_2^\circ - \widehat{w}_1^\circ), \qquad t_1 \leq t \leq t_2,$$

with $\widehat{w}_1^\circ := \widehat{w}^\circ(t_1; w)$ and $\widehat{w}_2^\circ := \widehat{w}^\circ(t_2; w)$. We have $\widehat{w}(t) \geq [w]_R - \delta t$ when $t = t_1$ and when $t = t_2$, and it then follows by linearity that this lower bound holds for every $t \in [t_1, t_2]$. Convexity of R implies that $\widehat{w}(t) \in R$ for $t \in [t_1, t_2]$, so that by the definition of w° in (5.27) we must have $\widehat{w}(t) \geq [[w]_R - \delta t]_R = \widehat{w}^\circ(t; w)$ for all $t \in [0, t_2]$. This is the desired convexity. $\square$

To prove the semigroup property (5.28) we establish upper and lower bounds. To obtain the upper bound note that $\widehat{w}^\circ(t_1; w) \geq [w]_R - \delta t_1$ for any t_1, so that

$$\widehat{w}^\circ(t_1 + t_2; w) := [[w]_R - \delta(t_1 + t_2)]_R \leq [\widehat{w}^\circ(t_1; w) - \delta t_2]_R.$$

To obtain the lower bound we note that convexity of $\widehat{w}^\circ$ on $(0, \infty)$ implies the bound

$$\widehat{w}^\circ(t_1 + t_2; w) - \widehat{w}^\circ(t_1; w) \geq \widehat{w}^\circ(t_2; w) - \widehat{w}^\circ(0+; w).$$

The right-hand side is bounded below by $-\delta t_2$, which gives

$$\widehat{w}^\circ(t_1 + t_2; w) \geq \widehat{w}^\circ(t_1; w) - \delta t_2,$$

so that $\widehat{w}^\circ(t_1 + t_2; w) \geq [\widehat{w}^\circ(t_1; w) - \delta t_2]_R$. We thereby obtain (5.28). $\square$

Proof of Theorem 5.3.7. If (i) holds, then for each $w \in W$ the vector $\widehat{w}(0+; w) \in R$ must be pointwise minimal, which implies (ii).

Conversely, if (ii) holds then the trajectory given by $\widehat{w}^\bullet(t; w) = [[w]_R - \delta t]_R$, $t \geq 0$, is a piecewise linear function of t for each initial condition w. The semigroup property (5.28) implies that $\frac{d}{dt}\widehat{w}^\bullet(t; w) \geq -\delta$ for all t, so that this trajectory is feasible for the relaxed fluid model, and hence (i) holds with $\widehat{w}^\circ$ defined in (5.27). $\square$

When the minimal process exists and can be expressed via (5.27) we find that it can also be represented as the solution to the ODE

$$\frac{d^+}{dt}\widehat{w}(t) = -\delta + \iota(t), \qquad t > 0, \ \widehat{w}(0+) = [w]_R, \tag{5.29}$$

where the idleness rate ι is expressed as the state-feedback law $\iota(t) = \phi(\widehat{w}(t))$, $t \geq 0$, with

$$
\begin{aligned}
\phi(w) &:= \delta - \delta(w); \\
\delta(w) &:= \lim_{t \downarrow 0} t^{-1}\{[w - t\delta]_\mathsf{R} - w\}, \quad w \in \mathsf{R}.
\end{aligned}
\tag{5.30}
$$

Proposition 5.3.9 summarizes some properties of the projection, and resulting properties of the feedback law ϕ. We assume that the region $\mathsf{R} \subset \mathsf{W}$ is polyhedral, of the form

$$
\mathsf{R} = \{w \in \mathbb{R}^n : \langle n^i, w \rangle \geq -\beta_i, \, 1 \leq i \leq \ell_\mathsf{R}\},
\tag{5.31}
$$

where $\{\beta_i\}$ are nonnegative constants, and $\{n_i\} \subset \mathbb{R}^n$. The ith face $F(i) \subset \mathsf{R}$ is defined by

$$
F(i) := \{w \in \mathsf{R} : \langle n^i, w \rangle = -\beta_i\}, \qquad 1 \leq i \leq \ell_\mathsf{R}.
\tag{5.32}
$$

We assume moreover that R has nonempty interior, and that each of the sets $F(i)$ is of dimension $n - 1$.

Proposition 5.3.9. *Suppose that the set R given in (5.31) has nonempty interior, and that the pointwise projection $[\,\cdot\,]_\mathsf{R} : \mathbb{R}^n \to \mathsf{R}$ exists. Then:*

(i) *For each $y \in \mathbb{R}^n$, the projection $[y]_\mathsf{R}$ is the unique optimizer $w^\circ \in \mathsf{R}$ of the linear program*

$$
\begin{aligned}
\textbf{min} \quad & w_1 + \cdots + w_n \\
\textbf{s.t.} \quad & \langle n^i, w \rangle \geq -\beta_i, \, 1 \leq i \leq \ell_\mathsf{R} \\
& w \geq y.
\end{aligned}
$$

(ii) *For each face $F(i)$ with $i \in \{1, \ldots \ell_\mathsf{R}\}$, there is a unique $j_i \in \{1, \ldots, n\}$ satisfying*

$$
n^i_{j_i} > 0, \quad \text{and} \quad n^i_j \leq 0, \, j \neq j_i.
$$

(iii) *The feedback law $\phi \colon \mathsf{W} \to \mathbb{R}^n_+$ defined in (5.30) satisfies $\phi(w) = 0$ for $w \in$ interior (R). Otherwise, if for some $i \in \{1, \ldots, \ell_\mathsf{R}\}$ we have $w \in F(i)$, and $w \notin F(i')$ for $i' \neq i$, then $\phi(w)_j = 0$ for $j \neq j_i$, where j_i is defined in (ii).*

Proof. Let $w^\circ \in \mathsf{R}$ be an optimizer of the linear program in (i). We have $w^\circ \geq [y]_\mathsf{R}$ since $w^\circ \geq y$, and hence also $\sum w^\circ_i \geq \sum([y]_\mathsf{R})_i$. It follows that $[y]_\mathsf{R}$ is the unique optimizer of this linear program.

We prove part (ii) by contradiction: Fix $1 \leq i \leq \ell_\mathsf{R}$, and suppose that in fact $n^i_j > 0$ and $n^i_k > 0$ for some $1 \leq j < k \leq \ell_\mathsf{R}$.

Consider $w \in F(i)$, with $w \notin F(i')$ for $i' \neq i$. For $\varepsilon > 0$ we consider the open ball centered at w given by $B(w, \varepsilon) = \{y \in \mathbb{R}^n : \|w - y\| < \varepsilon\}$, where $\|\cdot\|$ denotes the Euclidean norm.

Fix $w^0 \in B(w, \varepsilon)$ with $w^0 \notin \mathsf{R}$, and define

$$
s_j = -(n^i_j)^{-1}(\beta_i + \langle n^i, w^0 \rangle), \quad s_k = -(n^i_k)^{-1}(\beta_i + \langle n^i, w^0 \rangle).
$$

We have $s_j > 0$, $s_k > 0$ by construction since $w^0 \notin \mathsf{R}$, and the vectors $\{w^j := w^0 + s_j 1^j,\; w^k := w^0 + s_k 1^k\}$ satisfy $\langle n^i, w^j \rangle = \langle n^i, w^k \rangle = -\beta_i$. Hence, we may choose $\varepsilon > 0$ so that each of these vectors lies in $F(i) \subset \mathsf{R}$ for any $w^0 \in B(w, \varepsilon) \cap \mathsf{R}^c$. We conclude that $[w^0]_{\mathsf{R}} \leq \min(w^j, w^k)$, which is only possible if $[w^0]_{\mathsf{R}} = w^0$. This violates our assumptions, and completes the proof of (ii).

Part (iii) is immediate from (ii) and the definition (5.30). $\qquad\square$

We conclude this section with some results illustrating the structure of a minimal solution under the conditions of Theorem 5.3.7.

Resource s is said to be *satiated* at state x provided there exists $v \in \widehat{\mathsf{V}}$ satisfying $\langle \xi^s, v \rangle = -(1 - \rho_s)$, and $v_i \geq 0$ whenever $x_i = 0$.

Proposition 5.3.10. *Suppose that $\rho_\bullet < 1$. Then for any two states x^1, $x^2 \in \mathsf{X}$, the minimal time $\widehat{T}^*(x^1, x^2)$ to travel from x^1 to x^2 is finite. If $\widehat{T}^*(x^1, x^2) > 0$ then*

$$
\begin{aligned}
\widehat{T}^*(x^1, x^2) &= \max\left\{ \frac{\langle \xi^j, x^1 - x^2 \rangle}{1 - \rho_j} : 1 \leq j \leq n \right\} \\
&= \max\left\{ \frac{\langle \xi^j, x^1 - x^2 \rangle}{1 - \rho_j} : j \text{ is satiated by } x^1 \right\}.
\end{aligned}
$$

Proof. With $v = (x^2 - x^1)/\widehat{T}^*(x^1, x^2) \in \widehat{\mathsf{V}}$, the trajectory below is both feasible and time-optimal:

$$
\widehat{q}(t; x^1) = x^1 + vt, \qquad 0 \leq t \leq \widehat{T}^*(x^1, x^2).
$$

Moreover, simple dynamic programming arguments (along the lines of the proof of Proposition 4.3.4) ensure that

$$
\tfrac{d}{dt} \widehat{T}^*(\widehat{q}(t; x^1), x^2) = -1, \qquad 0 < t < \widehat{T}^*(x^1, x^2).
$$

Hence, whenever s^* is a maximizer, so that

$$
\widehat{T}^*(x^1, x^2) = \frac{\langle \xi^{s^*}, x^1 - x^2 \rangle}{1 - \rho_{s^*}},
$$

we must have $\langle \xi^{s^*}, v \rangle = -(1 - \rho_{s^*})$. This implies that resource s^* is satiated by x^1.

Satiated resources play a role analogous to dynamic bottlenecks in the construction of a time-optimal trajectory. The following result is an analog of Proposition 4.3.4. It is an easy corollary to Proposition 5.3.10. $\qquad\square$

Theorem 5.3.11 (Satiated resources and minimality). *Suppose that $\rho_\bullet < 1$. Let $\widehat{q}$ be any solution to the nth workload relaxation, starting at $x \in \mathsf{X}$, and let $w(t; x) = \widehat{\Xi}\widehat{q}(t; x)$, $t \geq 0$. We then have the following:*

(i) *If w is pointwise minimal, then each satiated resource is working at capacity for each $0 < t < \infty$. That is, if resource i is satiated at time t, then*

$$
\tfrac{d}{dt} w_i(t) = -(1 - \rho_i).
$$

(ii) *If each satiated resource works at capacity for all t, then the resulting workload trajectory w is pointwise minimal.*

5.3.2 Effective cost

We now turn to control solutions for a workload relaxation based on a given cost function for the ℓ-dimensional network. For simplicity we restrict to a linear cost function $c(x) = c^{\mathsf{T}}x$ with $c_i > 0$ for each $i \in \{1, \dots, \ell\}$.

Suppose that two states $x^0, x^1 \in \mathsf{X}$ are given with $\langle \xi^s, x^0 \rangle \leq \langle \xi^s, x^1 \rangle$ for all $1 \leq s \leq n$. For any $\kappa > 0$ let $v = \kappa(x^1 - x^0)$. This velocity vector lies in the set $\widehat{\mathsf{V}}$ defined in (5.22), and the feasible trajectory $\widehat{q}(t; x^0) = x^0 + tv, 0 \leq t \leq 1/\kappa$ reaches x^1 in $1/\kappa$ sec. It follows that the minimum time to reach x^1 starting from x^0 is *zero*. This leads to the following terminology.

Definition 5.3.12 (Effective cost for a workload relaxation). Suppose that the state space X is a convex polyhedron, and that $c \colon \mathsf{X} \to \mathbb{R}_+$ is a linear cost function. Then,

(i) The *effective cost* $\bar{c} \colon \mathsf{W} \to \mathbb{R}_+$ is defined for $w \in \mathsf{W}$ as the value of the linear program

$$\bar{c}(w) = \mathbf{min} \quad \langle c, x \rangle \tag{5.33}$$
$$\mathbf{s.t.} \quad \widehat{\Xi}x = w, \; x \in \mathsf{X}.$$

(ii) The region on which $\bar{c}$ is *monotone* is denoted W^+. That is,

$$\mathsf{W}^+ := \left\{ w \in \mathsf{W} : \bar{c}(w') \geq \bar{c}(w) \text{ whenever } w' \geq w \text{ and } w' \in \mathsf{W}. \right\}$$

(iii) For each $w \in \mathsf{W}$, an *effective state* $\mathcal{X}^*(w)$ is a vector $x^* \in \mathsf{X}$ that minimizes the linear program (5.33):

$$\mathcal{X}^*(w) \in \underset{x \in \mathsf{X}}{\arg\min} \left(c(x) : \widehat{\Xi}x = w \right). \tag{5.34}$$

(iv) For any $x \in \mathsf{X}$, an *optimal exchangeable state* $\mathcal{P}^*(x) \in \mathsf{X}$ is defined via

$$\mathcal{P}^*(x) = \mathcal{X}^*(\widehat{\Xi}x). \tag{5.35}$$

∎

The effective cost is easily computed for a one-dimensional relaxation. Consider the case in which there are no buffer constraints so that $\mathsf{X} = \mathbb{R}_+^\ell$. Hence the linear program (5.33) that defines the effective cost becomes

$$\bar{c}(w) = \mathbf{min} \quad c(x) \quad \mathbf{s.t.} \quad \langle \xi^1, x \rangle = w, \; x \in \mathbb{R}_+^\ell \,.$$

Proposition 5.3.13. *If $n = 1$ and there are no buffer constraints, then the effective cost is given by the linear function, $\bar{c}(w) = (c_{i_*}/\xi_{i_*}^1)w, \; w \in \mathbb{R}_+$, where the index i_* is any* *solution to* $c_{i_*}/\xi^1_{i_*} = \min_{1 \le i \le \ell}(c_i/\xi^1_i)$. *An effective state can be expressed as a linear function of w via*

$$\mathcal{X}^*(w) = \left(\frac{1}{\xi^1_{i_*}}1^{i_*}\right)w, \qquad w \in \mathbb{R}_+.$$

The conclusion of Proposition 5.3.13 is an example of *state space collapse*: Regardless of the size of ℓ, there exists an optimal solution $\widehat{q}^*$ that evolves on a one-dimensional region for all $t > 0$.

Throughout this book it is assumed that $\mathcal{X}^* \colon \mathsf{W} \to \mathsf{X}$ is defined to be a single-valued, continuous function on W. This can be assumed without loss of generality:

Proposition 5.3.14. *Suppose that c is a linear function on X. Then, there exists a continuous function $\mathcal{X}^* \colon \mathsf{W} \to \mathsf{X}$ such that $x^* = \mathcal{X}^*(w)$ is a minimizer of the linear program (5.33) for each $w \in \mathsf{W}$.*

Proof. The result follows from [136, Theorem II.1.4 and Theorem I.3.2]. The conditions of these theorems hold for the linear program defined in (5.34) since the feasible set has nonempty interior. □

Recall from Section 1.3.1 the definition of the dual of a linear program. The dual of (5.33) with $\mathsf{X} = \mathbb{R}^\ell_+$ can be expressed as

$$\begin{aligned} &\textbf{max} && \psi^\mathsf{T} w \\ &\textbf{s.t.} && \widehat{\Xi}^\mathsf{T}\psi \le c. \end{aligned} \tag{5.36}$$

The dual variable ψ in (5.36) is not sign-constrained since this corresponds to the equality constraint $\widehat{\Xi}x = w$ in (5.33). The following result follows from the fact that an optimal solution to (5.36) can be found among *basic feasible solutions*. We denote these by $\{\bar{c}^i : 1 \le i \le \ell_{\bar{c}}\}$.

Proposition 5.3.15. *Suppose that $\mathsf{X} = \mathbb{R}^\ell_+$. Then, the effective cost is piecewise linear:*

$$\bar{c}(w) = \max_{1 \le i \le \ell_{\bar{c}}}\langle \bar{c}^i, w\rangle, \qquad w \in \mathsf{W}, \tag{5.37}$$

where $\{\bar{c}^i : 1 \le i \le \ell_{\bar{c}}\}$ are extreme points of the constraint region in the linear program (5.36).

Example 5.3.16 (KSRS model). Consider the KSRS model with $\mathsf{X} = \mathbb{R}^4_+$ and cost function equal to the ℓ_1 norm. The effective cost for the two-dimensional relaxation is the solution to the linear program (5.44), where $\widehat{\Xi} = \Xi$ is the 4×2 matrix given in (4.34). We consider the following two special cases:

CASE 1 $\mu_2 = \mu_4 = 1/3$ and $\mu_1 = \mu_3 = 1$. Then

$$\{\bar{c}^i : i = 1, 2, 3\} = \{\tfrac{1}{3}\binom{1}{0}, \tfrac{1}{4}\binom{1}{1}, \tfrac{1}{3}\binom{0}{1}\}.$$

CASE 2 $\mu_2 = \mu_4 = 1$ and $\mu_1 = \mu_3 = 1/3$. Then

$$\{\bar{c}^i : i = 1, 2, 3\} = \{\binom{1}{-2}, \tfrac{1}{4}\binom{1}{1}, \binom{-2}{1}\}.$$

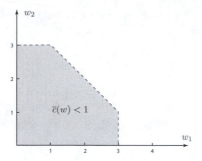
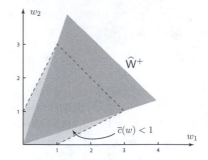

Figure 5.3. Level sets of the effective cost for the KSRS model in Cases 1 and 2 respectively.

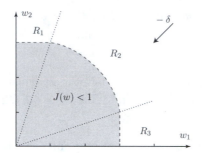
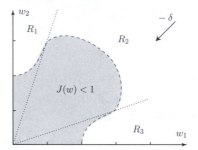

Figure 5.4. Level sets of the value function $\widehat{J}$ for the KSRS model in Cases 1 and 2 respectively under the work-conserving policy.

In each case $\mathsf{W} = \mathbb{R}_+^2$, and the effective cost $\bar{c} \colon \mathsf{W} \to \mathbb{R}_+$ is continuous, piecewise linear, and strictly linear on each of the three regions $\{R_i : i = 1, 2, 3\}$ defined by

$$R_1 = \{0 < 3w_1 < w_2 < \infty\}, \; R_2 = \{0 < w_1 < 3w_2 < 9w_1\}, \; R_3 = \{0 < 3w_2 < w_1\}.$$

Level sets of the effective cost in each case are shown in Fig. 5.3. It is evident that $\mathsf{W}^+ = \mathbb{R}_+^2$ so that the effective cost is monotone in Case 1; in Case 2 the monotone region is $\mathsf{W}^+ = \mathrm{closure}\,(R_2)$. Level sets of the corresponding value functions are shown in Fig. 5.4. We return to this numerical example in Example 5.3.22. ∎

The monotone region can be empty.

Example 5.3.17 (Effective cost for tandem queues). The effective cost for the two-dimensional relaxation is defined by the linear program

$$\bar{c}(w) = \mathbf{min} \quad c_1 x_1 + c_2 x_2$$

$$\begin{aligned} \mathbf{s.t.} \quad & x_1 = \mu_1 w_1 \\ & x_1 + x_2 = \mu_2 w_2 \\ & x \geq 0. \end{aligned} \tag{5.38}$$

This can be solved to give $\bar{c}(y) = (c_1 - c_2)\mu_1 w_1 + c_2\mu_2 w_2$. Hence the effective cost is linear, and it is monotone if and only if $c_1 \geq c_2$. If this inequality fails then $\mathsf{W}^+ = \emptyset$. ∎

Suppose now that X is restricted via buffer constraints. It is assumed that $0 < b_i \leq \infty$ for $1 \leq i \leq \ell$, and $\mathsf{X} := \{x \in \mathbb{R}_+^\ell : x \leq b\}$. In this case the effective cost $\bar{c} \colon \mathsf{W} \to \mathbb{R}_+$ is the solution to the linear program

$$\bar{c}(w) := \textbf{min} \quad c^\mathsf{T}x$$

$$\textbf{s.t.} \quad \Xi x = w, \tag{5.39}$$

$$x \leq b, \quad x \geq 0.$$

Its dual is expressed

$$\textbf{max} \quad \psi^\mathsf{T}w - \sigma^\mathsf{T}b$$

$$\tag{5.40}$$

$$\textbf{s.t.} \quad \widehat{\Xi}^\mathsf{T}\psi - \sigma \leq c, \quad \sigma \geq 0.$$

The variable ψ in (5.40) is again not sign-constrained. The dual variable σ corresponds to the *inequality* constraint $x \leq b$ in (5.39), which requires that $\sigma \geq 0$.

The extreme points in the constraint region of (5.40) are denoted $\{\bar{c}^i, \overline{d}^i : 1 \leq i \leq \ell_{\bar{c}}\}$. Note that the integer $\ell_{\bar{c}}$ defined here is in general larger than the integer used in (5.37).

The optimizers (ψ^*, σ^*) to (5.40) are Lagrange multipliers. Consequently, ψ^* provides sensitivity of the effective cost to workload, and σ^* provides sensitivity with respect to buffer constraints. The following result is a consequence of these observations. In particular, Proposition 5.3.18 (ii) quantifies how the effective cost $\bar{c}(w; b)$ decreases with increasing values of b.

Proposition 5.3.18. *For the nth relaxation, and each $w \in \mathsf{W}$, $b \in \mathbb{R}_+^\ell$:*

(i) *The effective cost $\bar{c}(w) = \bar{c}(w; b)$ is given by*

$$\bar{c}(w) = \max_{1 \leq i \leq \ell_{\bar{c}}} \left(\bar{c}^{i\mathsf{T}}w - \overline{d}^{i\mathsf{T}}b \right). \tag{5.41}$$

(ii) *Suppose that there is a unique maximizing index in (5.41), denoted $i^* = i^*(w, b)$. Then*

$$\frac{\partial}{\partial w_j}\bar{c}(w; b) = \bar{c}_j^{i^*}, \quad 1 \leq j \leq n,$$

$$\frac{\partial}{\partial b_j}\bar{c}(w; b) = -\overline{d}_j^{i^*}, \quad 1 \leq j \leq \ell, \quad \textit{if } b_j < \infty.$$

The definition of the index i_* in Proposition 5.3.13 is similar to the construction of the Klimov indices defined in the c–μ rule for the Klimov model in Section 2.2. Proposition 5.3.18 strengthens this analogy when specialized to a one-dimensional relaxation. In this special case the linear program that defines the effective cost is written

$$\textbf{min} \quad c^\mathsf{T}x$$

$$\textbf{s.t.} \quad \langle \xi^1, x \rangle = w$$

$$x \leq b, \quad x \geq 0.$$

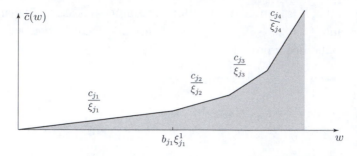

Figure 5.5. Effective cost for a one-dimensional relaxation with four buffer constraints. The cost is piecewise linear, with slopes $\{c_{j_i}/\xi_{j_i} : i \geq 1\}$ as indicated in the figure.

Let $\{\theta_1, \ldots, \theta_\ell\}$ denote a permutation of $\{1, \ldots, \ell\}$ such that

$$\frac{c_i}{\xi_i^1} \leq \frac{c_j}{\xi_j^1} \qquad \text{if } \theta_i < \theta_j. \tag{5.42}$$

The integers $\{\theta_1, \ldots, \theta_\ell\}$ are regarded as generalized Klimov indices.

Proposition 5.3.19. *Suppose that the state space includes buffer constraints, so that it can be expressed as (4.35) for constants $\{b_i\} \subset (0, \infty]$. Then, the effective cost $\bar{c} \colon \mathsf{W} \to \mathbb{R}_+$ for the one-dimensional relaxation is piecewise linear and convex, as illustrated in Fig. 5.5. For a given $w \in \mathbb{R}_+$, the effective state $x^* = \mathcal{X}^*(w)$ satisfies*

$$x_i^* = b_i \qquad \text{whenever} \qquad \sum_{j: \theta_j > \theta_i} x_j^* > 0, \tag{5.43}$$

where the generalized Klimov indices $\{\theta_j\}$ are defined in (5.42).

We next illustrate the impact of buffer constraints using a two-dimensional relaxation of the simple re-entrant line.

Example 5.3.20 (Relaxations for the simple re-entrant line). Consider the three buffer model shown in Fig. 2.9 with $\mathsf{X} = \mathbb{R}_+^3$, and $c(x) = c^{\mathsf{T}} x$, $x \in \mathsf{X}$. It is assumed that the model is homogeneous, so that $\mu_1 = \mu_3$. For a given $w \in \mathsf{W}$ the linear program (5.33) that defines the effective cost for the two-dimensional relaxation is expressed

$$\bar{c}(w) := \mathbf{min} \quad [c_1 x_1 + c_2 x_2 + c_3 x_3],$$

$$\mathbf{s.t.} \quad \begin{bmatrix} 2 & 1 & 1 \\ 1 & 1 & 0 \end{bmatrix} x = \begin{bmatrix} \mu_1 & 0 \\ 0 & \mu_2 \end{bmatrix} w, \ x \geq 0.$$

Its dual can be expressed

$$\bar{c}(w) = \mathbf{max} \quad [\psi_1 \mu_1 w_1 + \psi_2 \mu_2 w_2],$$

$$\mathbf{s.t.} \quad \begin{bmatrix} 2 & 1 \\ 1 & 1 \\ 1 & 0 \end{bmatrix} \psi \leq \begin{bmatrix} c_1 \\ c_2 \\ c_3 \end{bmatrix}. \tag{5.44}$$

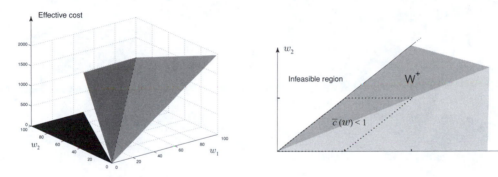

Figure 5.6. On the left is a plot of the effective cost for the simple re-entrant line. The monotone region W^+ is shown on the right. The effective cost is not monotone since feeding the second station conflicts with draining buffer 3.

Consider the special case in which the cost function is equal to the total inventory, $c(x) = |x| = x_1 + x_2 + x_3$. Below are some properties of the effective cost under these assumptions:

(i) The workload space is expressed

$$W = \left\{ w : 0 \leq \frac{w_2}{w_1} \leq 2\frac{\rho_2}{\rho_1} \right\}. \qquad (5.45)$$

(ii) The effective cost is given by

$$\bar{c}(w) = \max(\mu_2 w_2, \mu_1 w_1 - \mu_2 w_2). \qquad (5.46)$$

The effective state $x^* := \mathcal{X}^*(w)$ satisfies $x_2^* = 0$ when $w_2/w_1 \leq \rho_2/\rho_1$.

(iii) The monotone region for the effective cost is the positive cone

$$W^+ = \left\{ w : \frac{\rho_2}{\rho_1} \leq \frac{w_2}{w_1} \leq 2\frac{\rho_2}{\rho_1} \right\}. \qquad (5.47)$$

Figure 5.6 shows a sublevel set of $\bar{c}$. It is apparent that the effective cost is not monotone since W^+ is a strict subset of W. This is to be expected since when c is equal to the total inventory, reducing workload at Station 2 does not necessarily reduce $c(q(t))$. In particular, for the unrelaxed model, if $q_2(t) = 0$ and $q_3(t) > 0$ then the myopic policy will set $\zeta_3 = 1$, which results in starvation of the second resource.

Given a vector $b \geq 0$ of buffer constraints, the effective cost is the solution to the linear program

$$
\begin{aligned}
\bar{c}(w) = \ \mathbf{min} \quad & x_1 + x_2 + x_3 \\
\mathbf{s.t.} \quad 2x_1 + x_2 + x_3 &= \mu_1 w_1, \\
x_1 + x_2 &= \mu_2 w_2, \\
x &\geq 0, \\
x &\leq b.
\end{aligned}
$$

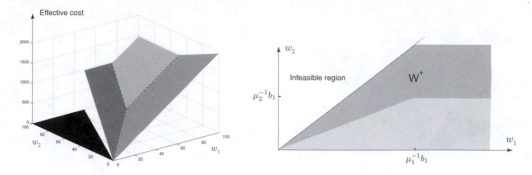

Figure 5.7. On the left is shown the effective cost for the three buffer model shown in Fig. 2.9 when the constraint $q_1 \leq b_1$ is imposed. The monotone region W^+ increases as the upper-bound b_1 decreases.

whose solution is given by

$$\bar{c}(w; b) = \max(\mu_2 w_2, \mu_1 w_1 - \mu_2 w_2, \mu_1 w_1 - b_1). \tag{5.48}$$

It is independent of b_2 and b_3, even though the workload space W depends upon the values of these parameters. The effective cost and the monotone region W^+ are shown in Fig. 5.7 when b_1 is finite and $b_2 = b_3 = \infty$.   ∎

5.3.3 Value functions

Recall that an application of the fundamental theorem of calculus gives the representation (3.31) for the derivative $\frac{d}{dt} J(q(t))$ in the fluid model. It follows that if J is smooth at $x \in \mathbb{R}_+^\ell$, and if $\zeta(t) = u$ (constant) on some nonempty time interval $[0, t_0)$, then

$$\langle \nabla J(x), Bu + \alpha \rangle = -c(x).$$

Similar reasoning can be applied to the fluid workload model with value function

$$\widehat{J}(w) := \int_0^\infty \bar{c}(\widehat{w}(t)) \, dt, \qquad w \in \mathsf{W}. \tag{5.49}$$

If $\widehat{J}$ is continuously differentiable then the fundamental theorem of calculus leads to the identity

$$\mathcal{D}_0 \widehat{J} = -\bar{c}, \tag{5.50}$$

where the differential generator for the fluid model is defined in analogy with (3.4) via

$$\mathcal{D}_0 f := -\delta^{\mathsf{T}} \nabla f, \tag{5.51}$$

where $\delta_s = 1 - \rho_s$, $1 \leq s \leq n$, is defined below (5.24).

We now establish general conditions under which the value function is smooth when $\widehat{w}$ is the R-minimal process on a polyhedral region of the form (5.31). To evaluate the gradient of $\widehat{J}$ we differentiate the workload process $\widehat{w}(t) = [w - \delta t]_{\mathsf{R}}$ with respect to the initial condition $w \in \mathsf{R}$. Let $\mathcal{O} \subset \mathbb{R}^n$ denote the maximal open set such that

$[o]_\mathsf{R}$ and $\bar{c}([o]_\mathsf{R})$ are each C^1 for $o \in \mathcal{O}$. Generally, since $\bar{c}$ and the projection are each piecewise linear on $\mathbb{R}^n$, it follows that the set $\mathcal{O}$ can be expressed as the union

$$\mathcal{O} = \bigcup_{1 \leq i \leq \ell_O} R_i, \qquad \bar{\mathcal{O}} = \mathbb{R}^n. \tag{5.52}$$

where each of the sets $\{R_i : 1 \leq i \leq \ell_O\}$ is an open polyhedron, and the functions $[\,\cdot\,]_\mathsf{R}$ and $\bar{c}([\,\cdot\,]_\mathsf{R})$ are each linear on R_i for each i.

Theorem 5.3.21 establishes several useful properties of the fluid value function.

Theorem 5.3.21. *Assumptions* (a)–(c) *given below refer to a convex polyhedral domain* $\mathsf{R} \subset \mathbb{R}^n_+$ *containing the origin:*

(a) *The set* $\mathsf{R} \subset \mathbb{R}^n_+$ *has nonempty interior, and the pointwise projection* $[\,\cdot\,]_\mathsf{R} \colon \mathbb{R}^n \to \mathsf{R}$ *exists.*

(b) $\int_0^\infty \mathbf{1}\{(w - \delta t) \in \mathcal{O}^c\}\, dt = 0$ *for each* $w \in \mathbb{R}^n$.

(c) $\delta \in$ interior (R).

If Assumptions (a) *and* (b) *hold, then*

(i) *The function* $\widehat{J} \colon \mathsf{R} \to \mathbb{R}_+$ *is piecewise-quadratic,* C^1, *and its gradient* $\nabla \widehat{J}$ *is globally Lipschitz continuous on* R.

(ii) *The dynamic programming equation* (5.50) *holds on* R.

(iii) *The following boundary conditions hold,*

$$\langle \phi(w), \nabla \widehat{J}(w) \rangle = 0, \qquad w \in \mathsf{R}, \tag{5.53}$$

where ϕ *is defined in* (5.30) *with respect to the region* R.

(iv) *If in addition Assumption* (c) *holds, then the function* ϕ *does not vanish on* $\partial \mathsf{R}$.

Before proceeding with the proof of Theorem 5.3.21 we again turn to the KSRS model to illustrate its assumptions and conclusions.

Example 5.3.22 (KSRS model). Consider first the case in which $\delta_1 = \delta_2 > 0$, the cost function is $\bar{c}(w) = \max(w_1, w_2)$, and the constraint region is $\mathsf{R} = \mathsf{W} = \mathbb{R}^2_+$. Assumption (b) does not hold: The integral in (b) is nonzero for any nonzero initial condition w on the diagonal in $\mathbb{R}^2_+$. The value function given by $\widehat{J}(w) = \frac{1}{2}\delta_1^{-1} \max(w_1^2, w_2^2)$ is not C^1 on $\mathbb{R}^2_+$. This explains why (b) is required in Theorem 5.3.21 (i).

To see why (c) is required in Part (iv) we take $\delta_1 = 4\delta_2$; $\bar{c}(w) = w_1 + w_2$; and $\mathsf{R} = \{0 \leq w_1 \leq 3w_2 \leq 9w_1\}$. Assumptions (a) and (b) hold, and the C^1 value function may be explicitly computed:

$$\widehat{J}(w) = \tfrac{1}{2} w^{\mathsf{T}} D w, \; w \in \mathsf{R}, \quad \text{with } D = \tfrac{1}{11}\delta_1^{-1} \begin{bmatrix} 3 & -1 \\ -1 & 15 \end{bmatrix}.$$

Although smooth, we have $\frac{\partial}{\partial w_2}\widehat{J}(w) \neq 0$ and $\phi(w) = 0$ along the lower boundary of R. This is possible since the model violates (c).

We now consider some positive results based on the parameter values specified in Example 5.3.16 with $\mathsf{R} = \mathsf{W} = \mathbb{R}^2_+$. The functions $[\,\cdot\,]_\mathsf{R}$ and $\bar{c}([\,\cdot\,]_\mathsf{R})$ are each linear on each of the sets $\{R_1, R_2, R_3\}$ shown in Fig. 5.4. The open polyhedron $R_4 = \{y \in \mathbb{R}^2 : y < 0\}$ is also contained in $\mathcal{O}$ since $[y]_\mathsf{R} = 0$ and $\bar{c}([y]_\mathsf{R}) = 0$ on R_4. This reasoning leads to a representation of the form (5.52).

Shown in Fig. 5.4 are level sets of the value function in Case 1 and Case 2. The level sets are smooth since $\widehat{J}$ is C^1 under the work-conserving policy by Theorem 5.3.21. It is also piecewise quadratic, and purely quadratic within each of the regions $\{R_i : i = 1, 2, 3\}$: $\widehat{J}(w) = \frac{1}{2} w^\mathsf{T} D^i w$ for $w \in R_i$, $i = 1, 2, 3$.

The C^1 property also facilitates computation of $\widehat{J}$: Exactly as shown in Example 4.3.9 and Example 4.3.10, the dynamic programming equation (5.50) is equivalent to the equations

$$-w^\mathsf{T} D^i \delta = -\langle \bar{c}^i, w \rangle, \quad w \in R_i, \ i = 1, 2, 3. \tag{5.54}$$

It follows that $D^i \delta = \bar{c}^i$ for each i. A single constraint on each of $\{D^1, D^3\}$ is obtained on considering w on the two boundaries of $\mathbb{R}_+^2$. Finally, the C^1 property implies that $D^1 \binom{1}{3} = D^2 \binom{1}{3}$ and $D^3 \binom{3}{1} = D^2 \binom{3}{1}$. These linear constraints can be solved to compute $\{D^1, D^2, D^3\}$. For example, $D^1 = \frac{1}{3} \delta_1^{-1} \begin{bmatrix} 1 & 0 \\ 0 & 0 \end{bmatrix}$ in Case 1. ∎

Proof of Theorem 5.3.21. The dynamic programming equation (ii) is simply the fundamental theorem of calculus: We have

$$\widehat{J}(w) = \int_0^t \bar{c}(\widehat{w}(s; w)) \, ds + \widehat{J}(\widehat{w}(t; w)), \quad t \geq 0.$$

Rearranging terms, dividing by t, and letting $t \to 0$ gives the result.

To establish the smoothness property in (i) we define for $t \geq 0$, $w \in \mathsf{R}$,

$$\Gamma(t; w) := \nabla_w \{\bar{c}(\widehat{w}(t; w))\} = \nabla_w \{\bar{c}([w - \delta t]_\mathsf{R})\}, \tag{5.55}$$

whenever the gradient with respect to w exists. Let $\Pi \colon \mathcal{O} \to \mathbb{R}^{n \times n}$ denote the $n \times n$ derivative of $[\,\cdot\,]_\mathsf{R}$ with respect to w. The matrix-valued function Π is constant on each of the connected components of $\mathcal{O}$. The gradient (5.55) exists whenever $w^\infty(t) := w - \delta t \in \mathcal{O}$ for a given $w \in \mathsf{R}$, $t > 0$, and by the chain rule,

$$\Gamma(t; w) = \Pi(w^\infty(t))^\mathsf{T} \left(\{\nabla_w \bar{c}\} ([w^\infty(t)]_\mathsf{R}) \right).$$

Under (b) we have $w^\infty(t) \in \mathcal{O}$ for each $w \in \mathsf{R}$, and a.e. $t \in \mathbb{R}_+$.

We conclude that, for each $w \in \mathsf{W}$, the gradient (5.55) exists for a.e. $t \geq 0$. Lipschitz continuity of $\bar{c}(\widehat{w}(t; w))$ with respect to (t, w) justifies exchanging differentiation and integration as follows:

$$\nabla_w \widehat{J}(w) = \int_0^\infty \nabla_w \{\bar{c}(\widehat{w}(s; w))\} \, ds.$$

This implies the representation

$$\nabla_w \widehat{J}(w) = \int_0^{\widehat{T}^*(w)} \Gamma(t; w) \, dt, \tag{5.56}$$

where the minimal draining time for the fluid model is the piecewise linear function

$$\widehat{T}^*(w) = \max_{1 \le i \le n} \delta_i^{-1} w_i, \qquad w \in \mathsf{W}.$$

The range of integration in (5.56) is finite since $\Gamma(t; w) = 0$ for $t > \widehat{T}^*(w)$.

The proof of (i) is completed on establishing Lipschitz continuity of the right-hand side of (5.56). For a given initial condition $w \in \mathsf{R}$, let $\{R_i^w : i = 1, \ldots, m\}$ denote the sequence of regions in $\{R_i\}$ (defined in (5.52)) so that $w^\infty(t) := w - \delta t$ enters regions $\{R_1^w, R_2^w, \ldots\}$ sequentially. We set $T_0 = 0$, let T_i denote the exit time from R_i^w, and set $\Gamma_i := \Gamma(t; w)$ for $t \in (T_{i-1}, T_i)$, $i \ge 1$. From the foregoing we see that the gradient can be expressed as the finite sum

$$\nabla_w \widehat{J}(w) = \sum_{i=1}^m (T_i(w) - T_{i-1}(w)) \Gamma_i.$$

Each of the functions $\{T_i(w)\}$ is piecewise linear and continuous on R under (a) and (b), and this implies that $\nabla \widehat{J}$ is Lipschitz continuous.

To see (iii), fix $w \in \partial \mathsf{R}$, and $w' \in$ interior (R). Then $w^\theta := (1 - \theta)w + \theta w' \in$ interior (R) for each $\theta \in (0, 1]$. Consequently, applying (ii) we have for $\theta \in (0, 1]$,

$$\langle \delta, \nabla J(w^\theta) \rangle = \bar{c}(w^\theta), \; and \; \langle \delta(w), \nabla \widehat{J}(w) \rangle = \bar{c}(w).$$

This combined with continuity of $\nabla \widehat{J}$ and $\bar{c}$ establishes the boundary property

$$\langle \phi(w), \nabla \widehat{J}(w) \rangle = \langle \delta - \delta(w), \nabla \widehat{J}(w) \rangle = 0.$$

It is obvious from (5.30) that $\phi(w) \ne 0$ on the boundary under (c), proving (iv). $\quad \square$

5.3.4 Optimization

We conclude this section with some examples of optimal control solutions for workload relaxations. Under appropriate conditions it can be shown that the respective optimal solutions for the relaxation and the original fluid model *couple* in the sense that $q^*(t; x) = \widehat{q}^*(t; x)$ following a relatively short transient period. Moreover, given the solution to the relaxation, we can construct a policy for the unrelaxed model so that q couples in this sense. The policy construction is based on the solution of finite-dimensional linear programs. We postpone this development to Section 6.6.3 where we treat more general models. Here, we illustrate the concepts through examples.

The following definition generalizes myopic policies for a workload model.

Definition 5.3.23 (Myopic policy for the fluid relaxation). For an n-dimensional workload model with effective cost $\bar{c}: \mathsf{W} \to \mathbb{R}_+$, suppose that the pointwise projection $[\cdot]_\mathsf{R}: \mathbb{R}^n \to \mathsf{R}$ exists with $\mathsf{R} = \mathsf{W}^+$. The *myopic policy for the fluid model* is defined so that

(i) $\widehat{w}$ is R-minimal, and
(ii) $\widehat{q}(t) = \mathcal{P}^*(\widehat{q}(t))$ for each t.   ∎

For small values of n, an optimal solution to the workload relaxation is frequently myopic, and also pathwise optimal. This is always the case in one dimension:

Proposition 5.3.24. *The myopic policy is pathwise optimal for a one-dimensional relaxation.*

Proof. The myopic solution $\widehat{q}^\circ$ satisfies $c(\widehat{q}^\circ(t)) = \bar{c}(\widehat{w}^\circ(t))$ for $t > 0$, and

$$\widehat{w}^\circ(t) = (\widehat{w}^\circ(0) - (1 - \rho_1)t)_+, \qquad t > 0.$$

Since $\bar{c}$ is a monotone function of $w \in \mathbb{R}_+$, it follows that for any other feasible state trajectory q,

$$c(\widehat{q}(t)) \geq \bar{c}(\widehat{w}(t)) \geq \bar{c}((\widehat{w}(0) - (1 - \rho_1)t)_+), \qquad t \geq 0.$$

That is, $c(\widehat{q}(t)) \geq c(\widehat{q}^\circ(t))$ for $t > 0$.

For relaxations of dimension greater than one it is not always so easy to identify an optimal solution, but we can obtain some qualitative results. The (optimal) fluid value function is defined as a function of $w \in \mathsf{W}$ by

$$\widehat{J}^*(w) = \min \int_0^\infty \bar{c}(\widehat{w}(s; w)) \, ds \,, \tag{5.57}$$

where the minimum is over all policies, subject to the conditions of Definition 5.3.1. ◻

The following result follows from the fact that the workload relaxation is a true relaxation of the fluid model, in the sense that it is subject to fewer constraints.

Proposition 5.3.25. *For any $1 \leq n \leq \ell_m$ the following hold for the nth relaxation:*

(i) *The value function $\widehat{J}^*$ is a convex, monotone function of w satisfying*

$$\widehat{J}^*(\widehat{\Xi}x) \leq J^*(x), \qquad x \in \mathsf{X}.$$

(ii) *Similarly, the minimal draining times satisfy $\widehat{T}^*(\widehat{\Xi}x) \leq T^*(x)$ for $x \in \mathsf{X}$.*
(iii) *If $n = \ell_m$ then $\widehat{T}^*(\widehat{\Xi}x) = T^*(x)$ for all $x \in \mathsf{X}$.*

Proof. The only part that requires proof is part (iii). However, this follows directly from the representation (4.29) for $T^*(x)$. ◻

Example 5.3.26 (Optimizing workload in the KSRS model). We now compute optimal solutions for the KSRS model in the two cases introduced in Example 5.3.16. One can show that a pathwise optimal solution exists in each case, and the optimal policy is myopic.

In Case 1 the effective cost is monotone, so that $\bar{c}^i \geq 0$ for each i. In this case the work-conserving policy coincides with the myopic policy. It follows that the value function $\widehat{J}$ for the work-conserving policy is convex in this case, which is seen in the plot on the left in Fig. 5.4.

A level set for the value function $\widehat{J}$ under the work-conserving policy is shown on the right in Fig. 5.4 in Case 2. Based on this plot it is clear that the value function is not convex. In Case 2 the monotone region for the effective cost is given by $\mathsf{W}^+ = $ closure (R_2), as shown on the right in Fig. 5.3. In this case, a pathwise optimal solution exists and coincides with the myopic policy. The value function is purely quadratic on R^*,

$$\widehat{J}^*(w) = \tfrac{1}{2} w^{\mathsf{T}} D w, \quad w \in \mathsf{R}^*, \qquad D = \tfrac{1}{8} \delta_1^{-1} \begin{bmatrix} 3 & -1 \\ -1 & 3 \end{bmatrix}. \tag{5.58}$$

In each case the boundary conditions (5.53) hold for $\widehat{J}$. In this two-dimensional example, this means that $\partial \widehat{J} / \partial w_2$ is zero along the lower boundary of R^*, and $\partial \widehat{J} / \partial w_1$ is zero along the upper boundary. ∎

The behavior of a workload relaxation may appear entirely unnatural since $\widehat{V}$ is unbounded. Nevertheless, for a properly chosen relaxation the gap between the value functions in the respective optimal control solutions can be tightly bounded. We illustrate this claim using the simple re-entrant line. Again, precise results for general models are developed in Section 6.6.

Example 5.3.27 (Optimizing workload in the simple re-entrant line). Consider the total cost optimization problem for the model considered in Example 5.3.20 without buffer constraints. Our goal is to achieve the minimal cost $\widehat{J}^*(w)$ defined in (5.57). Recall that the effective cost and monotone region are obtained in Example 5.3.20. In particular, the workload space and monotone region are given by

$$\mathsf{W} = \left\{ w : 0 \leq \frac{w_2}{w_1} \leq \overline{m} \right\}, \qquad \mathsf{W}^+ = \left\{ w : \tfrac{1}{2} \overline{m} \leq \frac{w_2}{w_1} \leq \overline{m} \right\},$$

with $\overline{m} = 2\rho_2 / \rho_1$. The optimal solution has the following properties:

(i) $\widehat{w}^*$ is R^*-minimal on the polyhedron

$$\mathsf{R}^* = \{ w : m_w^* w_1 \leq w_2 \leq \overline{m} w_1 \},$$

where $m_w^* = \mu_1 \mu_2^{-1} (1 + m_x^*)^{-1}$, with m_x^* is given in (4.53).

(ii) If $\rho_2 \leq \rho_1$ then $m_w^* = \tfrac{1}{2} \overline{m}$ so that $\mathsf{R}^* = \mathsf{W}^+$, and the resulting optimal policy coincides with the myopic policy. Moreover, it is pathwise optimal solution from each initial condition.

(iii) If $\rho_2 > \rho_1$ then the optimal policy is not myopic. Moreover, if the initial condition $\widehat{w}^*(0) = w$ satisfies

$$\frac{w_2}{w_1} < \frac{\rho_2}{\rho_1},$$

then a pathwise optimal solution starting from w cannot exist.

These conclusions follow from consideration of the fluid value function, exactly as in Example 4.3.10. The parameter m_w^* is consistent with the corresponding parameter m_x^* given in (4.53) for the unrelaxed model: If $x_2 = 0$ then $x_3 > m_x^* x_1$ if and only if $w_2 < m^*_w w_1$. Moreover, the effective state $x^* = \mathcal{X}^*(w)$ satisfies $x^*_2 = 0$ when $w_2 < m^*_w w_1$.

Consider, for example, the parameters defined in (2.26). The conditions of (iii) hold since $\rho_2 = 9/10 > 9/11 = \rho_1$, so we conclude that a pathwise optimal solution does not exist from certain initial conditions. The infinite-horizon optimal policy for $\widehat{q}$ is defined by the switching curve in workload space,

$$s_*(w_1) = m^*_w w_1 = \tfrac{11}{20} w_1, \qquad w_1 > 0.$$

That is, Station 2 is permitted to idle as long as $20w_2 < 11w_1$. In the relaxed model, if the initial workload is below this line, then the optimal state trajectory jumps upward to reach it at time $t = 0+$. Henceforth, the idleness process for the second station satisfies $\frac{d}{dt} I_2(t) = 0$ until the upper boundary of the workload space W is reached.

In Fig. 5.8, the optimal trajectory minimizing (4.37) is compared to the optimal solution for the two-dimensional relaxation. The triangular region shows the error introduced by relaxing the original network optimization problem. Note that the unrelaxed model catches up with the relaxed model after a short transient period, denoted $\widehat{T}^\circ$ in the figure. Consequently, although the optimal policy for the unrelaxed model incurs a higher cost when compared with the relaxation, this error is small relative to the total cost.

For comparison, shown in Fig. 5.9 are the optimal solutions for the first and second workload relaxations for this model. The two plots are very different since the loads at Stations 1 and 2 are nearly equal. In the one-dimensional relaxation only $\widehat{q}^{*1}_1(t)$ is ever nonzero in the optimal solution for $t > 0$, while in the two-dimensional relaxation $\widehat{q}^{*2}_2(t) = 0$ for all $t > 0$, but the other two buffers remain nonzero for a period of time. The optimal solution q^* couples with each relaxation, at times denoted $\widehat{T}^{\circ i}$, $i = 1, 2$. The coupling time $\widehat{T}^{\circ 2}$ is short relative to the draining time, while $\widehat{T}^{\circ 1}$ is far greater. ∎

5.4 Stochastic workload models

5.4.1 Relaxations for the CRW model

Relaxations of a stochastic workload model are constructed just as in the fluid model. The workload in units of time and the workload in units of inventory each evolve according to a CRW model, described by the respective recursions (5.17) and (5.9). For simplicity we consider here relaxations of the workload in units of inventory for a homogeneous model; the relaxation of workload in units of time is defined analogously based on the models introduced in Section 5.2.2.

Recall that the $n \times \ell$ matrix $\widehat{C}$ was introduced in Definition 5.3.1 to define the workload relaxation for the fluid model.

Definition 5.4.1 (Relaxation for the homogeneous CRW model). Assume that the following conditions hold:

(i) The CRW scheduling model is homogeneous, so that the workload process Y is expressed as (5.11).

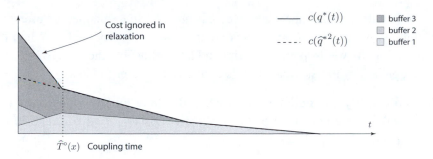

Figure 5.8. Optimal trajectories for the fluid model and its two-dimensional relaxation in the simple re-entrant line with $c(x) = |x|$. The dashed line shows the cost $c(\widehat{q}^*(t;x))$ for the optimized workload relaxation. The two solutions couple, $\widehat{q}^*(t;x) = q^*(t;x)$, for $t \geq \widehat{T}^\circ(x)$.

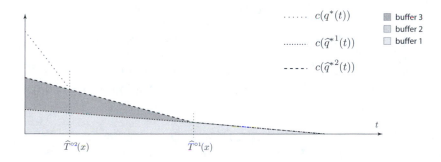

Figure 5.9. Optimal cost curves for the first and second workload relaxations in the simple re-entrant line. The plot of the cost $c(\widehat{q}^{*2}(t;x))$ coincides with the plot $c(\widehat{q}^*(t;x))$ shown in Fig. 5.8 for the two-dimensional relaxation. The plot of the cost $c(\widehat{q}^{*1}(t;x))$ is the optimal cost as a function of time for the one-dimensional relaxation.

(ii) The state process $\boldsymbol{Q}$ is restricted to $\mathsf{X}_\diamond := \mathsf{X} \cap \mathbb{Z}_+^\ell$, and the allocation process $\boldsymbol{U}$ is restricted to the set $\mathsf{U}_\diamond$ defined in (4.5).

Then, for a given integer $1 \leq n \leq \ell_m$, the *nth relaxation* of the CRW model is defined by the n-dimensional recursion

$$\widehat{Y}(t+1) = \widehat{Y}(t) + \widehat{S}(t+1)\widehat{\imath}(t) - \widehat{S}(t+1)\mathbf{1} + \widehat{L}(t+1), \qquad t \geq 0, \qquad (5.59)$$

where $\widehat{L}(t) := \widehat{C}[I - R^\mathsf{T}]^{-1} A(t)$, and $\widehat{S}$ is the $n \times n$ diagonal matrix sequence, $\{\widehat{S}(t) = \operatorname{diag}(S_1(t), \ldots, S_n(t)) : t \geq 1\}$. The idleness process $\widehat{\imath}$ is adapted to $\{\widehat{S}, \widehat{L}, \widehat{Y}\}$, and $\widehat{Y}$ is restricted to the *workload space*

$$\mathsf{Y}_\diamond = \{y = \widehat{C}[I - R^\mathsf{T}]^{-1} x : x \in \mathsf{X}_\diamond\}. \qquad \blacksquare$$

The process $\widehat{Y}$ is a relaxation of the workload in units of inventory. The relaxation of the (conditional) workload in units of time is defined by the simple scaling $\widehat{W}_i(t) := \mu_i^{-1} \widehat{Y}_i(t)$, $i = 1, \ldots, n$.

For a given cost function $c \colon \mathsf{X} \to \mathbb{R}_+$ the effective cost (5.33) defines a cost function $\bar{c}$ on W. This, along with the mapping $\mathcal{X}^*$ that defines the effective state can be defined for $\widehat{Y}$ based on the correspondence between $\widehat{Y}$ and $\widehat{W}$. In particular, the effective cost on Y is defined by $\bar{c}y(y) := \bar{c}(w)$, $y \in \mathsf{Y}$, with $w_i := \mu_i^{-1} y_i$ for each i. When there is no risk of ambiguity we drop the subscript and let $\bar{c}(y)$ denote the effective cost for $\widehat{Y}$.

The myopic policy for the relaxation is defined as in other models.

Definition 5.4.2 (Myopic policy for a workload relaxation). Suppose that $\widehat{Y}$ is the n-dimensional workload model (5.59) with effective cost $\bar{c}y \colon \mathsf{Y} \to \mathbb{R}_+$. The $\bar{c}y$-myopic policy is defined by the feedback law

$$\hat{\phi}(y) \in \arg\min \mathsf{E}[\bar{c}y(y + \widehat{S}(1)\iota - \widehat{S}(1)\mathbf{1} + \widehat{L}(1))], \qquad y \in \mathsf{Y}_\diamond,$$

where the minimum is over all $\iota \in \mathbb{R}_+^n$ such that $y + \widehat{S}(1)\iota - \widehat{S}(1)\mathbf{1} + \widehat{L}(1) \in \mathsf{Y}_\diamond$ with probability one. The queue length process $\widehat{Q}$ is defined as the effective state

$$\widehat{Q}(t) = \mathcal{X}^*(\widehat{W}(t)), \qquad t \geq 0,$$

where $\mathcal{X}^* \colon \mathsf{W} \to \mathsf{X}$ is defined in (5.34), and $\widehat{W}_i(t) := \mu_i^{-1} \widehat{Y}_i(t)$, $i = 1, \ldots, n$. ∎

The proof of Proposition 5.4.3 (i) is identical to the proof of Proposition 5.3.24, and (ii) is then immediate from Theorem 3.0.1 and the definition of the optimal average cost given in Section 4.1.3.

Proposition 5.4.3. *The following hold for a one-dimensional relaxation (5.59) based on Station s, for any $s \in \{1, \ldots, \ell_m\}$:*

(i) *The $\bar{c}y$-myopic policy is pathwise optimal.*

(ii) *The optimal average cost for the unrelaxed model satisfies $\eta_x^* \geq \widehat{\eta}^*$ for all $x \in \mathsf{X}_\diamond$, where*

$$\widehat{\eta}^* := \frac{c_{i_*}}{\mu_s \xi_{i_*}^s} \left(\frac{1}{2} \frac{\sigma_s^2}{\mu_s - \lambda_s} \right), \tag{5.60}$$

with i_* defined in Proposition 5.3.13, $\lambda_s := \mu_s \rho_s$, and

$$\sigma_s^2 := \rho_s \mathsf{E}[(\widehat{L}(1) - \widehat{S}(1))^2] + (1 - \rho_s)\mathsf{E}[(\widehat{L}(1))^2]$$

Example 5.4.4 (Dai–Wang model). We revisit the model of Dai and Wang under the assumptions of Example 4.6.4 to illustrate the application of Proposition 5.4.3.

The two workload processes in units of inventory for a CRW model are defined by

$$\begin{aligned} Y_1(t) &= 3Q_1(t) + 2Q_2(t) + Q_3(t) + Q_4(t) + Q_5(t) \\ Y_2(t) &= 2\big(Q_1(t) + Q_2(t) + Q_3(t)\big) + Q_4(t), \qquad t \geq 0. \end{aligned} \tag{5.61}$$

It was previously assumed that the second station is a bottleneck, so that $\rho_1 < \rho_2 = \rho_\bullet$, with $\rho_1 = 3\alpha_1/\mu_1$ and $\rho_2 = 2\alpha_1/\mu_2$.

Consider a one-dimensional relaxation based on the second workload process. Given any linear cost function on buffer levels, the effective cost on W can be expressed

$$\bar{c}(w) = \bar{c}_* w, \quad w \in \mathbb{R}_+, \qquad \text{with } \bar{c}_* = \min\left\{\tfrac{1}{2}c_1, \tfrac{1}{2}c_2, \tfrac{1}{2}c_3, c_4\right\}. \tag{5.62}$$

Based on the cost function $c(x) = x_1 + 2(x_2 + x_3 + x_4 + x_5)$ we have $c_1 < \min\{c_2, c_3, 2c_4\}$, and $\bar{c}_* = \frac{1}{2}c_1 = \frac{1}{2}$.

The other parameters in the bound (5.60) are obtained as follows: $\mu_s \xi_{i_*}^s = \mu_2 \xi_1^2 = 2$, $\lambda_s = 2\alpha_1$, $\rho_s = \lambda_2/\mu_2$, and based on the statistics (4.74),

$$\sigma_s^2 = \rho_2\left(\mu_2 + m_L^2\right) + (1 - \rho_2)m_L^2 = \lambda_2 + m_L^2,$$

where $m_L^2 = \mathsf{E}[(\widehat{L}(1))^2] = (\alpha_1/\kappa)(2\kappa)^2 = 2\lambda_2\kappa$. Hence (5.60) becomes

$$\widehat{\eta}^* = \frac{1 + 2\kappa}{4} \frac{\rho_2}{1 - \rho_2}.$$

This lower bound is indicated in Fig. 4.16 for $\kappa = 1, 3, 5$, and 9. The relative error in the bound (5.60) decreases as variability increases, corresponding to larger values of κ. ∎

5.4.2 Controlled Brownian model

Relaxations for the general CRW model can be constructed based on the recursion (5.17). In many cases the relaxed model is far simpler than the original model in "buffer coordinates." Here, we consider diffusion approximations for these discrete-time workload models.

The general controlled Brownian motion model defined in Definition 5.4.5 is obtained on adding a Brownian motion to the fluid model equation (5.29).

Definition 5.4.5 (Controlled Brownian motion). The controlled Brownian motion (CBM) model is the continuous-time controlled stochastic process $\widehat{W}$ on W subject to the dynamics

$$\widehat{W}(t; w) = w - \delta t + \hat{I}(t) + \widehat{N}(t), \qquad \widehat{W}(0) = w \in \mathsf{W}, \ t \geq 0, \tag{5.63}$$

where $\delta \in \mathbb{R}^n$ denotes the *drift vector* with components

$$\delta_s = 1 - \rho_s, \qquad 1 \leq s \leq n.$$

The state process $\widehat{W}$, the idleness process $\hat{I}$, and the disturbance process $\widehat{N}$ are subject to the following constraints:

(i) $\widehat{W}$ is constrained to the workload space W defined in (5.21).
(ii) The stochastic process $\widehat{N}$ is a drift-less n-dimensional Brownian motion, whose instantaneous covariance Σ is full rank, that is, $\Sigma > 0$.
(iii) The idleness process $\hat{I}$ is constrained to be adapted to the Brownian motion $\widehat{N}$, with $\hat{I}(0) = \mathbf{0}$, and

$$\hat{I}_s(t_1) \geq \hat{I}_s(t_0), \qquad 0 \leq t_0 \leq t_1 < \infty, \ 1 \leq s \leq n.$$

A process $\hat{I}$ satisfying these constraints is called *admissible*. ∎

Recall that the CBM model was constructed as an approximation for the single server queue based on the scaling (2.4) for a family of processes $\{Q^\kappa\}$ parameterized by κ with load $\rho^\kappa \uparrow 1$. A similar "heavy traffic" scaling can be used in the network setting. We do not consider this limit theory in this book, beyond what was already established for the single server queue.

We turn to the simple re-entrant line to illustrate how a CBM model is constructed.

Example 5.4.6 (Simple re-entrant line). Consider the simple re-entrant line under the homogeneity assumption, so that the CRW model can be expressed according to the recursion (4.10). It follows from (5.8) that the workload process in units of inventory is defined by the pair of equations

$$Y_1(t) = 2Q_1(t) + Q_2(t) + Q_3(t);$$
$$Y_2(t) = Q_1(t) + Q_2(t), \qquad t \geq 0. \tag{5.64}$$

The workload relaxation is defined as the controlled random-walk model

$$\widehat{Y}(t+1) = \widehat{Y}(t) + S(t+1)\widehat{\iota}(t) - S(t+1)\mathbf{1} + C[I - R^{\mathsf{T}}]^{-1}A(t+1), \qquad t \in \mathbb{Z}_+,$$

where S denotes the two-dimensional i.i.d. matrix sequence $\{S(t) = \mathrm{diag}\,(S_1(t), S_2(t)) : t \geq 1\}$. This can be expressed in a form similar to (5.63),

$$\widehat{Y}(t+1) = \widehat{Y}(t) - \delta + S(t+1)\widehat{\iota}(t) + \widehat{N}(t+1), \qquad t \geq 0, \tag{5.65}$$

where the drift vector is defined by the mean increment $\mathsf{E}[\widehat{Y}(t+1) - \widehat{Y}(t)]$ when $\widehat{\iota}(t) = 0$, given by

$$-\delta := \alpha_1 \binom{2}{1} - \mu_1 \binom{1}{0} - \mu_2 \binom{0}{1}, \tag{5.66}$$

and $\widehat{N}(t) := \delta - S(t) + C[I - R^{\mathsf{T}}]^{-1}A(t), t \geq 1$. Given this description, we can construct the CBM model so that the first- and second-order statistics in the two workload models agree, with drift $-\delta$ and covariance matrix is $\Sigma = \mathsf{Cov}(\widehat{N}(1))$.

For example, if the marginal distribution of the i.i.d. process (S_1, S_2, A_1) is as defined in (4.11), then the covariance is

$$\Sigma = \mathsf{Cov}(\widehat{N}(t)) = \alpha_1 \left(-\delta + \binom{2}{1}\right)\left(-\delta + \binom{2}{1}\right)^{\mathsf{T}} + \mu_1 \left(\delta + \binom{1}{0}\right)\left(\delta + \binom{1}{0}\right)^{\mathsf{T}}$$
$$+ \mu_2 \left(\delta + \binom{0}{1}\right)\left(\delta + \binom{0}{1}\right)^{\mathsf{T}}. \tag{5.67}$$

This model is considered in-depth in Section 5.6.2. ∎

The construction of an R-minimal solution for the CBM model is based on Definition 5.3.4 for the fluid model. It is assumed that the polyhedral region R has nonempty interior, and has the form (5.31), so that the existence of an R-minimal solution for the fluid model is characterized using Theorem 5.3.7, based on the projection $[\,\cdot\,]_{\mathsf{R}}$. We have seen that the existence of the projection requires some assumptions on the set R for dimensions 2 or higher. Figure 5.2 shows an example in which a minimal process does not exist from each initial condition in this three-dimensional fluid model, and the same conclusion will be reached in the three-dimensional CBM model.

We begin with the following refinement of the deterministic workload model through the introduction of an additive disturbance: Fix a continuous function $d\colon \mathbb{R}_+ \to \mathbb{R}^n$ satisfying $d(0) = 0$, and consider the controlled model

$$\widehat{w}(t) = w - t\delta + I(t) + d(t), \qquad t \geq 0. \tag{5.68}$$

An idleness process I is again called admissible if it has nondecreasing components, with $I(0) = 0$, and $\widehat{w}(t) \in \mathsf{W}$ for all $t > 0$.

Proposition 5.4.7 establishes the existence of an R-minimal solution for the model (5.68) whenever it exists for the disturbance-free model (with $d \equiv 0$.) The L_∞-norm in Proposition 5.4.7 (ii) is given by

$$\|d - d'\|_{[0,t]} := \max_{0 \leq s \leq t} \|d(s) - d'(s)\|, \qquad t \geq 0, \tag{5.69}$$

where $\| \cdot \|$ denotes the Euclidean norm.

The existence of a minimal solution can be regarded as a multidimensional generalization of the *Skorokhod map* described in Section 3.1.2.

Proposition 5.4.7. *Suppose that the set* R *is given in (5.31), and that the pointwise projection* $[\cdot]_\mathsf{R}\colon \mathbb{R}^n \to \mathsf{R}$ *exists. Then:*

(i) *An* R-*minimal solution* $\widehat{w}^\circ$ *exists for each continuous disturbance* d, *and each initial condition* $w \in \mathsf{W}$.

(ii) *There exists a fixed constant* $k_\mathsf{R} < \infty$ *satisfying the following: For any two initial conditions* $w, w' \in \mathsf{W}$, *and any two continuous disturbance processes* d, d', *the resulting workload processes* $\{\widehat{w}^\circ(\cdot\,;w), \widehat{w}^{\circ'}(\cdot\,;w')\}$ *obtained with the respective initial conditions and disturbances satisfy the following bounds for* $t \geq 0$:

$$\|\widehat{w}^\circ(t;w) - \widehat{w}^{\circ'}(t;w')\| \leq k_\mathsf{R}\big[\|w - w'\| + \|d - d'\|_{[0,t]}\big].$$

(iii) *For each* $\widehat{w}^\circ(0) \in \mathsf{W}$, $i \in \{1, \ldots, \ell_\mathsf{R}\}$, *and* $j \neq j_i$,

$$\int_0^\infty \mathbf{1}\big(\widehat{w}^\circ(t) \in F(i),\ \widehat{w}^\circ(t) \notin F(i') \text{ for } i' \neq i\big)\, dI_j(t) = 0.$$

Proof. Fix $r \geq 1$ and define $t_k = r^{-1}k$ for $k \geq 0$. We define the piecewise linear, continuous disturbance d^r as follows: It is linear on $[t_k, t_{k+1})$, and $d^r(t_k) = d(t_k)$ for each $k \geq 0$. The piecewise linear minimal process $\widehat{w}^r$ may be defined inductively by $\widehat{w}^r(t) = [[w]_\mathsf{R} - t\delta]_\mathsf{R}$ for $t \in (0, t_1]$, and for $k \geq 1$,

$$\widehat{w}^r(t) = [\widehat{w}(t_k) - (t - t_k)\delta + (t - t_k)(d(t_{k+1}) - d(t_k))]_\mathsf{R}, \qquad t \in [t_k, t_{k+1}].$$

$\square$

This is the minimal solution for the disturbance d^r. The continuity property Proposition 5.4.7 (ii) can be verified with k_R independent of $r \geq 1$, and (iii) is a direct consequence of Proposition 5.3.9 (iii) and the definition of $\widehat{w}^r$.

Since d is continuous it is uniformly approximated by the $\{d^r : r \geq 1\}$ in $L_\infty[0, t]$ for each $t > 0$. We conclude that $\{\widehat{w}^r : r \geq 1\}$ is a Cauchy sequence in $L_\infty[0, t]$, and hence it is convergent to some function $\widehat{w}^\circ$. The limiting process $\widehat{w}^\circ$ is the desired solution in (i), and this satisfies the properties stated in (ii) and (iii).

Proposition 5.4.7 leads to the following extension of Definition 5.3.4.

Definition 5.4.8 (**Minimal solution**). Suppose that the pointwise projection $[\,\cdot\,]_R$ exists. The R-minimal solution for the CBM model is the pair $(\widehat{W}, I)$ coinciding with the R-minimal solution for (5.68) with $d = N$. ∎

The resulting idleness process I is random since N is assumed to be a Gaussian stochastic process. The controlled process $\widehat{W}$ is a time homogeneous, strong Markov process. The strong Markov property follows from the sample-path construction of $\widehat{W}$ [469].

5.5 Pathwise optimality and workload

In this section we apply workload relaxation techniques to establish pathwise optimality in several special cases.

5.5.1 Klimov model

Recall that the Klimov model was defined in Section 2.2 as the single station scheduling problem based on a linear cost function $c\colon \mathsf{X} \to \mathbb{R}_+$. The c–μ rule was introduced, which is precisely the c-myopic policy, and we found that this defined a pathwise optimal solution for the fluid model. Similar conclusions can be reached for the CRW model. We begin with the simplest case in which pathwise optimality is feasible.

Homogeneous model: The homogeneous Klimov model can be expressed as follows,

$$Q_i(t+1) = Q_i(t) - S(t+1)U_i(t) + A_i(t+1), \qquad t \ge 0,\ 1 \le i \le \ell,$$

where S is a (scalar) i.i.d. Bernoulli sequence, and A is i.i.d. on $\mathbb{Z}_+^\ell$ with finite second moment. The workload process in units of inventory is defined as the sum $Y(t) = \sum Q_i(t)$, and evolves as a single server queue,

$$Y(t+1) = Y(t) - S(t+1)U(t) + L(t+1), \qquad t \ge 0,$$

where $L(t) = \sum A_i(t)$ and $U(t) = \sum U_i(t)$. Pathwise optimality requires that the policy be nonidling: $U(t) = 1$ if $Y(t) \ge 1$. This does hold for the c–μ rule, and in fact we do have the following result:

Proposition 5.5.1. *The c–μ rule is pathwise optimal for the homogeneous Klimov model.*

Proof. Assume without loss of generality that the cost parameters are ordered so that $c_1 \ge c_2 \ge \cdots \ge c_\ell$. The c–μ rule is then the priority policy satisfying for each $m \in \{1, \ldots, \ell\}$,

$$U^{[m]}(t) := \sum_{i=1}^m U_i(t) = 1 \quad \textit{whenever} \quad Y^{[m]}(t) := \sum_{i=1}^m Q_i(t) \ge 1. \qquad (5.70)$$

We interpret $Y^{[m]}(t)$ as the workload at time t for all of the m *most costly* unfinished jobs.

On letting $L^{[m]}(t) = \sum_{i=1}^{m} A_i(t)$ we find that the partial workload also evolves as a single server queue,

$$Y^{[m]}(t+1) = Y^{[m]}(t) - S(t+1)U^{[m]}(t) + L^{[m]}(t+1), \qquad t \geq 0, \qquad (5.71)$$

and the priority policy is precisely the nonidling policy for $Y^{[m]}$.

It follows that under the c–μ rule policy, $Y^{[m]}$ is pathwise optimal for each m. This completes the proof since

$$c(Q(t)) = c_\ell Y(t) + \sum_{m=1}^{\ell-1} (c_m - c_{m+1}) Y^{[m]}(t).$$

$\square$

Homogeneous cost: Suppose that $c(x) = |x| = \sum x_i$, but the service rates depend upon the customer class. The c–μ rule gives priority to the buffer with the largest service rate in this case. We assume that the buffers are ordered so that $\mu_1 \geq \mu_2 \geq \cdots \geq \mu_\ell$.

The CRW model is given by the ℓ-dimensional recursion (5.13) where A is again i.i.d. with finite second moment, and each M_i is Bernoulli with mean μ_i. Suppose that the following stochastic dominance condition holds: for each $m \in \{1, \ldots, \ell\}$, and any time t,

$$M_m(t) = 1 \Rightarrow M_i(t) = 1 \quad \text{for each } 1 \leq i \leq m. \qquad (5.72)$$

Then, under the c–μ rule, the proof of Proposition 5.5.1 can be adapted to show that the process $Y^{[m]}$ defined in (5.70) is again pathwise optimal for each m. This gives:

Proposition 5.5.2. *Suppose that (5.72) holds for the Klimov model, and that $c_i = c_1$ for each i. Then, the c–μ rule is pathwise optimal.*

Lastly, we turn to the general model.

Pathwise optimality in the mean: Consider the general Klimov model (5.13) with general linear cost function. As before it is assumed that the c–μ rule defines the priority policy (5.70), which means that the buffers have been ordered so that $c_1\mu_1 \geq c_2\mu_2 \geq \cdots \geq c_\ell\mu_\ell$.

We consider partial workload as before, but now measured in units of time,

$$W^{[m]}(t) := \sum_{i=1}^{m} \mu_i^{-1} Q_i(t), \qquad t \geq 1, \ m \in \{1, \ldots, \ell\}.$$

We have

$$c(Q(t)) = c_\ell\mu_\ell W(t) + \sum_{m=1}^{\ell-1} (c_m\mu_m - c_{m+1}\mu_{m+1}) W^{[m]}(t).$$

Hence, the expectation $\mathsf{E}[c(Q(t))]$ is minimized if the expectation of each of $\{W^{[m]}(t) : 1 \leq m \leq \ell\}$ are simultaneously minimized. Proposition 5.5.3 shows that the c–μ rule achieves this, and thereby proves that this policy is *pathwise optimal in the mean*, in that $E[c(Q(t))]$ is minimized over all policies for each time t.

Proposition 5.5.3. *For the general Klimov model,* $E[W^{[m]}(t)]$ *is minimized under the* c–μ *rule at each time* $t \geq 1$, *and each* $m \in \{1, \dots, \ell\}$.

Proof. Similar to the proof of Proposition 5.5.2, the proof here rests on a particular probabilistic representation of the queue. In this case we apply the workload in units of time for the Klimov model introduced in Section 5.2.2.1.

Suppose that for each $i \in \{1, \dots, \ell\}$, the jth customer to be serviced at buffer i requires $\mathcal{G}_{i,j}$ time units of service. The random variables $\{\mathcal{G}_{i,j} : 1 \leq i \leq \ell, \, j \geq 1\}$ are i.i.d. with geometric distribution satisfying

$$P\{\mathcal{G}_{i,j} = T + 1 \mid \mathcal{G}_{i,j} > T\} = \mu_j, \qquad T \geq 0.$$

Let $\mathcal{W}^{[m]}(t)$ denote the *actual time* required to clear all of the customers in the system at time t, ignoring future arrivals. This partial workload process evolves according to the recursion

$$\mathcal{W}^{[m]}(t+1) = \mathcal{W}^{[m]}(t) - U^{[m]}(t) + \mathcal{L}^{[m]}(t+1),$$

where $U^{[m]}(t) := \sum_{i=1}^{m} U_i(t)$ and $\{\mathcal{L}^{[m]}(t) : t \geq 1\}$ is i.i.d. with common distribution given by

$$\mathcal{L}^{[m]}(t) \approx \sum_{i=1}^{m} \mathbf{1}\{A_i(t) \geq 1\} \sum_{j=1}^{A_i(t)} \mathcal{G}_{i,j}(t).$$

When $m = \ell$ we obtain $\mathcal{L}^{[m]}(t) = \mathcal{L}(t)$ (defined in (5.14)).

Thus $\{\mathcal{W}^{[m]}(t) : t \geq 0\}$ evolves exactly as the CRW model for the single server queue. We conclude that the nonidling policy is optimal,

$$U^{[m]}(t) := \sum_{i=1}^{m} U_i(t) = 1 \quad \text{whenever} \quad \mathcal{W}^{[m]}(t) \geq 1.$$

The proof is completed on noting that $E[W^{[m]}(t)] = E[\mathcal{W}^{[m]}(t)]$ for each t. $\qquad \square$

The conclusion of Proposition 5.5.3 can also be reached by constructing a linear program that characterizes the optimal cost, and considering extreme points in the constraint set. Let $\bar{\mathsf{X}} \subset \mathbb{R}_+^\ell$ denote the set of all possible average-state values,

$$\bar{\mathsf{X}} = \{E^\phi[Q(t)] : \phi \text{ is stationary}\},$$

where the expectation is taken in steady state (given by infinity if the controlled process is not ergodic). Since the cost is assumed linear, for any stationary policy ϕ we must have, for some $\bar{x} \in \bar{\mathsf{X}}$,

$$E^\phi[c(Q(t))] = c^{\mathsf{T}}\bar{x}.$$

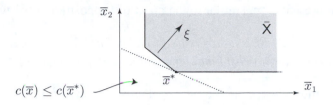

Figure 5.10. Extreme point $\bar{x}^*$ optimizing the linear program (5.73).

Consequently, the optimization problem can be cast as the linear program

$$\mathbf{min} \quad c^{\mathsf{T}} x$$

$$\mathbf{s.t.} \quad x \in \bar{\mathsf{X}}.$$

(5.73)

Without loss of generality the optimizer can be chosen as an extreme point in $\bar{\mathsf{X}}$. An extremal optimizer is illustrated in Fig. 5.10. It follows from Proposition 5.5.3 that the optimizing extreme point coincides with the c–μ priority policy.

This point of view is the basis of the linear programming approaches to optimization considered in Sections 9.2 and 9.3.

5.5.2 Single station re-entrant line

Consider the homogeneous re-entrant line consisting of a single station, defined by the ℓ-dimensional recursion

$$Q(t+1) = Q(t) + A_1(t+1)\mathbf{1}^1 + S(t+1)\sum_{i=1}^{\ell} U_i(t)[\mathbf{1}^{i+1} - \mathbf{1}^i],$$

(5.74)

where we define $\mathbf{1}^{\ell+1} := \mathbf{0}$. For this homogeneous model there is a pathwise optimal solution provided the cost is also homogeneous.

Suppose that $c(x) = |x| = \sum x_i$. The myopic policy for the fluid model coincides with LBFS in the simple re-entrant line, and the same holds true for any re-entrant line. The analysis of the CRW model under this policy is almost identical to the homogeneous Klimov model.

The quantity $Y^{[m]}(t)$ is now defined by

$$Y^{[m]}(t) := \sum_{i=m}^{\ell} (m-i+1)Q_i(t) \qquad t \geq 0.$$

This is interpreted as the workload in units of inventory, for all "down-stream" jobs, at time t. We have $Y(t) = \Xi_Y Q(t) = Y^{[1]}(t)$ using Definition 5.2.1. The LBFS policy can be described as follows: for each $m \in \{1, \ldots, \ell\}$,

$$U^{[m]}(t) := \sum_{i=m}^{\ell} U_i(t) = 1 \quad \text{whenever} \quad Y^{[m]}(t) \geq 1.$$

(5.75)

Based on this representation, similar to the proof of Proposition 5.5.1 we have:

Proposition 5.5.4. *The LBFS policy is pathwise optimal for the single station, homogeneous re-entrant line with cost function equal to the ℓ_1 norm, $c(\,\cdot\,) \equiv |\cdot|$.*

In moving to multiple-resource models pathwise optimality is typically infeasible. In such cases we consider a workload relaxation. Conditions for the existence of a pathwise optimal solution are transparent for the relaxation. When pathwise optimality is not feasible it is revealing to understand why.

5.5.3 Workload models

The simplest case tells most of the story: A criterion for pathwise optimality is easily obtained in a one-dimensional relaxation.

Proposition 5.5.5. *Consider the workload model (5.11) in one dimension,*

$$Y(t+1) = Y(t) - S(t+1) + S(t+1)\iota(t) + L(t+1), \qquad t \geq 0, \qquad (5.76)$$

where (S, L) is an i.i.d. process on $\mathbb{Z}_+^2$ and S is Bernoulli. If the cost function $\bar{c}y \colon Y \to \mathbb{R}_+$ is monotone, then the nonidling policy is pathwise optimal.

Monotonicity is of course critical. For example, in the simple inventory model introduced in Section 2.6 in which $Y = \mathbb{Z}$ the state process evolves as (5.76), yet pathwise optimality is impossible since the cost function cannot be monotone.

A single example illustrates how Proposition 5.5.5 can be generalized to multidimensional workload models.

Example 5.5.6 (KSRS model). Consider the KSRS model with $X = \mathbb{R}_+^4$, and cost function $c \colon \mathbb{R}^4 \to \mathbb{R}_+$ equal to the ℓ_1 norm. The effective cost for the two-dimensional relaxation was computed in two particular cases in Example 5.3.16. Figure 5.3 shows the sublevel sets of the effective cost $\bar{c} \colon \mathbb{R}_+^2 \to \mathbb{R}_+$. In Case 1 the effective cost is monotone, while in Case 2 it is not.

The nonidling policy is pathwise optimal in Case 1 since the resulting workload process is minimal. It is also easily seen that in Case 2 pathwise optimality cannot hold for any policy.

In conclusion, pathwise optimality holds if and only if the effective cost is monotone.

∎

For workload models of dimension 3 or higher the situation is more complex even when $\bar{c}$ is monotone since a minimal solution may not exist. It was shown in Theorem 5.3.7 that the existence of a minimal solution is equivalent to the existence of the projection operator $[\,\cdot\,]_Y$.

We now consider how to optimize when the effective cost is not monotone.

5.6 Hedging in networks

The main conclusion in Section 4.4 is that safety stocks are sometimes necessary to compensate for variability and delay in a stochastic network. Here, we describe a second important aspect of control in the face of variability. In a workload model the following class of policies generalizes the hedging points defined earlier in the one-dimensional inventory model.

This definition applies to any of the workload models defined in this chapter, both deterministic and stochastic. We use the notation $\mathbf{Y}$ for workload in this section since our main examples treat workload in units of inventory.

Definition 5.6.1 (Affine policies). Suppose that $\mathsf{R} \subset \mathsf{Y}$ is a polyhedral region of the form

$$\mathsf{R} = \{y \in \mathsf{Y} : \langle n^i, y \rangle \geq 0,\ 1 \leq i \leq \ell_r\}.$$

Given *affine-shift parameters* $\{\beta_i : 1 \leq i \leq \ell_R\} \subset \mathbb{R}_+$, we define the affine constraint region

$$\mathsf{R}(\beta) = \{y \in \mathsf{Y} : \langle n^i, y \rangle \geq -\beta_i,\ 1 \leq i \leq \ell_R\}. \tag{5.77}$$

The associated *affine policy* for an n-dimensional workload model is defined as the idleness process associated with the minimal process on $\mathsf{R}(\beta)$, assuming this exists. ∎

Typically, we define $\mathsf{R} = \mathsf{R}(0)$ based on the fluid model. It is assumed that a piecewise linear cost function is given,

$$\bar{c}(y) = \max \langle c^i, y \rangle, \qquad y \in \mathbb{R}^n. \tag{5.78}$$

The constraint region $\mathsf{R}(0)$ is either the optimal constraint region for the fluid model, or the region Y^+ that defines the myopic policy. The region $\mathsf{R}(\beta)$ is thus an affine enlargement of the constraint region for the fluid model when $\beta > 0$. The constants $\{\beta_i\}$ are interpreted as hedging points for the workload model.

We postpone analysis to Chapter 9, based on some preliminary results described in Chapter 7. In this section we introduce the main concept used to construct hedging points, and present a detailed example.

5.6.1 Height process

Here, we introduce a general two-dimensional workload model. For the purposes of optimization it is convenient to relax integer constraints: The workload process $\mathbf{Y}$ is simply constrained to a convex cone $\mathsf{Y} \subset \mathbb{R}^2$, and evolves in discrete time according to the recursion

$$Y(t+1) = Y(t) - \mu + \iota(t) + L(t+1), \qquad t \geq 0,\ Y(0) \in \mathsf{Y}. \tag{5.79}$$

It is assumed that $\mu \in \mathbb{R}_+^2$, and that L an i.i.d. sequence taking values in $\mathbb{R}_+^2 \cap \mathsf{Y}$, with mean $\lambda \in \mathbb{R}_+^2$ satisfying $\lambda < \mu$. The idleness increment process ι also takes on nonnegative values. This may be viewed as a relaxation of the workload in units of time defined in (5.17), in which case $\mu = 1$ and $L(t) = \mathcal{L}(t)$. The two-dimensional fluid model is defined as in (5.24) by

$$y(t) = y(0) - \delta t + I(t), \qquad t \geq 0, \; y(0) \in Y,$$

where $\delta := \mu - \lambda > 0$.

The cost function $\bar{c} \colon Y \to \mathbb{R}_+$ is piecewise linear. To describe structural properties of optimal control solutions in two dimensions we simplify discussion by assuming that the cost function is monotone in its first variable y_1. Under this assumption, the monotone region Y^+ can be expressed as the cone $Y^+ = \{y \in Y : y_2 \geq m_* y_1\}$ for some $m_* > 0$. This allows us to focus attention on a single switching curve that defines an optimal policy for the workload model.

In the approximations described here based on the affine region (5.77) we always choose $R(0) = Y^+$. Let $\bar{c}^+, \bar{c}^- \in \{c^i\}$ denote the cost vectors that define the lower boundary of Y^+. By definition of the monotone region and the constant m_* we must have

$$\bar{c}_1^+ \geq 0, \; \bar{c}_2^+ \geq 0, \; \bar{c}_1^- \geq 0, \quad \bar{c}_2^- < 0,$$

$$\text{and} \quad \bar{c}((1, m_*)^{\mathsf{T}}) = \bar{c}_1^+ + m_* \bar{c}_2^+ = \bar{c}_1^- + m_* \bar{c}_2^- .$$

$$(5.80)$$

We consider several separate cases, according to the structure of the effective cost:

CASE I   *The effective cost $\bar{c}$ is monotone on Y, that is, $Y^+ = Y$.*

CASE II   *The effective cost $\bar{c}$ is not monotone on Y, and $\delta \in Y^+$.*

CASE II* *The effective cost $\bar{c}$ is not monotone on Y, and the vector δ lies in the interior of Y^+.*

CASE III *The effective cost $\bar{c}$ is not monotone on Y, and $\delta \notin Y^+$.*

CASE IV $Y^+ = \emptyset$.

The following taxonomy is obtained for control solutions:

(i)   In Case I the work-conserving policy is optimal for the CBM and fluid models. For the CRW model the policy that makes Y minimal on Y is optimal, provided a minimal solution exists.

(ii)   In Case II the two-dimensional *fluid model* admits a pathwise optimal solution from each initial condition: The optimal trajectories evolve in Y^+ for $t > 0$, and y^* is the minimal process on Y^+.

(iii) In Case III pathwise optimality cannot hold for arbitrary initial conditions, even in the fluid model. Optimal trajectories for the fluid model evolve in a closed positive cone $R^*(0) \subset Y$ that is strictly larger than Y^+.

(iv)   Case IV is unusual, although we have seen in Example 5.3.17 that $Y^+ = \emptyset$ is possible for a workload relaxation of the tandem queues.

The structural properties of optimal policies described informally here, and strengthened in later chapters, are obtained through an associated *height process* defined with

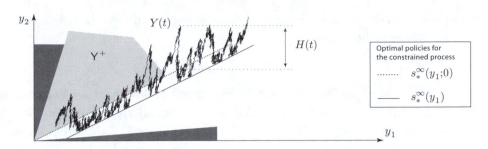

Figure 5.11. Workload model controlled using an affine policy and the associated height process. The dark grey region is infeasible for Y, but this constraint is disregarded in the unconstrained model with state process Y^∞. The light grey region indicates the monotone region for the effective cost $\bar{c}$. The height process H^∞ is positive recurrent in Case II* when the policy is affine.

respect to the half-line $\{y_2 = m_* y_1 : y_1 \geq 0\}$ defining the lower boundary of Y^+. It is formally defined as the difference

$$H(t) = Y_2(t) - m_* Y_1(t), \qquad t \geq 0, \tag{5.81}$$

and up until the first time τ_1 that $\iota_1(t)$ is nonzero it evolves as follows,

$$H(t+1) = H(t) - \mu_H + \iota_2(t) + L_H(t+1), \qquad 0 \leq t < \tau_1, \tag{5.82}$$

where $L_H(t) := L_2(t) - m_* L_1(t)$ and $\mu_H = (-m_*, 1)\mu = \mu_2 - m_* \mu_1$. A typical sample path of the height process is illustrated in Fig. 5.11, with Y controlled using an affine policy.

The process H is similar to a CRW model of a simple inventory model of the form introduced in Section 2.6. This similarity is strengthened on introducing a cost function $\bar{c}_H \colon \mathbb{R} \to \mathbb{R}_+$ for the height process

$$\bar{c}_H(x) = c_H^+ x_+ + c_H^- x_-, \qquad x \in \mathbb{R},$$

where $c_H^+ = c_2^+$ and $c_H^- = |c_2^-|$, based on (5.80).

In our approximation of the optimal policy we consider the following relaxation in which it is allowable to set $\iota_1(t) = 0$ for all t:

Definition 5.6.2 (Unconstrained process). The *unconstrained process* Y^∞ is the workload model satisfying (5.79), but with the state-space constraint $Y(t) \in Y$ relaxed, so that Y^∞ evolves in $\mathbb{R}^2$. The cost function for the unconstrained process is a relaxation of (5.78),

$$\bar{c}^\infty(y) := \max\{\langle c^+, y \rangle, \langle c^-, y \rangle\}, \qquad y \in \mathbb{R}^2. \tag{5.83}$$

The associated height process is denoted $H^\infty(t) := Y_2^\infty(t) - m_* Y_1^\infty(t)$, $t \geq 0$, with cost function $\bar{c}_H$. ∎

We will see that a discounted-cost optimal policy for Y^∞ is precisely affine: $\iota_1^{\infty*}(t) = 0$ for all t, and for some constant $\overline{y}_2^* \geq 0$,

$$\iota_2^{\infty*}(t) = \max(0, -\overline{y}_2^* - H_2^{\infty*}(t)), \qquad t \geq 0. \tag{5.84}$$

The resulting workload process is minimal on a half-space in $\mathbb{R}^2$. The height process evolves according to the recursion

$$H^\infty(t+1) = \max(H^\infty(t), -\overline{y}_2^*) - \mu_H + L_H(t+1). \tag{5.85}$$

For the two-dimensional model (5.79) the discounted-cost optimal policy is approximated by an affine policy of the form (5.84). Approximations are obtained for the average-cost optimal policy in Case II* only. The significance of Case II* is that the height process H^∞ satisfying (5.85) is positive recurrent if and only if $\delta_H := \delta_2 - m_*\delta_1 > 0$, so that its steady-state mean is finite (see Theorem 3.0.1.) It follows from the definitions that $\delta_H > 0$ precisely when the assumptions of Case II* are satisfied.

These remarks are intended to convey the main ideas. With a bit more work we can obtain an explicit expression for the value of the hedging point $\overline{y}_2^*$. We return to this in Section 7.5, where a precise statement is provided in Theorem 7.5.2.

5.6.2 Case study: Simple re-entrant line

Consider the simple re-entrant line shown in Fig. 2.9 with c the ℓ_1-norm, so that $c(x) = |x|$ equals the total inventory. It is assumed that the model is homogeneous, which in this example implies that $\mu_1 = \mu_3$. We consider several sets of parameters $\{\mu_i, \alpha_1\}$, and in each case construct policies for a CRW model and its relaxation.

We consider several models simultaneously:

(i) The CRW model that evolves according to the three-dimensional recursion (4.10).
(ii) The two-dimensional workload relaxation in units of inventory for the fluid and CRW models

$$\begin{aligned}
\tfrac{d^+}{dt}\widehat{y}_1(t) &= \mu_1\hat{\imath}(t) + 2\alpha_1 - \mu_1 \\
\tfrac{d^+}{dt}\widehat{y}_2(t) &= \mu_2\hat{\imath}(t) + \alpha_1 - \mu_2
\end{aligned} \tag{5.86}$$

$$\begin{aligned}
\widehat{Y}_1(t+1) - \widehat{Y}_1(t) &= -S_1(t+1) + S_1(t+1)\hat{\imath}_1(t) + 2A_1(t+1) \\
\widehat{Y}_2(t+1) - \widehat{Y}_2(t) &= -S_2(t+1) + S_2(t+1)\hat{\imath}_2(t) + A_1(t+1),
\end{aligned} \tag{5.87}$$

with initialization $\widehat{Y}(0), \widehat{y}(0) \in \mathsf{Y} := \{y \in \mathbb{R}_+^2 : y_2 \leq y_1\}$.

(iii) The CBM model described in Example 5.4.6 whose drift δ and covariance Σ are chosen to approximate the first- and second-order statistics in the CRW model (5.87).
(iv) The unconstrained process $\widehat{Y}^\infty$ that evolves as in (5.87) on $\mathsf{Y}^\infty := \mathbb{R}^2$, as well as associated height processes.

5.6.2.1 Hedging and workload

The two-dimensional relaxation of this model is examined in Example 5.3.20. With c the ℓ_1-norm, the expression (5.46) implies that the effective cost is given as a function of $y \in \mathsf{Y}$ by

$$\bar{c}(y) = \max(y_2, y_1 - y_2). \tag{5.88}$$

The plot of the level sets of $\bar{c}$ shown in Fig. 5.6 shows that the effective cost is not monotone in this example: The monotone region is a strict subset of Y, and is independent of the service and arrival rates,

$$\mathsf{Y}^+ := \{y \in \mathbb{R}_+^2 : \tfrac{1}{2}y_1 \le y_2 \le y_1\} .$$

An optimal solution for the fluid workload relaxation (5.86) is described by the *linear policy* obtained in Example 5.3.27,

$$\text{Work resource 2 at maximal rate if } y_2 > m_* y_1,$$

where $m_* \in [0, \tfrac{1}{2}]$ is a constant. For the CRW relaxation we consider affine policies of the specific form

$$\text{Work resource 2 at maximal rate if } y_2 > m_* y_1 - \bar{y}_2 . \tag{5.89}$$

For the three-dimensional CRW model we consider translations of (5.89) to buffer coordinates.

5.6.2.2 Policy translation

We now compare a family policy for the three-dimensional CRW model inspired by the form of the affine policy (5.89).

On substituting $y_1 = 2x_1 + x_2 + x_3$ and $y_2 = x_1 + x_2$ into the inequality in (5.89) we obtain $x_1 + x_2 > m_*(2x_1 + x_2 + x_3) - \bar{y}_2$, which we interpret as a warning that Station 2 should not be starved of work.

When $m_* = \tfrac{1}{2}$ this becomes $x_2 > x_3 - \bar{y}_1$, where $\bar{y}_1 := 2\bar{y}_2$. This motivates the following candidate policy for the three-dimensional model based on the hedging point $\bar{y}_1$ and an additional safety-stock $\bar{x}_2$,

$$\text{Serve buffer 1 if buffer 3 is zero, or}$$
$$x_3 < x_2 + \bar{y}_1, x_2 \le \bar{x}_2, \text{ and } x_1 > 0. \tag{5.90}$$

For the fluid model, the optimal parameters are $\bar{x}_2 = \bar{y}_1 = 0$. The parameter $\bar{y}_1$ is interpreted as a hedging point since it is defined with respect to the workload model. The safety-stock $\bar{x}_2$ is used to compensate for delay when transferring inventory from buffer 1 to buffer 2.

Consider the following two instances of Case II for this model:

Case II (a) $\mu_1 = \mu_3 = 20, \mu_2 = 10, \alpha_1 = 9.$ $\rho_1 = \rho_2 = 9/10.$
Case II (b) $\mu_1 = \mu_3 = 20, \mu_2 = 11, \alpha_1 = 9.$ $\rho_1 = 9/10$ and $\rho_2 = 9/11.$

In each instance the vector $\delta_Y = (\mu_1 - \lambda_1, \mu_2 - \lambda_2)^{\mathsf{T}}$ lies within the monotone region Y^+, where $\lambda_1 = 2\alpha_1$ and $\lambda_2 = \alpha_1$. The optimal policy for the fluid model is defined by the switching curve $y_2 = m_* y_1$ with $m_* = \tfrac{1}{2}$, since this defines the lower boundary of Y^+.

Figure 5.12 shows results from simulation of the controlled network for a range of parameters using the affine policy (5.90) in Case II (a). For each data-point the

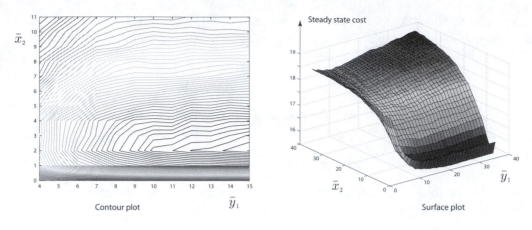

Figure 5.12. Contour and surface plots of average cost for Case II (a). The constant $\overline{y}_1 = 2\overline{y}_2$ is used in the affine policy (5.90).

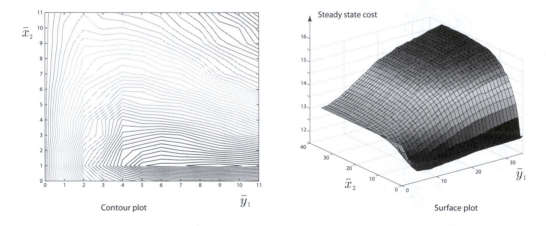

Figure 5.13. Contour and surface plots of average cost for Case II (b).

simulation was run for 5×10^6 time units, starting with an empty network. We see that the best affine shift $\overline{y}_1$ is very large. However, sensitivity with respect to this parameter is extremely small.

Figure 5.13 shows the results from simulation experiments in Case II (b). The sensitivity is much higher to the threshold $\overline{y}_1$ since the switching curve is attracting, in the sense that the height process is positive recurrent. The optimal values based on this simulation are $\overline{y}_1^* = 9$ and $\overline{x}_2^* = 1$.

We now consider an instance of Case III using the rates seen previously in (2.26). The loads at the two machines are $\rho_1 = \frac{9}{11}$ and $\rho_2 = \frac{9}{10}$ respectively. In this case one would expect lower sensitivity with respect to the policy since the mean drift forces the process away from the optimal switching curve for both stochastic and fluid models.

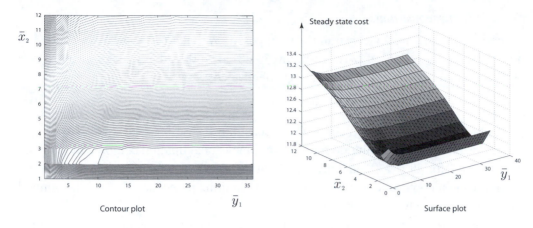

Figure 5.14. Contour and surface plots of average cost for affine policies in Case III.

The infinite-horizon optimal control for the fluid model is given by the switching curve illustrated in Fig. 2.10. For this set of parameters it is expressed

$$\text{Serve buffer 1 if buffer 3 is zero, or } x_1 > x_3 \text{ and } x_2 = 0. \tag{5.91}$$

The optimal policy (5.91) enforces the following condition for y:

$$\text{Work resource 2 at maximal rate if } \quad y_2 > \tfrac{1}{3} y_1. \tag{5.92}$$

An affine translation is implemented as follows: From the expressions for Y_1 and Y_2 given in (5.64) we write $y_2 > \tfrac{1}{3} y_1 - \overline{y}_2$ as

$$x_1 + x_2 > \tfrac{1}{3}(2x_1 + x_2 + x_3) - \overline{y}_2$$

which can be simplified to $2x_2 + x_1 > x_3 - \overline{y}_1$, where $\overline{y}_1 = 3\overline{y}_2$. For $\overline{y}_1 \geq 0$ and $\overline{x}_2 \geq 0$ we define the following policy for the CRW model:

$$\text{Serve buffer 1 if buffer 3 is zero, or } 2x_2 + x_1 > x_3 - \overline{y}_1 \text{ and } x_2 \leq \overline{x}_2, \tag{5.93}$$

where $\overline{x}_2$ is again interpreted as a static safety-stock. The average cost optimal policy is illustrated in Fig. 4.11. The policy (5.93) is a good fit with the optimal policy when $\overline{x}_2 = 2$ and $\overline{y}_1 = 10$.

Shown in Fig. 5.14 are the results of a simulation of the policies defined in (5.93) with $\overline{x}_2 = 1, 2, \ldots, 12$ and $\overline{y}_1 = 1, 2, \ldots, 36$. The simulation was run for $T = 5 \times 10^6$ simulation steps.

This example illustrates the relative sensitivity of cost to control parameters. On a fluid scale, the sensitivity of cost with respect to the threshold $x_2 \leq \overline{x}_2$ is strictly positive. On the other hand, the first-order sensitivity is zero for perturbations of the switching curve $y_2 = \tfrac{1}{3} y_1$. The contour plot given in Fig. 5.14 shows that this dichotomy is inherited by the stochastic model in this example. Sensitivity with respect to $\overline{y}_1$ is *extremely low*, while sensitivity with respect to $\overline{x}_2$ is *relatively high*.

5.6.2.3 Buffer constraints

Suppose now that buffer constraints are imposed. We take $b_2 = b_3 = \infty$, and vary the constraint b_1 on buffer 1. Transforming (5.48) to workload in units of inventory we arrive at the following expression for the effective cost,

$$\bar{c}(y) = \max(y_2, y_1 - y_2, y_1 - b_1). \tag{5.94}$$

The workload space is the same as in the unconstrained case, $\mathsf{Y} = \{y \in \mathbb{R}_+^2 : y_2 \leq y_1\}$.

In Case II an optimal solution for the two-dimensional fluid relaxation will maintain $\hat{y}^*(t) \in \mathsf{Y}^+$ for all $t > 0$, and any $y \in \mathsf{Y}$. As b_1 decreases, the idle time at Station 2 decreases in an optimal control solution. In the limiting case where $b_1 = 0$ the optimal solution $\hat{y}^*$ is pointwise minimal.

Computing the sensitivity of cost with respect to a buffer constraint $b_i < \infty$ in the fluid model is straightforward: Applying (5.39),

$$\frac{\partial}{\partial b_i} c(\hat{q}^*(t)) = -\psi_i \mathbf{1}(\hat{q}_i^*(t) = b_i), \tag{5.95}$$

where $\psi_i \geq 0$ is a Lagrange multiplier, and $\hat{q}^*$ is the optimal solution to the relaxation on the constrained state space.

Consider now the stochastic workload relaxation. Suppose that the policy is fixed, and that the state space Y and the workload process $\hat{Y}$ do not depend on b in a neighborhood of some value b_0 of interest. In this case we obtain an exact expression for first-order sensitivity since only the effective cost $\bar{c}$ is subject to variation. To obtain a simple form write $\bar{c}(\hat{Y}(t)) = c(\hat{Q}(t))$, where $\hat{Q}(t) = \mathcal{X}^*(\hat{Y}(t))$ is the effective state. Hence, exactly as in (5.95) we obtain

$$\frac{\partial}{\partial b_i} \bar{c}(\hat{Y}(t)) = -\psi_i \mathbf{1}(\hat{Q}_i(t) = b_i).$$

Remember, there is nothing probabilistic in this argument: The cost $\bar{c}(\hat{Y}(t))$ varies with b_i only because the cost function $\bar{c}$ depends on b_i. On taking expectations (and exchanging derivative and expectation) we obtain

$$\frac{\partial}{\partial b_i} \mathsf{E}[\bar{c}(\hat{Y}(t))] = -\psi_i \mathsf{P}(\hat{Q}_i(t) = b_i). \tag{5.96}$$

If the policy in workload space also changes with b_i then the formula (5.96) must be modified to take this into account.

We conclude with results from a single numerical experiment in Case II (a).

In the CRW model (4.10) it is impossible to maintain a strict upper bound on Q_1 – we simply impose the constraint that buffer 1 receives priority whenever $Q_1(t) \geq b_1$. A policy of this form will maintain a constraint of the form $Q_1(t) \leq b_1 + K$ with high probability (of order $1 - (9/20)^K$). The control laws for Q considered in the numerical experiments below are modifications of those given in (5.90):

> Serve buffer 1 if buffer 3 is zero, or
> $$x_3 < x_2 + \bar{y}_1 \text{ and } x_2 \leq \bar{x}_2 \text{ and } x_1 > 0, \text{ or} \tag{5.97}$$
> $$x_1 \geq b_1.$$

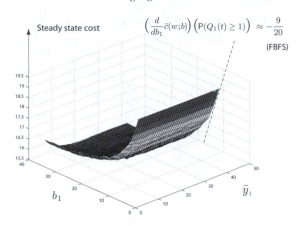

Figure 5.15. Surface plots of average cost for the CRW model in Case II (a) with buffer constraints. The sensitivity estimate (5.98) is nearly exact in this example. Sensitivity with respect to $\bar{y}_1$ remains very small.

For this policy we use the expression (5.96) as a guide in an approximation of sensitivity with respect to b_1. Applying (5.94) we obtain $\psi_1 = 1$, which suggests the sensitivity approximation

$$\left[\mathsf{E}^{b_1=k}[c(Q(t))] - \mathsf{E}^{b_1=k-1}[c(Q(t))]\right] \approx -[\mathsf{P}^{b_1=k}(Q_1(t) \geq k)], \qquad k \geq 1.$$

The right-hand side converges to zero geometrically fast as $k \to \infty$. The simplest case is $k = 1$ since the resulting policy is First-Buffer First-Served (FBFS). In this case Q_1 is equivalent to an M/M/1 queue with arrival rate α_1 and service rate μ_1. For the numerical values considered here it follows from Proposition 3.4.1 that

$$\mathsf{P}(Q_1(1) \geq 1) = \alpha_1/\mu_1 = 9/20,$$

and this gives the approximation

$$\mathsf{E}^{b_1=1}[c(Q(t))] - \mathsf{E}^{b_1=0}[c(Q(t))] \approx -9/20. \tag{5.98}$$

The surface plot shown in Fig. 5.15 shows the steady-state mean $\mathsf{E}^{b_1=k}[c(Q(t))]$, computed via simulation, for a range of k under the policy (5.97). Again, for each data point in this plot, the simulation was run for 5×10^6 time units, starting with an empty network. We see in this example that the approximation (5.98) is nearly exact.

The value of the workload relaxation in this example is its use in visualizing the impact of policy decisions. Almost any statistic of interest can be computed numerically for the full network model for a network with only three buffers and two stations. In more complex networks, nothing is explicitly computable for the full network model. In such cases we can again construct a workload relaxation to obtain sensitivity and performance estimates that will serve as a guide to policy selection, or design issues such as the choice of total buffer storage capacity.

5.7 Notes

Many of the concepts in this chapter, such as the *effective cost* and *state space collapse* are borrowed from the heavy-traffic literature on stochastic networks. The central idea is to construct a parameterized family of networks with increasing network load, as constructed for the single server queue in Section 3.2.2. Under appropriate conditions on the network and policy, the n-dimensional process $\{\widehat{\Xi}Q(t) : t \geq 0\}$ is approximated by a version of the CBM model.

Soon after the seminal work of Kingman [301, 302], heavy-traffic theory was developed by Iglehart and Whitt [276, 277, 495], Foschini and Salz [190], Harrison, Reiman, Williams, and others [229, 240, 407, 231, 244, 232, 327]. The network shown in Fig. 4.15 was introduced in [131] to show that a heavy-traffic limit may not exist in a network setting.

See Kimura's survey [300] for the state of the art in 1993. Harrison [236, 237] and Bramson and Williams [84] contain more up-to-date bibliographies, as well as [81, 503, 323, 498] and the monographs by Kleinrock [303], Asmussen [20], and Chen and Yao [96].

The construction of workload in units of time described in Section 5.2.2 is adapted from the classical construction in the M/G/1 queue [114].

The viewpoint of this chapter is similar to [292] where no attempt is made to justify a heavy-traffic limit. Rather, a candidate CBM model is considered so that "important features of good control policies are displayed in sharpest relief." The CRW workload model, like the CBM model, allows independent modeling of variability and drift. An advantage of the discrete-time workload model is that solutions always exist, and performance evaluation is straightforward through simulation.

The fact that one obtains a model of reduced dimension as a relaxation without constructing a Brownian model was first applied in Neil Laws' thesis [333]. Laws introduces a one-dimensional relaxation of a stochastic network to obtain a performance bound similar to Proposition 5.4.3. In this thesis and [292] the authors also consider the CBM model with applications to policy synthesis. Extensions are contained in [242, 205, 332, 491, 389, 393], and surveyed in [291].

The multidimensional workload relaxation for the fluid model was introduced in [361]. Multidimensional relaxations for the stochastic CRW model are considered in [103, 256, 362, 363], and an application to electric power networks is contained in [101]. Affine policies were proposed in [357, p. 194].

There are parallels between workload relaxations and the theory of two time-scale Markov processes, as described in books by Ethier and Kurtz [167] and Yin and Zhang [507]. Consider a family of Markov chains parameterized by $\varepsilon \in [0, 1]$. For each parameter the Markov chain is defined via a transition matrix of the form $P^\varepsilon = P^0 + \varepsilon\Delta$, where P^0 is itself a transition matrix, and the matrix Δ satisfies $\Delta 1 = 0$. The setting is analogous to the parameterization of a network by load, where P^0 corresponds to a network satisfying $\rho_\bullet = 1$. The concepts developed in [507] are themselves an extension of the singular perturbation technique for model reduction in deterministic dynamical systems, as described in Kokotovic et al. [307].

Also, similar to a workload relaxation is the use of *coarse variables* to approximate the dynamics of molecular structures. The origins of this approach lie in the work of Mori and Zwanzig [375, 509], which remains an important tool for model reduction in statistical mechanics [108, 295]. A version of this technique is summarized in Proposition A.1.1 in the appendix. Recently, attempts have been made to extend these techniques for application to the type of production systems considered in Chapter 7 – see [508].

The Skorokhod map is introduced in an analysis of Jackson networks in Harrison and Reiman [240], and applied to prove heavy-traffic limit theorems in Reiman [407]. The origin is of course Skorokhod [451]. The minimal process on R is the subject of the thesis of P. Yang [506]. Proposition 5.3.9 (ii) is given in [506, Lemma 4.4]. Theorem 5.3.21 is taken from [363]. Proposition 5.4.7 is a minor extension of [361, Theorem 3.10]. Far deeper and more general results can be found in work of Soner and Shreve [454] and Dupuis and Ramanan [158]. See also [156, 157, 22] for more results and more history.

The approaches to pathwise optimality contained in Section 5.5, and in particular the proof of optimality of the c–μ rule, are based on *conservation laws*. Little's law is another conservation law, and related invariance equations abound in the steady-state analysis of networks. See Kleinrock [304] for Little's law and its generalizations.

The seminal work of Coffman and Mitrani [112] extends the conservation law for a single class model to multi-class/single-station models by characterizing the set of achievable performance vectors. Such a characterization provides a proof of optimality of the c–μ rule in Federgruen and Groenevelt [172]. Tsoucas and Walrand [475], Sigman [449], and Green and Stidham [220] have shown that similar conservation laws hold pathwise in some models.

The structure of the achievable region in network models is typically defined by *polymatroid constraints*, as described in the book of Chen and Yao [96]. Related work and extensions are contained in Shanthikumar, J. G. and Sumita [440], Shanthikumar and Yao [441, 442], Dacre, Glazebrook and Niño-Mora [123], and Bertsimas and Niño-Mora [53]. Textbook treatments are contained in Gelenbe and Mitrani [200, Chapter 6], Heyman and Sobel [263, Chapter 8], Glasserman and Yao [208], Baccelli and Brémaud [29, Chapter 3], as well as Chen and Yao [96]. There are also fascinating parallel developments in information theory for multiple access communication networks [470].

Exercises

5.1   Compute the effective cost for three queues in tandem, with cost $c(x) = |x|$.

5.2   Figure 4.15 shows a re-entrant line with two stations. Assume that the model is homogeneous, so that $\mu_1 = \mu_2 = \mu_5$ and $\mu_3 = \mu_4$. Compute the following based on the cost function $c(x) = |x| = x_1 + \cdots + x_5$:

(a) the effective cost for each one-dimensional relaxation;

(b) the effective cost for each one-dimensional relaxation with the buffer constraints

$$q_i \leq 1 \text{ for each } i = 1, \ldots, 5.$$

5.3   The re-entrant line shown below is homogeneous, so that $\mu_1 = \mu_3 = \mu_5$ and $\mu_2 = \mu_4$.

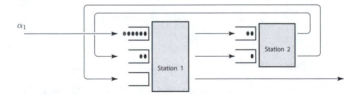

(a) Find the two workload vectors.

(b) Find the effective cost for a two-dimensional relaxation, with $c(x) = |x| = x_1 + \cdots + x_5$. Note that $\bar{c}$ is expressed more elegantly in terms of the workload in units of inventory, $\bar{c}(w) = \bar{c}_Y(y)$, with $y_i = \mu_i w_i$.

(c) How can your result be generalized to more complex re-entrant lines consisting of two stations?

5.4   Consider the criss-cross network shown in Fig. 4.18. The fluid model is three-dimensional with $\alpha = (\alpha_1, 0, \alpha_3)^\mathsf{T}$, and constituency matrix defined implicitly in (E4.1). It is assumed that the model is homogeneous, so that $\mu_1 = \mu_3$.

(a) Obtain the effective cost for the two-dimensional workload relaxation for an arbitrary linear cost function $c(x) = c_1 x_1 + c_2 x_2 + c_3 x_3$, $x \in \mathbb{R}^3_+$. Under what conditions is $\bar{c}$ monotone?

(b) Compute the infinite-horizon optimal policy for the workload model when $c_i = 1$ for each i.

(c) Repeat (ii) with the linear cost function $c_1 = 1$ and $c_2 = c_3 = 2$.

5.5   Consider again the criss-cross network. Compute the fluid value function $\widehat{J}^*$ for a workload relaxation, and based on this define

$$h_0(x) = \widehat{J}^*(\widehat{\Xi}x) + \tfrac{b}{2}\|x - \mathcal{P}^*(x)\|^2, \qquad (E5.1)$$

with $b > 0$ a constant, and $\mathcal{P}^*(x)$ the effective state (5.35). Simulate the resulting h-MaxWeight policy in a CRW model based on (4.94) for a range of b and θ (with θ defined in (4.94).)

5.6   Proposition 5.3.13 states that every buffer but one is empty in an effective state for a one-dimensional relaxation. This underlines the conflict that arises frequently in network optimization: Optimization of an idealized model dictates zero inventory at certain stations, while in a more realistic discrete/stochastic model, adopting a "zero-inventory policy" results in starvation of resources.

Suppose that $c\colon \mathsf{X} \to \mathbb{R}_+$ is quadratic, of the form $c(x) = \tfrac{1}{2}x^\mathsf{T} D x$, $x \in \mathbb{R}^\ell_+$, for a symmetric matrix D. For an appropriately chosen matrix D the effective state is far less "singular":

(a) Compute the effective state for a one-dimensional relaxation based on the first workload vector, and show that it is linear in the workload value $w \in \mathbb{R}_+$.

(b) Suppose that D^{-1} exists and has nonnegative entries. For what values of i is $x^*_i = 0$ (where $x^* = \mathcal{X}^*(w)$ with $w > 0$)?

(c) If D is diagonal, argue that the station is not starved of work:

$$w > 0 \Longrightarrow \sum_{i=1}^{\ell} C_{1i} x^*_i > 0.$$

(d) Consider now a general relaxation. Prove the following proposition:

Proposition 5.7.1. *Suppose that $D > 0$ is diagonal, and let $\bar{c}\colon \mathsf{W} \to \mathbb{R}_+$ denote the effective cost for the n-dimensional relaxation. Then the monotone region is given by*

$$\mathsf{W}^+ = \{\widehat{\Xi} D^{-1}\widehat{\Xi}^\mathsf{T}\psi : \psi \in \mathbb{R}^n_+\}.$$

For any given $w \in \mathsf{W}^+$, the effective state is given by

$$\mathcal{X}^*(w) = D^{-1}\widehat{\Xi}^\mathsf{T}\psi,$$

where ψ is the vector obtained in the representation $w = \widehat{\Xi} D^{-1}\widehat{\Xi}^\mathsf{T}\psi$.

6

Routing and Resource Pooling

In the preceding chapters we learned that resources in a network must be shared among different classes of customers. This sharing is performed through scheduling at each station in the network.

Scheduling is just one of many decision processes encountered in typical applications. In the Internet there are many paths between nodes, and hence protocols must be constructed to determine appropriate routes. In a power distribution system there may be many generators that can meet current demands distributed across the power grid. A manufacturing system may have redundant processing equipment, or multiple vendors, and this then leads to a network somewhat more complex than those considered in the previous two chapters.

Figure 6.1 shows a network with eight nodes, four arrival streams, and ten links. The high congestion between nodes 1 and 3 can be modeled through an additional linear constraint on the rate vector ζ in a fluid model. There are many routes from node 1 to node 8, even though node 4 is temporarily unavailable. This example demonstrates that there may be many equilibria in a routing model; the best route for a given user will depend upon the current environment.

Most of the concepts introduced in previous chapters, such as stabilizability and workload relaxations, will be extended to this more general setting. Consideration is largely restricted to a fluid model since the definitions are most transparent in this setting.

The state process q that defines the evolution of the vector of queue lengths in the fluid model again evolves on a convex, polyhedral state space denoted X, with $\mathsf{X} \subseteq \mathbb{R}_+^\ell$. As in (4.24), the state space is used to model hard constraints, such as finite buffers. The cumulative allocation process z evolves on $\mathbb{R}_+^{\ell_u}$ for some integer $\ell_u \geq 1$. The dynamics of the network are captured by the relationship between these two deterministic processes. These dynamics are again assumed linear, of the form

$$q(t) = x + Bz(t) + \alpha t, \qquad t \geq 0, \ x \in \mathbb{R}_+^\ell. \tag{6.1}$$

When $t = 0$ we have $q(0) = x \in \mathbb{R}_+^\ell$, so that x is the initial condition for the fluid model.

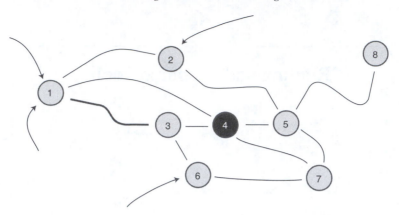

Figure 6.1. Network routing.

Each index $i \in \{1, \ldots, \ell_u\}$ is viewed as a particular *activity*. Activities may include a combination of scheduling of jobs at a particular station, and routing those jobs to other stations once service is completed.

The cumulative allocation process z indicates how much time has been devoted to each activity. Constraints on the allocation process arise from link constraints, processing constraints at nodes, and possibly also contention for service among customer classes. All of these constraints are assumed to be captured by the following set of linear inequality constraints:

(i) For each $i \in \{1, \ldots, \ell_u\}$ the deterministic process $z_i(t)$ is nondecreasing, with $z_i(0) = 0$. This expresses the constraint that the cumulative time that an activity is performed can never decrease.

(ii) There is an $\ell_m \times \ell_u$ *constituency matrix* C such that

$$C[z(t_1) - z(t_0)] \le (t_1 - t_0)\mathbf{1}, \qquad 0 \le t_0 \le t_1 < \infty, \qquad (6.2)$$

where $\mathbf{1}$ denotes a vector of ones. It is assumed that C has binary entries.

The ℓ_u-dimensional process z is called *admissible* if it satisfies these constraints, and in addition q does not leave the state space X. The vector of *allocation rates* is defined as the right derivative, $\zeta(t) = \frac{d^+}{dt} z(t)$, which is constrained to lie in the polyhedron U defined in (4.3) by $\mathsf{U} := \{u \in \mathbb{R}^{\ell_m} : u \ge 0,\ Cu \le \mathbf{1}\}$.

Each of the ℓ_m rows of C is regarded as a *resource*. We identify "resource r" with the set of corresponding activities:

$$\mathcal{I}_r := \{j \in \{1, \ldots, \ell_u : C_{rj} = 1\}. \qquad (6.3)$$

The definitions of *stabilizability*, and the *minimal draining times* T^* and W^* are stated exactly as in Definition 4.2.2. In particular, the minimal draining time for the arrival-free model is the solution to the linear program

$$W^*(x) = \min \quad T$$

$$\begin{aligned}
\text{s.t.} \quad Bz + x &= 0 \\
Cz &\leq T\mathbf{1} \\
z &\geq 0.
\end{aligned} \tag{6.4}$$

It is assumed throughout this chapter that the arrival-free model is stabilizable, so that $W^*(x) < \infty$ for $x \in \mathbb{R}_+^\ell$.

Also, as in the scheduling problem, stabilizability is characterized by a formulation of *network load*. This is formally defined as the solution to a linear program very similar to (6.4):

Definition 6.0.1 (Network load). The *network load* $\rho_\bullet$ is the value of the linear program

$$\min \quad \varrho$$

$$\begin{aligned}
\text{s.t.} \quad B\zeta + \alpha &= 0 \\
C\zeta &\leq \varrho\mathbf{1} \\
\zeta &\geq 0.
\end{aligned} \tag{6.5}$$

A vector $\zeta^e \in \mathbb{R}^{\ell_u}$ is called an *equilibrium* if it is feasible for the linear program (6.5):

$$B\zeta^e + \alpha = 0, \quad and \quad \zeta^e \in \mathsf{U}. \tag{6.6}$$

■

The idea behind Definition 6.0.1 is that we consider all allocation rates ζ^e that provide an equilibrium, and choose among these the one that has minimal overall impact on the system: For any equilibrium vector ζ^e, and any resource $r \in \{1, \ldots, \ell_m\}$, the value $(C\zeta^e)_r$ is the proportion of time that this resource will be busy in equilibrium. The worst case is given as the maximum

$$\varrho(\zeta^e) := \max_{r=1,\ldots,\ell_m} (C\zeta^e)_r,$$

and the network load as defined in (6.5) is $\rho_\bullet = \min \varrho(\zeta^e)$, where the minimum is over all equilibrium allocation rates.

The next section contains characterizations of load and workload in this general setting. In Section 6.3 we consider three very different approaches to choose a "good" equilibrium vector ζ^e when $\rho_\bullet < 1$:

(i) Section 6.3.1 develops the maximal-flow approach, leading to the *Min-Cut Max-Flow Theorem*.
(ii) Alternatively, on assigning linear cost on link utilization, an equilibrium is obtained via optimization. This approach leads to an equilibrium corresponding to a single optimal path from source to destination. This path can be computed using a recursive algorithm known as the *Bellman–Ford algorithm*. These ideas are explored in Section 6.3.2.

(iii) Section 6.4 contains a version of the MaxWeight algorithm to address routing in a distributed, multiuser setting.

In this chapter we come across phenomena that could not have arisen in the simpler scheduling model. A fundamental concept in routing models is *resource pooling*. An example is the way links in a communication system cooperate to send information across the network. In examples we find that resource pooling can reduce complexity dramatically. When several resources pool to form a *single bottleneck* we find that a one-dimensional relaxation can be justified for policy synthesis and performance approximation.

In communication, computer, and transportation systems we frequently see far less favorable behavior known as *simultaneous resource possession* in which a single "customer" can tie up several resources simultaneously. Some examples of simultaneous resource possession are treated in Section 6.5.

Section 6.6 concerns workload relaxations and related concepts for the fluid model. Here we improve the motivation for this approach by establishing results that bound the gap between the two models. A simple translation of a policy obtained from the relaxation is approximately optimal in heavy traffic.

This brings us to a frontier in the stochastic networks area. Bounds on the performance gap between a CRW model and its optimized relaxation are also possible, but these results are highly technical. In Section 6.7 we provide examples to illustrate the bounds that can be expected for stochastic models.

Section 6.7 also contains a few words on policy translation. In particular, we revisit MaxWeight and myopic policies from the point of view of workload, and show how to introduce hedging to reduce idling. However, there is not much more to say in this chapter since the translation techniques introduced in the previous two chapters carry over to this more general setting with few changes.

6.1 Workload in general models

Here we consider notions of workload for the general network model (6.1). The definition of workload vectors is more subtle in this general setting in which the matrix B is not square. In particular, we cannot define the workload matrix through a matrix inversion as in Definition 4.2.3. However, a complete generalization of the development of Section 4.2 is possible by re-examining the minimal draining time.

The main results of this section are summarized in Theorem 6.1.1. The representation (6.7) implies that the workload vectors can be interpreted as Lagrange multipliers since they define sensitivity of the minimal draining time with respect to the initial condition x.

The representation $\rho_\bullet = W^*(\alpha)$ provides the following interpretation of the network load: The vector α is equal to the total amount of exogenous arrivals in 1 sec. Consequently, the network load is equal to the amount of time required to process these arrivals, given that there are no additional arrivals to the network.

Theorem 6.1.1 (Geometric construction of workload). *Suppose that the arrival-free model is stabilizable. Then, there are vectors $\{\xi^s : 1 \leq s \leq \ell_r\} \subset \mathbb{R}^\ell$ such that the following hold for the model (6.1):*

(i) *The minimal draining time for the arrival-free model defined in (6.4) can be expressed as the maximum,*

$$W^*(x) = \max_{1 \leq s \leq \ell_r} \langle \xi^s, x \rangle, \qquad x \in \mathbb{R}^\ell_+. \tag{6.7}$$

(ii) *The network load defined in Definition 6.0.1 can be expressed*

$$\rho_\bullet = W^*(\alpha) = \max_{1 \leq s \leq \ell_r} \rho_s, \qquad \alpha \in \mathbb{R}^\ell_+,$$

where $\rho_s = \langle \xi^s, \alpha \rangle$.

(iii) *The model (6.1) is stabilizable if, and only if, $\rho_\bullet < 1$.*

(iv) *If the network is stabilizable then the minimal draining time can be expressed*

$$T^*(x) = \max_{1 \leq s \leq \ell_r} \frac{\langle \xi^s, x \rangle}{1 - \rho_s} < \infty, \qquad x \in \mathbb{R}^\ell_+. \tag{6.8}$$

Proof. Parts (i), (iii), and (iv) are contained in Proposition 6.1.5. The vectors $\{\xi^s\}$ are defined in Definition 6.1.2 below.

To see (ii) consider the linear program (6.5) that defines the network load. This is precisely (6.4) when $x = \alpha$, and hence (ii) follows from (i). $\qquad\square$

We consider both geometric and algebraic (linear programming) interpretations of workload vectors since both approaches have value for intuition, computation, and improving the theory.

6.1.0.4 Workload and the velocity space

The proof of Theorem 6.1.1 requires a closer look at the geometry of the velocity space $\mathsf{V} \subset \mathbb{R}^\ell$, defined by

$$\mathsf{V} := \{B\zeta + \alpha : \zeta \in \mathsf{U}\}. \tag{6.9}$$

Letting V_0 denote the velocity space for the arrival-free model,

$$\mathsf{V}_0 := \{v = B\zeta : \zeta \in \mathsf{U}\}, \tag{6.10}$$

one may conclude from the definition (6.9) that the set V is expressed as the translation, $\mathsf{V} = \{\mathsf{V}_0 + \alpha\} := \{v + \alpha : v \in \mathsf{V}_0\}$.

The polyhedron V_0 contains the origin in $\mathbb{R}^\ell$ since by definition we have $0 \in \mathsf{U}$. It follows that this set can be expressed as the intersection of half-spaces,

$$\mathsf{V}_0 = \{v \in \mathbb{R}^\ell : \langle \xi^s, v \rangle \geq -o_s, \qquad 1 \leq s \leq \ell_v\}, \tag{6.11}$$

where the constants $\{o_s : 1 \leq s \leq \ell_v\}$ take on values zero or one, and $\xi^s \in \mathbb{R}^\ell$ for each s. We assume that the vectors $\{\xi^s\}$ are minimally specified, in the sense that V_0 cannot be represented via (6.11) using a subset of the vectors $\{\xi^s\}$.

The set V_0 has nonempty interior since the arrival-free model is stabilizable: This follows from Proposition 6.1.4 with $\alpha = 0$, which implies that $\{v \in \mathbb{R}^\ell : -\varepsilon < v_i < 0\} \subset V_0$ for some $\varepsilon > 0$. Hence the set $\{v \in V_0 : \langle \xi^s, v \rangle = -o_s\}$ is an $(\ell - 1)$-dimensional face of the polyhedron V_0. This face passes through the origin if, and only if, $o_s = 0$.

Given this structure, we arrive at a geometric construction of workload vectors:

Definition 6.1.2 (Workload in units of time). Suppose that the arrival-free model is stabilizable. Then:

(i) The vector $\xi^s \in \mathbb{R}^\ell$ is called a *workload vector* if $o_s = 1$. The number of distinct workload vectors is denoted ℓ_r. By reordering, we assume that $o_s = 1$ if and only if $s \leq \ell_r$.

(ii) The ℓ_r-dimensional *vector load* is given by $\rho = (\rho_1, \ldots, \rho_{\ell_r})^\mathsf{T}$, where $\rho_s := \langle \xi^s, \alpha \rangle$ for $s \leq \ell_v$. ∎

The load parameters $\{\rho_s\}$ can be used to represent the velocity space V as follows:

Proposition 6.1.3. *The velocity space can be expressed*

$$V = \{v \in \mathbb{R}^\ell : \langle \xi^s, v \rangle \geq -(o_s - \rho_s),\ s = 1, \ldots, \ell_v\}.$$

Proof. We begin with the representation of the velocity space as a translation $V = \{V_0 + \alpha\}$. Hence, from (6.11), on writing $v' = v + \alpha$,

$$\begin{aligned}
V &= \{v + \alpha \in \mathbb{R}^\ell : \langle \xi^s, v \rangle \geq -o_s,\ s = 1, \ldots, \ell_v\} \\
&= \{v' \in \mathbb{R}^\ell : \langle \xi^s, v' - \alpha \rangle \geq -o_s,\ s = 1, \ldots, \ell_v\} \qquad (6.12) \\
&= \{v' \in \mathbb{R}^\ell : \langle \xi^s, v' \rangle \geq -o_s + \rho_s,\ s = 1, \ldots, \ell_v\}.
\end{aligned}$$
□

A geometric construction of the minimal draining time is possible based on consideration of the translation $\{V + x^0\}$ for a given initial condition $x^0 \in X$. If $x^1 \in X \cap \{V + x^0\}$ then $v = x^1 - x^0 \in V$, and it follows that the following trajectory is feasible, and travels from x^0 to x^1 in exactly 1 sec,

$$q(t) = x^0 + tv, \qquad 0 \leq t \leq 1.$$

This shows that $X \cap \{V + x^0\}$ is precisely the set of states reachable from x^0 in 1 sec or less. Similarly, on scaling V by a real number $T > 0$ we find that $X \cap \{TV + x^0\}$ is identified as the set of states reachable from x^0 in no more than T sec. This geometry is illustrated for a two-dimensional example in Fig. 6.2. From the initial condition $x^0 \in \mathbb{R}^2_+$ shown in the figure, the minimal draining time is $T^*(x^0) = 3$, and is expressed in terms of the first workload vector as follows:

$$T^*(x^0) = \min\{T : 0 \in \{TV + x^0\}\} = \max_{s=1,\ldots,6} \frac{\langle \xi^s, x^0 \rangle}{o_s - \rho_s} = \frac{\langle \xi^1, x^0 \rangle}{1 - \rho_1}.$$

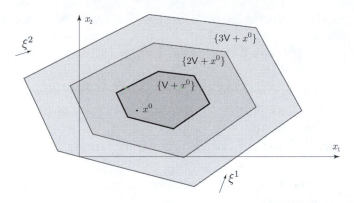

Figure 6.2. The minimal draining time T^* is piecewise linear since the velocity space V is a polyhedron. The minimal draining time from the initial state x^0 shown is $T^*(x^0) = 3$.

We thus arrive at the following representation for the time required to travel between two states:

Proposition 6.1.4. *Suppose that the arrival-free model is stabilizable, and that $0 \in V$. Then the following are equivalent for two states $x^1, x^2 \in X$ with $x^1 \neq x^2$, and $T > 0$:*

(i) *There exists an admissible allocation z such that $q(T) = x^2$ when $q(0) = x^1$.*
(ii) $T^{-1}(x^2 - x^1) \in V$.

Moreover, if either of these equivalent conditions hold, then the minimal time to reach x^2 from x^1 is finite, and given by

$$T^*(x^1, x^2) = \min\{T > 0 : T^{-1}(x^2 - x^1) \in V\} = \max_{s=1,\dots,\ell_v} \frac{\langle \xi^s, x^1 - x^2 \rangle}{o_s - \rho_s}. \quad (6.13)$$

Proof. Suppose that the conditions of (i) hold: q is a feasible trajectory starting from x^1 and satisfying $q(T) = x^2$. On denoting $v(t) := \frac{d^+}{dt} q(t) \in V$, $0 \leq t \leq T$, we obtain by convexity of V,

$$T^{-1}(q(T) - q(0)) = T^{-1} \int_0^T v(t)\, dt \in V.$$

That is, $v := T^{-1}(x^2 - x^1) \in V$.

Conversely, if $v := T^{-1}(x^2 - x^1) \in V$ then the trajectory

$$q(t) = x^1 + tv, \qquad 0 \leq t \leq T,$$

evidently reaches x^2 in precisely T sec. This proves the desired equivalence, and shows that $T^*(x^1, x^2) = \min\{T > 0 : T^{-1}(x^2 - x^1) \in V\}$.

The representation (6.13) is established as follows. First, note that since $0 \in V$ by assumption, it follows from Proposition 6.1.3 that $0 \geq -(o_s - \rho_s)$, or

$$o_s \geq \rho_s, \quad s = 1, \dots, \ell_v. \quad (6.14)$$

Consequently, again by Proposition 6.1.3, we have the following implications:

$$T^{-1}(x^2 - x^1) \in \mathsf{V} \iff T^{-1}\langle \xi^s, x^2 - x^1 \rangle \geq -(o_s - \rho_s), \quad s = 1, \ldots, \ell_v$$
$$\iff \langle \xi^s, x^1 - x^2 \rangle \leq (o_s - \rho_s)T, \quad s = 1, \ldots, \ell_v$$
$$\iff \frac{\langle \xi^s, x^1 - x^2 \rangle}{o_s - \rho_s} \leq T, \quad s = 1, \ldots, \ell_v,$$

where the division in the last equivalence is justified by (6.14).   □

Based on Proposition 6.1.4 we obtain representations for the minimal draining times in analogy with Proposition 4.2.5.

Proposition 6.1.5. *Suppose that the arrival-free model is stabilizable. Then the following hold for the model (6.1):*

(i) *For the arrival-free model, the minimal draining time is given by*

$$W^*(x) = \max_{1 \leq s \leq \ell_r} \langle \xi^s, x \rangle, \quad x \in \mathbb{R}_+^\ell.$$

(ii) *For the model with arrivals, provided $\rho_\bullet < 1$, the minimal draining time is finite, and can be expressed*

$$T^*(x) = \max_{1 \leq s \leq \ell_r} \frac{\langle \xi^s, x \rangle}{1 - \rho_s}, \quad x \in \mathbb{R}_+^\ell.$$

Proof. Consider first the arrival-free model in which $\alpha = 0$. Proposition 6.1.4 continues to hold in this situation, and setting $x^2 = 0$ gives

$$W^*(x) = \max_{1 \leq s \leq \ell_v} \frac{1}{o_s} \langle \xi^s, x \rangle, \quad x \in \mathbb{R}_+^\ell.$$

Recall that $o_s = 0$ for $s > \ell_r$. If for some such s we have $\langle \xi^s, x \rangle > 0$, we conclude that $W^*(x)$ cannot be finite. Since by assumption $W^*(x) < \infty$ for each $x \in \mathbb{R}_+^\ell$, it must follow that

$$\xi_j^s \leq 0 \text{ for each } s > \ell_r \text{ and each } 1 \leq j \leq \ell. \tag{6.15}$$

We thus arrive at the desired expression for $W^*(x)$ in (i).

The proof of Part (ii) is similar to (i): Proposition 6.1.4 gives

$$T^*(x) = \max_{1 \leq s \leq \ell_v} \frac{\langle \xi^s, x \rangle}{o_s - \rho_s}, \quad x \in \mathbb{R}_+^\ell.$$

The ratio $\langle \xi^s, x \rangle / (o_s - \rho_s)$ is nonpositive for $s > \ell_r$: (6.15) implies that $\langle \xi^s, x \rangle \leq 0$ and $o_s - \rho_s = |\rho_s|$ for such s. Hence the maximum can be restricted to $s \leq \ell_r$ as claimed.   □

6.1.0.5 Workload and linear programs

The minimal draining time $T^*(x)$ can be expressed as the value of a linear program analogous to (6.4). Combining this observation with Proposition 6.1.5 we obtain an alternate representation of workload.

Consider the computation of the minimal time to reach one state from another. For any two states $x^1, x^2 \in \mathsf{X}$, a feasible allocation process z taking q from x^1 to x^2 in T sec must satisfy $q(T) = x^1 + Bz + \alpha T = x^2$ and $T^{-1}z \in \mathsf{U}$. These constraints on z and T can be expressed as linear constraints, and we thereby obtain the linear program,

$$
\begin{aligned}
T^*(x^1, x^2) \;=\; \mathbf{min} \quad & T \\
\mathbf{s.t.} \quad x^1 + Bz + \alpha T \;&=\; x^2 \\
Cz \;&\leq\; \mathbf{1}T \\
z \;&\geq\; \mathbf{0}.
\end{aligned}
\tag{6.16}
$$

To construct a dual of (6.16) we express the constraints in matrix form,

$$
\begin{bmatrix} -B & -\alpha \\ -C & \mathbf{1} \end{bmatrix} \begin{bmatrix} z \\ T \end{bmatrix} \begin{array}{c} = \\ \geq \end{array} \begin{array}{c} x^1 - x^2 \\ \mathbf{0} \end{array}, \qquad z \geq \mathbf{0}, \; T \geq 0.
$$

We define two dual variables, denoted $\xi \in \mathbb{R}^\ell$ and $\nu \in \mathbb{R}_+^{\ell_m}$. The vector ξ is not sign constrained since it corresponds to the equality constraint above, and ν is nonnegative since it corresponds to the inequality constraint $Cz \leq \mathbf{1}T$. The dual of (6.16) takes the form

$$
\begin{aligned}
T^*(x^1, x^2) \;=\; \mathbf{max} \quad & \langle \xi, x^1 - x^2 \rangle \\
\mathbf{s.t.} \quad -B^\mathsf{T}\xi - C^\mathsf{T}\nu \;&\leq\; 0 \\
-\alpha^\mathsf{T}\xi + \mathbf{1}^\mathsf{T}\nu \;&\leq\; 1 \\
\nu \;&\geq\; \mathbf{0}.
\end{aligned}
\tag{6.17}
$$

The primal linear program (6.16) and its dual (6.17) have the same value, provided a bounded solution to (6.16) exists (see Theorem 1.3.2.)

Equation (6.17) can be applied to compute $W^*(x)$, which is defined to be $T^*(x^1, x^2)$ with $x^1 = x$, $x^2 = \mathbf{0}$, and $\alpha = \mathbf{0}$. In this special case the linear program (6.17) becomes

$$
\begin{aligned}
W^*(x) \;=\; \mathbf{max} \quad & \langle \xi, x \rangle \\
\mathbf{s.t.} \quad -B^\mathsf{T}\xi - C^\mathsf{T}\nu \;&\leq\; 0 \\
\mathbf{1}^\mathsf{T}\nu \;&\leq\; 1 \\
\nu \;&\geq\; \mathbf{0}.
\end{aligned}
\tag{6.18}
$$

Hence, provided the origin is reachable from the initial condition x, the minimal draining time can be expressed $W^*(x) = \max\langle \xi, x \rangle$, where the maximum is over all pairs (ξ, ν) satisfying the constraints in (6.18). This maximum can be taken over extreme points in the constraint region (6.18).

Recall that a workload vector ξ^s defines a face of V_0 that does not pass through the origin. Similarly, in Proposition 6.1.6 we see that $(\xi, \mathbf{0})$ is never an extreme point if ξ is nonzero.

Proposition 6.1.6. *Suppose that the arrival-free model is stabilizable. Then, the nonzero extreme points of the constraint set in (6.18) are of the form (ξ, ν) with $\langle \nu, 1 \rangle = 1$.*

Proof. We first establish that $1^{\mathsf{T}}\nu = 0$ or 1 if (ξ, ν) is extremal. This follows from the form of the constraint region (6.18): If $1^{\mathsf{T}}\nu \in (0, 1)$ then $(r\xi, r\nu)$ will be feasible for r in a neighborhood of 1.

Suppose now that $1^{\mathsf{T}}\nu = 0$. It follows that $\nu = 0$ since $\nu \geq 0$, and the constraint on ξ reduces to $-B^{\mathsf{T}}\xi \leq 0$. If $\xi \neq 0$ then $(r\xi, 0)$ is feasible for each $r > 0$ when $(\xi, 0)$ is feasible. This shows that $(\xi, 0)$ cannot be extremal if $\xi \neq 0$. $\square$

All of these results point to a relationship between extreme points of (6.18), and the workload vectors $\{\xi^s\}$ defined previously. The question we must answer is, *If ξ^s is a workload vector, can it be extended to form a feasible solution to (6.18)?* Proposition 6.1.7 provides an affirmative answer.

Proposition 6.1.7. *If ξ^s is a workload vector, then there exists $\nu^s \in \mathbb{R}_+^{\ell_m}$ such that (ξ^s, ν^s) is feasible for (6.18), with $\langle \nu^s, 1 \rangle = 1$.*

Proof. If ξ^s is a workload vector then $\langle \xi^s, v \rangle \geq -1$ for each $v \in \mathsf{V}_0$, and this lower bound is attained at some v^s. From the definition of V_0 we have $v^s = B\zeta^s$ for some $\zeta^s \in \mathsf{U}$. These observations can be expressed in a linear program,

$$1 = \mathbf{max} \; \left\{ \langle -B^{\mathsf{T}}\xi^s, \zeta \rangle \quad \mathbf{s.t.} \; C\zeta \leq 1, z \geq 0 \right\},$$

where the inequality $\langle \xi^s, v \rangle \geq -1$ has been replaced by the equivalent inequality $\langle -B^{\mathsf{T}}\xi^s, \zeta \rangle = \langle \xi^s, -B\zeta \rangle \leq 1$. Letting $\psi \in \mathbb{R}_+^{\ell_m}$ denote a Lagrange multiplier for the inequality constraint $C\zeta \leq 1$, we obtain from Theorem 1.3.2,

$$
\begin{aligned}
1 \;=\; &\mathbf{min} \quad \langle \psi, 1 \rangle \\
&\mathbf{s.t.} \qquad C^{\mathsf{T}}\psi \;\geq\; -B^{\mathsf{T}}\xi^s \\
&\qquad\qquad\;\; \psi \;\geq\; 0.
\end{aligned}
$$

Any optimizer ψ^* of this dual satisfies the desired conclusions: $\psi^* \geq 0$, $\langle \psi^*, 1 \rangle = 1$, and $C^{\mathsf{T}}\psi^* + B^{\mathsf{T}}\xi^s \geq 0$. Hence we can take $\nu^s := \psi^*$. $\square$

Proposition 6.1.7 is the basis of the construction of *pooled resources* in general network models.

6.2 Resource pooling

In the scheduling model each workload vector is associated with a station. Each workload vector may be associated with a *set of resources* in the more general model considered in this chapter.

Definition 6.2.1 (Resource pooling). Suppose that the arrival-free model is stabilizable. Fix $s \in \{1, \dots, \ell_r\}$, and let ν^s denote the dual vector constructed in Proposition 6.1.7. Then:

(i) The associated *pooled resource* is

$$\mathcal{R}_s^p := \{r \in \{1, \ldots, \ell_m\} : \nu_r^s > 0\}. \tag{6.19}$$

(ii) The pooled resource is called a *bottleneck* if $\rho_s = \rho_\bullet$. In this case, each resource $r \in \mathcal{R}_s^p$ is called a *bottleneck resource*.   ■

Proposition 6.2.2 implies that the set $\mathcal{R}_s^p$ is uniquely determined by the workload vector ξ^s, even if ν^s is not uniquely determined by ξ^s. For each $s \le \ell_r$, and any t, we interpret the inner product $\langle \xi^s, q(t; x) \rangle$ as the workload, measured in units of time, for the sth pooled-resource. Proposition 6.2.2 shows that this workload decreases at maximum rate $(1 - \rho_s)$ if and only if each associated resource cooperates to work at full capacity.

Proposition 6.2.2. *For any $1 \le s \le \ell_r$, and any $\zeta \in \mathsf{U}$, the following are equivalent:*

(i) $\langle \xi^s, B\zeta \rangle = -1$

(ii) $(C\zeta)_r = 1$ *for* $r \in \mathcal{R}_s^p$, *and* ζ *satisfies the complementary slackness condition,*

$$\zeta_r > 0 \implies [B^\mathsf{T}\xi^s + C^\mathsf{T}\nu^s]_r = 0, \quad 1 \le r \le \ell_m. \tag{6.20}$$

Proof. Suppose that (i) holds. Because (ξ^s, ν^s) is feasible for (6.18) we have the inequality $B^\mathsf{T}\xi^s + C^\mathsf{T}\nu^s \ge 0$. This combined with the assumption that $\zeta \in \mathsf{U} \subset \mathbb{R}_+^{\ell_u}$ implies the bound

$$-1 + \langle \nu^s, C\zeta \rangle = \langle \xi^s, B\zeta \rangle + \langle \nu^s, C\zeta \rangle = \langle \zeta, [B^\mathsf{T}\xi^s + C^\mathsf{T}\nu^s] \rangle \ge 0,$$

so that $\langle \nu^s, C\zeta \rangle \ge 1$. Since the reverse inequality also holds when $\zeta \in \mathsf{U}$ we must have equality:

$$-1 + \langle \nu^s, C\zeta \rangle = \langle \zeta, [B^\mathsf{T}\xi^s + C^\mathsf{T}\nu^s] \rangle = 0. \tag{6.21}$$

In fact, since ν^s is a probability distribution on $\{1, \ldots, \ell_u\}$ and $C\zeta \le 1$, the equality (6.21) implies that $(C\zeta)_r = 1$ for all $r \in \mathcal{R}_s^p$. The equation (6.21) also implies the complementary slackness condition (6.20) since $[B^\mathsf{T}\xi^s + C^\mathsf{T}\nu^s] \ge 0$, and $\zeta \ge 0$.

Conversely, if (ii) holds then the complementary slackness condition implies the identity, $\langle \xi^s, B\zeta \rangle + \langle \nu^s, C\zeta \rangle = \langle \zeta, [B^\mathsf{T}\xi^s + C^\mathsf{T}\nu^s] \rangle = 0$. This combined with the assumption in (ii) that $(C\zeta)_r = 1$ whenever $r \in \mathcal{R}_s^p$ (equivalently $\langle \nu^s, C\zeta \rangle = 1$), gives (i).   □

The following example illustrates the need for the complementary slackness condition (6.20).

Example 6.2.3 (Queue with redundant activities). The system shown in Fig. 6.3 is described by the fluid model equations

$$\frac{d^+}{dt}q(t) = -\mu_1\zeta_1(t) - \mu_2\zeta_2(t), \qquad t \ge 0.$$

It is assumed that $\zeta_1(t) + \zeta_2(t) \le 1$ for all t so that this system is of the form (6.1) with

$$B = [-\mu_1, -\mu_2], \quad C = [1, 1].$$

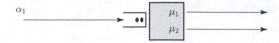

Figure 6.3. A queue with two similar activities.

Suppose that $\xi \in \mathbb{R}_+$ is a workload vector. Applying Proposition 6.1.7 we can find $\nu \in \mathbb{R}_+$ such that the pair (ξ, ν) is feasible for the LP (6.18). In this simple example feasibility means that $\nu \in [0, 1]$, and the inequality constraint is satisfied

$$\begin{pmatrix} -\mu_1 \\ -\mu_2 \end{pmatrix} \xi + \begin{pmatrix} 1 \\ 1 \end{pmatrix} \nu = B^{\mathsf{T}} \xi + C^{\mathsf{T}} \nu \geq \begin{pmatrix} 0 \\ 0 \end{pmatrix}$$

Proposition 6.1.7 also implies that we can take $\nu = 1$, so that the vector inequality above reduces to $\mu_i \xi \leq 1$ for $i = 1, 2$. Assuming that $\mu_1 > \mu_2$ it then follows that $\xi = \mu_1^{-1}$ is the unique workload vector, and the network load is given by $\rho_\bullet = \xi \alpha_1 = \alpha_1/\mu_1$.

Consider the allocation rate given by $\zeta = (0, 1)^{\mathsf{T}} \in \mathsf{U}$. This satisfies $C\zeta = 1$, yet

$$\langle \xi, B\zeta \rangle = -\mu_2/\mu_1 > -1.$$

It follows from Proposition 6.2.2 that the complementary slackness condition is violated. Indeed, (6.20) can be written for this ζ as

$$[B^{\mathsf{T}} \xi + C^{\mathsf{T}} \nu]_2 = 0,$$

which is *false* since the left-hand side is equal to $1 - \mu_2/\mu_1$.

Of course, if $\mu_1 > \mu_2$ then the allocation rate given by $\zeta^* = (1, 0)^{\mathsf{T}}$ minimizes the time to drain the queue. This satisfies $C\zeta^* = 1$ and does satisfy the complementary slackness condition (6.20). ∎

Example 6.2.4 (Simple routing model). Consider the simple routing model shown in Fig. 2.14. There are three buffers (one at the router), and four activities: processing at buffer 1 or buffer 2, as well as routing to buffer 1 or to buffer 2. The fluid model is expressed as the controlled ODE

$$\frac{d^+}{dt} q = \begin{bmatrix} -\mu_1 & 0 & \mu_r & 0 \\ 0 & -\mu_2 & 0 & \mu_r \\ 0 & 0 & -\mu_r & -\mu_r \end{bmatrix} \zeta + \begin{bmatrix} 0 \\ 0 \\ \alpha_r \end{bmatrix}. \tag{6.22}$$

The constituency matrix is given by

$$C = \begin{bmatrix} 1 & 0 & 0 & 0 \\ 0 & 1 & 0 & 0 \\ 0 & 0 & 1 & 1 \end{bmatrix}.$$

The third row of C expresses the constraint at the router, $\zeta_3 + \zeta_4 \leq 1$.

The workload vectors are constructed as the extreme points of the constraint region shown in (6.18). This can be expressed

$$
B^{\mathsf{T}}\xi + C^{\mathsf{T}}\nu = \begin{bmatrix} -\mu_1 & 0 & 0 \\ 0 & -\mu_2 & 0 \\ \mu_r & 0 & -\mu_r \\ 0 & \mu_r & -\mu_r \end{bmatrix} \xi + \begin{bmatrix} 1 & 0 & 0 \\ 0 & 1 & 0 \\ 0 & 0 & 1 \\ 0 & 0 & 1 \end{bmatrix} \nu \geq \mathbf{0},
$$

together with the additional constraints $\mathbf{1}^{\mathsf{T}}\nu \leq 1$ and $\nu \geq \mathbf{0}$. There are a total of six variables in (ξ, ν) and five constraints beyond the positivity constraints for ν.

Extreme points correspond to *basic feasible solutions* [342]. A systematic approach to computation is to first put the LP in the standard form (1.8), which requires that we expand the number of variables by writing $\xi = \xi^+ - \xi^-$ with ξ^+ and ξ^- each nonnegative vectors. We then have an LP of the form (1.8) with five constraints and *nine* variables. A basic solution is obtained on setting four of the nine variables equal to zero, setting all inequalities to equalities, and then attempting to solve the resulting system of equations. If a unique solution exists and it is feasible, then this is a basic feasible solution as well as an extreme point.

This process can be performed on a computer, with the following result. First suppose that the inequality constraints involving B and C are replaced by the equality constraint $B^{\mathsf{T}}\xi + C^{\mathsf{T}}\nu = \mathbf{0}$. This gives

$$
\begin{aligned}
\xi_1 &= \frac{\nu_1}{\mu_1} & \xi_2 &= \frac{\nu_2}{\mu_2} \\
\xi_1 - \xi_3 &= \frac{\nu_3}{\mu_r} & \xi_2 - \xi_3 &= \frac{\nu_3}{\mu_r}.
\end{aligned}
$$

The last two equalities imply $\xi_1 = \xi_2$, and the first two give $\nu_1\mu_2 = \nu_2\mu_1$. If $\nu_3 = 0$ these equations can be solved to give the extreme point

$$
\begin{aligned}
\nu^1 &= \left(\frac{\mu_1}{\mu_1 + \mu_2}, \frac{\mu_2}{\mu_1 + \mu_2}, 0 \right)^{\mathsf{T}} \\
\xi^1 &= \left(\frac{1}{\mu_1 + \mu_2}, \frac{1}{\mu_1 + \mu_2}, \frac{1}{\mu_1 + \mu_2} \right)^{\mathsf{T}}.
\end{aligned}
$$

Three other extreme points are

$$
\begin{aligned}
\nu^2 &= (1, 0, 0)^{\mathsf{T}}, & \xi^2 &= (\mu_1^{-1}, 0, 0)^{\mathsf{T}}, \\
\nu^3 &= (0, 1, 0)^{\mathsf{T}}, & \xi^3 &= (0, \mu_2^{-1}, 0)^{\mathsf{T}}, \\
\nu^4 &= (0, 0, 1)^{\mathsf{T}}, & \xi^4 &= (0, 0, \mu_r^{-1})^{\mathsf{T}}.
\end{aligned}
$$

The corresponding load parameters are

$$
\rho_1 = \frac{\alpha_r}{\mu_1 + \mu_2}, \ \rho_2 = \rho_3 = 0, \ \rho_4 = \frac{\alpha_r}{\mu_r}.
$$

Recall that the router is assumed to be *fast*. We interpret this to mean that $\mu_r > \mu_1 + \mu_2$, and that $x_3 \equiv 0$. The network load is then $\rho_\bullet = (\mu_1 + \mu_2)^{-1}\alpha_r$. Assume moreover that the model is heavily loaded, so that neither station on its own can handle

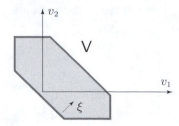

Figure 6.4. Velocity space for the simple routing model when $q_3 \equiv 0$.

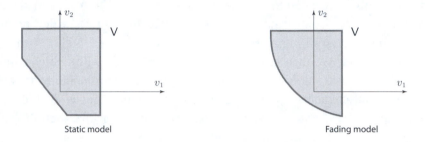

Figure 6.5. Velocity spaces in two versions of the ALOHA communication model.

the arrival rate. That is, $\max(\mu_1, \mu_2) < \alpha_r$. In this case the minimal draining time is a linear function of x,

$$T^*(x) = \frac{1}{1 - \rho_\bullet} \left(\frac{x_1 + x_2}{\mu_1 + \mu_2} \right), \qquad x \in \mathbb{R}_+^3, \ x_3 = 0. \qquad (6.23)$$

The extreme point (ξ^1, ν^1) shows resource pooling of the two downstream nodes. When (6.23) holds, then this resource pooling is interpreted as follows: a trajectory q is time optimal if and only if

$$\zeta_1(t) = 1 \quad \text{and} \quad \zeta_2(t) = 1 \qquad \text{whenever } q(t) \neq \mathbf{0}. \qquad (6.24)$$

This is illustrated in Fig. 6.4 where the workload vector ξ^1 defines a face of the velocity space V that nearly passes through the origin in $\mathbb{R}^2$.   ∎

Example 6.2.5 (ALOHA model). Consider the multiple access communications model illustrated in Fig. 2.4. Letting $\alpha \in \mathbb{R}_+^2$ denote the vector of arrival rates, the velocity space for the fluid model is expressed as the translation $V = \{-U + \alpha\}$, or

$$V = \{v \in \mathbb{R}^2 : v = \alpha - u \text{ for some } u \in U\}.$$

The velocity space is illustrated in Fig. 6.5 for each of the two cases shown in Fig. 2.5.

In the static model (without fading) the velocity space is polyhedral, as shown on the left in Fig. 6.5. The set V is very similar to that obtained in the simple routing model. In particular, when the network load is high then there is a single dominant face that nearly meets the origin in $\mathbb{R}^2$.

Although the model is not naturally described through the state space equations (6.1), we can construct workload vectors to represent both V and the minimal

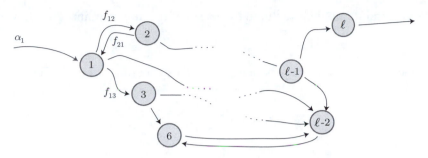

Figure 6.6. Single-user routing model.

draining time for the static model. First, the set U is polyhedral, and has the simple form

$$\mathsf{U} = \{u \in \mathbb{R}^2 : u \geq 0, \ \langle \xi^s, u \rangle \leq 1, \ s = 1, 2, 3\},$$

where the vectors $\{\xi^s : s = 1, 2, 3\}$ define the three faces of U shown in Fig. 2.5 that do not meet the origin. Consequently, the velocity space for the arrival-free model is given by

$$\mathsf{V}_0 := -\mathsf{U} = \{v \in \mathbb{R}^2 : v \leq 0, \ \langle \xi^s, v \rangle \geq -1, \ s = 1, 2, 3\}.$$

In the fading model the situation is more subtle. The velocity space is not polyhedral, as shown on the right in Fig. 6.5. This makes some physical sense, given the identification of faces of V with workload vectors, and the correspondence of workload vectors with *pooled resources*: In a wireless communication system resources may include multiple paths, multiple frequency bands, and a range of coding algorithms. Loosely speaking, the number of resources is infinite in wireless communications. ∎

6.3 Routing and workload

We now consider how to choose efficient routes for a network as complex as the subset of the Internet shown in Fig. 1.2. It is assumed that constraints are imposed only on links in the network. This is a reasonable simplification in the Internet and in wireless networks.

We begin with the single-user model illustrated in Fig. 6.6, consisting of ℓ nodes, various links between them, and a single-arrival stream. We let (i, j) denote the link from node i to node j, for any pair $i, j \in \{1, \ldots, \ell\}$, and assume that the total rate along this link is bounded by its capacity μ_{ij}.

We do not require that $\mu_{ij} = \mu_{ji}$; that is, the capacity values are *directional*. The directional flow on link (i, j) is denoted f_{ij}, and the corresponding allocation rate is defined by the ratio

$$\zeta_{ij} := \frac{f_{ij}}{\mu_{ij}}, \qquad 1 \leq i, j \leq \ell.$$

The link constraints are then captured by the set of linear constraints $\zeta \in U$, where

$$U := \{0 \le \zeta_{ij} \le 1, \quad 1 \le i, j \le \ell\}.$$

The ℓ-dimensional vector of queue lengths evolves according to the differential equation

$$\tfrac{d^+}{dt} q_i(t) = \alpha_i + \sum_{k=1}^{\ell} \zeta_{ki} \mu_{ki} - \sum_{j=1}^{\ell} \zeta_{ij} \mu_{ij}. \tag{6.25}$$

In the single-user model there is a single exogenous arrival stream. Without loss of generality we assume that this flow enters buffer 1, so that

$$\alpha = (\alpha_1, 0, \ldots, 0)^{\mathsf{T}}.$$

In this model each link is viewed as a resource, and the (i, j)th *activity* is interpreted as the process of routing packets along the corresponding link.

One goal is to obtain intuition regarding the nature of workload and network load in a link-constrained network. In view of the definition of network load given in (6.5), to obtain an expression for the network load $\rho_{\bullet}$ it suffices to restrict attention to allocation-rate vectors $\zeta^{\mathrm{e}} \in \mathbb{R}^{\ell^2}$ that define an equilibrium,

$$0 = \alpha_i + \sum_{k=1}^{\ell} \zeta^{\mathrm{e}}_{ki} \mu_{ki} - \sum_{j=1}^{\ell} \zeta^{\mathrm{e}}_{ij} \mu_{ij}, \quad 1 \le i, j \le \ell, \tag{6.26}$$

or, using the equivalent flow notation,

$$\alpha_i + \sum_{k=1}^{\ell} f_{ki} = \sum_{j=1}^{\ell} f_{ij}, \quad 1 \le i, j \le \ell,$$

where $\alpha_i := 0$ for $i \ge 2$.

In Section 6.4 we consider a general model with multiple arrivals.

6.3.1 Resource pooling and maximal flows

Here we interpret the maximal admissible arrival rate α_1^* in equilibrium as a *maximal flow*. In this way we build a bridge between the notions of workload introduced in this chapter, and the *Min-Cut Max-Flow theorem* of graph theory. This classical result implies a characterization of network load based on binding paths for a network in equilibrium.

Definition 6.3.1 (Binding link constraints). For the link-constrained network:

(i) For a given arrival rate vector $\alpha_1 \in \mathbb{R}_+^{\ell}$ and equilibrium rate vector $\zeta^{\mathrm{e}} \in U$ satisfying (6.26), the link-constraint μ_{ij} is called *binding* if $\zeta^{\mathrm{e}}_{ij} = 1$ and $\zeta^{\mathrm{e}}_{ji} = 0$. Otherwise, the constraint is *nonbinding*.

(ii) A path between two nodes i_1, i_2 is called *nonbinding* if each constraint along each link along the path is nonbinding. Otherwise, the path is called *binding*.

(iii) Let α_1^* denote the maximum over all arrival rates $\alpha_1 \geq 0$ such that $\zeta^e \in U$ exists satsifying (6.26). The associated vector of flows f is then called *maximal*. ∎

A maximal flow may have associated with it several paths that are binding, which is a particular instance of resource pooling: The resources in the networks (the links) cooperate to send material from link 1 to link ℓ at the fastest possible rate.

Calculation of a maximal flow is closely related to the calculation of network load as defined in Definition 6.0.1:

Proposition 6.3.2. *The arrival-free single-user model is stabilizable, and the network load is given by* $\rho_\bullet = \alpha_1/\alpha_1^*$.

Proof. It is obvious that $\rho_\bullet = 1$ if $\alpha_1 = \alpha_1^*$, and that $\rho_\bullet = 0$ if $\alpha_1 = 0$. The general formula given in the proposition then follows from the following homogeneity property:

$$\rho_\bullet(\kappa\alpha) = \kappa\rho_\bullet(\alpha), \qquad \kappa \geq 0.$$

Homogeneity is a consequence of the structure of the linear program (6.5) that defines the network load. □

Suppose that an inflow rate $\alpha_1 > 0$ is given, and suppose that there is a nonbinding path from node 1 to node ℓ. Then, there exists $\varepsilon > 0$ such that $f_{ij} \leq \mu_{ij} - \varepsilon$, or $f_{ji} \geq \varepsilon$ for each link $l = (i, j)$ on this path. We can then define a new flow $(\bar{\alpha}, \bar{f})$ as follows: $\bar{\alpha}_1 = \alpha_1 + \varepsilon$, and for all (i, j) on the path, set

$$\begin{aligned}
\bar{f}_{ij} &= f_{ij} + \varepsilon, \quad \bar{f}_{ji} = f_{ji} & \text{if } f_{ij} \leq \mu_{ij} - \varepsilon \\
\bar{f}_{ji} &= f_{ji} - \varepsilon, \quad \bar{f}_{ij} = f_{ij} & \text{otherwise.}
\end{aligned}$$

Thus, if there is a nonbinding path from node 1 to node ℓ, then α_1 is not maximal.

To find a maximal flow we formulate a test to see if a nonbinding path exists.

Definition 6.3.3 (Cuts and flows). For the single-user network

(i) A *cut* is any set $A \subset \{1, 2, \ldots, \ell\}$, satisfying $1 \in A$ and $\ell \in A^c$.
(ii) The (maximal) *flow from A to A^c* is the sum

$$\mu(A, A^c) = \sum \{\mu_{ij} : i \in A, j \in A^c\}.$$
∎

Theorem 6.3.4 (Min-Cut Max-Flow theorem). *The maximal flow* α_1^* *is equal to*

$$\min \mu(A, A^c),$$

where the minimum is over all cuts $A \subset \{1, 2, \ldots, \ell - 1\}$.

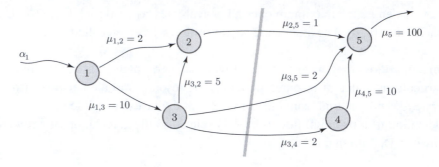

Figure 6.7. Single-user routing model with min-cut $A = \{1, 2, 3\}$.

Proof. Suppose that f is a maximal flow. Since no material is lost in an equilibrium, we must have $\mu(A, A^c) \geq \alpha_1^*$ for any cut A, which implies

$$\min \mu(A, A^c) \geq \alpha_1^*.$$

To prove the theorem we construct a cut A that achieves this lower bound based on a flow f^* that achieves the maximum flow α_1^*.

The set $A \subset \{1, \ldots, \ell\}$ is defined to be the set of all $1 \leq j \leq \ell$ such that there exists a nonbinding path from node 1 to node j for the flow f^*. It is assumed that $1 \in A$, and we must have $\ell \in A^c$ since the flow f^* is assumed to be maximal. Any link (i, j) with $i \in A$, $j \in A^c$ must be binding by definition. Hence, for such links, $f_{ij}^* = \mu_{ij}$ and $f_{ji}^* = 0$. The result then follows.   □

The theorem is a foundation for constructing fast algorithms to compute efficient routes as well as network load.

Example 6.3.5 (Simple link-constrained model). Consider the single-user routing model shown in Fig. 6.7. The capacities on each of seven links are as indicated. The links are unidirectional so, for example, we have $\mu_{5,4} = 0$. The imposition of the rate $\mu_5 = 100$ at node 5 is for the purposes of constructing a linear fluid model of the form (6.1).

Listed below are four candidate cuts for this model, and the corresponding maximal flow from A to A^c:

$$
\begin{aligned}
A &= \{1, 2\} & \mu(A, A^c) &= \mu_{1,3} + \mu_{2,5} = 11 \\
A &= \{1, 3\} & \mu(A, A^c) &= \mu_{1,2} + \mu_{3,2} + \mu_{3,4} + \mu_{3,5} = 11 \\
A &= \{1, 2, 3\} & \mu(A, A^c) &= \mu_{2,5} + \mu_{3,4} + \mu_{3,5} = 5 \\
A &= \{1, 3, 4\} & \mu(A, A^c) &= \mu_{1,2} + \mu_{3,2} + \mu_{3,5} + \mu_{4,5} = 19.
\end{aligned}
$$

There are of course other cuts, but the choice $A = \{1, 2, 3\}$ is minimal, and hence by the Min-Cut Max-Flow theorem,

$$\alpha_1^* = \mu(A, A^c) = 5.$$

The fluid model is stabilizable for any arrival rate satisfying $\alpha_1 < 5$, and the network load is then $\rho_\bullet = \alpha_1/5$.

We now consider the fluid model (6.1) to show that the conclusion $\rho_\bullet = \alpha_1/5$ can also be reached by constructing the workload vectors. There is a workload vector ξ satisfying $\langle \xi, \alpha \rangle = \rho_\bullet = \alpha_1/\alpha_1^*$, and the associated pooled resources are consistent with the minimal cut $A = \{1, 2, 3\}$.

To construct a fluid model we first order the vector of activity rates as follows:

$$\zeta = (\zeta_{1,2},\ \zeta_{1,3},\ \zeta_{2,5},\ \zeta_{3,2},\ \zeta_{3,4},\ \zeta_{3,5},\ \zeta_{1,2},\ \zeta_5)^\mathsf{T}$$

so that (6.25) is of the form (6.1) with $\alpha = (\alpha_1, 0, 0, 0, 0)^\mathsf{T}$, and

$$B = \begin{bmatrix} -2 & -10 & 0 & 0 & 0 & 0 & 0 & 0 \\ 2 & 0 & -1 & 5 & 0 & 0 & 0 & 0 \\ 0 & 10 & 0 & -5 & -2 & -2 & 0 & 0 \\ 0 & 0 & 0 & 0 & 2 & 0 & -10 & 0 \\ 0 & 0 & 1 & 0 & 0 & 2 & 10 & -100 \end{bmatrix}.$$

Several workload vectors can be constructed, following the approach illustrated in Example 6.2.4. There is a unique extreme point (ξ, ν) of the constraint set in (6.18) satisfying $\xi_i = 0$ for $i \in A^c$ and $\nu_{ij} = 0$ for $(i, j) \notin A \times A^c$:

$$\begin{aligned} \xi &= (\xi_1 \quad \xi_2 \quad \xi_3 \quad \xi_4 \quad \xi_5)^\mathsf{T} \\ &= (1/5 \quad 1/5 \quad 1/5 \quad 0 \quad 0)^\mathsf{T} \end{aligned}$$

$$\begin{aligned} \nu &= (\nu_{1,2} \quad \nu_{1,3} \quad \nu_{2,5} \quad \nu_{3,2} \quad \nu_{3,4} \quad \nu_{3,5} \quad \nu_{1,2} \quad \nu_5)^\mathsf{T} \\ &= (0 \quad 0 \quad 1/5 \quad 0 \quad 2/5 \quad 2/5 \quad 0 \quad 0)^\mathsf{T}. \end{aligned}$$

In particular, the variables $\{\nu_{2,5},\ \nu_{3,4},\ \nu_{3,5}\}$ are positive and sum to one, which identifies resource pooling among the three corresponding links. The network load is indeed given by $\rho_\bullet = \langle \xi, \alpha \rangle = \alpha_1/5$. ∎

6.3.2 Bellman–Ford algorithm

The development of Section 6.3.1 is motivated by a system-level desire to utilize all available resources. We now consider routing from the point of view of an individual who wishes to find the "best" single route to node ℓ starting from node 1. On constructing an appropriate cost function one can model mathematically the desire to avoid routing traffic through Bangalore, when it is intended to travel from Athens to Chicago.

The optimization problem considered here is similar to the *traveling salesman problem* in which a traveler must choose a path that passes through each node exactly once, and returns to the start. The routing problems considered here are far less complex, and there are many tractable algorithms to obtain efficient or optimal solutions. We present here the most famous technique, known as the Bellman–Ford algorithm.

Several restrictions are imposed to simplify the solution:

(i) The cost function c depends only on ζ. This is justified when considering equilibrium rates so that we may take $q \equiv 0$.
(ii) The cost is linear, so that $c(\zeta) = \langle c, \zeta \rangle$ for some vector c.
(iii) Node and link constraints are ignored (although the cost parameters can be designed to impose soft constraints on utilization rates on links).

A common choice is to identify $c_{(i,j)}$ with the delay on link (i,j). We set $c_{(i,j)} = \infty$ if there is no link between nodes i and j.

For a given arrival rate vector $\alpha \in \mathbb{R}_+^\ell$, paths are then found by solving the linear program

$$L^*(\alpha) := \quad \textbf{min} \quad \langle c, \zeta^e \rangle$$
$$\textbf{s.t.} \quad B\zeta^e + \alpha = 0 \tag{6.27}$$
$$\zeta^e \geq 0.$$

Its dual is written

$$\textbf{max} \quad \langle \alpha, \psi \rangle$$
$$\textbf{s.t.} \quad B^\mathsf{T}\psi + c \geq 0. \tag{6.28}$$

The dual variable $\psi \in \mathbb{R}^\ell$ corresponds to the equality constraint in (6.27). If the optimizer ψ^* is unique then $L^*(\alpha') = \langle \alpha', \psi^* \rangle$ for all α' in a neighborhood of α.

It is assumed that $\alpha_i = 0$ for $i \geq 2$ in the single-user model. In this special case we have the following simple representation for the unique optimizer ψ^*.

Proposition 6.3.6. *Consider the single-user model without link constraints. Then:*

(i) *The unique optimizer $\psi^* \in \mathbb{R}^\ell$ of the linear program (6.28) is independent of $\alpha_1 > 0$ and can be expressed as the cost to go,*

$$\psi_m^* = \textbf{min} \left(\sum_{(i,j)\in\mathcal{P}_k} c_{(i,j)} \right), \qquad 1 \leq m \leq \ell - 1,$$

where the minimum is over all paths $\mathcal{P}_k$ from k to ℓ.
(ii) *The optimizer satisfies the following dynamic programming equation,*

$$\psi_i^* = \textbf{min}\left\{c_{ij} + \psi_j^*\right\}, \qquad 1 \leq i \leq \ell - 1, \tag{6.29}$$

where the minimum is over all links (i,j).
(iii) *For a given i, if an optimal path between nodes 1 and ℓ uses link (i,j), then this link is an optimizer of the minimum in (6.29).*

The dynamic programming equation leads to the Bellman–Ford algorithm. The recursion (6.30) is an example of *value iteration*, to be explored in a different context in Chapter 9.

Definition 6.3.7 (Bellman–Ford algorithm). The algorithm is initialized with any $\psi^0 \in \mathbb{R}_+^\ell$ satisfying $\psi_\ell^0 = 0$. For $n \geq 1$ the sequence of vectors $\{\psi^i : i \geq 1\}$ is defined recursively by

$$\psi_i^{n+1} = \min\{c_{(i,j)} + \psi_i^n\}, \qquad 1 \leq i \leq \ell - 1, \ n \geq 0, \tag{6.30}$$

where the minimum is over all links. ∎

The proof of Theorem 6.3.8 (and also Proposition 6.3.6) can be found in [43].

Theorem 6.3.8 (Convergence of the Bellman–Ford algorithm). *For any $\psi^0 \in \mathbb{R}_+^\ell$ satisfying $\psi_\ell^0 = 0$, the sequence of vectors $\{\psi^i : i \geq 1\}$ defined by the Bellman–Ford algorithm satisfy*

$$\psi^n \to \psi^*, \qquad n \to \infty,$$

where the convergence is geometrically fast.

6.4 MaxWeight for routing and scheduling

In his 1995 paper introducing the back-pressure algorithm, Tassiulas argues that dynamic control achieves better utilization of resources than can be achieved using a static algorithm [467]. That is, the static viewpoint adopted in the previous section may not be effective in a network consisting of thousands of competing flows generated by different users, new flows joining the system, and old flows expiring.

Moreover, so far we have not addressed the critical issue of information: Is it realistic to assume that each node in the network shown in Fig. 1.2 has complete information regarding every flow in the system? It is unimaginable that so much information could be distributed across the network, or that so much redundant information would be useful.

While distributed versions of the Bellman–Ford algorithm are available and implemented in practice, for many reasons it is helpful to move away from the equilibrium model.

We first consider the MaxWeight policy for an extension of the routing model introduced in Section 6.3.

6.4.1 MaxWeight and back-pressure routing

It may appear that the addition of the queue-length process to the routing problem will add substantial complexity to the problem of choosing routes. Remarkably, in some sense the opposite is true. The queue process captures global information regarding network behavior that can be exploited to obtain highly decentralized routing policies for complex networks.

Consider the following extension of the routing model (6.25) consisting of ℓ_f different users, along with the topology and link constraints imposed in Section 6.3. User m wishes to send packets to a destination $d_m \in \{1, \ldots, \ell\}$. For each node i there are

ℓ_f different buffers corresponding to the different users. In a fluid model the buffer contents of user m at node i evolves according to the ODE

$$\tfrac{d^+}{dt} q_i^m(t) = \sum_{j=1}^{\ell} \big(-\mu_{ij}^m \zeta_{ij}^m(t) + \mu_{ji}^m \zeta_{ji}^m(t)\big), \qquad 1 \le i \le \ell,\ 1 \le m \le \ell_f.$$

The allocation rates are subject to the constraints $\zeta \ge 0$, $\sum_m \zeta_{ij}^m \le 1$ for each m, and we also impose the constraint that $q_j^m = 0$ and $\zeta_{ij}^m = 0$ if starting from node j there is no route to the desired destination d_m. Based on these constraints we can construct the set U of all allowable allocation rates.

The MaxWeight policy for this model is defined exactly as in (4.85),

$$\phi^{\text{MW}}(x) \in \underset{u \in \mathsf{U}(x)}{\arg\min} \langle Bu + \alpha, Dx \rangle, \qquad x \in \mathsf{X}. \tag{6.31}$$

The MaxWeight policy has an appealing representation that is similar to (4.88). The importance of this policy in applications is its simple form, and the fact that it can be implemented at each node in the network based solely on buffer levels at the node and its neighbors.

This is called the back-pressure policy in the special case $D = I$. For each link (i, j) the *maximal back-pressure* is defined by $\overline{\Theta}_{ij} := \max \mu_{ij}^m(x_i^m - x_j^m)$, where the maximum is over $m \in \{1, \dots, \ell_f\}$ such that node d_m is reachable from node j. If $\overline{\Theta}_{ij}(x) \ge 0$, then the back-pressure policy gives strict priority to buffers achieving the maximal back-pressure:

$$\sum_{m=1,\dots,\ell_f} \{\zeta_{ij}^m : \mu_{ij}^m(x_i^m - x_j^m) = \overline{\Theta}_{ij}(x)\} = 1, \qquad x \in \mathsf{X}, \tag{6.32}$$

and subject to the constraint that $\zeta_{ij}^m = 0$ if node d_m is not reachable from node j.

Example 6.4.1 (Simple routing model: MaxWeight policy). Consider the MaxWeight policy defined in (6.31) applied to the routing model introduced in Section 2.10. Assuming that the buffer at the router remains empty we obtain a two-dimensional network model. For a given diagonal matrix $D > 0$ the MaxWeight policy can be expressed

$$\phi(x) \in \underset{u \in \mathsf{U}(x)}{\arg\min} \big\{ D_{11}(\alpha_r u_1^r - \mu_1 u_1) x_1 + D_{22}(\alpha_r u_2^r - \mu_2 u_2) x_2 \big\}, \qquad x \in \mathbb{Z}_+^2.$$

On minimizing over u_1^r (with $u_2^r = 1 - u_1^r$) we obtain the linear switching curve

$$u_1^r = 1 \text{ whenever } D_{11}x_1 < D_{22}x_2.$$

A simulation of a CRW model controlled using this policy with $D_{ii} = \mu_i^{-1}$ is shown on the right in Fig. 2.16.   ∎

6.4.2 Simultaneous routing and scheduling

The MaxWeight policy for a general stochastic or fluid network model is defined precisely as for the scheduling model in Definition 4.8.1. In the fluid model it coincides with the h-myopic policy, where $h(x) = \frac{1}{2}x^{\mathsf{T}}Dx$ with D diagonal. Consequently, it can be described using a feedback law ϕ as in (4.88) or (6.31).

For the general model (6.1) we denote

$$\Theta_j(x) := -\sum_{i=1}^{\ell} x_i D_{ii} B_{ij}, \qquad 1 \leq j \leq \ell_u, \ x \in \mathsf{X}. \tag{6.33}$$

The MaxWeight policy for the fluid model can be expressed

$$\phi^{\text{MW}}(x) \in \arg\max \left\{ \sum_{j=1}^{\ell_u} \Theta_j(x)u_j \quad \text{s.t. } u \in \mathsf{U}(x) \right\}, \qquad x \in \mathsf{X}. \tag{6.34}$$

Proposition 6.4.3 implies that the MaxWeight policy for the fluid model defines a stationary deterministic policy for a CRW model.

A key assumption is that each activity can cause at most one queue to drain. Equivalently, the matrix $-B$ is Leontief:

Definition 6.4.2 (Leontief matrix). A matrix is *Leontief* if each column of the matrix contains at most one strictly positive element. ∎

Proposition 6.4.3. *Suppose that the arrival-free model is stabilizable and that $\rho_\bullet < 1$. Suppose moreover that the matrix $-B$ is Leontief. That is, for each $j \in \{1, \ldots, \ell_u\}$ there exists a unique value $i_j \in \{1, \ldots, \ell\}$ satisfying*

$$B_{ij} \geq 0 \qquad i \neq i_j. \tag{6.35}$$

For a given diagonal matrix $D > 0$ consider the MaxWeight policy defined for the fluid model via (6.34).

(i) *For each $x \in \mathsf{X}$ the allocation vector $\phi(x)$ is a solution u^* to the linear program*

$$\arg\max \left\{ \sum_{j=1}^{\ell_u} \Theta_j(x)u_j \quad \text{s.t. } Cu \leq 1, \ u \geq 0 \right\}. \tag{6.36}$$

(ii) *The optimizer u^* of the linear program (6.36) can be chosen so that $u_j^* \in \{0,1\}$ for each j, and $u_j^* = 0$ whenever $x_{i_j} = 0$.*

Proof. To establish (6.36) it suffices to demonstrate that an optimizer of (6.36) can be constructed with $u^* \in \mathsf{U}(x)$. For this we show that an optimizer exists such that $u_j^* = 0$ whenever $x_{i_j} = 0$. Any $u^* \in \mathsf{U}$ satisfying this condition also lies in $\mathsf{U}(x)$. That is, (i) will follow from (ii).

To prove (ii), suppose that u^* is an extremal optimizer for (6.36). For any j, if $x_{i_j} > 0$ and $u_j^* \in (0,1)$ then the allocation vector $u^\theta := (1-\theta)u^* + \theta\mathbf{1}^j$ is feasible whenever $\theta \leq 1$ and $u_j^\theta \geq 0$. That is, feasibility holds for $\theta \in [\theta^-, \theta^+]$ where $\theta^- = -u_j^*/(1-u_j^*)$ and $\theta^+ = 1$. Hence u^* can be written as a convex combination of u^{θ^+} and u^{θ^-}, which is a contradiction to the assumption that u^* is extremal. We conclude that an extremal optimizer exists, and it must satisfy $u_j^* \in \{0,1\}$ for each j for which $x_{i_j} > 0$.

Suppose now that $x_{i_j} = 0$. In this case $\Theta_j(x) \leq 0$, and it follows that we can take $u_j^* = 0$ without loss of generality for any optimizer of (6.36). □

The description of the MaxWeight policy can be simplified further under a mild additional restriction: Suppose that the sets $\{\mathcal{I}_r : 1 \leq r \leq \ell_m\}$ are disjoint. In this case the maximization (6.36) can be decomposed into ℓ_m separate optimization problems that can be solved explicitly. The MaxWeight policy can be expressed as follows under this assumption and the assumptions of Proposition 6.4.3:

(a) If $\overline{\Theta}_r(x) := \max_{j \in \mathcal{I}_r} \Theta_j(x) < 0$ then

$$\phi_j(x) = 0 \text{ for each } j \in \mathcal{I}_r(x). \tag{6.37}$$

(b) If $\overline{\Theta}_r(x) > 0$ then

$$\sum_{j \in \mathcal{I}_r} \{\phi_j(x) : \Theta_j(x) = \overline{\Theta}_r\} = 1. \tag{6.38}$$

(c) If $\overline{\Theta}_r(x) = 0$ then the choice is not unique: Either of the decision rules (6.37) or (6.38) is consistent with the MaxWeight policy.

If $\mathcal{I}_r \cap \mathcal{I}_{r'}$ is not empty for some $r \neq r'$ this means that the corresponding rows of the constituency matrix are not orthogonal. In this case a representation of the MaxWeight policy may be far more complex. We consider examples of these more exotic networks in the next section.

6.5 Simultaneous resource possession

Simultaneous resource possession arises in computer and communication systems, and many other applications. For example, a computer process may require the use of multiple resources simultaneously including access to memory, disk-drive search, and processor utilization. Audio or video communication may rely on simultaneous utilization of shared links in a network.

The following example concerns air traffic control, but the model could also describe Internet congestion control for delay sensitive traffic.

Example 6.5.1 (Air traffic scheduling). Fig. 6.8 is a schematic showing four classes of customers who wish to travel from source to destination. Class 1 and 2 customers

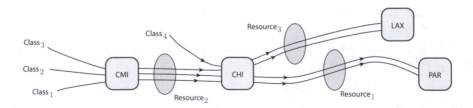

Figure 6.8. Simultaneous resource possession in aircraft scheduling: The resources highlighted in the figure indicate air-space between flights.

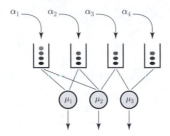

Figure 6.9. Customers of Class 1, 2, and 3 each require two resources simultaneously to complete service. Class 4 customers require service at resource 3 only.

each travel from Champaign–Urbana to Paris, but Class 1 customers pay a higher fare. Class 3 customers travel from Champaign–Urbana to Los Angeles, and Class 4 customers travel from Chicago to Los Angeles. The resources highlighted in the figure indicate air-space between flights. *Scheduling* amounts to determining which customers can make use of these resources at a given time.

A fluid model is used to obtain estimates of the maximal traffic flow and to obtain initial intuition for decision making over, say, a 12-h period. To refine this intuition requires consideration of variability and the finite number of flights between destinations. For example, the common practice of over-booking flights is a form of safety-stock to avoid starvation of resources. In this case the resources are planes and airspace; an example of *starvation* is a half-full airplane in flight.

Constraints on the scheduling policy are imposed to ensure customer satisfaction and avoid congestion at the airport in Chicago. For example, we must ensure that customers leaving Champaign–Urbana destined for Los Angeles will have a reasonably short wait for their flight in Chicago.

This scheduling problem can be placed within the framework of this chapter using the network model shown in Fig. 6.9, in which the four allocation rates $(\zeta_1, \ldots, \zeta_4)$ correspond to the four customer classes shown in the figure. The constituency matrix is given by

$$C = \begin{bmatrix} 1 & 1 & 0 & 0 \\ 1 & 1 & 1 & 0 \\ 0 & 0 & 1 & 1 \end{bmatrix}.$$

The respective rows of C correspond to airspace between (1) CHI and PAR; (2) CMI and CHI; (3) CHI and LAX. Simultaneous resource possession is captured by the fact that the rows of C are not orthogonal.   ∎

The effect of simultaneous resource possession is essentially the opposite of resource pooling. A seemingly innocuous system can explode in complexity. This is evident in the next example.

Example 6.5.2 (Input-queued switch). Figure 6.10 shows a particular architecture known as an *input-queued switch* used in Internet routers. In the example shown there

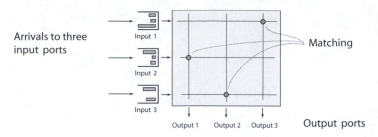

Figure 6.10. Input-queued switch showing a matching three selected inputs to their corresponding outputs.

are three arrival streams to the three *input-queues*, and three *output ports* that lead to three destinations. There are nine customer classes in this model, where a customer is a packet of data, and the customer class is determined by its point of origin and desired destination. This can be modeled using a nine-dimensional CRW or fluid model.

Assuming that the service rates are unity, the fluid model equations become

$$\tfrac{d^+}{dt} q_{ij}(t) = -\zeta_{ij}(t) + \alpha_{ij}, \qquad i,j = 1,2,3, \tag{6.39}$$

where $\zeta_{ij}(t)$ denotes the processing rate of Class (i,j) customers (i.e., customers at buffer i with destination j). Viewed as a 3×3 matrix, the constraint $\zeta \in \mathsf{U}$ is equivalent to the assumption that this matrix is "doubly substochastic":

$$\zeta \ge 0, \qquad \sum_{k_{\mathrm n}=1}^{3} \zeta_{k_{\mathrm n} j} \le 1, \qquad \sum_{k_{\mathrm o}=1}^{3} \zeta_{i k_{\mathrm o}} \le 1, \qquad i,j = 1,2,3. \tag{6.40}$$

The constituency matrix has six rows corresponding to these six constraints, and nine columns corresponding to the nine activities with rates $\{\zeta_{ij}\}$.

Simultaneous resource possession can be interpreted as blocking, exactly as in the air traffic scheduling problem. For example, there are two active constraints while data is routed from input queue 2 to output 1 as shown in the figure. That is, if $\zeta_{2,1} = 1$ then two of the constraints in (6.40) are binding,

$$\sum_{k_{\mathrm n}=1}^{3} \zeta_{k_{\mathrm n} 1} = 1, \qquad \sum_{k_{\mathrm o}=1}^{3} \zeta_{2 k_{\mathrm o}} = 1.$$

The first equality is a consequence of the constraint that no data can be sent to output 1 from any other input queue while $\zeta_{2,1} = 1$. The second equality follows from the fact that the second input queue is busy processing packets when $\zeta_{2,1} = 1$.

The MaxWeight policy for this model is defined exactly as in other models. Since q is nine-dimensional, we choose nine positive constants, denoted $\{D_{ij}\}$, and choose $\zeta(t)$ at time t to minimize the derivative

$$\tfrac{d^+}{dt} \left(\tfrac{1}{2} \sum_{ij} D_{ij} [q_{ij}(t)]^2 \right)$$

subject to the constraint that $\zeta(t) \in \mathsf{U}(x)$. From (6.39) and the representation of the MaxWeight policy in (6.34) we conclude that the policy can be expressed as the feedback law

$$\phi^{\text{MW}}(x) \in \arg\max_{u \in \mathsf{U}(x)} \sum D_{ij} u_{ij} x_{ij}, \qquad x \in \mathbb{R}_+^9. \tag{6.41}$$

The assumptions of Proposition 6.4.3 hold, so that we can assume that the maximum is achieved to give $\phi^{\text{MW}}(x) \in \mathsf{U}_\diamond(x)$ for each $x \in \mathsf{X}_\diamond$.

There is no simple description of the policy of the form (6.37), (6.38). In the $N \times N$ switch the set U is N^2-dimensional, so that this is also the dimension of the search space for computing ϕ^{MW}.

The literature surrounding the input-queued switch has settled on a name for this policy in a special case: If the constants $\{D_{ij}\}$ are all equal to one, then (6.41) is known as the *MaxWeight matching* policy, so that $\phi^{\text{MW}}(x) \in \arg\max_{u \in \mathsf{U}(x)} \sum u_{ij} x_{ij}$. The matching shown in Fig. 6.10 is a MaxWeight matching for the value of x indicated in the nine queues.   ∎

6.6 Workload relaxations

Most of the concepts in this section are straightforward extensions of those described for the scheduling model in Chapter 5. We also expand on the results of the previous chapter to obtain a better understanding of the potential "gap" in behavior between a network model and its relaxation.

6.6.1 Relaxations of the fluid model

The definition of a workload relaxation is essentially identical to Definition 5.3.1. Recall that $\{\xi^s : 1 \le s \le \ell_r\}$ are the workload vectors introduced in Definition 6.1.2.

Definition 6.6.1 (Workload relaxation). Suppose that $\rho_\bullet < 1$. Then, for a given integer $1 \le n \le \ell_r$:

(i) The *nth relaxation* of (6.1) is defined to be the differential inclusion based on the larger velocity space defined by

$$\widehat{\mathsf{V}} = \{v : \langle \xi^s, v \rangle \ge -(1 - \rho_s), 1 \le s \le n\}. \tag{6.42}$$

(ii) A feasible state trajectory is denoted $\widehat{q}$, and satisfies

$$\widehat{q}(t) \in \mathsf{X} \quad \text{and} \quad \frac{\widehat{q}(t_1) - \widehat{q}(t_0)}{t_1 - t_0} \in \widehat{\mathsf{V}}, \qquad t \ge 0,\ 0 \le t_0 < t_1 < \infty.$$

(iii) The *n*-dimensional *workload process* is defined by

$$\widehat{w}(t) = \widehat{\Xi}\widehat{q}(t), \qquad t \ge 0,$$

where the $n \times \ell$ *workload matrix* is given by $\widehat{\Xi} := [\xi^1 \mid \cdots \mid \xi^n]^\mathsf{T}$.

(iv) The workload process is constrained to the *workload space*

$$\mathsf{W} := \{\widehat{\Xi}x : x \in \mathsf{X}\}. \tag{∎}$$

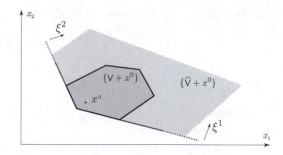

Figure 6.11. The velocity space $\widehat{V}$ is an unbounded relaxation of the original velocity space V.

A particular example illustrating the form of $\widehat{V}$ is shown in Fig. 6.11. The inclusion $V \subset \widehat{V}$ is a consequence of Theorem 6.1.1 (iv).

For each $t \geq 0$, $1 \leq s \leq n$, the sth component of the workload process is again subject to the simple constraint (5.23)

$$\tfrac{d^+}{dt}\widehat{w}_s(t) \geq -(1 - \rho_s), \qquad 1 \leq s \leq n.$$

These constraints are *decoupled* if the workload vectors $\{\xi^s : 1 \leq s \leq n\}$ are linearly independent.

The state space for the workload process is W, which is always a convex polyhedron (recall that X is a product of intervals, $X = \{x \in \mathbb{R}^\ell_+ : x_i \leq b_i, \quad 1 \leq i \leq \ell\}$.) It is a *positive cone* when $X = \mathbb{R}^\ell_+$. That is, if $w \in W$, then $\kappa w \in W$ for each $\kappa \geq 0$.

Convexity of X and the form of $\widehat{V}$ imply that the minimal draining time for the relaxation has a simple form, analogous to that obtained in Proposition 6.1.4. Note that the representation in Proposition 6.6.2 (i) uses an infimum, rather than a minimum over T, to reflect the fact that $\widehat{V}$ is unbounded and hence it is possible that $\widehat{T}^*(x^1, x^2) = 0$.

Proposition 6.6.2. *Suppose that $\rho_s < 1$ for each $1 \leq s \leq n$. Then, for any two states $x^1, x^2 \in X$:*

(i) *The minimal time to reach x^2 from x^1 for the n-dimensional relaxation is given by*

$$\widehat{T}^*(x^1, x^2) = \inf\{T > 0 : T^{-1}(x^2 - x^1) \in \widehat{V}\}.$$

(ii) $\widehat{T}^*(x^1, x^2) = \max\left(0, \; \max_{1 \leq s \leq n}\left(\dfrac{\langle \xi^s, x^1 - x^2 \rangle}{1 - \rho_s}\right)\right).$

Proof. Note that we must have $\widehat{T}^*(x^1, x^2) < \infty$ since 0 is an interior point of $\widehat{V}$.

The proof of (i) is identical to that given for the unrelaxed model in Proposition 6.1.4.

To see (ii), note first that if $\langle \xi^s, x^1 - x^2 \rangle \leq 0$ for each $1 \leq s \leq n$ then $T^{-1}(x^2 - x^1) \in \widehat{V}$ for each $T > 0$. Consequently, the infimum in (i) is precisely zero: In this nth relaxation the process can jump instantly to x^2 starting from x^1.

If $\langle \xi^s, x^1 - x^2 \rangle > 0$ for some s then $v^* := (x^2 - x^1)/\widehat{T}^*(x^1, x^2) \in \widehat{V}$ from the definition of the relaxed velocity space (6.42). The resulting state trajectory using this constant value reaches x^2 starting from x^1 in minimal time. $\qquad\square$

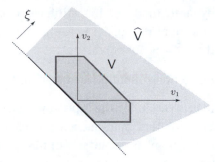

Figure 6.12. Velocity space for the simple routing model and its one-dimensional relaxation.

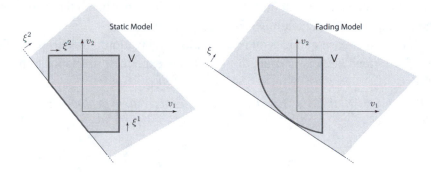

Figure 6.13. One-dimensional relaxations of the velocity space in two versions of the ALOHA communication model.

Proposition 6.6.2 implies that $\widehat{T}^*(x) = T^*(x)$ for any $x \in \mathbb{R}_+^\ell$ when $n = \ell_r$. For the two-dimensional model shown in Fig. 6.11 it is apparent that $\widehat{T}^*(x) = T^*(x)$ for any $x \in \mathbb{R}_+^2$ in a two-dimensional relaxation.

In the following two examples a one-dimensional relaxation is justified for sufficiently high load.

Example 6.6.3 (Simple routing model). Recall that for sufficiently high load, the minimal draining time for the simple routing model is given by the *linear* function T^* shown in (6.23). Figure 6.12 shows a one-dimensional relaxation of the velocity space V in the simple routing model when the expression (6.23) is valid.

Proposition 6.1.4 and Proposition 6.6.2 imply that the minimal draining times coincide for $x \in X = \mathbb{R}_+^3$,

$$T^*(x) = \widehat{T}^*(x) = \min\{T > 0 : T^{-1}x \in \widehat{V}\} = \frac{1}{1 - \rho_\bullet}\left(\frac{x_1 + x_2}{\mu_1 + \mu_2}\right). \qquad \blacksquare$$

Example 6.6.4 (ALOHA model). The velocity set V and its one-dimensional relaxation are shown in Fig. 6.13 for the two multiple-access communication models. In the static ALOHA model the set V and its relaxation are similar to the simple routing model.

A fluid model for the dynamic model (2.15) will have a convex, nonpolyhedral velocity space as shown on the right in Fig. 6.13. A one-dimensional relaxation can be constructed even though the set V is not polyhedral.   ∎

We now turn to the more complex input-queued switch.

Example 6.6.5 (Workload relaxation for the input-queued switch). Relaxation techniques are especially useful for the input-queued switch because of the explosive growth in complexity with the number of buffers. In the example shown in the figure there are three input buffers and three output ports, and six possible matchings between inputs and outputs. For a general $N \times N$ switch there are $N!$ possibilities, so that the number of possible matchings grows *super-exponentially* with N. It is very fortunate that a workload relaxation of modest dimension can be constructed that captures many important aspects of this scheduling problem.

In fact, the worst-case complexity in a workload relaxation is linear in N since the number of workload vectors is equal to $2N$. This is explained in Proposition 6.6.6 for $N = 3$.   ∎

Proposition 6.6.6. *In the 3×3 switch there are exactly six workload vectors, denoted $\{\xi^{n,i}, \xi^{o,j} : i, j = 1, 2, 3\} \subset \{0,1\}^{3 \times 3}$, and defined so that for each $x \in \mathbb{R}_+^{3 \times 3}$,*

$$\langle \xi^{n,i}, x \rangle = \sum_{k_o} x_{ik_o}, \qquad \langle \xi^{o,j}, x \rangle = \sum_{k_n} x_{k_n j}, \qquad i, j = 1, 2, 3 \,.$$

The minimal draining time for the arrival-free model is the maximum

$$W^*(x) = \max_{i=1,2,3} \left(\max\left(\langle \xi^{n,i}, x \rangle, \langle \xi^{o,i}, x \rangle \right) \right), \qquad x \in \mathbb{R}_+^{3 \times 3}. \qquad (6.43)$$

For a given arrival rate matrix α the network load is given by the maximum of all row-sums and column-sums of α,

$$\rho_\bullet = \max_{i,j} \left(\sum_{k_o} \alpha_{ik_o}, \sum_{k_n} \alpha_{k_n j} \right). \qquad (6.44)$$

Proof. In words, $\langle \xi^{n,i}, x \rangle := \sum_{kl} \xi_{kl}^{n,i} x_{kl}$ denotes the total amount of processing time required by customers waiting in the input buffer i, and $\langle \xi^{o,j}, x \rangle$ is the total amount of processing time required by customers destined for output j. This implies the representation (6.43) for $W^*(x)$.

The vectors $\{\xi^{n,i}, \xi^{o,j} : i, j = 1, 2, 3\}$ are the complete set of workload vectors since it is not possible to express $W^*(x)$ for all x using a maximum over some fixed subset of these six vectors.

Applying Definition 6.0.1 we conclude that the load is given by (6.44).   □

6.6.2 Effective cost

Suppose now that a cost function is given $c \colon \mathsf{X} \to \mathbb{R}_+$. It is assumed that c is convex, continuous, and vanishes only at the origin.

Definition 6.6.7 (Effective cost for a workload relaxation). Suppose that the state space X is a convex polyhedron, and that $c: X \to \mathbb{R}_+$ is a convex cost function that vanishes only at the origin. Then:

(i) The *effective cost* $\bar{c}: W \to \mathbb{R}_+$ is defined for each $w \in W$ as the value of the convex program,

$$\bar{c}(w) = \mathbf{min} \quad c(x)$$
$$\text{s.t.} \quad \widehat{\Xi}x = w, \quad x \in X. \tag{6.45}$$

(ii) The region where $\bar{c}$ is *monotone* is denoted W^+. That is,

$$W^+ := \{w \in W : \bar{c}(w') \geq \bar{c}(w) \text{ whenever } w' \geq w \text{ and } w' \in W\}.$$

(iii) For any $w \in W$, an *effective state* $\mathcal{X}^*(w)$ is defined to be any vector $x^* \in X$ that minimizes the convex program (6.45):

$$\mathcal{X}^*(w) \in \arg\min_{x \in X}\left(c(x) : \widehat{\Xi}x = w\right). \tag{6.46}$$

(iv) For any $x \in X$, an *optimal exchangeable state* $\mathcal{P}^*(x) \in X$ is defined via

$$\mathcal{P}^*(x) = \mathcal{X}^*(\widehat{\Xi}x). \tag{6.47}$$

∎

Just as in Proposition 5.3.15 we have the following:

Proposition 6.6.8. *For any workload relaxation, and any convex cost function:*

(i) *The effective cost $\bar{c}: W \to \mathbb{R}_+$ is convex.*
(ii) *If the cost function $c: X \to \mathbb{R}_+$ is linear, then the effective cost is piecewise linear,*

$$\bar{c}(w) = \max_i \langle \bar{c}^i, w \rangle, \qquad w \in W. \tag{6.48}$$

The vectors $\{\bar{c}^i\} \in \mathbb{R}^\ell$ are the extreme points obtained in the dual of the linear program (6.45).

Consider a one-dimensional relaxation. Just as in the scheduling problem, the effective cost is easily computed in this case. For simplicity we ignore buffer constraints in Proposition 6.6.9. The proof is left as an exercise – see Exercise 6.6 and Exercise 6.8.

Proposition 6.6.9. *For the one-dimensional workload relaxation on $X = \mathbb{R}_+^\ell$:*

(i) *If $c: \mathbb{R}_+^\ell \to \mathbb{R}_+$ is linear then an effective state is given by*

$$\mathcal{X}^*(w) = \left(\frac{1}{\xi_{i_*}^1} 1^{i_*}\right) w, \qquad w \in \mathbb{R}_+,$$

where the index i_* is any solution to $c_{i_*}/\xi_{i_*}^1 = \min_{1 \leq i \leq \ell}(c_i/\xi_i^1)$. The effective cost is given by the linear function

$$\bar{c}(w) = c(\mathcal{X}^*(w)) = (c_{i_*}/\xi_{i_*}^1)w, \qquad w \in \mathbb{R}_+.$$

(ii) *Suppose that* $c: \mathbb{R}_+^{\ell} \to \mathbb{R}_+$ *is the quadratic* $c(x) = \frac{1}{2}x^{\mathsf{T}}Dx$, $x \in \mathbb{R}_+^{\ell}$, *where* D *is an invertible* $\ell \times \ell$ *matrix, and suppose that* $x^{*1} \geq 0$, *where*

$$x^{*1} := \left(\xi^{1^{\mathsf{T}}}D^{-1}\xi^1\right)^{-1}D^{-1}\xi^1.$$

Then the effective state is again linear in the workload value, with

$$\mathcal{X}^*(w) = wx^{*1}, \qquad w \in \mathbb{R}_+.$$

The effective cost is the one-dimensional quadratic

$$\bar{c}(w) = \frac{1}{2}\left(\xi^{1^{\mathsf{T}}}D^{-1}\xi^1\right)^{-1}w^2, \qquad w \in \mathbb{R}_+.$$

Example 6.6.10 (Simple routing model). Returning to the simple routing model considered in Example 6.6.3, in the one-dimensional relaxation considered previously we have $\xi = (\mu_1 + \mu_2)^{-1}\mathbf{1} \in \mathbb{R}^3$.

If $c: \mathbb{R}_+^3 \to \mathbb{R}_+$ is a linear cost function then the effective state and effective cost are given by

$$\mathcal{X}^*(w) = (\mu_1 + \mu_2)w\mathbf{1}^{i_*}, \quad \bar{c}(w) = c_{i_*}(\mu_1 + \mu_2)w, \qquad w \in \mathbb{R}_+, \qquad (6.49)$$

where $i_* \in \arg\min_{1 \leq i \leq 3} c_i$.

Consider a quadratic cost function of the form

$$c(x) = \frac{1}{2}(c_1 x_1^2 + c_2 x_2^2 + c_3 x_3^2). \qquad (6.50)$$

That is, D is the diagonal matrix $D = \operatorname{diag}(c_i)$, and hence by Proposition 6.6.9 the effective state is given by $\mathcal{X}^*(w) = wx^{*1}$ for $w \in \mathbb{R}_+$, where

$$x_i^{*1} = \frac{\mu_1 + \mu_2}{c_1^{-1} + c_2^{-1} + c_3^{-1}}(c_1^{-1}, c_2^{-1}, c_3^{-1})^{\mathsf{T}}.$$

The effective cost is

$$\bar{c}(w) = \frac{1}{2}\frac{(\mu_1 + \mu_2)^2}{c_1^{-1} + c_2^{-1} + c_3^{-1}}w^2, \qquad w \in \mathbb{R}_+. \qquad \blacksquare$$

We now consider a high-dimensional relaxation for the input-queued switch.

Example 6.6.11 (Monotone region for the input-queued switch). Consider a balanced model in which all six of the workload vectors defined in Proposition 6.6.6 are used to construct a six-dimensional workload relaxation. The workload vectors are not linearly independent since $\sum_i \xi^{n,i} = \sum_j \xi^{o,j}$. They span five dimensions, and hence the workload space W lies in a five-dimensional subspace of $\mathbb{R}_+^6$.

Suppose that c is the quadratic $c(x) = \frac{1}{2}\|x\|$, $x \in \mathbb{R}^3 \times \mathbb{R}^3$. For given vectors $(w^n, w^o) \in$ W, the effective cost is the solution to the quadratic program

$$\bar{c}(w) = \mathbf{min} \quad \frac{1}{2}\sum_{ij} x_{ij}^2$$

$$\mathbf{s.t.} \quad \sum_k x_{ik} = w_i^n, \quad \sum_l x_{lj} = w_j^o, \quad i, j = 1, 2, 3 \qquad (6.51)$$

$$x \geq 0.$$

A Lagrangian is constructed based on Lagrange multipliers $\psi^{\mathrm{n}}, \psi^{\circ} \in \mathbb{R}^3$,

$$\mathcal{L}(x, \psi) = \tfrac{1}{2} \sum_{ij} x_{ij}^2 + \sum_i \psi_i^{\mathrm{n}} \left(w_i^{\mathrm{n}} - \sum_k x_{ik} \right) + \sum_j \psi_j^{\circ} \left(w_j^{\circ} - \sum_l x_{lj} \right).$$

Assuming that the Lagrangian is minimized with $x^* > 0$, the first-order optimality conditions give

$$0 = \frac{\partial}{\partial x_{ij}} \mathcal{L}\left(x^*, \psi\right) = x_{ij}^* - \psi_i^{\mathrm{n}} - \psi_j^{\circ}, \qquad i, j = 1, 2, 3.$$

The optimizer can thus be expressed in matrix form as the sum of two "outer products," $x^* = \psi^{\mathrm{n}} \mathbf{1}^{\mathsf{T}} + \mathbf{1} \psi^{\circ\mathsf{T}}$. The Lagrange multipliers and workload vectors are related by

$$w_i^{\mathrm{n}} = 3\psi_i^{\mathrm{n}} + \sum_j \psi_j^{\circ} \quad \text{and} \quad w_j^{\circ} = 3\psi_j^{\circ} + \sum_i \psi_i^{\mathrm{n}}.$$

The monotone region is precisely those $(w^{\mathrm{n}}, w^{\circ}) \in \mathsf{W}$ for which the Lagrange multipliers are nonnegative:

$$\mathsf{W}^+ = \left\{ (w^{\mathrm{n}}, w^{\circ}) = (3\psi^{\mathrm{n}} + |\psi^{\circ}| \mathbf{1}, \, 3\psi^{\circ} + |\psi^{\mathrm{n}}| \mathbf{1}) : \psi^{\mathrm{n}}, \psi^{\circ} \in \mathbb{R}_+^3 \right\}. \qquad \blacksquare$$

Proposition 6.6.12 extends the conclusions of this example to the general model. The assumption that D^{-1} has nonnegative entries is a relaxation of the assumption that $D > 0$ is diagonal.

Proposition 6.6.12. *Suppose that $\widehat{\Xi}_{ij} \geq 0$ for each i, j in the nth relaxation, and suppose that D is a positive definite matrix satisfying $D_{ij}^{-1} \geq 0$ for each i, j. Suppose finally that $D^{-1}\widehat{\Xi}^{\mathsf{T}}\mathbf{1} > 0$. Let $c(x) = \tfrac{1}{2} x^{\mathsf{T}} D x$, and let $\bar{c}$ denote the effective cost. Then the monotone region is given by*

$$\mathsf{W}^+ = \{ \widehat{\Xi} D^{-1} \widehat{\Xi}^{\mathsf{T}} \psi : \psi \in \mathbb{R}_+^n \}.$$

Proof. The effective cost is the solution to the quadratic program

$$\bar{c}(w) = \min \quad \tfrac{1}{2} x^{\mathsf{T}} D x$$
$$\text{s.t.} \quad \widehat{\Xi} x = w, \; x \geq 0. \tag{6.52}$$

Let $\psi \in \mathbb{R}^d$ denote a Lagrange multiplier, and define the Lagrangian as a function of (x, ψ) via

$$\mathcal{L}(x, \psi) = \tfrac{1}{2} x^{\mathsf{T}} D x + \psi^{\mathsf{T}} [w - \widehat{\Xi} x].$$

If x^* minimizes (6.52) then there exists a Lagrange multiplier ψ^* such that x^* also minimizes $\mathcal{L}(x, \psi^*)$ over $x \geq 0$. Moreover, ψ^* defines a gradient for the effective cost at w. Hence $w \in \mathsf{W}^+$ if and only if $\psi^* \geq 0$.

Suppose that x^* is an interior minimizer of $\mathcal{L}(x, \psi^*)$ so that we can apply the first-order necessary conditions for optimality,

$$0 = \nabla_x \mathcal{L}\left(x^*, \psi^*\right) = D x^* - \widehat{\Xi}^{\mathsf{T}} \psi^*,$$

giving $x^* = D^{-1}\widehat{\Xi}^\mathsf{T}\psi^*$. If $\psi^* > 0$ then $x^* > 0$ under the assumptions on D and $\widehat{\Xi}$. Finally, writing

$$w = \widehat{\Xi}x^* = \widehat{\Xi}D^{-1}\widehat{\Xi}^\mathsf{T}\psi^*$$

we conclude that w is a typical element of the interior of W^+. This completes the proof since

$$\text{closure } \{\widehat{\Xi}D^{-1}\widehat{\Xi}^\mathsf{T}\psi : \psi > 0\} = \{\widehat{\Xi}D^{-1}\widehat{\Xi}^\mathsf{T}\psi : \psi \geq 0\}.$$

$\square$

6.6.3 Coupling the relaxation

We have seen through examples that it is not difficult to devise efficient policies for a workload model of moderate dimension. The question then arises, *how can this be adapted to provide a policy for the original complex network of interest?* Here we show how to translate a policy from the workload relaxation to the ℓ-dimensional fluid model.

Suppose that $[\widehat{q}^*(t;x), \widehat{\zeta}^*(t;x)]$ is any solution to the relaxed control problem, and that $c\colon \mathsf{X} \to \mathbb{R}_+$ is a convex cost function. For simplicity we take c linear, and $\mathsf{X} = \mathbb{R}_+^\ell$. A policy for the unrelaxed model is defined as follows.

Definition 6.6.13 (Translation of a policy for the relaxation). For each initial condition x, and time $t \geq 0$, the allocation rate $\zeta(t;x)$ for the unrelaxed model is defined to be a function of $[\widehat{q}^*(t;x), \widehat{\zeta}^*(t;x), q(t;x)]$ as follows: Given the current states $x^1 = q(t;x)$, $\widehat{x}^1 = \widehat{q}^*(t;x)$, let $\zeta(t;x)$ be the optimizer $\zeta^* \in \mathsf{U}$ in the linear program

$$\begin{aligned}
\textbf{min} \quad & \langle c, B\zeta \rangle \\
\textbf{s.t.} \quad C\zeta &\leq 1 \\
\zeta &\geq 0 \\
(B\zeta + \alpha)_i &\geq 0 \quad \text{if } x_i^1 = 0 \\
\langle \xi^s, (B\zeta + \alpha) \rangle &\leq \langle \xi^s, (B\widehat{\zeta}^* + \alpha) \rangle, \quad \text{whenever } s \leq n, \\
& \qquad\qquad\qquad\qquad\qquad \text{and } \langle \xi^s, x^1 \rangle = \langle \xi^s, \widehat{x}^1 \rangle. \quad (6.53)
\end{aligned}$$

$\blacksquare$

The last constraint in (6.53) ensures that $w_s(t;x) \leq \widehat{w}_s^*(t;x)$ for all $s \leq n$ and all t. To analyze this policy we consider the error process and "coupling time" defined by

$$e(t;x) = q(t;x) - \widehat{q}^*(t;x), \ t > 0, \qquad \underline{T}^\circ(x) = \min\{t : e(t;x) = 0\}. \quad (6.54)$$

First we consider its application using the simple routing model.

Example 6.6.14 (Coupling in the simple routing model). In the simple routing model considered in Example 6.6.3 we have seen that a one-dimensional relaxation is well motivated, and in Example 6.6.10 we computed the effective cost when c is linear or quadratic. In the former case with $c(x) = c_1x_1 + c_2x_2 + c_3x_3$, $x \in \mathbb{R}_+^3$, $c_1 < c_2 < c_3$, the effective state is given by $\mathcal{X}^*(w) = (\mu_1 + \mu_2)w\mathbf{1}^1$ for $w \geq 0$ (see (6.49)).

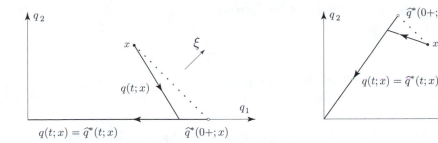

Figure 6.14. Coupling of optimal solutions in the simple routing model. When the cost function is linear, the policy defined in Definition 6.6.13 couples with the optimal relaxation at time $\underline{T}^\circ(x) = x_2/\mu_2$ as shown on the left. The figure on the right illustrates this coupling when the cost is quadratic.

Consider an initial condition of the form $q(0) = \widehat{q}(0) = x$ with $x_3 = 0$. In an optimal solution for the relaxation we have $\widehat{q}^*(0+) = \mathcal{P}^*(x) = (\mu_1 + \mu_2)\widehat{w}(0)\mathbf{1}^1 = (x_1 + x_2)\mathbf{1}^1$, and

$$\widehat{q}^*(t) = \widehat{q}(0+) - (\mu_1 + \mu_2 - \alpha_r)t \quad \text{for } 0 \le t \le T^*(x).$$

As illustrated on the left in Fig. 6.14, the norm of the error $\|e(t; x)\|$ is decreasing with t, and the state trajectory q defined in Definition 6.6.13 couples with the optimal relaxation at time $\underline{T}^\circ(x) = x_2/\mu_2$.

Similarly, if the cost function is the quadratic shown in (6.50) then the optimal trajectory $\widehat{q}^*$ is again restricted to a line segment for $t > 0$, and q couples with the relaxation in finite time. This coupling is illustrated on the right in Fig. 6.14. ∎

So far we have taken for granted the feasibility of the linear program defining ζ. In Theorem 6.6.15 sufficient conditions are formulated based on the assumption that the load is sufficiently large, and that the n workload vectors capture all of the bottlenecks. In addition to verifying feasibility, in this heavy-traffic setting we can bound the gap in performance between the fluid model and its relaxation.

It is simplest to formulate these conditions using a family of networks with increasing load. We let $\kappa \ge 1$ denote a scaling parameter, and construct for each κ a stabilizable network with $\rho_\bullet^\kappa < 1$ for $\kappa < \infty$ and $\rho_\bullet^\infty = 1$. It is assumed that B and C are independent of κ, and that the arrival-free model is stabilizable. Two arrival-rate vectors α^1, α^∞ are given, and for arbitrary $\kappa \ge 1$ we define

$$\alpha^\kappa := \alpha^\infty - \kappa^{-1}(\alpha^\infty - \alpha^1). \tag{6.55}$$

The respective load parameters can be expressed

$$\rho_\bullet^1 := W^*(\alpha^1), \quad \rho_\bullet^\infty := W^*(\alpha^\infty). \tag{6.56}$$

The choice of a perturbation in the arrival stream is for the sake of convenience since we can then take a fixed set of workload vectors. Let $\mathcal{R}_b \subset \{1, \dots, \ell_m\}$ denote the index set of bottleneck resources for the model with arrival rate α^∞. That is, those $s \in \{1, \ldots, \ell_r\}$ satisfying

$$\rho_s^\infty = \langle \xi^s, \alpha^\infty \rangle = 1.$$

By reordering, we can assume, without loss of generality, that $\mathcal{R}_b = \{1, \ldots, \ell_b\}$ for some integer $\ell_b \geq 1$. *Throughout the remainder of this section we consider the nth workload relaxation with $n = \ell_b$.*

Theorem 6.6.15 shows that little is lost when optimizing the ℓ_bth relaxation. Let $J^{\kappa*}$, $\widehat{J}^{\kappa*}$ denote the value functions for the infinite-horizon optimal control problems. We always have

$$\widehat{J}^{\kappa*}(x) \leq J^{\kappa*}(x), \qquad x \in \mathsf{X},$$

and Theorem 6.6.15 provides bounds in the reverse direction.

Let e^κ and $\underline{T}^{\kappa\circ}(x)$ denote the error process and hitting time defined in (6.54) for the κth network. The following result provides uniform bounds on $\underline{T}^{\kappa\circ}$, and justifies the term *coupling time*.

Theorem 6.6.15 (Coupling the relaxation). *Suppose that the following hold:*

(a) *The cost function $c\colon \mathsf{X} \to \mathbb{R}_+$ is linear, and the optimizer $\mathcal{X}^*(w)$ of (6.45) is unique for each $w \in \mathsf{W}$.*

(b) *For $x \in \mathsf{X}$, and with $\kappa = 1$, the ℓ_bth workload relaxation admits a solution $\widehat{q}^{1*}$ that minimizes the total cost*

$$\widehat{J}(x) = \int_0^\infty c(\widehat{q}(t))\, dt, \qquad \widehat{q}(0) = x,$$

and this solution is unique.

(c) *The workload vectors $\{\xi^s : 1 \leq s \leq \ell_b\}$ are linearly independent.*

Then, there exists $\kappa_0 \geq 1$ such that the linear program (6.53) is feasible for each $\kappa \geq \kappa_0$. The following hold for the resulting state trajectory and its relaxation:

(i) $w_s(t; x) \leq \widehat{w}_s^{\kappa*}(t; x),\, t \geq 0,\, 1 \leq s \leq \ell_b.$

(ii) *The time $\underline{T}^{\kappa\circ}$ is uniformly bounded: For some $b_0 < \infty$ independent of κ,*

$$\underline{T}^{\kappa\circ}(x) \leq b_0 \|e^\kappa(0+; x)\|, \qquad x \in \mathsf{X}.$$

(iii) $q(t; x) = \widehat{q}^{\kappa*}(t; x)$ *for all $t \geq \underline{T}^{\kappa\circ}(x).$*

(iv) *There is a constant b_1 such that the value function satisfies*

$$J^\kappa(x) := \int_0^\infty c(q(t; x)) \ \leq \ (1 + \kappa^{-1}b_1)\widehat{J}^{\kappa*}(x)$$

$$\leq \ (1 + \kappa^{-1}b_1)J^{\kappa*}(x), \qquad x \in \mathsf{X}$$

(v) *Suppose that $\widehat{q}^{1*}$ is a pathwise optimal solution. Then, for $\kappa \geq \kappa_0$,*

$$c(q(t; x)) = \underline{c}^*(t; x), \qquad t \geq \underline{T}^{\kappa\circ}(x),$$

where $\underline{c}^*$ is the lower bound on the cost defined in the linear program (4.40).

The proof of Theorem 6.6.15 is based on the following three lemmas.

Lemma 6.6.16. *Under the assumptions of Theorem 6.6.15 the workload relaxation satisfies the independent differential constraints*

$$\frac{d^+}{dt}\widehat{w}_s(t;x) \geq -\kappa^{-1}\delta_s, \qquad 1 \leq s \leq \ell_b, \ \kappa \geq 1, \ t > 0, \tag{6.57}$$

where $\delta_s = \langle \xi^s, \alpha^\infty - \alpha^1 \rangle = 1 - \rho_s^1$.

Proof. From the definitions we have $1 - \rho_s^\kappa = \kappa^{-1}\langle \xi^s, \alpha^\infty - \alpha^1 \rangle$, $s \in \mathcal{R}_b$. $\qquad\square$

Next we obtain a scaling property that relates the optimal solutions $\{\widehat{q}^{\kappa*} : \kappa \geq 1\}$.

Lemma 6.6.17. *Suppose that the assumptions of Theorem 6.6.15 hold, so that in particular a solution to the infinite-horizon control problem for the ℓ_bth relaxation is unique for each initial condition when $\kappa = 1$. Letting $\widehat{J}^{1*}$ denote the value function when $\kappa = 1$, for each finite $\kappa \geq 1$ the optimal solution for the relaxation satisfies*

$$\begin{aligned}
\widehat{q}^{\kappa*}(t;x) &= \widehat{q}^{1*}(\kappa^{-1}t;x), & t &> 0, \\
\widehat{J}^{\kappa*}(x) &= \kappa\widehat{J}^{1*}(x), & x &\in \mathsf{X}.
\end{aligned} \tag{6.58}$$

Proof. For each $\kappa \geq 1$, there is a one to one mapping between feasible $\boldsymbol{w}^1$ and feasible $\boldsymbol{w}^\kappa$. This is defined by the simple scaling $\widehat{w}^1(t) \mapsto \widehat{w}^\kappa(t) = \widehat{w}^1(\kappa^{-1}t)$.

Given the optimal workload process $\widehat{\boldsymbol{w}}^{1*}$, the scaled process does satisfy the required lower bounds

$$\frac{d^+}{dt}\widehat{w}_s^{1*}(\kappa^{-1}t;x) \geq -\kappa^{-1}\delta_s, \qquad 1 \leq s \leq \ell_b, \ t > 0,$$

and from this and uniqueness of the optimal solution we can conclude that $\widehat{w}^{\kappa*}(t;x) = \widehat{w}^{1*}(\kappa^{-1}t;x)$ for all t. Since $\mathcal{X}^*$ is uniquely defined, we conclude that a similar correspondence holds between $\widehat{q}^{1*}$ and $\widehat{q}^{\kappa*}$,

$$\widehat{q}^{\kappa*}(t;x) = \mathcal{X}^*\big(\widehat{w}^{\kappa*}(t;x)\big) = \mathcal{X}^*\big(\widehat{w}^{1*}(\kappa^{-1}t;x)\big) = \widehat{q}^{1*}(\kappa^{-1}t;x), \qquad t > 0.$$

This proves (i).

Part (ii) follows from (i) and the change of variables $s = \kappa^{-1}t$:

$$\begin{aligned}
\widehat{J}^{\kappa*}(x) &= \int_0^\infty c(\widehat{q}^{\kappa*}(t;x))\, dt \\
&= \int_0^\infty c(\widehat{q}^{1*}(\kappa^{-1}t;x))\, dt \\
&= \kappa \int_0^\infty c(\widehat{q}^{1*}(s;x))\, ds = \kappa\widehat{J}^{1*}(x). \qquad\square
\end{aligned}$$

The next result shows that, relative to the system load, exchangeable states for the ℓ_bth workload relaxation are almost exchangeable for the original fluid model when κ is large.

Lemma 6.6.18. *Suppose that the assumptions of Theorem 6.6.15 hold. Then, there exists $b_0 < \infty$ such that for any $x^1, x^2 \in X$ satisfying $\widehat{\Xi}x^2 \geq \widehat{\Xi}x^1$, the time to reach x^2 from x^1 is uniformly bounded in κ as follows,*

$$T^{\kappa*}(x^1, x^2) \leq b_0 \|x^1 - x^2\|, \qquad \kappa \geq 1.$$

Proof. If $\langle \xi^s, x^2 - x^1 \rangle \geq 0$ for $1 \leq s \leq \ell_b$ it then follows from Proposition 6.1.4 that the draining time can be expressed as the maximum

$$T^{\kappa*}(x^1, x^2) = \max_{s \geq 1} \frac{\langle \xi^s, x^1 - x^2 \rangle}{o_s - \langle \xi^s, \alpha^\kappa \rangle} = \max_{s > \ell_b} \frac{\langle \xi^s, x^1 - x^2 \rangle}{o_s - \langle \xi^s, \alpha^\kappa \rangle}, \qquad 1 \leq \kappa < \infty.$$

The right-hand side is bounded in κ by construction of α^κ, and the definition of ℓ_b. This gives the required bound, with $b_0 = \max_{\kappa \geq 1, s > \ell_b} \|\xi^s\| / |o_s - \rho_s^\kappa|$. □

Proof of Theorem 6.6.15. We first establish feasibility of the linear program (6.53) that defines the policy. For any $t > 0$ such that $w_s(t) \leq \widehat{w}_s(t)$ for $s \leq \ell_b$ we define

$$\hat{v}^\perp = -\gamma \frac{e^\kappa(t; x)}{\|e^\kappa(t)\|}.$$

The constant $\gamma > 0$ is chosen so that $\hat{v}^\perp$ is a boundary point of the velocity space $V_\kappa := \{V_0 + \alpha^\kappa\}$. We have the explicit formula $\gamma^{-1} = \frac{T^{\kappa*}(x^1, x^2)}{\|x^2 - x^1\|}$, with $x^1 = q(t)$, $x^2 = \widehat{q}^{\kappa*}(t)$.

A candidate velocity vector satisfying the constraints in (6.53) is given as follows:

$$v = \hat{v}^* + \left(1 - \tfrac{1}{2}\kappa^{-1}\kappa_0\right)\hat{v}^\perp, \tag{6.59}$$

where $\hat{v}^* = \frac{d}{dt}\widehat{q}^{\kappa*}(t)$, and κ_0 is a constant. Note that the final inequality in (6.53) is satisfied with equality: $\langle \xi^s, v \rangle = \langle \xi^s, \hat{v}^* \rangle$ for $s \leq \ell_b$ whenever $w_s(t) = \widehat{w}_s^*(t)$. We now construct $\kappa_0 \geq 1$ such that $v \in V_\kappa$ for $\kappa \geq \kappa_0$. For such κ we conclude that v is feasible at time t for the policy defined by (6.53).

For $1 \leq s \leq \ell_b$ we have $\langle \xi^s, \hat{v}^\perp \rangle \geq 0$ since $w_s(t) \leq \widehat{w}_s^{\kappa*}(t)$ by assumption. Hence, for $\kappa \geq \kappa_0$,

$$\langle \xi^s, v \rangle = \langle \xi^s, \hat{v}^* + (1 - (2\kappa)^{-1}\kappa_0)\hat{v}^\perp \rangle \geq \langle \xi^s, \hat{v}^* \rangle \geq -(1 - \rho_s^\kappa).$$

For $s > \ell_b$ we can reason as follows: The identity (6.58) implies that $\|\frac{d}{dt}\widehat{w}^{\kappa*}(t)\| \leq b_3/\kappa$ for some $b_3 < \infty$, and all $t > 0$. Since $\mathcal{X}^*$ is continuous, we must have a similar bound for $\widehat{q}^{\kappa*}$, so that $\|\hat{v}^*\| \leq b_4/\kappa$ for some finite b_4. Then, for $s > \ell_b$ and $\kappa \geq \kappa_0$, equation (6.59) and the assumption that $v^\perp \in V_\kappa$ gives

$$\begin{aligned}
\langle \xi^s, v \rangle &\geq \langle \xi^s, \hat{v}^* \rangle - (1 - \tfrac{1}{2}\kappa^{-1}\kappa_0)(1 - \rho_s^\kappa) \\
&\geq \kappa^{-1}\left(\tfrac{1}{2}\kappa_0(1 - \rho_s^\kappa) - b_4\|\xi^s\|\right) - (1 - \rho_s^\kappa).
\end{aligned}$$

Hence, to ensure feasibility of v, it is sufficient to choose $\kappa_0 > 2b_4 \max_{s > \ell_b}\left(\|\xi^s\|(1 - \rho_s^\infty)^{-1}\right)$.

This establishes feasibility and also the bounds in (i).

To establish the remaining conclusions we now obtain bounds on the coupling time. Lemma 6.6.18 implies that $\gamma = \gamma(\kappa)$ is uniformly bounded from below in κ. Applying this and Assumption (a) in the theorem, we conclude that there is some fixed $\varepsilon_c > 0$, independent of $x \in \mathsf{X}$ and $\kappa \geq 1$, such that for all $0 \leq t < \underline{T}^{\kappa\circ}(x)$, and sufficiently small $\varepsilon > 0$,

$$c(q(t) + \varepsilon\hat{v}^\perp) - c(q(t)) \leq -\varepsilon_c\varepsilon. \tag{6.60}$$

Define the error process $e_c(t) := c(q(t)) - c(\hat{q}^{\kappa*}(t))$, $t \geq 0$. Using (6.60), feasibility of the velocity vector v defined in (6.59), and minimality of $\langle c, \frac{d}{dt}q \rangle$ we obtain the following bound on the derivative of the error under the policy (6.53): whenever $e_c(t) \neq 0$ and $\kappa \geq \kappa_0$,

$$\tfrac{d}{dt}e_c(t) = \tfrac{d}{dt}\Big(c(q(t)) - c(\hat{q}^{\kappa*}(t))\Big) \leq -\Big(1 - \tfrac{1}{2}\kappa^{-1}\kappa_0\Big)\varepsilon_c \leq -\tfrac{1}{2}\varepsilon_c.$$

The error is nonnegative, so this bound implies that once the error reaches zero it stays at zero. Moreover, we have $e_c(0+) = c(x) - c(\mathcal{P}^*(x))$, where $\mathcal{P}^*$ is defined in (6.47), so that on integrating we obtain

$$e_c(t) \leq \max(0, e_c(0+) - \tfrac{1}{2}\varepsilon_c t), \qquad t \geq 0.$$

In particular, $e_c(t) = 0$ for $t \geq 2e_c(0+)/\varepsilon_c$.

Assumption (b) then implies that $q(t) = \hat{q}^{\kappa*}(t)$ for such t, and hence

$$\underline{T}^{\kappa\circ}(x) < \frac{2}{\varepsilon_c}e_c(0+) = \frac{2}{\varepsilon_c}\Big(c(x) - c(\mathcal{P}^*(x))\Big).$$

This proves (ii) and (iii) since c linear, and results (iv) and (v) follow immediately. $\qquad\square$

6.7 Relaxations and policy synthesis for stochastic models

We close this chapter with some remarks on policy synthesis techniques for stochastic models. A general model is the generalization of the CRW model (4.2),

$$Q(t+1) = Q(t) + \sum_{i=1}^{\ell_u}\sum_{j=1}^{\ell} M_i(t+1)[-\mathbf{1}^i + R_{ij}(t+1)\mathbf{1}^j]U_i(t) + A(t+1), \tag{6.61}$$

where as usual $Q(0) = x$ is given as initial condition, M_i is Bernoulli, A takes values in $\mathbb{Z}_+^\ell$, and (M, A, R) is i.i.d. The entries of the matrix sequence R are Bernoulli, satisfying

$$\sum_j R_{ij}(t) \in \{0, 1\}, \qquad i = 1, \ldots, \ell_u, \ t \geq 1.$$

We will not repeat all of the translation techniques that have been introduced for the simpler scheduling model in previous chapters. For example, we leave it to the reader to contemplate how to formulate a discrete-review policy for a real-life network based on a policy for the fluid model.

A question we would like to answer is, *how can Theorem 6.6.15 be generalized to the CRW model?* Unfortunately, this question is far too big for this book. We consider three aspects of this question in this section. For more precise answers the reader is referred to the research literature, starting with the Notes section of this chapter.

(i) If we are very optimistic we can hope for approximate sample path coupling that mirrors the conclusion of Theorem 6.6.15. Results of this kind can be found in the literature on networks in heavy traffic. In Section 6.7.1 we use a single example to illustrate the sample path behavior of Q and a relaxation $\widehat{Q}$ under a myopic policy.

(ii) Alternatively, suppose we have optimized a workload relaxation. We then face two questions: How can we translate the policy for the relaxation to obtain an effective policy for Q? And, in what sense can we compare Q with its optimized relaxation? When there is a single bottleneck, a policy can be obtained based on an h-myopic policy or through logarithmic safety stocks that is approximately average-cost optimal in heavy traffic. This is again a technical topic, so we present the ideas using a single example in Section 6.7.2.

(iii) Extensions to networks with multiple bottlenecks is more complex precisely because the relaxation may have a complex optimal policy. The hedging techniques described in Section 5.6 can be used to construct a policy that mirrors the structure of the optimal policy. Quantifying the gap in performance between the CRW model and its relaxation when there are multiple bottlenecks remains a significant open problem.

6.7.1 State-space collapse under a myopic policy

To illustrate the nature of state-space collapse in a stochastic model we consider a setting somewhat simpler than Theorem 6.6.15. Rather than the optimal policy, we apply a particular myopic policy to the CRW model, and consider the resulting sample path behavior. The myopic policy for a relaxation is defined exactly as for the unrelaxed model using (4.15). Definition 5.4.2 gives a formal definition for the workload in units of inventory for the homogeneous scheduling model.

We might expect (or hope) that the resulting workload processes will evolve in a neighborhood of the monotone region. For the relaxation we *define* $\widehat{Q}(t) := \mathcal{X}^*(\widehat{W}(t))$ so that $\widehat{Q}(t) = \mathcal{P}^*(\widehat{Q}(t))$ holds by definition. This may only hold after a long transient period in the unrelaxed model if $W(0)$ is a long distance from W^+. With a bit more optimism we will expect to have $Q(t) \approx \mathcal{X}^*(W(t))$ following a transient period (again depending on how far $Q(0)$ is from the optimal exchangeable state $\mathcal{P}^*(Q(0))$).

If these expectations hold true, then we will not be surprised to observe $Q(t) \approx \widehat{Q}(t)$ under a myopic policy.

Example 6.7.1 (Coupling in the input-queued switch CRW model). We now return to the setting of Example 6.6.11. The network is balanced, and we consider the six-dimensional workload model. The MaxWeight matching policy is applied to the CRW

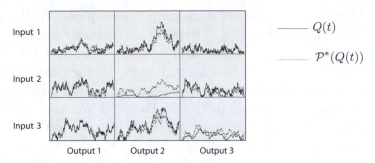

Figure 6.15. Evolution of the queue length process along with the projection onto the effective state in a 3×3 switch running MaxWeight matching. The queue size $Q_{ij}(t)$ tracks the effective state closely in most cases.

model. This is a special case of the MaxWeight policy in which the diagonal entries of D are all equal to unity.

In a relaxation we have $\widehat{Q}(t) = \mathcal{X}^*(\widehat{W}(t))$ for all $t \geq 1$ under the MaxWeight matching policy, and moreover $\widehat{W}(t)$ evolves in a neighborhood of the monotone region W^+. In the unrelaxed model we expect that these conclusions should hold approximately, and the approximations should become increasingly accurate for increasing load.

Figure 6.15 illustrates this state-space collapse in a particular numerical example. Service is taken deterministic with rate one, and the arrival process is Bernoulli with arrival-rate matrix

$$\alpha = \begin{bmatrix} 0.143 & 0.435 & 0.417 \\ 0.435 & 0.002 & 0.558 \\ 0.417 & 0.558 & 0.020 \end{bmatrix}$$

The network is balanced with $\rho_\bullet = 0.995$.

Each of the nine cells in Fig. 6.15 shows a trajectory of $Q_{ij}(t)$ and its projection $\mathcal{P}^*(Q(t))$. In each case except the $2, 2$-cell the approximation $Q_{ij}(t) \approx [\mathcal{P}^*(Q(t))]_{ij}$ is accurate over the length of the run, $t = 0, \ldots, 5{,}000$.

We are *not* saying that $Q(t) \approx \widehat{Q}(t)$, where $\widehat{Q}$ is the relaxation controlled using the MaxWeight policy. This approximation may fail because the workload for the unrelaxed model may idle significantly under the MaxWeight matching policy. ∎

6.7.2 Coupling mean behavior

Sample path coupling is well beyond our needs in the context of policy synthesis. What we would find very useful is a set of results stating that a policy based on a relaxation is nearly optimal. Failing this, we would like techniques to obtain useful upper bounds on the performance of the policy.

The following example shows how tight bounds can be obtained in a heavy traffic setting for a network possessing a single bottleneck.

Example 6.7.2 (Logarithmic regret in the simple routing model). We return to the two-dimensional model in the setting of Example 6.6.14, except that now we consider the CRW model given in (2.34). Recall that it is assumed (without loss of generality) that $c_1 \leq c_2$, and in "Case 2" we have equality. In the fluid model the myopic policy based on a linear cost function with $c_1 < c_2$ requires that all arrivals be routed to buffer 1, up until the first time that $q_2(t) = 0$. From this time up until the emptying time for the network, the policy maintains $q_2(t) = 0$, but sends material to this buffer at rate $\min(\alpha_r, \mu_2)$, so that $\zeta_1 + \zeta_2 = 1$ for all $0 \leq t < (x_1 + x_2)/(\mu_1 + \mu_2 - \alpha_r)$. This policy is pathwise optimal for the fluid model.

Theorem 6.6.15 states that a policy can be designed to enforce $q(t) = \widehat{q}^*(t)$ in the fluid model (and hence also $q(t) = \mathcal{P}^*(q(t))$) following a transient period. In Case 1 the projection is uniquely defined by $\mathcal{P}^*(x) = |x|\mathbf{1}^1$ since the inequality $c_1 \leq c_2$ is strict. For the CRW model we now ask, in what sense can we enforce an approximation of the form $Q(t) = \widehat{Q}^*(t)$?

To obtain a true relaxation in the CRW model we consider workload in units of time, $Y(t) = Q_1(t) + Q_2(t)$. As in the scheduling model the relaxation is denoted $\widehat{Y}$, and is defined so that its dynamics mimic those of Y,

$$\widehat{Y}(t+1) = \widehat{Y}(t) + A(t+1) - (S_1(t+1) + S_2(t+1)) + (S_1(t+1) + S_2(t+1))\widehat{I}(t).$$

The idleness process $\widehat{I}$ is constrained to evolve on $\{0, 1, 2, \dots\}$. The effective cost is monotone, so the optimal policy is nonidling, $\widehat{I}^*(t) = \mathbf{1}\{\widehat{Y}(t) \geq 1\}$, $t \geq 0$. Under any policy, the two-dimensional state process is defined as the effective state $\widehat{Q}(t) := \mathcal{X}^*(\widehat{Y}(t))$, which in Case 1 becomes $\widehat{Q}(t) = (Y_1(t) + Y_2(t))\mathbf{1}^1$.

To obtain $Q(t) = \widehat{Q}^*(t)$ requires the approximation $Q(t) \approx \mathcal{P}^*(Q(t))$ *and* approximate minimality of $Y(t) = Q_1(t) + Q_2(t)$. To achieve this, consider the following dynamic safety-stock policy based on the logarithmic switching curve (4.71): Customers are routed to buffer 2 whenever

$$Q_2(t) \leq s_\theta(Q_1(t)), \qquad t \geq 0. \tag{6.62}$$

For comparison we also revisit the "direct translation" of the fluid policy,

$$U_2^r(t) = \mathbf{1}(Q_2(t) = 0), \qquad t \geq 0. \tag{6.63}$$

If the policy (6.63) is applied we have $Q_2(t) \leq 1$ for all t provided $Q_2(0) \leq 1$. It then follows that $\|Q(t) - \mathcal{P}^*(Q(t))\| \leq 1$ for all t, regardless of load. However, the simulation shown on the left in Fig. 2.16 (and repeated in Fig. 6.16) shows that Y is far from minimal.

Under the policy (6.62) based on the switching curve s_θ we have

$$Q(t) - \mathcal{P}^*(Q(t)) = Q_2(t)(\mathbf{1}^2 - \mathbf{1}^1),$$

and hence $\|Q(t) - \mathcal{P}^*(Q(t))\| \leq \sqrt{2}(1 + s_\theta(Q_1(t))) = \sqrt{2}(1 + \theta \log(1 + \theta^{-1}Q_1(t)))$.

If idling is rare under the dynamic safety-stock policy this suggests that we can expect approximate optimality with error that grows logarithmically in $\mathsf{E}[\log(1 + Q_1(t))]$.

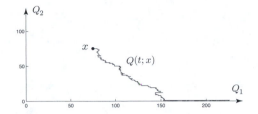 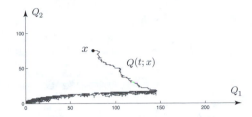

Figure 6.16. The simple routing model was simulated using the CRW model satisfying (2.36). In the simulation shown on the left the priority policy (6.63) was used, and the resulting queue length process $\mathbf{Q}$ explodes along the Q_1-axis. The simulation on the right shows results obtain using the policy (6.62) with $\theta = 5$.

In fact, this is true: If the parameter $\theta > 0$ is chosen sufficiently large then the dynamic safety-stock policy is approximately optimal with *logarithmic regret*. This means that for some constant K_1, independent of load,

$$\widehat{\eta}_* \leq \eta_* \leq \eta_\theta \leq \widehat{\eta}_* + K_1 |\log(1 - \rho_\bullet)|, \tag{6.64}$$

where η_* denotes the optimal steady-state cost, η_θ denotes the steady-state cost using the policy (6.62), and $\widehat{\eta}_*$ is the optimal steady-state cost for the relaxation. Based on the effective cost $\bar{c}(y) = c_1 y$, $y \in \mathbb{R}_+$, we obtain from Theorem 3.0.1 or Proposition 5.4.3,

$$\widehat{\eta}_* = \tfrac{1}{2} c_1 \frac{\sigma^2}{\mu_1 + \mu_2 - \alpha_r}, \tag{6.65}$$

where $\sigma^2 = \rho_2 m^2 + (1 - \rho_2) m_A^2$, with $m^2 = \mathsf{E}[(S_1(1) + S_2(1) - A(1))^2]$ and $m_A^2 = \mathsf{E}[(A(1))^2]$.

In Case 2 where the cost parameters are equal then *any* time-optimal policy is optimal for the fluid model. In particular, the policy defined by the linear switching curve $s(x_1) = \varpi x_1$, $x_1 \geq 0$ is optimal, where the constant $\varpi > 0$ is fixed, but arbitrary. In this case we find that insensitivity in the fluid model is reflected in the CRW model.

To make these statements precise consider a one-dimensional family of models defined with increasing load. Suppose that $(\mathbf{A}^1, \mathbf{S}_1, \mathbf{S}_2)$ defines a nominal model for which $\rho_\bullet = 1$ (i.e., $\mathsf{E}[A^1(k)] = \mu_1 + \mu_2$). For each $\kappa \geq 1$ we define a thinning of this arrival stream by $A^\kappa(k) = T(k) A^1(k)$, where $\mathbf{T}$ is an i.i.d. Bernoulli process that is indpendent of $(\mathbf{A}^1, \mathbf{S}_1, \mathbf{S}_2)$ with $\mathsf{P}\{T(k) = 1\} = 1 - \kappa^{-1}$. The system load is given by $\rho_\bullet^\kappa = \kappa$ for each $\kappa \geq 1$.

Proposition 6.7.3. *Suppose that $\theta > 3/\beta_2$, where $\beta_2 > 0$ solves*

$$\mathsf{E}[\exp(\beta_2(S_2(k) - A^1(k)))] = 1. \tag{6.66}$$

Suppose moreover that $\mathsf{E}[(S_1(1) + S_2(1) - A^1(1))^2] > 0$. Then, the following hold for the controlled network:

(i) *The bounds (6.64) hold for the steady state cost, with $\widehat{\eta}_*$ is given in (6.65). Equivalently, for some fixed constant K_1,*

$$\eta_\theta \leq \widehat{\eta}_* + K_1 \log(\kappa), \qquad \kappa \geq 1.$$

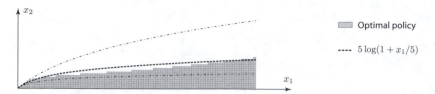

Figure 6.17. Optimal policy for the simple routing model. The grey region shows the optimal switching curve obtained using value iteration for the simple routing model. Within this region, arriving packets are routed to buffer 2. The three concave curves are plots of the switching curve $s_\theta(x_1) = \theta \log(1 + x_1/\theta)$ for $\theta = 2, 5, 20$. The value $\theta = 5$ yields a remarkably accurate approximation to the average-cost optimal policy.

(ii) *If $c_1 = c_2$ and the linear switching curve is used, then for some $K_2 < \infty$,*

$$\widehat{\eta}_* \le \eta \le \widehat{\eta}_* + K_2, \qquad \kappa \ge 1. \tag{6.67}$$

Proof. The proof contained in [363] is based on the construction of a function h solving a bound similar to the Poisson inequality (4.13) with $\overline{\eta} \approx \widehat{\eta}_*$. It is a perturbation of the fluid value function of the form $h = J^* + b$ where

$$b(x) = \frac{c_1}{\mu_1 + \mu_2 - \alpha_r}\Big((\mu_1 x_2 + m_A^2)k_1 e^{-\beta_1 x_1} + (\mu_2 x_1 + m_A^2)k_2 e^{-\beta_2 x_2}\Big), \tag{6.68}$$

with β_2 defined in (6.66) and β_1 defined analogously. $\qquad\square$

Shown on the right in Fig. 6.17 is the average-cost optimal policy for this stochastic model, as well as several instances of the logarithmic switching curve (4.71). It is apparent that the optimal policy can be closely approximated by a switching curve of this simple logarithmic form. In this experiment the statistics for the CRW model are defined in (2.36), so that in particular A is a Bernoulli process. The network parameters are $\alpha_r = 9/19$, and $\mu_1 = \mu_2 = 5/19$ so that $\rho_\bullet = \alpha_r/(\mu_1 + \mu_2) = 9/10$. The linear cost function is defined by $c(x) = 2x_1 + 3x_2$. $\qquad\blacksquare$

6.7.3 Hedging for MaxWeight

Hedging was introduced in Section 5.6 as a way to reduce the risk of high costs and idleness when constructing control solutions based on a workload model. Here we specialize this technique to obtain refinements of the MaxWeight policy.

As in Section 5.6, hedging will be defined narrowly as a policy based on an enlargement of the monotone region for the effective cost. Since the MaxWeight policy is an h-MaxWeight policy, we expect that the workload process will evolve in a neighborhood of the monotone region for the effective cost obtained from h. If this monotone region is small, this means that the workload process will be far from minimal.

Proposition 6.6.12 contains a characterization of the effective cost when the cost function is quadratic. For example, if $n = 2$ then the proposition implies that W^+ is the convex cone generated by the two vectors

$$w^i = \widehat{\Xi}D^{-1}\widehat{\Xi}^\mathsf{T}\mathbf{1}^i = \widehat{\Xi}D^{-1}\xi^i, \qquad i = 1, 2.$$

That is, $\mathsf{W}^+ = \{\psi_1 w^1 + \psi_2 w^2 : \psi \in \mathbb{R}^2_+\}$. If $D = I$ then these vectors become,

$$w^1 = \begin{pmatrix} \langle \xi^1, \xi^1 \rangle \\ \langle \xi^2, \xi^1 \rangle \end{pmatrix}, \qquad w^2 = \begin{pmatrix} \langle \xi^1, \xi^2 \rangle \\ \langle \xi^2, \xi^2 \rangle \end{pmatrix}$$

The monotone region is small if w^1 and w^2 are nearly aligned. In this case the myopic policy can result in large average workload values.

In this case performance can be improved with hedging.

Definition 6.7.4 (MaxWeight policy with hedging). Suppose that $D > 0$ is a diagonal $\ell \times \ell$ matrix, fix $\bar{\psi} > 0$, and let $\bar{x} = D^{-1}\widehat{\Xi}^{\mathsf{T}}\bar{\psi}$. The *MaxWeight policy with hedging* is the h-MaxWeight policy using,

$$h(x) = \tfrac{1}{2}(x + \bar{x})^{\mathsf{T}} D(x + \bar{x}), \qquad x \in \mathbb{R}^\ell. \tag{6.69}$$

∎

For the CRW model (6.61) we have for any $u \in \mathsf{U}$, with $\{\Theta_j\}$ defined in (6.33),

$$\langle \nabla h\,(x), Bu \rangle = \sum_{j=1}^{\ell_u} \Theta_j (x + \bar{x}) u_j.$$

If simultaneous resource possession is absent, that is, the rows of C are orthogonal, then the description of the policy in (6.37, 6.38) is modified as follows: For each resource r (corresponding to a row of C), given $Q(t) = x$,

(i) if $\overline{\Theta}_r(x + \bar{x}) < 0$ then $U_j(t) = 0$ for each $j \in \mathcal{I}_r(x)$;
(ii) if $\overline{\Theta}_r(x + \bar{x}) > 0$ then

$$\sum_{i \in \mathcal{I}_r} \{\phi_i(x) : \Theta_i(x + \bar{x}) = \overline{\Theta}_r,\ x_i \geq 1\} = 1.$$

The introduction of hedging does not have any impact on stability. The proof of Theorem 6.7.5 is very similar to the proof of Theorem 4.8.4. We leave this as an exercise (see Exercise 6.9).

Theorem 6.7.5. *Suppose that $\rho_\bullet < 1$, $\mathsf{E}[\|A(1)\|^2] < \infty$ in the CRW model (6.61), and that the MaxWeight policy with hedging is applied for some diagonal matrix $D > 0$, and n-dimensional vector $\bar{x} > 0$. Then the pair of functions $V(x) = bh(x)$, $c(x) = |x|$ solve the Poisson inequality (4.13) for some sufficiently large $b > 0$.*

In a workload relaxation the policy introduced in Definition 6.7.4 results in $\widehat{W}(t) \in \mathsf{W}^+$ for all t, with W^+ the monotone region with respect to the "cost function" h. Proposition 6.7.6 compares the monotone region obtained with and without the vector $\bar{x}$. Based on this result we conclude that the monotone region increases when the components of the vector $\bar{\psi}$ are increased.

The reason for the special form of the vector $\bar{x}$ assumed in Definition 6.7.4 is illustrated in Fig. 6.18. The vector $\bar{w} = \widehat{\Xi}\bar{x}$ lies in the interior of the monotone region constructed in Proposition 6.6.12. A simulation of the MaxWeight policy with and without hedging is shown in Fig. 6.19.

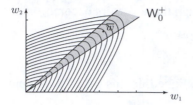
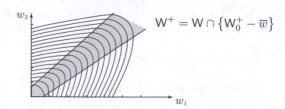

Figure 6.18. On the left is the monotone region for a two-dimensional relaxation under the MaxWeight policy. On the right is the monotone region obtained with hedging. The curves indicate level sets of the effective cost $\bar{h}$ corresponding to the particular quadratic "cost" h used in the definition of the policy. On the left this is $h(x) = \frac{1}{2}x^{\mathrm{T}}Dx$, and on the right $h(x) = \frac{1}{2}(x+\overline{x})^{\mathrm{T}}D(x+\overline{x})$.

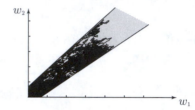

Figure 6.19. Two simulations of a CRW workload model: Shown on the left is a trajectory obtained using the MaxWeight policy, and on the right the MaxWeight policy with hedging. The mean workload is reduced significantly with the introduction of hedging.

Figure 6.18 also contains the essentials of the proof of Proposition 6.7.6:

Proposition 6.7.6. *Suppose that D, $\bar{\psi}$, and h are given in Definition 6.7.4. Let W_0^+ denote the monotone region for the effective cost with respect to $c_0(x) = \frac{1}{2}x^{\mathrm{T}}Dx$, and let W^+ denote the monotone region with respect to the cost function $c \equiv h$. Then*

$$\mathsf{W}^+ = \mathsf{W} \cap \{\mathsf{W}_0^+ - \overline{w}\}$$

where $\overline{w} = \widehat{\Xi}D^{-1}\widehat{\Xi}^{\mathrm{T}}\bar{\psi}$.

Proof. We have $w \in \mathsf{W}^+$ if and only if $w \in \mathsf{W}$ and $w + \overline{w} \in \mathsf{W}_0^+$. □

6.8 Notes

The material in Sections 6.1 and 6.2 is adapted from [361], which again borrows heavily from the literature on networks in heavy traffic, in particular [292, 241, 239, 236, 238, 83, 84, 237]. The linear program (6.5) that defines load is the *static planning problem* of Harrison [236, 238]. In this work and in several sequels it is assumed that the load is unity and the system is balanced, in the sense that the value of the linear program below is equal to one,

$$\begin{aligned} \min \ \varrho \quad \text{s.t.} \quad B\zeta + \alpha &= 0 \\ C\zeta &= \varrho\mathbf{1} \\ \zeta &\geq 0. \end{aligned}$$

This is simply (6.5) with the inequality constraint $C\zeta \leq \varrho1$ replaced by equality. In this special setting the analysis is simplified, but the story remains subtle. For example, in a simple example introduced in [84], the constraint region in this LP contains no extreme points. The lack of extreme points in the dual is related to the fact that the primal may be unbounded: The arrival-free model is not stabilizable in their example.

A central question in [292, 236, 238, 84] is the number of workload vectors corresponding to $\rho_s = 1$. Many interesting examples are presented in these papers, and in Laws' thesis [333]. In particular, examples in [292] show that pooled resources may overlap, and they cannot in general be identified via cut constraints.

The matrix classification Definition 6.4.2 is taken from the economics literature. The name honors Wassily Leontief, the creator of input–output analysis in economics. Bramson and Williams [84] call a network *unitary* if $-B$ is Leontief, and in addition simultaneous resource possession is absent.

The optimal routing problem described in Section 6.3.2 is known as the *single-source shortest path problem* in graph theory. A generalization is the *all-pairs shortest path problem* in which optimal paths are sought for each pair of vertices.

The Bellman–Ford algorithm is just one of many existing algorithms for solving shortest the path problem. Dijkstra's algorithm is probably more popular in practice; another is Johnson's algorithm [43, 485]. The *OSPR* protocol, also known as *shortest path first*, is a variant of the Bellman–Ford algorithm which is implemented in a distributed fashion. Algorithms for multicommodity flow problems and further references are contained in [352, 181, 180].

Pathwise optimality for *join the shortest queue* in routing was established by Winston [505] and Weber [486]. Ephrimedes, Varaiya, and Walrand [164] extend these results to investigate the role of information. If the queue-length process is not observed then it is sensible to distribute work more evenly to guard against starvation of resources.

There is an extensive literature on optimal routing for Markovian models. See Hariharan, Kulkarni, and Stidham [227] and Stidham and Weber [460] for extensive surveys. The first heavy-traffic analysis of routing is contained in the 1978 paper of Foschini and Salz [190].

In this chapter it is assumed that arrival rates to the system are uncontrolled. There is a growing body of work on congestion control for networks with *elastic traffic* consisting of flows competing for access to communication links. The models considered in the next chapter allow admission control so that arrival rates are determined by the policy employed.

The landmark work of Kelly and Gibbens [206] introduces economic intuition to allocate resources "fairly" and efficiently in an equilibrium model. Suppose that there are ℓ_f flows accessing the system with equilibrium rates $\{r_m\}$. It is assumed that the mth flow measures the value of its rate via a *utility function* $\mathcal{U}_m \colon \mathbb{R}_+ \to \mathbb{R}_+$, and the overall utility is defined as the sum

$$\mathcal{U}(r) := \sum_{m=1}^{\ell_f} \mathcal{U}_m(r_m).$$

On maximizing U over $r \in \mathbb{R}_+^{\ell_f}$ one obtains a so-called *fair allocation*. The utility functions can be designed to provide differentiated service performance to different classes of traffic.

This allocation rule is very similar in spirit to the MaxWeight and back-pressure algorithms. And, like these earlier methods, the utility-function approach leads to decentralized algorithms for flow control. It is possible to combine MaxWeight and utility function algorithms to perform simultaneous scheduling and flow control in a dynamic setting [165]. Srikant's monograph [459] contains much more on resource sharing in the Internet.

Wireless models related to the ALOHA model are considered in a heavy-traffic setting in [462, 166, 58]. In [58] a two-user wireless *downlink* model is considered. Under certain conditions a one-dimensional relaxation similar to that described in Example 6.2.5 is justified through a heavy-traffic limit. If the capacity region contains a "corner" near the arrival rate vector α, then a natural *two-dimensional* relaxation is obtained in this limit.

In the context of computer systems, models to analyze performance of networks subject to simultaneous resource possession began in the late seventies [421, 281], and one of the earliest papers on diffusion approximations for networks concerns precisely this topic [113]. The "dead-lock" that can occur in computer systems in which competing jobs require common resources is a central topic in the field of discrete-event systems [88].

The theory behind the input-queued switch described in Example 6.5.2 and Example 6.6.5 is largely taken from recent work of D. Shah and D. Wischik. In particular, Fig. 6.15 is based on a figure in [435]. In the influential paper of Karol et al. [284] it is shown that certain simple scheduling policies, including a priority policy and a certain greedy policy, are destabilizing for sufficiently large load in this model.

Exercises

6.1   The schematic shown below is a very simple instance of the network models used to describe call centers. Many other examples can be found in [199].

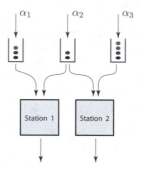

This network is similar to the processor sharing model described in Section 2.5. It is *not* similar to the model shown in Fig. 6.9 since simultaneous resource possession is absent.

Either station can process the contents of buffer 2; the other two buffers are served only by the stations shown. The total rate of service at Station i is limited by $\zeta_{ii} + \zeta_{i2} \leq 1$, where ζ_{ij} denotes the rate at which material at buffer j is processed at Station i. Hence in a fluid model $\frac{d}{dt} q_1(t) = \alpha_1 - \mu_1 \zeta_{11}$, $\frac{d}{dt} q_2(t) = \alpha_2 - \mu_1 \zeta_{12} - \mu_2 \zeta_{22}$, etc.

(a) Write down the 3×4 matrix B, and the 2×4 constituency matrix C.

(b) Compute the worklad vectors $\{\xi^i\}$, and the network load.

(c) Compute the c-myopic policy with $c(x) = x_1 + 2x_2 + x_3$.

(d) Construct a time-optimal policy.

(e) Find conditions on the service and arrival rates so that resource pooling is critical. That is, the load $\rho_\bullet$ is associated with a pooled-resource.

(f) Obtain a policy for q based on the one-dimensional relaxation under the conditions of (v).

You will return to this example in Exercise 6.6.

6.2   Repeat the previous exercise for the scheduling/routing model shown below:

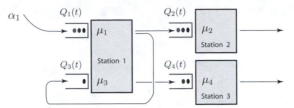

The routing at buffer 1 is controlled. You may take the cost function of your choice in Part (iii).

6.3   The example shown below is taken from Bramson and Williams [84].

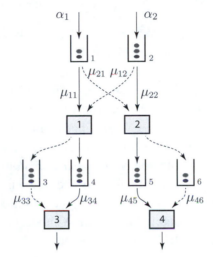

There are four stations, each of which can work on one of two buffers at any time. Consider the evolution of the contents of the first buffer. There is a constant arrival rate α_1, and it can be drained through activities at servers 1 and 2. Processing at server 1 results in material being sent to buffer 4, and processing at server 2 results in material being sent to buffer 6. The fluid model dynamics at this buffer are expressed $\frac{d^+}{dt} q_1(t) = \alpha_1 - \mu_{11}\zeta_{11}(t) - \mu_{21}\zeta_{21}(t)$.

Note that if the "cross-paths" are never used (shown as dashed lines, with rates μ_{12}, μ_{21}, μ_{33}, and μ_{46}) then the system reduces to two independent tandem queues.

Consider $\alpha_1 = \alpha_2$, $\mu_{12} = \mu_{21} = \frac{1}{2}$, and $\mu_{ij} = 1$ for all other i, j.

(a) Show that $\rho_\bullet = \alpha_1$ by solving the LP (6.5). Find the unique optimizer ζ^*.

(b) The dual LP has several optimal solutions (ξ, ν). Show that one is given by

$$\xi = \tfrac{1}{2}(1, 1, -1, 0, 0, -1)^\mathsf{T} \quad \nu = \tfrac{1}{2}(1, 1, 0, 0)^\mathsf{T}.$$

Note that ξ has negative entries.

(c) Find (ξ^*, ν^*) satisfying the complementary slackness condition (6.20). *Note*: that this does not require a computer. Two solutions can be obtained by realizing that certain activities do not pay off.

6.4   Consider the single-user routing model shown below. Suppose that each link
       has capacity 2, except for link $(1, 3)$ which has capacity 1.

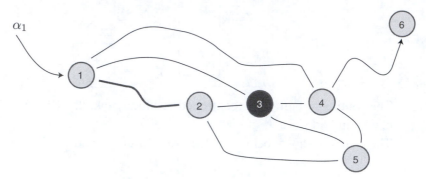

       What is the maximal rate α_1 that data can be sent from node 1 to node 6?
       Which link constraints are binding?
       How does your answer change if node 3 has finite capacity: the sum of traffic
       passing through this node is subject to the strict bound $\mu < \infty$.

6.5   How would you generalize the Min-Cut Max-Flow theorem beyond the single-
       user model? For example, what happens if a second user requires a link which
       is a bottleneck for the first user? Should the second user get none of the network
       capacity? How would you define a *maximal flow* for a two-user model?

6.6   Compute the MaxWeight policy for the following models: (a) the simple rout-
       ing model, (b) the processor sharing model introduced in Section 2.5. Are the
       resulting policies stabilizing in a CRW model?

6.7   Consider the ALOHA communications model described in Section 2.4 and
       Example 6.2.5. For the velocity set V you can take any region that is consistent
       with the set shown on the left in Fig. 6.5, but assume that the workload vector
       ξ^2 shown in Fig. 6.13 satisfies $\xi_1^2 > \xi_2^2$. For a cost function take $c(x) = |x|$,
       $x \in \mathbb{R}_+^2$.

       (a) Compute the optimal value function $\widehat{J}^*$ for the one-dimensional relaxation
           of the fluid model.
       (b) Obtain a policy for q based on the relaxation that couples with $\widehat{q}$ (see
           Example 6.6.14).
       (c) Construct a two-dimensional CRW model that is consistent with the fluid
           model.
       (d) Construct a policy for Q based on a translation of the policy obtained
           for the fluid model. One possibility is the h-MaxWeight policy based on
           (4.94), using the function h_0 defined in (E5.1):
$$h_0(x) = \widehat{J}^*(\xi^2 \cdot x) + \tfrac{b}{2}\|x - \mathcal{P}^*(x)\|^2.$$

       Note: In Exercise 6.7 you are asked to simulate this policy, and in Exercise 6.5
       you are asked to improve this policy using value or policy iteration.

6.8   Prove Proposition 6.6.9.

6.9   Prove Theorem 6.7.5.

6.10 The network model shown on the right was introduced in [334] to illustrate a potential source of performance loss in the MaxWeight policy. With the exception of Station 3, each queue shown is a version of the single-server queue. The third queue $Q_3(t)$ evolves as the simple routing model, $Q_3(t+1) - Q_3(t) = S_2(t+1)U_2(t) - S_{3,4}(t+1)U_{3,4}(t) - S_{3,5}(t+1)U_{3,5}(t)$ with the constituency constraint that $U_{3,4}(t) + U_{3,5}(t) \leq 1$.

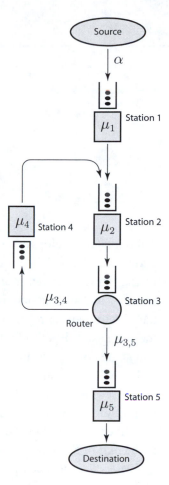

A centralized routing algorithm intended to minimize delay would never route any packets to Station 4. Though stabilizing, the MaxWeight or back-pressure routing algorithm will route customers to buffer 4 for certain values of (x_3, x_4, x_5).

(i) Obtain an expression for the network load, compute all workload vectors, and describe a time-optimal policy.

(ii) Describe the general MaxWeight policy.

(iii) Simulate the back-pressure policy with the following specifications:

$$\mu_1 = \mu_{3,5}, \mu_4 = \mu_{3,4} = 2\mu_1$$

and with α chosen so that $\rho_\bullet < 1$.

Experiment with different statistics for the arrival process. How does the percentage of traffic routed to buffer 4 change as you increase the variance of A? How is this influenced by network load?

7

Demand

This chapter develops extensions of the fluid and stochastic network models to capture a wider range of activities. As in the previous chapters we allow scheduling and routing. In the demand-driven models considered in this chapter we also permit "admission control" of raw material arriving to the network so that the total amount of material in the system can be regulated. Although manufacturing systems will motivate most of the discussion in this chapter, power distribution systems as described in Section 2.7 and some communication systems can be modeled as demand-driven networks.

Figure 7.1 illustrates a typical example of the class of models to be investigated. In this 16-buffer network there are two sources of exogenous demand, and the release of two different types of raw material into the system is controlled. At two of the five stations there are multiple buffers so that scheduling is required, and routing is controlled at the exit of Station 3.

This is the most complex example considered in any detail in this book, although it is far simpler than a typical semiconductor wafer fab as described in Section 1.1.1. The International Semiconductor Roadmap for Semiconductors (ITRS) provides an annual assessment of the challenges facing the semiconductor industry [279]. In recent years their reports have contained some recurring themes:

Contention for resources There may be dozens of different product flows in a single factory. Occasionally a manager will demand a "hot-lot" or "super hot-lot." This is a group of products to be given special priority due to the whim of a customer who is willing to pay, or to test a new prototype.

Uncertainty The impact of failure of a single resource in the system has increasing impact in a highly integrated factory and supply chain. The ITRS roadmap also highlights uncertainty in demand due to rapid changes in the market, as well as complete factory restructuring following changes in business needs.

Information Communication among disparate factories may be insufficient. Conversely, information may be *overwhelming* due to the explosive growth in inexpensive data collection technology.

In addition, *complexity* is a theme that appears throughout the ITRS website.

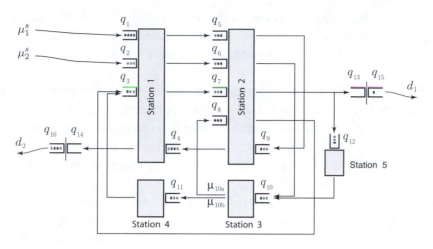

Figure 7.1. Demand-driven model with routing, scheduling, and rework.

In early 2000, unscheduled down-time was ranked as the most significant cause of capacity loss in semiconductor fabs according to [411]. Large deficits may be generated while a critical resource is temporarily unavailable, especially when upstream or downstream resources are forced into idleness due to an interruption of material flow. By using the available resources wisely the system can be placed in a position to return to its normal operating state quickly once the resource is restored. This is especially true in the case of planned maintenance where there is advanced notice of system down-time.

The ITRS website contains much discussion of some of the basic issues addressed in this book. The industry wishes to operate factories at very high loads; the figure $\rho_\bullet = 94\%$ is frequently cited, where the bottleneck resource in semiconductor manufacturing is lithography. It is recognized that monitoring workload so that bottleneck resources work almost continuously is essential to achieve such high throughput.

Nowhere in the ITRS roadmap is a declaration that the semiconductor industry requires more efficient techniques to solve the average-cost optimality equations, or new methods to verify ergodicity of their manufacturing models! Rather, the control issues cited by ITRS concern primarily response to crisis: How should the system respond when a resource is temporarily unavailable? When should preventative maintenance be performed? How can a hot-lot be processed quickly in a complex factory with many competing products? How can the impact of these disruptions be minimized?

In this chapter we begin with the fluid model in which we can construct first-order answers to these questions in the simplest setting. Following some discussion on modeling in Section 7.1, we describe in Section 7.2 how the GTO algorithm can be adapted to account for disruptions due to maintenance, unanticipated breakdown, or an unanticipated surge in demand.

Demand-driven models are also called *pull models*. A virtual queue is used to model inventory for each product: A negative value indicates deficit (unsatisfied demand), and a positive value indicates excess inventory. Our standing assumption that the arrival-free model is stabilizable is *violated* in a typical pull model. If there is excess inventory, then with no demand there is no mechanism to shed inventory unless an additional control is introduced.

To define workload we revisit the construction in the previous chapter. Recall that the workload vectors constructed in Theorem 6.1.1 were interpreted as Lagrange multipliers since they determine sensitivity of the minimal draining time with respect to buffer levels. To understand the structure of workload vectors in the setting of this chapter, suppose for example that the inventory at buffer 4 is increased in the network shown in Fig. 7.1. Will the minimal draining time increase? The answer is just the opposite: Increasing inventory at buffer 4 means that the second source of demand can be satisfied more quickly. For this reason we conclude in Section 7.1 that workload can be negative or positive. In Section 7.3 we investigate some of the unique features of workload relaxations for demand-driven models.

Once we have understood control solutions on a fluid scale we must consider how to improve reliability in the face of uncertainty. For this we take a closer look at the application of hedging. In manufacturing it appears that this technique was introduced by Henry Ford. In his 1922 book [187] he writes,

> We buy only enough to fit into the plan of production, taking into consideration the state of transportation at the time. If transportation were perfect and an even flow of materials could be assured, it would not be necessary to carry any stock whatsoever. The carloads of raw materials would arrive on schedule and in the planned order and amounts, and go from the railway cars into production. That would save a great deal of money, for it would give a very rapid turnover and thus decrease the amount of money tied up in materials. With bad transportation one has to carry larger stocks.

Note that Ford's emphasis is uncertainty of delivery. In the 16-buffer manufacturing example this is indicated by the two controlled arrival streams at buffers 1 and 2 shown in Fig. 7.1. Transportation remains a source of uncertainty in manufacturing applications among many other factors.

We begin in Section 7.4 with an analysis of a one-dimensional stochastic inventory model of the form introduced in Section 2.6. Computation of an optimal hedging-point policy amounts to a relatively simple calculus exercise in this model.

In moving to more complex networks we consider workload relaxations. We find in Section 7.5.1 that a one-dimensional relaxation of a demand-driven model coincides with the simple inventory model treated in Section 7.4. For models with a single bottleneck the identification of this hedging point combined with safety stocks to avoid idleness (as described in Section 4.6) provides a solution to the overall control problem.

For models with multiple bottlenecks a one-dimensional relaxation may not tell the entire story. In this case we construct a relaxation of higher dimension and consider how to construct multiple hedging points based on the structure of the average-cost or discounted-cost optimal control solution.

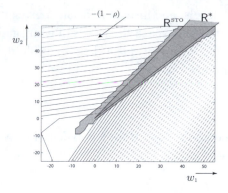
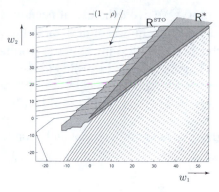

Figure 7.2. Optimal policies for two instances of the network shown in Fig. 7.1. In each figure the optimal stochastic control region R^{STO} is compared with the optimal region R^* obtained for the two-dimensional fluid model.

Section 7.5.2 describes approximation theory for optimal control of two-dimensional workload models based on a version of the *height process* introduced in Section 5.6. The control synthesis problem is reduced to the one-dimensional inventory model corresponding to the height process relative to a boundary of the monotone region of the effective cost.

While this sounds like a modest advancement over the one-dimensional case, in fact the main ideas extend to workload models of arbitrary dimension.

The approximations constructed in Section 7.5.2 are illustrated in the plots shown in Fig. 7.2. In each of these two plots the average-cost optimal policy for a CRW workload model is defined by the constraint region R^{STO} – the optimized workload process is minimal on this region. Also shown are level sets of the effective cost, and the constraint region R^* that defines the total-cost optimal solution for the fluid model. The numerical values for the effective cost and statistical assumptions for the CRW model are specified in Section 7.5.2, based on a two-dimensional relaxation of the network shown in Fig. 7.1.

In each case the region R^{STO} is approximated by an affine enlargement of the fluid-optimal policy. The parameters that define this affine enlargement are interpreted as hedging points at the two virtual queues.

In the next section we specify the class of network models to be considered in this chapter and take a look at workload for demand-driven models.

7.1 Network models

Here we construct models for networks of the form shown in Fig. 7.1 and begin to investigate the nature of workload for demand-driven models. We can restrict to the fluid model since variability plays no role at this stage.

Recall that two virtual queues are used in the 16-buffer model. A virtual queue takes on a negative value when demand cannot be met, and it is positive if there is surplus inventory. We avoid the use of negative buffer levels, and instead translate a given pull model into an equivalent push model in which the state space X consists of nonnegative buffer levels. Following this transformation, the vector of queue lengths in the fluid model again evolves according to the familiar equation

$$q(t) = x + Bz(t) + \alpha t, \qquad t \geq 0, \ x \in \mathbb{R}^\ell_+, \tag{7.1}$$

or the equivalent differential inclusion representation $\frac{d^+}{dt}q \in V$ for a polyhedral velocity space $V \subset \mathbb{R}^\ell$.

7.1.1 Pull to push translation

To obtain a cohesive theory we avoid the use of negative buffer levels, and instead interpret a virtual queue as two separate buffers. To place the resulting model in the standard form (7.1) we introduce a "virtual server" that combines completed work with current demand: An *exit resource* is a single-station model with $n \geq 2$ buffers, and a fluid model of the form

$$\frac{d^+}{dt}q_i(t) = -\mu\zeta(t) + \mu_{i-}\zeta_{i-}(t), \qquad t \geq 0, \ 1 \leq i \leq n, \tag{7.2}$$

where $\mu_{i-}\zeta_{i-}(t)$ represents the cumulative inflow. The single allocation process z satisfies $\zeta(t) = \frac{d^+}{dt}z(t) \in [0,1]$ for $t \geq 0$. Note that the single allocation rate $\zeta(t)$ is used for each of the n buffers. No buffer at the exit resource can be drained unless an equal amount is drained from each of the other buffers at the server.

Exit resources will form a part of the overall network. An example in manufacturing is the final assembly stage on a production line, where several components are combined to produce the finished product.

To model the release of work to the network we introduce the *supply resource*. This allows controlled output of raw material to the network, but has no buffer.

The construction of a fluid model in the following example can be generalized to any of the demand-driven models envisioned in this chapter.

Example 7.1.1 (Single-station demand-driven model). Consider the simple inventory model illustrated in Fig. 2.7. An equivalent push model is obtained by replacing the virtual queue with a single exit resource with two buffers whose contents are denoted q_s and q_d. These buffers are interpreted as surplus and deficit buffers, respectively. The surplus buffer is fed by the production resource, and the deficit buffer is fed by the demand process with instantaneous rate d. In this transformed model the arrival process is actually exogenous demand. A single supply resource is used to model controlled release of raw material to the system. This construction is illustrated on the right in Fig. 7.3.

Letting (q_p, z_p) denote the inventory and cumulative allocation at the production resource, z_n the cumulative allocation at the supply resource, z_d the cumulative allocation

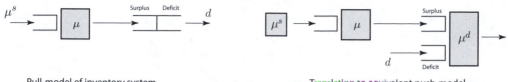

Pull-model of inventory system   Translation to equivalent push-model

Figure 7.3. Conversion of a pull model to a push model. The network on the right shows a supply resource, a production station, and an exit resource. If $\mu^d > \max(\mu, d)$ then the two models shown are equivalent.

at exit resource, and setting $q = (q_p, q_s, q_d)^{\mathsf{T}}$, $z = (z_p, z_d, z_n)^{\mathsf{T}}$, the resulting fluid model equations are given by (7.1) with

$$
B = \begin{bmatrix} -\mu & 0 & \mu^s \\ \mu & -\mu^d & 0 \\ 0 & -\mu^d & 0 \end{bmatrix}, \quad
\alpha = \begin{bmatrix} 0 \\ 0 \\ d \end{bmatrix}, \quad
C = \begin{bmatrix} 1 & 0 & 0 \\ 0 & 1 & 0 \\ 0 & 0 & 1 \end{bmatrix}. \tag{7.3}
$$

Thus, after the push-to-pull conversion, a single-resource queue becomes a three-resource system consisting of a production station, an exit resource, and a supply resource. The three resources are represented by the three rows of the constituency matrix C.

It is assumed that μ^d is much greater than any other rate parameters. It is also assumed that the exit resource is nonidling, so that $\zeta_d(t) = 1$ when $q_s(t) > 0$ and $q_d(t) > 0$. Consequently, if $\min(q_s(0), q_d(0)) = 0$, then $\min(q_s(t), q_d(t)) = 0$ for all $t \geq 0$. In this case, the two systems shown in Fig. 7.3 are equivalent.

The deficit buffer will accumulate deficit if the demand rate exceeds the arrival rate to the surplus buffer, and $q_s = 0$ so that no surplus material is present. This captures the output behavior of the pull model: The server combines purchases (demand) with product (the network output). ∎

7.1.2 What is workload?

Consider a general network model described by the differential equation (7.1), where ζ is constrained to a polyhedral set $\mathsf{U} \subset \mathbb{R}_+^{\ell_u}$ containing the origin. Some components of the vector α may represent rates of demand for goods produced by the system, as in the example considered in Section 2.6, or the model shown in Fig. 7.1.

The velocity space is again defined as $\mathsf{V} = \{\mathsf{V}_0 + \alpha\}$, where V_0 is the velocity space for the arrival-free model as defined in (6.10). The set V_0 is a polyhedron, and $0 \in \mathsf{V}_0$ since $0 \in \mathsf{U}$ by assumption. Consequently, the set V_0 can again be expressed as the intersection of half-spaces as in (6.11) for a finite collection of vectors $\{\xi^s : 1 \leq s \leq \ell_v\} \subset \mathbb{R}^\ell$, and constants $\{o_s : 1 \leq s \leq \ell_v\} \subset \{0, 1\}$.

The terminology used in Definition 6.1.2 is maintained in this general model. In particular, the vector $\xi^s \in \mathbb{R}^\ell$ is called a *workload vector* if $o_s = 1$. We denote by ℓ_r the number of distinct workload vectors, and assume that the indices are ordered so that $\{\xi^s : 1 \leq s \leq \ell_r\}$ are workload vectors.

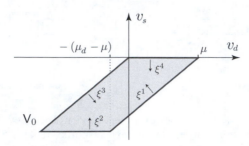

Figure 7.4. Velocity space V_0 in the single-station demand-driven model: arrival-free case, restricted to the v_s–v_d axis. This describes the projected dynamics with $q_p(t) \equiv 0$.

We begin with the following observations regarding the minimal draining time.

Proposition 7.1.2. *For any fluid model of the form (7.1):*

(i) *For any $x \in \mathbb{R}_+^\ell$, if $\langle \xi^s, x \rangle > 0$ for some $s > \ell_r$ (so that $o_s = 0$) then $W^*(x) = \infty$. Otherwise, the minimal draining time for the arrival-free model is expressed*

$$W^*(x) = \max_{1 \le s \le \ell_r} \langle \xi^s, x \rangle < \infty.$$

(ii) *For the model with arrivals, provided it is stabilizable, then the minimal draining time is expressed*

$$T^*(x) = \max_{1 \le s \le \ell_v} \frac{\langle \xi^s, x \rangle}{o_s - \rho_s}, \qquad x \in \mathbb{R}_+^\ell. \tag{7.4}$$

Proof. The proof of Proposition 6.1.5 can be adapted. It is necessary to take the maximum over all of the faces of V, rather than just those corresponding to $o_s = 1$, since we do not know that $\langle \xi^s, x \rangle \le 0$ and $o_s - \rho_s = |\rho_s|$ for $s > \ell_r$. $\qquad \square$

The following examples illustrate the construction of a fluid model and associated workload vectors for demand-driven models.

Example 7.1.3 (Single-station demand-driven model). Consider the simple network shown in Fig. 7.3. For simplicity we project onto the two-dimensional subspace $\{x_p = 0\}$. That is, the production resource is empty, $q_p(t) = 0$ for all $t \ge 0$, which is feasible provided $q_p(0) = 0$ and $\mu^s \ge \mu$. We thereby obtain a two-dimensional model with state $(q_s, q_d)^\mathsf{T}$ by ignoring the empty buffer at the production resource. The fluid model equations are of the form (7.1), with

$$B = \begin{bmatrix} \mu & -\mu^d \\ 0 & -\mu^d \end{bmatrix}, \quad C = \begin{bmatrix} 1 & 0 \\ 0 & 1 \end{bmatrix}, \quad \alpha = \begin{bmatrix} 0 \\ d \end{bmatrix}.$$

The velocity space V_0 is illustrated in Fig. 7.4. There are four faces, so that $\ell_v = 4$, and two of these faces pass through the origin. The vectors $\{\xi^s : 1 \le s \le \ell_v\}$ and

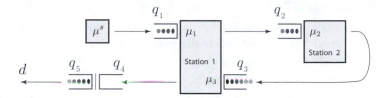

Figure 7.5. Simple re-entrant line with demand.

constants $\{o_s : 1 \leq s \leq \ell_v\}$ are given by

$$\xi^1 = \mu^{-1}\begin{pmatrix} -1 \\ 1 \end{pmatrix} \quad \xi^2 = (\mu^d)^{-1}\begin{pmatrix} 0 \\ 1 \end{pmatrix} \quad \xi^3 = \begin{pmatrix} 1 \\ -1 \end{pmatrix} \quad \xi^4 = \begin{pmatrix} 0 \\ -1 \end{pmatrix}$$

$$o_1 = 1 \qquad o_2 = 1 \qquad o_3 = 0 \qquad o_4 = 0.$$

We have $o_3 = o_4 = 0$ since the corresponding faces pass through the origin in $\mathbb{R}^2$.

Based on the notation introduced in Definition 6.1.2, for a given demand rate $d < \mu^s$, we have

$$(\rho_1, \ldots, \rho_4) = (\mu^{-1}d, (\mu^d)^{-1}d, -d, -d).$$

Whenever the origin is reachable from an initial condition $x \in \mathbb{R}_+^2$, the minimal draining time is expressed

$$T^*(x) = \max_{1 \leq s \leq \ell_r} \frac{\langle \xi^s, x \rangle}{o_s - \rho_s}$$

$$= \max\left\{ \frac{x_2 - x_1}{\mu - d}, \frac{x_2}{\mu^d - d}, \frac{x_1 - x_2}{d}, \frac{-x_2}{d} \right\}.$$

The arrival-free model is not stabilizable in any of the demand-driven models considered in this chapter. In this example it is evident that if there is no demand, then there is no mechanism to discard any excess inventory at the surplus buffer. Loss of stabilizability is also clear from the form of the velocity space V_0: If the arrival-free model is stabilizable, then there is an $\varepsilon > 0$ such that $-x \in V_0$ whenever $x \in \mathbb{R}^n$ with $\|x\| \leq \varepsilon$. ∎

Example 7.1.4 (Simple re-entrant line with demand). The network shown in Fig. 7.5 is similar to the simple re-entrant line shown in Fig. 2.9. It consists of two stations, with two buffers at the first station. The input is regulated by the policy, with an upper bound of $\mu^s > 0$ on the rate at which raw material can be brought into buffer 1.

This model is represented as the fluid model (7.1) with system parameters

$$B = \begin{bmatrix} -\mu_1 & 0 & 0 & 0 & \mu^s \\ \mu_1 & -\mu_2 & 0 & 0 & 0 \\ 0 & \mu_2 & -\mu_3 & 0 & 0 \\ 0 & 0 & \mu_3 & -\mu^d & 0 \\ 0 & 0 & 0 & -\mu^d & 0 \end{bmatrix}, \quad \alpha = \begin{bmatrix} 0 \\ 0 \\ 0 \\ 0 \\ d \end{bmatrix}, \quad C = \begin{bmatrix} 0 & 1 & 0 & 1 & 0 \\ 0 & 0 & 1 & 0 & 0 \\ 0 & 0 & 0 & 1 & 0 \\ 0 & 0 & 0 & 0 & 1 \end{bmatrix}.$$

$$(7.5)$$

Recall that the four rows of C correspond to the four resources in this model. The first two rows of the constituency matrix shown in (7.5) correspond to Stations 1 and 2, respectively. The third row corresponds to the exit resource with rate μ^d, and the fourth row of C corresponds to the supply resource with rate μ^s. Each of the parameters μ^d and μ^s is assumed to be very large.

Consider an initial condition x and allocation z such that the resulting state process q reaches the origin at some time T_0:

$$0 = q(T_0) = x + Bz(T_0) + \alpha T_0.$$

The allocation at time T_0 can be computed since the matrix B is square and invertible $z(T_0) = -B^{-1}[x + \alpha T_0]$, and feasibility then dictates the following vector bound:

$$T_0 \mathbf{1} \geq Cz(T_0) = -CB^{-1}[x + \alpha T_0].$$

As in the scheduling model, a *workload matrix* is defined by $\Xi = -CB^{-1}$, the four rows are denoted $\{\xi^i : i = 1, \ldots, 4\}$, and $\rho_i = \langle \xi^i, \alpha \rangle$. The bound above can be expressed as four scalar equations,

$$T_0 \geq \langle \xi^i, x \rangle + \rho_i T_0, \qquad 1, \ldots, 4.$$

Provided $\rho_i < 1$ for each i, we can subtract and divide to obtain the lower bound $T_0 \geq \max \langle \xi^i, x \rangle / (1 - \rho_i)$. Hence the vectors $\{\xi^i\}$ do play the role of workload vectors corresponding to the four resources in this network.

The first two workload vectors, corresponding to the two physical stations, correspond to the first two rows of Ξ:

$$\begin{aligned}
\xi^1 &= (0, -\mu_1^{-1}, -\mu_1^{-1}, -(\mu_1^{-1} + \mu_3^{-1}), \mu_1^{-1} + \mu_3^{-1})^{\mathsf{T}}, \\
\xi^2 &= (0, 0, -\mu_2^{-1}, -\mu_2^{-1}, \mu_2^{-1})^{\mathsf{T}}.
\end{aligned} \tag{7.6}$$

For a given demand rate d, the first two load parameters are given by

$$\rho_1 = (\mu_1^{-1} + \mu_3^{-1})d, \qquad \rho_2 = \mu_2^{-1}d. \tag{7.7}$$

Suppose that μ^d and μ^s are very large, and that the surplus buffer is empty ($q_4(0) = 0$). Then from the foregoing we obtain

$$T^*(x) = \max_{s=1,2} \frac{\langle \xi^s, x \rangle}{1 - \rho_s}.$$

As in the previous example we find negative entries in the workload vectors that define this minimal draining time. This is physically reasonable given the expression (7.4) for the minimal draining time T^*. The processing time required to meet demand is *reduced* if the quantity of material at buffers $2, 3$, or 4 is increased.

From the representation of T^* above it follows that $\xi_j^1 < 0$ for buffers $j = 2, 3, 4$, and $\xi_j^2 < 0$ for $j = 3, 4$. Similarly, $\xi_1^s = 0$, $s = 1, 2$, since the addition of material at buffer 1 will not reduce the time required to meet demand.   ■

7.2 Transients

We now describe policies that address a range of issues in a dynamic environment. In particular, we find that the GTO policy can be adapted to provide effective policies in the following circumstances:

(i) A *transient* demand is imposed, and is given priority over other products in the system.
(ii) Some component in the network is temporarily in-operable. During repair it is still necessary to choose allocations at those resources in the network that are functioning.
(iii) Some resources require preventative maintenance, so that some portion of the network is disabled for a period of time in the future. Decisions regarding allocations prior to maintenance will be made subject to the knowledge of approaching down-time, and subsequent maintenance.

The impact of these disturbances can be reduced if the control synthesis problem is solved using all relevant information. This is especially true in the case of scheduled maintenance. Given prior knowledge of station down-time, the network can be positioned such that starvation of active resources during the maintenance period is minimized.

7.2.1 GTO policy

In the general model (6.1) considered in this chapter the definition of the GTO policy is essentially unchanged: The allocation rate $\zeta(t)$ at time t is defined so that $\frac{d^+}{dt}c(q(t; x^0))$ is minimized, subject to state space constraints, and nonidling constraints to ensure time optimality.

We begin with some examples to illustrate the application of the GTO policy in a demand-driven model.

Example 7.2.1 (GTO policy for the simple re-entrant line with demand). The following service rates are used for the five-buffer demand-driven model shown in Fig. 7.5:

$$\mu_1 = \mu_3 = 22, \quad \mu_2 = 10, \quad \mu^s = \mu^d = 100. \tag{7.8}$$

With $d = 9.5$ we obtain from (7.7)

$$\rho_1 = 2\mu_1^{-1}d = 9.5/11, \qquad \rho_2 = \mu_2^{-1}d = 9.5/10.$$

Consider the initial condition $x = (10, 25, 55, 0, 100)^\mathsf{T}$, which corresponds to an initial deficit of 100 units. The internal buffers contain 90 units of inventory, so in addition to the input necessary to meet the normal demand of 9.5 units per unit time, an additional 10 units must be brought into the system.

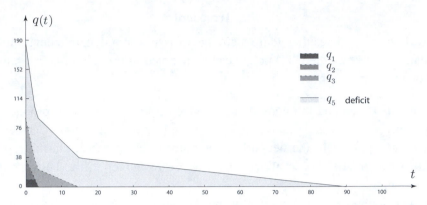

Figure 7.6. Buffer levels versus time for the simple re-entrant line with demand under the GTO policy. The y-axis indicates the *cumulative* buffer levels defined by $q_1(t)$, $q_1(t) + q_2(t)$, ..., $q_1(t) + \cdots + q_5(t)$.

Based on (7.6) we obtain the following expression for the minimal draining time from this initial condition,

$$T^*(x) = \max_{1 \le s \le 2} \frac{\langle \xi^s, x \rangle}{1 - \rho_s} = \max(40, 90) = 90. \tag{7.9}$$

Note that the index set used in the maximum in (7.9) must be enlarged if there is initial surplus, so that $x_4 > x_5 \ge 0$.

Recall the definition of a *dynamic bottleneck* for a scheduling model in Definition 4.3.3. The definition is maintained for the networks considered in this section. Since the second station is a dynamic bottleneck at time 0, it follows that $\zeta_2(t) \equiv 1$ for $0 \le t < 90$ under any time-optimal policy.

In formulating control solutions based on a cost function, one typically assigns cost to each internal buffer as well as virtual queues, and the cost of deficit is significantly higher than the cost assigned to excess completed inventory or internal buffers. Following this convention, a reasonable cost function for this model is linear, with

$$c(x) = \langle c, x \rangle, \qquad c = (1, 1, 1, 5, 10)^{\mathsf{T}}, \quad x \in \mathbb{R}_+^5. \tag{7.10}$$

That is, the holding cost for each unit of completed inventory is 5; the cost for each unit of unsatisfied demand is 10; and the holding cost for each unit of unfinished work is equal to 1. The evolution of buffer levels versus time under the GTO policy based on this cost function is illustrated in Fig. 7.6.

State space collapse is observed for $t > 13$ (approximately), when all buffers in the network are empty except the deficit buffer.   ∎

Example 7.2.2 (Multiple demands). Consider the network shown in Fig. 7.7 consisting of two stations and two separate demand processes.

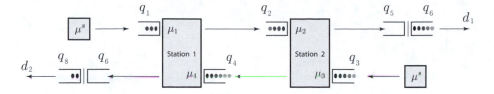

Figure 7.7. Demand-driven network with two sources of demand.

A fluid model can be constructed using the transformation outlined in Section 7.1.1. The resulting network model is of the form (7.1) with

$$
B = \begin{bmatrix}
-\mu_1 & 0 & 0 & 0 & 0 & 0 & \mu^s & 0 \\
\mu_1 & -\mu_2 & 0 & 0 & 0 & 0 & 0 & 0 \\
0 & 0 & -\mu_3 & 0 & 0 & 0 & 0 & \mu^s \\
0 & 0 & \mu_3 & -\mu_4 & 0 & 0 & 0 & 0 \\
0 & \mu_2 & 0 & 0 & -\mu^d & 0 & 0 & 0 \\
0 & 0 & 0 & 0 & -\mu^d & 0 & 0 & 0 \\
0 & 0 & 0 & \mu_4 & 0 & -\mu^d & 0 & 0 \\
0 & 0 & 0 & 0 & 0 & -\mu^d & 0 & 0
\end{bmatrix}
\tag{7.11}
$$

$$
\alpha = (\, 0 \quad 0 \quad 0 \quad 0 \quad 0 \quad d_1 \quad 0 \quad d_2 \,)^{\mathsf{T}}.
$$

As in Example 7.2.1, the matrix B is square and invertible, so that the workload vectors corresponding to Stations 1 and 2 are given by $\xi^i = -(C^i B^{-1})^{\mathsf{T}}$, with

$$
C^1 = (1, 0, 1, 0, 0, 0, 0, 0), \quad C^2 = (0, 1, 0, 1, 0, 0, 0, 0).
$$

We thereby obtain the two workload vectors

$$
\begin{aligned}
\xi^1 &= (0, -\mu_1^{-1}, 0, 0, -\mu_1^{-1}, \mu_1^{-1}, -\mu_2^{-1}, \mu_2^{-1})^{\mathsf{T}} \\
\xi^2 &= (0, 0, 0, -\mu_1^{-1}, -\mu_2^{-1}, \mu_2^{-1}, -\mu_1^{-1}, \mu_1^{-1})^{\mathsf{T}}.
\end{aligned}
\tag{7.12}
$$

For a given pair of demand rates d_1, d_2, the associated load parameters are given by

$$
\rho_1 = \mu_1^{-1} d_1 + \mu_2^{-1} d_2, \qquad \rho_2 = \mu_2^{-1} d_1 + \mu_1^{-1} d_2.
\tag{7.13}
$$

For the parameter values

$$
\mu^s = \mu^d = 100, \quad \mu_1 = \mu_3 = 12, \quad \mu_2 = \mu_4 = 9, \quad d_1 = d_2 = 5,
$$

the network is balanced with $\rho_\bullet = \rho_1 = \rho_2 = 0.972$. The cost structure imposed on this network is identical in form to that used in (7.10) for the previous example,

$$
c = c(x) = \langle c, x \rangle, \qquad c = (1, 1, 1, 1, 5, 10, 5, 10)^{\mathsf{T}}, \quad x \in \mathbb{R}_+^8.
$$

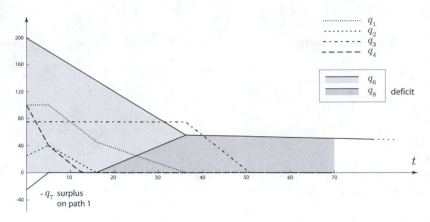

Figure 7.8. Individual buffer levels under the GTO policy.

We refer to the route consisting of μ_1^s, μ_1, and μ_2 as *path 1*, and the route consisting of μ_2^s, μ_3, and μ_4 as *path 2*. These are the paths raw material must follow to produce the two final products.

The initial condition chosen below results in initial deficit along path 1, while path 2 has an initial surplus:

$$x = (100, 25, 75, 100, 0, 200, 25, 0)^{\mathsf{T}}.$$

The minimum emptying time from this initial condition can be expressed in terms of the two workload vectors given in (7.12),

$$T^*(x) = \max_{s=1,2} \frac{\langle \xi^s, x \rangle}{1 - \rho_s} = 425.$$

Resources 1 and 2 are each dynamic bottlenecks at time $t = 0$, and hence any time-optimal solution satisfies $\frac{d}{dt} w_s(t) = -(1 - \rho_s)$ for $s = 1, 2$ and $0 \leq t < T^*(x)$.

Figure 7.8 shows the buffer levels under the GTO policy for $0 \leq t \leq 70$. The controlled system has some interesting characteristics:

(i) The virtual buffer on path 1 initially starts in surplus, but is quickly drained on the time interval $[0, 4)$. Simultaneously, buffer 2 is loaded utilizing the available capacity at resource 1.

(ii) Buffers 2 and 4 drain slowly on the time interval $[5, 13)$, and both are empty by time $t = 15$.

(iii) The behavior on the time interval $[16, 35.23)$ may at first appear counterintuitive: The virtual buffer on the second path was empty at time $t = 16$, but goes into deficit and remains in deficit until the draining time $T^* = 425$. This is required to ensure time optimality: As noted previously, during the entire time interval $[0, T^*)$, the dynamic bottleneck resources operate at capacity. At $t = 35$ (approximately) the two deficit buffers equalize and remain equal throughout the remainder of the run.

(iv) State space collapse is evident on the time interval $[50.69, 425)$ on which just *two* of eight buffers in the system are nonempty. This is consistent with a two-dimensional relaxation of this model. ∎

Example 7.2.3 (Routing, scheduling, and exogenous demand). The 16-buffer model shown in Fig. 7.1 has two virtual buffers since there are two sources of demand. Activities include processing of customers at the stations, routing customers exiting Station 3, and workload release of material to buffers 1 and 3. This model also contains uncontrolled routing at the exit of buffer 7 where, for a given allocation rate ζ_7, the inflow rate to buffer 12 is $0.2\mu_7\zeta_7$, and the inflow rate to buffer 13 is $0.8\mu_7\zeta_7$. The uncontrolled routing represents reworking a flawed product, and the controlled routing reflects the availability of alternate resources.

Calculation of workload vectors in Examples 7.1.4 and 7.2.2 was straightforward due to invertibility of the matrix B in (7.1). In the model considered here the matrix B is not square, and hence not invertible, due to the controlled routing at Station 3. Nevertheless, it is possible to compute the workload vectors defined in Definition 6.1.2 by considering the extreme points in the constraint set in (6.18). To construct supporting vectors for every face of V, including those corresponding to $o_s = 0$, it is necessary to compute extreme points of the LP (6.17).

To construct the workload vectors note that the dimension of the matrix B is 16×17. Letting $B^{[i]}$ denote the 16×16 matrix obtained by removing the ith column, and defining $C^{[i]}$ similarly, with C the 7×17-dimensional constituency for the model, a potential extreme point is obtained on consideration of the following inequality obtained from (6.18):

$$-(B^{[i]})^{\mathsf{T}}\xi - (C^{[i]})^{\mathsf{T}}\eta \leq 0.$$

If $B^{[i]}$ is invertible, then for any $1 \leq s \leq 7$ we obtain

$$\eta^{i,s} := 1^s, \qquad \xi^{i,s} := \left((B^{[i]})^{\mathsf{T}}\right)^{-1}(C^{[i]})^{\mathsf{T}}1^s.$$

The pair $(\xi^{i,s}, \eta^{i,s})$ is an extreme point provided it is feasible for (6.18). Repeating this construction for each i results in potentially $7 \cdot 17 = 119$ extreme points, and there may be others if resource pooling is possible for some extreme point.

Consider the following parameters for the fluid model: The demand rate for each of the two products is equal to $d_1 = d_2 = 19/75$, and the 12-dimensional vector of service rates is given by

$$\mu = (13/15, 26/15, 13/15, 26/15, 1, 2, 1, 2, 1, 1/3, 1/2, 1/10)^{\mathsf{T}}. \tag{7.14}$$

The rates for supply resources and exit resources are taken to be 100 in the numerical examples below.

The network load obtained from (6.5) is $\rho_{\bullet} = 0.95$, and there are three extreme points $\{(\xi^s, \eta^s) : s = 1, 2, 3\}$ satisfying $\rho_s := \langle \xi^s, \alpha \rangle = 0.95$, expressed in matrix form by

$$
\widehat{\Xi}^{\mathsf{T}} = \begin{bmatrix}
0.00 & 0.00 & 0.00 \\
0.00 & 0.00 & 0.00 \\
-0.58 & -0.50 & -3.03 \\
-1.15 & -2.00 & 0.00 \\
-1.15 & 0.00 & 0.00 \\
-0.58 & 0.00 & 0.00 \\
-1.73 & -0.50 & -3.03 \\
-0.58 & -0.26 & -3.03 \\
-1.15 & -1.00 & 0.00 \\
-0.58 & -0.50 & 0.00 \\
-0.58 & -0.50 & -3.03 \\
-0.58 & -0.50 & 0.00 \\
-2.02 & -1.75 & -3.79 \\
-1.73 & -2.00 & 0.00 \\
2.01 & 1.75 & 3.79 \\
1.73 & 2.00 & 0.00
\end{bmatrix} . \tag{7.15}
$$

The remaining load parameters are substantially smaller. We have $\eta^s = 1^s$ for $s = 1, 2, 3$, so that resource pooling is not significant in this example, and we can identify the first three workload vectors as describing workload for the first three stations shown in Fig. 7.1.

Just as in previous examples we find that some of the entries in the workload vectors are negative, and that entries corresponding to deficit buffers are nonnegative.

As argued in Example 7.2.1, a reasonable cost function for this model is linear, $c(x) = \langle c, x \rangle$ for $x \in \mathbb{R}_+^{16}$, with

$$
c = (1, 1, 1, 1, 1, 1, 1, 1, 1, 1, 1, 1, 1, 5, 5, 10, 10)^{\mathsf{T}}, \tag{7.16}
$$

so that cost is uniform across internal buffers, and the cost of unsatisfied demand is twice the cost of completed inventory.

Consider the initial condition given by

$$
x^0 = (10, 5, 8, 3, 4, 4, 6, 5, 6, 7, 2, 2, 0, 0, 15, 10)^{\mathsf{T}}, \tag{7.17}
$$

so that initially there are 15 units of unsatisfied demand for the first product, and 10 units of unsatisfied demand for the second. When the service rates are given in (7.14), the minimal draining time is determined by the second workload vector as follows:

$$
\begin{aligned}
T^*(x^0) &= \max_{s=1,\dots,\ell_v} \frac{\langle \xi^s, x \rangle}{o_s - \rho_s} \\
&= \max \left\{ \frac{5.88}{1 - \rho_1}, \frac{20.45}{1 - \rho_2}, \frac{-6.78}{1 - \rho_3}, \cdots \right\} = \frac{20.45}{1 - 0.95} = 409.
\end{aligned}
$$

Figure 7.9 shows the buffer levels as a function of time for this network with this initial condition controlled using the GTO policy based on the cost function given in (7.16). Initially, the system behavior is dominated by the draining of internal buffers in

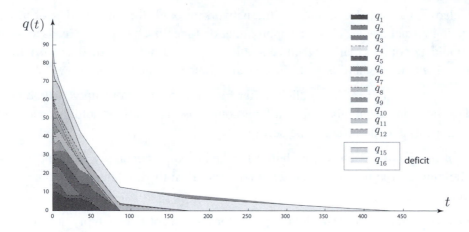

Figure 7.9. Cumulative buffer levels vs. time for the 16-buffer model under the GTO policy.

the network. After a transient period we once again observe state space collapse: The deficit buffers drain slowly once all internal buffers have emptied. ∎

7.2.2 Hot-lots

In production parlance, a *lot* is a group of products in the system. A *hot-lot* is a lot (or group of lots) that is given high priority. Hot-lots are frequently seen in semiconductor manufacturing facilities for various reasons:

(i) Some customers are more valuable than others, and hence may receive high priority even with a short order.

(ii) In evaluating a prototype for a new product, batches may be given preference to speed the testing process.

(iii) The factory may produce standard "make to stock" products in large quantities, as well as more expensive "make to order" products. The latter will be given higher priority due to their greater value.

(iv) A sudden change in priorities may be imposed because some products are behind schedule due to, for example, a surge in demand. This results in a short-term shock to the system that can be modeled as a hot-lot.

The ITRS roadmap [279] classifies hot-lots in a semiconductor factory into "regular hot-lots" that comprise 5% of total workload, and "super hot-lots" that comprise just 1%.

In her newsletter on semiconductor fab management, Jennifer Robinson lists suggestive terms to describe a hot-lot, *Ambulance Lots, Racetrack Lots, Screamer Lots, Lightning Lots, Platinum Lots, Priority1 Lots, Nuclear Lots, Rocket Lots, Turbo Lots, and CEO Lots.* When a manager declares that a favored product is an ambulance lot or CEO lot, the message is clear!

In most cases the (long-run) demand rate for a hot-lot is essentially zero, which justifies the following modeling step: for the purpose of policy synthesis, a hot-lot is modeled in the initial condition of the network through the introduction of a corresponding virtual queue. This is an approximation to reality, but quickly leads to a range of candidate policies that can then be fine-tuned given a more detailed network model.

Priorities may be hard or soft. We consider three general classes below:

Normal-priority hot-lot The deficit buffer for the hot-lot is nonempty at time $t = 0$, and the holding cost at this deficit buffer is commensurate with the holding cost at other (non-hot-lot) deficit buffers in the network.

High-priority hot-lot The deficit buffer for the hot-lot is nonempty at time $t = 0$, and the holding cost at this deficit buffer is far larger than the holding cost at other deficit buffers.

Critical-priority hot-lot The deficit buffer for the hot-lot is nonempty at time $t = 0$, and the allocation is chosen to meet this demand in the shortest time possible.

The high-priority hot-lot is used to model a finite size customer order where, due to the premium paid by the customer, the system is realigned to meet the order quickly; a critical-priority hot-lot receives the highest priority possible.

Given the convention that a hot-lot is modeled by specifying the initial condition of the network, the GTO policy can be applied without modification in systems with normal-priority or high-priority hot-lots since positive demand rates are not required in definition (4.46).

Policy synthesis for a critical hot-lot is more complex. In this section we consider a two-layer generalization of the GTO policy: First, the minimal clearing time of the hot-lot over all policies is computed. Based on this, the minimal draining time for the overall system is computed, given that the critical hot-lot is cleared in minimal time. Each of these computations can be cast as a finite-dimensional linear program.

Example 7.2.4 (Hot-lots for the complex demand-driven model). To illustrate the impact of a hot-lot on overall system performance we consider a version of the network model shown in Fig. 7.1 in which the demand rate d_2 is set to *zero*. However, the corresponding buffer q_{16} is nonzero since we assume that there is a transient demand for this product. In the numerical results below the initial buffer levels are given by

$$x^0 = (0, 5, 8, 0, 0, 4, 6, 6, 0, 7, 2, 2, 0, 0, 15, 20)^{\mathsf{T}}. \tag{7.18}$$

Note that all of the buffers in the hot-lot path are initially empty. Consequently, to meet the hot-lot demand $q_{16} = 20$ it is necessary to bring into the entrance buffers all of the required raw material.

We first establish a baseline for comparison. The hot-lot is absent, so that $q_{16}(0) = 0$, and the remaining initial-condition values are unchanged in (7.18). Results from the GTO policy are shown on the left in Fig. 7.10. This illustrates normal operation of the network under the GTO policy when there is a single recurrent demand with rate d_1.

Consider now a normal-priority hot-lot with initial condition (7.18), and linear cost function defined in (7.16). Numerical results obtained using the GTO policy are shown

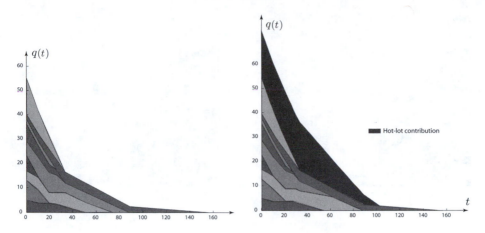

Figure 7.10. The plot on the left-hand side shows cumulative buffer levels vs. time for normal operation under the GTO policy. The plot on the right-hand side shows cumulative buffer levels vs. time for a normal-priority hot-lot of size 20.

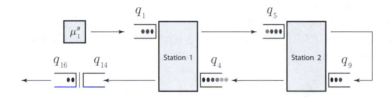

Figure 7.11. Subsystem serving the hot-lot.

on the right in Fig. 7.10. The emptying time $T^*(x^0) \approx 160$ is approximately equal to the emptying time for the baseline system. The hot-lot is not cleared until late into the time horizon $[0, T^*(x^0)]$.

The plot on the left-hand side of Fig. 7.12 shows results obtained for a high-priority hot-lot under the GTO policy in which the holding cost at buffer 16 is increased from 10 to 10^4, with initial condition again given in (7.18). The resulting state trajectory clears the hot-lot demand in approximately 60 time units as opposed to about 100 time units for the normal-priority hot-lot. Of course, because the GTO policy imposes a global time-optimality constraint, the system drains at time $T^*(x^0) \approx 160$, exactly as seen for the normal-priority hot-lot.

Finally, we consider a critical-priority hot-lot. To determine the minimal time required to clear the hot-lot we examine the model shown in Fig. 7.11, obtained by ignoring any buffers that are not on the path leading to the hot-lot deficit buffer. The workload vectors for this six-buffer model are given by

$$\xi^{\text{HL 1}} = (0, -\mu_1^{-1}, -\mu_1^{-1}, -\mu_1^{-1}, -(\mu_1^{-1} + \mu_4^{-1}), \mu_1^{-1} + \mu_4^{-1})^{\text{T}},$$
$$\xi^{\text{HL 2}} = (0, 0, -\mu_5^{-1}, -(\mu_5^{-1} + \mu_9^{-1}), -(\mu_5^{-1} + \mu_9^{-1}), \mu_5^{-1} + \mu_9^{-1})^{\text{T}}.$$

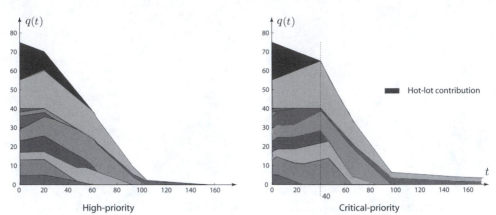

Figure 7.12. On the right is a plot of cumulative buffer levels vs. time under the GTO policy for the same network when a critical hot-lot is present. The system clearing time was found to be $T^* = 320$ in this case, which is approximately twice the minimal clearing time observed in the previous three experiments.

The initial condition for the subsystem is given by $x^{\text{HL}} = (0, 0, 0, 0, 0, 20)^{\text{T}}$. Hence, on applying Proposition 7.1.2, the minimal time required to clear the hot-lot deficit buffer can be computed as follows:

$$W^{\text{HL}*}(x^{\text{HL}}) = \max_{s=1,2} \frac{\langle \xi^{\text{HL}\,s}, x^{\text{HL}} \rangle}{o_s} = 20 \max(15/26 + 15/13, 1 + 1) = 40.$$

The minimal draining time for the network, subject to the constraint that the critical-priority deficit is cleared at time $t = 40$, is equal to approximately 325. This is approximately twice the clearing time seen for the normal or high-priority model. Numerical results are shown on the right in Fig. 7.12. ∎

7.2.3 Unanticipated breakdown

We introduce here an extension of the GTO policy for network control during a period in which some resource is unavailable. Our goal is to obtain allocations that minimize the impact of these gross disturbances.

By normalization we may assume that the breakdown occurs at time $t = 0$, and take x^0 as the state at this time. It is assumed that the time to repair, denoted T_{MTTR}, is known exactly, and on the time interval $[0, T_{\text{MTTR}})$ the evolution of the state process is described by a fluid model of the form

$$q(t; x) = x + B^{\text{down}} z(t) + \alpha t, \qquad t \geq 0, \ x \in \mathbb{R}_+^\ell, \tag{7.19}$$

where $\zeta \in \mathsf{U}^{\text{down}}$ is set of feasible allocation rates during the time period $[0, T_{\text{MTTR}})$. For $t > T_{\text{MTTR}}$ the system description reverts to the form given in (7.1). Note that the model (7.19) is typically not stabilizable.

The acronym MTTR stands for *mean time to repair*, reflecting the fact that in practice only mean values, and perhaps some higher order statistics, will be available.

We now consider the evolution of q on the time interval $[0, T_{\text{MTTR}}]$ on which repair is taking place. The state process evolves according to one set of dynamics for $0 \leq t < T_{\text{MTTR}}$, and the ordinary fluid model dynamics after this period. To compute the minimum time required to empty the system we consider a cumulative allocation z^1 on $[0, T_{\text{MTTR}}]$, and a cumulative allocation $z^2 - z^1$ on $[T_{\text{MTTR}}, T]$ satisfying $q(T) = 0$. Minimizing over all such (z^1, z^2, T) we obtain the minimal draining time $T^*(x^0; T_{\text{MTTR}})$ as the value of the following linear program:

$$\textbf{min } T$$
$$\begin{aligned}
\textbf{s.t.} \quad x^1 &= x^0 + B^{\text{down}} z^1 + \alpha T_{\text{MTTR}} \\
0 &= x^1 + B(z^2 - z^1) + \alpha(T - T_{\text{MTTR}})
\end{aligned} \tag{7.20}$$
$$Cz^1 \leq \mathbf{1} T_{\text{MTTR}}, \; Cz^2 \leq \mathbf{1} T, \; z^i \geq 0, \; x^1 \in \mathsf{X}.$$

This is a piecewise linear function of (x, T_{MTTR}).

Suppose that q is trajectory from the initial condition $q(0) = x^0$, fix $t \in [0, T^*(x; T_{\text{MTTR}}))$, and consider the *time remaining to drain the network at time t*. From the definitions this is given by $T^*(q(t); T_{\text{MTTR}} - t)$. If q is time optimal, then we must have $T^*(q(t); T_{\text{MTTR}} - t) = T^*(x^0; T_{\text{MTTR}}) - t$; this can be interpreted as a dynamic programming equation, similar to (3.53). On taking the derivative of each side with respect to t, and noting that $T^*(x^0; T_{\text{MTTR}})$ does not depend upon t, we obtain for any $0 \leq t < T^*(x; T_{\text{MTTR}})$, with $x^\bullet = q(t)$ and $v^\bullet = \frac{d^+}{dt} q(t)$,

$$\langle \nabla_x T^*(x^\bullet; T_{\text{MTTR}} - t), v^\bullet \rangle - \frac{\partial}{\partial T_{\text{MTTR}}} T^*(x^\bullet; T_{\text{MTTR}} - t) = -1. \tag{7.21}$$

This can be expressed as a linear constraint on $\zeta(t)$ since $v^\bullet = B\zeta(t) + \alpha$.

The *GTO (with-breakdowns) policy* (GTO-B) defined next is a refinement of the GTO policy in which the dynamic programming constraint (7.21) is imposed to ensure time optimality.

Definition 7.2.5 (GTO-B policy). Given $0 \leq t < T_{\text{MTTR}}$ and $q(t) = x^1$, the allocation rate $\zeta(t)$ is an optimizer of the linear program

$$\textbf{min } \langle c, v \rangle$$
$$\textbf{s.t.} \quad \text{Eq. (7.21) holds with } v = v^\bullet, \text{ and}$$
$$\begin{aligned}
v_i &\geq 0, && \text{if } x_i^1 = 0, 1 \leq i \leq \ell, \\
v_i &\leq 0, && \text{if } x_i^1 = b_i, 1 \leq i \leq \ell, \\
v &= B^{\text{down}} \zeta + \alpha, \\
\zeta &\in \mathsf{U}^{\text{down}}.
\end{aligned}$$

$\blacksquare$

7.2.4 Preventative maintenance

We now suppose that advance warning is provided regarding system down-time. Our main conclusion is that the impact of such disruptions is reduced significantly with advanced planning.

For the purposes of control synthesis, we again assume that this information is exact. The time at which the resource is lost is denoted T_{MTTF}, and we continue to denote by T_{MTTR} the time required to bring the resource back into service. The acronym MTTF stands for *mean time to failure*, which again reflects the fact that this time may not be known exactly in practice.

Given a time period $[T_{\text{MTTF}}, T_{\text{MTTF}} + T_{\text{MTTR}}]$ when a certain resource is not operational, the minimal draining time from the initial state x^0 is denoted $T^*(x) = T^*(x; T_{\text{MTTF}}, T_{\text{MTTR}})$. This can be found through an obvious modification of (7.20), and we again obtain a dynamic programming equation: For $0 \le t < T^*(x; T_{\text{MTTF}}, T_{\text{MTTR}})$, $x^\bullet = q(t)$, and $v^\bullet = \frac{d^+}{dt} q(t)$,

$$\langle \nabla_x T^*(x^\bullet; T_{\text{MTTF}}, T_{\text{MTTR}} - t), v^\bullet \rangle - \frac{\partial}{\partial T_{\text{MTTR}}} T^*(x^\bullet; T_{\text{MTTF}}, T_{\text{MTTR}} - t) = -1. \tag{7.22}$$

The *greedy-time-optimal-with-maintenance* (GTO-M) policy is given in Definition 7.2.6. We present the algorithm only for $0 \le t < T_{\text{MTTF}}$. For $t \in [T_{\text{MTTF}}, T_{\text{MTTF}} + T_{\text{MTTR}})$ the GTO-B policy is used to define the allocation rate $\zeta(t)$.

As with the GTO-B policy, the GTO-M policy empties the system in minimal time.

Definition 7.2.6 (GTO-M policy). Given $0 \le t < T_{\text{MTTF}}$ and $q(t) = x^1$, the allocation rate $\zeta(t)$ is an optimizer of the linear program

$$\textbf{min} \ \langle c, v \rangle$$

$$\textbf{s.t.} \quad \text{Eq. (7.22) holds with } v = v^\bullet, \text{ and}$$

$$v_i \ge 0, \qquad \text{if } x_i^1 = 0, 1 \le i \le \ell,$$
$$v_i \le 0, \qquad \text{if } x_i^1 = b_i, 1 \le i \le \ell,$$
$$v = B\zeta + \alpha,$$
$$\zeta \in \mathsf{U}.$$
∎

Example 7.2.7 (Maintenance and breakdown). We illustrate application of the GTO-B and GTO-M policies for the network shown in Fig. 7.5, with rates $d_1 = 9$, $\mu_1 = \mu_3 = 22$, and $\mu_2 = 10$. Applying (7.7), the vector load is given by

$$\rho = (9/11, 9/10, 9/\mu^d)^\mathsf{T},$$

and the network load is $\rho_\bullet = \rho_2 = 0.9$ since μ^d is assumed large. The loss of either resource will cause significant disruption. A breakdown at Station 2 is particularly significant due to its higher load.

Consider first the application of the GTO-B policy. It is assumed that resource 2 is in-operable during the time interval $[0, T_{\text{MTTR}}]$ with $T_{\text{MTTR}} = 10$, and the initial condition is $q(0) = [10, 15, 20, 0, 0]^\mathsf{T}$. The plot on the left-hand side of Fig. 7.13 shows the buffer trajectories versus time under the GTO-B policy.

The unscheduled breakdown generated tremendous deficit: Assuming that the initial 20 units in buffer 3 were used to meet some of the demand, an additional 70 units of deficit accrued during repair. The inventory in buffer 2 is quickly cleared once repair

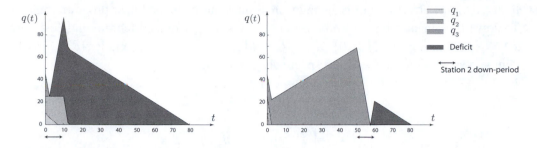

Figure 7.13. The GTO-B and GTO-M policies for the network shown in Fig. 7.5. The first plot shows cumulative buffer levels vs. time under the GTO-B policy when resource 2 is inoperable for $0 \leq t < 10$. The plot on the right shows cumulative buffer levels vs. time for the GTO-M policy applied to the same network, with identical initial conditions, when resource 2 is inoperable for $50 \leq t < 60$. The impact of resource down-time is reduced significantly with advanced planning.

has been completed, after which the system works at capacity until the deficit created by the temporary loss of resource 2 has been cleared.

Consider now the GTO-M policy with $T_{\text{MTTR}} = 10$ and $T_{\text{MTTF}} = 50$. The plot on the right-hand side of Fig. 7.13 shows the resulting buffer trajectories. At the point where maintenance begins, there are about 70 units of inventory in q_3 that is used to feed demand during the maintenance period $50 \leq t < 60$. The surplus staged at q_3 is completely consumed by demand time $t = 57$, and an additional 20 units of deficit are incurred before repair is complete at time $t = 60$.   ∎

7.3 Workload relaxations

The general theory of workload relaxations carries over to the present setting. In the nth relaxation we have a workload process $\widehat{w}$ evolving on a polyhedral set $\mathsf{W} \subset \mathbb{R}^n$, subject to the independent linear constraints

$$\tfrac{d^+}{dt}\widehat{w}_s(t) \geq -(1 - \rho_s), \qquad 1 \leq s \leq n. \tag{7.23}$$

The structure of the effective cost again provides significant insight in models with multiple bottlenecks in which customers are competing for service.

The model (7.23) is obtained as a workload relaxation of $\widehat{\Xi}q(t)$; the rows of $\widehat{\Xi}$ constituting n workload vectors are assumed to be linearly independent, so that the n constraints in (7.23) are independent.

The key difference between the workload models considered here and those of Chapter 4 is that workload vectors typically have negative entries, and hence $\mathsf{W} \not\subset \mathbb{R}^n_+$.

7.3.1 One-dimensional relaxations

Consider a one-dimensional relaxation of a general network model in which some entries of $\xi^1 \in \mathbb{R}^\ell$ are negative. If X is restricted via buffer constraints, so that $\mathsf{X} := \{x \in \mathbb{R}^\ell_+ : x \leq b\}$ for constants $\{0 < b_i \leq \infty : 1 \leq i \leq \ell\}$, and if $c \colon \mathsf{X} \to \mathbb{R}_+$ is linear, then the effective cost is again given as the solution to the linear program (5.39). To solve this linear program we consider *two* sets of generalized Klimov indices. Let $\mathcal{I}^+ = \{i \in \{1, \ldots, \ell\} : \xi_i^1 \geq 0\}, \mathcal{I}^- = \{i \in \{1, \ldots, \ell\} : \xi_i^1 < 0\}$, and let $\{\theta_i^+\}, \{\theta_j^-\}$ denote permutations of $\mathcal{I}^+, \mathcal{I}^-$, respectively, satisfying

$$\frac{c_i}{\xi_i^1} \leq \frac{c_j}{\xi_j^1} \qquad \text{if } \theta_i^+ < \theta_j^+, i \in \mathcal{I}^+,$$

$$\frac{c_i}{|\xi_i^1|} \leq \frac{c_j}{|\xi_j^1|} \qquad \text{if } \theta_i^- < \theta_j^-, j \in \mathcal{I}^-.$$

We then arrive at the following extension of Proposition 5.3.19. We leave the proof to the reader as Exercise 5.

Proposition 7.3.1. *The effective cost* $\bar{c} \colon \mathsf{W} \to \mathbb{R}_+$ *for the one-dimensional relaxation is piecewise linear and convex. For a given* $w \in \mathsf{W}$, *the effective state* $x^* = \mathcal{X}^*(w)$ *satisfies*

$$x_i^* = b_i \qquad \text{if } w > 0 \text{ and } \sum_{j:\theta_j^+ > \theta_i^+} x_j^* > 0,$$

$$x_i^* = b_i \qquad \text{if } w < 0 \text{ and } \sum_{j:\theta_j^- > \theta_i^-} x_j^* > 0.$$

Given two indices i, j, if $\theta_j^+ > \theta_i^+$ then buffer j is "more costly." Proposition 7.3.1 asserts that buffer j should be empty whenever buffer i is not full in this one-dimensional relaxation.

Note that since $\bar{c}$ is convex with $\bar{c}(0) = 0$, and $\bar{c}(w) > 0$ for $w \neq 0$, it follows that the optimal solution for a one-dimensional relaxation can be expressed in workload space via

$$\widehat{w}^*(t) = [\widehat{w}^*(0) - (1 - \rho_1)t]_+, \qquad t \geq 0. \tag{7.24}$$

For $\widehat{w}^*(0) \leq 0$ this gives $\widehat{w}^*(t) = 0$ for $t > 0$.

Example 7.3.2 (Simple re-entrant line with demand). Consider a one-dimensional relaxation based on $\xi^2 \in \mathbb{R}^5$ for the homogeneous model, with cost given in (7.10). We first construct the effective cost when there are no buffer constraints. Based on the expression for ξ^2 given in (7.6), the effective cost as a function of w is the solution to the linear program

$$\bar{c}(w) := \begin{array}{ll} \min & x_1 + x_2 + x_3 + 5x_4 + 10x_5 \\ \text{s.t.} & -\mu_2^{-1}(x_3 + x_4) + \mu_2^{-1}x_5 = w, \\ & x \geq 0, \end{array}$$

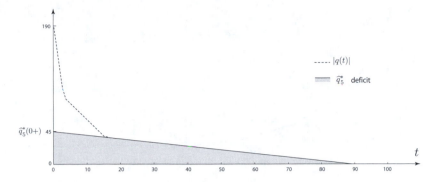

Figure 7.14. Buffer levels vs. time under the optimal policy for a one-dimensional relaxation of the five-buffer demand-driven network. Only $\widehat{q}_5^*(t)$ is nonzero for $t > 0$ in the relaxed model. The dashed line shows the sum $|q(t)| := q_1(t) + \cdots + q_5(t)$ for the unrelaxed model under the GTO policy.

which can be solved to give the following formulae for the effective cost and effective state:

$$\bar{c}(w) = \mu_2 \max(-w, 10w), \tag{7.25}$$

$$\mathcal{X}^*(w) = \begin{cases} \mu_2|w|\mathbf{1}^3 & w < 0 \\ \mu_2 w \mathbf{1}^5 & w \geq 0, \end{cases} \qquad w \in \mathbb{R}. \tag{7.26}$$

Inventory is stored in buffer 3 rather than buffer 4 when $w < 0$ due to the higher cost at buffer 4.

The form of the effective state is consistent with the results of the numerical experiment described in Example 7.2.1. In this example $\rho_1 = 9.5/11 = 0.8636 < 0.95 = \rho_2$, so that a one-dimensional relaxation is justifiable. For the initial condition $x = (10, 25, 55, 0, 100)^\mathsf{T}$ we have

$$w = \langle \xi^2, x \rangle = (-55 + 100)\mu_2^{-1} = 4.5 > 0,$$

so that $\mathcal{P}^*(x) = \mathcal{X}^*(w) = \mu_2 w \mathbf{1}^5 = 45\,\mathbf{1}^5$, and the optimal solution for the one-dimensional relaxation is given by

$$\widehat{q}^*(t; x) = \mathcal{P}^*(x) - (\mu_2 - d)\mathbf{1}^5 t, \qquad 0 < t \leq T^*(x).$$

The plot contained in Fig. 7.6 shows that under the GTO policy the state process satisfies $q(t) = \mathcal{P}^*(q(t)) = \widehat{q}^*(t)$ for $t > 13$ (approximately). Shown in Fig. 7.14 is a comparison of $|q(t)|$ and $\widehat{q}^*(t)$ as a function of time under the GTO policy, and the optimal policy for the one-dimensional relaxation, respectively.

If buffer constraints are imposed at buffer 3 or 5, then the effective cost and effective state will change. In this case we apply Proposition 7.3.1 to obtain

$$\mathcal{I}^+ = \{1, 2, 5\}, \quad \mathcal{I}^- = \{3, 4\},$$

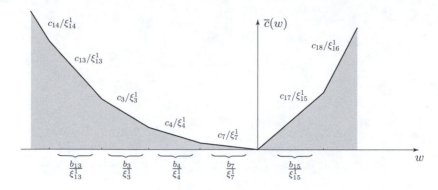

Figure 7.15. Effective cost for a one-dimensional relaxation of the 16-buffer inventory model.

and the two permutations are given by

$$\{\theta_i^+\} = \{5, 2, 1\}, \qquad \{\theta_i^-\} = \{3, 4\}.$$

In particular, if $\mu_2 w < -b_3$ then $(\mathcal{X}^*(w))_4 > 0$.   ∎

Example 7.3.3 (Complex demand-driven model). Consider the one-dimensional relaxation of the model shown in Fig. 7.1, based on the workload vector ξ^1 shown in (7.15). We have

$$\mathcal{I}^+ = \{1, 2, 15, 16\}, \quad \mathcal{I}^- = \{3, \ldots, 14\},$$

and we can choose the two permutations as

$$\{\theta_i^+\} = \{15, 16, 1, 2\}, \quad \{\theta_i^-\} = \{7, 4, 5, 9, 3, 6, 8, 10, 11, 12, 13, 14\},$$

to satisfy the conditions of Proposition 7.3.1.

Figure 7.15 shows a plot of the resulting piecewise linear effective cost function on $\mathbb{R}$. Indicated on the plot is the slope on each interval on which $\bar{c}$ is linear (actually, affine). In the relaxation the system "cheats" by storing inventory in buffer 7 rather than buffer 13 since storage is cheaper there. In the relaxation, material can be transferred from buffer 7 to buffer 13 instantaneously.   ∎

7.3.2 Multidimensional relaxations and control

This section contains some examples to illustrate the construction of the effective cost, as well as policy synthesis for workload relaxations of dimension 2 or higher.

Example 7.3.4 (Simple re-entrant line with demand). The effective cost for a two-dimensional relaxation of the model shown in Fig. 7.5 is defined by the linear program

$$
\begin{aligned}
\bar{c}(w) := \quad \mathbf{min} \qquad & x_1 + x_2 + x_3 + 5x_4 + 10x_5 \\
\mathbf{s.t.} \qquad & -\mu_1^{-1}x_2 - \mu_1^{-1}x_3 - 2\mu_1^{-1}x_4 + 2\mu_1^{-1}x_5 = w_1, \\
& -\mu_2^{-1}(x_3 + x_4) + \mu_2^{-1}x_5 = w_2, \quad x \in \mathbb{R}_+^5.
\end{aligned}
$$

The dual can be expressed in terms of the variables $y_i = \mu_i w_i$, $i = 1, 2$, via

$$\bar{c}(w) = \mathbf{max} \ [\psi_1 y_1 + \psi_2 y_2],$$

$$\mathbf{s.t.} \begin{bmatrix} 0 & 0 \\ -1 & 0 \\ -1 & -1 \\ -2 & -1 \\ 2 & 1 \end{bmatrix} \psi \leq \begin{bmatrix} 1 \\ 1 \\ 1 \\ 5 \\ 10 \end{bmatrix}.$$

To find an extreme point in the constraint set of the dual we choose two rows and attempt an inversion of the resulting 2×2 matrix. This yields a *basic solution*, which is called a *basic feasible solution* if it is indeed feasible.

For example, choosing the second and fifth rows of the 2×5 matrix above gives

$$\begin{bmatrix} -1 & 0 \\ 2 & 1 \end{bmatrix} \psi = \begin{bmatrix} 1 \\ 10 \end{bmatrix}.$$

This matrix equation can be solved to give $\psi = \left(\begin{smallmatrix} -1 \\ 12 \end{smallmatrix} \right)$. This solution is feasible for the dual, and is hence an extreme point. When this value of ψ is the optimizer for the dual, then the effective state is supported on $\{2, 5\}$, so that $\mathcal{X}^*(w) = (-y_1 + 2y_2)\mathbf{1}^2 + y_2\mathbf{1}^5$ if $\bar{c}(w) = \psi^\mathsf{T} y$ (recall $y_i = \mu_i w_i$, $i = 1, 2$).

These calculations can be repeated for each pair of rows, and one thereby obtains three extreme points given by $\psi^1 = (-1, 12)^\mathsf{T}$, $\psi^2 = (11, -12)^\mathsf{T}$, $\psi^3 = (-1, 0)^\mathsf{T}$, so that the vectors $\{\bar{c}^i\}$ that define the effective cost on $\mathsf{W} = \mathbb{R}_+^2$ are given by

$$\bar{c}^1 = \begin{pmatrix} -\mu_1 \\ 12\mu_2 \end{pmatrix}, \quad \bar{c}^2 = \begin{pmatrix} 11\mu_1 \\ -12\mu_2 \end{pmatrix}, \quad \bar{c}^3 = \begin{pmatrix} -\mu_1 \\ 0 \end{pmatrix}.$$

The corresponding effective state is given by, with $y_i = \mu_i w_i$,

$$\mathcal{X}^*(w) = \begin{cases} (2y_2 - y_1)\mathbf{1}^2 + y_2\mathbf{1}^5 & \text{if } \bar{c}(w) = \langle \bar{c}^1, w \rangle, \\ (y_1 - 2y_2)\mathbf{1}^3 + (y_2 - y_1)\mathbf{1}^5 & \text{if } \bar{c}(w) = \langle \bar{c}^2, w \rangle, \\ (y_2 - y_1)\mathbf{1}^2 - y_2\mathbf{1}^3 & \text{if } \bar{c}(w) = \langle \bar{c}^3, w \rangle. \end{cases} \tag{7.27}$$

Figure 7.16 shows a level set of the effective cost $\bar{c}$. The monotone region is a line segment in this example, $\mathsf{W}^+ = \{w \in \mathbb{R}_+^2 : \langle \bar{c}^1 - \bar{c}^2, w \rangle = 0\}$.

The myopic policy for this two-dimensional model is again defined so that $\frac{d^+}{dt}\bar{c}(\widehat{w}(t; w^0))$ is minimized. For each initial condition $w^0 \in \mathsf{W}$, the workload trajectory satisfies $\widehat{w}(t; w^0) \in \mathsf{W}^+$, $t > 0$. In particular, if $w_i^0 \leq 0$ for $i = 1, 2$, then $\widehat{w}(t; w^0) = \mathbf{0}$ for all $t > 0$.

Suppose that the initial condition w^0 also lies strictly above W^+. In this case we have $\widehat{w}(0+; w^0) \in \mathsf{W}^+$ with $\widehat{w}_1(0+; w^0) > w_1^0$ and $\widehat{w}_2(0+; w^0) = w_2^0$. That is, resource 1 will idle at time $0+$ just enough so that the workload process reaches the monotone region. If the line segment $\{(1 - \rho)r : r \geq 0\}$ lies strictly above the line segment

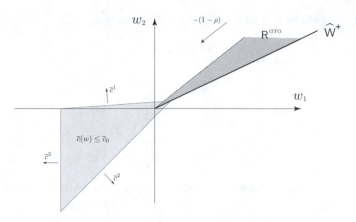

Figure 7.16. Level set of the effective cost $\bar{c}$ for the two-dimensional relaxation of the network shown in Fig. 7.5. The monotone region $\mathsf{W}^+ \subset \mathbb{R}_+^2$ is the line segment shown.

W^+, then resource 1 is a dynamic bottleneck at time $t = 0$, in the sense that any time-optimal solution is nonidling at this resource. Consequently, under the myopic policy the workload trajectory is not time optimal from this initial condition.

One obtains a time-optimal allocation if the region W^+ is expanded to the set R^{GTO} shown in Fig. 7.16 since we then have $(1 - \rho) \in \mathsf{R}^{\text{GTO}}$. This is an instance of the GTO policy for the relaxation.   ∎

Example 7.3.5 (Complex demand-driven model). Consider again the operation of the GTO policy illustrated in Fig. 7.9. After a short transient period the deficit buffers drain slowly once all internal buffers have been emptied. This can be interpreted as an instance of state space collapse. In a workload relaxation, one will observe this behavior instantaneously since $\widehat{q}(t) = \mathcal{X}^*(\widehat{w}(t))$ for $t > 0$.

Here we consider two- and three-dimensional relaxations for this model with parameters defined in (7.14), and with $d_1 = d_2 = 19/75$, so that $\rho_s := \langle \xi^s, \alpha \rangle = 0.95$, $s = 1, 2, 3$. The cost function is given in (7.16). In a relaxation it is evident that the fast initial draining of internal buffers is at the expense of starvation of bottleneck resources. Increasing some components of the workload $w(t)$ will initially reduce the effective cost $\bar{c}(w(t))$ since the effective cost is not monotone.

Consider first a two-dimensional relaxation in which there are no buffer constraints on any of the 16 buffers. In this case the workload space W is equal to all of $\mathbb{R}^2$. The graph on the left-hand side of Fig. 7.17 shows the contour plot of the resulting effective cost $\bar{c}(w)$ on $\mathsf{W} = \mathbb{R}^2$. The figure shows that the monotone region W^+ is a small subset of W.

The form of the effective state in a two-dimensional relaxation is consistent with the behavior of q shown in Fig. 7.9 under the GTO policy from the initial condition $x^0 \in \mathbb{R}_+^{16}$ given in (7.17). After a transient period, the deficit buffers drain slowly once all internal buffers have been emptied, and we have $q(t) \approx \mathcal{P}^*(q(t))$ for $t \geq 80$ (approximately), where $\mathcal{P}^*: \mathbb{R}_+^2 \rightarrow \mathbb{R}_+^{16}$ is the projection operator defined for the

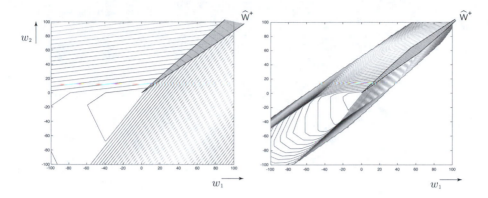

Figure 7.17. Two instances of the effective cost $\bar{c}$ for the two-dimensional relaxation of the network shown in Fig. 7.1. On the left is shown contour plots of the effective cost when no buffer constraints are imposed; on the right is shown contour plots of the effective cost with buffer constraints given in (7.28). In both figures, the shaded regions are monotone.

two-dimensional relaxation. The approximation becomes exact after $t = 175$, which is substantially less than the network draining time $T^*(x) \approx 430$.

The level sets of $\bar{c}$ change dramatically when buffer constraints are imposed. Consider the vector of constraints given by

$$b = (10, 10, 20, 20, 20, 10, 10, 10, 10, 10, 10, 10, 30, 30, 30, 30)^{\mathsf{T}}, \qquad (7.28)$$

where the effective cost is now computed based on the compact state space

$$\mathsf{X} = \{x \in \mathbb{R}_+^\ell : x_i \le b_i, \quad 1 \le i \le \ell\}.$$

The level sets of $\bar{c}$ are shown on the right-hand side of Fig. 7.17.

A three-dimensional relaxation may be more appropriate for this model since the first three resources have equal, and relatively high loads. Fig. 7.19 shows contour plots of the effective cost for the three-dimensional relaxation, with w_3 fixed at the values $w_3 = 30$, $w_3 = 60$, and $w_3 = 90$, respectively.

The myopic, GTO, and infinite-horizon optimal policies can be constructed easily for a two-dimensional relaxation. Consider for simplicity the case without buffer constraints as illustrated on the right in Fig. 7.17, based on the linear cost function $c \colon \mathbb{R}_+^{16} \to \mathbb{R}_+$ with parameters given in (7.16). It is assumed that the vector $-(1 - \rho)$ lies above W^+, as illustrated in Fig. 7.18.

Myopic policy Under the myopic policy the workload trajectory satisfies $\widehat{w}(t; w^0) \in \mathsf{W}^+$ for each initial condition $w^0 \in \mathsf{W}$ and $t > 0$.

GTO policy Under the GTO policy the workload process is constrained for $t > 0$ to the region

$$\mathsf{R}^{\text{GTO}} := \{w \in \mathbb{R}_+^2 : aw_1 < w_2 < bw_1\}, \qquad (7.29)$$

where $b = (1 - \rho_2)/(1 - \rho_1)$ and a is the slope of the lower boundary of the monotone region W^+.

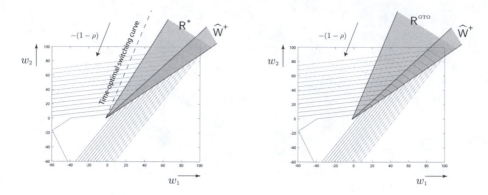

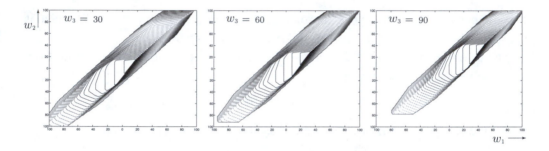

Figure 7.18. On the left is shown the switching region for the infinite-horizon optimal policy, R*, and on the right is shown the switching region R^{GTO} for the GTO policy. In this numerical example, the vector $(1 - \rho)$ does not lie in W⁺. Consequently, the region R* is strictly larger than the monotone region W⁺, and R^{GTO} is strictly larger than R*.

Figure 7.19. Contour plots of effective cost $\bar{c}$ for the three-dimensional relaxation of the 16-buffer model, with the buffer constraints given in (7.28). The projections shown are based on fixing w_3 at the values 30, 60, and 90.

Infinite-horizon optimal policy Exactly as in the analysis of the simple re-entrant line in Example 5.3.20, it can be shown that the infinite-horizon optimal policy is defined by a smaller region $\mathsf{R}^* = \{w \in \mathbb{R}^2_+ : aw_1 < w_2 < b_* w_1\} \subset \mathsf{R}^{\text{GTO}}$, where a is as in (7.29), and $b_* < b$. Consequently, for initial conditions lying above the infinite-horizon optimal switching curve with slope b_*, the infinite-horizon optimal control is not time optimal.   ∎

We now consider the impact of variability.

7.4 Hedging in a simple inventory model

We consider in this section a version of the inventory model described in Section 2.6. As in the treatment of the two-dimensional workload model (5.79) we relax the integer constraints on Y so that the one-dimensional process evolves on $\mathsf{Y} = \mathbb{R}$. It evolves according to a one-dimensional version of (5.79),

$$Y(t + 1) = Y(t) - \mu + \iota(t) + L(t + 1), \qquad t \geq 0, \tag{7.30}$$

in which L is a nonnegative i.i.d. process that models increments in new demand, and μ is the maximal rate at which the desired product can be manufactured. Further assumptions on the model are imposed in Theorem 7.4.2 that follows.

We also consider the fluid model

$$y(t) = y(0) - \delta t + I(t), \qquad t \geq 0, \; y(0) \in \mathbb{R}, \tag{7.31}$$

with $\delta = \lambda - \mu$ and $\lambda = \mathsf{E}[L(t)]$. This can be viewed as a workload relaxation for the fluid model constructed in Example 7.1.1 in which $\mathsf{X} = \mathbb{R}_+ \times \mathbb{R}$, the two workload vectors are $\{\xi^1 = \mu^{-1}(0,1)^{\mathsf{T}}, \; \xi^2 = \lambda^{-1}(-1,1)^{\mathsf{T}}\}$, and the vector load is $\rho = d(\mu^{-1}, \lambda^{-1})^{\mathsf{T}}$. Suppose that $\lambda \gg \mu > d$. In this case a one-dimensional workload relaxation is justified based on the first workload vector. The relaxation in units of inventory is simply the contents of the virtual queue, for which (7.31) is a reasonable model. If a cost function $c \colon \mathsf{X} \to \mathbb{R}_+$ is given for the two-dimensional model, of the form $c(x) = c_1 x_1 + c^+ x_{2+} + c^- x_{2-}$, then the effective cost for the relaxation is given by $\bar{c}_Y(y) = c^+ y_+ + c^- y_-$ for $y \in \mathsf{Y} = \mathbb{R}$.

In this section it is assumed that a piecewise linear cost function $c \colon \mathsf{Y} \to \mathbb{R}_+$ of this form is given,

$$c(y) = c^+ y_+ + c^- y_- = c^+ \mathbf{1}\{y \geq 0\} y + c^- \mathbf{1}\{y < 0\} |y|. \tag{7.32}$$

Optimization of the resulting one-dimensional fluid model is trivial: A pathwise optimal solution drives the workload process to the origin in minimal time, $y^*(t) = [y^*(0) - \delta t]_+, t \geq 0$. For the stochastic model control is more subtle.

7.4.1 CRW model

We will see in Section 9.7 that an optimal policy is defined by a hedging point $\bar{y} \geq 0$ such that the idleness process can be represented as a maximum,

$$\iota(t) = \max(0, -\bar{y} - Y(t)), \tag{7.33}$$

so that (7.30) becomes

$$Y(t+1) = \max(Y(t), -\bar{y}) - \mu + L(t+1), \qquad t \geq 0. \tag{7.34}$$

We characterize the optimal hedging-point value $\bar{y}^*$ as the solution to a nonlinear equation in $\bar{y}$. Unfortunately, the equation cannot be solved in closed form except in special cases.

Section 7.4.2 contains a treatment of the CBM model. In this case the optimal hedging point has an explicit and very intuitive form given in Theorems 7.4.7 and 7.4.6 for discounted cost and average cost, respectively.

The conclusions obtained in Section 7.4.2 motivate the following heuristic for hedging points in the CRW inventory model. Note that the hedging-point value given in (7.35) grows linearly with $(1 - \rho)^{-1}$ and *not logarithmically* as found in Section 4.6 when considering appropriate safety-stock values (see also the discussion surrounding (10.84)).

Definition 7.4.1 (Diffusion heuristic for hedging). Consider the simple inventory model (7.30) with first- and second-order statistics

$$\delta = \mu - \lambda, \quad \sigma^2 = \mathsf{Var}(L(1)).$$

The *diffusion heuristic* is to choose the hedging point

$$\bar{y}^{\mathrm{D}} = \frac{1}{\Theta} \log\left(1 + \frac{c^+}{c^-}\right) \tag{7.35}$$

where, under the discounted-cost optimality criterion, $\Theta = \Theta(\gamma)$ is the positive root of the quadratic equation

$$\tfrac{1}{2}\sigma^2\Theta^2 - \delta\Theta - \gamma = 0. \tag{7.36}$$

To approximate the average-cost optimal policy we take $\Theta = \Theta(0) = \left(\tfrac{1}{2}\sigma^2/\delta\right)^{-1}$. ∎

Throughout the discussion here we assume that Y is controlled using the hedging-point policy (7.33). When necessary we use the notation $Y^{\bar{y}}(t)$ to emphasize the hedging point chosen. The processes are related via

$$Y^{\bar{y}}(t) = Y^0(t) - \bar{y}, \quad t \geq 0, \quad \text{provided} \quad Y^0(0) = Y^{\bar{y}}(0) + \bar{y}. \tag{7.37}$$

Theorem 7.4.2 asserts that Y^0 possesses a unique invariant measure. Although the proof appeals to methods outside of the scope of this book, we can give the main idea here. Note firstly that the process "regenerates" on reaching the set $[-\mu, 0]$: If $Y^0(t) \in [-\mu, 0]$ then $Y^0(t+1) = L(t+1) - \mu$, which is independent of $Y^0(0), \ldots, Y^0(t)$. Based on this observation, the invariant measure π^0 can be written down explicitly as the "mean number of visits over a cycle,"

$$\pi^0(A) = \frac{1}{\mathsf{E}[\tau_0 \mid Y^0(0) = 0]} \mathsf{E}\left[\sum_{t=1}^{\tau_0} \mathbf{1}\{Y^0(t) \in A\} \mid Y^0(0) = 0\right], \tag{7.38}$$

where

$$\tau_0 = \min\{t \geq 1 : Y^0(t) \leq 0\} = \min\{t \geq 1 : Y^{\bar{y}}(t) \leq -\bar{y}\}. \tag{7.39}$$

The representation (7.38) is the generalized Kac's Theorem A.2.2.

The density assumption in Theorem 7.4.2 is imposed to enable derivative calculations in the proof of the following.

Theorem 7.4.2 (Optimizing the simple inventory model). *Consider the simple inventory model (7.30) subject to the following constraints: L is a nonnegative i.i.d. process whose marginal distribution possesses a bounded continuous density p_L. There exists $\bar{L} \in (\mu, \infty)$ such that*

$$p_L(x) > 0, x \in (0, \bar{L}) \quad \text{and} \quad p_L(x) = 0, x \in [0, \bar{L}]^c.$$

Assume moreover that $\lambda := \mathsf{E}[L(t)] < \mu$. Then:

(i) *Y^0 is positive recurrent: There exists a unique invariant measure π^0 whose continuous density satisfies $p^0(y) > 0$ for all $y > -\mu$.*

(ii) *For any hedging point $\bar{y} \in \mathbb{R}$ there exists a unique invariant measure π, expressed for any bounded function $f: \mathbb{R} \to \mathbb{R}$ via*

$$\int f(x)\,\pi(dx) = \frac{1}{\mathsf{E}[\tau_0 \mid Y^0(0) = 0]}\mathsf{E}\left[\sum_{t=1}^{\tau_0} f(Y^0(t) - \bar{y}) \mid Y^0(0) = 0\right].$$

The densities for different values of $\bar{y}$ can be expressed in terms of the density obtained when $\bar{y} = 0$,

$$p^{\bar{y}}(y) = p^0(y + \bar{y}), \qquad y \in \mathbb{R}. \tag{7.40}$$

(iii) *The average cost $\eta(\bar{y})$ is finite for each hedging point $\bar{y}$ when c is given in (7.32), where*

$$\eta(\bar{y}) = \int_{-\infty}^{\infty} c(y)p^{\bar{y}}(y)\,dy. \tag{7.41}$$

Moreover, the unique parameter $\bar{y}^*$ minimizing the average cost solves the equation

$$P_{\pi^0}\{Y^0(t) \geq \bar{y}^*\} = \int_{\bar{y}^*}^{\infty} p^0(y)\,dy = \left(1 + \frac{c^+}{c^-}\right)^{-1}. \tag{7.42}$$

We first establish the existence of a density. The proof follows from (7.34) with $\bar{y} = 0$.

Lemma 7.4.3. *Under the assumptions of Theorem 7.4.2 the invariant measure π^0 has a continuous density satisfying the invariance equation*

$$p^0(y) = \pi\{[-\mu, 0]\}p_L(y + \mu) + \int_0^{\infty} p_L(-x + y + \mu)p^0(x)\,dx.$$

To show that $\bar{y}^*$ is unique we demonstrate that $\eta(\bar{y})$ is strictly convex. We let χ_c denote the "sensitivity function,"

$$\chi_c(y) = \tfrac{d^+}{dy}c(y) = c^+\mathbf{1}\{y \geq 0\} - c^-\mathbf{1}\{y < 0\}, \qquad y \in \mathbb{R}. \tag{7.43}$$

Lemma 7.4.4. *The steady-state mean $\eta(\bar{y})$ is strictly convex and C^1 as a function of $\bar{y} \in (0, \infty)$. Its first and second derivatives can be expressed*

$$\frac{d}{d\bar{y}}\eta(\bar{y}) = -\int_{-\infty}^{\infty} \chi_c(y - \bar{y})p^0(y)\,dy \tag{7.44a}$$

$$\frac{d^2}{d\bar{y}^2}\eta(\bar{y}) = (c^- + c^+)p^0(\bar{y}). \tag{7.44b}$$

Proof. Given (7.40), we have

$$\eta(\bar{y}) = \int_{-\infty}^{\infty} c(y)p^0(y + \bar{y})\,dy = \int_{-\infty}^{\infty} c(y - \bar{y})p^0(y)\,dy.$$

Either of these two expressions implies that $\eta(\bar{y})$ is differentiable with respect to $\bar{y}$. Expression (7.44a) is obtained from the final expression by taking the derivative under the integral sign. This is justified by the Dominated Convergence Theorem.

To establish (7.44b) we express (7.44a) as follows:

$$\int_{-\infty}^{\infty} \chi_c(y - \bar{y}) p^0(y)\, dy = c^- \int_{-\infty}^{\bar{y}} p^0(y)\, dy + (-c^+) \int_{\bar{y}}^{\infty} p^0(y)\, dy.$$

Hence (7.44b) follows from the fundamental theorem of calculus.

Strict convexity follows since the second derivative is strictly positive for any $\bar{y} > 0$. ☐

Proof of Theorem 7.4.2. The unique invariant measure is given in (7.38) when $\bar{y} = 0$, and the translation given in (ii) is obvious from (7.37). Moreover, since the distribution of $L(t)$ possesses a continuous density, the representation in Lemma 7.4.3 implies that π^0 also admits a continuous density supported on $(-\mu, \infty)$.

The remaining results follow from Lemma 7.4.4: Eq. (7.42) follows from the first-order necessary condition for optimality,

$$\frac{d}{d\bar{y}} \eta(\bar{y}) = 0.$$

This is also sufficient since $\eta(\,\cdot\,)$ is convex, and the optimizer is unique since $\eta(\,\cdot\,)$ is strictly convex on $(-\mu, \infty)$. ☐

The density assumption is imposed so that we can obtain the derivative formulae given in Lemma 7.4.4. In a discrete state-space model we use the diffusion heuristic introduced in Definition 7.4.1, or alternatively we can use (7.42) as a heuristic.

These two approaches are illustrated in the following example.

Example 7.4.5 (Inventory model with Bernoulli supply and demand). Consider the CRW workload model,

$$Y(t+1) = Y(t) - S(t+1) + S(t+1)\iota(t) + L(t+1), \qquad t \geq 0,$$

where L and S are Bernoulli processes satisfying $\mathsf{P}\{(S(t), L(t))^{\mathsf{T}} = \mathbf{1}^i\} = \mu$ if $i = 1$, and $\lambda = 1 - \mu$ for $i = 2$. This is a variation of (7.30) that can be analyzed using the same techniques to obtain a formula similar to (7.42) for the optimal hedging point.

If $\bar{y} = 1$ then the controlled process Y^1 is a version of the M/M/1 queue for which we can apply Proposition 3.4.1: Provided $\rho = \lambda/\mu < 1$, its unique invariant measure is supported on $\mathbb{Z}_+$ with $\pi_1(y) = (1 - \rho)\rho^y$ for $y \in \mathbb{Z}_+$. For $\rho \approx 1$ we have the approximation $\mathsf{P}_{\pi_0}\{y \geq \bar{y}\} \approx \mathsf{P}_{\pi_1}\{y \geq \bar{y}\} \approx \rho^{\bar{y}}$. Substituting this into the formula (7.42) we obtain the approximation

$$\bar{y}^* \log(\rho) \approx -\log\left(1 + \frac{c^+}{c^-}\right).$$

This is itself a heuristic approximation since (7.42) is based on the density assumption on L. Using the approximation $-\log(\rho) = \log(\rho^{-1}) \approx \rho^{-1} - 1$ for $\rho \approx 1$ this then gives

$$\bar{y}^* \approx \frac{\rho}{1 - \rho} \log\left(1 + \frac{c^+}{c^-}\right). \tag{7.45}$$

For the diffusion heuristic we take $\delta = \mu - \lambda$ and $\sigma^2 = \mathsf{Var}(L(1) - S(1))$. Under the assumed statistics this gives $\sigma^2 = 1$ so that (7.35) suggests the approximation

$$\overline{y}^* \approx \tfrac{1}{2} \frac{1}{\mu - \lambda} \log\left(1 + \frac{c^+}{c^-}\right). \tag{7.46}$$

The difference between the two estimates (7.45) and (7.46) remains bounded as $\rho \uparrow 1$.

∎

We now explain the diffusion heuristic by solving the optimal control problem for the CBM model.

7.4.2 CBM model

A CBM model for the simple inventory model can be analyzed exactly as in its discrete-time counterpart. Consider the one-dimensional model

$$Y(t) = y - \delta t + I(t) + N(t), \qquad Y(0) = y \in \mathbb{R}, \ t \geq 0, \tag{7.47}$$

where N is a driftless Brownian motion with instantaneous variance σ^2, and $\delta > 0$. We find that the optimal policy is a hedging-point policy, defined so that Y is a one-dimensional reflected Brownian motion (RBM) on $[-\overline{y}^*, \infty)$ for some hedging point $\overline{y}^* \in \mathbb{R}_+$.

Theorem 7.4.6 (Optimizing the brownian inventory model). *Consider the one-dimensional inventory model (7.47) and cost function $c \colon \mathbb{R} \to \mathbb{R}$ defined in (7.32). The average cost $\eta(\overline{y})$ is finite for each hedging point $\overline{y}$ and can be expressed*

$$\eta(\overline{y}) = \int_0^\infty c(y - \overline{y})\Theta e^{-\Theta y} \, dy, \tag{7.48}$$

with $\Theta^{-1} = \tfrac{1}{2}\sigma^2/\delta$. Moreover, the unique parameter $\overline{y}^*$ minimizing the average cost is

$$\overline{y}^* = \frac{1}{\Theta} \log\left(1 + \frac{c^+}{c^-}\right). \tag{7.49}$$

Proof. Computation of the optimal hedging-point value is possible since the invariant measure of Y is exponential, with parameter Θ [72, p. 250]. Consequently, for any hedging point, the average cost is given by (7.48). Differentiating this expression with respect to $\overline{y}$ and setting the resulting expression to zero gives the optimal value (7.49), exactly as in the derivation of (7.42). $\square$

It is possible to use the same techniques to address the discounted-cost optimization problem. For a given policy denote the discounted cost with discount rate $\gamma > 0$ by

$$h_\gamma(y) = \int_0^\infty e^{-\gamma t} \mathsf{E}_y[c(Y(t))] \, dt.$$

Theorem 7.4.7 (Optimization of the discounted Brownian model). *Suppose that the assumptions of Theorem 7.4.6 hold. Then, the optimal hedging point is again of the form (7.49), but with $\Theta(\gamma) > 0$ the positive root of the quadratic equation*

$$\tfrac{1}{2}\sigma^2\Theta^2 - \delta\Theta - \gamma = 0. \tag{7.50}$$

Proof. To compute the optimal hedging-point value we express the discounted cost as the expectation

$$h_\gamma(y) = \gamma^{-1}\mathsf{E}[c(Y(T)) \mid Y(0) = y], \tag{7.51}$$

where T is an exponential random variable with parameter γ that is independent of N. Consider a hedging-point policy such that Y is an RBM on $[-\overline{y}, \infty)$, and fix an initial condition $Y(0) = y < -\overline{y}$, so that $Y(0+) = -\overline{y}$. In this case the random variable $Y(T)$ has an exponential distribution, with parameter Θ (this follows from identities in [72, p. 250]). That is, for any $\overline{y}$ and any y satisfying $y < -\overline{y}$,

$$h_\gamma^{\overline{y}}(y) = \int_0^\infty c(y - \overline{y})\Theta e^{-\Theta y}\, dy.$$

The formula for $\overline{y}^*(\gamma)$ is obtained exactly as in the average-cost case by differentiating this expression with respect to $\overline{y}$, and setting the derivative equal to zero. $\square$

7.5 Hedging in networks

We close this chapter with consideration of workload relaxations for the CRW model. For a homogeneous model, a relaxation is constructed exactly as in previous chapters.

7.5.1 One-dimensional relaxation

In the stochastic scheduling models considered in Chapter 4, and any of the stochastic models considered in Chapter 6, it was not necessary to make any policy considerations for a one-dimensional relaxation. The cost function is always monotone when the workload space is $\mathbb{R}_+$, and in this case the nonidling policy is pathwise optimal by Proposition 5.4.3.

In the demand-driven models considered here a one-dimensional relaxation is similar to the simple inventory model considered in Section 7.4.

Example 7.5.1 (Simple re-entrant line with demand: one-dimensional relaxation). Consider a CRW model for the network shown in Fig. 7.5 based on the fluid model with parameters given in (7.5). It is assumed that the model is homogeneous, so that the CRW model can be expressed $Q(t + 1) = Q(t) + B(t + 1)U(t) + A(t + 1)$ for $t \geq 0$, with

$$B(t) = \begin{bmatrix} S_s(t) & -S_1(t) & 0 & 0 & 0 \\ 0 & S_1(t) & -S_2(t) & 0 & 0 \\ 0 & 0 & S_2(t) & -S_1(t) & 0 \\ 0 & 0 & 0 & S_1(t) & -S_d(t) \\ 0 & 0 & 0 & 0 & -S_d(t) \end{bmatrix} \quad A(t) = \begin{bmatrix} 0 \\ 0 \\ 0 \\ 0 \\ A_5(t) \end{bmatrix}. \tag{7.52}$$

The process A_5 models the demand process for this network, and S_s models the supply resource.

The workloads in units of inventory for Stations 1 and 2 are given by

$$Y_1(t) = \mu_1 \langle \xi^1, Q(t) \rangle = -Q_2(t) - Q_3(t) - 2Q_4(t) + 2Q_5(t),$$
$$Y_2(t) = \mu_2 \langle \xi^2, Q(t) \rangle = -Q_3(t) - Q_4(t) + Q_5(t), \qquad t \geq 0. \tag{7.53}$$

Consider the relaxation of the second workload process. If the cost function for Q is given by (7.10), then from (7.26) it follows that the optimal solution for the relaxation satisfies $\widehat{Q}(t) = |\widehat{Y}(t)| \mathbf{1}^{i_*}$ where $i_* = 3$ if $\widehat{Y}(t) < 0$, and $i_* = 5$ if $\widehat{Y}(t) \geq 0$. The effective cost is of the form (7.32) with $c^+ = 10$ and $c^- = 1$.

This model is similar to the one considered in Example 7.4.5. To adapt the diffusion heuristic we take $\delta = \mu_2 - \alpha_5$ and $\sigma^2 = \mathsf{Var}(A_5(1) - S_2(1)) = \mu_2 + \alpha_5$. Taking (7.35) for granted we obtain

$$\overline{y}^* \approx \frac{1}{2} \frac{\mu_2 + \alpha_5}{\mu_2 - \alpha_5} \log(11). \tag{7.54}$$

If $\rho_2 = \alpha_5 / \mu_2 \approx 1$, then this is approximated by $\overline{y}^* \approx \log(11)/(1-\rho_2) \approx 2.4/(1-\rho_2)$.

To obtain a policy for the five-dimensional CRW network will require the application of safety stocks at buffers 4 and 2. The idealized model takes $\widehat{Q}(t) = |\widehat{Y}(t)| \mathbf{1}^3$ when $\widehat{Y}(t) < 0$, while in the CRW model this might be replaced by the target state $q^* = \vartheta |Y(t)| \mathbf{1}^3 + (1 - \vartheta) |Y(t)| \mathbf{1}^4$ for some $\vartheta \in (0, 1)$ to ensure that inventory is available at buffer 4 when required. In Exercise 7.3 you are asked to obtain a full description of a policy based on the one-dimensional relaxation. ∎

7.5.2 Two-dimensional relaxation

We now return to the "height process analysis" described at the start of Section 5.6 for a two-dimensional workload model. As in Section 7.4 we omit the "hat" in Y to simplify notation. The workload process evolves in discrete time with

$$Y(t+1) = Y(t) - \mu + \iota(t) + L(t+1), \qquad t \geq 0, \ Y(0) \in \mathsf{Y}. \tag{7.55}$$

Recall that the one-dimensional height processes each evolve as the simple inventory model

$$H(t+1) = H(t) - \mu_H + \iota_2(t) + L_H(t+1), \qquad 0 \leq t < \tau_1, \tag{7.56}$$
$$H^\infty(t+1) = H^\infty(t) - \mu_H + \iota_2^\infty(t) + L_H(t+1), \qquad t \geq 0, \tag{7.57}$$

where $\{L_H(t)\}$ is a nonnegative sequence, and τ_1 denotes the first time that $\iota_1(t)$ is nonzero. The cost function for the height process H^∞ is defined by

$$\overline{c}_H(x) = c_H^+ x_+ + c_H^- x_-, \qquad x \in \mathbb{R}, \tag{7.58}$$

where $c_H^+ = c_2^+$ and $c_H^- = |c_2^-|$ (see (5.80)).

We obtain the following conclusions for the two-dimensional CRW model.

Theorem 7.5.2 (Hedging in the workload model). *Suppose that the following hold for the two-dimensional CRW model described by (7.55):*

(a) *The process L is i.i.d. with mean λ satisfying $\delta = \mu - \lambda > 0$. The distribution of $L(t)$ has a continuous density that is supported on a domain $O_L \subset \mathsf{Y} \cup \mathbb{R}^2_+$ that is bounded, convex, and with $0 \in$ closure (O_L).*

(b) *The piecewise linear cost function $\bar{c} : \mathbb{R}^2 \to \mathbb{R}_+$ defines a norm on $\mathbb{R}^2$. When restricted to Y, the cost function $\bar{c}$ is monotone in y_1.*

(c) *The workload space Y is a positive cone, the monotone region Y^+ has nonempty interior, and the inclusion $\mathsf{Y}^+ \subset \mathsf{Y}$ is strict.*

Then:

(i) *Under the discounted-cost optimal control criterion the optimal policy for the unconstrained process Y^∞ is uniquely specified by (5.84) for some $\bar{y}_2^* = \bar{y}_2^*(\gamma) > 0$. Equivalently, the idleness process under the optimal policy is described by a feedback law of the form $\iota^{\infty*}(t) = \phi^{\infty*}(Y^{\infty*}(t)), t \in \mathbb{Z}_+$. Its first component is identically zero, and the second component is expressed*

$$\phi_2^{\infty*}(y) = (-y_2 + s_*^\infty(y_1))_+,$$

where the switching curve s_*^∞ is linear,

$$s_*^\infty(y_1) = m_* y_1 - \bar{y}_2^* \qquad y_1 \in \mathbb{R}. \tag{7.59}$$

(ii) *Under the discounted-cost or average-cost control criteria, the optimal policy for the two-dimensional workload model is described by a feedback law of the form $\iota^*(t) = \phi^*(Y^*(t)), t \in \mathbb{Z}_+$. There exists $\underline{y}_1 \geq 0$ so that the first component is zero whenever $\phi_2(y) > 0$ and $y_1 \geq \underline{y}_1$, and the second component is described by a switching curve $s_* : \mathbb{R}_+ \to \mathbb{R}$ so that*

$$\phi_2^*(y) = (-y_2 + s_*(y_1))_+, \qquad y_1 \geq \underline{y}_1. \tag{7.60}$$

(iii) *Under the discounted-cost optimality criterion the optimal policy is approximately affine:*

$$\lim_{y_1 \to \infty} |s_*(y_1) - s_*^\infty(y_1)| = 0. \tag{7.61}$$

(iv) *In Case II* the hedging points $\{\bar{y}_2^*(\gamma) : \gamma > 0\}$ defined in (i) converge as $\gamma \downarrow 0$ to a finite limit $\bar{y}_2^*(0)$. The convergence (7.61) holds under the average-cost criterion, where the switching curve (7.59) is defined using the hedging point $\bar{y}_2^*(0)$.*

Although the proof is based on concepts that will not be revealed until Section 9.8, we can give the main ideas in the case of the discounted-cost criterion.

The first step in the proof of (7.61) is to obtain an explicit construction of the unconstrained process using a state transformation. Define for each $r \geq 0$,

$$\begin{aligned} \mathsf{Y}_r &= \{\mathsf{Y} - r(1, m_*)^\mathsf{T}\} \subset \mathbb{R}^2, \\ \bar{c}^r(y) &= \bar{c}(y + r(1, m_*)^\mathsf{T}) - r\bar{c}((1, m_*)^\mathsf{T}), \quad y \in \mathsf{Y}_r. \end{aligned} \tag{7.62}$$

For each $y^0 \in \mathsf{Y}_r$ we associate $y^r \in \mathsf{Y}$ defined by

$$y^r = y^0 + r(1, m_*)^{\mathsf{T}}. \qquad (7.63)$$

Under the assumptions of the theorem the cost function $\bar{c}$ is not monotone, and $(1, m_*)^{\mathsf{T}}$ lies in the interior of Y. It then follows that, as $r \to \infty$,

$$\mathsf{Y}_r \uparrow \mathbb{R}^2 \quad \text{and} \quad \bar{c}^r(y) \downarrow \bar{c}^{\infty}(y), \qquad y \in \mathbb{R}^2.$$

We let $h_\gamma^{r*} : \mathsf{Y}_r \to \mathbb{R}_+$ denote the optimal discounted-cost value function based on the state space Y_r and cost function $\bar{c}^r$. The key step in the proof of parts (i)–(iii) of Theorem 7.5.2 is convergence of the normalized value functions. This follows virtually by definition:

Proposition 7.5.3. *For each $\gamma > 0$ and initial condition $y \in \mathsf{Y}$, the discounted-cost value functions for the rth model converge monotonically to the value function for the unconstrained process: as $r \uparrow \infty$,*

$$h_\gamma^{r*}(y) \downarrow h_\gamma^{\infty *}(y). \qquad (7.64)$$

Moreover, for each $y^0 \in \mathsf{Y}$ we have

$$h_\gamma^{r*}(y^0) = h_\gamma^*(y^r) - r\gamma^{-1}\bar{c}((1, m_*)^{\mathsf{T}}). \qquad (7.65)$$

Proof. The result (7.64) is a consequence of three observations:

(a) $\bar{c}^{r_2}(y) \leq \bar{c}^{r_1}(y) \leq \bar{c}(y)$ for all $y \in \mathsf{Y}$ and all $0 \leq r_1 < r_2 \leq \infty$,
(b) $\bar{c}^r(y) \downarrow \bar{c}^{\infty}(y)$ as $r \uparrow \infty$, and
(c) the process $\mathbf{Y}^{r_2}$ is subject to fewer constraints than $\mathbf{Y}^{r_1}$ for each $0 \leq r_1 < r_2 \leq \infty$.

The identity (7.65) follows from the form of $\bar{c}^r$ in (7.62). $\qquad \square$

Example 7.5.4 (Hedging in a power distribution system). To illustrate a different kind of application of hedging we return to the electric power distribution system described in Section 2.7.

Consider first a model with a single generator and consumer. This is a simplification of (2.17) since the ancillary generation is absent. Recall that this is not a queueing system since electricity is not stored.

At time t the amount of power the generator can supply is denoted $G(t)$, and the demand for power in watts is denoted $D(t)$. The excess capacity is the difference, denoted

$$Q(t) = G(t) - D(t), \quad t \geq 0. \qquad (7.66)$$

The constraints on generation are as in (2.20): For the discrete time model it is assumed that $U(t) := G(t) - G(t - 1) \in \mathsf{U}$ for $t \geq 0$ where

$$\mathsf{U} := \{u \in \mathbb{R} : -\zeta^{p-} \leq u \leq \zeta^{p+}\}. \qquad (7.67)$$

In a CBM model in which D is Brownian motion we relax the lower rate constraint, so that $\zeta^{p-} = \infty$.

Recall that demand can be negative since this is *normalized demand*: current demand minus an earlier forecast.

We now construct a cost function from the point of view of the consumer. First note that demand for power is not flexible. Moreover, if demand is not met, so that $Q(t)$ is negative, then an emergency is declared. Blackouts may be imposed, resulting in high cost. A second cost is the price paid for *reserves*: If the generator is asked to have on hand $G(t)$ units of power, then some compensation is reasonable, even if $G(t) > D(t)$. Finally, the consumer finds value from consuming power, which can be modeled as a utility or "negative cost." A simple cost function capturing these features is piecewise linear,

$$c(G(t), Q(t)) = c^p G(t) + c^{bo} Q_-(t) - v \min(D(t), G(t)), \qquad (7.68)$$

where c^{bo} is the marginal cost of an increment of unfulfilled demand, c^p is the marginal price paid for power reserve, and v is the marginal value of consumption.

This can be converted to a cost function of $Q(t)$ alone as follows. On substituting Eq. (7.66) to eliminate G on the right-hand side we obtain

$$\begin{aligned}
c(G(t), Q(t)) &= c^p(Q(t) + D(t)) + c^{bo} Q_-(t) - v \min(D(t), Q(t) + D(t)) \\
&= c^p Q(t) + c^{bo} Q_-(t) - v \min(0, Q(t)) + (c^p - v)D(t) \\
&= c^p Q(t) + (c^{bo} + v)Q_-(t) + (c^p - v)D(t).
\end{aligned}$$

Since D is not controlled, the following function of $Q(t)$ serves as an equivalent cost function for the purposes of control:

$$c(Q(t)) = c^p Q(t) + (c^{bo} + v)Q_-(t) = \max(c^p Q(t), (c^p - c^{bo} - v)Q(t)). \quad (7.69)$$

We conclude that the model (7.66) with this cost function is identical to the simple inventory model considered in Section 7.4.

We now consider the model with ancillary service whose state process X is defined in Section 2.7. The excess capacity at time $t \geq 1$ is defined by (2.17), where $D(t)$ is the demand at time t, and $(G^p(t), G^a(t))$ are current capacity levels from primary and ancillary service.

The state process X is constrained to the state space $X = \mathbb{R} \times \mathbb{R}_+$, and obeys the recursion (2.19),

$$X(t+1) = X(t) + BU(t) + A(t+1), \qquad t = 0, 1, \dots,$$

where B is defined in (2.18), the two-dimensional "arrival process" is defined as $A(t) := -(\mathcal{E}(t), 0)^{\mathsf{T}}$, and the inputs are increments of capacity, $U^p(t) := G^p(t+1) - G^p(t)$, $U^a(t) := G^a(t+1) - G^a(t)$. For simplicity we relax the ramp-down constraints $\{\zeta^{a-}, \zeta^{p-}\}$ in (2.20): It is assumed that $U(t) \in U(X(t))$ for all $t \in \mathbb{Z}_+$, where

$$U := \{u = (u^p, u^a)^{\mathsf{T}} \in \mathbb{R}^2 : -\infty \leq u^p \leq \zeta^{p+}, \ -\infty \leq u^a \leq \zeta^{a+}\},$$

$$U(x) := \{u \in U : x + Bu \in X\}, \qquad x \in X.$$

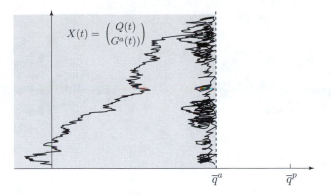

Figure 7.20. Ancillary service ramps up at maximal rate when $Q(t) < \overline{q}^a$.

A hedging-point policy is defined by a pair of thresholds $\overline{q}^a, \overline{q}^p$. If $Q(t) < \overline{q}^p$ then primary generation ramps up at maximal rate, subject to the constraint that $Q(t)$ does not exceed this threshold. That is,

$$U^p(t) = \max(\overline{\zeta}^{p+}, \overline{q}^p - Q(t)).$$

If $Q(t) \geq \overline{q}^p$ then primary generation ramps down, $U^p(t) = \overline{q}^p - Q(t) < 0$.

The specification of the policy for ancillary service also has two cases. If $Q(t) + U^p(t) < \overline{q}^a$ then ancillary generation ramps up,

$$U^a(t) = \max(\overline{\zeta}^{a+}, \overline{q}^a - Q(t) - U^p(t)).$$

If $Q(t) + U^p(t) > \overline{q}^a$ then ancillary generation ramps down, but not too fast,

$$U^a(t) = \max(-G^a(t), \overline{q}^a - Q(t) - U^p(t)).$$

A typical sample path of X under a hedging-point policy is shown in Fig. 7.20.

The cost function is an elaboration of (7.69) to include the cost $c^a > c^p$ for ancillary service,

$$c(Q(t)) = c^a G^a(t) + c^p Q(t) + (c^{bo} + v)Q_-(t). \tag{7.70}$$

Performance analysis is based upon a pair of height processes with respect to each of the thresholds $\overline{q}^p, \overline{q}^a$,

$$H^p(t) := -Q(t) + G^a(t), \quad H^a(t) := -Q(t), \qquad t \geq 0.$$

In the CRW model, the first process evolves as the simple inventory model with hedging point equal to $\overline{q}^p$. The process H^a has the same behavior, but only up to the first time that $G^a(t)$ reaches zero. The threshold for H^a is $\overline{q}^a$. Another difference between the two processes is that H^p has mean drift ζ^{p+}, while H^a has mean drift $\zeta^{p+} + \zeta^{p+}$,

$$H^p(t+1) = \max(H^p(t) - \zeta^{p+}, -\overline{q}^p) + \mathcal{E}(t+1) \qquad t \geq 0,$$
$$H^a(t+1) = \max(H^a(t) - \zeta^{p+} - \zeta^{a+}, -\overline{q}^a) + \mathcal{E}(t+1) \quad \text{while } G^a(t) > 0.$$

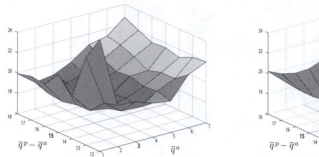 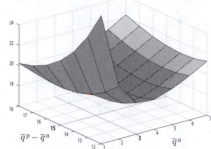

Figure 7.21. Shown on the left is the average cost obtained using simulation for the network with demand increments supported on $\{-1, 1\}$ using various hedging-point values. On the right is the average cost obtained for the CBM model.

Cost functions consistent with the two-dimensional model are of the form (7.58) with

$$c_H^+ = c^a - c^p, \quad c_H^- = c^p \qquad \text{for } H^p$$
$$c_H^+ = c^{bo} + v - c^a, \quad c_H^- = c^a \qquad \text{for } H^a.$$

Based on these cost functions and these dynamics, we obtain from the diffusion heuristic the two parameters

$$\overline{q}^{a*} = \frac{1}{2} \frac{\sigma_D^2}{\zeta^{p+} + \zeta^{a+}} \log\left(\frac{c^{bo} + v}{c^a}\right) \qquad \overline{q}^{p*} = \overline{q}^{a*} + \frac{1}{2} \frac{\sigma_D^2}{\zeta^{p+}} \log\left(\frac{c^a}{c^p}\right). \qquad (7.71)$$

Consider the following special case in which the marginal distribution of the increment process $\mathcal{E}$ is symmetric on $\{\pm 1\}$. The model parameters are $c^p = 1$, $c^a = 20$, and $v + c^{bo} = 400$; the ramp-up rates are $\zeta^{p+} = 1/10$ and $\zeta^{a+} = 2/5$.

Shown on the right in Fig. 7.21 is the average cost for the CBM model with first- and second-order statistics identical to those of the CRW model. For these numerical values, the diffusion heuristic (7.71) gives $(\overline{q}^{p*}, \overline{q}^{a*}) = (14.974, 2.996)$.

Affine policies for the CRW model were constructed based on threshold values $\{\overline{q}^p, \overline{q}^a\}$. The average cost was approximated at several values of $(\overline{q}^p, \overline{q}^a)$ based on the (unbiased) *smoothed estimator* described in Chapter 11. In the simulation shown on the left in Fig. 7.21 the time horizon was $n = 8 \times 10^5$. Among the affine parameters considered, the best policy for the discrete time model is given by $(\overline{q}^{p*}, \overline{q}^{a*}) = (19, 3)$, which almost coincides with the values obtained using (7.71).

In conclusion, in spite of the drastically different demand statistics, the best affine policy for the discrete-time model is remarkably similar to the average-cost optimal policy for the continuous-time model with Gaussian demand. Moreover, the optimal average costs for the two models are in close agreement.

To see the impact of variability we now consider three models in which the increment process $\mathcal{E}$ is again symmetric, and supported on the finite set $\{0, \pm 3, \pm 6\}$.

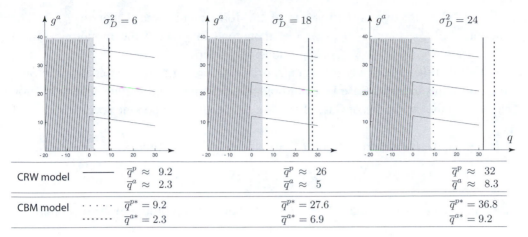

CRW model	——	$\bar{q}^p \approx 9.2$ $\bar{q}^a \approx 2.3$	$\bar{q}^p \approx 26$ $\bar{q}^a \approx 5$	$\bar{q}^p \approx 32$ $\bar{q}^a \approx 8.3$
CBM model	· · · · · - - - - - -	$\bar{q}^{p*} = 9.2$ $\bar{q}^{a*} = 2.3$	$\bar{q}^{p*} = 27.6$ $\bar{q}^{a*} = 6.9$	$\bar{q}^{p*} = 36.8$ $\bar{q}^{a*} = 9.2$

Figure 7.22. Optimal policies in a power distribution system. The policy for the CRW model was computed using value iteration. The grey region indicates those states for which ancillary service ramps up at maximum rate, and the constant $\bar{q}^p$ is the value such that primary ramps up at maximum rate when $Q(t) < \bar{q}^p$. The optimal policy for the CBM model closely matches the optimal policy for the discrete-time model in each case.

In each case the marginal distribution of $\mathcal{E}$ is uniform on its respective support. The support and respective variance values are given by

(a) $\{-3, 0, 3\}$, $\sigma_a^2 = 6$ (b) $\{-6, -3, 0, 3, 6\}$, $\sigma_b^2 = 18$ (c) $\{-6, 0, 6\}$, $\sigma_c^2 = 24$.

The marginal distribution has zero mean since the support is symmetric in each case.

The cost parameters are taken to be $c^{bo} + v = 100$, $c^a = 10$, $c^p = 1$, and we take $\zeta^{p+} = 1$, $\zeta^{a+} = 2$. The state process X is restricted to an integer lattice to facilitate computation of an optimal policy using value iteration.

Optimal policies are illustrated in Fig. 7.22: The constant $\bar{q}^p$ is defined as the maximum of $q \geq 0$ such that $U^p(t) = 1$ when $X(t) = (q, 0)^{\mathsf{T}}$. The grey region represents R^a, and the constant $\bar{q}^a$ is an approximation of the value of q for x on the right-hand boundary of R^a.

Also shown in Fig. 7.22 is a representation of the optimal policy for the CBM model with first- and second-order statistics consistent with the CRW model. That is, the demand process D was taken to be a drift-less Brownian motion with variance σ_D^2 equal to 6, 18, or 24 as shown in the figure.

The constants $\bar{q}^{p*}$, $\bar{q}^{a*}$ indicated in the figure are the optimal parameters for the CBM model obtained using the diffusion heuristic (7.71). The optimal policy for the CBM model closely matches the optimal policy for the discrete-time model in each case. ∎

Theorem 7.5.2 and the approximations obtained through the diffusion heuristic predict the numerical results previously described in Section 5.6.2.

Example 7.5.5 (Simple re-entrant line without demand). Consider the three-dimensional CRW model (4.10) for the simple re-entrant line shown in Fig. 2.9. We maintain the assumptions of Section 5.6.2: The model is homogeneous, and c is the

ℓ_1 norm. The effective cost (5.88) is of the form required in Theorem 7.5.2: The function $\bar{c}(y) = \max(y_2, y_1 - y_2)$ is monotone in y_1, and the monotone region $Y^+ := \{y \in \mathbb{R}_+^2 : \frac{1}{2}y_1 \leq y_2 \leq y_1\}$ has nonempty interior.

Recall that the effective cost was categorized according to four cases listed in Section 5.6. In this example Case I cannot hold since $\bar{c}$ is never monotone. The relaxation may satisfy the conditions of Case II or Case III, depending upon the specific values of $\{\mu_i, \alpha_1\}$.

Based on the lower boundary of Y^+, we define the height process

$$H(t) = \widehat{Y}_2(t) - \tfrac{1}{2}\widehat{Y}_1(t), \qquad t \geq 0.$$

Applying (5.87), we find that while $\widehat{\iota}_1(t) = 0$ it evolves according to the recursion

$$H(t+1) = H(t) - S_2(t+1) + S_2(t+1)\widehat{\iota}_2(t) + \tfrac{1}{2}S_1(t+1), \qquad t \geq 0.$$

The cost function for the height process is defined by

$$\bar{c}_H(r) = \max(-c_H^- r, c_H^+ r) \qquad r \in \mathbb{R}.$$

We have $c_H^+ = \bar{c}_2^+ = 1$ and $c_H^- = |\bar{c}_2^-| = 1$, giving $\bar{c}_H(r) = |r|$.

The height process $\boldsymbol{H}^\infty$ for the unconstrained process $\widehat{\boldsymbol{Y}}^\infty$ satisfies the same recursion. Under an affine policy for $\widehat{\boldsymbol{Y}}^\infty$ defined by the switching curve (5.89), we have $\widehat{\iota}_1 \equiv 0$ and $\widehat{\iota}_2$ is defined so that the height process evolves as a reflected random walk,

$$H^\infty(t+1) = \max(H^\infty(t), -\bar{y}_2) - S_2(t+1) + \tfrac{1}{2}S_1(t+1), \qquad t \geq 0. \quad (7.72)$$

We have under any policy

$$\bar{c}^\infty(\widehat{Y}^\infty(t)) = \bar{c}_H(H^\infty(t)) + (\bar{c}_1^+ + \bar{c}_2^+/m_*)\widehat{Y}_1^\infty(t), \quad (7.73)$$

and applying (5.88) we have $\bar{c}_1^+ + \bar{c}_2^+/m_* = 1/m* = 2$ with $m_* = \tfrac{1}{2}$. The discounted-cost optimal policy for $\widehat{\boldsymbol{Y}}^\infty$ is obtained on solving the discounted-cost optimal control problem for the height process $\boldsymbol{H}^\infty$ since $\widehat{\boldsymbol{Y}}_1^\infty$ is uncontrolled.

To estimate the threshold $\bar{y}_2^*$ that solves the average-cost optimal control problem in Case II* we take two different paths. First we compute the invariant measure for $\boldsymbol{H}^\infty$ when $\bar{y}_2 = 0$ and apply formula (7.42). This is only a heuristic since the density assumption imposed to obtain (7.42) is not satisfied for $\boldsymbol{H}^\infty$. Next, we construct a CBM model with identical first- and second-order statistics and apply the diffusion heuristic Definition 7.4.1 to obtain a second estimate. Based on these two approaches we obtain the two approximations

$$\bar{y}_2^*(\text{CRW}) \approx \tfrac{1}{2}\frac{\varsigma}{1-\varsigma}\log\left(1 + \frac{c_H^+}{c_H^-}\right), \quad (7.74)$$

$$\bar{y}_2^*(\text{CBM}) \approx \tfrac{1}{2}\frac{\sigma_H^2}{\delta_H}\log\left(1 + \frac{c_H^+}{c_H^-}\right), \quad (7.75)$$

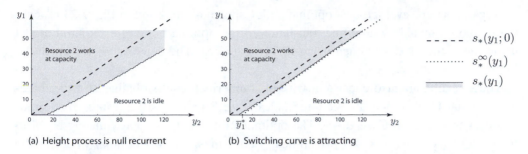

(a) Height process is null recurrent

(b) Switching curve is attracting

Figure 7.23. Optimal switching curves for the CRW workload relaxation in Case II. On the left is Case II (a) where the network is balanced, and on the right is Case II (b). The difference between the fluid and stochastic switching curves is significant in Case II (a). This corresponds to null recurrence of the associated height process. The optimal policy in Case II (b) is accurately approximated by an affine policy whose offset is determined by an associated height process with threshold $\overline{y}_2^*$ obtained using the diffusion heuristic. The parameter shown in the figure is $\overline{y}_1^* := 2\overline{y}_2^* \approx 11$.

where $c_H^- = c_H^+ = 1$, and

$$\varsigma = \tfrac{1}{2}\left(1 + 4\frac{\mu_1}{\mu_2}\right)^{\frac{1}{2}} - \tfrac{1}{2}. \tag{7.76}$$

$$\delta_H = \mu_2 - \mu_1/2, \quad \sigma_H^2 \approx \mu_2 + \mu_1/4. \tag{7.77}$$

The parameter ς is strictly less than 1 and δ_H is strictly positive since $\mu_1 < 2\mu_2$. We leave the derivation of (7.74) as a guided homework problem in the exercises at the end of this chapter.

Optimal policies for $\widehat{Y}$ are shown in Fig. 7.23 based on the two instances of Case II described in Section 5.6.2.

Case II (a) is a marginal case since the drift vector δ_Y is exactly aligned with the lower boundary of Y^+. The height process H^∞ is *not* positive recurrent in this case (see Theorem 7.4.2). Consequently, we would expect the optimal policy for the probabilistic workload relaxations to differ significantly from the fluid relaxation. The plot on the left in Fig. 7.23 shows that the difference between the fluid and stochastic switching curves is significant.

Case II (b) is an instance of Case II*, which is the assumption required in Theorem 7.5.2 (iv). The height process for the CRW model is positive recurrent, and the parameter ς defined in (7.76) is given by

$$\varsigma = \tfrac{1}{2}\left(1 + 4\frac{20}{11}\right)^{\frac{1}{2}} - \tfrac{1}{2} \approx 0.94.$$

On applying (7.74) we obtain $\overline{y}_2^* \approx \tfrac{1}{2}(\varsigma/(1-\varsigma))\log(2) \approx 5.25$. The approximation (7.75) gives a similar value, $\overline{y}_2^* \approx \tfrac{1}{2}(\sigma_H^2/\delta_H)\log(2) \approx 5.5$. These are simply estimates of the optimal parameter for the height process H^∞ – at this stage it is not clear that they have any relevance to the two-dimensional CRW model.

Regardless, the average-cost optimal policy shown on the right in Fig. 7.23 is accurately approximated by an affine translation of the optimal policy for the fluid model using the value of $\overline{y}_2^*$ obtained using (7.74) or (7.75). The constant $\overline{y}_1^*$ shown in the figure is $\overline{y}_1^* := 2\overline{y}_2^* \approx 11$.

These results are also consistent with the numerical results obtained for the three-dimensional CRW model in Section 5.6.2. Figure 5.13 shows the results from simulation experiments in Case II (b). The optimal value of $\overline{y}_1$ is approximately equal to $\overline{y}_1 = 9$, or $\overline{y}_2 = \frac{1}{2}\overline{y}_1 = 4.5$, which nearly coincides with the optimal value for the CRW model $\widehat{Y}$ shown in Fig. 7.23 (b).   ∎

Example 7.5.6 (Complex demand-driven model). The 16-buffer model with cost parameters specified in Example 7.3.5 illustrates how the application of hedging points generalizes to a complex network setting. We consider a two-dimensional relaxation of the form considered in the previous example. In our earlier consideration of the 16-buffer model we considered workload in units of time. We maintain this convention here.

The myopic, GTO, and optimal policies for the fluid workload model are described at the end of Example 7.3.5. They each are defined by some constraint region $\mathsf{R} \subset \mathsf{W}$ containing W^+. Just as in the one-dimensional inventory model, the region R obtained for the fluid model is not a suitable constraint region to construct a policy for the stochastic model. To see this, observe that for any $w \in \mathsf{W}^+$ the effective state $\mathcal{X}^*(w)$ corresponds to a set of buffer levels for which there is deficit at some virtual queue. Consequently, as in the simple inventory model (7.30), if $W(t)$ is constrained to W^+ then high costs will be incurred due to unsatisfied demand.

A second fundamental difficulty in this multidimensional setting was described in Section 5.6. If W is constrained to W^+ as dictated by the myopic policy, this will result in excessively large values of $W(t)$ due to "chattering" near the boundaries of W^+. This is illustrated for a generic example in Fig. 6.19. Note that the constraint region shown on the left in Fig. 6.19 is similar to the monotone region for the present example shown on the left in Fig. 7.17.

In the numerical results described here the CRW workload process is defined in discrete time as in (5.59),

$$\widehat{W}(t+1) = \widehat{W}(t) + \widehat{S}(t+1)\widehat{\iota}(t) - \widehat{S}(t+1)\mathbf{1} + \widehat{L}(t+1).$$

The process $\{\widehat{L}(t) - \widehat{S}(t) : t \geq 1\}$ is i.i.d., and its marginal distribution is supported on $\{-(2,0)^{\mathsf{T}},\ -(0,2)^{\mathsf{T}},\ (5,5)^{\mathsf{T}}\}$. Two cases are considered, where in each case the mean drift is defined by $-\delta := -\mathsf{E}[\widehat{L}(t) - \widehat{S}(t)]$.

Case II* The marginal distribution is uniform (each possible value occurs with probability $1/3$.) This results in a mean drift given by $-\delta = -[1,1]^{\mathsf{T}}$, so that δ lies within the interior of the monotone region shown on the left in Fig. 7.2 (where $\mathsf{R}^* = \mathsf{W}^+$).

Case III The probabilities are chosen so that $-\delta = -[1/10, 1]^{\mathsf{T}}$, which is consistent with the drift vector shown in Fig. 7.18 and on the right in Fig. 7.2. The conditions of Case III hold since $\delta \notin \mathsf{W}^+$.

Figure 7.2 illustrates the average-cost optimal policy for a CRW workload model in these two cases. The optimal idleness process $\widehat{\iota}^*$ was obtained using value iteration. The light grey regions indicate values of w such that $\widehat{\iota}^*(t) = 0$ when $\widehat{W}^*(t) = w$. Note that in each case, the boundaries of the optimal stochastic control region R^{STO} shown are approximated by an affine enlargement of the fluid-optimal policy. ∎

7.6 Summary of steady-state control techniques

In the last two sections we have emphasized the application of hedging points to improve performance in the control of workload.

The numerical results obtained in Example 7.5.5 and previously in Section 5.6.2 show that hedging combined with safety stocks can approximate the structure of the optimal policy for the simple re-entrant line. Similarly, the optimal policy for the 16-buffer example treated in Example 7.5.6 can be approximated by an affine policy obtained using two hedging points in workload space. This can be interpreted as hedging at the two virtual queues.

In summary, here is a two-step program for policy synthesis based on what we have learned so far:

1. Construct a workload relaxation of suitable dimension to obtain a policy in workload space, described as state feedback via $\iota(t) = \hat{\phi}(\widehat{W}(t))$, $t \geq 0$. For example, this can be the optimal policy for the relaxation, or an affine policy based on restricting $\widehat{W}$ to an affine enlargement of the monotone region.
2. Devise a policy for Q to satisfy $W(t) \approx \widehat{W}(t)$. Equally important is the desire to maintain $Q(t) \approx \mathcal{P}^*(Q(t))$ for all t, subject to this workload constraint.

Two techniques introduced in Chapter 4 can be adapted to achieve the two approximations in (2):

(i) A discrete-review policy of the form described in Section 4.7 can be formulated based on safety stocks as described in Section 4.6. Exactly as in the myopic-DR and GTO-DR policies, a vector of $\overline{x} \in \mathbb{R}^\ell_+$ is constructed to avoid starvation of resources so that the goals of (ii) are achieved.

   Based on the numerical results treated in Section 4.6 and in Example 6.7.2 (see also Proposition 6.7.3), the following parametric form is suggested: For a constant vector $\theta \in \mathbb{R}^{\ell_r}$, the vector $\overline{x}$ should grow with the total population so that

$$\sum_{j \geq 1} C_{sj} \overline{x}_j = \theta_s \log(1 + |x|/\theta_s), \qquad 1 \leq s \leq \ell_r, \qquad (7.78)$$

   where C is the constituency matrix, and $x \in \mathsf{X}$ is the initial state at the start of the planning horizon.

(ii) If a stationary policy is preferable then this can be constructed using a particular instance of the h-MaxWeight or h-myopic policy. Suppose that the policy $\hat{\phi}$ constructed in (1) is described as a myopic policy,

$$\hat{\phi}(w) = \arg\min \mathsf{E}[\hat{h}(\widehat{W}(t+1)) \mid \widehat{W}(t) = w, \ \iota(t) = \iota],$$

where the minimum is over $\iota \geq 0$ such that $W(t+1) \in \mathsf{W}$ with probability 1. We then define for a given constant $b > 0$,

$$h(x) := \hat{h}(\widehat{\Xi}\tilde{x}) + \tfrac{b}{2}[c(\tilde{x}) - \bar{c}(\widehat{\Xi}\tilde{x})]^2, \qquad x \in \mathbb{R}_+^\ell, \qquad (7.79)$$

where $\tilde{x}$ is defined in (4.93).

A stationary policy is obtained using the h-myopic policy. Alternatively, if $\hat{h}$ is smooth then the simpler h-MaxWeight policy can be applied.

Conditions for stability of the resulting policy and methods to evaluate performance are developed in the chapters that follow – see in particular Theorem 8.4.1 for general conditions ensuring stability of the h-MaxWeight policy.

7.7 Notes

The material in Sections 7.1 and 7.2 is taken from [151, 102]. In particular, all of the figures in Section 7.2 are based on numerical work performed by Dubrawskii [151]. The focus on transient issues is common to the first part of Newell's monograph [385].

In manufacturing models an alternative approach to managing complexity is based on decomposition techniques introduced in the seminal paper of Clark and Scarf [110]. Related hierarchical decomposition techniques are contained in the work of Gershwin [202, 134], Perkins and Kumar [396, 397, 395], Sethi [434, 433], and Muharremoglu and Tsitsiklis [377]. These approaches are similar in flavor to workload relaxations, and they may be equivalent for certain models.

The construction and analysis of policies to determine work release of raw material to a production system has been the subject of some research [336, 267]. A well-known approach is the CONWIP (constant work-in-process) policy in which new material is released to the system just as completed jobs depart [456, 457, 64, 4, 429]. Hence, under the CONWIP policy the total customer population remains constant. Research has focused on the determination of an optimal value for the overall customer population [243, 320]. Refinements are described in [489, 152]. Scheduling and work-release policies for demand-driven models of specific structure are developed in, for instance, [204] based on the construction of hedging points. Admission control is also an essential part of flow control in computer networks [43, 459, 8].

Uncertainty in demand presents a new dimension in constructing efficient policies. The issues are summarized in the *news-vendor problem*: A news vendor must decide how many papers to purchase from a supplier to be delivered the next morning. In deciding how many papers to order, the new vendor must balance two conflicting costs: Any extra papers will have no value at the end of the day, but of course she will miss sales if she does not purchase newspapers in sufficient quantities. Practical solutions require the use of hedging as described in Section 7.4.

The mathematical foundations of hedging began in the early fifties in the work of Arrow, Harris, and Marschak [17], and Dvoretzky, Kiefer, and Wolfowitz [161]. These ideas expanded in the work of Clark and Scarf [18, 422, 110, 423]. Existence and uniqueness of optimal policies for a version of the simple inventory model was established in 1955 by Bellman et al. in [39].

Federgruen and Zipkin [177, 176] provide conditions under which a base-stock policy is optimal, based on existence and structural theory for average-cost optimal control of MDP models obtained in [174], following Clark and Scarf [110]. This remains an area of active research [175, 203, 394, 490, 398, 262, 77, 377].

The CBM inventory model (7.47) is considered in [488] where the optimal hedging-point value (7.49) is derived.

The development in Section 7.5 is based on [103], where the structure of optimal policies for multidimensional workload models was investigated for the first time. In particular, Theorem 7.5.2 that describes the emergence of hedging points for the CRW workload model is based on results from [103] for the CBM model. The numerical results in Example 7.5.5 are also taken from [103].

The numerical plots in Example 7.5.4 are taken from Chen's thesis [100]. Similar solidarity between the CRW and diffusion models is found when $\mathcal{E}$ is Markov rather than i.i.d.. To apply the diffusion heuristic, the variance σ_D^2 is taken to be the asymptotic variance appearing in the Central Limit Theorem for $\mathcal{E}$ (described in Chapter 11). It is shown in [107] that the diffusion heuristic is exact for the CBM model. That is, the affine policy described in Example 7.5.4 is average-cost optimal.

Exercises

7.1   Compute the workload vectors for the network in Example 7.2.1 by computing (numerically or analytically) the extreme points in the linear program (6.18).

7.2   Repeat Exercise 7.1 for the model in Example 7.2.2.

7.3   Devise a policy for the simple re-entrant line with demand based on the hedging point constructed in Example 7.5.1. The policy can be based on safety stocks at buffers 2 and 4. Or, you might construct a function $\hat{h}\colon \mathbb{R}_+ \to \mathbb{R}_+$ such that the $\hat{h}$-MaxWeight policy coincides with the hedging-point policy for the one-dimensional relaxation constructed in in Example 7.5.1. Based on this, construct a function $h\colon \mathbb{R}_+^5 \to \mathbb{R}_+$ using (7.79) to define an h-MaxWeight policy for Q. Simulate the CRW model to find the best parameters in your policy.

7.4   Repeat Exercise 7.3 using the model in Example 7.2.2.

7.5   Prove Proposition 7.3.1.

7.6   Consider the fluid model of the simple inventory model (7.31) with piecewise linear cost function $c\colon \mathbb{R} \to \mathbb{R}$. Suppose that y is controlled using a hedging-point policy with *nonzero* value $\bar{y} > 0$. Compute the fluid value function J and its normalization K,

$$J(y) = \int_0^T c(y(t; y))\, dt$$

$$K(y) = \int_0^T \big(c(y(t; y)) - c(-\bar{y}) \big)\, dt, \qquad y \in \mathbb{R},$$

where $T = T(y) = \min(t : y(t) = -\bar{y})$. Sketch J and K as a function of y, and verify the dynamic programming equations

$$-\delta \tfrac{d}{dy} J(y) = -c(y), \qquad -\delta \tfrac{d}{dy} J(y) = -c(y) + c(-\bar{y}), \qquad y > -\bar{y}.$$

Are these functions smooth? In Exercise 11.10 these functions are used to construct a simulation algorithm.

7.7   In this exercise you will derive (7.76) based on the height process defined in (7.72).

(a) When $\bar{y}_2 = 1$ the height process for the unconstrained process is a countable state-space Markov chain on $\mathsf{X}_H = \{0, \tfrac{1}{2}, 1, \tfrac{3}{2}, \dots\}$. Show that it is positive recurrent if and only if $\mu_1 < 2\mu_2$. That is, $\rho_1 > \rho_2$, which is precisely Case II*.

(b) Show that the invariant measure for $\boldsymbol{H}^\infty$ with $\bar{y}_2 = 0$ is given by

$$\pi_H^0(0) = \pi_H^0(1) = \frac{1}{\varsigma}\frac{1-\varsigma}{2-\varsigma} \tag{E7.1}$$

$$\pi_H^0(r) = \pi_H^0(0)\varsigma^{2r-1}, \qquad r \in \mathsf{X}_H, \ r \geq 2,$$

where ς is given in (7.76).

(c) The process $\boldsymbol{H}^\infty$ is similar to the example considered in Example 7.4.5. To optimize $\pi_H^{\bar{y}}(\bar{c}_H)$ over $\bar{y}_2$ we apply (7.42) as a heuristic to obtain

$$\left(1 + \frac{c_H^+}{c_H^-}\right)^{-1} = \mathsf{P}_{\pi_H^{\bar{y}}}\{y \geq 0\} \approx \mathsf{P}_{\pi_H^0}\{y \geq -\bar{y}_2^*\} \approx \frac{1}{\varsigma}\frac{1}{2-\varsigma}\varsigma^{2\bar{y}_2^*}.$$

Taking logarithms gives

$$\bar{y}_1^* := 2\bar{y}_2^* \approx \frac{1}{\log(\varsigma^{-1})}\log\left(1 + \frac{c_H^+}{c_H^-}\right) + \frac{1}{\log(\varsigma^{-1})}\Big(\log(\varsigma^{-1}) - \log(2-\varsigma)\Big).$$

Using the same sequence of bounds used to obtain the approximation (7.45), obtain the analogous approximation (7.74).

Part III

Stability and Performance

8

Foster–Lyapunov Techniques

So far we have visited several approaches to policy synthesis for networks, including a survey of control techniques for the scheduling model in Chapter 4. The focus of much of Chapter 7 was transient performance metrics such as the impact of hot-lots on congestion, or the recovery time following the failure of some resource. We have considered several techniques to account for variability, including hedging points, safety stocks, and discrete-review policies. This chapter provides foundations for evaluating these methods. We concentrate primarily on steady-state performance criteria, and the surrounding theory of Lyapunov functions for stochastic networks.

To address questions regarding finer performance properties we assume that the queue length process can be modeled as a Markov chain, or as a function of a Markov chain denoted $\boldsymbol{X} = \{X(0), X(1), \dots\}$. We let $\mathsf{X}_\diamond$ denote the state space which is always assumed countable, and P the transition matrix defined by

$$P(x,y) = \mathsf{P}[X(t+1) = y \mid X(t) = x], \qquad x, y \in \mathsf{X}_\diamond.$$

Just as is standard in the theory of deterministic dynamical systems, stability of a Markov model is characterized by a version of the *direct method of Lyapunov* [59, 296]. Section 8.1 highlights some Lyapunov stability theory for Markov chains based on the survey contained in the Appendix, and this is specialized to the CRW model in Section 8.2. This theory is largely based on inequalities involving the *generator*, defined as the difference operator

$$\mathcal{D} := P - I. \tag{8.1}$$

The generator acts on functions $h \colon \mathsf{X}_\diamond \to \mathbb{R}$ via

$$\mathcal{D}h\,(x) = Ph\,(x) - h(x) = \mathsf{E}[h(X(t+1)) - h(X(t)) \mid X(t) = x], \qquad x \in \mathsf{X}_\diamond,$$

whenever the expectation is well defined.

The general form of the Lyapunov condition considered here is condition (V3), or the special case known as Foster's criterion. If either of these bounds hold, then V is called a *Lyapunov function*.

The Markov chain X is called x^*-*irreducible* if the given state $x^* \in X_\diamond$ is reachable with positive probability from each initial condition:

$$\sum_{t=0}^{\infty} P^t(x, x^*) > 0, \qquad x \in X_\diamond.$$

Definition 8.0.1 (Foster–Lyapunov drift conditions). The chain satisfies *Foster's criterion* if there is a function $V \colon X_\diamond \to \mathbb{R}_+$, a constant $b < \infty$, and a finite set $S \subset X_\diamond$ such that

$$\mathcal{D}V(x) \leq -1 + b\mathbf{1}_S(x), \qquad x \in X_\diamond. \tag{V2}$$

The chain satisfies condition (V3) if for a function $f \colon X_\diamond \to [1, \infty)$,

$$\mathcal{D}V(x) \leq -f(x) + b\mathbf{1}_S(x), \qquad x \in X_\diamond, \tag{V3}$$

where again $b < \infty$ and $S \subset X_\diamond$ is a finite set. If this is the case, and in addition X is x^*-irreducible for some $x^* \in X_\diamond$, then the chain is called f-*regular*. ∎

Written as an expectation, Condition (V3) becomes

$$\mathsf{E}[V(X(t+1)) - V(X(t)) \mid X(t) = x] \leq \begin{cases} -f(x) & x \in S^c \\ -f(x) + b & x \in S. \end{cases}$$

Hence $V(X(t))$ is decreasing on average until the first time that $X(t)$ reaches the set S. The function f modulates the rate of decrease.

The cost functions considered in this chapter are assumed to admit an extension to all of $\mathbb{R}^\ell$ to define a norm. Examples are the ℓ_1-norm $c(x) = |x|$, or any piecewise linear function that is strictly positive on $\mathbb{R}_+^\ell \setminus \{0\}$.

More generally, c is called *near monotone* if the sublevel set $S_c(r) := \{x \in \mathbb{R}^\ell : c(x) \leq r\}$ is bounded for each $r < \sup_{x \in \mathbb{R}^\ell} c(x)$. If c is also unbounded, then the cost function is called *coercive*.

Typically, a solution to (V3) is sought for which the function V can be viewed as an approximate solution to *Poisson's equation* with *forcing function c*,

$$\mathcal{D}h = -c + \eta. \tag{8.2}$$

We noted in the Introduction that Poisson's equation is the basis of much of the ergodic theory of Markov chains, and this equation is also central to average-cost optimal control theory. When h is nonnegative valued, and the set $S := \{c(x) \leq \eta + 1\}$ is finite, then (8.2) implies (V3) with $V = h$, $b = \eta + 1$, and

$$f = (c - \eta)\mathbf{1}_{S^c} + \mathbf{1}_S.$$

Conversely, Proposition 8.1.6 asserts that Poisson's equation (8.2) has a solution h that is bounded by a constant times V, provided (V3) holds with c upper bounded by a constant multiple of f.

Definition 8.0.2 (Regularity). Given a cost function $c \colon \mathbb{R}_+^\ell \to \mathbb{R}_+$, a policy is called *regular* if the controlled process is an f-regular Markov chain with $f = 1 + c$. ∎

Recall that a policy is called *stabilizing* if η_x is finite for each initial condition, where the *average cost* is defined in (4.14) for the scheduling model,

$$\eta_x := \limsup_{n \to \infty} \frac{1}{n} \sum_{t=0}^{n-1} \mathsf{E}\big[c(X(t))\big]. \tag{8.3}$$

We will see that a regular policy is always stabilizing, and moreover the average cost is independent of the initial condition x.

Following the general discussion in Section 8.1 we specialize to network models. We have seen in Theorem 4.8.4 that the MaxWeight policy satisfies a bound similar to (V3) for the CRW scheduling model. Under general conditions the controlled network is also 0-irreducible, so that the MaxWeight policy is regular when c is a norm on $\mathbb{R}_+^\ell$. In this chapter we generalize these results to a wider class of networks. We also provide extensions to the h-MaxWeight policies introduced in Chapter 4.

Most of the concepts developed in these final chapters of the book require very little structure for the CRW model. As in Parts I and II the state process Q evolves on $\mathbb{Z}_+^\ell$ according to the recursion

$$Q(t+1) = Q(t) + B(t+1)U(t) + A(t+1), \qquad t \geq 0,\ Q(0) = x. \tag{8.4}$$

The allocation sequence U evolves on $\mathbb{Z}_+^{\ell_u}$ for some integer ℓ_u; the arrival sequence A is ℓ-dimensional, and B is an $\ell \times \ell_u$ matrix sequence, each with integer-valued entries. We continue to assume that $U(t) \in \mathsf{U}_\diamond$ where as in (4.3),

$$\mathsf{U} := \{u \in \mathbb{R}^{\ell_u} : u \geq 0, \quad Cu \leq 1\}. \tag{8.5}$$

The constituency matrix is an $\ell_m \times \ell$ matrix with binary entries.

The fluid model is expressed as the ODE model

$$\frac{d^+}{dt} q(t) = B\zeta(t) + \alpha, \qquad t \geq 0, \tag{8.6}$$

where $\zeta(t) \in \mathbb{R}_+^{\ell_u}$ denotes the allocation rate vector at time t, also constrained via $\zeta(t) \in \mathsf{U}$ for each t. The velocity set for the fluid model is defined as in Part II by

$$\mathsf{V} := \{v = B\zeta + \alpha : \zeta \in \mathsf{U}\}. \tag{8.7}$$

In this general setting, the "load condition" $\rho_\bullet < 1$ translates to the following:

$$\textit{The origin is an interior point of } \mathsf{V}. \tag{8.8}$$

Further discussion on this interpretation is contained in Section 6.1.

The following result follows from Theorem A.2.3.

Theorem 8.0.3 (**Ergodicity and regularity**). *Consider the CRW model (8.4) subject to the following conditions:*

(a) *The allocation sequence U is defined by the stationary policy*

$$U(t) = \phi(Q(t)), \qquad t \geq 0, \tag{8.9a}$$

where $\phi \colon \mathbb{Z}_+^\ell \to \mathsf{U}_\diamond$ with $\mathsf{U}_\diamond = \{u \in \{0,1\}^\ell : Cu \leq 1\}$.

(b) *Condition (V3) holds for the controlled network with* $f = 1 + c$, *where the cost function c defines a norm on* $\mathbb{R}^\ell$.

(c) *The set $S \subset \mathsf{X}_\diamond$ given in (V3) is "small": There exists $x^* \in S$ satisfying, for any other $x \in S$,*

$$\sum_{t=0}^{\infty} P^t(x, x^*) > 0. \tag{8.9b}$$

Moreover, $P(x^*, x^*) > 0$.

Then:

  (i) *The policy is regular: Q is an aperiodic, f-regular Markov chain with $f = 1 + c$ and $\eta_x = \pi(c) < \infty$ for each $x \in \mathsf{X}_\diamond$, where π is the invariant measure.*

 (ii) *The Strong Law of Large Numbers holds: For each initial condition,*

$$\eta(n) := n^{-1} \sum_{t=0}^{n-1} c(Q(t)) \to \eta, \qquad n \to \infty, \text{ a.s.}$$

(iii) *The mean ergodic theorem holds: For each initial condition,*

$$\mathsf{E}_x[c(Q(t))] \to \eta, \qquad t \to \infty. \tag{8.10}$$

To verify the assumptions of the theorem for a given stationary policy we must establish the Lyapunov condition (b) and the "small set" condition (c). General conditions for the latter are given in Proposition 8.2.1; in particular, it is assumed that the policy is weakly nonidling.

The Lyapunov condition (b) holds for the MaxWeight policy under *very general conditions*. Of course, stabilizability is required. In addition, it is assumed in Theorem 8.0.4 that the matrix sequence B satisfies a condition similar to the Leontief assumption (6.35).

Theorem 8.0.4 is a substantial generalization of Theorem 4.8.4. Further generalizations are obtained in Section 8.4. Note that (8.11) is automatic in the scheduling model.

Theorem 8.0.4 (Stability of MaxWeight). *Consider the CRW model (8.4) satisfying the following conditions:*

(a) *The i.i.d. process (A, B) has integer entries, and a finite second moment. Moreover, $-1 \leq B_{ij}(t) \leq \ell_u$ for each i, j, and t.*

(b) *For each $j \in \{1, \ldots, \ell_u\}$ there exists a unique value $i_j \in \{1, \ldots, \ell\}$ satisfying*

$$B_{ij}(t) \geq 0 \qquad a.s. \ \ i \neq i_j. \tag{8.11}$$

(c) *The stabilizability condition (8.8) holds.*

Then (V3) holds for the MaxWeight policy with $f = 1 + \|x\|$ and $V = \frac{1}{2} b_0 x^\mathsf{T} D x$ for sufficiently large $b_0 > 0$.

Proof. Recall that Proposition 4.8.3 was the basis of Theorem 4.8.4 that established (V3) for the scheduling model under the MaxWeight policy. The proof in this more general setting is almost identical, based on Proposition 6.4.3, which generalizes Proposition 4.8.3 to the network model considered in this chapter.

Since the stabilizability condition (8.8) holds, there exists $\varepsilon > 0$ such that the vector v with coefficients $v_i = -\varepsilon$, $i \geq 1$, lies in V for each x. By definition there exists $u \in \mathsf{U}$ such that $Bu + \alpha = v$, so that by Proposition 6.4.3

$$\langle B\phi^{\mathrm{MW}}(x) + \alpha, \nabla h(x) \rangle = \langle B\phi^{\mathrm{MW}}(x) + \alpha, Dx \rangle \leq v^{\mathsf{T}}Dx = -\varepsilon \sum D_{ii}x_i \leq -\varepsilon_0|x|,$$

with $\varepsilon_0 = \varepsilon(\min_i D_{ii})$.

Note that u may not be feasible if $x_i = 0$ for some i. Thanks to Proposition 6.4.3 this is irrelevant since we are only seeking bounds on the value of (4.89).

We thus arrive at a version of condition (V3) in the form of Poisson's inequality,

$$\mathcal{D}_{\mathrm{MW}}h(x) := \mathsf{E}_{\mathrm{MW}}[h(Q(t+1)) - h(Q(t)) \mid Q(t) = x] \leq -\varepsilon_0|x| + b_D,$$

with b_D the finite constant

$$\frac{1}{2}\max_{x' \in \mathbb{Z}_+^\ell, u \in \mathsf{U}_\diamond(x')} \mathsf{E}[(Q(t+1) - Q(t))^{\mathsf{T}}D(Q(t+1) - Q(t)) \mid Q(t) = x', \, U(t) = u].$$

$\square$

Stability is a basic issue in any network application: We need to know whether a queue ever empties, whether a computer network jams, or if products will be delivered in finite time. Moreover, the stability theory for stochastic networks described in this chapter is a foundation for much of the material in this book.

Beginning in Section 8.6 we show how Lyapunov theory developed for stability can be refined to provide methods for performance evaluation. The following techniques can be used to compute or approximate the average cost, each generalizing a technique used in Chapter 3 in our analysis of the single-server queue:

(i) One can take a Markovian model, solve the equations that define the invariant measure, and then compute the average cost $\eta = \pi(c)$. Unfortunately, the complexity of the invariance equations is formidable. In a few cases the Markov chain is *reversible*, so that a generalization of the detailed balance equations (3.44) holds. However, this is possible only under very special assumptions on the policy [290], so we do not pursue this topic in this book.

(ii) Upper and lower bounds can be obtained based on the Comparison Theorem (see Section 8.1.2 and Theorem A.4.3). In particular, to obtain an upper bound, suppose that we can construct a function $h \colon \mathsf{X}_\diamond \to (0, \infty)$ and a constant $\overline{\eta} < \infty$ that solve the Poisson inequality introduced in (4.13),

$$\mathcal{D}h \leq -c + \overline{\eta}. \tag{8.12}$$

On applying the Comparison Theorem A.4.3 we can conclude that $\eta_x \leq \overline{\eta}$ for each x.

Two general approaches are considered to construct a solution to (8.12). It is possible to automate the search among quadratic functions. Alternatively, sometimes the fluid value function is a solution, just as in the single-server queue. We have seen in several examples in Part I that the fluid value function is piecewise quadratic when c is linear, and frequently also continuously differentiable (see also Theorem 5.3.21).

(iii) Suppose that the bound (8.12) is replaced by the identity

$$\mathcal{D}h = -f + g \tag{8.13}$$

where g is a bounded function on $\mathsf{X}_\diamond$, and $f \geq c$. Subject to a growth condition on the function h, it is shown in Proposition 8.2.5 that the invariant measure π satisfies $\pi(f) = \pi(g)$. This result is the basis of algorithmic approaches to performance approximation developed in Section 8.6.

Refinements and extensions of these methods are developed for workload models in Section 8.7.2.

(iv) Finally, of course, performance estimates can be obtained through simulation. This topic is postponed to Chapter 11. Remarkably, a solution to Poisson's inequality can be used to construct an algorithm for simulation with reduced variance.

We begin with some stability theory for Markov chains based on the survey contained in the Appendix.

8.1 Lyapunov functions

In this section we collect together some consequences of the drift criterion (V3) and the related bound (8.12).

First we take a brief look at the x^*-irreducibility assumption.

8.1.1 Irreducibility

For a set $S \subset \mathsf{X}_\diamond$ the definitions of the first hitting time and the first return time are recalled here,

$$\sigma_S = \min(t \geq 0 : X(t) \in S), \quad \tau_S = \min(t \geq 1 : X(t) \in S). \tag{8.14}$$

The chain is x^*-irreducible if $\mathsf{P}_x\{\tau_{x^*} < \infty\} > 0$ for each $x \in \mathsf{X}_\diamond$.

Theorem 8.1.1 is a generalization of Kac's Theorem A.2.2 in which the set S is a singleton.

Theorem 8.1.1 (Regenerative representations). *Suppose that X is x^*-irreducible and that an invariant measure π exists. Then, for any π-integrable function f and set $S \subset \mathsf{X}_\diamond$ satisfying $\pi(S) > 0$,*

$$\pi(f) = \sum_{x \in S} \pi(x)\mathsf{E}_x \left[\sum_{t=0}^{\tau_S - 1} f(X(t)) \right]. \tag{8.15}$$

In complex models verification of x^*-irreducibility can be difficult, but condition (V3) can be used to simplify this step. Proposition 8.1.2 is the motivation for assumption (c) of Theorem 8.0.3.

Proposition 8.1.2. *Suppose that (V2) holds for some (V, S, b), and that there exists a state $x^* \in \mathsf{X}_\diamond$ satisfying*

$$\sum_{t=0}^{\infty} P^t(x, x^*) > 0, \qquad x \in S. \tag{8.16}$$

Then $\mathbf{X}$ is x^*-irreducible.

Proof. Applying Theorem A.4.1 we have $\mathsf{E}_x[\tau_S] < \infty$ for each initial condition, and hence for each $x \in \mathsf{X}_\diamond$ we can find $x^1 \in S$ and $t_0 \geq 1$ such that $P^{t_0}(x, x^1) > 0$. By assumption there exists $t_1 \geq 1$ such that $P^{t_1}(x^1, x^*) > 0$. Hence we have $P^{t_0+t_1}(x, x^*) > 0$. $\qquad\square$

If (8.16) cannot be verified, then we can still use the techniques of the Appendix for a Markov chain restricted to a subset of $\mathsf{X}_\diamond$. This follows from Doeblin's famous decomposition theorem for Markov chains. For a proof see [368].

Theorem 8.1.3 (Doeblin decomposition). *Any (time-homogeneous) Markov chain on the countable state space $\mathsf{X}_\diamond$ admits a Doeblin decomposition: There exists a collection of disjoint sets $\{\mathsf{X}_i : i \in \mathcal{I}\}$ where $\mathcal{I}$ is finite or countable, such that for each i the chain can be restricted to X_i, and the restricted process is irreducible and recurrent. Moreover, the following holds with probability 1 for each $x \in \mathsf{X}_\diamond$ and finite set $S \subset \mathsf{X}_\diamond$:*

$$\mathsf{P}\left\{ \sum_{t=1}^{\infty} \mathbf{1}\{X(t) \in S\} < \infty \,\Big|\, X(r) \notin \bigcup \mathsf{X}_i \text{ for all } r \right\} = 1.$$

Note that the final conclusion of Theorem 8.1.3 can be equivalently expressed as follows: For any coercive function $c\colon \mathsf{X}_\diamond \to \mathbb{R}_+$,

$$\mathsf{P}\left\{ \lim_{t \to \infty} c(X(t)) = \infty \,\Big|\, X(r) \notin \bigcup \mathsf{X}_i \text{ for all } r \right\} = 1.$$

8.1.2 Comparison Theorem

A simple, but central result in this chapter is the Comparison Theorem A.4.3. If V, f, g are nonnegative-valued functions satisfying $\mathcal{D}V \leq -f + g$, then we have for each $x \in \mathsf{X}_\diamond$ and any stopping time τ,

$$\mathsf{E}_x\left[V(X(\tau)) + \sum_{t=0}^{\tau-1} f(X(t)) \right] \leq V(x) + \mathsf{E}_x\left[\sum_{t=0}^{\tau-1} g(X(t)) \right]. \tag{8.17}$$

This bound and refinements were used in our treatment of invariance equations for the single-server queue in Section 3.4.

Given a function $V: \mathsf{X} \to [1, \infty)$ we let L_∞^V denote the set of functions $g: \mathsf{X} \to \mathbb{R}$ for which $|g|/V$ is bounded on X. This is viewed as a normed linear space with norm defined as the supremum

$$\|g\|_V := \sup_{x \in \mathsf{X}} \frac{|g(x)|}{V(x)}. \tag{8.18}$$

In applications the function V is typically taken to be a Lyapunov function, or the function f appearing in (V3).

We can choose τ in (8.17) equal to the first return time to the set S appearing in (V3) to obtain

Proposition 8.1.4. *Under (V3) we have for each initial condition,*

$$V_f(x) := \mathsf{E}_x \left[\sum_{t=0}^{\tau_S - 1} f(X(t)) \right] \leq V(x) + b\mathbf{1}_S(x), \tag{8.19}$$

where $\tau_S := \min\{t \geq 1 : X(t) \in S\}$. *Consequently,* $V_f \in L_\infty^V$.

As another corollary to the Comparison Theorem we obtain bounds on the average and the discounted cost based on a solution to Poisson's inequality (8.12).

Proposition 8.1.5. *Suppose that there exists a function* $h: \mathsf{X}_\circ \to \mathbb{R}_+^\ell$ *satisfying (8.12). Then the following transient bounds hold for each initial condition* $x \in \mathsf{X}_\circ$:

$$r^{-1} \sum_{t=0}^{r-1} \mathsf{E}_x\big[c(X(t))\big] \ \leq \ r^{-1}h(x) + \overline{\eta}, \qquad r \geq 1, \tag{8.20a}$$

$$\gamma \sum_{t=0}^{\infty} (1+\gamma)^{-t-1} \mathsf{E}_x\big[c(X(t))\big] \ \leq \ \gamma h(x) + \overline{\eta}, \qquad \gamma > 0. \tag{8.20b}$$

Proof. The proof follows from the Comparison Theorem, which gives for any $r \geq 1$, $x \in \mathsf{X}_\circ$,

$$\mathsf{E}_x \left[h(X(r)) + \sum_{i=0}^{r-1} c(X(t)) \right] \leq h(x) + \overline{\eta}r. \tag{8.21}$$

This implies the first bound since $h \geq 0$.

The second bound follows from the first by multiplying each side of (8.20a) by $r(1+\gamma)^{-r}$ and summing from $r = 0$ to ∞. To obtain (8.20b) we change the order of summation as follows:

$$\sum_{r=0}^{\infty} \gamma(1+\gamma)^{-r-1} \sum_{t=0}^{r-1} c(X(t)) = \sum_{t=0}^{\infty} c(X(t)) \sum_{r=t+1}^{\infty} \gamma(1+\gamma)^{-r-1} = \sum_{t=0}^{\infty} (1+\gamma)^{-t-1} c(X(t)).$$

Taking expectations of both sides and applying (8.20a) to the left-hand side completes the proof. $\qquad\square$

8.1.3 Poisson's equation

Recall that h solves Poisson's equation with forcing function c if $Ph = h - c + \eta$, with $\eta = \pi(c)$. If X is x^*-irreducible with $\eta = \pi(c) < \infty$, then the function h given below is finite valued for a.e. $x \in X_\diamond$ $[\pi]$ by Proposition A.3.1,

$$h(x) = \mathsf{E}_x \left[\sum_{t=0}^{\tau_{x^*}-1} [c(X(t)) - \eta] \right], \qquad x \in X_\diamond. \tag{8.22}$$

Proposition 8.1.6. *Suppose that X is an x^*-irreducible Markov chain and that the function c is coercive. Suppose moreover that an invariant measure π exists with $\pi(c) < \infty$. Then the function h defined in (8.22) satisfies the following:*

(i) *The support $X_\pi = \{x : \pi(x) > 0\}$ is full and absorbing. That is,*

$$\pi(X_\pi) = 1, \quad and \quad P(x, X_\pi) = 1 \quad for\ all\ x \in X_\pi.$$

(ii) *The set X_h on which the function h is finite valued is also full and absorbing, with $X_\pi \subseteq X_h$.*

(iii) *The function h is uniformly bounded from below, $\inf_x h(x) > -\infty$. If h' is another solution to Poisson's equation that is bounded from below, then for some constant b,*

$$h'(x) = h(x) + b,\ x \in X_h, \quad h'(x) \geq h(x) + b,\ for\ all\ x \in X_\diamond.$$

(iv) *If the policy is regular, then $h \in L_\infty^{1+V}$, where V is any solution to (V3) satisfying $c \in L_\infty^f$.*

Proof. Part (i) follows from Proposition A.3.1, parts (ii) and (iii) are given in Proposition A.3.12, and (iv) follows from Theorem A.4.6 (iii) or Theorem A.2.3. □

8.1.4 Ergodicity

The bound (8.20a) in Proposition 8.1.5 implies the ergodic bound

$$\lim_{N \to \infty} N^{-1} \sum_{t=0}^{N-1} \mathsf{E}_x \left[c(X(t)) \right] \leq b,$$

with b the constant appearing in (V3). This can be refined for an aperiodic, x^*-irreducible Markov chain through coupling techniques described in the Appendix, and described for the single-server queue in Chapter 3. Theorem A.2.3 contains a survey of consequences of (V3).

8.2 Lyapunov functions for networks

What is special about a network? First, for most policies the transition probability $P(x, \cdot)$ is piecewise constant as a function of x. A second basic property is that the state cannot jump too far in one time-step: Under any policy the state process Q is *skip free in mean square*,

$$\sup_{x \in \mathsf{X}_\circ, u \in \mathsf{U}_\circ(x)} \mathsf{E}[\|Q(t+1) - Q(t)\|^2 \mid Q(t) = x, \ U(t) = u] < \infty. \tag{8.23}$$

This property enables a geometric approach to constructing Lyapunov functions that is developed in Section 8.2.4.

Network structure can also simplify the verification of irreducibility.

8.2.1 Irreducibility

Regularity requires two conditions: (V3) and x^*-irreducibility. Although the latter is a mild property, it is not easy to obtain completely general conditions for the CRW model (8.4).

In Proposition 8.2.1 we illustrate the verification of x^*-irreducibility for the scheduling model. Recall from (4.4) that the state process can be described by the recursion

$$Q(t+1) = Q(t) + \sum_{i=1}^{\ell} M_i(t+1)[-1^i + 1^{i+}]U_i(t) + A(t+1), \quad Q(0) = x, \tag{8.24}$$

where M is the i.i.d. diagonal matrix sequence whose diagonal elements satisfy $\mathsf{P}\{M_i(t) = 1\} = 1 - \mathsf{P}\{M_i(t) = 0\} = \mu_i$ for each $t \geq 0$ and $1 \leq i \leq \ell$. The key assumption on the policy is that it is *weakly nonidling*: For any t,

$$\sum_{i=1}^{\ell} U(t) \geq 1 \ \text{ whenever } Q(t) \neq 0. \tag{8.25}$$

In Proposition 8.2.1 we do not even require the load condition $\rho_\bullet < 1$. The condition (8.26) was introduced in Section 4.1 as a consequence of uniformization. It is a bit stronger than necessary for our purposes here, but (8.26) will find other uses later in this chapter.

Proposition 8.2.1. *Consider the CRW scheduling model (8.24) subject to the following conditions:*

(a) *The sequence (M, A) is i.i.d., M_i is Bernoulli, and A takes values in $\mathbb{Z}_+^\ell$.*
(b) *The joint process (A, M) satisfies*

$$A(t)M_i(t) = 0, \quad and \quad M_i(t)M_j(t) = 0 \quad for\ each\ i, j, t. \tag{8.26}$$

(c) *The network is controlled using a stationary, weakly nonidling policy (see (8.25)).*
(d) *$\mu_i > 0$ for each i.*

Then the controlled process Q is an x^*-irreducible Markov chain with $x^* = 0$, and it is also aperiodic. That is, for each $x \in \mathsf{X}_\circ$ there exists $T(x) < \infty$ such that

$$P^t(x, 0) = \mathsf{P}_x\{Q(t) = 0\} > 0, \qquad t \geq T(x).$$

Proof. Under these assumptions we have for some $\delta > 0$, and any nonzero $x \in X_\diamond$,

$$P\{A \text{ service is completed at time } t \text{ and } A(t) = 0 \mid Q(t-1) = x\} \geq \delta, \qquad t \geq 1.$$

Since each customer in the network requires service at most ℓ times, it follows that $P^T(x, 0) \geq \delta^T$ for $T = \ell|x|$. This establishes 0-irreducibility.

Aperiodicity also follows from (8.26) since $P(0, 0) = P\{A(t) = 0\} > 0$. □

8.2.2 Value functions

Define the polynomial weighting functions

$$V_0 \equiv 1, \quad V_p = 1 + \|x\|^p, \qquad x \in \mathbb{R}^\ell, \ p \geq 1. \tag{8.27}$$

The following result is used to obtain bounds on solutions to Poisson's equation, or solutions to average-cost dynamic programming equations.

Proposition 8.2.2. *Suppose that assumptions (a)–(c) of Theorem 8.0.4 hold, and that an invariant measure π exists with $\pi(V_{p-1}) < \infty$ for some $p \geq 1$. Then, with*

$$V_{p*}(x) := \mathsf{E}_x\left[\sum_{t=0}^{\sigma_0} V_{p-1}(Q(t))\right], \qquad x \in X_\diamond,$$

the following conclusions hold:

(i) $V_p \in L_\infty^{V_{p*}}$ *and the set* $X_p := \{x \in X_\diamond : V_{p*}(x) < \infty\}$ *satisfies* $\pi(X_p) = 1$.
(ii) *For* $x \in X_p$,

$$\lim_{t \to \infty} \frac{1}{t} \mathsf{E}_x[V_{p*}(Q(t))] = \lim_{t \to \infty} \mathsf{E}_x[V_{p*}(Q(t))\mathbf{1}\{\tau_0 > t\}] = 0.$$

(iii) *The following identity holds on* $X_\diamond$,

$$PV_{p*} = V_{p*} - V_{p-1} + b_p \mathbf{1}_0, \tag{8.28}$$

where $b_p := (\pi(0))^{-1}\pi(V_{p-1})$.

Proof. Parts (ii) and (iii) and the conclusion that $\pi(X_p) = 1$ are given in Proposition A.6.2.

To complete the proof we demonstrate that $V_p \in L_\infty^{V_{p*}}$, so that (i) holds. Note that at most ℓ_u services can be completed in each time slot, where ℓ_u denotes the dimension of U. Consequently, we have the strict lower bound $|Q(t)| \geq |Q(0)| - t\ell_u$ for each $t \geq 0$ and $Q(0) = x \in X_\diamond$. This implies the following lower bound for $x \neq 0$:

$$V_{p*}(x) \geq \sum_{t \leq |x|/\ell_u} (|x| - t\ell_u)^{p-1}$$

$$\geq \int_0^{|x|/\ell_u} (|x| - t\ell_u)^{p-1}\, dt = \frac{1}{p\ell_u}|x|^p.$$

It follows that $V_p \in L_\infty^{V_{p*}}$, as claimed. □

As a direct corollary we obtain the following refinement of Proposition 8.1.6.

Proposition 8.2.3. *Suppose that the network is controlled using a stationary policy, and that the resulting queue-length process Q is x^*-irreducible. Suppose moreover that an invariant measure π exists with $\pi(c) < \infty$, where c is a norm on $\mathbb{R}^\ell$. Then the function h defined in (8.22) satisfies the following lower bound: For some constant $\varepsilon_0 > 0$,*

$$h(x) \geq -\varepsilon_0^{-1} + \varepsilon_0 \|x\|^2, \qquad x \in \mathsf{X}_\diamond.$$

Example 8.2.4 (Poisson's equation for the M/M/1 queue revisited). In Theorem 3.0.1 we constructed a solution to Poisson's equation with forcing function $c(x) \equiv x$ for the M/M/1 queue. Provided $\rho_\bullet < 1$, a solution is the quadratic $h(x) = \frac{1}{2}(x^2 + x)/(\mu - \alpha)$, $x \in \mathbb{Z}_+$.

Consider now the forcing function $g = \mathbf{1}_0$ with mean $\eta_g := \pi(g) = \pi(0) = 1 - \rho_\bullet$. A solution to Poisson's equation is $h(x) = -\mu^{-1}x$: Writing $u = \mathbf{1}\{x \geq 1\} = 1 - g(x)$ we have

$$\mathcal{D}h(x) = -\mu^{-1}\mathsf{E}[Q(t+1) - Q(t) \mid Q(t) = x] = -\mu^{-1}(-u\mu + \alpha) = -g(x) + \eta_g.$$

The solution to Poisson's equation is not unique since we can always add a constant to h to obtain another solution. We now ask, *are there others?*

Applying the transition matrix to the function $h_0(x) = \rho_\bullet^{-x}$ gives

$$Ph_0(x) = \alpha\rho_\bullet^{-x-1} + \mu\rho_\bullet^{-x+u}, \qquad x \in \mathbb{Z}_+.$$

We have $\alpha\rho_\bullet^{-x-1} = \mu\rho_\bullet^{-x}$ and $\mu\rho_\bullet^{-x+1} = \alpha\rho_\bullet^{-x}$, so that

$$\mathcal{D}h_0(x) = -2\mu\mathbf{1}_0(x).$$

That is, with $h = (2\mu)^{-1}h_0$ we have $\mathcal{D}h = -g$. *Can we conclude that h solves Poisson's equation with forcing function g?* Obviously not, since $\pi(g) \neq 0$. ∎

This example shows that even if the function h solves the linear equation $\mathcal{D}h = -g + b$ with b constant, a growth condition on h is necessary to conclude that $b = \pi(g)$ so that h solves Poisson's equation. In particular, for a given function h the function

$$g := -\mathcal{D}h \tag{8.29}$$

does not necessarily have zero mean, even if g is π-integrable. Subject to growth conditions on g and h we can resolve this ambiguity, and conclude that $\pi(g) = 0$. Proposition 8.2.5 can be viewed as a refinement of the Comparison Theorem (note that the positivity assumption on V is relaxed). This result is an extension of Proposition 3.4.6 that was the key step used in the proof of Theorem 3.0.1.

Proposition 8.2.5. *Suppose that assumptions (a)–(c) of Theorem 8.0.4 hold, and that an invariant measure π exists with $\pi(V_{p-1}) < \infty$. Suppose that $h \in L_\infty^{V_p}$ is given, and the function given in (8.29) satisfies $g \in L_\infty^{V_{p-1}}$. Then $\pi(g) = 0$, h solves Poisson's* *equation with forcing function g, and moreover for a.e. $x \in X_\diamond [\pi]$,*

$$h(x) - h(0) = \mathsf{E}_x \left[\sum_{t=0}^{\tau_0 - 1} g(Q(t)) \right].$$

Proof. This follows from Proposition A.6.2 combined with Proposition 8.2.2 (i). □

8.2.3 Geometric ergodicity

In network models the state process typically satisfies a form of ergodicity far stronger than the mean ergodic theorem given in Theorem 8.0.3.

A formal definition of geometric ergodicity is given in Section A.5.3. The definition is easy to guess: The convergence in the mean ergodic theorem holds at a geometric rate. However, as in the ordinary ergodic theorem expressed in limit (8.10) in Theorem 8.0.3, it is necessary to impose bounds on the class of allowable functions $c \colon X \to \mathbb{R}$.

It is most convenient to express this in terms of the operator theoretic framework of Section A.5.3, based on the norm (8.18). A Markov chain X is called V-uniformly ergodic if there exist $d_0 > 0$ and $b_0 < \infty$ such that for any $g \in L_\infty^V$,

$$|\mathsf{E}[g(X(t)) \mid X(0) = x] - \pi(g)| \leq b_0 \|g\|_V e^{-d_0 t} V(x), \qquad t \in \mathbb{Z}_+, \ x \in X_\diamond. \quad (8.30)$$

Writing $\tilde{g} = g - \pi(g)$, this bound can be expressed $\|P^t \tilde{g}\|_V \leq b_0 \|g\|_V e^{-d_0 t}$.

It is shown in Section A.5.3 that this form of ergodicity is a consequence of the following Foster–Lyapunov drift condition: The chain satisfies condition (V4) if for a function $V \colon X_\diamond \to [1, \infty), \varepsilon > 0, b < \infty$, and a finite set $S \subset X_\diamond$,

$$\mathcal{D}V(x) \leq -\varepsilon V(x) + b\mathbf{1}_S(x), \qquad x \in X_\diamond. \quad \text{(V4)}$$

Perhaps surprisingly, (V4) follows from (V2) under general conditions. The following result is a version of Proposition A.5.7.

Theorem 8.2.6 (Strengthening ergodicity to geometric ergodicity). *Consider the CRW model (8.4) subject to the following conditions:*

(a) *The allocation sequence U is defined by a stationary policy.*
(b) *Condition (V2) holds for the controlled network with $V \colon X \to \mathbb{R}_+^\ell$ Lipschitz continuous.*
(c) *With $S \subset X_\diamond$ in (V2), there exists $x^* \in S$ satisfying (8.16) and $P(x^*, x^*) > 0$.*

Suppose moreover that the arrival process possesses an exponential moment: For some $\varepsilon_0 > 0$,

$$\mathsf{E}[e^{\varepsilon_0 \|A(t)\|}] < \infty, \qquad t \geq 0. \quad (8.31)$$

Then there exists $\vartheta > 0$ such that the controlled process is V_ϑ-uniformly ergodic with $V_\vartheta = \exp(\vartheta V)$.

Theorem 8.2.7 is a useful corollary. The Lipschitz condition on V_0 holds when V is quadratic.

Theorem 8.2.7. *Suppose that the assumptions of Theorem 8.0.3 hold, and that $V_0 := \sqrt{1+V}$ is Lipschitz continuous, where V is the solution to (V3). Suppose moreover that A satisfies the exponential bound (8.31) for some $\varepsilon_0 > 0$. Then there exists $\vartheta > 0$ such that the controlled process is V_ϑ-uniformly ergodic with $V_\vartheta = \exp(\vartheta\sqrt{1+h})$.*

Proof. Since V_0 is Lipschitz we have $V_0 \in L_\infty^{V_1}$. Since c is a norm it follows that $V_0(x) \le rc(x)$ for some $r < \infty$ and all x outside of a finite set. For a possibly larger finite set S we can conclude that

$$PV(x) - V(x) = \mathcal{D}V(x) \le -r^{-1}V_0(x) + b\mathbf{1}_S(x), \qquad x \in \mathsf{X}_\diamond.$$

Jensen's inequality then provides an upper bound on PV_0,

$$PV_0 \le \sqrt{PV} \le \sqrt{1+V-r^{-1}V_0+b\mathbf{1}_S}.$$

From concavity of the square root the right-hand side is bounded as follows,

$$\begin{aligned}
\sqrt{1+V-r^{-1}V_0+b\mathbf{1}_S} \\
\le \sqrt{1+V} + \tfrac{1}{2}\frac{1}{\sqrt{1+V}}\left(-r^{-1}V_0+b\mathbf{1}_S\right) \\
\le V_0 - r^{-1} + b\mathbf{1}_S,
\end{aligned}$$

so that

$$\mathcal{D}V_0 \le -r^{-1} + b\mathbf{1}_S. \tag{8.32}$$

Hence the function $V_{00}:=rV_0$ satisfies Foster's criterion (V2). This function is Lipschitz continuous by assumption, so the conclusions follow from Theorem 8.2.6. $\qquad\square$

8.2.4 Drift vector field

Propositions A.4.2 and A.4.7 describe geometric conditions ensuring that the respective drift conditions (V2) or (V3) hold for a continuously differentiable function $V\colon \mathbb{R}^\ell \to \mathbb{R}_+$. These results are based on bounds assumed on the inner product $\langle \Delta(y), \nabla V(x)\rangle$ where ∇V denotes the gradient of V, and the *drift vector field* $\Delta\colon \mathsf{X}_\diamond \to \mathbb{R}^\ell$ is defined by

$$\Delta(x) := \mathsf{E}[Q(t+1) - Q(t) \mid Q(t) = x] = B\phi(x) + \alpha, \qquad x \in \mathsf{X}_\diamond. \tag{8.33}$$

The following is an application of Propositions A.4.2 and A.4.7 to the CRW model (8.4).

Proposition 8.2.8. *Suppose that $V\colon \mathbb{R}^\ell \to \mathbb{R}_+$ is a C^1 function. Then:*

(i) *If there exists $\varepsilon_0 > 0$, $b_0 < \infty$, such that*

$$\langle \Delta(y), \nabla V(x)\rangle \le -(1+\varepsilon_0) + b_0(1+\|x\|)^{-1}\|x-y\|, \qquad x, y \in \mathbb{R}_+^\ell, \tag{8.34}$$

then the function V solves Foster's criterion (V2).

(ii) *If there exists $b_0 < \infty$ such that for each $x, y \in \mathbb{R}_+^\ell$,*

$$\langle \Delta(y), \nabla V(x) \rangle \le -\|x\| + b_0 \|x - y\|^2, \tag{8.35}$$

then the function V solves (V3) with $f(x) = 1 + \frac{1}{2}\|x\|$.

In this section we illustrate this approach using the pair of tandem queues shown in Fig. 4.3. We take the CRW model (4.58) with S_i Bernoulli for each i, and we assume that A has an exponential moment in the sense of (8.31).

It is not surprising that the system is in some sense stable under the nonidling policy when $\rho_\bullet < 1$, and there are many simple ways to verify stability of this model. Conversely, even though this is a simple network model, computation of π is nontrivial except in very special cases.

The purpose of this section is to illustrate a valuable geometric point of view. The ODE methods developed in Chapter 10 are based on the theory developed in this chapter. The resulting techniques provide much more efficient methods for verifying stability.

8.2.4.1 Foster's criterion

To show that Q is positive Harris recurrent when controlled using the nonidling policy we apply Proposition 8.2.8 to construct a solution to Foster's criterion. The Lyapunov function will be Lipschitz continuous, so that we conclude that Q is V-uniformly ergodic by Theorem 8.2.6.

Under the nonidling policy the drift vector (8.33) can take on one of four values,

$$\Delta(x) = \begin{cases} (\alpha_1 - \mu_1, \mu_1 - \mu_2)^\mathsf{T} & \text{if } x_1 \ge 1 \text{ and } x_2 \ge 1 \\ (\alpha_1 - \mu_1, \mu_1)^\mathsf{T} & \text{if } x_1 \ge 1 \text{ and } x_2 = 0 \\ (\alpha_1, -\mu_2)^\mathsf{T} & \text{if } x_1 = 0 \text{ and } x_2 \ge 1 \\ (\alpha_1, 0)^\mathsf{T} & \text{if } x_1 = x_2 = 0. \end{cases} \tag{8.36}$$

To construct a solution to Foster's criterion (V2) consider the minimal draining time for the fluid model. This is expressed in terms of the two workload vectors

$$\xi^1 = (\mu_1^{-1}, 0)^\mathsf{T}, \quad \xi^2 = (\mu_2^{-1}, \mu_2^{-1})^\mathsf{T}.$$

On defining $d^s = (1 - \rho_s)^{-1}\xi^s \in \mathbb{R}^2$, $s = 1, 2$, the minimal draining time is expressed

$$T^*(x) = \max_{s=1,2} = \max \langle d^s, x \rangle, \qquad x \in \mathbb{R}_+^2.$$

Unfortunately, because T^* is piecewise linear it is difficult to compute the conditional expectation $\mathsf{E}[T^*(Q(t+1)) \mid Q(t) = x]$ for x near the boundary $\langle \xi^1, x \rangle = \langle \xi^2, x \rangle$, $x \in \mathsf{X}_\diamond$. To construct a solution to Foster's criterion we smooth the function T^* by replacing the maximum with an exponential approximation as follows:

$$V_0(x) = \log\Big(\exp(\langle d^1, x \rangle) + \exp(\langle d^2, x \rangle)\Big), \qquad x \in \mathbb{R}_+^2. \tag{8.37}$$

This is a continuously differentiable, convex function of x.

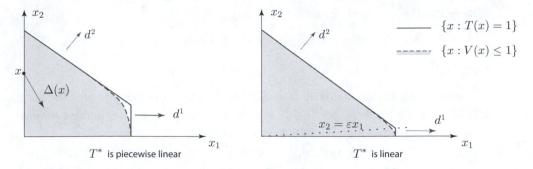

Figure 8.1. Smoothed piecewise linear Lyapunov function. The level sets of T^* are polyhedral, and the level sets of V_0 are a smooth approximation.

We have $V_0(x) \geq T^*(x)$ for each $x \in \mathbb{R}_+^2$. Moreover, V_0 is an approximation to T^* for large $x \in \mathbb{R}_+^2$ in the following sense: For any nonzero $x \in \mathbb{R}_+^2$ satisfying $\langle d^1, x \rangle \neq \langle d^2, x \rangle$,

$$V_0(\kappa x) - T^*(\kappa x) \downarrow 0, \qquad \kappa \uparrow \infty.$$

To prove this assume for definiteness that $\langle d^1, x \rangle > \langle d^2, x \rangle$. We can then write

$$T^*(x) = \langle d^1, x \rangle = \log\big(\exp(\langle d^1, x \rangle)\big),$$

and by subtraction,

$$
\begin{aligned}
V_0(\kappa x) - T^*(\kappa x) &= \log\Big(1 + \exp\big(-\kappa \langle d^1, x \rangle\big) \exp\big(\kappa \langle d^2, x \rangle\big)\Big) \\
&= \log\Big(1 + \exp\big(-\kappa \langle d^1 - d^2, x \rangle\big)\Big).
\end{aligned}
$$

This decreases to zero exponentially fast as $\kappa \uparrow \infty$ since $\langle d^1 - d^2, x \rangle > 0$.

We consider two cases. In the first case the draining time is piecewise linear, and we apply Proposition 8.2.8 (i) to show that $V := (1 + \varepsilon_0)V_0$ is a solution to (V2) for any $\varepsilon_0 > 0$. In the second we have $T^*(x) = \langle d^2, x \rangle$ for all $x \in \mathbb{R}_+^2$ so that the draining time is linear. In this case a small modification of the definition (8.37) is required to ensure that (V2) is satisfied.

CASE 1: $\mu_1 < \mu_2$ Shown on the left in Fig. 8.1 are level sets of the functions T^* and V_0 in Case 1. Note that in Cases 1 or 2 we have the following individual drift conditions:

$$\langle \Delta(x), d^1 \rangle = -1, \ \text{ if } x_1 \geq 1; \qquad \langle \Delta(x), d^2 \rangle = -1, \ \text{ if } x_2 \geq 1. \qquad (8.38)$$

This geometry will ultimately lead to a proof that (V2) holds.

The function V_0 is differentiable, and on defining

$$\vartheta(x) = \Big(\exp(\langle d^1, x \rangle) + \exp(\langle d^2, x \rangle)\Big)^{-1} \exp(\langle d^1, x \rangle),$$

the gradient can be expressed

$$\nabla V_0(x) = \vartheta(x)d^1 + (1 - \vartheta(x))d^2, \qquad x \in \mathbb{R}_+^2.$$

It is easily verified that $\vartheta(x) \in (0,1)$ for each $x \in \mathbb{R}_+^2$, and as $\kappa \to \infty$,

$$\vartheta(\kappa x) \to 1 \quad \text{if } \langle d^1, x \rangle > \langle d^2, x \rangle$$
$$\vartheta(\kappa x) \to 0 \quad \text{if } \langle d^1, x \rangle < \langle d^2, x \rangle.$$

It then follows from (8.36) that the conditions of Proposition 8.2.8 (i) hold with $V = (1 + \varepsilon_0)V_0$ for any given $\varepsilon_0 > 0$, and we conclude that (V2) holds.

Note that the condition $\mu_1 < \mu_2$ is critical here so that $\vartheta(x) \to 1$ as $x \to \infty$ along the lower boundary of $\mathbb{R}_+^2$. That is,

$$\lim_{r \to \infty} \vartheta(r\mathbf{1}^1) = 1.$$

This limit follows from the dominance of d^1 in the level set of T^* when x is near the lower boundary, as shown on the left in Fig. 8.1.

CASE 2: $\mu_1 \geq \mu_2$ In this case the minimal draining time is linear in x, $T^*(x) = (\mu_2 - \alpha_1)^{-1}(x_1 + x_2)$ for $x \in \mathbb{R}_+^2$. One might hope to use T^* as a Lyapunov function, but the drift condition fails when $x_2 = 0$ since we have

$$\mathsf{E}[T^*(Q(t+1)) \mid Q(t) = x] = -\left(\frac{\phi_2(x) - \rho_2}{1 - \rho_2}\right), \qquad x \in \mathbb{Z}_+^2,$$

where $\phi_2(x) = \mathbf{1}\{x_2 \geq 1\}$ since the policy is nonidling.

Consider then a small perturbation of T^*: Maintain the definition $d^2 = (1 - \rho_2)^{-1}\xi^2$, but redefine the second vector as

$$d^1 = (1 + \varepsilon)(1 - \rho_2)^{-1}(\mu_2^{-1}, 0)^{\mathsf{T}}. \tag{8.39}$$

With this new definition we do have $\max\langle d^s, x \rangle = \langle d^1, x \rangle$ when $x_2 = 0$, so that the function $T(x) := \max\langle d^s, x \rangle$ is piecewise linear (and not linear). Note that T is the minimal draining time for the tandem queue model in which the service rate μ_1 is reduced to $\mu_1' = (1 + \varepsilon)^{-1}(\mu_2 + \varepsilon\alpha_1)$. With this new model we return to Case 1 since

$$\mu_1' := \frac{\mu_2 + \varepsilon\alpha_1}{1 + \varepsilon} < \frac{\mu_2 + \varepsilon\mu_2}{1 + \varepsilon} = \mu_2.$$

The following two drift conditions hold in Case 2, analogous to (8.38):

$$\langle \Delta(x), d^1 \rangle = -(1+\varepsilon)\frac{\mu_1 - \alpha_1}{\mu_2 - \alpha_1} < -1, \quad \text{if } x_1 \geq 1; \qquad \langle \Delta(x), d^2 \rangle = -1, \quad \text{if } x_2 \geq 1.$$

The level sets of the resulting function V_0 are as shown on the right in Fig. 8.1. We can use identical reasoning to verify that (V2) holds with Lyapunov function $V = (1 + \varepsilon_0)V_0$ for any given $\varepsilon_0 > 0$.

8.2.4.2 Condition (V3)

To construct a solution to (V3) consider the fluid value function

$$J(x) = \int_0^\infty c(q(t))\, dt, \qquad q(0) = x,$$

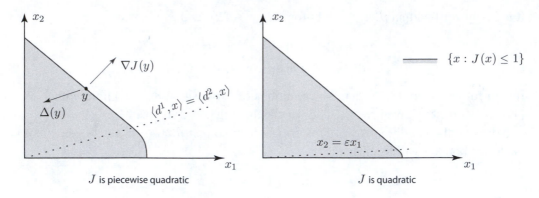

Figure 8.2. Fluid value functions for the tandem queue. In Case 1 the value function is piecewise quadratic, and the level set is as shown on the left. In Case 2 the value function is purely quadratic. The plot on the right shows a level set of the modified value function.

where $c(x) = x_1 + x_2$. We find that J is always continuously differentiable, so the smoothing used in Section 8.2.4.1 to apply Proposition 8.2.8 is not required here.

We again consider the two cases in which the minimal draining time T^* is linear or not. As in Section 8.2.4.1, we find that the value function must be modified slightly in Case 2 to ensure that the desired drift condition holds along the lower boundary of $\mathbb{R}_+^2$.

CASE 1: $\mu_1 < \mu_2$ On the left in Fig. 8.2 is an illustration of a level set of the function J in Case 1, where the value function is piecewise quadratic:

$$J(x) = \begin{cases} \frac{1}{2}x^{\mathsf{T}}D^1x & \text{if } \langle d^1, x \rangle \geq \langle d^2, x \rangle \\ \frac{1}{2}x^{\mathsf{T}}D^2x & \text{otherwise,} \end{cases}$$

$$\text{(8.40)}$$

$$\text{with} \qquad D^1 = \begin{bmatrix} (\mu_1 - \alpha_1)^{-1} & 0 \\ 0 & (\mu_2 - \mu_1)^{-1} \end{bmatrix}, \quad D^2 = \frac{1}{\mu_2 - \alpha_1}\begin{bmatrix} 1 & 1 \\ 1 & 1 \end{bmatrix}.$$

The function J is convex and continuously differentiable. This is illustrated in Fig. 8.2 where it is evident that the level set $\{x \in \mathbb{R}_+^2 : J(x) = 1\}$ is smooth, and the sublevel set $\{x \in \mathbb{R}_+^2 : J(x) \leq 1\}$ is convex. The gradient of J is given by $\nabla J(x) = D^i x$, where $i = 1$ when $\langle d^1, x \rangle \geq \langle d^2, x \rangle$. Moreover, we have the following individual drift conditions, exactly as in (8.38):

$$\langle \Delta(x), D^1 x \rangle = -c(x), \quad \text{if } x_1 \geq 1; \qquad \langle \Delta(x), D^2 x \rangle = -c(x), \quad \text{if } x_2 \geq 1.$$

$$\text{(8.41)}$$

It follows that $\langle \Delta(x), \nabla J(x) \rangle = -c(x)$ for all x, and by Proposition 8.2.8 (ii) we conclude that (V3) holds with $V = J$.

CASE 2: $\mu_1 \geq \mu_2$ In this case J is the pure quadratic,

$$J(x) = \tfrac{1}{2}x^{\mathsf{T}}D^2x, \quad x \in \mathsf{X}, \qquad D^2 = \frac{1}{\mu_2 - \alpha_1}\begin{bmatrix} 1 & 1 \\ 1 & 1 \end{bmatrix}.$$

Unfortunately the drift condition (V3) is violated for $\{x_2 = 0\}$, so we perturb the value function as before: Define the new piecewise quadratic function $J \colon \mathbb{R}_+^2 \to \mathbb{R}_+$ exactly as in (8.40), but with D^1 redefined as follows:

$$D^1 = \begin{bmatrix} (\mu_1' - \alpha_1)^{-1} & 0 \\ 0 & (\mu_2 - \mu_1)^{-1} \end{bmatrix},$$

where again $\mu_1' = (1 + \varepsilon)^{-1}(\mu_2 + \varepsilon\alpha_1)$.

The function J coincides with the value function J' for the model with this reduced value of μ_1. Hence this function is continuously differentiable, and it also satisfies the inequality $\langle \Delta(x), \nabla J(x) \rangle \le -c(x)$ for all $x \in \mathbb{R}_+^2$. By Proposition 8.2.8 (ii) we again conclude that (V3) holds with $V = J$.

The next two sections describe classes of policies for which a solution to (V3) is automatic.

8.3 Discrete review

Here we show how to establish stability of the discrete-review (DR) policies introduced in Section 4.7.

In general, the resulting queue-length process is not a Markov chain since the DR policy is not stationary. By expanding the state space we do obtain a Markovian model under mild conditions on the review times.

8.3.1 Markovian realizations

To construct a Markovian state process $\boldsymbol{X}$ we appeal to the particular form of the sequence of review times (4.75) introduced in Section 4.7 for the scheduling model. We maintain the same assumption for the CRW model (8.4): The review times are defined inductively by

$$T_0 = 0, \quad T_{m+1} = T_m + \mathcal{T}(Q(T_m)), \qquad m \ge 0, \tag{8.42}$$

where $\mathcal{T} \colon \mathsf{X}_\diamond \to \{1, 2, 3, \dots\}$.

The following results are related to the state-dependent drift criteria considered in Section A.4.3.

Proposition 8.3.1. *Consider the CRW model (8.4) controlled by a randomized policy of the following form: There exists an i.i.d. process $\{\Gamma(t) : t \ge 0\}$ evolving $\mathbb{R}^\ell$, and a family of functions $\{\phi^t \colon \mathsf{X}_\diamond^{t+1} \times \mathbb{R}^\ell \to \mathsf{U}_\diamond : t \ge 0\}$ satisfying*

$$\phi^t(x_0, \dots, x_t, \Gamma(t)) \in \mathsf{U}_\diamond(x_t), \qquad 0 \le t \le \mathcal{T}(x_0) - 1.$$

The policy ϕ is defined for $0 \le t \le T_1 - 1$ via $U(t) := \phi^t(Q(0), \dots, Q(t), \Gamma(t))$, and for arbitrary $m \ge 1$ and $t \in \{T_m, \dots, T_{m+1} - 1\}$ by

$$U(t) := \phi^{t-T_m}(Q(T_m), \dots, Q(t), \Gamma(t)). \tag{8.43}$$

Then:

(i) *The sampled process $\{Q(T_m) : m \geq 0\}$ is a Markov chain on $\mathsf{X}_\diamond$.*

(ii) *The process*

$$X(t)^{\mathsf{T}} = [Q(T_m), \dots, Q(t)], \qquad T_m \leq t \leq T_{m+1} - 1,\ m \geq 0$$

is a countable state space Markov chain.

(iii) *Suppose that X is 0-irreducible and aperiodic, and that $\{Q(T_m) : m \geq 0\}$ possesses an invariant measure π_0 satisfying*

$$\bar{\mathcal{T}} := \sum_{x \in \mathsf{X}_\diamond} \pi_0(x)\, \mathcal{T}(x) < \infty.$$

Then X is positive recurrent, and for a.e. initial condition $[\pi_0]$,

$$\lim_{t \to \infty} \mathsf{P}_x\{Q(t) = y\} = \pi(y), \quad \lim_{t \to \infty} \mathsf{E}_x[c(Q(t)] = \pi(c), \qquad (8.44)$$

where

$$\pi(y) = \frac{1}{\bar{\mathcal{T}}} \mathsf{E}_{\pi_0} \left[\sum_{t=0}^{T_1 - 1} \mathbf{1}\{Q(t) = y\} \right], \qquad y \in \mathsf{X}_\diamond, \qquad (8.45)$$

with $T_1 = \mathcal{T}(Q(0))$.

Proof. We provide a sketch since the main ideas are contained in Theorem 8.3.2 that follows.

Observe that $X(T_m) = Q(T_m)$. Based on this observation, one can check directly that the measure defined for arbitrary sets A by

$$\pi(A) = \frac{1}{\bar{\mathcal{T}}} \mathsf{E}_{\pi_0} \left[\sum_{t=0}^{T_1 - 1} \mathbf{1}\{X(t) \in A\} \right]$$

is invariant for X. The normalization by $\bar{\mathcal{T}}$ ensures that π has total mass 1.

Theorem A.2.3 implies that the mean ergodic theorem holds for X, which implies the two limits in (8.44). □

8.3.2 Stability of DR policies

To address stability we apply the fluid model approximation described in Section 4.7. The "target process" q is assumed to be stable: There is a smooth function $V_f \colon \mathbb{R}^\ell \to \mathbb{R}_+$ satisfying

$$V_f(q(T; x)) \leq V_f(x) - \int_0^T c(q(t; x))\, dt, \qquad T \geq 0,\ x \in \mathsf{X}, \qquad (8.46)$$

where c is a norm on $\mathbb{R}^\ell$. It is also assumed that the approximations (4.78a), (4.78b) hold in the following L_2 sense: There is a positive sequence $\{\varepsilon(n) : n \geq 1\}$ satisfying $\varepsilon(n) \downarrow 0$ as $n \to \infty$, and

$$T^{-2} \mathsf{E}[\|Q(T; x) - q(T; x)\|^2] \leq \varepsilon(T), \qquad x \in \mathsf{X}_\diamond,\ T \geq 1. \qquad (8.47)$$

Theorem 8.3.2 (Stability of discrete review policies). *Suppose that the following hold for the CRW model (8.4):*

(a) *The policy is defined in (8.43), and the resulting state process* $\mathbf{X}$ *is* 0*-irreducible and aperiodic.*
(b) *There exists a function* $V_f \colon \mathbb{R}_+^\ell \to \mathbb{R}_+$ *satisfying (8.46), with* ∇V_f *Lipschitz continuous on* $\mathbb{R}_+^\ell$.

Then there exists $\underline{T}_1 < \infty$ *such that the following conclusions hold whenever the function* $\mathcal{T} \colon \mathsf{X}_\diamond \to \mathbb{Z}_+$ *defined in (8.42) satisfies*

$$\liminf_{r \to \infty} \inf_{|x| \ge r} \mathcal{T}(x) \ge \underline{T}_1, \quad \text{and} \quad \limsup_{r \to \infty} \sup_{|x|=r} r^{-1} \mathcal{T}(x) = 0.$$

(i) *The sampled chain* $\{Q(T_m) : m \ge 0\}$ *is* c_0*-regular with* $c_0(x) = \|x\| \mathcal{T}(x)$.
(ii) $\mathbf{X}$ *is* f*-regular and aperiodic with* $f = 1 + c_{DR}$, *and* $c_{DR}(X(t)) := c(Q(t))$. *Consequently, for each initial condition,*

$$\lim_{t \to \infty} \mathsf{E}[c(Q(t; x))] = \pi(c) < \infty.$$

Proof. The proof of (i) is based on the mean value theorem, following the approach used in Section 8.2.4. There is a vector $\bar{Q} \in \mathsf{X}$ on the line connecting $Q(T_1)$ and $q(T_1)$ such that

$$V_f(Q(T_1)) - V_f(q(T_1)) = \langle \nabla V_f(\bar{Q}), Q(T_1) - q(T_1) \rangle. \tag{8.48}$$

The expectation of the right-hand side is bounded using the Cauchy–Schwartz inequality,

$$\left| \mathsf{E}[\langle \nabla V_f(\bar{Q}), Q(T_1) - q(T_1) \rangle] \right|^2 \le \mathsf{E}[\|\nabla V_f(\bar{Q})\|^2] \mathsf{E}[\|Q(T_1) - q(T_1)\|^2] \\ \le \mathsf{E}[\|\nabla V_f(\bar{Q})\|^2] \varepsilon(T_1) T_1^2, \tag{8.49}$$

where in the last inequality we have used (8.47). Under the Lipschitz condition on ∇V_f we have for some constant b_1

$$\mathsf{E}[\|\nabla V_f(\bar{Q})\|^2] \le b_1(1 + \mathsf{E}[\|\bar{Q}\|^2]),$$

which is bounded by a constant times $(\mathcal{T}(x) + c(x))^2$. Combining these bounds with (8.49), we obtain for some constant b_1

$$\mathsf{E}[V_f(Q(T_1)) - V_f(q(T_1))] \le b_1(c(x) + T_1) T_1 \sqrt{\varepsilon(T_1)}.$$

We then have, based on our assumptions on $\mathbf{q}$,

$$\mathsf{E}[V_f(Q(T_1)) - V_f(x)] \le -\int_0^{T_1} \left[c(q(t)) - b_1(c(x) + T_1) \sqrt{\varepsilon(T_1)} \right] dt. \tag{8.50}$$

The constant $\underline{T}_1$ is chosen so that $b_1 \sqrt{\varepsilon(\underline{T}_1)} < 1$. We thus obtain for finite constants $\varepsilon_2 > 0$ and b_2,

$$\mathsf{E}[V_f(Q(T_1)) - V_f(x)] \le -\varepsilon_2 c(x) \mathcal{T}(x) + b_2, \qquad x \in \mathsf{X}_\diamond, \tag{8.51}$$

and this implies (i).

To prove (ii) we construct a Lyapunov function for X as a conditional expectation: For each $m \geq 0$ and $t \in \{T_m, \ldots, T_{m+1} - 1\}$,

$$V(X(t)) := \mathsf{E}\left[\sum_{k=t}^{T_{m+1}-1} c(X(k)) + b_3 V_f(Q(T_{m+1})) \,\middle|\, X(t)\right]. \qquad (8.52)$$

The finite constant b_3 is chosen so that

$$\mathsf{E}_x\left[\sum_{k=0}^{T(x)-1} c(X(k))\right] \leq b_3 \varepsilon_2 c(x) T(x), \qquad 0 \leq t \leq T(x) - 1, \; x \in \mathsf{X}_\diamond. \qquad (8.53)$$

If $t \in \{T_m, \ldots, T_{m+1} - 2\}$ we have

$$\mathsf{E}\big[V(X(t+1)) \mid X(t)\big] = \mathsf{E}\left[\sum_{k=t+1}^{T_{m+1}-1} c(X(k)) + b_3 V_f(Q(T_{m+1})) \,\middle|\, X(t)\right]$$
$$= V(X(t)) - c(X(t)). \qquad (8.54)$$

In the remainder of the proof we restrict to $t = T_{m+1} - 1$. We have in this case

$$\mathsf{E}\big[V(X(t+1)) \mid X(t)\big] = \mathsf{E}\left[\sum_{k=T_{m+1}}^{T_{m+2}-1} c(X(k)) + b_3 V_f(Q(T_{m+2})) \,\middle|\, X(t)\right]. \qquad (8.55)$$

Applying (8.51) we obtain the following bound on the conditional expectation of $V_f(Q(T_{m+2}))$:

$$\mathsf{E}\big[V_f(Q(T_{m+2})) \mid X(t), Q(T_{m+1})\big]$$
$$= \mathsf{E}\big[V_f(Q(T_{m+2})) \mid Q(T_{m+1})\big]$$
$$\leq V_f(Q(T_{m+1})) - \varepsilon_2 c(Q(T_{m+1})) T(Q(T_{m+1})) + b_2.$$

By the smoothing property of the conditional expectation this gives

$$\mathsf{E}\big[V_f(Q(T_{m+2})) \mid X(t)\big] \leq \mathsf{E}\big[V_f(Q(T_{m+1})) - \varepsilon_2 c(Q(T_{m+1})) T(Q(T_{m+1})) + b_2 \mid X(t)\big].$$

Applying (8.53) we see that the conditional expectation (8.55) is bounded as follows:

$$\mathsf{E}\big[V(X(t+1)) \mid X(t)\big]$$
$$\leq \mathsf{E}\left[\sum_{k=T_{m+1}}^{T_{m+2}-1} c(X(k)) + b_3\big(V_f(Q(T_{m+1})) - \varepsilon_2 c(Q(T_{m+1})) T(Q(T_{m+1})) + b_2\big) \,\middle|\, X(t)\right]$$
$$\leq \mathsf{E}\big[b_3 V_f(Q(T_{m+1})) \mid X(t)\big] + b_2 b_3.$$

The right-hand side is $V(X(t)) - c(Q(t)) + b_2 b_3$ from the definition (8.52) and since $t = T_{m+1} - 1$. Combining this with (8.54) shows that the Poisson inequality holds for X, which is (ii).   □

8.4 MaxWeight

Recall that the h-MaxWeight policy was defined in (4.92) with h a given C^1 function $h \colon \mathbb{R}^\ell \to \mathbb{R}_+$. The same definition is used here,

$$\phi^{\mathrm{MW}}(x) \in \arg\min_{u \in \mathsf{U}_\diamond(x)} \langle Bu + \alpha, \nabla h(x) \rangle, \qquad x \in \mathsf{X}_\diamond. \tag{8.56}$$

This is the MaxWeight policy if h is quadratic, $h(x) = V_D(x) = \frac{1}{2} x^\mathsf{T} D x$, with D diagonal.

In this section we establish several generalizations of Theorem 8.0.4 for the perturbed Lyapunov functions introduced in Section 4.9. Recall that for a given function $h_0 \colon \mathbb{R}_+^\ell \to \mathbb{R}_+$ we define

$$h(x) = h_0(\tilde{x}), \qquad x \in \mathbb{R}_+^\ell, \tag{8.57}$$

where $\tilde{x}_i := x_i + \theta(e^{-x_i/\theta} - 1)$, $1 \le i \le \ell$, with $\theta > 0$ a fixed constant.

Perhaps the most important result of this section is Theorem 8.4.1 in which the function h_0 is assumed to serve as a Lyapunov function for the fluid model. We suppose that c is a norm on $\mathbb{R}^\ell$, and that $h_0 \colon \mathbb{R}^\ell \to \mathbb{R}_+$ is any C^1 function that satisfies the following dynamic programing inequality for the fluid model:

$$\min_{u \in \mathsf{U}(x)} \langle \nabla h_0(x), Bu + \alpha \rangle \le -c(x), \qquad x \in \mathbb{R}_+^\ell. \tag{8.58}$$

For example, we might take $h_0 = J^*$, the optimal fluid value function (4.37). We can also take for h_0 a constant multiple of *any* strictly convex quadratic function.

Theorem 8.4.1 (Stability of h-MaxWeight). *Suppose that assumptions (a)–(c) of Theorem 8.0.4 hold, and that the function $h_0 \colon \mathbb{R}^\ell \to \mathbb{R}_+$ satisfies the following:*

(a) *Smoothness: The gradient ∇h_0 is Lipschitz continuous.*
(b) *Monotonicity: $\nabla h_0(x) \in \mathbb{R}_+^\ell$ for $x \in \mathbb{R}_+^\ell$.*
(c) *The dynamic programing inequality (8.58) holds, with c a norm on $\mathbb{R}^\ell$.*

Then there exist $\theta_0 < \infty$ and $b_h < \infty$ such that (V3) holds for each $\theta \ge \theta_0$, with $V = h$ and $f = 1 + \frac{1}{2} c$.

8.4.1 Roadmap

The analysis in this section is based on the drift vector field (8.33) and bounds from elementary calculus similar to the arguments used to prove Propositions A.4.2 and A.4.7 in the Appendix. We recall the critical derivative condition (4.91) here:

$$\frac{\partial}{\partial x_j} h(x) = 0 \quad \text{when } x_j = 0. \tag{8.59}$$

The first step in establishing stability of the h-MaxWeight policy is the following generalization of Proposition 6.4.3. The proof is left as an exercise – it is identical to the proof of Proposition 6.4.3.

Proposition 8.4.2. *Suppose that h is any C^1 monotone function satisfying the derivative conditions (8.59). Then, for each $x \in \mathbb{Z}_+^\ell$, the allocation $\phi^{\text{MW}}(x)$ defined by the h-MaxWeight policy can be expressed as a solution to the linear program*

$$\phi^{\text{MW}}(x) \in \mathbf{arg\,min} \quad \langle Bu, \nabla h(x) \rangle$$

$$\text{s.t.} \quad Cu \leq 1, \quad u \geq 0. \tag{8.60}$$

Proposition 4.9.1 is valid in this general setting since this result only involves the relationship between h and h_0. That is, we have the derivative identities

$$\nabla h(x) = [I - M_\theta] \nabla h_0(\tilde{x}), \tag{8.61}$$

$$\nabla^2 h(x) = [I - M_\theta] \nabla^2 h_0(\tilde{x})[I - M_\theta] + \theta^{-1} \text{diag}(M_\theta \nabla h_0(\tilde{x})), \tag{8.62}$$

with $M_\theta = \text{diag}(e^{-x_i/\theta})$ for $x \in \mathbb{R}^\ell$. In particular, the derivative conditions (8.59) do hold:

Proposition 8.4.3. *For any C^1 function h_0, the function h defined in (8.57) satisfies the derivative conditions (8.59).*

To apply these results to the CRW model we use the mean value theorem, which implies the following representation for any $Q(t) \in \mathbb{Z}_+^\ell$, and any $t \geq 0$,

$$\begin{aligned} h(Q(t+1)) - h(Q(t)) &= \langle \nabla h(\bar{Q}), \Delta(t+1) \rangle \\ &= \langle \nabla h(Q(t)), \Delta(t+1) \rangle \\ &\quad + \langle \nabla h(\bar{Q}) - \nabla h(Q(t)), \Delta(t+1) \rangle \end{aligned} \tag{8.63}$$

where $\Delta(t+1) := Q(t+1) - Q(t)$, and $\bar{Q} \in \mathbb{R}_+^\ell$ lies on the line connecting $Q(t+1)$ and $Q(t)$. Consequently,

$$\mathcal{D}h(x) = \langle \nabla h(x), \Delta(x) \rangle + b_h, \tag{8.64}$$

where

$$b_h(x) = \mathsf{E}[\langle \nabla h(\bar{Q}) - \nabla h(Q(t)), \Delta(t+1) \rangle \mid Q(t) = x]. \tag{8.65}$$

To deduce stability of the h-MaxWeight policy based on (8.64) we obtain a bound on $\langle \nabla h(x), \Delta(x) \rangle$, and we then show that the second term $b_h(x)$ is relatively small in magnitude.

In Section 8.4.2 we follow this approach to establish stability of the h-MaxWeight policy based on the perturbed linear function defined in (4.98). This is precisely (8.57) with h_0 linear. In Section 8.4.3 we prove Theorem 8.4.1.

8.4.2 Perturbed linear function

Suppose that $h_0(x) = c^{\mathsf{T}} x$ with $c \in \mathbb{R}_+^\ell$. In this case the first and second derivatives of the function h defined in (4.98) are given in (4.99). The function h is monotone, convex, and satisfies the derivative condition (8.59):

$$\nabla h\left(x\right) \geq \mathbf{0}, \quad \nabla^2 h\left(x\right) > \mathbf{0}, \quad \frac{\partial}{\partial x_j} h\left(x\right) = 0 \quad \text{when } x_j = 0. \tag{8.66}$$

Figure 4.17 shows a plot of the sublevel sets of this function in a typical two-dimensional example.

In Example 4.9.2 we found that the h-MaxWeight policy resembles a static safety stock policy with threshold value $\bar{x}_2$ given in (4.101). Similarly, Proposition 8.4.4 can be interpreted as an indirect approach to the construction of safety stocks to ensure stability.

Proposition 8.4.4. *Suppose that assumptions (a)–(c) of Theorem 8.0.4 hold. Then there exists $\theta_0 > 0$ such that the following hold for all $\theta \geq \theta_0$:*

(i) *The controlled network satisfies Foster's criterion. The function V in (V2) can be taken as a constant multiple of h.*

(ii) *Condition (V3) holds: There exist $\varepsilon_2 > 0$, $b_2 < \infty$, and a finite set S satisfying*

$$\mathcal{D}V \leq -f + b_2 \mathbf{1}_S,$$

with $V = 1 + \frac{1}{2} h^2$, $f = 1 + \varepsilon_2 h$.

(iii) *Suppose that for some $\varepsilon > 0$ the arrival process satisfies $\mathsf{E}[\exp(\varepsilon \|A(t)\|)] < \infty$. Then condition (V4) holds: For some $\varepsilon_e > 0$, $\delta_e > 0$, $b_e < \infty$, and a finite set S,*

$$\mathcal{D}V \leq -\delta_e V + b_e \mathbf{1}_S,$$

with $V = \exp(\varepsilon_e h)$.

Proof. We apply the second-order mean value theorem to obtain

$$\begin{aligned} h(Q(t+1)) - h(Q(t)) &= \langle \nabla h\left(Q(t)\right), \Delta(t+1) \rangle \\ &\quad + \tfrac{1}{2} \Delta(t+1)^{\mathsf{T}} \big[\nabla^2 h\left(\bar{Q}\right) \big] \Delta(t+1), \end{aligned} \tag{8.67}$$

where again $\bar{Q} \in \mathbb{R}_+^\ell$ lies on the line connecting $Q(t+1)$ and $Q(t)$. This implies the identity (8.64) with b_h redefined as

$$b_h(x) = \tfrac{1}{2} \mathsf{E}\big[\Delta(t+1)^{\mathsf{T}} (\nabla^2 h\left(\bar{Q}\right)) \Delta(t+1) \mid Q(t) = x \big].$$

The expression for the second derivative in (4.99) then gives

$$\mathcal{D}h\left(x\right) = \langle \nabla h\left(x\right), v \rangle + \theta^{-1} b_\Delta,$$

where $v = \Delta(x) \in \mathsf{V}$ and

$$b_\Delta = \tfrac{1}{2} \|c\| \sup_{x' \in \mathbb{Z}_+^\ell, u \in \mathsf{U}_\circ(x')} \mathsf{E}[\|\Delta(t+1)\|^2 \mid Q(t) = x', U(t) = u] < \infty.$$

We now obtain an upper bound on $\langle \nabla h\left(x\right), v \rangle$ under the h-MaxWeight policy. The expression for the first derivative in (4.99) implies the bound

$$\frac{\partial}{\partial x_i} h\left(x\right) = c_i(1 - e^{-x_i/\theta}) \geq \underline{c}(1 - e^{-x_i/\theta}), \quad 1 \leq i \leq \ell,$$

with $\underline{c} := \min_j c_j$. Exactly as in the proofs of Theorems 4.8.4 and 8.0.4 we can consider arbitrary $v \in V$ to obtain bounds on the value of (8.60). This is justified by Proposition 8.4.2. The stabilizability condition (8.8) implies that there exists $\varepsilon > 0$ such that the vector with components $v_i = -\varepsilon$, $1 \leq i \leq \ell$, lies in V for each $x \in \mathbb{R}_+^\ell$. By definition, there exists $u \in U$ satisfying $Bu + \alpha = v$. Consequently, under the h-MaxWeight policy,

$$\mathcal{D}h\,(x) \leq -\varepsilon \underline{c} \max_i (1 - e^{-x_i/\theta}) + \theta^{-1} b_\Delta.$$

Suppose that $|x| \geq \ell\theta$. Then $x_i \geq \theta$ for at least one i, and we obtain the bound

$$\mathcal{D}h\,(x) \leq -\tfrac{1}{2}\varepsilon \underline{c} + \theta^{-1} b_\Delta, \qquad \text{if } |x| \geq \ell\theta. \tag{8.68}$$

The right-hand side is negative provided $\theta > 2b_\Delta/(\varepsilon \underline{c})$. Fixing θ satisfying this bound we obtain the desired solution to (V2) with $V = 2(\varepsilon \underline{c})^{-1}h$, and $S = \{x : |x| < \ell\theta\}$. This establishes (i).

To establish (ii) we begin with the identity

$$\tfrac{1}{2}[h(Q(t+1))]^2 - \tfrac{1}{2}[h(Q(t))]^2 = h(Q(t))(h(Q(t+1) - h(Q(t)))$$
$$+ \tfrac{1}{2}[h(Q(t+1)) - h(Q(t))]^2.$$

On taking conditional expectations of both sides we obtain $\mathcal{D}V \leq h[\mathcal{D}h] + b_{\Delta 2}$, where $b_{\Delta 2}$ is the constant

$$b_{\Delta 2} = \tfrac{1}{2} \sup_{x' \in \mathbb{Z}_+^\ell, u \in U_\circ(x')} \mathsf{E}[[h(Q(t+1)) - h(Q(t))]^2 \mid Q(t) = x', U(t) = u] < \infty.$$

Applying (i) we obtain a version of the Poisson inequality (4.13) with this V, which implies that (V3) also holds.

Part (iii) follows from (i) combined with Theorem 8.2.6. $\square$

8.4.3 Perturbed value function

We conclude this section with the proof of Theorem 8.4.1. It is organized in the following two lemmas.

Lemma 8.4.5. *Under the assumptions of Theorem 8.4.1 we have under the h-MaxWeight policy, for some constant k_1,*

$$\langle \nabla h\,(x), v^{\mathrm{MW}} \rangle \leq -c(x) + k_1 \log(1 + \|x\|), \qquad x \in \mathbb{Z}_+^\ell,$$

where $v^{\mathrm{MW}} = B\phi^{\mathrm{MW}}(x) + \alpha$.

Proof. Fix a constant $\beta_- \geq \theta$, and define

$$s_-(r) = \beta_- \log(1 + r/\beta_-), \qquad r \geq 0.$$

We impose the following constraint on the velocity vector v:

$$v_i \geq 0 \quad \text{whenever} \quad x_i < s_-(\|x\|), \qquad i = 1, \ldots, \ell. \tag{8.69}$$

The minimum of $\langle \nabla h(x), v \rangle$ over v satisfying these constraints provides a bound under the h-MaxWeight policy. Proposition 8.4.2 is critical here so that we can ignore lattice constraints and boundary constraints as we search for bounds on this inner product.

The purpose of (8.69) is to obtain the following bound:

$$-e^{-x_i/\theta} v_i \leq |v_i| \left(1 + \|x\|/\beta_- \right)^{-\beta_-/\theta}, \qquad i = 1, \ldots, \ell. \tag{8.70}$$

Since h_0 is assumed monotone we have $\nabla h_0 \colon \mathbb{R}_+^\ell \to \mathbb{R}_+^\ell$, and applying (8.61) we obtain

$$\langle \nabla h(x), v \rangle \leq \langle \nabla h_0(\tilde{x}), v \rangle + \|v\| \|\nabla h_0(\tilde{x})\| \left(1 + \|x\|/\beta_- \right)^{-\beta_-/\theta}.$$

Since ∇h_0 is also Lipschitz and $\beta_- \geq \theta$, this gives for some constant k_0

$$\langle \nabla h(x), v \rangle \leq \langle \nabla h_0(\tilde{x}), v \rangle + k_0. \tag{8.71}$$

To bound (8.71) we shift $\tilde{x}$ as follows: Let $\tilde{x}^- \in \mathbb{Z}_+^\ell$ denote the vector with components

$$\tilde{x}_i^- = \lfloor (\tilde{x}_i - s_-(\|x\|))_+ \rfloor, \qquad i = 1, \ldots, \ell,$$

where $\lfloor \cdot \rfloor$ denotes the integer part. In view of (8.58), there exists $u \in \mathsf{U}(x)$ such that with $v = Bu + \alpha$,

$$\langle \nabla h_0(\tilde{x}^-), v \rangle \leq -c(\tilde{x}^-).$$

Moreover, we have $\tilde{x}_i^- = 0$ whenever $x_i < s_-(\|x\|)$. Since $u \in \mathsf{U}(x)$, this implies that the vector $v = Bu + \alpha$ satisfies $v_i \geq 0$. That is, v satisfies the constraint (8.69).

Using this v in (8.71) gives

$$\begin{aligned}
\langle \nabla h(x), v^{\text{MW}} \rangle &\leq \langle \nabla h(x), v \rangle \\
&\leq \langle \nabla h_0(\tilde{x}^-), v \rangle + \langle \nabla h_0(\tilde{x}) - \nabla h_0(\tilde{x}^-), v \rangle + k_0 \\
&\leq -c(\tilde{x}^-) + \|v\| \|\nabla h_0(\tilde{x}) - \nabla h_0(\tilde{x}^-)\| + k_0 \\
&\leq -c(x) + |c(x) - c(\tilde{x}^-)| + \|v\| \|\nabla h_0(\tilde{x}) - \nabla h_0(\tilde{x}^-)\| + k_0.
\end{aligned}$$

This completes the proof since c and ∇h_0 are each Lipschitz. $\qquad \square$

Lemma 8.4.6. *Under the assumptions of Theorem 8.4.1 we have under the h-MaxWeight policy, for some constant k_2,*

$$\mathcal{D}h(x) \leq \langle \nabla h(x), v^{\text{MW}} \rangle + k_2(1 + \theta^{-1} \|x\|), \qquad x \in \mathbb{Z}_+^\ell,$$

where $v^{\text{MW}} = B\phi^{\text{MW}}(x) + \alpha$.

Proof. The first-order mean value theorem (8.63) results in the representation (8.64) with b_h defined in (8.65). Applying the Cauchy–Schwartz inequality we obtain $\mathcal{D}h(x) \leq \langle \nabla h(x), v^{\text{MW}} \rangle + b(x)$ with

$$b(x) := \mathsf{E}\big[\|\nabla h(\bar{Q}) - \nabla h(Q(t))\|^2 \mid Q(t) = x\big]^{\frac{1}{2}} \mathsf{E}\big[\|\Delta(t+1)\|^2 \mid Q(t) = x\big]^{\frac{1}{2}}. \tag{8.72}$$

It remains to bound this function of x.

Given $Q(t) = x$, an application of the derivative identities (8.59) gives

$$\nabla h\left(\bar{Q}\right) - \nabla h\left(x\right) = [I - M_\theta(\bar{Q})](\nabla h_0\left(\bar{Q}\right) - \nabla h_0\left(\tilde{x}\right)) + [M_\theta(x) - M_\theta(\bar{Q})]\nabla h_0\left(\tilde{x}\right),$$

and applying the triangle inequality, the conditional expectation in (8.72) is bounded as follows:

$$\mathsf{E}\big[\|\nabla h\left(\bar{Q}\right) - \nabla h\left(Q(t)\right)\|^2 \mid Q(t) = x\big]^{\frac{1}{2}}$$

$$\leq \mathsf{E}\big[\|[I - M_\theta(\bar{Q})](\nabla h_0\left(\bar{Q}\right) - \nabla h_0\left(\tilde{x}\right))\|^2 \mid Q(t) = x\big]^{\frac{1}{2}}$$

$$+ \mathsf{E}\big[\|[M_\theta(x) - M_\theta(\bar{Q})]\nabla h_0\left(\tilde{x}\right)\|^2 \mid Q(t) = x\big]^{\frac{1}{2}}.$$

$$(8.73)$$

To bound the first term on the right-hand side of (8.73) we apply the Lipschitz condition on h_0: For some constant k_1,

$$\|\nabla h_0\left(\tilde{\bar{Q}}\right) - \nabla h_0\left(x\right)\| \leq k_1\|\Delta(t+1)\|, \qquad x \in \mathbb{Z}_+^\ell.$$

Hence the first term is bounded over x.

The second term in (8.73) is bounded using the mean value theorem. The ith diagonal element of $[M_\theta(x) - M_\theta(\bar{Q})]$ admits the bound

$$|e^{-x_i/\theta} - e^{-\bar{Q}_i/\theta}| = e^{-x_i/\theta}|1 - e^{-(\bar{Q}_i - x_i)/\theta}|$$

$$\leq e^{-x_i/\theta}(1 - e^{-\overline{\Delta}_i/\theta})\mathbf{1}\{\bar{Q}_i > x_i\}$$

$$+ e^{-x_i/\theta}(e^{\ell_u/\theta} - 1)\mathbf{1}\{\bar{Q}_i < x_i\}$$

where $\overline{\Delta}_i = A_i(1) + \sum_j |B_{ij}(1)|$, and we have used the fact that $\sum_j B_{ij}(1) \geq -\ell_u$ under (8.11). The right-hand side can be bounded through a second application of the mean value theorem, giving

$$|e^{-x_i/\theta} - e^{-\bar{Q}_i/\theta}| \leq e^{-x_i/\theta}(e^{\ell_u/\theta} - e^{-\overline{\Delta}_i/\theta}) \leq \theta^{-1}e^{\ell_u/\theta}(\ell_u + \overline{\Delta}_i).$$

The Lipschitz condition on ∇h_0 and second moment conditions on $(\boldsymbol{A}, \boldsymbol{B})$ then imply that for some $k_3 < \infty$,

$$\mathsf{E}\big[\|[M_\theta(x) - M_\theta(\bar{Q})]\nabla h_0\left(\tilde{x}\right)\|^2 \mid Q(t) = x\big]^{\frac{1}{2}}$$

$$\leq \theta^{-1}e^{\ell_u/\theta}(\sqrt{\ell}\ell_u + \mathsf{E}[\|\overline{\Delta}\|^2]^{\frac{1}{2}})\|\nabla h_0\left(\tilde{x}\right)\|$$

$$\leq k_3\theta^{-1}\|x\|.$$

This combined with (8.72) and (8.73) completes the proof that $b(x)$ is a bounded function of x.   □

Proof of Theorem 8.4.1. Combining the bounds given in Lemmas 8.4.5 and 8.4.6 gives, under the h-MaxWeight policy,

$$\mathcal{D}h\left(x\right) \leq -c(x) + k_1\log(1 + \|x\|) + k_2(1 + \theta^{-1}\|x\|), \qquad x \in \mathbb{Z}_+^\ell.$$

Choosing $\theta > 0$ sufficiently large, we obtain a version of the Poisson inequality (4.13).   □

8.5 MaxWeight and the average-cost optimality equation

We now consider the MaxWeight policy within the context of average-cost optimality for the CRW scheduling model (8.24). We find that the MaxWeight policy solves the average-cost optimality equation (ACOE) for a particular cost function.

However, since it is derived from the particular matrix used in the construction of the policy, the cost function may have little to do with any true "cost" associated with the network. Our motivation is not optimization: The dynamic programming equations obtained in this section imply performance bounds for the MaxWeight policy, and these results will be applied in the next section to obtain performance bounds for more general classes of policies.

We consider a general quadratic form, which we denote

$$V_D(x) = \tfrac{1}{2}x^{\mathsf{T}}Dx, \qquad x \in \mathbb{R}^\ell. \tag{8.74}$$

We relax the assumption that D is diagonal in the MaxWeight policy introduced in Definition 4.8.1. Note however that this assumption is critical in the proof of Theorem 4.8.4 or Theorem 8.0.4. In general, the h-MaxWeight policy with $h(x) = \tfrac{1}{2}x^{\mathsf{T}}Dx$, $x \in \mathbb{R}^\ell_+$, is destabilizing when D is not diagonal – see Exercise 8.

It is assumed that D is a symmetric $\ell \times \ell$ matrix (i.e., $D_{ij} = D_{ji}$ for each i and j), and that $V_D \colon \mathbb{R}^\ell_+ \to \mathbb{R}_+$. Under the *assumption* that the V_D-MaxWeight policy is regular, we can conclude that this policy is optimal with respect to a certain cost function on $\mathbb{R}^\ell_+$.

For any stationary policy ϕ we define

$$c_D(x) = c_D^\phi(x) = -(B\phi(x) + \alpha)^{\mathsf{T}}Dx, \tag{8.75}$$

and we let c_D^{MW} denote the maximum over all stationary policies,

$$c_D^{\mathrm{MW}}(x) = \max_{u \in \mathsf{U}_\circ(x)} \left[-(Bu + \alpha)^{\mathsf{T}}Dx \right], \qquad x \in \mathsf{X}_\circ. \tag{8.76}$$

Applying Proposition 4.8.3 we can conclude that c_D^{MW} is piecewise linear when D is diagonal:

Proposition 8.5.1. *Suppose that $D > 0$ is diagonal. Then c_D^{MW} can be extended to all of $\mathbb{R}^\ell_+$ to form a piecewise linear function via*

$$c_D^{\mathrm{MW}}(x) = \quad \textbf{max} \quad x^{\mathsf{T}}D(I - R^{\mathsf{T}})Mu$$

$$\textbf{s.t.} \quad Cu \le 1, \quad u \ge 0.$$

Moreover, this function is linearly unbounded: For some $\varepsilon_0 > 0$,

$$c_D^{\mathrm{MW}}(x) \ge \varepsilon_0|x|, \qquad x \in \mathsf{X}_\circ. \tag{8.77}$$

Although c_D depends upon the particular policy employed, its *steady-state mean* is the same for any regular policy.

Theorem 8.5.2. *Suppose that the assumptions of Proposition 8.2.1 hold for the CRW scheduling model (8.24). Then, for any symmetric matrix D and any norm c:*

(i) *For any stationary policy ϕ for which the chain is positive recurrent with $\pi(c) < \infty$, the steady-state mean of c_D is finite, and given by*

$$\eta_D = \tfrac{1}{2}\left(\mathrm{trace}\,(\Sigma_A D) + \|\alpha\|_D^2 + \sum_{i=1}^{\ell} \bar{U}_i \mu_i \|1^{i+} - 1^i\|_D^2\right), \qquad (8.78)$$

where $\eta_D := \pi(c_D)$, and $\bar{U} \in \mathsf{U}$ denotes the steady-state mean utilization vector

$$\bar{U} := \sum \pi(x)\phi(x) = -B^{-1}\alpha. \qquad (8.79)$$

(ii) *With ϕ in (i), the quadratic function defined by*

$$h_D(x) := V_D(x) + \sum_{i=1}^{\ell} \|1^{i+} - 1^i\|_D^2 \left([I - R^{\mathsf{T}}]^{-1}x\right)_i, \qquad x \in \mathsf{X}_\diamond, \qquad (8.80)$$

solves Poisson's equation $P_\phi h_D = h_D - c_D + \eta_D$.

(iii) *The ACOE holds,*

$$\eta_D + h_D(x) = c_D^{\text{MW}}(x) + \min_{u \in \mathsf{U}(x)} P_u h_D(x), \qquad (8.81)$$

and the minimum is achieved using the MaxWeight policy ϕ^{MW}.

(iv) *If $V_D \colon \mathbb{R}_+^{\ell} \to \mathbb{R}_+$ and c_D^{MW} is linearly unbounded, in the sense that (8.77) holds for some $\varepsilon_0 > 0$, then the MaxWeight policy is regular and optimal over all stationary policies with respect to the cost function c_D^{MW}.*

To prove Theorem 8.5.2 we first show that the quadratic form V_D satisfies an identity similar to the Poisson equation.

Lemma 8.5.3. *Suppose that the assumptions of Proposition 8.2.1 hold for the CRW scheduling model (8.24). Then for any stationary policy,*

$$\mathcal{D}V_D = -c_D + b_D, \qquad (8.82)$$

where, for $x \in \mathsf{X}_\diamond$,

$$b_D(x) = \tfrac{1}{2}\left(\mathrm{trace}\,(\Sigma_A D) + \|\alpha\|_D^2 + \sum_{i=1}^{\ell} \phi_i(x)\mu_i \|1^{i+} - 1^i\|_D^2\right),$$

and $\Sigma_A := \mathsf{E}[(A(1) - \alpha)(A(1) - \alpha)^{\mathsf{T}}]$.

Proof. We first write

$$V_D(Q(1)) = \tfrac{1}{2}(Q(0) + (Q(1) - Q(0)))^{\mathsf{T}} D(Q(0) + (Q(1) - Q(0)))$$

$$= V_D(Q(0)) + (Q(1) - Q(0))^{\mathsf{T}} DQ(0) + \tfrac{1}{2}(Q(1) - Q(0))^{\mathsf{T}} D(Q(1) - Q(0)).$$

Using (8.24) and the expression $\mathsf{E}_x[Q(1) - Q(0)] = B\phi(x) + \alpha$ we obtain

$$PV_D(x) = V_D(x) + (B\phi(x) + \alpha)^{\mathsf{T}} Dx + \tfrac{1}{2}\mathsf{E}_x[\|Q(1) - Q(0)\|_D^2],$$

and this is precisely (8.82) with

$$b_D(x) = \tfrac{1}{2} \left(\mathsf{E}[\|A(1)\|_D^2] + \sum_{i=1}^{\ell} \phi_i(x)\mu_i \|\mathbf{1}^{i+} - \mathbf{1}^i\|_D^2 \right).$$

To complete the proof we obtain an alternative expression for the expectation: Since $\mathsf{E}[A(1)] = \alpha$ we may write

$$\mathsf{E}[\|A(1)\|_D^2] = \mathsf{E}[(A(1) - \alpha)^\mathsf{T} D(A(1) - \alpha)] + \alpha^\mathsf{T} D\alpha.$$

Moreover, for any vectors $X, Y \in \mathbb{R}^\ell$ we have $X^\mathsf{T} Y = \operatorname{trace} Y X^\mathsf{T}$. Setting $X = A(1) - \alpha$ and $Y = DX$ we thus obtain

$$\mathsf{E}[(A(1) - \alpha)^\mathsf{T} D(A(1) - \alpha)] = \operatorname{trace} \mathsf{E}[D(A(1) - \alpha)(A(1) - \alpha)^\mathsf{T}] = \operatorname{trace} (D\Sigma_A),$$

and this establishes the desired form for b_D.   □

Lemma 8.5.4. *Suppose that the assumptions of Proposition 8.2.1 hold for the CRW scheduling model (8.24). Then for any stationary policy for which the chain is positive recurrent with $\pi(c) < \infty$ we have:*

(i) *The steady-state mean utilization vector $\bar{U} = \pi(\phi)$ is given by (8.79). That is, $\bar{U} = -B^{-1}\alpha$.*

(ii) *The vector-valued function $g(x) = -B^{-1}x$, $x \in \mathsf{X}_\diamond$, solves*

$$Pg = g - \phi + \bar{U}.$$

Proof. We have from the definition of the CRW model

$$\mathsf{E}_x[Q(1) - Q(0)] = B\phi(x) + \alpha, \qquad x \in \mathsf{X}_\diamond. \tag{8.83}$$

Letting $h(x) = x_i$ and $f(x) = [B\phi(x) + \alpha]_i$ for $1 \le i \le \ell$ we obtain $\mathcal{D}h = f$. Moreover, we have $h \in L_\infty^{V_p}$ and $f \in L_\infty^{V_{p-1}}$ with $p = 1$. From Proposition 8.2.5 we obtain $\pi(f) = -(B\bar{U} + \alpha)_i = 0$ for each i, or in vector form,

$$B\bar{U} + \alpha = 0.$$

This completes the proof of (i) since B is invertible.

Postmultiplying (8.83) by $-B^{-1}$ gives

$$\mathsf{E}_x[g(Q(1)) - g(Q(0))] = -B^{-1}[B\phi(x) + \alpha] = -\phi(x) - B^{-1}\alpha,$$

which is (ii).   □

Proof of Theorem 8.5.2. The identity $\pi(c_D) = \pi(b_D)$ follows from Lemma 8.5.3 combined with Proposition 8.2.5.

With g defined in Lemma 8.5.4 we have by definition of h_D

$$h_D(x) = V_D(x) + \sum_{i=1}^{\ell} \|\mathbf{1}^{i+} - \mathbf{1}^i\|_D^2 (Mg(x))_i, \qquad x \in \mathsf{X}_\diamond.$$

Lemma 8.5.4 then gives

$$Ph_D = PV_D + \sum_{i=1}^{\ell} \|\mathbf{1}^{i+} - \mathbf{1}^i\|_D^2 \big(M[-\phi + \bar{U}] \big)_i, \qquad x \in \mathsf{X}_\diamond.$$

This identity combined with Lemma 8.5.3 completes the proof of (ii).

We can now prove (iii). Based on (ii) we have

$$\min_u P_u h_D (x) = \min_u \{ h_D(x) - c_D(x, u) + \eta_D \},$$

where $c_D(x, u) = -(Bu + \alpha)^\mathsf{T} Dx$. That is,

$$\min_u P_u h_D (x) = h_D(x) + \eta_D + \min_u \{ (Bu + \alpha)^\mathsf{T} Dx \},$$

and this minimum is achieved using the MaxWeight policy to give (8.81).

To prove (iv) first note that regularity follows from (iii) and the assumptions on V_D and c_D^{MW}, so that they each are unbounded. Suppose that ϕ is a stationary policy satisfying $\pi^\phi(c_D^\phi) < \infty$. We then have

$$\pi^{\mathrm{MW}}(c_D^{\mathrm{MW}}) = \pi^\phi(c_D^\phi) \le \pi^\phi(c_D^{\mathrm{MW}})$$

where the equality holds by (i), and the inequality follows from maximality of c_D^{MW}. Hence the MaxWeight policy solves the ACOE as claimed. $\square$

8.6 Linear programs for performance bounds

We now extend the results of Section 8.5 to construct performance bounds and approximate solutions to Poisson's equation through linear programming methods.

We restrict to the CRW scheduling model (8.24). Several simplifying assumptions will be imposed so that the main ideas can be clearly exposed.

Definition 8.6.1 (CRW model assumptions for LP bounds).

(a) For each $t \ge 1$, $1 \le i \le \ell$, the distribution of $M_i(t)$ is Bernoulli, the distribution of $A(t)$ is supported on $\mathbb{Z}_+^\ell$, and satisfies $\mathsf{E}[\|A(t)\|^2] < \infty$, and the joint process (A, M) satisfies (8.26).

(b) The queue length process Q is constrained to the integer lattice,

$$Q(t) \in \mathsf{X}_\diamond, \qquad t \ge 0, \tag{8.84a}$$

where $\mathsf{X}_\diamond = \mathbb{Z}_+^\ell$.

(c) The allocation sequence U is defined as a stationary policy that is weakly non-idling,

$$U(t) = \phi(Q(t)), \qquad t \ge 0, \tag{8.84b}$$

where $\phi \colon \mathbb{Z}_+^\ell \to \mathsf{U}_\diamond$.

If the chain is positive recurrent, then we let π denote its stationary measure, and we let $\mathsf{X}_\pi \subset \mathsf{X}_\diamond$ denote the support of π. That is,

$$\mathsf{X}_\pi = \{ x \in \mathsf{X}_\diamond : \pi(x) > 0 \}. \qquad \blacksquare$$

Buffer constraints can be included in the linear programs described here. In this case the state space is redefined as $X_\diamond = X \cap \mathbb{Z}_+^\ell, X = \{x \in \mathbb{R}_+^\ell : x_i < \overline{x}_i + 1, \quad 1 \le i \le \ell\}$, with $0 < \overline{x}_i \le \infty$ for each i.

8.6.1 Linear test for stability

In Section 8.5 we found that the MaxWeight policy is regular, and optimal for the cost function defined in (8.76), provided V_D is a solution to (V3). There are two properties that must be verified: V_D is nonnegative, and $c_D \ge c$, with c a given cost function. In this section we construct an algebraic test to determine if V_D satisfies these bounds. Throughout this section it is assumed that c is linear.

We first take a short detour to survey some concepts from linear algebra concerning an $\ell \times \ell$ matrix D:

(i) It is called *copositive* if the quadratic form V_D defined in (8.74) is nonnegative on the positive orthant in $\mathbb{R}_+^\ell$, i.e., $V_D \colon \mathbb{R}_+^\ell \to \mathbb{R}_+$.

(ii) A *principal submatrix* is a $k \times k$ matrix formed by deleting $\ell - k$ columns and the same $\ell - k$ rows from D. The kth *leading principal submatrix* is the special case in which the last $\ell - k$ rows and columns are deleted.

(iii) The *leading principal minor* is the determinant of the leading principal submatrix.

(iv) The (i, j)th cofactor is $(-1)^{i+j} \det(D^{ij})$, where D^{ij} is the $(\ell-1) \times (\ell-1)$ matrix obtained by deleting the ith row and jth column.

Recall that a matrix is *positive definite* if it is *symmetric* $(D = D^\mathsf{T})$, and also $x^\mathsf{T} D x \ge 0$ for any nonzero x. The symmetric matrix D is positive definite if and only if all ℓ leading principal minors are positive. Similarly, Keller's Theorem asserts that copositive matrices are characterized by the signs of certain determinants:

Theorem 8.6.2 (Keller's Theorem). *A symmetric matrix is copositive if and only if each principal submatrix for which the cofactors of the last row are nonnegative has a nonnegative determinant.*

Algorithms for testing copositivity can be found in the literature, but the determination of copositivity is very difficult for large matrices [379, 14]. To avoid this complexity we usually confine our attention to matrices satisfying $D_{ij} \ge 0$ for each i and j. Such matrices are obviously copositive.

Alternatively, under certain conditions we find that copositivity is automatic when V_D satisfies Poisson's inequality (8.12) for all nonidling policies. See in particular Proposition 8.6.8.

The function c_D appearing in (8.82) can be expressed

$$-c_D(x) = \alpha^\mathsf{T} D x + \sum_{j=1}^{\ell} x_j \sum_{i=1}^{\ell} \mu_i (D_{ji_+} - D_{ji}) \phi_i(x), \qquad x \in X.$$

So, to enforce the condition $c_D \geq c$ we impose the following bounds:

$$x_j \left((\alpha^\mathsf{T} D)_j + \sum_{i=1}^{\ell} \mu_i (D_{ji_+} - D_{ji}) \phi_i(x) \right) \leq -c_j x_j, \qquad 1 \leq j \leq \ell, \ x \in \mathsf{X}.$$

Suppose that the policy is nonidling: $\sum_{i \in \mathcal{I}_{s(j)}} \phi_i(x) = 1$ whenever $x_j \geq 1$. Then considering the worst case over all policies gives the following sufficient condition to guarantee that Poisson's inequality (8.12) holds: For each $1 \leq j \leq \ell$,

$$\max_{i \in \mathcal{I}_{s(j)}} \left(\mu_i (D_{ji_+} - D_{ji}) \right) + \sum_{\substack{s \neq s(j)}} \max_{i \in \mathcal{I}_s} \left(\mu_i (D_{ji_+} - D_{ji})_+ \right) \leq -c_j - (\alpha^\mathsf{T} D)_j. \quad (8.85)$$

The notation $(z)_+$ indicates the positive part of $z \in \mathbb{R}$, i.e., $(z)_+ = \max(z, 0)$. The use of the positive part in (8.85) is required since we do not know if station s is idling for $s \neq s(j)$.

Proposition 8.6.3. *Suppose that there exists a symmetric copositive matrix D satisfying (8.85). Then the function V_D defined in (8.74) satisfies (8.12) for any policy that is nonidling, and*

$$\pi(c) \leq \pi(c_D) = \pi(b_D).$$

Proof. Poisson's inequality (8.12) follows from the fact that the function b_D defined below (8.82) is a bounded function of x. The identity $\pi(c_D) = \pi(b_D)$ then follows from Theorem 8.5.2. □

Example 8.6.4 (A quadratic Lyapunov function may not exist). The simple re-entrant line is shown in Fig. 2.9. The nonidling assumption is given by the three constraints $(u_1 + u_3)x_1 = x_1$; $u_2 x_2 = x_2$, and $(u_1 + u_3)x_3 = x_3$ when $\phi(x) = u \in \mathsf{U}_\circ(x)$.

With the cost function given by $c(x) = x_1 + x_2 + x_3$, the quadratic form V_D will satisfy Poisson's inequality (8.12) for every nonidling policy provided the matrix D is copositive and satisfies the following six inequalities:

$$\mu_1(D_{12} - D_{11}) + \mu_2(D_{13} - D_{12})_+ \leq -c_1 - D_{11}\alpha_1$$
$$\mu_3(-D_{13}) + \mu_2(D_{13} - D_{12})_+ \leq -c_1 - D_{11}\alpha_1 \tag{8.86a}$$

$$\mu_2(D_{23} - D_{22}) + \mu_1(D_{22} - D_{21})_+ \leq -c_2 - D_{21}\alpha_1$$
$$\mu_2(D_{23} - D_{22}) + \mu_3(-D_{23})_+ \leq -c_2 - D_{21}\alpha_1 \tag{8.86b}$$

$$\mu_1(D_{32} - D_{31}) + \mu_2(D_{33} - D_{32})_+ \leq -c_3 - D_{31}\alpha_1$$
$$\mu_3(-D_{33}) + \mu_2(D_{33} - D_{32})_+ \leq -c_3 - D_{31}\alpha_1. \tag{8.86c}$$

Each pair of inequalities corresponds to the drift constraints on each of the three buffers $j = 1, 2, 3$ in (8.85).

The linear constraints (8.86a)–(8.86c) can be tested using simple computer code. In a homogeneous model in which $\mu_1 = \mu_3$ and $(\rho_1, \rho_2) = (2\alpha_1/\mu_1, \alpha_1/\mu_2)$, it is found that for *most* service rates satisfying $\rho_\bullet < 1$ a quadratic Lyapunov function satisfying (8.12) does exist, so the CRW model (8.24) is stable under any nonidling policy.

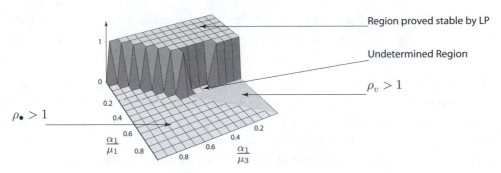

Figure 8.3. Feasibility of (8.85) for the KSRS model. If the virtual load condition (2.30) is not satisfied, then there is a priority policy that is destabilizing. Hence, in this case the linear test cannot be feasible.

However, there is a small set of parameters for which the test fails. An example is the value $\rho_1 = 0.95$ and $\rho_2 = 0.6$. In this case the test is inconclusive, so we do not know if all nonidling policies are stabilizing. ∎

Example 8.6.5 (KSRS model). Consider the system shown in Fig. 2.12 in which $\mu_1 = \mu_4$, $\mu_2 = \mu_3$, and $\alpha_1 = \alpha_4$. The network load is given by $\rho_\bullet = \rho_1 = \rho_2 = \alpha_1(\mu_1^{-1} + \mu_3^{-1})$.

The constraint region (8.85) to test for stability over all nonidling policies is expressed as follows:

$$\max((\alpha_1 - \mu_1)D_{11} + \alpha_1 D_{13} + \mu_1 D_{12}, \alpha_1 D_{11} + \alpha_1 D_{13} - \mu_4 D_{14})$$
$$+ \max(0, -\mu_1 D_{13} + \mu_3 D_{14}) \le 1$$
$$\alpha_1 D_{12} + \alpha_1 D_{23} - \mu_2 D_{22} + \max(0, -\mu_1 D_{12} + \mu_1 D_{22}) \le 1$$
$$\max(\alpha_1 D_{13} + \alpha_1 D_{33} - \mu_2 D_{23}, \alpha_1 D_{13} + \alpha_1 D_{33} - \mu_3 D_{33} + \mu_3 D_{34})$$
$$+ \max(0, -\mu_1 D_{13} + \mu_1 D_{23}) \le 1$$
$$\alpha_1 D_{14} + \alpha_1 D_{34} - \mu_4 D_{44} + \max(0, -\mu_3 D_{34} + \mu_3 D_{44}) \le 1.$$
$$(8.87)$$

Figure 8.3 shows results for the stability test for a range of values of $(\alpha_1 \mu_1^{-1}, \alpha_1 \mu_3^{-1}) \in [0,1] \times [0,1]$. The region $\alpha_1 \mu_1^{-1} + \alpha_1 \mu_3^{-1} > 1$ should be disregarded as this corresponds to $\rho_\bullet > 1$.

We have seen that there exist nonidling policies that are not stabilizing even when $\rho_\bullet < 1$. Suppose for example that priority is given to buffer 3 at Station 1 and to buffer 2 at Station 2. This policy was analyzed in Section 2.9 where the virtual load condition (2.30) was introduced, and it was shown that the fluid limit q is stable only if $\alpha_1 < \frac{1}{2}\mu_1$. The simulation shown in Section 2.9 shows how instability can arise when this condition is violated.

Consequently, for parameter values satisfying $\alpha_1 \ge \frac{1}{2}\mu_1$ the set of inequalities (8.87) cannot be feasible with D copositive. The results illustrated in Fig. 8.3 are consistent with this observation.

In *most* of the rest of the capacity region the system is stable, since (8.87) is feasible. However, as shown in the figure, there is a small region in which (8.87) is infeasible; thus the stability remains unresolved there based on these methods.   ∎

Recall that a priority policy for the scheduling model was defined in Section 4.4.1 based on a permutation $\{\theta_1, \dots, \theta_\ell\}$ of the buffer indices $\{1, \dots, \ell\}$ that determines a rank ordering of the buffers in the network. Since the policy can be represented by the set of linear constraints (4.55) it is possible to refine the constraint region (8.85) to obtain a test for a specific buffer priority policy. We illustrate the construction of an LP in the simple re-entrant line.

Example 8.6.6 (Simple re-entrant line: linear test for priority policy). Recall from the discussion in Example 8.6.4 that there are network parameters satisfying $\rho_\bullet < 1$ for which the test for a quadratic Lyapunov function fails. We now consider the two buffer priority policies: In LBFS buffer 3 received priority over buffer 1, and in FBFS these priorities are reversed.

LBFS The linear program to test for stability is almost identical to (8.86a)–(8.86c): The two constraints in (8.86c) are relaxed to the single constraint

$$-\mu_3 D_{33} + \mu_2 (D_{33} - D_{32})_+ \le -c_3 - D_{31}\alpha_1.$$

FBFS The linear program is again almost identical to (8.86a)–(8.86c). We remove one of the two constraints in (8.86a) to obtain the single constraint

$$\mu_1 (D_{12} - D_{11}) + \mu_2 (D_{13} - D_{12})_+ \le -c_1 - D_{11}\alpha_1.$$

In either case, when $\rho_\bullet < 1$ and $\mu_1 = \mu_3$, then these linear constraints are feasible. We conclude that each policy is stabilizing, in the sense that a quadratic Lyapunov function can be constructed for each priority policy.   ∎

8.6.2 The Drift and Performance LPs

The linear test described in Proposition 8.6.3 tells us that an explicit performance bound is obtained provided a copositive matrix D satisfies a linear test. The Drift LP is a refinement of this result in which this bound is optimized over all matrices D.

Definition 8.6.7 (Drift LP). This is the linear program in the variables $\{D_{ij}\} \subset \mathbb{R}$,

$$
\begin{aligned}
\textbf{min} \quad & \tfrac{1}{2}\left(\text{trace}\,(\Sigma_A D) + \alpha^{\mathsf{T}} D\alpha + \sum_{i=1}^{\ell} \bar{U}_i \mu_i (\mathbf{1}^{i+} - \mathbf{1}^i)^{\mathsf{T}} D (\mathbf{1}^{i+} - \mathbf{1}^i) \right) \\
\textbf{s.t.} \quad & D = D^{\mathsf{T}}
\end{aligned}
\tag{8.88}
$$

The linear constraints (8.85).   ∎

Note that the constraints (8.88) *do not* include a copositivity condition on D. Copositivity is in fact a consequence of feasibility:

Proposition 8.6.8. *Suppose that $\rho_\bullet < 1$ and that the Drift LP is feasible. Then it is bounded, and the following conclusions hold:*

(i) *The optimizer D^* is copositive.*

(ii) *The quadratic form V_{D^*} is a solution to Poisson's inequality (8.12) for any non-idling policy.*

(iii) *For any nonidling policy, the value of the Drift LP provides an upper bound on the average cost $\pi(c)$.*

Proof. Part (i) is a consequence of Proposition 8.2.5 combined with Theorem 8.5.2: Given any regular policy, the function h_{D^*} defined in Theorem 8.5.2 is the unique solution to Poisson's equation satisfying $h_{D^*} \in L_\infty^{V_2}$ and $h_{D^*}(0) = 0$, and can be expressed

$$h_{D^*}(x) = \mathsf{E}_x\left[\sum_{t=0}^{\tau_0-1}[c_{D^*}(Q(t)) - \eta_{D^*}]\right].$$

The function h_{D^*} must be bounded from below since the constraint (8.85) implies that $c_{D^*} \geq c$, and Proposition 8.2.3 asserts that the solution to Poisson's equation is bounded from below. It follows that the matrix D^* is copositive.

Part (ii) is by definition, and (iii) follows from Proposition 8.6.3. □

An alternative approach to obtain bounds on the steady-state cost is to consider the constraint $\mathsf{E}_\pi[c_D(Q(t))] = \mathsf{E}_\pi[b_D(Q(t))]$ for *each* $\ell \times \ell$ matrix D, and attempt to find a maximum or minimum of $\mathsf{E}_\pi[c(Q(t))]$ subject to these abstract constraints.

Consider the simplest quadratic satisfying $\frac{1}{2}x^\mathsf{T}Dx = x_ix_j$ for some indices i,j satisfying $1 \leq i \leq j \leq \ell$. For each $i \in \{1,\dots,\ell\}$ denote by i_- the unique $i' \in \{1,\dots,\ell\}$ satisfying $R_{i',i} = 1$ if such an integer exists. That is, $j = i_-$ if and only if $j_+ = i$. If $R_{i',i} = 0$ for all i' then we set $\mu_{i_-} = 0$. Under these conventions we obtain the following expression for the conditional expectation,

$$\begin{aligned}
\mathsf{E}[Q_i(t+1)Q_j(t+1) \mid Q(t) = x, U(t) = u] = {}& x_ix_j \\
&+ x_i(\alpha_j - \mu_ju_j + \mu_{j_-}u_{j_-}) \quad\quad (8.89)\\
&+ x_j(\alpha_i - \mu_iu_i + \mu_{i_-}u_{i_-}) + E_{ij}(u),
\end{aligned}$$

where

$$E_{ij}(u) := \mathsf{E}[(A_i(t) - u_iM_i(t) + u_{i_-}M_{i_-}(t))(A_j(t) - u_jM_j(t) + u_{j_-}M_{j_-}(t))].$$

Due to the statistical assumptions imposed on the service processes in (8.26), the expectation can be expressed

$$E_{ij}(u) = \mathsf{E}[A_i(t)A_j(t)] + \mu_iu_i(\mathbf{1}_{i=j} - \mathbf{1}_{i=j_-}) + \mu_{i_-}u_{i_-}(\mathbf{1}_{i_-=j_-} - \mathbf{1}_{i_-=j}).$$

For each i,j we define $\overline{\Gamma}_{ij} = \mathsf{E}_\pi[U_i(t)Q_j(t)]$. We view $\overline{\Gamma}$ as a vector in $\mathbb{R}^N$, with $N = \ell^2$, and obtain linear constraints on this vector.

When $\mathsf{E}_\pi[c(Q(t))]$ is finite and U is defined using a nonidling policy, it then follows from (8.89) and Proposition 8.2.5 that the following identity holds for each i and j:

$$\left(\mu_j\overline{\Gamma}_{ji} - \mu_{j_-}\overline{\Gamma}_{j_-,i} - \alpha_j\mathsf{E}_\pi[Q_i(t)]\right) + \left(\mu_i\overline{\Gamma}_{ij} - \mu_{i_-}\overline{\Gamma}_{i_-,j} - \alpha_i\mathsf{E}_\pi[Q_j(t)]\right) = \mathsf{E}_\pi[E_{ij}(U(t))].$$
$$(8.90)$$

Under the nonidling assumption (8.91) we also have

$$Q_j(t) = \sum_{i \in \mathcal{I}_{s(j)}} U_i(t) Q_j(t),$$
(8.91)

and hence the identity (8.90) can be expressed

$$
\left(\mu_j \overline{\Gamma}_{ji} - \mu_{j_-} \overline{\Gamma}_{j_-,i} - \alpha_j \left(\sum_{i' \in \mathcal{I}_{s(i)}} \overline{\Gamma}_{i'i} \right) \right)
$$
$$
+ \left(\mu_i \overline{\Gamma}_{ij} - \mu_{i_-} \overline{\Gamma}_{i_-,j} - \alpha_i \left(\sum_{i' \in \mathcal{I}_{s(j)}} \overline{\Gamma}_{i'j} \right) \right) = \mathsf{E}_\pi [E_{ij}(U(t))].
$$
(8.92)

The nonidling assumption also provides the following representation for the steady-state cost:

$$
\mathsf{E}_\pi [c(Q(t))] = \sum_{j=1}^{\ell} c_j \left(\sum_{i \in \mathcal{I}_{s(j)}} \overline{\Gamma}_{ij} \right).
$$

Let $\langle \Gamma^1, \Gamma^2 \rangle$ denote the usual inner product,

$$
\langle \Gamma^1, \Gamma^2 \rangle = \sum_{i,j \in \{1,\dots,\ell\}} \Gamma^1_{ij} \Gamma^2_{ij}, \qquad \Gamma^1, \Gamma^2 \in \mathbb{R}^N.
$$

In this notation we may express the steady-state cost as the inner product $\mathsf{E}_\pi [c(Q(t))] = \langle c^{\bullet}, \overline{\Gamma} \rangle$, where $c^{\bullet} \in \mathbb{R}^N_+$ is defined consistently with $c \in \mathbb{R}^\ell_+$ as

$$
c^{\bullet}_{ij} := c_j \mathbf{1}(i \in s(j)) \qquad \text{for } i,j \in \{1,\dots,\ell\}.
$$
(8.93)

For each pair $i,j \in \{1,\dots,\ell\}$, the constraint (8.92) can be expressed as follows. Define $a^{ij} \in \mathbb{R}^N$ so that for any $\Gamma \in \mathbb{R}^N$,

$$
\langle a^{ij}, \Gamma \rangle = -\alpha_j \left(\sum_{i' \in \mathcal{I}_{s(i)}} \Gamma_{i'i} \right) + \mu_j \Gamma_{ji} - \mu_{j_-} \Gamma_{j_-,i} - \alpha_i \left(\sum_{i' \in \mathcal{I}_{s(j)}} \Gamma_{i'j} \right) + \mu_i \Gamma_{ij} - \mu_{i_-} \Gamma_{i_-,j},
$$

and define $b \in \mathbb{R}^N$ by

$$
b_{ij} = \pi(E_{ij}) = \mathsf{E}[A_i(t) A_j(t)] + \mu_i \bar{U}_i (\mathbf{1}_{i=j_-} - \mathbf{1}_{i=j}) + \mu_{i_-} \bar{U}_{i_-} (\mathbf{1}_{i_-=j} - \mathbf{1}_{i_-=j_-}).
$$

In this notation, the identity (8.92) becomes the linear constraint $\langle a^{ij}, \overline{\Gamma} \rangle = b_{ij}$.

We arrive at the second linear program to obtain bounds on the steady-state cost:

Definition 8.6.9 (Performance LP). This is the linear program in the variables $\{\Gamma_{ij}\} \in \mathbb{R}^N_+$,

$$
\begin{aligned}
\textbf{max} \quad & \langle c^{\bullet}, \Gamma \rangle \\
\textbf{s.t.} \quad & \langle a^{ij}, \Gamma \rangle = b_{ij}, \qquad i,j \in \{1,\dots,\ell\}, \\
& \Gamma \geq 0.
\end{aligned}
$$
(8.94)

∎

We also consider (8.94) with the max replaced by a min to obtain lower bounds on steady-state performance.

If the Performance LP is bounded then any nonidling policy is stabilizing. The proof of the following proposition will be obtained in Section 8.6.3 by establishing duality between the two linear programs.

Proposition 8.6.10. *Suppose that the Performance LP is bounded. Then:*

(i) *All nonidling policies are stabilizing.*
(ii) *The value of the Performance LP provides an upper bound on $\pi(c)$ for any non-idling policy.*
(iii) *If the "max" is replaced by a "min" in (8.94), then the value of the linear program is again bounded, and its value provides a lower bound on $\pi(c)$ for any nonidling policy.*

8.6.3 Duality

We now show that the Performance and Drift LPs are dual. Applying the Duality Theorem of Linear Programming Theorem 1.3.2 we conclude that the Drift LP has a feasible solution whenever the Performance LP is bounded. Consequently, there exists a single quadratic function h that solves Poisson's inequality (8.12) for every nonidling policy.

Consider how the dual of a linear program such as (8.94) is typically constructed: Take any constants $\{G_{ij} : i, j \in \{1, \ldots, \ell\}\}$, and consider the vector $a^G \in \mathbb{R}^N$ defined as the linear combination

$$a^G := \tfrac{1}{2} \sum_{i,j \in \{1,\ldots,\ell\}} G_{ij} a^{ij},$$

where the constants $\{a^{ij}\}$ are defined in the Performance LP. If we choose these coefficients so that $a^G \geq c^\bullet$ pointwise, then from the equality constraint in (8.94) we obtain an upper bound on the value of (8.94) as follows, whenever $\Gamma \geq 0$:

$$\langle c^\bullet, \Gamma \rangle \leq \langle a^G, \Gamma \rangle = \tfrac{1}{2} \sum G_{ij} \langle a^{ij}, \Gamma \rangle = \tfrac{1}{2} \sum G_{ij} b_{ij} = \tfrac{1}{2} \langle G, b \rangle. \tag{8.95}$$

The dual of (8.94) is defined as the linear program obtained on minimizing the upper bound (8.95) over all $G \in \mathbb{R}^N$. This is precisely the Drift LP:

Proposition 8.6.11. *The dual of the Performance LP is the linear program in the N variables $\{G_{ij}\} \subset \mathbb{R}$,*

$$\begin{aligned} \mathbf{min} \quad & \tfrac{1}{2} \langle G, b \rangle \\ \mathbf{s.t.} \quad & a^G \geq c^\bullet, \quad G \in \mathbb{R}^N. \end{aligned} \tag{8.96}$$

Consequently:

(i) *The linear program (8.96) is precisely the Drift LP.*

(ii) *If the Performance LP is bounded, then the Drift LP is feasible, and the values of the two LPs are identical.*

Proof. It follows from the definition of b and (8.88) that for any matrix $G = (G_{ij})$,

$$\langle G, b \rangle = \tfrac{1}{2} \left(\sum_{i,j=1}^{\ell} G_{ij} \mathsf{E}[A_i(t) A_j(t)] + \sum_{i=1}^{\ell} \bar{U}_i \mu_i (\mathbf{1}^{i-} - \mathbf{1}^i)^{\mathsf{T}} G_{i-,i} (\mathbf{1}^{i-} - \mathbf{1}^i) \right)$$

$$= \tfrac{1}{2} \left(\operatorname{trace}(\Sigma_A G) + \alpha^{\mathsf{T}} G \alpha + \sum_{i=1}^{\ell} \bar{U}_i \mu_i (\mathbf{1}^{i-} - \mathbf{1}^i)^{\mathsf{T}} G (\mathbf{1}^{i-} - \mathbf{1}^i) \right).$$

Hence the objective function in (8.96) is precisely the objective function in the Drift LP. Similarly, the constraint set for D in (8.85) for the Drift LP is precisely the constraint set on G given in (8.96), which proves (i).

Part (ii) follows from the Duality Theorem of Linear Programming stated in Theorem 1.3.2. $\qquad\square$

8.7 Brownian workload model

We now extend the Lyapunov theory developed in this chapter to the CBM model. We return to the setting of Section 5.4.2 where $\widehat{W}$ denotes an R-minimal solution on the domain R defined in (5.31),

$$\mathsf{R} = \{w \in \mathbb{R}^n : \langle n^i, w \rangle \geq -\beta_i, \ 1 \leq i \leq \ell_R\}, \tag{8.97}$$

where $\{n^i\}$ and $\{\beta_i\}$ are fixed vectors and nonnegative constants, respectively.

Lyapunov theory requires a generator for the process.

8.7.1 The extended generator

The generator for the CBM model is defined as for the RBM model in Section 3.3.3.

A stochastic process M adapted to some filtration $\{\mathcal{F}_t : t \geq 0\}$ is called a martingale if $\mathsf{E}[M(t + s) \mid \mathcal{F}_t] = M(t)$ for each $t, s \in \mathbb{R}_+$ (see Section 1.3.5). It is called a *local martingale* if there exists an increasing sequence of stopping times $\{\varsigma_n\}$ such that $\{M(t \wedge \varsigma_n) : t \in \mathbb{R}_+\}$ is a martingale, for each $n \geq 1$, and $\varsigma_n \uparrow \infty$ a.s. as $n \to \infty$. Some background in discrete time is surveyed in Section 1.3.5.

For the CRW model controlled using a stationary policy, whenever the function $g := \mathcal{D}h = Ph - h$ is finite valued, the following process is a discrete-time local martingale with respect to $\mathcal{F}_t = \sigma\{Q(k) : k \leq t\}$:

$$M_h(t) := h(Q(t)) - \left\{ h(Q(0)) + \sum_{k=0}^{t-1} g(Q(k)) \right\}, \qquad t \geq 0.$$

The extended generator for the CBM model is meant to capture this property of the generator in discrete time.

Definition 8.7.1 (Generators for the CBM model). The *differential generator* $\mathcal{D}$ is defined by the second-order differential operator, defined for C^2 functions on $\mathbb{R}^n$ by

$$\mathcal{D}h := -\delta^{\mathsf{T}}\nabla h + \tfrac{1}{2}\Delta h, \qquad (8.98)$$

where Δ denotes the weighted Laplacian,

$$\Delta h := \sum_{i,j}\Big(\Sigma_{i,j}\frac{\partial^2}{\partial x_i \partial x_j}h\Big).$$

The *extended generator* $\mathcal{A}$ is defined as follows: A measurable function $h\colon \mathbb{R} \to \mathbb{R}$ is in the domain of $\mathcal{A}$ if there is a measurable function $g\colon \mathbb{R} \to \mathbb{R}$ such that, for each initial condition $\widehat{W}(0) \in \mathsf{R}$, the stochastic process defined below is a local martingale with respect to the filtration $\{\mathcal{F}_t\}$ generated by $\boldsymbol{N}$,

$$M_h(t) := h(\widehat{W}(t)) - \Big\{ h(\widehat{W}(0)) + \int_0^t g(\widehat{W}(s))\,ds \Big\}, \qquad t \geq 0. \qquad (8.99)$$

In this case, we write $g = \mathcal{A}h$. ∎

The following version of Itô's formula (8.100) is used to identify a large class of functions within the domain of $\mathcal{A}$. This result will be applied repeatedly in the treatment of value functions that follows.

Theorem 8.7.2. *Suppose that the set R is given in (8.97), and that the pointwise projection $[\,\cdot\,]_\mathsf{R}\colon \mathbb{R}^n \to \mathsf{R}$ exists. Suppose that $h\colon \mathbb{R}^n \to \mathbb{R}$ is continuously differentiable, that ∇h is Lipschitz continuous on bounded subsets of $\mathbb{R}^n$, and that Δh exists for a.e. $w \in \mathsf{R}$. Then:*

(i) *For each initial condition $\widehat{W}(0) \in \mathsf{R}$,*

$$h(\widehat{W}(t)) = h(\widehat{W}(0)) + \int_0^t [\mathcal{D}h](\widehat{W}(s))\,ds$$
$$+ \int_0^t \langle \nabla h(\widehat{W}(s)), dI(s)\rangle \qquad (8.100)$$
$$+ \int_0^t \langle \nabla h(\widehat{W}(s)), dN(s)\rangle.$$

(ii) *Suppose that $\delta \in$ interior (R), and in addition the following boundary conditions hold,*

$$\langle \phi(w), \nabla h(w)\rangle = 0, \qquad w \in \mathsf{R}, \qquad (8.101)$$

where ϕ is defined in (5.30). Then h is in the domain of $\mathcal{A}$, and $\mathcal{A}h = \mathcal{D}h$.

(iii) *If the conditions of (ii) hold, and in addition,*

$$\mathsf{E}\Big[\int_0^t \|\nabla h(\widehat{W}(s;w))\|^2\,ds\Big] < \infty, \qquad w \in \mathsf{R},\ t \geq 0,$$

then $\boldsymbol{M}_h$ is a martingale for each initial condition.

Proof. When h is C^2, then (i) is given in [167, Theorem 2.9, p. 287]. In the more general setting given in the theorem, the function h and its derivatives can be approximated uniformly on compacta by C^2 functions so that $\mathcal{D}h$ is simultaneously approximated in $L_2(C)$ for any compact set $C \subset \mathsf{R}$. See [314, Theorem 1, p. 122], and the extensions in [315, 106, 185].

Suppose now that the assumptions of (ii) hold. The assumption $\delta \in \text{interior}\,(\mathsf{R})$ is imposed to ensure that ϕ does not vanish on $\partial \mathsf{R}$. Itô's formula (8.100) then gives the following representation:

$$M_h(t) = \int_0^t \langle \nabla h(\widehat{W}(s)), dN(s)\rangle, \qquad t \geq 0. \tag{8.102}$$

The local-martingale property is immediate since N is driftless Brownian motion (we can take $\varsigma_n := \min\{t \geq 0 : \|\nabla h(\widehat{W}(t))\| \geq n\}$ in the definition above).

We have for each $0 \leq s < t$, and each initial w,

$$\mathsf{E}[(M_h(t) - M_h(s))^2] = \int_s^t \mathsf{E}[(\nabla h(\widehat{W}(s)))^\mathsf{T} \Sigma \nabla h(\widehat{W}(s))].$$

The right-hand side is finite under the conditions of (iii), and the local-martingale property can be strengthened to the ordinary martingale based on this bound (see [167]). ∎

Poisson's inequality for the CBM model is defined exactly as in discrete time: For a constant $\overline{\eta} < \infty$ and function $h\colon \mathsf{X} \to \mathbb{R}_+$ in the domain of $\mathcal{A}$,

$$\mathcal{A}h \leq -c + \overline{\eta}. \tag{8.103}$$

Under the assumptions of Theorem 5.3.21 we find that the fluid value function is one solution. We note that it follows directly from the definition (8.98) that a version of this bound holds for the differential generator

$$\mathcal{D}\widehat{J} = -\overline{c} + b_{\text{CBM}}$$

where b_{CBM} is the piecewise constant function defined by

$$b_{\text{CBM}}(w) := \tfrac{1}{2}\Delta\widehat{J}(w). \tag{8.104}$$

Proposition 8.7.3. *Suppose that assumptions (a)–(c) of Theorem 5.3.21 hold. Then, with $\widehat{W}$ equal to the minimal process on R, the following hold:*

(i) *The fluid value function $\widehat{J}$ is in the domain of the extended generator, and $\mathcal{A}\widehat{J} = \mathcal{D}\widehat{J} = -\overline{c} + b_{\text{CBM}}$, where b_{CBM} is defined a.e. on R by (8.104).*
(ii) *The function $V = \sqrt{1 + \widehat{J}}$ is in the domain of the extended generator, and for some $\varepsilon_0 > 0$, $b_0 < \infty$, and a compact set $C_0 \subset \mathsf{R}$,*

$$\mathcal{A}V = \mathcal{D}V \leq -\varepsilon_0 + b_0 \mathbf{1}_{C_0}.$$

(iii) *The function $V_\vartheta = e^{\vartheta V}$ is in the domain of the extended generator for each $\vartheta > 0$. There exists $\vartheta_0 > 0$ such that the following bound holds for each $\vartheta \in (0, \vartheta_0]$: For finite constants $\varepsilon_\vartheta > 0$, $b_\vartheta > 0$,*

$$\mathcal{A}V_\vartheta = \mathcal{D}V_\vartheta \le -\varepsilon_\vartheta V_\vartheta + b_\vartheta.$$

Proof. Theorem 5.3.21 (iv) implies that $\widehat{J}$ satisfies the boundary conditions required in Theorem 8.7.2 (ii), and hence $\widehat{J}$ is in the domain of $\mathcal{A}$. Each of the functions considered in (ii) and (iii) also satisfies the conditions of Theorem 8.7.2 (ii) since these properties are inherited from $\widehat{J}$.

The bounds on $\mathcal{D}$ in (ii) and (iii) follow from the identity $\mathcal{D}\widehat{J} = -\overline{c} + b_{\text{CBM}}$, and straightforward calculus. $\square$

Just as in the discrete-time definition (8.30), we say that $\widehat{W}$ is V-*uniformly ergodic* if there exist $d_0 > 0$ and $b_0 < \infty$ such that for any $g \in L_\infty^V$,

$$|\mathsf{E}[g(\widehat{W}(t)) \mid \widehat{W}(0) = w] - \pi(g)| \le b_0 \|g\|_V e^{-d_0 t} V(x), \qquad t \in \mathbb{R}_+, \ w \in \mathsf{W}.$$

Theorem 8.7.4. *Suppose that assumptions (a)–(c) of Theorem 5.3.21 hold. Then the steady-state mean satisfies $\widehat{\eta} = \pi(\overline{c}) = \pi(b_{\text{CBM}})$, and the minimal process $\widehat{W}$ is V_ϑ-uniformly ergodic for each $\vartheta \in (0, \vartheta_0]$.*

Proof. V_ϑ-uniform ergodicity follows from [369, Theorem 6.1] and Proposition 8.7.3 (iii). The identity $\widehat{\eta} = \pi(b_{\text{CBM}})$ is obtained as follows. We have from Theorem 8.7.2 the representation

$$\widehat{J}(\widehat{W}(t)) = \widehat{J}(\widehat{W}(0)) + \int_0^t \left(-\overline{c}(\widehat{W}(s)) + b_{\text{CBM}}(\widehat{W}(s)) \right) ds$$

$$+ \int_0^t \langle \nabla \widehat{J}(\widehat{W}(s)), dN(s) \rangle.$$

We necessarily have $\pi(\widehat{J}) < \infty$ since $\pi(V_\vartheta) < \infty$. Setting $\widehat{W}(0) \sim \pi$, taking expectations, and canceling the common terms $\mathsf{E}_\pi[\widehat{J}(\widehat{W}(t))] = \mathsf{E}_\pi[\widehat{J}(\widehat{W}(0))]$ gives

$$t\pi(\overline{c}) = \mathsf{E}\left[\int_0^t \overline{c}(\widehat{W}(s)) \right] = \mathsf{E}\left[\int_0^t b_{\text{CBM}}(\widehat{W}(s)) \right] = t\pi(b_{\text{CBM}}). \qquad \square$$

8.7.2 Linear programs

Here we consider extensions of the linear programming theory to stochastic workload models. The main complication is that the effective cost $\overline{c}$ defined in Definition 5.3.12 is in general piecewise linear when the cost function for $\mathbf{Q}$ is linear. In spite of this apparent complexity, the linear programming approach has a direct and elegant extension to the CBM model based on Proposition 8.7.3.

Suppose that assumptions (a)–(c) of Theorem 5.3.21 hold, and let $\{R_j : i = 1, \ldots, \ell_O\}$ denote open, connected polyhedral regions satisfying the following: The function b_{CBM} given in (8.104) is constant on each R_j, $\overline{c}$ is linear on R_j, and $\mathsf{R} = \text{closure}(\cup R_j)$.

We also consider a family of auxiliary functions $\{\overline{c}^{ai} : 1 \leq i \leq \ell_a\}$ that are compatible with $\overline{c}$, in the sense that each of these functions is continuous, piecewise linear, and linear on each of the sets $\{R_j\}$. Consequently, the assumptions of Theorem 8.7.4 hold: Letting $\{\widehat{J}^i : 1 \leq i \leq \ell_a\}$ denote the associated C^1 value functions, and setting $b^{ai}_{\mathrm{CBM}} = \mathcal{D}\widehat{J}^i + \overline{c}^{ai}$, we obtain the identity $\pi(\overline{c}^{ai}) = \pi(b^{ai}_{\mathrm{CBM}})$ for each i. These identities are interpreted as equality constraints below.

The variables in the linear program are defined for $1 \leq i \leq n$, $1 \leq j \leq \ell_O$, by

$$P_j = \pi(R_j), \qquad \Gamma_{ij} = \mathsf{E}_\pi[\widehat{W}_i(t)\mathbf{1}_{R_j}].$$

We have several constraints:

(a) *Mass constraints*: $P_j \geq 0$ for each j, and $\sum P_j = 1$.
(b) *Region constraints*: For example, $\Gamma_{1j} \geq \Gamma_{2j}$ if $\widehat{w}_1 \geq \widehat{w}_2$ within region R_j.
(c) *Value function constraints*: For some constants $\{a_{ij}\} \subset \mathbb{R}$ and vectors $\{\varpi^{ij} : 1 \leq i \leq \ell_a, \ 1 \leq j \leq n\} \subset \mathbb{R}^n$ we have the representations for any $1 \leq j \leq \ell_O$, $1 \leq i \leq \ell_a$,

$$b^{ai}_{\mathrm{CBM}}(w) = a_{ij}; \quad \overline{c}^{ai}(w) = \langle \varpi^{ij}, w \rangle, \qquad w \in R_j.$$

Letting $\Gamma^j = (\Gamma_{1j}, \dots, \Gamma_{nj})^{\mathsf{T}} \in \mathbb{R}^n$, $1 \leq j \leq \ell_O$, we obtain from Theorem 8.7.4, for each $i \in \{1, \dots, \ell_a\}$,

$$\sum_{j=1}^{\ell_O} \langle \varpi^{ij}, \Gamma^j \rangle = \pi(\overline{c}^{ai}) = \pi(b^{ai}_{\mathrm{CBM}}) = \sum_{j=1}^{\ell_O} a_{ij} P_j. \qquad (8.105)$$

(d) *Objective function*: There is $d \in \mathbb{R}^{n \times \ell_O}$ such that $\widehat{\eta} := \pi(\overline{c}) = \sum d_{ij}\Gamma_{ij}$.

We illustrate this construction in a two-dimensional example.

Example 8.7.5 (KSRS model). We return to the two cases introduced in Example 5.3.16 in the setting of Example 5.3.26. In the workload model we have, by the symmetry assumptions imposed in these examples,

$$\delta = (\delta_1, \delta_1)^{\mathsf{T}},$$

and we assume in the CBM model that the covariance matrix satisfies $\Sigma_{11} = \Sigma_{22} > 0$.

Consider first the policy defined by the constraint region $\mathsf{R} = \{w \in \mathsf{W} : w_1/3 \leq w_2 \leq 3w_1\}$. This coincides with the monotone region $\mathsf{W}^+ = \mathrm{closure}\,(\mathsf{R}_2)$ shown on the right in Fig. 5.3 in Case II.

The cost function restricted to R is the same in Cases I and II, and the common value function shown in (5.58) is purely quadratic on R. Consequently, in this case we have $h = \widehat{J}$, and

$$\widehat{\eta} = \pi(b_{\mathrm{CBM}}) = b_{\mathrm{CBM}} = \tfrac{1}{8}\delta_1^{-1}(3\Sigma_{11} - \Sigma_{12}). \qquad (8.106)$$

Consider now the minimal process on $\mathsf{W} = \mathbb{R}_+^2$ in Case I. The function b_{CBM} is not constant, so it is not obvious that we can compute $\widehat{\eta}$ exactly using these techniques when $\mathsf{R} = \mathsf{W}$.

To construct an LP we restrict to the following specifications: $\ell_O = 3$, with $\{R_i : i = 1, 2, 3\}$ as shown in Fig. 5.4, and $\ell_a = 2$, with $\overline{c}^{a1}(w) = w_1 + w_2$ and $\overline{c}^{a2}(w) = \max(\frac{1}{3}w_1, \frac{1}{3}w_2, \frac{1}{4}(w_1 + w_4))$.

We thus obtain the following constraints:

(a) *Mass constraints*: $P_1 + P_2 + P_3 = 1$.

(b) *Region constraints*: We have $\Gamma_{ij} \geq 0$ for all i, j since $W = \mathbb{R}_+^2$. Moreover, on considering the structure of the sets $\{R_i\}$ we obtain $3\Gamma_{21} \leq \Gamma_{11}$ and $3\Gamma_{13} \leq \Gamma_{23}$. In addition, there are numerous symmetry constraints. For example, $P_1 = P_3$, and $\Gamma_{12} = \Gamma_{22}$ since $\delta_1 = \delta_2$ and $\Sigma_{11} = \Sigma_{22}$.

(c) *Value function constraints*: The value function $\widehat{J}^1$ is a pure quadratic. In fact, if $k(w) = \langle \overline{c}^{a1}, w \rangle$ is any linear function on W, then

$$\widehat{J}(w) = \tfrac{1}{2}\delta_1^{-1}(\overline{c}_1^{a1}w_1^2 + \overline{c}_2^{a1}w_2^2), \quad w \in W.$$

We conclude from Theorem 8.7.4 that $\mathsf{E}_\pi[\widehat{W}_1(t)] = \mathsf{E}_\pi[\widehat{W}_2(t)] = \tfrac{1}{2}\delta_1^{-1}\Sigma_{11}$. The identity (8.105) then implies the equality constraint, $\frac{1}{3}\Gamma_{11} + \frac{1}{4}(\Gamma_{12} + \Gamma_{22}) + \frac{1}{3}\Gamma_{23} = \delta_1^{-1}\big[\frac{1}{6}\Sigma_{11}P_1 + \frac{1}{8}(3\Sigma_{11} - \Sigma_{12})P_2 + \frac{1}{6}\Sigma_{22}P_3\big]$.

(d) *Objective function*: In Case I we have

$$\mathsf{E}_\pi[\overline{c}(\widehat{W}(t))] = \tfrac{1}{3}\Gamma_{11} + \tfrac{1}{4}(\Gamma_{12} + \Gamma_{22}) + \tfrac{1}{3}\Gamma_{23}.$$

We conclude with results from one numerical experiment in Case I, using parameters consistent with the values used in the simulation illustrated in Fig. 4.13 for the controlled random walk model (4.67). The first-order parameters were scaled as follows:

$$\mu = K[1, 1/3, 1, 1/3], \quad \alpha = \tfrac{1}{4}K\rho_\bullet[1, 0, 1, 0],$$

where K is chosen so that $\sum(\mu_i + \alpha_i) = 1$, and $\rho_\bullet = 0.9$. The effective cost was similarly scaled, $\overline{c}(w) = K\max(w_1/3, w_2/3, (w_1 + w_4)/4)$ for $w \in \mathbb{R}_+^2$.

The random variables $\{S_i(t), A_i(t)\}$ used in (4.67) were taken mutually independent, with the variance of each random variable equal to its mean.

To construct a CBM model we choose the first- and second-order statistics to approximate this CRW model. Suppose that the CRW model is in steady state, so that the mean of $U(t)$ is constant with $\mathsf{E}[U(t)] \equiv [0.25, 0.75, 0.25, 0.75]^\mathsf{T}$. In this case the steady-state covariance is approximated by

$$\mathsf{Cov}[W(t+1) - W(t)] \approx \begin{bmatrix} 13.8 & 4.21 \\ 4.21 & 13.8 \end{bmatrix},$$

where $W(t) := \Xi Q(t)$. In the CBM model we take Σ equal to the covariance matrix given above, and $\delta_1 = \delta_2 = 1 - \rho_\bullet = 0.1$.

Solving the resulting linear program then gives bounds on the steady-state cost, as well as bounds on the probabilities of each region:

$$11.07 \leq \mathsf{E}_\pi[\overline{c}(\widehat{W}(t))] \leq 14.76$$

$$0.489 \leq \mathsf{P}_\pi(\widehat{W}(t) \in R_2) \leq 0.979. \tag{8.107}$$

Although these bounds apply to the CBM model, they roughly approximate the estimated value of $\mathsf{E}_\pi[c(Q(t))] \approx 18$ for the CRW model obtained in the simulation illustrated in Fig. 4.13.

For the same parameters with the region $\mathsf{R} := R_2 = \{w_1/3 \leq w_2 \leq 3w_1\}$ we obtain from (8.106)

$$\mathsf{E}_\pi[\bar{c}(\widehat{W}(t))] = 14.92\,.$$

The steady-state mean for the process restricted to the region R_2 is strictly greater than the upper bound (8.107) obtained for the minimal process on W. This is consistent with the fact that the minimal process on W is optimal whenever $\bar{c}$ is monotone.   ∎

8.8 Notes

Stochastic Lyapunov theory appears in many different contexts in many different academic areas. Foster's criterion and its variants are the focus of Meyn and Tweedie [368] for general Markov chains, and for countable state-space chains in [170]. The latter emphasizes application to queueing networks. A more complete history can be found in [368].

Borovkov's 1986 paper [73] stimulated several papers on *positive Harris recurrence* of special classes of stochastic networks or reflected random walks [169, 447, 448, 287, 76, 93, 171, 366, 146, 30]. The main results in later papers make use of stochastic Lyapunov functions, such as a version of the Poisson inequality (8.12). Propositions 8.2.2 and A.6.2 are based on Theorem 5.2 of [356].

Stochastic Lyapunov theory plays a role in the large-deviation analysis of networks in the work of Foley and MacDonald [183, 184], and Hordijk and Popov [271].

It appears that the operator norm introduced in (8.18) was introduced by Veinott in a study of MDPs [483]. With a particular choice of V it is known that the dynamic programming operator is a contraction in this norm [339, 478, 493, 47, 471, 48, 44]. See also Denardo's original 1967 approach to dynamic programing via contraction [141], and the monographs [259, 45].

Completely independently, the weighted norm (8.18) has found application in the ergodic theory of Markov chains. Kartashov was the first to use the weighted norm in this context [285, 286], and it was applied to characterize geometric ergodicity for countable state-space Markov chains in [458, 272]. Generalizations to general state spaces and to other aspects of Markov chain theory have appeared in several sequels [148, 343, 369, 370, 416, 415, 310, 34].

Tsitsiklis in [472] introduces a DR policy to obtain approximate optimality in a very general version of the simple inventory model with a fixed "set-up cost" for any batch of orders. A version of this policy was used in [346, 345] to obtain stabilizing policies and "fluid-scale asymptotic optimality" (see the Notes section in Chapter 10). Harrison applied this technique in [235] to obtain a policy that is approximately optimal in heavy traffic in a particular example. This result was generalized in work of the author [361] and Ata and Kumar [21].

For a history on the MaxWeight policy see the Notes section in Chapter 4. The stability analysis in Section 8.4 is adapted from [365]. Part (iii) of Proposition 8.4.4 is taken from [359, Theorem 4], which is based on [368, Theorem 16.3.1].

The material in Section 8.5 is new, but is motivated by the techniques introduced by Kumar et al. [319, 317, 316] and Bertsimas et al. [54] for performance approximation in Markovian network models obtained through uniformization. The linear programs in Section 8.6 generalize these techniques to the bulk-arrival CRW model. See also [111, 147, 51, 52, 376].

Also related are approaches to approximate dynamic programming via linear programming techniques [137, 1, 482, 248].

Performance approximation is commonly approached through analytic techniques [280, 32, 290, 245, 246] or numerical computation [425, 383, 126, 443, 63]. These approaches are generally intractable in complex network models. The situation is more favorable in workload models due to the significant reduction in dimension. Analytic techniques specialized to work-load models are contained in [240, 245, 246, 247] and numerical methods are developed in [322, 325, 326, 126].

The material in Section 8.7 is largely taken from [363]. More on Lyapunov functions for the CBM model can be found in the work of Dupuis, Williams, Atar, Ramanan, and Budhiraja [160, 23, 159, 405]. Linear programs for performance bounds in the CBM model were introduced in [426, 427] based on pure quadratic functions, similar to the development for the CRW model in Section 8.6.

Exercises

8.1   Consider the two-buffer model shown below.

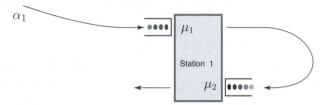

Assume that the arrival stream is Poisson, that services are exponentially distributed, and that the Last Buffer-First Served policy is used.

(a) Compute the associated rate matrix for this Markov process in continuous time, and construct a CRW model via sampling (see Section 2.1.1).

(b) Compute the draining time T^* for the associated fluid model, and show that $V = T^*$ is a Lyapunov function.

(c) Construct a Lyapunov function to approximate Poisson's equation, with $c(x) = |x| = x_1 + x_2$. Using this, apply the Comparison Theorem to obtain a bound on the steady-state mean of $c(Q(t))$.

8.2   Ten tellers at a bank perform services with a common mean of 10 min per customer. Assume that the distribution of services is exponential. Let $\alpha > 0$ be the rate at which customers arrive to the bank (customers/minute), and assume that the arrival process is Poisson.

(a) What is the largest value of α that the bank can handle without hiring new tellers?

(b) Assume that there is a seperate line for each teller. Compute the mean waiting time for any given α.

(c) Repeat (b) when there is a single line at the bank: When customers arrive, they enter the end of the queue if every teller is busy.

(d) Any suggestions for future supermarket design?

8.3   Consider the Klimov model considered in Section 5.5.1 with just two customer classes. Suppose that $c_1\mu_1 = c_2\mu_2$. Show that any nonidling policy is optimal by following these steps:

(a) There is a constant $\beta > 0$ such that $\mathsf{E}[c(Q(t))] = \beta\mathsf{E}[W(t)]$ for any t and any nonidling policy, where $W(t) = Q_1(t)/\mu_1 + Q_2(t)/\mu_2$.

(b) There exist a single quadratic function $h^*\colon \mathbb{R}_+ \to \mathbb{R}_+$ and a constant $\widehat{\eta}^{\mathrm{w}}$ such that the following identity holds for any nonidling policy,

$$\mathsf{E}[h^*(W(t+1)) \mid W(t) = w] = h^*(w) - w + \widehat{\eta}^{\mathrm{w}},$$

where w is restricted to the lattice, $\{x_1/\mu_1 + x_2/\mu_2 : x \in \mathbb{Z}_+^2\}$.

(c) Explain why $\widehat{\eta}^{\mathrm{w}}$ must be the steady-state mean of $W(t)$, so that the equation in (b) is Poisson's equation.

8.4   A *leaky bucket* regulation scheme is designed to "smooth" an arrival processes. It can be described in continuous time as follows: There is a token pool, that holds up to B tokens. When a packet arrives, if there is a token in the token pool, then the packet instantly passes through the regulator, and one of the tokens is removed from the token pool. On the other hand, if the token pool is empty, then the packet is lost. (Notice that packets are never queued.) New tokens are generated periodically, with one time unit between successive generation times. If a token is generated when the token pool is full, the token is lost.

Suppose that the packet stream to be regulated is modeled as a Poisson stream with arrival rate α.

(a) Identify an embedded discrete-time, discrete-state Markov process, and describe the one-step transition probabilities of the chain.

(b) Express the fraction of packets lost (long-term average) in terms of α, B, and π, where π denotes the equilibrium probability vector for your Markov chain. (You do *not* need to compute π.)

(c) As an approximation, suppose that the times between new token generations are independent and exponentially distributed with common mean 1. Find a fairly simple expression for the loss probability.

8.5   Find conditions on (A, S_1, S_2) such that the CRW model (4.58) for the tandem queues is reversible, in the sense that the detailed balance equations hold:

$$\pi(x)P(x,y) = \pi(y)P(y,x), \qquad x, y \in \mathsf{X}.$$

8.6   This problem is designed as an introduction to the operator-theoretic techniques described in the Appendix. Consider a Markov chain on $\mathsf{X}_\diamond = \{1,2\}$ with transition matrix

$$P = \begin{bmatrix} \frac{1}{2} & \frac{1}{2} \\ \frac{3}{4} & \frac{1}{4} \end{bmatrix}.$$

(a) Compute the *resolvent matrix* $R_z = [I - zP]^{-1}$ for $z \in \mathbb{C}$.

(b) Find a general expression for P^n based on the representation

$$R_z = \sum_{t=0}^{\infty} z^t P^t, \qquad |z| < 1.$$

(c) Compute the *fundamental matrix* $Z = [I - (P - s \otimes \nu)]^{-1}$, where s, ν are any two-dimensional nonzero vectors for which

$$P(x, y) \geq s(x)\nu(y), \qquad x, y \in \{1, 2\}.$$

Verify that the row vector

$$\mu := \nu[I - (P - s \otimes \nu)]^{-1}$$

is invariant for P, so that $\mu P = \mu$.

(d) Show that the function $h = Z\tilde{c}$ solves Poisson's equation.

8.7 Consider the simple routing model shown in Fig. 2.14. As usual it is assumed that $\mu_3 \gg \mu_1 + \mu_2$, and it is assumed that $\rho_\bullet = \alpha/(\mu_1 + \mu_2) \approx 1$.

(a) For the fluid model, with cost $c(x) = x_1 + 2x_2$, find the pathwise optimal policy.

(b) Apply the policy in (a) to the stochastic model in which service times are exponentially distributed, and the arrival stream is Poisson. Construct a Markov process for the controlled system, and an associated rate matrix, with state space $\mathsf{X}_\diamond = \mathbb{Z}_+ \times \mathbb{Z}_+$.

(c) *Estimate* the set of $\{\alpha, \mu_1, \mu_2\}$ for which the process constructed in (b) is positive recurrent. *Hint: Estimate the busy time for the second queue.*

8.8 Find a network and a symmetric matrix D satisfying $D_{ij} > 0$ for each i, j, such that the V_D-myopic policy is not stabilizing. *Hint*: See Example 4.4.5.

8.9 Consider the tandem queues in Case 2, where $\mu_1 \geq \mu_2$. Compute the h-MaxWeight policy for $h(x) = T^*(\tilde{x})$ with $T^*(x) = (\mu_2 - \alpha_1)^{-1}(x_1 + x_2)$ and $\tilde{x}_i := x_i + \theta_i(e^{-x_i/\theta_i} - 1)$ for $x \in \mathbb{R}_+^2$ and $\theta \in \mathbb{R}_+^2$. Does the policy satisfy Foster's criterion?

8.10 Consider the h-MaxWeight policy (8.56) with

$$h(x) = \log\left(\sum_{i=1}^\ell e^{c_i x_i} + e^{-c_i x_i}\right), \qquad x \in \mathbb{R}_+^\ell, \qquad \text{(E8.1)}$$

with $c > 0$. Stability is established following the steps used in the proof of Proposition 8.4.4.

(a) Show that the function $h \colon \mathbb{R}_+^\ell \to \mathbb{R}_+$ is convex and monotone.

(b) The gradient ∇h is bounded over $\mathbb{R}_+^\ell$, so that h is Lipschitz continuous.

(c) The derivative condition (8.59) holds.

(d) For any $x \in \mathbb{R}_+^\ell$,

$$\lim_{r \to \infty} r^{-1} h(rx) = c(x) := \max(c_1 x_1, \ldots, c_\ell x_\ell).$$

Based on this structure, prove the following theorem:

Theorem 8.8.1. *Suppose that assumptions (a)–(c) of Theorem 8.0.4 hold. Consider the h-MaxWeight policy with $h \colon \mathbb{R}_+^\ell \to \mathbb{R}_+$ defined in (E8.1). Then:*

(a) *The minimization (4.92) admits a solution satisfying $\phi^{\text{MW}}(x) \in \mathsf{U}_\diamond(x)$ for each $x \in \mathsf{X}_\diamond$.*

(b) *The controlled network satisfies Foster's criterion, $PV \leq V - \varepsilon_0 + b\mathbf{1}_S$, where $V = 1 + h$, $\varepsilon_0 > 0$, $b < \infty$, and S is a finite set.*

(c) *Condition (V3) holds,*

$$PV \leq V - f + b\mathbf{1}_S,$$

with $V = 1 + \frac{1}{2}h^2$, $f = 1 + \frac{1}{2}h$, $b < \infty$, and S is a finite set.

8.11 *Phase-type arrivals.* Consider the simple queue in which A is not i.i.d., but is independent of the past given the current state of a certain *phase process* denoted I. The phase process is an irreducible Markov chain on a finite set $\{1, \dots, \ell_I\}$, and for transition functions p_A, P_I, given $I(t) = i$,

$$\mathsf{P}\{A(t+1) = m,\ I(t+1) = j \mid A(r), S(r), I(r) : r \leq t\} = p_A(i, m) P_I(i, j),$$

$$\mathsf{E}\{(A(t+1))^2 \mid A(r), S(r), I(r) : r \leq t\} = \sum_m p_A(i, m) m^2 < \infty,$$

for all $t \geq 0$, $i, j \in \{1, \dots, \ell_I\}$ and $m \in \mathbb{Z}_+$. Assume moreover that $S(t)A(t) = 0$ a.s., and that S is Bernoulli. This model is a special case of the queue introduced in Exercise 4.10.

(a) Verify that under the nonidling policy the bivariate process (Q, I) is a Markov chain.

(b) Compute the conditional expectation $\mathsf{E}[Q(t+1) - Q(t) \mid Q(t) = x, I(t) = i]$ for each i and x, and conclude that if the steady-state queue length is finite then in steady state $\mathsf{P}\{Q(t) = 0\} = 1 - \rho$ where $\rho = \alpha/\mu = (\sum \varpi_i \alpha_i)/\mu$, with ϖ the invariant measure for I and $\alpha_i = \mathsf{E}[A(t+1) \mid I(t) = i]$.

(c) Consider the function of $x \in \mathbb{Z}_+$ and $i \in \{1, \dots, \ell_I\}$ given by

$$h(x, i) = \frac{1}{2} \frac{x^2 + b_i x}{\mu - \alpha}$$

where $\{b_i\}$ are constants. Choose these constants so that h solves the following version of the Poisson inequality,

$$\mathsf{E}[h(Q(t+1), I(t+1)) - h(Q(t), I(t)) \mid Q(t) = x, I(t) = i] = -x + b(x),$$

where b is a bounded function of x. This is only possible if $\rho < 1$. *Note*: Your solution will involve a version of the Poisson equation for I.

In Exercise 10.3 an alternative route to establishing stability is described, based on the fluid limit model.

8.12 Extend the previous exercise to the MaxWeight policy for the CRW scheduling model. Assume that there exists a Markov chain I such that A is conditionally independent of the past with finite second moment: As above, given $I(t) = i$,

$$P\{A(t+1) = a,\ I(t+1) = j \mid A(r), S(r), I(r) : r \le t\} = p_A(i,a)P_I(i,j),$$
$$E\{\|A(t+1)\|^2 \mid A(r), S(r), I(r) : r \le t\} = \sum_a p_A(i,a)\|a\|^2 < \infty,$$

with $i, j \in \{1, \ldots, \ell_I\}, a \in \mathbb{Z}_+^\ell$. Of course, you must also assume that the natural load conditions are satisfied. Show that the MaxWeight policy is stabilizing by constructing a Lyapunov function (solution to (V3)) of the form

$$V(x,i) = b_0 + \frac{b}{2}x^\mathsf{T}Dx + b_i^\mathsf{T}x, \qquad x \in \mathbb{Z}_+^\ell,\ i \in \{1, \ldots, \ell_I\},$$

for suitable $b, b_0 > 0$ and $\{b_i : i \ge 1\} \subset \mathbb{R}^\ell$ (the constant b_0 is only used to ensure positivity of V).

8.13 Consider the Dai–Wang model shown in Fig. 4.15 with the following parameters:

$$\mu_1 = 10,\ \mu_2 = 20,\ \mu_3 = 10/9,\ \mu_4 = 20,\ \mu_5 = 5/4. \tag{E8.2}$$

Verify that the network is balanced, with $\rho_1 = \rho_2 = \alpha_1(19/20)$. Numerically solve the Drift LP for a range of values in $[0, 20/19)$. What is the maximum value for which you can assert that the system is stable under any nonidling policy? Repeat your experiment for a few priority policies; then simulate your favorite using a CRW model.

9

Optimization

Optimization concepts have appeared in nearly every chapter of this book. In Section 4.1.3 we saw how the CRW model can be cast as a particular Markov decision process model for the purposes of policy synthesis, and Section 4.5 contains several examples illustrating the structure of optimal policies. The Bellman–Ford algorithm was realized as a particular version of a dynamic programming equation in Section 6.3. The optimality equations developed in Section 8.5 were the basis of the Stability and Performance LPs described in Section 8.6.

In this chapter we consider a general MDP model:

Definition 9.0.1 (Markov Decision Process). A *Markov Decision Process* or MDP consists of a state space $\mathsf{X}_\diamond$, an *action space* $\mathsf{U}_\diamond$, and the *controlled transition matrix*

$$P_u(x,y) := \mathsf{P}\{Q(t+1) = y \mid Q(t) = x, U(t) = u\}, \qquad x,y \in \mathsf{X}_\diamond, \ u \in \mathsf{U}_\diamond. \quad (9.1)$$

We adopt the same definitions and conventions introduced in Section 4.1.2 for the scheduling model:

 (i) For each $x \in \mathsf{X}_\diamond$ there is a set $\mathsf{U}_\diamond(x) \subseteq \mathsf{U}_\diamond$ whose elements are admissible actions when the state process $X(t)$ takes the value x.
 (ii) A (deterministic) *policy* ϕ is a sequence of functions $\{\phi^t : t \in \mathbb{Z}_+\}$, from $\mathsf{X}_\diamond^{t+1}$ to $\mathsf{U}_\diamond$ such that the allocation sequence defined by

$$U(t) = \phi^t(X(0), \ldots, X(t-1), X(t))$$

 satisfies $U(t) \in \mathsf{U}_\diamond(X(t))$ for each t.
 (iii) The Markov property for the controlled process is expressed as follows: For each $x^0, x^1 \in \mathsf{X}_\diamond$, $u \in \mathsf{U}_\diamond$, $t \geq 0$,

$$\mathsf{P}\{X(t+1) = x^1 \mid (X(0),U(0)), \ldots, (X(t), U(t)) ; U(t) = u, X(t) = x^0\}$$

$$= \mathsf{P}\{X(1) = x^1 \mid U(0) = u, X(0) = x^0\}$$

$$= P_u(x^0, x^1).$$

$$(9.2)$$

(iv) A randomized stationary policy is a mapping $\phi\colon \mathsf{X}_\diamond \to \mathcal{P}(\mathsf{U}_\diamond)$, where $\mathcal{P}(\mathsf{U}_\diamond)$ denotes the set of probability measures on $\mathsf{U}_\diamond$. For each $x \in \mathsf{X}_\diamond$, $\phi(x) = \{\phi_u(x) : u \in \mathsf{U}_\diamond\}$ satisfies $\phi_u(x) = 0$ for $u \notin \mathsf{U}_\diamond(x)$.

(v) If $\phi\colon \mathsf{X}_\diamond \to \mathsf{U}_\diamond$ defines a stationary policy (deterministic or randomized), then we let P_ϕ denote the resulting transition matrix,

$$P_\phi(x, y) = P_{\phi(x)}(x, y), \qquad x, y \in \mathsf{X}_\diamond.$$

(vi) Given a cost function $c\colon \mathbb{R}_+^\ell \to \mathbb{R}_+$, a stationary policy is called *regular* if the controlled process is an f-regular Markov chain with $f = 1 + c$ (recall Definition 8.0.2). ∎

The general results developed in this chapter are framed in this general MDP setting so that we are not forced to assume any special form for the state or control processes.

Until Section 9.7 we assume that the state space is discrete; typically the state space $\mathsf{X}_\diamond$ is taken countably infinite. When analyzing algorithms to construct an optimal policy this is not a very practical setting. However, the main goal of this chapter is not computation. We will explore the structure of optimality equations and value functions to expand the theory developed in previous chapters. This structure is most apparent when the state space is allowed to be unbounded.

When $\mathsf{X}_\diamond$ is discrete we always assume that $\mathsf{U}_\diamond$ is a finite set.

Throughout this chapter it is assumed that a cost function is given that is nonnegative and coercive. Based on this cost function we consider several different optimization criteria. The average-cost optimality equation defined in (4.21a) for the scheduling model and the associated myopic policy (4.21b) remain unchanged:

Definition 9.0.2 (Average-cost optimality). The average cost η_x^ϕ is defined in (8.3), and the infimum over all policies is denoted

$$\eta_x^* := \inf \eta_x^\phi \tag{9.3}$$

The policy ϕ^* is called (average-cost) *optimal* if it achieves (9.3) for each x.

The *average-cost optimality equation* (ACOE) is defined by (9.4a) where $h^*\colon \mathsf{X}_\diamond \to \mathbb{R}$ is a function and η^* is a constant. Equation (9.4b) defines the h^*-myopic policy:

$$\eta^* + h^*(x) = \min_{u \in \mathsf{U}_\diamond(x)} [c(x) + P_u h^*(x)] \tag{9.4a}$$

$$\phi^*(x) \in \operatorname*{arg\,min}_{u \in \mathsf{U}_\diamond(x)} P_u h^*(x), \qquad x \in \mathsf{X}_\diamond. \tag{9.4b}$$

The function h^* is called the *relative value function*. ∎

For example, for the single-server queue with $c(x) = x$ we have seen in Example 4.5.1 that the nonidling policy is optimal. A solution to the ACOE is given in Theorem 3.0.1,

$$\eta^* = \frac{1}{2}\frac{\sigma^2}{\mu - \alpha}, \qquad h^*(x) = J^*(x) + \frac{1}{2}\mu^{-1}\left(\frac{m^2 - m_A^2}{\mu - \alpha}\right)x, \qquad x \in \mathbb{Z}_+.$$

We might also impose a cost on service so that c is a function of two variables, say

$$c(x, u) = x + ru, \qquad x \in \mathsf{X}_\circ, \ u \in \mathsf{U}_\circ.$$

We can construct an MDP model using $X(t) = (Q(t), U(t-1))^{\mathsf{T}}$, $t \geq 0$, where $X(0) = (Q(0), U(-1))$ is given as an initial condition.

The ACOE is interpreted as a fixed point equation in the two unknowns (h^*, η^*). Under general conditions η^* coincides with the optimal average cost η_x^* for each x. Under these assumptions, the stationary policy ϕ^* defined in (9.4b) is optimal.

Theorem 9.0.3 provides one set of sufficient conditions. If the state space is finite then assumption (c) is automatic.

Theorem 9.0.3. *Suppose that the following conditions hold:*

(a) *The pair (h^*, η^*) solves the optimality equation (9.4a).*
(b) *The stationary policy ϕ^* is h^*-myopic. That is, it satisfies (9.4b).*
(c) *For any $x \in \mathsf{X}_\circ$, and any policy ϕ satisfying $\eta_x^\phi < \infty$,*

$$\frac{1}{n} \mathsf{E}_x^\phi [h^*(X(n))] \to 0, \qquad n \to \infty. \tag{9.5}$$

Then the stationary policy with feedback law ϕ^* is optimal with average cost η^*.

Proof. For any policy the optimality equation (9.4a) implies the bound $\mathsf{E}[h^*(X(n)) \mid \mathcal{F}_{n-1}] \geq \eta^* + h^*(X(n-1)) - c(X(n-1))$ with $\mathcal{F}_t = \sigma\{X(k) : k \leq t\}$, and hence on iterating this bound,

$$\mathsf{E}[h^*(X(n)) \mid \mathcal{F}_{n-2}] \geq \mathsf{E}[\eta^* + h^*(X(n-1)) - c(X(n-1)) \mid \mathcal{F}_{n-2}]$$
$$\geq h^*(X(n-2)) - \big(c(X(n-2)) - \eta^*\big)$$
$$- \mathsf{E}[\big(c(X(n-1)) - \eta^*\big) \mid \mathcal{F}_{n-2}].$$

Repeating this step n times gives

$$\mathsf{E}[h^*(X(n))] \geq h^*(x) - \mathsf{E}\left[\sum_{t=0}^{n-1}\big(c(X(t-1)) - \eta^*\big)\right], \qquad X(0) = x, \ n \geq 1.$$

Dividing by n and letting $n \to \infty$ shows that $\eta_x^\phi \geq \eta^*$.

When ϕ is defined using the stationary policy ϕ^* we then have equality $\mathsf{E}[h^*(X(n)) \mid \mathcal{F}_{n-1}] = \eta^* + h^*(X(n-1)) - c(X(n-1))$. Repeating the same arguments shows that $\eta_x^{\phi^*} = \eta^*$. $\qquad\square$

A partial converse is obtained under an irreducibility condition.

Theorem 9.0.4. *Suppose that the following conditions hold:*

(a) *The cost function $c \colon \mathsf{X}_\circ \to \mathbb{R}_+$ is coercive.*
(b) *For each $x^1 \in \mathsf{X}_\circ$ there exists a regular policy that is also x^1-irreducible.*

Then there exists a solution (h^*, η^*) to the optimality equation (9.4a), the h^*-myopic policy defined in (9.4b) is regular with average cost η^*, and for any other policy and initial condition we have

$$\liminf_{n \to \infty} \frac{1}{n} \sum_{t=0}^{n-1} \mathsf{E}\big[c(X(t))\big] \geq \eta^* \tag{9.6a}$$

and

$$\liminf_{n \to \infty} \frac{1}{n} \sum_{t=0}^{n-1} c(X(t)) \geq \eta^*, \quad a.s. \tag{9.6b}$$

Hence ϕ^* is average-cost optimal.

Proof. Proposition 9.1.4 implies that η^* is independent of x, and Theorem 9.2.1 shows that there exists a stationary policy with invariant measure π^* satisfying $\pi^*(c) = \eta^*$. The sample-path optimality (9.6b) is established in Lemma 9.2.8. Sample path arguments form an important part of the convex-analytic techniques introduced in Section 9.2.2.

The above conclusions are summarized in Proposition 9.2.10 at the close of Section 9.2.

The proof is completed in Section 9.4, where an explicit construction of the relative value function is contained in Section 9.4.3,

$$h^*(x) = \inf \mathsf{E}_x \left[\sum_{t=0}^{\tau_{x^*}-1} \big(c(X(t)) - \eta^* \big) \right], \quad x \in \mathsf{X}_\diamond,$$

where x^* is a particular state in $\mathsf{X}_\diamond$, and the infimum is over all policies. It is shown that h^* is bounded from below, and it follows that the h^*-myopic policy is both regular and optimal. $\square$

Using similar techniques we can conclude that an optimal policy exists for the general CRW model (8.4). It is not known whether assumption (ii) of Theorem 9.0.4 holds for the general model (8.4). We can adopt a weaker assumption in the CRW model: We suppose that there exists a policy ϕ that is irreducible in the usual sense,

$$\sum_{t=0}^{\infty} P_\phi^t(x, y) > 0, \quad x, y \in \mathsf{X}_\diamond. \tag{9.7}$$

Theorem 9.0.5. *Consider the CRW model (8.4) satisfying the assumptions of Theorem 8.0.4. Suppose moreover that there exists a stationary policy satisfying (9.7), and that the cost function c can be extended to define a norm on $\mathbb{R}^\ell$. Then,*

(i) *The optimal average cost $\eta_x^* = \eta^*$ is independent of x.*
(ii) *There exists a solution (h^*, η^*) to the dynamic programming equation (9.4a) satisfying $h^* \geq 0$ and $h^* \in L_\infty^{V_2}$, with $V_2(x) = 1 + \|x\|^2$, $x \in \mathbb{R}^\ell$.*
(iii) *The h^*-myopic policy is regular and average-cost optimal.*

Proof. The irreducibility assumption (9.7) is introduced to strengthen the conclusion of Theorem 8.0.4. It is shown in Proposition 9.6.1 that there exists a policy obtained as a perturbation of MaxWeight that is regular and satisfies (9.7). Hence the assumptions of Theorem 9.0.4 hold, which establishes (i) and (iii).

Part (ii) is given in Proposition 9.6.3.   □

In most applications it is not realistic to characterize system performance using a single objective function.[1] Rather, the MDP framework is primarily a tool for policy construction and analysis. Fortunately, multiobjective optimization also has an elegant formulation.

Consider for example a network with two different classes of customers. The goal is to minimize the cost (or maximize the offered performance) for the first class of customers, subject to a strict constraint on the cost for the second class. Given two cost functions c and c_a for the respective customer classes, and a prescribed bound $\overline{\eta}_a > 0$, an average-cost formulation of this optimization problem is expressed

$$\mathbf{min} \ \ \mathsf{E}[c(Q(t))], \qquad \mathbf{s.t.} \ \ \mathsf{E}[c_a(Q(t))] \leq \overline{\eta}_a,$$

where each expectation is taken with respect to the stationary version of the process, and the minimum is over all policies such that a stationary version exists.

Solution techniques for multiobjective optimization are described in Section 9.3 based on linear programming formulations of optimal control surveyed in Section 9.2.

In the final sections of this chapter we apply these techniques to stochastic networks. Theory surrounding the value iteration algorithm is applied in Section 9.6 to obtain structural results for value functions and policies in the CRW model.

Throughout the book we have stressed that the CRW network model is far too detailed to be useful in optimization except in the simplest examples. Fortunately, a lower dimensional CRW workload model may be entirely tractable. For workload models of moderate dimension an optimal policy can be constructed using the methods developed in Section 9.5.

In Section 9.7 we return to the simple inventory model. Recall that a hedging point was constructed in Section 7.4 such that the resulting policy is optimal *over all hedging-point policies*. Based on the general results in this chapter we can now establish optimality of this policy. This is summarized in Theorem 9.7.1.

In Section 9.8 we reconsider the affine approximations surveyed in Sections 5.6 and 7.5.2. Based on Theorem 9.7.1 we can prove that an optimal switching curve for a two-dimensional workload model is approximated by an affine policy. The approximating policy is defined by the optimal height process introduced in Section 7.5.2.

9.1 Reachability and decomposibility

We begin with an investigation of the "reachability" assumption in Theorem 9.0.4. The basic results obtained here will be combined with convex-analytic techniques developed in Section 9.2 to show that an optimal policy exists.

---

[1] *There is more to life than increasing its speed.* –Gandhi

We begin with the following corollary to the Doeblin Decomposition Theorem 8.1.3 that relates η_x^ϕ to the mean of c with respect to an invariant measure.

Proposition 9.1.1. *For any stationary policy ϕ and any x we have*

$$\eta_x \geq \min \pi(c),$$

where the minimum is over all invariant measures for P_ϕ.

Proof. Since c is coercive, the Doeblin Decomposition combined with the Strong Law of Large Numbers Theorem A.5.8 gives

$$\lim_{n\to\infty} \frac{1}{n} \sum_{t=0}^{n-1} c(X(t)) = \sum_{i\in\mathcal{I}} \pi_i(c)\mathbf{1}\{\boldsymbol{X} \text{ enters } \mathsf{X}_i\} + \infty\mathbf{1}\{\boldsymbol{X} \text{ does not enter any } \mathsf{X}_i\},$$

where $\pi_i(c) := \infty$ if the chain restricted to X_i is not positive recurrent.

On applying Fatou's Lemma we obtain

$$\eta_x^\phi \geq \liminf_{n\to\infty} \mathsf{E}_x\left[\frac{1}{n}\sum_{t=0}^{n-1} c(X(t))\right]$$

$$\geq \mathsf{E}_x\left[\liminf_{n\to\infty} \frac{1}{n}\sum_{t=0}^{n-1} c(X(t))\right] \geq \min_i \pi_i(c).$$

$\square$

We now explain why η_x^* can be taken independent of x. For two states $x^1, x^0 \in \mathsf{X}_\diamond$ we write $x^0 \rightsquigarrow x^1$ if x^1 is reachable from x^0 in the sense that there exists a policy ϕ such that

$$\mathsf{E}_{x^0}^\phi\left[\sum_{t=0}^{\tau_{x^1}-1} c(X(t))\right] < \infty. \tag{9.8}$$

We write $x^0 \overset{\phi}{\rightsquigarrow} x^1$ when we wish to stress the particular policy used.

Lemma 9.1.2 (Transitivity in MDPs). *If $x^0 \overset{\phi^0}{\rightsquigarrow} x^1$ and $x^1 \overset{\phi^1}{\rightsquigarrow} x^2$, then $x^0 \rightsquigarrow x^2$.*

Proof. Define a policy ϕ^2 that coincides with ϕ^0 before $X(t)$ reaches x^1,

$$U(t) = \phi^0(X(0), \dots, X(t)), \qquad 0 \leq t \leq \tau_{x^1} - 1.$$

Thereafter it is defined using ϕ^1,

$$U(t) = \phi^1(X(\tau_{x^1}), \dots, X(t)), \qquad t \geq \tau_{x^1}.$$

We obtain from the definitions

$$\mathsf{E}_{x^0}^{\phi^2}\left[\sum_{t=0}^{\tau_{x^2}-1} c(X(t))\right] = \mathsf{E}_{x^0}^{\phi^0}\left[\sum_{t=0}^{\tau_{x^1}-1} c(X(t))\right] + \mathsf{E}_{x^1}^{\phi^1}\left[\sum_{t=\tau_{x^2}}^{\tau_{x^1}} c(X(t))\right] < \infty.$$

$\square$

Lemma 9.1.3. *Suppose that $x^1, x^0 \in \mathsf{X}_\diamond$ and that $x^0 \rightsquigarrow x^1$. Then $\eta_{x^0}^* \leq \eta_{x^1}^*$.*

Proof. The proof is similar to the proof of Lemma 9.1.2. Let ϕ be an arbitrary policy, and let $\eta_{x^1}^\phi$ denote the resulting average cost starting from x^1. Define a new policy ϕ^1 as follows: The sequence U is defined by the policy ϕ^0 satisfying (9.8) before reaching x^1, and then is defined using ϕ,

$$U(t) = \phi(X(\tau_{x^1}), \dots, X(t)), \qquad t \geq \tau_{x^1}.$$

Since c is nonnegative valued we obtain the bound, for any $n \geq 1$,

$$\sum_{t=0}^{n-1} \mathsf{E}_{x^0}^{\phi^1}[c(X(t))] \leq \mathsf{E}_{x^0}^{\phi^1}\left[\sum_{t=0}^{\tau_{x^1}-1} c(X(t))\right] + \mathsf{E}_{x^0}^{\phi^1}\left[\sum_{t=\tau_{x^1}}^{\tau_{x^1}+n-1} c(X(t))\right]$$

$$= \mathsf{E}_{x^0}^{\phi^0}\left[\sum_{t=0}^{\tau_{x^1}-1} c(X(t))\right] + \mathsf{E}_{x^1}^{\phi}\left[\sum_{t=0}^{n-1} c(X(t))\right].$$

Dividing each side by n and letting $n \to \infty$ gives

$$\eta_{x^0}^* \leq \eta_{x^0}^{\phi^1} \leq \eta_{x^1}^\phi.$$

Minimizing over ϕ we conclude that $\eta_{x^0}^* \leq \eta_{x^1}^*$. $\qquad\square$

The following corollary proves a small part of Theorem 9.0.4.

Proposition 9.1.4. *Under the assumptions of Theorem 9.0.4 the optimal average cost η_x^* does not depend upon x.*

Proof. Under the conditions of Theorem 9.0.4, we can find, for any state x^1, a stationary policy ϕ^0 such that the controlled chain is f-regular and x^1-irreducible, where $f = 1 + c$. Proposition A.4.5 implies that $x^0 \rightsquigarrow x^1$ for each x^0, x^1. The result then follows from Lemma 9.1.3. $\qquad\square$

In the following section we establish that an optimal policy does exist, and that it can be taken to be stationary.

9.2 Linear programming formulations

Here we recast the average-cost optimal control problem within the framework of linear programming. This point of view has many advantages:

(i) The LP formulation leads to new intuition, and sometimes simpler derivations of theoretic results.

(ii) This setting suggests new algorithms based on linear programming, or more general convex-analytic techniques.

(iii) Throughout the book we have stressed that in typical applications, system performance cannot be characterized by a single performance objective. The LP formulation is the most elegant setting for multiobjective control problems.

(iv) The LP approach suggests other extensions, such as approximate dynamic programming based on a parameterized class of candidate value functions. Some extensions will be developed in Chapter 11.

Randomization is required to create a linear program that captures the ACOE. Throughout this section we restrict to randomized stationary policies denoted ϕ, with transition matrix P_ϕ, and invariant measure denoted π_ϕ (when it exists).

An elegant setting is obtained by expanding the decision space to capture both the policy and its associated invariant measure. A similar idea was used in the construction of the Performance LP, where the variables $\{\overline{\Gamma}_{ij} = \mathsf{E}_\pi[U_i(t)Q_j(t) : 1 \leq i, j \leq \ell]\}$ were introduced, even though the optimization problem of interest only depended on the steady-state queue lengths $\{\mathsf{E}_\pi[Q_j(t)]; 1 \leq j \leq \ell\}$.

The set of *occupation measures* $\mathcal{G}$ is defined as the collection of all probability measures on $\mathsf{X}_\diamond \times \mathsf{U}_\diamond$ of the form

$$\Gamma(x, u) = \pi_\phi(x)\phi_u(x), \qquad x \in \mathsf{X}_\diamond, \ u \in \mathsf{U}_\diamond. \tag{9.9}$$

We have the interpretation

$$\langle \Gamma, c \rangle := \sum_{x,u} \Gamma(x, u)c(x) = \mathsf{E}_{\pi_\phi}[c(X)].$$

We show in Lemma 9.2.4 that the set $\mathcal{G}$ is convex, and we thereby obtain a linear program over the set of occupation measures that captures the *value* of the average-cost optimal control problem.

Theorem 9.2.1 (Linear program for optimal average cost). *Under the assumptions of Theorem 9.0.4 the optimal average cost η^* is independent of x and can be expressed as the value of the linear program*

$$\mathrm{min} \quad \langle \Gamma, c \rangle \qquad \text{s.t.} \quad \Gamma \in \mathcal{G}. \tag{9.10}$$

Proof. Proposition 9.1.4 states that η_x^* is independent of x. Proposition 9.2.10 implies the conclusion that η^* is the value of the linear program (9.10). $\qquad \square$

Theorem 9.2.1 is a surprisingly strong result given the mild assumptions. In fact, it is only the coerciveness condition that is critical to establish the existence of a minimizer in (9.10). *Throughout this section the reachability assumption (ii) of Theorem 9.0.4 is relaxed.* In Proposition 9.2.10 we obtain the conclusions of Theorem 9.2.1 with η^* redefined as the minimum of η_x^* over all policies.

Theorem 9.2.1 and Proposition 9.2.10 also contain a major weakness. Because we are optimizing over occupation probabilities, we can say nothing about the MDP for initial conditions lying outside the support of π^*. To understand the optimization problem globally we must combine the results here with those of Section 9.1 to establish the existence of a regular optimal policy. This is done in Section 9.4.3.

To begin to understand the LP formulation we begin with the uncontrolled case.

9.2.1 Linear programming and the Comparison Theorem

The Poisson inequality and the Comparison Theorem A.4.3 lead to a linear program to characterize the steady-state cost. To emphasize its role as a variable, we denote $z = \bar{\eta}$, so that the Poisson inequality (8.12) becomes

$$Ph\,(x) := \sum P(x,y)h(y) \le h(x) - c(x) + z, \qquad x \in \mathsf{X}_\diamond, \qquad (9.11)$$

where $h \colon \mathsf{X}_\diamond \to \mathbb{R}_+$ and $z \ge 0$.

Suppose that the chain is x^*-irreducible and that c is coercive. Then (9.11) implies that the chain is f-regular with $f = 1 + c$, and applying the Comparison Theorem we obtain the bound $\pi(c) \le z$. Proposition 8.1.6 implies that (9.11) has a solution with $z = \eta = \pi(c)$ and with h nonnegative. Hence the steady-state mean η is the solution to a linear program. We summarize this conclusion in Theorem 9.2.2.

Even in the countable state-space case, we view h as a column vector: The linear program characterizing η is expressed in (9.12) in the variable $\binom{z}{h}$.

Theorem 9.2.2. *Suppose that X is an uncontrolled, x^*-irreducible Markov chain, and that $c \colon \mathsf{X}_\diamond \to \mathbb{R}_+$ is a coercive cost function with steady-state mean η. Then the steady-state mean can be expressed as the solution to the linear program*

$$\mathbf{min} \quad \left\langle \begin{pmatrix} 1 \\ 0 \end{pmatrix}, \begin{pmatrix} z \\ h \end{pmatrix} \right\rangle$$

$$\mathbf{s.t.} \quad [I \mid I - P] \begin{pmatrix} z \\ h \end{pmatrix} \ge c \qquad\qquad (9.12)$$

$$z \ge 0, \; h \ge 0.$$

The dual of (9.12) can be expressed

$$\mathbf{max} \quad \nu^\mathsf{T} c$$

$$\mathbf{s.t.} \quad [I \mid I - P]^\mathsf{T} \nu \le \begin{bmatrix} 1 \\ 0 \end{bmatrix} \qquad\qquad (9.13)$$

$$\nu \ge 0.$$

Proof. The representation of the primal (9.12) is merely notational. The feasibility constraint

$$[I \mid I - P] \begin{pmatrix} z \\ h \end{pmatrix} \ge c$$

means that $z + (I - P)h \ge c$, which is precisely the Poisson inequality (9.11). The dual (9.13) is then standard (see (1.9)).   □

It is worthwhile to take a closer look at the dual. The inequality constraint in (9.13) amounts to the pair of constraints

$$\sum_{x \in \mathsf{X}_\diamond} \nu(x) \le 1 \quad \text{and} \quad \nu(y) \le \sum_{x \in \mathsf{X}_\diamond} \nu(x) P(x,y), \qquad y \in \mathsf{X}_\diamond.$$

To maximize we will always take equality in the first constraint. Since we also have $\nu \geq 0$, it follows that ν is a probability measure on $\mathsf{X}_\diamond$, and the second inequality constraint shows that ν is subinvariant. Since ν is finite it must be invariant, so that $\nu = \pi$. In summary,

Proposition 9.2.3. *The invariant measure π is a maximizer of the dual (9.13) when c is coercive.*

To generalize these results to controlled Markov chains it is convenient to express duality through the following *Lagrangian relaxation*. Let $\mathcal{M}_+$ denote the set of all nonnegative functions from $\mathsf{X}_\diamond$ to $\mathbb{R}_+$. We view $\nu \in \mathcal{M}_+$ as a dual variable, and the Lagrangian is defined as a function of all variables,

$$\mathcal{L}(z, h, \nu) = z - \langle \nu, [z\mathbf{1} + (I - P)h - c] \rangle.$$

The dual functional $\Psi^* \colon \mathcal{M}_+ \to \mathbb{R} \cup \{-\infty\}$ is the infimum,

$$\Psi^*(\nu) = \inf_{z \geq 0, \, h \geq 0} \mathcal{L}(z, h, \nu).$$

We have $\mathcal{L}(z, h, \nu) \leq z$ whenever (z, h) is feasible for the primal (9.12). Hence $\Psi^*(\nu) \leq z^* = \eta$ for any $\nu \in \mathcal{M}_+$. The convex dual is expressed as the maximum,

$$\max_{\nu \in \mathcal{M}_+} \Psi^*(\nu). \tag{9.14}$$

The convex program (9.14) can be reduced to the dual (9.13). This is derived in the more complex controlled case in Section 9.2.3.

Next we consider in more detail the linear program (9.10), and in Section 9.2.3 we construct its dual.

9.2.2 Convex-analytic and sample-path approaches

There is much more to be done in the controlled model. Because we plan to establish optimality over *all policies*, not just those that are stationary, the theory in the appendix for time-homogeneous Markov chains can take us only so far.

In this section we introduce powerful sample-path techniques to treat nonstationary policies. This approach is used to verify the sample-path form of optimality expressed in Theorem 9.0.4. In Section 9.4.3 this technique is extended to establish the existence of a solution to the ACOE. The basic ideas are contained in Section A.1, where it is claimed that every process is *approximately* Markov.

We begin with a proof that the constraint set in the primal LP (9.10) is convex.

Lemma 9.2.4. *The set $\mathcal{G}$ of occupation measures defined in (9.9) is convex as a subset of $\mathcal{P}(\mathsf{X}_\diamond \times \mathsf{U}_\diamond)$.*

Proof. The inclusion $\Gamma \in \mathcal{G}$ is equivalent to the linear constraints

$$\sum_{u,x} \Gamma(x, u) = 1, \qquad \sum_{u,x} \Gamma(x, u) P_u(x, y) = \sum_{u} \Gamma(y, u), \; y \in \mathsf{X}_\diamond, \tag{9.15}$$

which implies convexity. □

It follows that the infimization in (9.10) is a linear program. We let η° denote its value:

$$\eta^\circ := \inf_{\Gamma \in \mathcal{G}} \langle \Gamma, c \rangle. \tag{9.16}$$

Throughout this section it is assumed that this value is finite.

An *extreme point* of $\mathcal{G}$ is any element that *cannot* be expressed as a convex combination of distinct elements of $\mathcal{G}$. This definition is illustrated in Fig. 9.1.

We considered extreme points in the finite-dimensional region $\bar{\mathsf{X}}$ in our treatment of the Klimov model, and the resulting geometry shown in Fig. 5.10 supported the conclusion that a priority policy is optimal in this example.

Extreme points for $\mathcal{G}$ have a simple characterization.

Proposition 9.2.5 (Extreme points over occupation measures). *For the MDP with countable state space, an occupation measure Γ is an extreme point if and only if*

(i) *ϕ is not randomized;*
(ii) *π_ϕ is ergodic under ϕ. That is, the chain restricted to the support of π_ϕ is irreducible for P_ϕ.*

Proof. The occupation measure Γ is *not* extremal if and only if it can be expressed as a convex combination of two distinct occupation measures.

Suppose that $\Gamma \in \mathcal{G}$ is not extremal. It follows that there exist $\theta \in (0, 1)$ and distinct $\{\Gamma^i\}$ for which

$$\Gamma = \theta \Gamma^1 + (1 - \theta) \Gamma^0.$$

From the definition (9.9) we can write the feedback law and invariant measure for Γ in terms of those for $\{\Gamma^i\}$ as follows,

$$\pi(x) = \theta \pi^1(x) + (1 - \theta) \pi^0(x)$$
$$\phi_u(x) = \Theta(x) \phi_u^1(x) + (1 - \Theta(x)) \phi_u^0(x)$$

where $\Theta(x) := \theta \pi^1(x) / \pi(x)$. There are precisely two ways in which Γ can be represented in this form:

(i) The feedback law ϕ_u is not deterministic.
(ii) ϕ_u is a deterministic policy, in which case we must have $\phi_u^1(x) = \phi_u^0(x)$ for all x and u. Since $\Gamma^0 \neq \Gamma^1$ it follows that $\pi^0 \neq \pi^1$. In this case each of the $\{\pi^i\}$ is invariant under the common transition law $P_\phi = P_{\phi^0} = P_{\phi^1}$. That is, π is not ergodic.

These two cases cover those declared in the proposition. $\square$

From the basic theory of linear programs, if the state space is finite then there exists an optimizer of (9.10) that is an extreme point in $\mathcal{G}$. In this case, Proposition 9.2.5 implies that an optimal policy exists that is not randomized. The finiteness assumption can be relaxed since c is coercive. It is simplest to prove this fact based on the dynamic programming equations, so we postpone the proof to Section 9.4.

Next we show that the infimum is achieved:

Proposition 9.2.6. *If c is coercive then the infimum defined in (9.16) is achieved by some $\Gamma^\circ \in \mathcal{G}$.*

Proof. Let $\{\Gamma^r\} \subset \mathcal{G}$ satisfy $\langle \Gamma^r, c \rangle \leq \eta^\circ + r^{-1}, r \geq 1$. By choosing a subsequence we can assume that the sequence is pointwise convergent to some $\Gamma^\infty \in \mathcal{G}$. For any constant $m < \infty$,

$$\langle \Gamma^\infty, m \wedge c \rangle = \lim_{r \to \infty} \langle \Gamma^r, m \wedge c \rangle \leq \eta^\circ.$$

Letting $m \uparrow \infty$ we obtain $\langle \Gamma^\infty, c \rangle \leq \eta^\circ$ by the Monotone Convergence Theorem. $\quad\square$

The next step in the proof of Theorem 9.2.1 is to show that $\eta^\circ = \eta^*$. An essential ingredient is to consider the sample-path averages rather than expectations. For each $n \geq 1$ the *empirical measure* on $\mathsf{X}_\diamond \times \mathsf{U}_\diamond$ is a random variable $\widetilde{\Gamma}_n$ taking values in $\mathcal{P}(\mathsf{X}_\diamond \times \mathsf{U}_\diamond)$. It is defined as the average

$$\widetilde{\Gamma}_n(x, u) = \frac{1}{n} \sum_{t=0}^{n-1} \mathbf{1}\{X(t) = x, \, U(t) = u\}, \qquad x \in \mathsf{X}_\diamond, \, u \in \mathsf{U}_\diamond(x). \tag{9.17}$$

Any limiting measure must be an occupation measure. The sample space notation $(\Omega, \mathcal{F}, \mathsf{P})$ is taken from Section 1.3.2.

Lemma 9.2.7. *For any policy, there exists a set of full probability $\Omega_0 \subset \Omega$ such that for each $\omega \in \Omega_0$, and any subsequence $\{n_i\}$ of $\mathbb{Z}_+$ for which $\{\widetilde{\Gamma}_{n_i}\}$ is convergent along this sample to a probability measure Γ, we must have $\Gamma \in \mathcal{G}$.*

Proof. For a given $x^1 \in \mathsf{X}_\diamond$ define $g(x, u) = g(x) = \mathbf{1}\{x = x^1\}$ and $g^1(x, u) = P_u(x, x^1), (x, u) \in \mathsf{X}_\diamond \times \mathsf{U}_\diamond$. Our goal is to show that there exists a full set $\Omega_0 \subset \Omega$ such that for any $\omega \in \Omega_0$ and any pointwise limit Γ for this ω we must have

$$\langle \Gamma, g \rangle = \langle \Gamma, g^1 \rangle. \tag{9.18}$$

This will establish (9.15), proving that $\Gamma \in \mathcal{G}$ whenever $\Gamma(\mathsf{X}_\diamond \times \mathsf{U}_\diamond) = 1$.

To establish (9.18) we begin with the following application of the LLN for martingales Theorem 1.3.9. We have with probability 1

$$\lim_{n \to \infty} \frac{1}{n} \sum_{t=1}^{n} \big(g(X(t)) - \mathsf{E}[g(X(t)) \mid \mathcal{F}_{t-1}]\big) = 0.$$

Since the state space is countable, there exists a single set Ω_0 such that this limit holds for $\omega \in \Omega_0$ and all x^1. We also have

$$\mathsf{E}[g(X(t)) \mid \mathcal{F}_{t-1}] = g^1(X(t-1), U(t-1)),$$

so that on Ω_0,

$$\lim_{n\to\infty}\left(\langle\widetilde{\Gamma}_n,g\rangle-\langle\widetilde{\Gamma}_n,g^1\rangle\right)$$

$$=\lim_{n\to\infty}\frac{1}{n}\sum_{t=0}^{n-1}\left(g(X(t))-g^1(X(t),U(t))\right)$$

$$=\lim_{n\to\infty}\frac{1}{n}\sum_{t=1}^{n}\left(g(X(t))-g^1(X(t-1),U(t-1))\right)=0.$$

$\square$

The lemma implies a uniform sample-path lower bound over all policies:

Lemma 9.2.8. *Suppose that $c\colon\mathsf{X}_\diamond\to\mathbb{R}_+$ is coercive. Then for each policy and initial condition,*

$$\liminf_{n\to\infty}\frac{1}{n}\sum_{t=0}^{n-1}c(X(t))\ge\eta^\circ,\quad a.s.$$

Proof. Take any sample $\omega\in\Omega_0$ such that the limit infimum is finite, where Ω_0 is the set defined in Lemma 9.2.7.

Suppose that $\{n_i\}$ is a subsequence of $\mathbb{Z}_+$ such that $\{\widetilde{\Gamma}_{n_i}\}$ is convergent along this sample,

$$\widetilde{\Gamma}_\infty(x,u):=\lim_{i\to\infty}\widetilde{\Gamma}_{n_i}(x,u),\qquad(x,u)\in\mathsf{X}_\diamond\times\mathsf{U},$$

with $\limsup_{i\to\infty}\widetilde{\Gamma}_{n_i}(c)<\infty$. The limiting measure Γ must be a probability measure since c is coercive.

Pointwise convergence implies that for each $m\ge0$,

$$\liminf_{i\to\infty}\langle\widetilde{\Gamma}_{n_i},c\rangle\ge\lim_{i\to\infty}\langle\widetilde{\Gamma}_{n_i},m\wedge c\rangle=\langle\Gamma,m\wedge c\rangle.$$

Since m is arbitrary the limit infimum is bounded below by $\langle\Gamma,c\rangle$, and Lemma 9.2.7 states that $\langle\Gamma,c\rangle$ is bounded below by η°. $\square$

The previous lemma implies that η° provides a uniform *lower* bound on the average cost. Recall that nowhere have we assumed the reachability condition imposed in Theorem 9.0.4, assumption (ii).

Proposition 9.2.9. *If $c\colon\mathsf{X}_\diamond\to\mathbb{R}_+$ is coercive, then $\eta^\circ=\inf_x\eta_x^*$.*

Proof. Lemma 9.2.8 and Fatou's Lemma give, for any policy,

$$\liminf_{n\to\infty}\mathsf{E}\left[\frac{1}{n}\sum_{t=0}^{n-1}c(X(t))\right]\ge\mathsf{E}\left[\liminf_{n\to\infty}\frac{1}{n}\sum_{t=0}^{n-1}c(X(t))\right]\ge\eta^\circ.$$

Infimizing over all policies we obtain the bound $\eta^\circ\le\eta_x^*$ for any x. This is achieved using the policy ϕ° and any x within the support of π°. $\square$

We now summarize these results. Proposition 9.2.10 establishes many of the conclusions of Theorem 9.0.4.

Proposition 9.2.10. *Suppose that $c\colon X_\diamond \to \mathbb{R}_+$ is coercive. Let $\eta^* = \inf_x \eta_x^*$, and assume this is finite. Then there exists a stationary policy ϕ^* with invariant measure π^* satisfying $\pi^*(c) = \eta^*$. Moreover, for any other policy ϕ and every initial condition the lower bounds (9.6a), (9.6b) hold.*

Proof. Proposition 9.2.9 implies that η^* coincides with η°. The optimizer $\Gamma^\circ \in \mathcal{G}$ constructed in Proposition 9.2.6 defines the optimal policy ϕ^* and the invariant measure π^*.

Proposition 9.2.9 and Lemma 9.2.8 imply (9.6a), (9.6b). □

9.2.3 Duality

To construct the dual for (9.10) we introduce a Lagrangian relaxation based on a pair of dual variables $z \in \mathbb{R}$, $h\colon X_\diamond \to \mathbb{R}$,

$$\mathcal{L}(\Gamma, z, h) = \langle \Gamma, c \rangle + \left\{ z\left(1 - \sum_{u,x} \Gamma(x,u) \right) \right.$$

$$\left. + \sum_{u,y} h(y)\left(-\Gamma(y,u) + \sum_x \Gamma(x,u)P_u(x,y) \right) \right\}.$$

We express this more compactly as

$$\mathcal{L}(\Gamma, z, h) = z + \langle \Gamma, c - z\mathbf{1} + h^1 - h^0 \rangle, \tag{9.19}$$

where

$$h^0(x,u) := h(x), \quad h^1(x,u) := \sum_{y \in X_\diamond} P_u(x,y)h(y), \qquad x \in X_\diamond,\ u \in U_\diamond. \tag{9.20}$$

The dual functional is the infimum

$$\Psi^*(z,h) = \inf_{\Gamma \geq 0} \mathcal{L}(\Gamma, z, h).$$

The LP dual of (9.10) is obtained upon maximizing the dual functional,

$$\max_{z,h} \Psi^*(z,h). \tag{9.21}$$

This is similar to the dual (9.14) constructed in Section 9.2.1. Note however that in the uncontrolled case our starting point was the Comparison Theorem, and the dual LP was the invariance equation $\pi = \pi P$. Here our starting point is the generalized invariance equation (9.15) and the desire to minimize $\langle \Gamma, c \rangle$ subject to this constraint.

To compute the functional Ψ^* we note that the infimum is frequently $-\infty$. Suppose that there is one $(x^0, u^0) \in X_\diamond \times U_\diamond$ satisfying $-\mathcal{L}_0 := c(x^0) - z + h^1(x^0, u^0) - h^0(x^0, u^0) < 0$. On setting $\Gamma = n\delta_{x^0, u^0}$ with $n \geq 0$ we obtain $\mathcal{L}(\Gamma, z, h) = -n\mathcal{L}_0$. From the definition of the dual functional this implies that $\Psi^*(z,h) = -\infty$. Conversely, the dual is finite valued when this term is nonnegative, and in this case the optimizing Γ is zero. This gives

$$\Psi^*(z,h) = \begin{cases} z & \text{if } c - z + h^1 - h^0 \geq 0 \text{ for all } (x,u) \\ -\infty & \text{otherwise.} \end{cases} \tag{9.22}$$

Hence (9.21) is equivalently expressed as the maximum of z subject to $c - z + h^1 - h^0 \geq 0$. Substituting the definitions of $\{h^0, h^1\}$ the dual becomes

max z

s.t. $c(x) - z + \sum_{y \in \mathsf{X}_\diamond} P_u(x, y) h(y) - h(x) \geq 0,$ $x \in \mathsf{X}_\diamond,\ u \in \mathsf{U}_\diamond.$ (9.23)

Proposition 9.2.11 states that the dual is essentially the ACOE under the assumptions of Theorem 9.0.4. The existence of a nonnegative optimizer h^* is established in Section 9.4.3.

Proposition 9.2.11 (ACOE and duality). *Suppose that the assumptions of Theorem 9.0.4 hold. Let Γ^* denote an optimizer of the primal (9.10), and (z^*, h^*) an optimizer of the dual LP (9.23). Assume moreover that h^* takes values in $\mathbb{R}_+$. We then have $z^* = \eta^* = \pi^*(c)$ where $\pi^*(x) = \sum_{u \in \mathsf{U}_\diamond(x)} \Gamma(x, u), x \in \mathsf{X}_\diamond$. The function $h^*: \mathsf{X}_\diamond \to \mathbb{R}$ satisfies*

$$\eta^* + h^*(x) \leq \min_{u \in \mathsf{U}_\diamond(x)} [c(x) + P_u h^*(x)], \qquad x \in \mathsf{X}_\diamond,$$

with equality whenever $\pi^*(x) := \sum_u \Gamma^*(x, u) > 0$.

These results imply that the simplex method or other LP techniques can be applied to solve the ACOE when the state space is finite. Moreover, this approach generalizes naturally to the multiobjective control problems considered next.

9.3 Multiobjective optimization

It is frequently desirable to construct a policy that minimizes one cost criterion, subject to inequality constraints on other criteria. In this section we extend the linear programming formulation to characterize solutions to such multiobjective optimization problems.

To avoid technicalities we restrict to a finite state space throughout this section.

The cost function to be minimized is again denoted $c: \mathsf{X}_\diamond \to \mathbb{R}_+$. In addition, it is assumed we are given ℓ_a "auxilliary cost functions," denoted $\{c^{ai} : 1 \leq i \leq \ell_a\}$, along with upper bounds $\{\overline{\eta}_i\}$. Our aim is to minimize the average cost subject to the constraints

$$\langle \Gamma, c^{ai} \rangle \leq \overline{\eta}_i, \qquad 1 \leq i \leq \ell_a.$$

On setting $\mathcal{H} = \{\Gamma \in \mathcal{P}(\mathsf{X}_\diamond \times \mathsf{U}_\diamond) : \langle \Gamma, c^{ai} \rangle \leq \overline{\eta}_i, 1 \leq i \leq \ell_a\}$, the resulting optimal control problem is expressed as the LP

min $\langle \Gamma, c \rangle$ **s.t.** $\Gamma \in \mathcal{G} \cap \mathcal{H}.$ (9.24)

We will see that an optimizer is described as a stationary policy that is typically *not* deterministic. This departure from the conclusions in the previous section has a geometric interpretation. First, consider the representation of an optimizer for the single-objective problem shown in Fig. 9.1. The extreme points in the polyhedral region $\mathcal{G}$

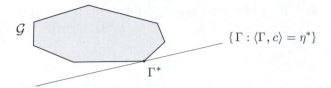

Figure 9.1. Extreme point Γ^* optimizing the linear program (9.10).

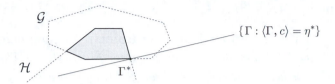

Figure 9.2. Extreme point Γ^* optimizing the linear program (9.24). The corners of the polyhedron $\mathcal{G}$ correspond to deterministic policies and ergodic invariant measures. The extreme points of $\mathcal{G} \cap \mathcal{H}$ allow randomization.

shown in the figure correspond to nonrandomized policies with ergodic invariant measures, as demonstrated in Proposition 9.2.5. Fig. 9.2 shows the extreme points for the optimization problem considered here. The occupation measure Γ^* may be randomized since it is not an extreme point of $\mathcal{G}$.

The following result of Dubins [149] implies an explicit bound on the degree of randomization: If Γ^* defined using the randomized policy ϕ^* is extremal, then $\phi_u^*(x)$ is zero for all but at most $\ell_a + 1$ values of $u \in \mathsf{U}_\diamond(x)$.

Proposition 9.3.1 (Extreme points with auxiliary constraints). *Any extreme point $\Gamma \in \mathcal{G} \cap \mathcal{H}$ can be expressed as the convex combination of $\ell_a + 1$ extreme points in $\mathcal{G}$.*

To construct the dual of (9.24) we can repeat the steps used to construct the dual of (9.10). We take a pair of dual variables $z \in \mathbb{R}_+$, $h \colon \mathsf{X}_\diamond \to \mathbb{R}$ as before, and another dual variable $\beta \in \mathbb{R}_+^{\ell_a}$ to capture the inclusion $\Gamma \in \mathcal{H}$. The Lagrangian is defined by $\mathcal{L}(\Gamma, z, h, \beta) =$

$$
\langle \Gamma, c \rangle + \left\{ z\Big(1 - \sum_{u,x} \Gamma(x, u)\Big) + \sum_{u,y} h(y)\left(-\Gamma(y, u) + \sum_x \Gamma(x, u) P_u(x, y)\right) \right.
$$
$$
\left. + \sum_{u,x,i} \beta_i \Gamma(x, u)\big(c^{ai}(x) - \overline{\eta}_i\big) \right\},
$$

which can be expressed

$$
\mathcal{L}(\Gamma, z, h, \beta) = z + \langle \Gamma, c_\beta - z\mathbf{1} + h^1 - h^0 \rangle,
$$

where h^0, h^1 are defined in (9.20), and

$$
c_\beta = c + \sum_{i=1}^{\ell_a} \beta_i (c^{ai} - \overline{\eta}_i). \tag{9.25}
$$

For fixed β the Lagrangian is identical to (9.19) with the new cost function c_β. Hence the dual functional defined as the infimum $\Psi^*(z, h, \beta) = \inf_{\Gamma \geq 0} \mathcal{L}(\Gamma, z, h, \beta)$ can be expressed using (9.22),

$$\Psi^*(z, h, \beta) = \begin{cases} z & \text{if } c_\beta - z + h^1 - h^0 \geq 0 \text{ for all } (x, u) \\ -\infty & \text{otherwise.} \end{cases}$$

In the finite state space case there is no duality gap, so that the value η^* of (9.24) is given by $\eta^* = \max\{\Psi^*(z, h, \beta) : z \in \mathbb{R}, \ h : \mathsf{X}_\circ \to \mathbb{R}, \ \beta \in \mathbb{R}_+^{\ell_a}\}$. This proves

Proposition 9.3.2 (ACOE with auxiliary constraints). *Suppose that $\mathbf{X}$ is a finite state space MDP, and let Γ^* denote an optimizer of the primal (9.24). Then, there exists a vector $\beta^* \in \mathbb{R}_+^{\ell_a}$ such that the solution η^* to (9.24) can be expressed as the linear program,*

$$\eta^* = \quad \max \quad z$$

$$\text{s.t.} \quad \sum_{y \in \mathsf{X}_\circ} P_u(x, y) h(y) \geq h(x) - c_{\beta^*}(x) + z \qquad (9.26)$$

$$x \in \mathsf{X}_\circ, \ u \in \mathsf{U}_\circ,$$

where c_{β^*} is defined in (9.25). Hence (η^*, h^*, β^*) solves the average-cost optimality inequality,

$$\eta^* + h^*(x) \leq \min_{u \in \mathsf{U}_\circ(x)} [c_{\beta^*}(x) + P_u h^*(x)]. \qquad (9.27)$$

The inequality (9.27) is an equality for any x satisfying $\pi^*(x) := \sum_u \Gamma^*(x, u) > 0$. Moreover, the complementary slackness conditions hold: For each $1 \leq i \leq \ell_a$,

$$\beta_i^* > 0 \Longrightarrow \langle \Gamma^*, c^{ai} \rangle = \bar{\eta}_i. \qquad (9.28)$$

Example 9.3.3 (Klimov model with hard moment constraints). The Klimov model illustrated in Fig. 2.3 is specialized here to the case of two queues,

Consider the constrained optimization problem,

$$\min \quad \mathsf{E}[Q_1(t)]$$

$$\text{s.t.} \quad \mathsf{E}[Q_2(t)] \leq \bar{\eta}_2,$$

with all expectations in steady state. Applying Theorem 9.3.2, there exists a Lagrange multiplier $\beta^* \in \mathbb{R}_+$ such that the primal can be expressed as the standard average-cost optimization problem,

$$\min \quad \mathsf{E}[Q_1(t) + \beta^*(Q_2(t) - \bar{\eta}_2)].$$

We have seen that the solution to this optimization problem can be taken to be a priority policy defined by the c–μ rule, with $c_\beta^*(x) = x_1 + \beta^* x_2$ (see Section 5.5.1). If the optimal solution is not a priority policy, it follows that the c–μ rule is degenerate in the sense that $\mu_1 = \beta^* \mu_2$.

Assuming that this is the case, we obtain $\beta^* = \mu_1/\mu_2 > 0$, and hence by the complementary slackness condition (9.28) the optimal policy satisfies $\mathsf{E}[Q_2(t)] = \bar{\eta}_2$. It turns out that *any* nonidling policy meeting this constraint is optimal, with value

$$\eta^* = \mu_1(\mathsf{E}[W(t)] - \bar{\eta}_2/\mu_2)$$

where $W(t) = Q_1(t)/\mu_1 + Q_2(t)/\mu_2$, and the expectation is with respect to the non-idling policy (see Exercise 3). One optimal policy is defined using a hedging point,

$$U_2(t) = \begin{cases} 1 & \text{if } Q_2(t) > -\bar{x}, \text{ or } Q_1(t) = 0, \\ 0 & \text{if } Q_2(t) < -\bar{x} \text{ and } Q_1(t) \geq 1, \end{cases}$$

with $U_1(t) = 1 - U_2(t)$ when $Q_1(t) > 0$. The threshold $\bar{x}$ is chosen to meet the inequality constraint with equality $\mathsf{E}[Q_2(t)] = \bar{\eta}_2$. It may be necessary to use randomization when $Q_2(t) = \bar{x}$ to achieve this. ∎

9.4 Optimality equations

Up to now we have settled the existence of a stationary policy that is optimal in the restrictive sense of Proposition 9.2.10: We can be assured that the policy ϕ^* achieves the minimal cost $\eta^* = \inf_x \eta_x^*$, but only for initial conditions in the support of π^*.

To complete the proof of Theorem 9.0.4 it remains to establish that there exists a policy ϕ^* with these properties that is deterministic, stationary, and regular.

We begin with a transient control problem that will serve to unify the remaining development.

9.4.1 Shortest path

Suppose that a distinguished state $x^\ddagger$ is given. For each initial condition x we seek a policy that minimizes the cost along the path from x to $x^\ddagger$ defined by

$$h_\ddagger(x) := \mathsf{E}_x\left[\sum_{t=0}^{\tau_\ddagger - 1} c(X(t))\right], \tag{9.29}$$

where $\tau_\ddagger := \min\{t \geq 1 : X(t) = x^\ddagger\}$ is the first time this distinguished state is visited. This is known as the (stochastic) *shortest path problem*, or SPP.

The SPP can be motivated by the desire to recover gracefully from a transient event: The state x was reached through some fault in the system, or other bad luck, and we would like to recover by the most efficient means possible. We will see that the SPP is sufficiently flexible to cover almost any other dynamic optimization problem we care to consider.

The analysis is simplified by the introduction of a modified MDP whose state process is denoted $X_{\ddagger}$, evolving on a state space denoted $\mathsf{X}_{\ddagger}$. The controlled transition law is denoted ${}_{\ddagger}P_u$. We generalize the SPP slightly by allowing $x^{\ddagger}$ to lie outside of the original state space. In this case we define $\mathsf{X}_{\ddagger} = \mathsf{X}_{\diamond} \cup \{x^{\ddagger}\}$.

Two critical conventions are imposed:

(i) The state $x^{\ddagger}$ is absorbing: ${}_{\ddagger}P_u(x^{\ddagger}, x^{\ddagger}) = 1$ for any u.
(ii) The cost function on $\mathsf{X}_{\ddagger}$ satisfies $c(x^{\ddagger}) = 0$.

These conventions are enforced through a modification of c and P_u. In this way the SPP objective function becomes the *total cost*,

$$h_{\ddagger}(x) = \mathsf{E}_x\left[\sum_{t=0}^{\infty} c(X_{\ddagger}(t))\right]. \qquad (9.30)$$

Let $h_{\ddagger}^*(x)$ denote the infimum of (9.30) over all policies. This solves the optimality equation (9.31a),

$$h_{\ddagger}^*(x) = \min_{u \in \mathsf{U}_{\diamond}(x)} [c(x) + P_u h_{\ddagger}^*(x)] \qquad (9.31a)$$

$$\phi^*(x) \in \arg\min_{u \in \mathsf{U}_{\diamond}(x)} P_u h_{\ddagger}^*(x), \qquad x \in \mathsf{X}_{\diamond}. \qquad (9.31b)$$

Proposition 9.4.1. *Suppose that the value function $h_{\ddagger}^*$ is finite valued. Then the dynamic programming equation (9.31a) holds, and the $h_{\ddagger}^*$-myopic policy ϕ^* is deterministic, stationary, and optimal.*

Proof. We express the infimum in the definition of $h_{\ddagger}^*(x)$ as the infimum over admissible sequences $\{U(1), U(2), \dots\}$ for a *given* $U(0) \in \mathsf{U}_{\diamond}(x)$, and then minimize over $U(0)$: We have for any fixed $U(0)$,

$$\min_{U(1), U(2),\dots} \mathsf{E}\left[\sum_{t=1}^{\infty} c(X_{\ddagger}(t)) \mid \mathcal{F}_1\right]$$

$$= \min_{U(1), U(2),\dots} \mathsf{E}_{X_{\ddagger}(1)}\left[\sum_{t=0}^{\infty} c(X_{\ddagger}(t))\right] = h_{\ddagger}^*(X_{\ddagger}(1)).$$

Consequently, for each $x \in \mathsf{X}_{\diamond}$,

$$h_{\ddagger}^*(x) = c(x) + \min_{U(0)} \mathsf{E}_x\left[\min_{U(1), U(2),\dots} \mathsf{E}\left[\sum_{t=1}^{\infty} c(X_{\ddagger}(t)) \mid \mathcal{F}_1\right]\right]$$

$$= c(x) + \min_{u \in \mathsf{U}_{\diamond}(x)} {}_{\ddagger}P_u h_{\ddagger}^*(x). \qquad \square$$

Next we establish uniqueness.

Proposition 9.4.2. *Suppose that the value function $h_{\ddagger}^*$ is finite for each x. If $h_{\ddagger}^{\circ} \in L_{\infty}^V$ is any other nonnegative solution to (9.31a), with $V = 1 + h_{\ddagger}^*$, then $h_{\ddagger}^{\circ} = h_{\ddagger}^*$.*

Proof. We first obtain a bound via iteration, exactly as in the proof of Theorem 9.0.3: For each $x \in X_\diamond$, any policy, and any $N \geq 1$,

$$\mathsf{E}_x[h_{\ddagger}^\circ(X(N))] \geq h_{\ddagger}^\circ(x) - \sum_{t=0}^{N-1} \mathsf{E}[c(X(t))]. \tag{9.32}$$

On applying the optimal policy ϕ^* and letting $N \to \infty$ we obtain

$$\limsup_{N \to \infty} \mathsf{E}_x^{\phi^*}[h_{\ddagger}^\circ(X(N))] \geq h_{\ddagger}^\circ(x) - h_{\ddagger}^*(x). \tag{9.33}$$

We claim that the limit supremum is zero. To see this write

$$\mathsf{E}_x^{\phi^*}[h_{\ddagger}^*(X(N))] = h_{\ddagger}^*(x) - \sum_{t=0}^{N-1} \mathsf{E}_x^{\phi^*}[c(X(t))],$$

and observe that the right-hand side vanishes as $N \to \infty$ since ϕ^* is optimal. We conclude that

$$\limsup_{N \to \infty} \mathsf{E}_x^{\phi^*}[h_{\ddagger}^\circ(X(N))] \leq \|h_{\ddagger}^\circ\|_V \limsup_{N \to \infty} \mathsf{E}_x^{\phi^*}[1 + h_{\ddagger}^*(X(N))] = 0,$$

and by (9.33) this shows that $h_{\ddagger}^\circ \leq h_{\ddagger}^*$.

To obtain a bound in the reverse direction, first note that the bound (9.32) is an equality under the $h_{\ddagger}^\circ$-myopic policy. Consequently,

$$\mathsf{E}_x^{\phi^\circ}[h_{\ddagger}^\circ(X(N))] = h_{\ddagger}^\circ(x) - \sum_{t=0}^{N-1} \mathsf{E}_x^{\phi^\circ}[c(X(t))].$$

This gives the desired inequality since the function $h_{\ddagger}^\circ$ takes on nonnegative values:

$$0 \leq \liminf_{N \to \infty} \mathsf{E}_x^{\phi^\circ}[h_{\ddagger}^\circ(X(N))]$$

$$= h_{\ddagger}^\circ(x) - \limsup_{N \to \infty} \sum_{t=0}^{N-1} \mathsf{E}_x^{\phi^\circ}[c(X(t))]$$

$$\leq h_{\ddagger}^\circ(x) - h_{\ddagger}^*(x), \qquad x \in X_\diamond. \qquad \square$$

The SPP formulation is applied in the next subsections to obtain representations for the solutions to the DCOE and the ACOE through a particular construction of $X_{\ddagger}$.

9.4.2 Discounted cost

The *discounted-cost value function* is defined as in the scheduling model,

$$h_\gamma^*(x) := \inf \sum_{t=0}^\infty (1+\gamma)^{-t-1} \mathsf{E}_x[c(X(t))], \qquad X(0) = x \in X_\diamond, \tag{9.34}$$

and the *discounted-cost optimality equations* are given by

$$(1 + \gamma)h_\gamma^*(x) \quad = \quad \min_{u \in U_\circ(x)} [c(x) + P_u h_\gamma^*(x)] \tag{9.35a}$$

$$\phi^*(x) \quad \in \quad \arg\min_{u \in U_\circ(x)} P_u h_\gamma^*(x), \qquad x \in X_\circ. \tag{9.35b}$$

The value function h_γ^* defined in (9.34) can be expressed as the solution to a SPP for some Markov chain $X_\ddagger$, and the associated dynamic programming equation (9.35a) is a special case of (9.31a). To make this claim precise we consider the augmented state space $X_\ddagger = X_\circ \cup \{x^\ddagger\}$ and define a particular MDP on $X_\ddagger$.

Consider an enlarged probability space that includes a random variable T that is independent of (U, X). The random time T is geometrically distributed with parameter γ,

$$P\{T = k\} = \gamma P\{T > k\} = \gamma(1 + \gamma)^{-k-1}, \qquad k \ge 0. \tag{9.36}$$

We define $X_\ddagger(t) = X(t)$ for $t \le T$, while $X(t) = x^\ddagger$ for $t > T$.

The statistics of $(X_\ddagger, U)$ are identical to those obtained with a particular controlled transition matrix denoted $_\ddagger P_u$: For each $x \ne x^\ddagger$, $u \in U_\circ(x)$, and $y \in X_\ddagger$ this is defined by

$$_\ddagger P_u(x, y) = \frac{1}{1 + \gamma} P_u(x, y) + \frac{\gamma}{1 + \gamma} \mathbf{1}_{\{x^\ddagger\}}(y), \tag{9.37}$$

with $_\ddagger P(x^\ddagger, x^\ddagger) = 1$, as always.

Since X and T are independent we can write

$$E_x[\mathbf{1}\{T = k\}c(X(T))] = \gamma(1 + \gamma)^{-k-1}E_x[c(X(T))]$$

$$E_x[\mathbf{1}\{T \ge k\}c(X(T))] = (1 + \gamma)^{-k}E_x[c(X(T))] \tag{9.38}$$

$$E_x[\mathbf{1}\{T > k\}c(X(T))] = (1 + \gamma)^{-k-1}E_x[c(X(T))], \qquad k \ge 1.$$

The following representations then follow on summing over k:

$$h_\gamma(x) = \gamma^{-1}E_x[c(X(T))] = (1 + \gamma)^{-1}E_x\left[\sum_{k=0}^{T} c(X(k))\right] = E_x\left[\sum_{k=0}^{T-1} c(X(k))\right]. \tag{9.39}$$

In the final expression the sum is interpreted as zero when $T = 0$.

Appealing to our convention that c vanishes at $x^\ddagger$ we obtain for any policy

$$E_x\left[\sum_{t=0}^{\infty} c(X_\ddagger(t))\right] = E_x\left[\sum_{t=0}^{T} c(X(t))\right] = (1 + \gamma)h_\gamma(x), \qquad x \in X_\circ. \tag{9.40}$$

Hence $h_\ddagger^* = (1 + \gamma)h_\gamma^*$.

Proposition 9.4.3 uses this correspondence to generalize Theorem 4.1.5 from the scheduling model to the more general setting of this chapter. Part (i) follows from Proposition 9.4.1 and Part (ii) follows from Proposition 9.4.2, based on the definition (9.37) and the correspondences given above.

Proposition 9.4.3. *Suppose that the value function h_γ^* defined in (9.34) is finite for each x. Then:*

(i) *h_γ^* solves the DCOE (9.35a).*
(ii) *Suppose that $h_\gamma^\circ \in L_\infty^V$ is any other nonnegative solution to (9.35a), with $V = 1 + h_\gamma^*$. Then $h_\gamma^\circ = h_\gamma^*$.*

9.4.3 Average cost

Under the assumptions of Theorem 9.0.4 we now construct a solution to the ACOE as the solution to the SPP for a particular MDP model. Under these conditions we show that this solution is essentially unique. We begin with the problem of existence.

9.4.3.1 Existence

To place this problem within the context of the SPP we do not augment the state space. Instead, we take a fixed state $x^* \in \mathsf{X}_\diamond$ to play the role of $x^\ddagger$. The modified MDP model has transition law consistent with $P(x, y)$ when $x \neq x^\ddagger$,

$$_\ddagger P_u(x, y) = \begin{cases} P_u(x, y) & y \neq x^*, \\ P_u(x, x^*) & y = x^\ddagger, \end{cases} \tag{9.41}$$

and again $_\ddagger P_u(x^\ddagger, x^\ddagger) = 1$.

Under our convention that $c(x^\ddagger) = 0$ we have for $x \in \mathsf{X}_\diamond \setminus \{x^*\}$

$$_\ddagger P^t c(x) = \mathsf{E}[c(X(t))\mathbf{1}\{\tau_{x^*} > t\}], \quad t \geq 0,$$

and the value function for the associated SPP can be expressed

$$\mathsf{E}_x\left[\sum_{t=0}^\infty c(X_\ddagger(t))\right] = \mathsf{E}_x\left[\sum_{t=0}^{\tau_{x^*}-1} c(X(t))\right], \qquad x \in \mathsf{X}_\diamond, \ x \neq x^*. \tag{9.42}$$

If the policy is regular, and the function c is replaced by $c - \pi(c)$ in (9.42), then the resulting function of x is a solution to Poisson's equation (see (8.22) or (A1.16)). That is, Poisson's equation is the value function for the SPP with *relative cost* $c - \eta$.

Similarly, a representation for the solution to the ACOE can be constructed as the minimum of (9.42) over all policies with c replaced by $c - \eta^*$. For any constant $\eta \in \mathbb{R}$ and $x \in \mathsf{X}_\diamond$ we denote

$$h^*(x; \eta) := \inf \mathsf{E}_x\left[\sum_{t=0}^{\tau_{x^*}-1} (c(X(t)) - \eta)\right], \tag{9.43}$$

where the infimum is over all policies satisfying $\tau_{x^*} < \infty$ with probability 1 from the initial condition x.

The function $c - \eta$ is no longer positive valued, so in general the infimum can be negatively infinite. However, we do have some useful structure, and under the assumptions of Theorem 9.0.4 we will see that $h^*(x; \eta)$ is finite valued.

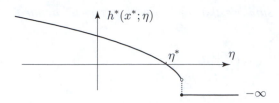

Figure 9.3. Construction of the relative value function as the solution to a SPP.

For each fixed x and fixed policy ϕ the expectation $\mathsf{E}_x^{\phi}\left[\sum_{t=0}^{\tau_{x^*}-1}(c(X(t)) - \eta)\right]$ is linear as a function of η. Consequently, the infimum over all ϕ is a concave function of η, as illustrated in Fig. 9.3. In particular, $h^*(x;\eta)$ is continuous and nonincreasing on an interval of the form $(-\infty, \overline{\eta}(x))$ for some $\overline{\eta}(x) \leq \infty$, and equal to $-\infty$ for $\eta > \overline{\eta}(x)$ when $\overline{\eta}(x)$ is finite.

In the uncontrolled model, the solution to Poisson's equation in (A1.16) (this is (9.43) without the infimum) satisfies $h(x^*) = 0$. This motivates our definition of $\eta^{\bullet}$ as the solution to

$$h^*(x^*, \eta^{\bullet}) = 0. \tag{9.44}$$

We will see shortly that $\eta^{\bullet} = \eta^*$ provided the state x^* is chosen in the support of π^*, with $\eta^* = \pi^*(c)$ given in Proposition 9.2.10.

The function $h^*(\,\cdot\,, \eta^{\bullet})$ does solve the ACOE:

Proposition 9.4.4. *Suppose that there exists $\eta^{\bullet} \in \mathbb{R}$ satisfying (9.44), and that $h^*(x) :=$ $h^*(x; \eta^{\bullet})$ is finite for each x, and uniformly bounded from below,*

$$\inf_{x \in \mathsf{X}_{\circ}} h^*(x; \eta^{\bullet}) > -\infty. \tag{9.45}$$

Then the pair $(h^*, \eta^{\bullet})$ solves the ACOE.

Proof. With $_{+}P_u$ defined in (9.41) we have by (9.31a) for each η and x,

$$h^*(x; \eta) = c(x) - \eta + \min_{u \in \mathsf{U}_{\circ}(x)} \left\{ \sum_{y} {_{+}P_u}(x,y) h^*(y; \eta) \right\}$$

$$= c(x) - \eta + \min_{u \in \mathsf{U}_{\circ}(x)} \left\{ \sum_{y \neq x^*} P_u(x,y) h^*(y; \eta) \right\}.$$

We obtain the desired equation on substituting $\eta = \eta^{\bullet}$ and applying (9.44),

$$h^*(x; \eta^{\bullet}) = c(x) - \eta^{\bullet} + \min_{u \in \mathsf{U}_{\circ}(x)} \left\{ \sum_{y} P_u(x,y) h^*(y; \eta^{\bullet}) \right\}. \qquad \square$$

Questions remain: Is the function $h^*(x; \eta)$ finite valued? Under what conditions is the existence of a solution to (9.44) guaranteed? Under the assumptions of

Theorem 9.0.4 we can apply sample-path arguments similar to those used in Section 9.2.2 to provide positive answers:

Proposition 9.4.5. *The following hold under the assumptions of Theorem 9.0.4, where the state $x^* \in \mathsf{X}_\diamond$ is used in the definition of $h^*(\,\cdot\,;\eta)$:*

(i) *Suppose that x^* belongs to the support of π^*, with π^* the invariant measure given in Proposition 9.2.10. Then $\eta^\bullet = \eta^*$.*

(ii) *For any $x^* \in \mathsf{X}_\diamond$ we have*

$$h^*(x^*;\eta^*) = \inf \mathsf{E}_{x^*}\left[\sum_{t=0}^{\tau_{x^*}-1}\big(c(X(t)) - \eta^*\big)\right] \geq 0.$$

This lower bound is achieved if x^* belongs to the support of π^*.

(iii) *For any $x^* \in \mathsf{X}_\diamond$, the uniform bound (9.45) holds with h^* defined in (9.43).*

Proof. To prove (i) and (ii) we first show that $h^\phi(x^*;\eta^*) \geq 0$ under any policy ϕ for which $\tau_{x^*} < \infty$ with probability 1 from the initial condition x^*. We assume without loss of generality that this value is finite:

$$h^\phi(x^*;\eta^*) := \mathsf{E}_{x^*}^\phi\left[\sum_{t=0}^{\tau_{x^*}-1}\big(c(X(t)) - \eta^*\big)\right] < \infty.$$

Also without loss of generality, we assume that the policy regenerates when x^* is reached, so that the samples $\{X(t) : \tau_{x^*}^{n-1} \leq t \leq \tau_{x^*}^n - 1\}$ are i.i.d., with $\{\tau_{x^*}^n\}$ the successive return times to x^*. Hence, the sequence $\{\mathcal{C}_n : n \geq 1\}$ is i.i.d., where

$$\mathcal{C}_n := \sum_{t=\tau_{x^*}^{n-1}}^{\tau_{x^*}^n-1}\big(c(X(t)) - \eta^*\big).$$

By the Strong Law of Large Numbers we have

$$\lim_{N\to\infty}\frac{1}{N}\sum_{n=1}^{N}\mathcal{C}_n = \mathsf{E}_{x^*}^\phi[\mathcal{C}_1] = h^\phi(x^*;\eta^*),$$

and

$$\lim_{N\to\infty}\frac{\tau_{x^*}^N}{N} = \lim_{N\to\infty}\frac{1}{N}\sum_{n=1}^{N}(\tau_{x^*}^n - \tau_{x^*}^{n-1}) = \mathsf{E}_{x^*}^\phi[\tau_{x^*}^2 - \tau_{x^*}^1] = \mathsf{E}_{x^*}^\phi[\tau_{x^*}].$$

Proposition 9.2.10 then gives the lower bound required in (ii):

$$h^\phi(x^*;\eta^*) = \lim_{N\to\infty}\frac{1}{N}\sum_{n=1}^{N}\mathcal{C}_n = \lim_{N\to\infty}\left(\frac{\tau_{x^*}^N}{N}\right)\frac{1}{\tau_{x^*}^N}\sum_{t=0}^{\tau_{x^*}^N-1}\big(c(X(t)) - \eta^*\big)$$

$$\geq \mathsf{E}_{x^*}^\phi[\tau_{x^*}]\liminf_{n\to\infty}\frac{1}{n}\sum_{t=0}^{n-1}\big(c(X(t)) - \eta^*\big) \geq 0.$$

It follows that $\eta^\bullet \geq \eta^*$. Conversely, Proposition 9.2.10 and Kac's Theorem in the form (8.15) imply that $h^{\phi^*}(x^*, \eta^*) = 0$ provided x^* is in the support of π^*. This establishes (i) and (ii).

To prove (iii) fix $x^1 \in \mathsf{X}_\circ$ and a policy ϕ for which τ_{x^*} is a.s. finite starting from x^1. Let ϕ^0 denote a stationary policy that is regular and x^1-irreducible, whose existence is required in Theorem 9.0.4. Define a new policy ϕ^1 that coincides with ϕ^0 until x^1 is reached, with $U(t) = \phi(X(\tau_{x^1}), \ldots, X(t))$ for $t \geq \tau_{x^1}$. Applying (ii) we have $h^{\phi^1}(x^*; \eta^*) \geq 0$, and this gives

$$0 \leq h^{\phi^1}(x^*; \eta^*) = \mathsf{E}_{x^*}^{\phi^0} \left[\sum_{t=0}^{\tau_{x^1} \wedge \tau_{x^*} - 1} \left(c(X(t)) - \eta^* \right) \right]$$

$$+ \, \mathsf{P}_{x^*}^{\phi^0} \{ \tau_{x^1} < \tau_{x^*} \} \mathsf{E}_{x^1}^{\phi} \left[\sum_{t=0}^{\tau_{x^*} - 1} \left(c(X(t)) - \eta^* \right) \right].$$

That is,

$$h^{\phi}(x^1; \eta^*) \geq - \left(\frac{1}{\mathsf{P}_{x^*}^{\phi^0} \{ \tau_{x^1} < \tau_{x^*} \}} \right) \mathsf{E}_{x^*}^{\phi^0} \left[\sum_{t=0}^{\tau_{x^1} \wedge \tau_{x^*} - 1} \left(c(X(t)) - \eta^* \right) \right] > -\infty.$$

$$(9.46)$$

Observe that this bound is *independent* of the particular policy ϕ, though it does depend upon x^1.

To obtain (9.45) we use the fact that c is coercive: Let $S_* := \{ x : c(x) \leq \eta^* \}$, so that $c(x) > \eta^*$ for $x \in S_*^c$. This gives the bound

$$h^{\phi}(x; \eta^*) \geq \mathsf{E}_x^{\phi} \left[\mathbf{1}\{ \tau_{S_*} < \tau_{x^*} \} \sum_{t=\tau_{S_*}}^{\tau_{x^*} - 1} (c(X(t)) - \eta^*) \right] = \mathsf{E}_x^{\phi} \left[\mathbf{1}\{ \tau_{S_*} < \tau_{x^*} \} h^{\phi}(X(\tau_{S_*}); \eta^*) \right]$$

where the equality follows from the Markov property. This implies the uniform bound

$$h^{\phi}(x; \eta^*) \geq - \sup_{x^1 \in S_*} |h^{\phi}(x^1; \eta^*)|,$$

which is finite by (9.46).   □

9.4.3.2 Uniqueness

Suppose that (h°, η°) is a solution to the ACOE. Under a growth condition on h°, similar to what was used in Propositions 9.4.2 and 9.4.3, we can establish optimality of the h°-myopic policy.

The growth bound is based on the value function for a particular SPP. However, instead of a single state, we fix a finite set denoted S_θ to serve as the final destination.

Define for $\theta > 0$

$$S_\theta = \{ x : c(x) \leq \theta \}, \tag{9.47}$$

$$\tau_\theta = \min\{ t \geq 1 : X(t) \in S_\theta \}, \quad \sigma_\theta = \min\{ t \geq 0 : X(t) \in S_\theta \}, \tag{9.48}$$

and consider the *minimal cost to reach* S_θ,

$$V_\theta^*(x) = \inf \mathsf{E}_x \left[\sum_{t=0}^{\sigma_\theta} c(X(t)) \right], \tag{9.49}$$

where the infimum is over all policies. Note that S_θ is a finite set under the assumption that c is coercive. Also, if under a regular policy ϕ we have $\pi(c) \le \theta$, then necessarily $\pi(S_\theta) > 0$, and hence by regularity the cost is finite,

$$\mathsf{E}_x^\phi \left[\sum_{t=0}^{\tau_\theta} c(X(t)) \right] < \infty, \qquad x \in \mathsf{X}_\diamond.$$

Proposition 9.4.6. *Suppose that the assumptions of Theorem 9.0.4 are satisfied, and that (h°, η°) is a solution to the ACOE with $h^\circ \colon \mathsf{X}_\diamond \to \mathbb{R}_+$ and $h^\circ \in L_\infty^{V_\theta^*}$ for some $\theta \ge \eta^\circ$. Then the h°-myopic policy is optimal, with average cost $\eta^\circ = \eta^*$.*

For the proof we require the following extension of Theorem A.6.1.

Lemma 9.4.7. *Suppose that X is an f-regular Markov chain, with $f = 1 + c$. Fix $\theta \ge \pi(c)$ and let V_θ denote the cost to reach S_θ,*

$$V_\theta(x) := \mathsf{E}_x \left[\sum_{t=0}^{\sigma_\theta} c(X(t)) \right], \qquad x \in \mathsf{X}_\diamond. \tag{9.50}$$

Then:

(i) *$V_\theta(x) < \infty$ for each x.*
(ii) *The following identity holds,*

$$PV_\theta = V_\theta - c + b_\theta,$$

with $b_\theta(x) := \mathbf{1}_{S_\theta}(x) \mathsf{E}_x \left[\sum_{t=1}^{\tau_\theta} c(X(t)) \right].$

(iii) *For each initial condition,*

$$\lim_{t\to\infty} \frac{1}{t} \mathsf{E}_x[V_\theta(X(t))] = \lim_{t\to\infty} \mathsf{E}_x[V_\theta(X(t))\mathbf{1}\{\tau_\theta > t\}] = 0.$$

Proof. Part (i) follows from the definition of regularity and the fact that $\pi(S_\theta) > 0$. Applying the Markov property, we obtain

$$PV_\theta(x) = \mathsf{E}_x \left[\mathsf{E}_{X(1)} \left[\sum_{t=0}^{\sigma_\theta} c(X(t)) \right] \right]$$

$$= \mathsf{E}_x \left[\mathsf{E} \left[\sum_{t=1}^{\tau_\theta} c(X(t)) \mid X(0), X(1) \right] \right]$$

$$= \mathsf{E}_x \left[\sum_{t=1}^{\tau_\theta} c(X(t)) \right] = \mathsf{E}_x \left[\sum_{t=0}^{\tau_\theta} c(X(t)) \right] - c(x), \qquad x \in \mathsf{X}_\diamond.$$

On noting that $\sigma_\theta = \tau_\theta$ for $x \in S_\theta^c$, the identity above implies (ii).

To prove the first limit in (iii) we iterate the identity in (ii) to obtain

$$\mathsf{E}_x[V_\theta(X(t))] = P^t V_\theta(x) = V_\theta(x) + \sum_{k=0}^{t-1}[-P^k c(x) + P^k b_\theta(x)], \quad t \geq 1.$$

Dividing by t and letting $t \to \infty$ we obtain from the LLN

$$\lim_{t\to\infty} \frac{1}{t}\mathsf{E}_x[V_\theta(X(t))] = -\eta + \pi(b_\theta),$$

and the right-hand side is zero by the generalized Kac's Theorem 8.1.1,

$$\eta = \pi(c) = \sum_{x\in S_\theta} \pi(x)\mathsf{E}_x\left[\sum_{t=0}^{\tau_\theta-1} c(X(t))\right] = \pi(b_\theta).$$

To prove the second limit we apply the Markov property to obtain, for each $m \geq 1$,

$$V_\theta(X(m)) = \mathsf{E}_{X(m)}\left[\sum_{t=0}^{\sigma_\theta} c(X(t))\right] = \mathsf{E}\left[\sum_{t=m}^{\tau_\theta} c(X(t)) \mid \mathcal{F}_m\right], \quad \text{on } \{\tau_\theta > m\}.$$

Note that $\tau_\theta = \sigma_\theta$ if $\sigma_\theta \geq 1$.

The event $\{\tau_\theta > m\}$ is $\mathcal{F}_m$ measurable. Consequently, by the smoothing property of the conditional expectation,

$$\mathsf{E}_x[V_\theta(X(m))\mathbf{1}\{\tau_\theta > m\}] = \mathsf{E}\left[\mathbf{1}\{\tau_\theta > m\}\mathsf{E}\left[\sum_{t=m}^{\tau_\theta} c(X(t)) \mid \mathcal{F}_m\right]\right]$$

$$= \mathsf{E}\left[\mathbf{1}\{\tau_\theta > m\}\sum_{t=m}^{\tau_\theta} c(X(t))\right] \leq \mathsf{E}\left[\sum_{t=m}^{\tau_\theta} c(X(t))\right]$$

The right-hand side vanishes as $m \to \infty$ by the Dominated Convergence Theorem. This proves the second limit in (iii). $\qquad\square$

Proof of Proposition 9.4.6. If ϕ is any regular policy, then by Lemma 9.4.7 we have for any x in the support of its invariant measure π,

$$\limsup_{t\to\infty} \frac{1}{t}\mathsf{E}_x^\phi[|h^\circ(X(t))|] \leq \|h^\circ\|_{V_\theta^*} \limsup_{t\to\infty} \frac{1}{t}\mathsf{E}_x^\phi[V_\theta^*(X(t))] = 0.$$

We conclude as in the proof of Theorem 9.0.3 that

$$0 = \lim_{n\to\infty} \frac{1}{n}\mathsf{E}_x^\phi[h^\circ(X(n))] \geq -\lim_{n\to\infty} \frac{1}{n}\mathsf{E}_x^\phi\left[\sum_{t=0}^{n-1}\big(c(X(t)) - \eta^\circ\big)\right] = -\eta_x^\phi + \eta^\circ,$$

with $\eta_x^\phi = \pi(c)$. That is, $\eta_x^\phi \geq \eta^\circ$.

Repeating the same steps using ϕ° we conclude as in the proof of Theorem 9.0.3 that $\eta_x^{\phi^\circ} \equiv \eta^\circ$. Hence ϕ° is optimal over all stationary policies. Optimality (over all policies) follows from Proposition 9.2.10. $\qquad\square$

9.5 Algorithms

The inventory model and the single-server queue are among the few examples for which an optimal policy can be obtained explicitly. In more interesting examples an optimal policy can be computed numerically using some algorithm.

The definition of the value iteration algorithm, or VIA, is precisely successive approximation. The policy iteration algorithm is in fact a version of the Newton–Raphson method. Exercise 8 is a guided exercise to prove this correspondence. Policy iteration is sometimes called *policy improvement* – fortunately, the acronym PIA covers either term.

In this section we introduce these algorithms and describe some basic properties. As with many recursive algorithms, a careful choice of initialization can lead to improved convergence. We illustrate this in Example 9.5.7.

9.5.1 Value iteration

Value iteration is actually a family of techniques based on successive approximation. One example is the Bellman–Ford algorithm described in Section 6.3.2.

Definition 9.5.1 (Value iteration). Given the *terminal penalty function* $V_0 \colon \mathsf{X}_\diamond \to \mathbb{R}_+$, the sequence of functions $\{V_n^* \colon \mathsf{X}_\diamond \to \mathbb{R}_+\}$ is defined recursively via $V_0^* = V_0$, and

$$V_{n+1}^*(x) = c(x) + \min_{u \in \mathsf{U}_\diamond(x)} P_u V_n^*(x), \qquad x \in \mathsf{X}_\diamond,\ n \geq 0. \qquad (9.51)$$

The feedback law $\phi_n^* \colon \mathsf{X}_\diamond \to \mathsf{U}_\diamond$ is defined to be any minimizer,

$$\phi_n^*(x) \in \arg\min_{u \in \mathsf{U}_\diamond(x)} P_u V_n^*(x), \qquad x \in \mathsf{X}_\diamond. \qquad\blacksquare$$

Proposition 9.5.2 shows that the VIA does solve a basic optimization problem. However, the utility of this result is largely theoretical. In practice, a single feedback law ϕ_n^* is obtained after running the algorithm for many steps, say, $n = 10{,}000$. The *stationary policy* with feedback law ϕ_n^* is then used for control.

Proposition 9.5.2. *For each $n \geq 1$ the function V_n^* can be expressed*

$$V_n^*(x) = \min \mathsf{E}\left[V_0(X(n)) + \sum_{t=0}^{n-1} c(X(t)) \right], \qquad (9.52)$$

where the minimum is over all policies. Moreover, for fixed $n \geq 1$ an optimal policy ϕ^n achieving V_n^* is Markov, but not necessarily stationary, with

$$U^*(t) = \phi_{n-1-t}^*(Q(t)), \qquad t = 0, \dots, n-1. \qquad (9.53)$$

Proof. The proof is by induction on n: For $n = 1$ the result holds by the definition of the VIA,

$$V_1^*(x) = \min_{U(0)} \mathsf{E}[V_0^*(X(1)) + c(X(0))],$$

and the minimizing allocation vector can be expressed $U^*(0) = \phi_0^*(Q(0))$.

Suppose that the statement of the proposition is true for a given $n \geq 1$. That is, the representation (9.52) holds, and the minimum is attained using the stationary policy defined in (9.53). Following the proof of Proposition 9.4.3, for the $(n+1)$-period problem we express the minimum over all admissible input sequences as the minimum first over admissible sequences $\{U(1), U(2), \dots\}$ for a given $U(0) \in \mathsf{U}_\circ(x)$, and then minimize over $U(0)$. This gives

$$\min_\phi \mathsf{E}_x \left[V_0^*(X(n+1)) + \sum_{t=0}^{n} c(X(t)) \right]$$

$$= c(x) + \min_\phi \mathsf{E}_x \left[V_0^*(X(n+1)) + \sum_{t=1}^{n} c(X(t)) \right]$$

$$= c(x) + \min_{U(0)} \mathsf{E}_x \left[\min_{U(1),U(2),\dots} \mathsf{E} \left[V_0^*(X(n+1)) + \sum_{t=1}^{n} c(X(t)) \mid \mathcal{F}_1 \right] \right].$$

We have by the structure of the MDP model,

$$\min_{U(1),U(2),\dots} \mathsf{E} \left[V_0^*(X(n+1)) + \sum_{t=1}^{n} c(X(t)) \mid \mathcal{F}_1 \right] = \min_\phi \mathsf{E}_{X(1)}^\phi \left[V_0^*(X(n)) + \sum_{t=0}^{n-1} c(X(t)) \right]$$

and the right-hand side is precisely $V_n^*(X(1))$ by the induction hypothesis. Moreover, by the induction hypothesis the minimum on the right-hand side is achieved using $U(1) = \phi_{n-1}^*(X(1))$, $U(2) = \phi_{n-2}^*(X(2))$, ...

We thus obtain the desired representation,

$$\min_\phi \mathsf{E}_x \left[V_0^*(X(n+1)) + \sum_{t=0}^{n} c(X(t)) \right] = c(x) + \min_{u \in \mathsf{U}_\circ(x)} P_u V_n^*(x) = V_{n+1}^*(x).$$

Moreover, the policy achieving the minimum is Markov as claimed. $\qquad\square$

We obtain a suggestive corollary on applying the bound,

$$\inf \left(\limsup_{n \to \infty} \frac{1}{n} \mathsf{E}_x \left[\sum_{t=0}^{n-1} c(X(t)) \right] \right) \geq \limsup_{n \to \infty} \frac{1}{n} \left(\min \mathsf{E}_x \left[\sum_{t=0}^{n-1} c(X(t)) \right] \right) \quad (9.54)$$

where the infimum and the minimum are over all policies. Exercise 7 contains a generalization of Corollary 9.5.3 to allow nonzero initialization in the VIA.

Corollary 9.5.3. *If $V_0 \equiv 0$, then for every x,*

$$\limsup_{n \to \infty} \frac{1}{n} V_n^*(x) \leq \eta_x^*.$$

Proof. We obtain the lower bound (9.54) since the policy can depend upon n in the minimum on the right-hand side, while this is not true for the infimum on the left-hand side. The corollary follows since the minimum is achieved to give $V_n^*(x)$, and the infimum is by definition η_x^*. $\qquad\square$

This gives hope that $\{\phi_n^*\}$ will converge to an average-cost optimal policy.

In spite of the positive message conveyed in Corollary 9.5.3, the initialization V_0 should be chosen with care.

Proposition 9.5.4. *If $V_0 \equiv 0$, then the policy ϕ_0^* obtained from value iteration is the c-myopic policy.*

Consequently, when $V_0 \equiv 0$ it is possible that the controlled chain is transient under ϕ_n^* for some n, such as $n = 0$; we have seen in several examples, such as Example 4.4.5, that the c-myopic policy may not be stabilizing in network models.

There is one special case under which the policy ϕ^n defined in Proposition 9.5.2 is stationary.

Proposition 9.5.5. *If (h^*, η^*) solves the ACOE (9.4a) and $V_0 = h^*$, then for each $n \geq 1$,*

$$V_n^* = h^* + n\eta^*.$$

Hence, there exists a single stationary policy minimizing the right-hand side of (9.52) for each n.

Proof. The proof is by induction: This holds by definition when $n = 1$, and then for arbitrary n since η^* is constant. □

Proposition 9.5.5 suggests that an initialization approximating the solution to the ACOE might result in quickened convergence. We assume in Theorem 9.5.6 below that at least one regular policy ϕ_{-1} exists, and that the function V_0 solves a relaxation of the ACOE,

$$\min_{u \in \mathsf{U}_\diamond(x)} P_u V_0 (x) \leq V_0(x) - c(x) + \overline{\eta},$$

where $\overline{\eta}$ is a constant. Equivalently, the *Poisson inequality* holds for *some policy ϕ_{-1}*,

$$P_{\phi_{-1}} V_0 \leq V_0 - c + \overline{\eta}. \tag{9.55}$$

The existence of a pair (V_0, ϕ_{-1}) satisfying (9.55) is a natural stabilizability assumption on the model, and we find below that this initialization ensures that the VIA generates stabilizing policies, in the sense that Poisson's inequality holds for each of the feedback laws $\{\phi_n^*\}$ obtained from the algorithm.

To simplify notation, for each n we denote $P_n = P_{\phi_n^*}$, and we let E^n denote the expectation operator induced by the stationary policy with feedback law ϕ_n^*. Let $b_n := V_{n+1}^* - V_n^*$ denote the incremental value, and $\overline{\eta}_n = \sup_x b_n(x)$.

Theorem 9.5.6 (Performance bounds for VIA). *Suppose the initialization V_0 satisfies (9.55). Then the upper bounds $\{\overline{\eta}_n\}$ are finite and nonincreasing:*

$$\overline{\eta}_0 \geq \overline{\eta}_1 \geq \cdots \geq \overline{\eta}_n \geq \cdots.$$

For each n, the average cost under the policy ϕ_n^* satisfies

$$\eta_x^{\phi_n^*} \leq \overline{\eta}_n, \qquad x \in \mathsf{X}_\diamond.$$

Proof. From the definitions, for each n we have the familiar looking identity $P_n V_n^* = V_n^* - c + b_n$, which implies a version of (9.55),

$$P_n V_n^* \leq V_n^* - c + \overline{\eta}_n. \tag{9.56}$$

The Comparison Theorem A.4.3 then implies that the average cost using policy ϕ_n^* is bounded by $\overline{\eta}_n$ for each initial condition. It remains to show that the $\{\overline{\eta}_n\}$ are finite and nonincreasing.

The minimization in the value-iteration algorithm leads to the bound

$$P_n b_n = P_n V_{n+1}^* - P_n V_n^* \geq P_{n+1} V_{n+1}^* - P_n V_n^* = (V_{n+1}^* - c + b_{n+1}) - (V_n^* - c + b_n),$$

giving $P_n b_n \geq b_{n+1}$. From this we deduce by induction that the upper bounds $\{\overline{\eta}_n\}$ are finite and decreasing,

$$\overline{\eta}_{n+1} := \sup_x b_{n+1}(x)$$

$$\leq \sup_x \left(\sum_y P_n(x,y) b_n(y) \right)$$

$$\leq \sum_y P_n(x,y) \left(\sup_{y'} b_n(y') \right) = \overline{\eta}_n.$$

Finiteness of $\overline{\eta}_0$ follows from the assumption that the initial condition V_0 satisfies (9.55). $\square$

To illustrate the role of the initial condition V_0 we turn to the simple re-entrant line.

Example 9.5.7 (Simple re-entrant line: initialization and convergence). We return to the homogeneous CRW model in Case III as defined in Section 5.6.2 using the rates given in (2.26), and with $c(\,\cdot\,) = |\,\cdot\,|$ the ℓ_1 norm.

To implement the VIA it is necessary to truncate the state space. Choosing $Q_i \in \{0, \ldots, 44\}$ for $i = 1, 2, 3$ we arrive at a finite state space consisting of $45^3 = 91,125$ states.

We shall see that the standard VIA in which $V_0 \equiv 0$ requires thousands of iterations for convergence, while the VIA implemented with an appropriate initial value function converges *much* more quickly.

Applying Theorem 8.0.4 we see that the MaxWeight policy satisfies the conditions of Theorem 9.5.6. This requires $V_0(x) = \frac{1}{2} x^{\mathsf{T}} D x$ with D diagonal.

Alternatively, we can obtain an initial quadratic to define ϕ_{-1} using the Performance LP introduced in Definition 8.6.9, resulting in

$$V_{D^1}(x) = \tfrac{1}{2} x^{\mathsf{T}} D^1 x; \qquad D^1 = \begin{bmatrix} 15.9 & 9.0 & 9.0 \\ 9.0 & 9.0 & 0 \\ 9.0 & 0 & 7.5 \end{bmatrix}. \tag{9.57}$$

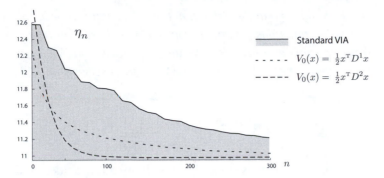

Figure 9.4. Convergence of the VIA with V_0 quadratic. The quadratic V_{D^1} was found using the Performance LP, and the second quadratic V_{D^2} was found through direct calculation.

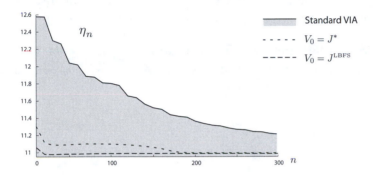

Figure 9.5. Convergence of the VIA with V_0 a value function for the fluid model. The value function for the optimal fluid policy and that for the LBFS priority policy were each used as initialization in two experiments.

Another quadratic that solves the Poisson inequality for some policy is

$$V_{D^2}(x) = \tfrac{1}{2}x^{\mathsf{T}}D^2 x; \qquad D^2 = \begin{bmatrix} 56 & 31.6 & 24.3 \\ 31.6 & 31.6 & 0 \\ 24.3 & 0 & 20.9 \end{bmatrix}. \tag{9.58}$$

The results are shown in Fig. 9.4. In each case the rate of convergence is significantly faster than the standard algorithm using $V_0 \equiv 0$.

Consider now initialization using the optimal value function J^* computed in Example 4.3.10. We also initialize using the fluid value function J^{LBFS} obtained using the LBFS priority policy. Results from the two experiments are shown in Fig. 9.5. The convergence is exceptionally fast in both experiments. Perhaps surprisingly, the "suboptimal" choice using LBFS leads to the fastest convergence to the optimal cost $\eta^* \approx 10.9$. ∎

9.5.2 VIA and shortest path formulations

The SPP setting can be used to unify analysis of algorithms. The following corollary to Proposition 9.5.2 settles convergence.

Proposition 9.5.8. *Suppose that the value function $h_{\ddagger}^*$ is finite valued. Let $\{V_n^*\}$ denote the sequence of value functions obtained from the VIA based on a nonnegative cost function c using the transition law $_{\ddagger}P_u$. Assume moreover that $V_0 \in L_\infty^V$ with $V = 1 + h_{\ddagger}^*$. Then the sequence of value functions $\{V_n^*\}$ is convergent to the solution of the SPP,*

$$\lim_{n\to\infty} V_n^*(x) = h_{\ddagger}^*(x) := \min \mathsf{E}_x \left[\sum_{t=0}^{\infty} c(X(t)) \right]. \tag{9.59}$$

As one simple application, we see that Proposition 9.5.8 provides an algorithm and convergence proof for constructing the optimal value function h_γ^* for the discounted-cost optimization problem.

The following version of the VIA is precisely the algorithm specified in Definition 9.5.1 using the controlled transition matrix $_{\ddagger}P_u$ defined in (9.37).

Definition 9.5.9 (Value iteration for discounted cost). For an initial condition $h_0 \colon \mathsf{X}_\diamond \to \mathbb{R}_+$, the sequence of functions $\{h_n \colon \mathsf{X}_\diamond \to \mathbb{R}_+\}$ is defined recursively via

$$h_{n+1}(x) = c(x) + \frac{1}{1+\gamma} \min_{u \in \mathsf{U}_\diamond(x)} P_u h_n(x), \qquad x \in \mathsf{X}_\diamond, \; n \geq 0. \tag{9.60}$$

∎

Theorem 9.5.10. *Suppose that the value function h_γ^* is finite valued, and that $V_0 \in L_\infty^V$ with $V = 1 + h_\gamma^*$. Then the sequence of functions $\{h_n\}$ obtained using the VIA for discounted cost is convergent to $(1+\gamma)h_\gamma^*$. The value function h_γ^* is a solution to the discounted-cost optimality equation (9.35a), and (9.35b) defines a stationary policy achieving h_γ^*.*

There is much more that can be said about the VIA and its convergence properties. The algorithm is also a powerful theoretical tool to establish existence of solutions to the dynamic programming equations, and to derive properties of their solutions. We provide just a few examples in Section 9.6.3.

9.5.3 Policy iteration

Theorem 9.5.6 provides performance guarantees for each of the feedback laws obtained from the VIA. We now ask for some improvement. At stage n we have a solution to Poisson's inequality, but there are many others, and some may give a better bound.

Suppose that at stage n on computing V_n^* and ϕ_n^* we can solve Poisson's equation,

$$P_n h_n = h_n - c + \eta_n, \tag{9.61}$$

where η_n is the *actual* steady-state cost using the policy ϕ_n^*. We obtain the PIA by repeating this step for each n:

Definition 9.5.11 (Policy iteration). Given an initial stationary policy ϕ_0, the sequence of stationary policies $\{\phi_n^*\}$ is defined recursively via

$$\phi_{n+1}^*(x) \in \arg\min_{u \in \mathsf{U}_\diamond(x)} P_u h_n(x), \qquad x \in \mathsf{X}_\diamond, \tag{9.62}$$

where h_n is any solution to Poisson's equation (9.61). ∎

Stability follows as in Theorem 9.5.6, provided there is some irreducibility so that solutions to Poisson's equation can be constructed.

Theorem 9.5.12 (Performance bounds for PIA). *Suppose that the cost function c is coercive, that the initial policy ϕ_0 is regular, and that each subsequent policy induces a Markov chain that is x^*-irreducible for some $x^* \in \mathsf{X}_\diamond$. Then each of the policies $\{\phi_n^*\}$ is regular, and the average cost $\{\eta_n\}$ is nonincreasing in n.*

Proof. We prove by induction that for each n there exists a solution (h_n, η_n) to (9.61) with $h_n \geq 0$ and an invariant measure π_n satisfying $\pi_n(c) = \eta_n \leq \eta_{n-1}$.

Suppose that the induction hypothesis is true for n. By the definition of the algorithm we have

$$P_{n+1} h_n \leq P_n h_n = h_n - c + \eta_n.$$

Positivity of h_n and the coercive assumption on c then implies that ϕ_{n+1}^* is regular. The Comparison Theorem A.4.3 gives the bound $\eta_{n+1} \leq \eta_n$, and Proposition A.3.12 implies that a nonnegative solution to Poisson's equation exists for P_{n+1}. □

9.6 Optimization in networks

In the remainder of this chapter we specialize to network models. We consider the general CRW model (8.4) satisfying the assumptions of Theorem 8.0.4. There are no buffer constraints, so that $\mathsf{X}_\diamond = \mathbb{Z}_+^\ell$.

Throughout this section we restrict to a cost function c that defines a norm on $\mathbb{R}^\ell$.

9.6.1 Reachability in networks

So far the only general result establishing x^*-irreducibility for network models is Proposition 8.1.2, and this result was established in a very special setting. We now construct a set $\mathsf{X}_\diamond^{\mathrm{M}}$ that is *maximally irreducible* for the general CRW model.

Proposition 9.6.1. *Suppose that the assumptions of Theorem 8.0.4 hold for the CRW model (8.4). Then there exists a regular (randomized) policy ϕ^{M} with invariant measure π^{M} whose support $\mathsf{X}_\diamond^{\mathrm{M}}$ is maximal in the sense that the following hold:*

(i) *The solution to (V3) is obtained with V a quadratic function of x, and with $f = 1 + c$.*

(ii) *For any (possibly randomized) stationary policy that is regular with invariant measure π, the support of π is contained in $\mathsf{X}_\diamond^{\mathrm{M}}$.*

(iii) *If the irreducibility condition (9.7) holds then $\mathsf{X}_\diamond^{\mathrm{M}} = \mathsf{X}_\diamond$.*

Proof. Let ϕ^{MW} denote the MaxWeight policy for a given diagonal matrix D. Let ϕ^e denote a randomized policy satisfying for some $\varepsilon > 0$,

$$\phi^e_u(x) \geq \varepsilon, \qquad x \in \mathsf{X}_\diamond,\ u \in \mathsf{U}_\diamond(x). \tag{9.63}$$

We then define $\phi^{\text{M}} = \delta\phi^e + (1-\delta)\phi^{\text{MW}}$, where $\delta \in (0,1)$ is to be determined. By construction we have $\phi^{\text{M}}_u(x) \geq \varepsilon\delta\phi_u(x)$ for *any other* randomized stationary policy ϕ, and all u, x.

Theorem 8.0.4 gives

$$P_{\phi^{\text{MW}}} V \leq V - f + b\mathbf{1}_S$$

where V is quadratic, $f = 1 + \|x\|$, S is a finite set, and $b < \infty$. Also, since $A(1)$ has a second moment and $B(1)$ is bounded we have for some constant k

$$P_{\phi^e} V \leq V + kf.$$

The randomized policy ϕ^{M} satisfies the bound

$$P_{\phi^{\text{M}}} V \leq (1-\delta)[V - f + b\mathbf{1}_S] + \delta[V + kf].$$

Hence (V3) holds with this V provided $\delta < 1/k$.

Consider now any randomized policy ϕ. For each $x \in \mathsf{X}_\diamond$ we have $P_\phi(x,y) > 0 \implies P_{\phi^{\text{M}}}(x,y) > 0$. That is, $\text{supp}\left(P_\phi(x,\cdot)\right) \subset \text{supp}\left(P_{\phi^{\text{M}}}(x,\cdot)\right)$. By induction it follows that the same holds for the iterates,

$$\text{supp}\left(P^t_\phi(x,\cdot)\right) \subset \text{supp}\left(P^t_{\phi^{\text{M}}}(x,\cdot)\right), \qquad t \geq 1,\ x \in \mathsf{X}_\diamond.$$

The set $\mathsf{X}^{\text{M}}_\diamond = \text{supp}\left(\pi^{\text{M}}\right)$ is absorbing by Proposition 8.1.6, and hence the inclusion above gives

$$\text{supp}\left(P^t_\phi(x,\cdot)\right) \subset \mathsf{X}^{\text{M}}_\diamond, \qquad t \geq 1,\ x \in \mathsf{X}^{\text{M}}_\diamond. \tag{9.64}$$

If ϕ is regular then for *any* x and any $y \in \text{supp}\left(\pi_\phi\right)$ there exists $t > 0$ such that $y \in \text{supp}\left(P^t_\phi(x,\cdot)\right)$. The conclusion $\text{supp}\left(\pi_\phi\right) \subseteq \mathsf{X}^{\text{M}}_\diamond$ then follows. $\square$

9.6.2 Optimality equations

Here we consider consequences of the general theory presented in Section 9.4 when specialized to the CRW scheduling model. The growth condition imposed on the value function is of the form $h^* \in L^{V_p}_\infty$ where V_p is defined in (8.27). In the discounted-cost problem we take $p = 1$ and for the average-cost problem $p = 2$.

9.6.2.1 Discounted cost

The following result is a refinement of Proposition 9.4.3: The skip-free property of the network implies growth bounds on the value function.

Proposition 9.6.2. *Suppose that $\rho_\bullet < 1$ and c is a norm on $\mathbb{R}^\ell$. Then:*

(i) *The value function (9.34) is finite valued with $h^*_\gamma \in L^{V_1}_\infty$ and solves the DCOE (9.35a).*

(ii) *Suppose that $h^\circ_\gamma \in L^{V_1}_\infty$ is any other solution to (9.35a). Then $h^\circ_\gamma = h^*_\gamma$.*

Proof. The result is given in Proposition 9.4.3, except for the conclusion that $h_\gamma^* \in L_\infty^{V_1}$.

In fact, the functions are equivalent in the sense that $h_\gamma^* \in L_\infty^{V_1}$ and $V_1 \in L_\infty^{V_\gamma^*}$ with $V_\gamma^* = 1 + h_\gamma^*$. This follows from the skip-free nature of the CRW model, as formalized in (8.23). □

9.6.2.2 Average cost

An extension of Proposition 9.6.2 to the average-cost setting is made possible by applying the results of Section 9.4.3. Without conditions ensuring some form of irreducibility (such as the reachability condition in Theorem 9.0.4) it is necessary to restrict to the state space $X_\diamond^M$. In Proposition 9.6.3 we adopt the assumptions of Theorem 9.0.5 to avoid this complication.

Proposition 9.6.3. *Suppose that the assumptions of Theorem 9.0.5 hold. Then:*

(i) *There exists a solution to the ACOE (9.4a) satisfying $h^* \in L_\infty^{V_2}$ and $\eta^* = \eta_x^*$ independent of x.*

(ii) *The solution in (i) is unique (up to an additive constant): Suppose that $h^\circ \in L_\infty^{V_2}$ and η° solve (9.4a),*

$$\eta^\circ + h^\circ(x) = \min_{u \in U_\diamond(x)} [c(x) + P_u h^\circ(x)], \qquad x \in X_\diamond.$$

Then $\eta^\circ = \eta^*$, and $h^\circ(x) - h^\circ(0) = h^*(x) - h^*(0)$ for each $x \in X_\diamond$.

Proof. To prove (i) we note that the assumptions of Theorem 9.0.4 hold: The cost function is coercive since it defines a norm on $\mathbb{R}^\ell$, and the reachability assumption holds due to Proposition 9.6.1. Hence a solution to the ACOE equation exists as claimed. Moreover, applying Proposition 9.6.1 once more we can conclude that the assumptions of Proposition 9.4.5 hold and that the minimal solution defined there satisfies $h^* \in L_\infty^{V_2}$.

We now prove (ii). By normalization we assume that $h^*(0) = h^\circ(0) = 0$.

Proposition 9.4.6 implies that $\eta^\circ = \eta^*$. Hence using the policy ϕ^* we have

$$P_{\phi^*} h^\circ \geq h^\circ - c + \eta^*.$$

Using familiar arguments (e.g., the proof of Proposition 9.4.5) we obtain for each $n \geq 1$

$$\mathsf{E}_x^{\phi^*}[h^\circ(Q(n \wedge \tau_0))] \geq h^\circ(x) - \mathsf{E}^{\phi^*}\left[\sum_{t=0}^{n \wedge \tau_0 - 1} (c(Q(t)) - \eta^*)\right].$$

Under our convention that $h^\circ(0) = 0$ we have $h^\circ(Q(n \wedge \tau_0)) = h^\circ(Q(n))\mathbf{1}\{\tau_0 > n\}$. Applying Proposition 8.2.2 we conclude that

$$\lim_{n \to \infty} \mathsf{E}_x^{\phi^*}[h^\circ(Q(n \wedge \tau_0))] = \lim_{n \to \infty} \mathsf{E}_x^{\phi^*}[h^\circ(Q(n))\mathbf{1}\{\tau_0 > n\}] = 0,$$

so that for each x,

$$h^*(x) = \mathsf{E}^{\phi^*}\left[\sum_{t=0}^{\tau_0-1}(c(Q(t)) - \eta^*)\right] \geq h^\circ(x).$$

Using similar arguments we obtain under the h°-myopic policy

$$\mathsf{E}_x^{\phi^\circ}\left[\sum_{t=0}^{\tau_0-1}(c(Q(t)) - \eta^*)\right] \leq h^\circ(x), \qquad x \in \mathsf{X}_\diamond.$$

Minimality of h^* implies that the left-hand side is greater than or equal to $h^*(x)$, which shows that $h^\circ = h^*$. □

Next we investigate the structural insight that can be obtained from the value-iteration algorithm.

9.6.3 Monotonicity and irreducibility

To illustrate how the VIA can be used as an analytical tool we consider the scheduling model (8.24) under the assumptions of Section 8.5. This is an MDP model with controlled transition matrix defined for any $x \in \mathsf{X}_\diamond$, $u \in \mathsf{U}_\diamond(x)$, and any function h on $\mathsf{X}_\diamond$ by

$$P_u h(x) = \mathsf{E}[h(x + A(1))] + \sum_{i=1}^{\ell}\mu_i h(x - (1^i - 1^{i+})u_i). \tag{9.65}$$

The value-iteration algorithm is expressed

$$V_{n+1}^*(x) = \mathsf{E}[V_n^*(x + A(1))] + c(x) + \min_{u \in \mathsf{U}_\diamond(x)}\left(\sum_{i=1}^{\ell}\mu_i V_n^*(x - (1^i - 1^{i+})u_i)\right).$$
$$\tag{9.66}$$

We have remarked that the algorithm can be used to obtain insight regarding the structure of value functions or the structure of optimal policies. Proposition 9.6.4 is one example in which properties of the algorithm are used to establish monotonicity of each value function, and hence any possible limiting function. Monotonicity is used to verify irreducibility of the process controlled using ϕ_n^* for $n \geq 0$.

Proposition 9.6.4. *Suppose that the cost function* $c\colon \mathsf{X}_\diamond \to \mathbb{R}_+$ *and the function* $V_0\colon\ \colon \mathsf{X}_\diamond \to \mathbb{R}_+$ *satisfy the following: For each* $i \in \{1, \ldots, \ell\}$ *and* $x \in \mathsf{X}_\diamond$,

$$\begin{aligned} c(x) &\leq c(x + 1^i), & c(x + 1^{i+} - 1^i) &\leq c(x), \text{ if } x_i \geq 1. \\ V_0(x) &\leq V_0(x + 1^i), & V_0(x + 1^{i+} - 1^i) &\leq V_0(x), \text{ if } x_i \geq 1. \end{aligned} \tag{9.67}$$

Then the sequence of value functions $\{V_n^*\}$ *obtained using the VIA satisfies for each* $n \geq 1$, $i \in \{1, \ldots, \ell\}$, *and* $x \in \mathsf{X}_\diamond$,

$$V_n^*(x) \leq V_n^*(x + 1^i), \quad and \quad V_n^*(x + 1^{i+} - 1^i) \leq V_n^*(x), \quad if\ x_i \geq 1. \tag{9.68}$$

It follows that there is no reason to idle:

Corollary 9.6.5. *Under the assumptions of Proposition 9.6.4, the stationary policies $\{\phi_n^*\}$ obtained using the VIA can be chosen so that each is nonidling, and the resulting controlled process is 0-irreducible.*

Proof. The form of the recursion (9.66) combined with Proposition 9.6.4 implies the nonidling property, and Proposition 8.2.1 implies that each policy is 0-irreducible. $\square$

Proof of Proposition 9.6.4. We prove the result by induction on n. For $n = 0$ there is nothing to prove. Assuming the value function V_n^* satisfies (i) and (ii) we write for any j, any x satisfying $x_j \geq 1$, and any $u \in \mathsf{U}_\diamond(x - 1^j + 1^{j+})$,

$$V_{n+1}^*(x - 1^j + 1^{j+}) \leq \mathsf{E}[V_n^*(x - 1^j + 1^{j+} + A(1))] + c(x - 1^j + 1^{j+})$$
$$+ \left(\sum_{i=1}^{\ell} \mu_i V_n^*(x - (1^j - 1^{j+}) - (1^i - 1^{i+})u_i) \right).$$

Hence by the induction hypothesis and the assumptions on the cost function,

$$V_{n+1}^*(x - 1^j + 1^{j+}) \leq \mathsf{E}[V_n^*(x + A(1))] + c(x)$$
$$+ \left(\sum_{\substack{i=1 \\ j \neq i}}^{\ell} \mu_i V_n^*(x - (1^i - 1^{i+})u_i) \right)$$
$$+ \mu_j V_n^*((x - 1^j + 1^{j+}) - (1^j - 1^{j+})u_j).$$

If $x_j = 1$ then $\mathsf{U}_\diamond(x - 1^j + 1^{j+})$ is a strict subset of $\mathsf{U}_\diamond(x)$. Nevertheless, we have by the induction hypothesis

$$V_n^*((x - 1^j + 1^{j+}) - (1^j - 1^{j+})u_j) \leq V_n^*(x - (1^j - 1^{j+})u_j'),$$

whenever $u \in \mathsf{U}_\diamond(x - 1^j + 1^{j+})$ and $u' \in \mathsf{U}_\diamond(x)$. Thus, for any $u \in \mathsf{U}_\diamond(x)$,

$$V_{n+1}^*(x - 1^j + 1^{j+}) \leq \mathsf{E}[V_n^*(x + A(1))] + c(x)$$
$$+ \left(\sum_{i=1}^{\ell} \mu_i V_n^*(x - (1^i - 1^{i+})u_i) \right).$$

Minimizing the right-hand side over all $u \in \mathsf{U}_\diamond(x)$ gives $V_{n+1}^*(x - 1^j + 1^{j+}) \leq V_{n+1}^*(x)$. The proof that $V_{n+1}^*(x + 1^j) \geq V_{n+1}^*(x)$ is identical. $\square$

9.7 One-dimensional inventory model

In Section 7.4 we computed a hedging point for the inventory model such that the resulting policy is optimal over all hedging-point policies. The proof was based on the representation (7.38) for the average cost. Under the same assumptions, we now construct the solution to the average-cost dynamic programming equations to establish that the hedging-point policy is average-cost optimal over *all* policies.

Recall that the state space is not countable, but the existence of a regeneration time is enough to generalize the methods obtained in the countable state-space case. For example, the representation for the invariant measure given in (7.38) is identical to the countable state-space formula given in the Appendix.

The state process for the simple inventory model is denoted $\boldsymbol{Y}$, evolving on $\mathbb{R}$. For a given hedging point $\overline{y}$, recall from (7.34) that the evolution of $\boldsymbol{Y}$ is given by $Y(t+1) = \max(Y(t), -\overline{y}) - \mu + L(t+1)$, $t \geq 0$. The relative value function can be expressed in a form similar to (9.43) or the invariant measure (7.38) as

$$h^{\overline{y}}(y) = \mathsf{E}\left[\sum_{t=0}^{\sigma_0}\left(\overline{c}(Y(t)) - \eta(\overline{y})\right)\right], \qquad Y(0) = y, \qquad (9.69)$$

where the stopping time σ_0 is defined in (8.14),

$$\sigma_0 = \min\{t \geq 0 : Y^0(t) \leq 0\} = \min\{t \geq 0 : Y(t) \leq -\overline{y}\}. \qquad (9.70)$$

Recall that τ_0 is defined similarly in (7.39), but with the minimum over $t \geq 1$.

We write $h(y) = h^{\overline{y}}(y)$ when there is no risk of confusion, and we let h^* denote the function (9.69) in the special case $\overline{y} = \overline{y}^*$ given in Theorem 7.4.2. Finally, denote

$$h^{*1}(y) := \mathsf{E}[h^*(y - \mu + L(1))], \qquad y \in \mathbb{R}. \qquad (9.71)$$

Theorem 9.7.1. *The function h^* solves the following dynamic programming equation for the simple inventory model (7.30),*

$$\min_{\iota \geq 0} h^{*1}(y + \iota) = h^*(y) - \overline{c}(y) + \eta(\overline{y}^*), \qquad y \in \mathbb{R}. \qquad (9.72)$$

The minimum in (9.72) is achieved using the hedging-point policy $\iota^*(t) = \max(0, -\overline{y}^* - Y^*(t))$.

This section provides a proof of Theorem 9.7.1. We state without proof the analogous result for the discounted-cost criterion.

Theorem 9.7.2. *For any $\gamma > 0$, the simple inventory model (7.30) admits a discounted-cost optimal policy. The optimal policy is unique: It is a hedging-point policy $\iota^*(t) = \max(0, -\overline{y}^* - Y^*(t))$ with $\overline{y}^* \in \mathbb{R}$ chosen so that*

$$\tfrac{d^+}{dy}h_\gamma^{*1}(-\overline{y}^*) = 0.$$

First we verify that $h^{\overline{y}}$ solves Poisson's equation.

Proposition 9.7.3. *The function defined in (9.69) solves Poisson's equation for the simple inventory model (7.30) controlled using a hedging-point policy. That is,*

$$Ph^{\overline{y}}(y) = \mathsf{E}[h^{\overline{y}}(Y(1)) \mid Y(0) = y] = h^{\overline{y}}(y) - \overline{c}(y) + \eta(\overline{y}), \qquad y \in \mathbb{R}. \qquad (9.73)$$

Proof. The function $h = h^{\overline{y}}$ can be expressed in terms of the process $\boldsymbol{Y}^0$ as follows:

$$h(y) = \mathsf{E}\left[\sum_{t=0}^{\sigma_0}\left(\overline{c}(Y^0(t) - \overline{y}) - \eta(\overline{y})\right) \mid Y^0(0) = y + \overline{y}\right].$$

Consequently, we have

$$Ph\,(y) = \mathsf{E}[h(Y(1)) \mid Y(0) = y] = \left[\sum_{t=1}^{\tau_0}(\overline{c}(Y^0(t) - \overline{y}) - \eta(\overline{y})) \mid Y^0(0) = y + \overline{y}\right].$$

Moreover, $\sigma_0 = \tau_0$ for $Y^0(0) > 0$, and $\sigma_0 = 0$ otherwise, so that this identity becomes

$$Ph\,(y) = h(y) - (c(y) - \eta(\overline{y}))$$

$$+ \mathbf{1}\{Y^0(0) \leq 0\}\mathsf{E}\left[\sum_{t=1}^{\tau_0}(\overline{c}(Y^0(t) - \overline{y}) - \eta(\overline{y})) \mid Y^0(0) = y + \overline{y}\right].$$

Hence to prove the proposition it is enough to show that the expectation on the right-hand side is zero when $Y^0(0) \leq 0$.

If $Y^0(0) \leq 0$ then $Y^0(1) = -\mu + L(1)$. Consequently, when $y + \overline{y} \leq 0$,

$$\mathsf{E}\left[\sum_{t=1}^{\tau_0}(\overline{c}(Y^0(t) - \overline{y}) - \eta(\overline{y})) \mid Y^0(0) = y + \overline{y}\right]$$

$$= \mathsf{E}\left[\sum_{t=1}^{\tau_0}(\overline{c}(Y^0(t) - \overline{y}) - \eta(\overline{y})) \mid Y^0(0) = 0\right].$$

This completes the proof since by Theorem 7.4.2 the right-hand side vanishes:

$$\mathsf{E}\left[\sum_{t=1}^{\tau_0}(\overline{c}(Y^0(t) - \overline{y}) - \eta(\overline{y})) \mid Y^0(0) = 0\right] = \mathsf{E}[\tau_0 \mid Y^0(0) = 0]\pi^{\overline{y}}(\overline{c} - \eta(\overline{y})) = 0.$$

$\square$

To prove (9.72) we require representations for the derivatives of h^*. The following result is a refinement of Lemma 7.4.4.

The expression "Ph" appearing in the lemma is the conditional expectation using the $\overline{y}$-hedging-point policy. Poisson's equation (9.73) holds, so that $Ph\,(y) = h(y) - \overline{c}(y) + \eta(\overline{y})$. Recall the definition of the sensitivity function $\chi_{\overline{c}}$ in (7.43).

Proposition 9.7.4. *For any hedging point $\overline{y} \geq 0$, initial condition $Y(0) = y > -\overline{y}$, and any stopping time satisfying $\tau \leq \tau_0$,*

$$\frac{d}{dy}Ph\,(y) \;=\; \mathsf{E}\left[\sum_{t=1}^{\tau}\chi_{\overline{c}}(Y(t)) + \frac{d}{dy}Ph\,(Y(\tau))\right] \tag{9.74a}$$

$$\frac{d^2}{dy^2}Ph\,(y) \;=\; |c|\mathsf{E}\left[\sum_{t=0}^{\tau-1}p_L(\mu - Y(t)) + \frac{d^2}{dy^2}Ph\,(Y(\tau))\right], \tag{9.74b}$$

where $|c| = c^- + c^+$.

The following lemma will be used to justify the Dominated Convergence Theorem in the proof of Proposition 9.7.4.

Lemma 9.7.5. *For any stopping time $\tau \leq \tau_0$ and each initial condition $y^0 = Y(0)$, the expectation $\mathsf{E}[|Y^0(\tau)|]$ is finite provided $\mathsf{E}[\tau]$ is finite.*

Proof. This follows from the skip-free property (8.23) for the inventory model. Representing $|Y(\tau)|$ as the sum of its increments we obtain

$$\mathsf{E}[|Y^0(\tau)|] = \mathsf{E}\left[\sum_{t=0}^{\tau-1}(|Y^0(t+1)| - |Y^0(t)|)\right] + |y^0|$$

$$\leq \mathsf{E}\left[\sum_{t=0}^{\infty}(\mu + L(t+1))\mathbf{1}\{\tau > t\}\right] + |y^0|$$

$$= (\mu + \lambda)\mathsf{E}[\tau] + |y^0| < \infty,$$

where in the final equation we have used independence of $L(t+1)$ and the event $\{\tau > t\}$; independence holds because the event $\{\tau > t\}$ is $\mathcal{Y}_t$-measurable, with $\mathcal{Y}_t = \sigma\{Y(0), \ldots, Y(t)\}$. $\qquad\square$

The second-derivative formula in Proposition 9.7.4 is based on the following expression for the derivative of the smoothed version of $\chi_{\bar{c}}$ defined by

$$\chi_{\bar{c}}^1(y) = \mathsf{E}[\chi_{\bar{c}}(y - \mu + L(1))], \qquad y \in \mathbb{R}. \tag{9.75}$$

Lemma 9.7.6. *The function $\chi_{\bar{c}}^1$ is C^1, with*

$$\tfrac{d}{dy}\chi_{\bar{c}}^1(y) = (c^- + c^+)p_L(\mu - y).$$

Proof. This follows from the representation

$$\chi_{\bar{c}}^1(y) = \int_0^\infty \chi_{\bar{c}}(y - \mu + r)\, p_L(r)\, dr = (-c^-)\int_0^{\mu-y} p_L(r)\, dr + (c^+)\int_{\mu-y}^\infty p_L(r)\, dr,$$

and the Fundamental Theorem of Calculus. $\qquad\square$

Proof of Proposition 9.7.4. Poisson's equation (9.73) and the martingale property imply that

$$h(y) = \mathsf{E}\left[\sum_{t=0}^{\tau-1}(\bar{c}(Y(t)) - \eta(\bar{y})) + h(Y(\tau))\right],$$

where we again write $h = h^{\bar{y}}$ to simplify the notation. With the stopping time τ and $y > -\bar{y} + \mu$ fixed, we consider two initial conditions: $Y(0) = y$, and $Y^\varepsilon(0) = y + \varepsilon$, with $\varepsilon > 0$ a small constant. We then have

$$Y^\varepsilon(t) = Y(t) + \varepsilon, \qquad 0 \leq t \leq \tau,$$

and consequently,

$$\frac{h(y + \varepsilon) - h(y)}{\varepsilon}$$

$$= \mathsf{E}\left[\sum_{t=0}^{\tau-1}\varepsilon^{-1}(\bar{c}(Y(t) + \varepsilon) - \bar{c}(Y(t))) + \varepsilon^{-1}(h(Y(\tau) + \varepsilon) - h(Y(\tau)))\right].$$

Since $\bar{c}$ is Lipschitz and τ has a first moment, the Dominated Convergence Theorem implies that

$$\lim_{\varepsilon \to 0} \mathsf{E}\left[\sum_{t=0}^{\tau-1} \varepsilon^{-1}\left(\bar{c}(Y(t)+\varepsilon) - \bar{c}(Y(t))\right)\right] = \mathsf{E}\left[\sum_{t=0}^{\tau-1} \chi_{\bar{c}}(Y(t))\right].$$

From the definition (9.69) we also have for some constant k_0,

$$\varepsilon^{-1}\big|h(Y(\tau)+\varepsilon) - h(Y(\tau))\big| \le k_0 |Y(\tau)|,$$

and the right-hand side has finite mean by Lemma 9.7.5. The Dominated Convergence Theorem again implies convergence,

$$\lim_{\varepsilon \to 0} \mathsf{E}\left[\varepsilon^{-1}\left(h(Y(\tau)+\varepsilon) - h(Y(\tau))\right)\right] = \mathsf{E}\left[\tfrac{d}{dy} h\left(Y(\tau)\right)\right].$$

We thereby obtain a result similar to (9.74a),

$$\tfrac{d}{dy} h\left(y\right) = \mathsf{E}\left[\sum_{t=0}^{\tau-1} \chi_{\bar{c}}(Y(t)) + \tfrac{d}{dy} h(Y(\tau))\right]. \tag{9.76}$$

To transform (9.76) into (9.74a) first note that Poisson's equation (9.73) implies that

$$\tfrac{d}{dy} h\left(y\right) = \tfrac{d}{dy} Ph\left(y\right) + \chi_{\bar{c}}(y) \quad \text{and} \quad \tfrac{d}{dy} h\left(Y(\tau)\right) = \tfrac{d}{dy} Ph\left(Y(\tau)\right) + \chi_{\bar{c}}(Y(\tau)).$$

Substituting these expressions into (9.76) gives (9.74a).

To prove (9.74b) we write, for $t \ge 1$,

$$\mathsf{E}[\chi_{\bar{c}}(Y(t))\mathbf{1}\{\tau \ge t\}] = \mathsf{E}[\chi_{\bar{c}}^{1}(Y(t-1))\mathbf{1}\{\tau \ge t\}].$$

This follows from the fact that $\{\tau \ge t\}$ is $\mathcal{Y}_t$-measurable (i.e., measurable with respect to the observations $\{Y(0), \ldots, Y(t-1)\}$). Consequently,

$$\mathsf{E}\left[\sum_{t=1}^{\tau} \chi_{\bar{c}}(Y(t))\right] = \sum_{t=1}^{\infty} \mathsf{E}[\chi_{\bar{c}}(Y(t))\mathbf{1}\{\tau \ge t\}]$$

$$= \sum_{t=1}^{\infty} \mathsf{E}[\chi_{\bar{c}}^{1}(Y(t-1))\mathbf{1}\{\tau \ge t\}]$$

$$= \mathsf{E}\left[\sum_{t=0}^{\tau-1} \chi_{\bar{c}}^{1}(Y(t))\right].$$

Applying (9.74a) we see that the first derivative can be written

$$\tfrac{d}{dy} Ph\left(y\right) = \mathsf{E}\left[\sum_{t=0}^{\tau-1} \chi_{\bar{c}}^{1}(Y(t)) + \tfrac{d}{dy} Ph\left(Y(\tau)\right)\right]. \tag{9.77}$$

In the special case $\tau = \tau_0$ we have $\frac{d}{dy} Ph\left(Y(\tau)\right) = 0$ since Ph is constant on $(-\infty, \overline{y}]$ and $Y(\tau) \in (-\infty, \overline{y})$ a.s. Hence (9.77) becomes in this case

$$\frac{d}{dy} Ph\left(y\right) = \mathsf{E}\left[\sum_{t=0}^{\tau_0 - 1} \chi_{\overline{c}}^{1}(Y(t))\right].$$

This shows that $\frac{d}{dy} Ph$ is a smooth, Lipschitz function of y on $(-\overline{y}, \infty)$.

On differentiating each side of (9.77) (applying the Dominated Convergence Theorem once more) we obtain (9.74b) from the derivative formula Lemma 9.7.6. $\square$

We now prove Theorem 9.7.1

Proof of Theorem 9.7.1. We have already established Poisson's equation. To establish the dynamic programming equation in the form (9.72) we demonstrate the following bounds:

$$\frac{d}{dy} h_\gamma^{*1}(y) \begin{cases} \leq 0 & y \leq \overline{y}^* \\ > 0 & y > \overline{y}^*. \end{cases} \tag{9.78}$$

Consider an initial condition satisfying $y > -\overline{y}^*$. We apply Proposition 9.7.4 with the stopping time τ_0. This obviously depends upon the initial condition, while Proposition 9.7.4 requires independence. Here we fix the random variable τ_0 based on the given initial condition y. We thereby obtain the identity

$$\frac{d}{dy} h^{*1}(y) = \mathsf{E}\left[\sum_{t=1}^{\tau_0} \chi_{\overline{c}}(Y^0(t) - \overline{y}^*) \mid Y^0(0) = y + \overline{y}^*\right], \qquad y > -\overline{y}^*.$$

The right-hand side is nondecreasing in y, and as $y \downarrow -\overline{y}^*$ is convergent:

$$\lim_{y \downarrow -\overline{y}^*} \frac{d}{dy} h^{*1}(y) = \mathsf{E}\left[\sum_{t=1}^{\tau_0} \chi_{\overline{c}}(Y^0(t) - \overline{y}^*) \mid Y^0(0) = 0\right].$$

The right-hand side is zero by Theorem 7.4.2, which implies that

$$\frac{d}{dy} h^{*1}(y) \geq \frac{d}{dy} h^{*1}(-\overline{y}^*) = 0, \qquad y > -\overline{y}^*.$$

Note that (9.74b) implies that h^{*1} is strictly convex on this domain since the second derivative is strictly positive. Hence the inequality above is strict, which proves the lower bound in (9.78).

To establish the upper bound for $y \leq -\overline{y}^*$ it is enough to establish convexity of h^{*1} on all of $\mathbb{R}$. By Poisson's equation we have

$$h^*(y) = Ph^*(y) + \overline{c}(y) - \eta(\overline{y}^*) = h^{*1}(\max(-\overline{y}^*, y)) + \overline{c}(y) - \eta(\overline{y}^*).$$

The right-hand side is an affine function of y for $y \leq -\overline{y}^*$, convex for $y > -\overline{y}^*$, and it is smooth at $-\overline{y}^*$ since $h^{*1}(\max(-\overline{y}^*, y))$ is a smooth function of y, with zero derivative at $\overline{y}^*$. This shows that h^* is convex, and it follows that h^{*1} is also convex. $\square$

9.8 Hedging and workload

We now have sufficient ammunition to seriously address the results surveyed in Sections 5.6 and 7.5.2.

It is assumed throughout this section that the assumptions of Theorem 7.5.2 hold. In particular, the distribution of $L(t)$ possesses a continuous density whose support is bounded, convex, and contained in $\mathsf{Y} \cup \mathbb{R}_+^2$. Its mean is denoted $\lambda \in \mathbb{R}_+^2$.

The cost function is monotone in y_1, so that the monotone region has the form

$$\mathsf{Y}^+ = \{y \in \mathsf{Y} : y_2 \geq m_* y_1\}. \tag{9.79}$$

Under these assumptions we focus our attention on the height process H and its relaxation H^∞ defined in (7.57). Its cost function $\bar{c}_H$ is defined in (7.58), and an associated sensitivity function is defined as in (7.43),

$$\bar{c}_H(x) = c_H^+ x_+ + c_H^- x_-$$
$$\chi_{\bar{c}_H}(x) = c_H^+ \mathbf{1}\{x \geq 0\} - c_H^- \mathbf{1}\{x < 0\}, \qquad x \in \mathbb{R}, \tag{9.80}$$

with $c_H^+ = c_2^+$ and $c_H^- = |c_2^-|$. The relative value function based on this cost function is denoted h_H^* (defined in (9.69) with $\bar{y} = \bar{y}^*$ given in Theorem 7.4.2).

Proof of Theorem 7.5.2. The key elements of the proof are distributed across this section:

(i) Part (i) asserts that the unconstrained process admits an optimal policy that is affine. This is established in Proposition 9.8.3.

(ii) The existence of a switching curve defining the optimal policy is established in Proposition 9.8.2 for the discounted-cost criterion, and in Proposition 9.8.7 for average cost.

(iii) Proposition 9.8.5 establishes the convergence (7.61) of the switching curves defining the discounted-cost optimal policy,

$$\lim_{y_1 \to \infty} |s_*(y_1) - s_*^\infty(y_1)| = 0,$$

where s_*^∞ is defined in (7.59), which we copy here:

$$s_*^\infty(y_1) = m_* y_1 - \bar{y}_2^*, \qquad y_1 \in \mathbb{R}. \tag{9.81}$$

(iv) The proof of convergence of policies for the average-cost problem in Case II* is contained in Lemmas 9.8.12 and 9.8.13. □

We require the following notation in this section. Let $\mathsf{Y}_{\bar{c}^\infty} \subset \mathsf{Y}$ denote an open convex cone that lies in the interior of $\mathsf{Y} \cap \mathbb{R}_+^2$ and satisfies $\bar{c}(y) = \bar{c}^\infty(y)$ for $y \in \mathsf{Y}_{\bar{c}^\infty}$. For $\varepsilon > 0$ sufficiently small we can take

$$\mathsf{Y}_{\bar{c}^\infty} := \{y \in \mathbb{R}_+^2 : 0 < (m_* - \varepsilon)y_1 < y_2 < (m_* + \varepsilon)y_1\}.$$

For any policy and any initial condition we denote the first exit time from $\mathsf{Y}_{\bar{c}^\infty}$ by

$$\tau_\bullet := \inf\{t \geq 1 : Y(t) \notin \mathsf{Y}_{\bar{c}^\infty}\}, \tag{9.82}$$

and the two "first idling times,"

$$\tau_i = \inf\{t \geq 1 : \iota_i(t) > 0\}, \qquad i = 1, 2. \tag{9.83}$$

9.8.1 Discounted cost

We let h_γ^* denote the optimal value function, extended to all of $\mathbb{R}^2$ via $h_\gamma^*(y) := \infty$ if $y \notin \mathsf{Y}$, and define

$$h_\gamma^{*1}(y) = \mathsf{E}[h^*(y - \mu + L(1))], \qquad y \in \mathsf{Y}. \tag{9.84}$$

The function h_γ^{*1} is necessarily infinite if $\mathsf{P}\{y - \mu + L(1) \in \mathsf{Y}^c\} > 0$.

The h_γ^*-myopic policy can be expressed

$$\phi^*(y) = \arg\min_\iota \{h_\gamma^{*1}(y + \iota) : \iota \geq 0\}. \tag{9.85}$$

We let P_* denote the transition *kernel* under the h_γ^*-myopic policy. That is, for any $y \in \mathsf{Y}$ and $A \in \mathcal{B}(\mathsf{Y})$, $P_*(y, A) = \mathsf{P}\{Y(t+1) \in A \mid Y(t) = y, \iota(t) = \phi^*(y)\}$.

Proposition 9.8.1. *The following hold for the discounted-cost optimization problem under the assumptions of Theorem 7.5.2:*

(i) *The value function $h_\gamma^* \colon \mathsf{Y} \to \mathbb{R}_+$ and the function $h_\gamma^{*1} \colon \mathsf{Y} \to \mathbb{R}_+$ are each convex. Moreover, the function $P_* h_\gamma^*$ is monotone,*

$$P_* h_\gamma^* (y + v) \geq P_* h_\gamma^* (y), \qquad y \in \mathsf{Y}, \ v \in \mathbb{R}_+^2 \cap \mathsf{Y}.$$

(ii) *The value function can be expressed as the infimum,*

$$h_\gamma^*(y) = \gamma^{-1} \inf \mathsf{E}_y[\bar{c}(Y(T))], \tag{9.86}$$

where T is a geometrically distribution random variable with mean γ^{-1} that is independent of L, and the infimum is over all admissible idleness processes.

Proof. Convexity of h_γ^* follows from convexity of $\bar{c}$ and convexity of the set of possible idleness sequences ι. Convexity of h_γ^{*1} follows from convexity of h_γ^*.

To establish monotonicity let $Y(0) = y \in \mathsf{Y}$ and $Y^v(0) = y + v$, with $v \in \mathbb{R}_+^2 \cap \mathsf{Y}$. We have $Y^v(0) \in \mathsf{Y}$ since Y is a convex cone. If ι^v is any feasible idleness process starting from $Y^v(0)$, then the following is feasible from $Y(0)$: $\iota(0) = \iota^v(0) + v$, and $\iota(t) = \iota^v(t)$ for $t \geq 1$ since this gives $Y(t) = Y^v(t)$ for $t \geq 1$. Consequently, using this policy we obtain the upper bound

$$h_\gamma^*(y) \leq \sum_{t=0}^\infty (1 + \gamma)^{-t-1} \mathsf{E}[\bar{c}(Y(t))]$$

$$= (1 + \gamma)^{-1} (\bar{c}(y) - \bar{c}(y + v)) + \sum_{t=0}^\infty (1 + \gamma)^{-t-1} \mathsf{E}[\bar{c}(Y^v(t))].$$

Minimizing over all ι^v gives

$$h_\gamma^*(y) - (1+\gamma)^{-1}\overline{c}(y) \leq h_\gamma^*(y+v) - (1+\gamma)^{-1}\overline{c}(y+v).$$

We also have the dynamic programming equation

$$(1+\gamma)^{-1}P_*h_\gamma^*(y) = h_\gamma^*(y) - (1+\gamma)^{-1}\overline{c}(y), \qquad y \in \mathsf{Y},$$

so that the previous inequality completes the proof of (i).

Part (ii) follows from arguments used elsewhere in the book – see for example (9.39).

$\square$

A proof of Theorem 7.5.2 was begun in Section 7.5.2 based on the translated optimization problem with state space Y_r and cost function $\overline{c}^r$ defined by

$$\begin{aligned} \mathsf{Y}_r &= \{\mathsf{Y} - r(1, m_*)^{\mathsf{T}}\} \subset \mathbb{R}^2, \\ \overline{c}^r(y) &= \overline{c}(y + r(1, m_*)^{\mathsf{T}}) - r\overline{c}((1, m_*)^{\mathsf{T}}), \quad y \in \mathsf{Y}_r. \end{aligned} \tag{9.87}$$

We let $h_\gamma^{r*}\colon \mathsf{Y}_r \to \mathbb{R}_+$ denote the optimal value function based on the state space Y_r and cost function $\overline{c}^r$ defined in (9.87).

Proposition 7.5.3 states that value functions converge monotonically as $r \to \infty$ to the value function for the analogous optimization problem for the unconstrained process $\boldsymbol{Y}^\infty$ with state space $\mathbb{R}^2$, and cost function $\overline{c}^\infty$. Below we refine the conclusion of Proposition 7.5.3 to establish convergence of the optimal policies.

We let $\mathcal{E}_\gamma(y)$ denote an error term of the form, for some positive constants $d_i = d_i(\gamma)$, $i = 0, 1$,

$$\mathcal{E}_\gamma(y) = O\big(\exp(-d_0 y_1 + d_1|y_2 - m_* y_1|)\big), \qquad y \in \mathsf{Y}. \tag{9.88}$$

We also let $\mathcal{E}_\gamma(y_1)$ denote an error term of the form $\mathcal{E}_\gamma(y_1) = O(\exp(-d_0 y_1))$.

The switching curve s_* described in Theorem 7.5.2 is defined as follows,

$$s_*(y_1) := \inf\Big\{y_2 : \frac{\partial}{\partial y_2}h_\gamma^{*1}(y) > 0\Big\}, \tag{9.89}$$

defined for $y_1 \in \mathbb{R}$ such that $(y_1, y_2)^{\mathsf{T}} \in \mathsf{Y}$ for some y_2.

Based on Proposition 9.8.1 we have

Proposition 9.8.2. *Suppose that $y \in \mathsf{Y}$ is given, and that $y^0 := (y_1, s_*(y_1))^{\mathsf{T}}$ is in the interior of Y and satisfies*

$$\frac{\partial}{\partial y_2}h_\gamma^{*1}(y^0) = 0.$$

Then from this initial condition $\phi_1^*(y) = 0$ and $\phi_2^*(y) = (s_*(y_1) - y_2)_+$.

Proof. Let $v \in \mathsf{Y} \cap \mathbb{R}_+^2$ satisfy $v_1 > 0$. Observe that $h_\gamma^{*1}(x) = P_*h_\gamma^*(x)$ whenever $\iota(0) = \mathbf{0}$. Applying the monotonicity result in Proposition 9.8.1 gives $v^{\mathsf{T}}\nabla h_\gamma^{*1}(y^0) \geq 0$ for any $v \in \mathbb{R}_+^2 \cap \mathsf{Y}$, and hence under the assumptions of the proposition we have

$$\frac{d}{dy_1}h_\gamma^{*1}(y^0) \geq 0, \qquad \frac{d}{dy_2}h_\gamma^{*1}(y^0) = 0.$$

Since h_γ^{*1} is convex, this shows that y^0 is a minimum of h_γ^{*1} on the region $\{y \in \mathsf{Y} : y_1 \geq y_1^0\}$. The conclusions follow from the expression for ϕ^* in (9.85). $\qquad \square$

For the unconstrained model we define a switching curve $s_*^\infty \colon \mathbb{R} \to \mathbb{R}$ in the same way. We establish in Proposition 9.8.3 that it is affine, of the form (7.59).

Proposition 9.8.3. *Suppose that the assumptions of Theorem 7.5.2 are satisfied. Then the optimal policy for $\boldsymbol{Y}^\infty$ is unique: It is affine,*

$$\phi_1^{\infty*}(y) = 0, \quad \phi_2^{\infty*}(y) = (s_*^\infty(y_1) - y_2)_+, \qquad y \in \mathbb{R}^2, \tag{9.90}$$

where s_*^∞ is defined in (9.81) with $\overline{y}_2^*$ given in Theorem 9.7.2, based on the height process (7.57).

Proof. Fix $y^0 \in \mathbb{R}^2$, and recall the definition of $y^r = y^0 + r(1, m_*)^\mathsf{T}$ in (7.63). If $Y^\infty(t; y^0)$ is any feasible trajectory starting from y^0, then

$$Y^\infty(t; y^r) := Y^\infty(t; y^0) + r(1, m_*)^\mathsf{T}, \qquad t \geq 0,$$

is a feasible trajectory starting from y^r. Moreover, by definition of the cost function,

$$\overline{c}^\infty(Y^\infty(t; y^r)) = \overline{c}^\infty(Y^\infty(t; y^0)) + r\overline{c}((1, m_*)^\mathsf{T}), \qquad t \geq 0, \; r \in \mathbb{R}.$$

Optimizing over all feasible $\boldsymbol{Y}^\infty$, it follows that

$$h_\gamma^{\infty*}(y^r) = h_\gamma^{\infty*}(y^0) + r\gamma^{-1}\overline{c}((1, m_*)^\mathsf{T}), \qquad t \geq 0, \; r \in \mathbb{R}. \tag{9.91}$$

An optimal policy for $\boldsymbol{Y}^\infty$ is defined by a dynamic programming equation analogous to that given in (9.85):

$$\phi^{\infty*}(y) = \arg\min_\iota \{h_\gamma^{\infty*1}(y + \iota) : \iota \geq 0\}.$$

The derivative $\frac{\partial}{\partial y_2} h_\gamma^{\infty*1}(y^r)$ does not depend upon r, by (9.91), which implies that the policy is affine.

Uniqueness follows by observing that the optimal idleness process is obtained as the solution to a one-dimensional inventory model with state process $\boldsymbol{H}^\infty$ and cost function given in (9.80). The optimal policy is determined uniquely by the hedging-point policy given in Theorem 9.7.2. $\qquad \square$

Proposition 9.8.4 gives upper and lower bounds on the value function, which is a refinement of Proposition 7.5.3.

Proposition 9.8.4. *Suppose that the affine policy s_*^∞ is applied to the CRW model $\boldsymbol{Y}$, and let h_γ denote the resulting value function. Then:*

(i) $\mathsf{P}\{\tau_\bullet < T \mid Y(0) = y\} = \mathcal{E}_\gamma(y)$, *where $\mathcal{E}_\gamma$ is an error term satisfying (9.88), and T is the geometric random variable introduced in Proposition 9.8.1.*

(ii) $h_\gamma^*(y) \leq h_\gamma(y) \leq h_\gamma^{\infty*}(y) + \mathcal{E}_\gamma(y)$ *for each $y \in \mathsf{Y}$.*

Proof. We leave the proof of (i) as an exercise. A formal proof is given in Lemma 9.8.9 below in a slightly different setting. The proof is based on the height process: Fig. 5.11 illustrates that if $Y_2(0) = m_* Y_1(0)$ so that $H(0) = 0$, then with high probability $Y(T)$ will be far from the boundary of $\mathsf{Y}_{\bar{c}^\infty}$ when $Y_1(0) \gg 1$.

To establish (ii) we apply the following identity for the workload process under the affine policy:

$$\gamma h_\gamma(y) = \mathsf{E}_y[\bar{c}(Y(T))]$$

$$= \mathsf{E}_y[\bar{c}^\infty(Y^{\infty*}(T))\mathbf{1}\{\tau_\bullet > T\}] + \mathsf{E}_y[\bar{c}(Y(T))\mathbf{1}\{\tau_\bullet \le T\}] \qquad (9.92)$$

$$\le \gamma h_\gamma^{\infty*}(y) + \mathsf{E}_y[\bar{c}(Y(T))\mathbf{1}\{\tau_\bullet \le T\}].$$

An application of the Cauchy–Schwartz inequality to the right-hand side, combined with the bound in (i), implies the desired upper bound in (ii). $\qquad\square$

Given these results, it is not surprising that the optimal policy for Y is approximated by an affine policy:

Proposition 9.8.5. *Under the assumptions of Theorem 7.5.2,*

$$\lim_{y_1 \to \infty} \left| s_*(y_1) - s_*^\infty(y_1) \right| = 0.$$

Proof. For each $r \ge 0$ we consider the optimal process Y^{r*}, with arrival process L independent of r. The initial condition is fixed, of the form $y^0 = (0, y_2^0)^\mathsf{T}$. We have $y^0 \in \mathsf{Y}_r$ for all $r \ge 1$ sufficiently large since $\mathsf{Y}_r \uparrow \mathbb{R}^2$ as $r \uparrow \infty$.

The function h_γ^{r*1} is defined in analogy with (9.84) by $h_\gamma^{r*1}(y) = \mathsf{E}[h_\gamma^{r*}(y - \mu + L(1))]$. It is convex and nondecreasing as a function of y_2, and for all sufficiently large r,

$$\frac{\partial}{\partial y_2} h_\gamma^{r*1}(y^0) > 0 \iff y_2^0 > s_*(r) - r m_*. \qquad (9.93)$$

From Proposition 7.5.3 we have $h_\gamma^{r*}(y^0) \downarrow h_\gamma^{\infty*}(y^0)$ as $r \uparrow \infty$.

To obtain a complementary lower bound and complete the proof we consider possible weak limits of the cumulative idleness process. This and the cumulative disturbance are defined by

$$I^{r*}(t) = \sum_{k=0}^{t-1} \iota^{r*}(k), \quad N(t) = \sum_{k=1}^{t} L(k), \qquad t \ge 1.$$

Let T denote the geometrically distributed random variable introduced in (9.36). To establish the existence of a weak limit for $\{Y^{r*}(T), I^{r*}(T) : r \ge 1\}$ it is enough to obtain a bound on the mean of the norm of each of these sequences of random variables.

It is assumed that $\bar{c}$ can be extended to define a norm on $\mathbb{R}^2$. Since all norms are equivalent, we can find $\varepsilon > 0$ such that $\bar{c}(y) \ge \varepsilon|y|$ for all y, where $|\cdot|$ denotes the ℓ_1-norm. Using the fact that the discounted cost is finite,

$$\mathsf{E}[\bar{c}^r(Y^{r*}(T))] = \gamma h_\gamma^{r*}(y^0) < \infty,$$

we obtain an upper bound on $\boldsymbol{I}^{r*}$ for any y^0,

$$
\begin{aligned}
\limsup_{r\to\infty} \mathsf{E}[\|I^{r*}(T)\|] &= \limsup_{r\to\infty} \mathsf{E}[\|Y^{r*}(T) - y^0 + \mu T - N(T)\|] \\
&\leq \limsup_{r\to\infty} \mathsf{E}[\|Y^{r*}(T)\| + |y^0| + |\mu T| + |N(T)|] \\
&\leq \varepsilon^{-1}\gamma h_\gamma^{\infty*}(y^0) + \varepsilon^{-1}\bar{c}^\infty(y^0) + \gamma^{-1}|\mu| + \mathsf{E}[|N(T)|] < \infty.
\end{aligned}
$$

We conclude that the associated distributions of $\{Y^{r*}(T), I^{r*}(T) : r \geq 1\}$ are *tight*. Let $\{Y^\circ(T), I^\circ(T)\}$ denote any weak limit: For some sequence $\{r_i\}$ we have

$$
\{I^{r_i*}(T)\} \xrightarrow{\text{w}} \{I^\circ(T)\}, \qquad i \to \infty.
$$

The processes $(\boldsymbol{L}, Y^\circ(T), Y^{\infty*}(T))$ are defined on a common probability space with

$$
Y^\circ(T) = y^0 - \mu T + I^\circ(T) + N(T).
$$

Weak convergence and Proposition 9.8.1 imply that for any constant $m < \infty$,

$$
\begin{aligned}
\gamma h_\gamma^{r_i*}(y^0) &\geq \mathsf{E}[\bar{c}^\infty(Y^{r_i*}(T))] \\
&\geq \mathsf{E}[m \wedge \bar{c}^\infty(Y^{r_i*}(T))] \\
&\to \mathsf{E}[m \wedge \bar{c}^\infty(Y^\circ(T))], \qquad i \to \infty.
\end{aligned} \tag{9.94}
$$

Letting $m \to \infty$ we obtain from the Dominated Convergence Theorem,

$$
\liminf_{i\to\infty} \gamma h_\gamma^{r_i*}(y^0) \geq \mathsf{E}[\bar{c}^\infty(Y^\circ(T))],
$$

and hence from Proposition 9.8.4 we conclude that $\mathsf{E}[\bar{c}^\infty(Y^\circ(T))] \leq \gamma h_\gamma^{\infty*}(y^0)$. That is, $I^\circ(T)$ is an optimal allocation. This is only possible if the processes $I^\circ(T)$ and $I^*(T)$ are identical (see Proposition 9.8.3). $\qquad\square$

Note that the Dominated Convergence Theorem and the final limit in (9.94) all depend on bounds on moments of $\{Y^{r_i*}(T)\}$. The verification of such bounds is carried out in a similar setting in the next section – see in particular Lemma 9.8.9.

9.8.2 Average cost

We now consider approximations of the average-cost optimal policy in Case II*. Under the assumptions of Case II* we obtain a uniform bound on the exit time $\tau_\bullet$, which leads to stronger solidarity between the workload model and its unconstrained relaxation. Recall that $\tau_\bullet$ is defined in (9.82), and here we let $\tau_\bullet^+$ denote the hitting time to the region *above* $\mathsf{Y}_{\bar{c}^\infty}$,

$$
\tau_\bullet^+ := \inf\{t \geq 1 : Y(t) \notin \mathsf{Y}_{\bar{c}^\infty} \text{ and } Y_2(t) \geq m_* Y_1(t)\}, \tag{9.95}
$$

Although we make reference to the discounted-cost optimization problem, the bounds obtained in the remainder of this section are independent of $\gamma > 0$. For this reason, the subscript γ in the definition of the error term $\mathcal{E}(y)$ is omitted. It satisfies (9.88) with $\{d_i\}$ independent of γ.

We let $\overline{\eta}^*$ denote the optimal average cost,

$$\overline{\eta}^* := \inf\left(\limsup_{n\to\infty} \frac{1}{n}\sum_{t=0}^{n-1} \mathsf{E}[\overline{c}(Y(t))]\right). \tag{9.96}$$

We shall take for granted that $h^* \in L_\infty^{V_2}$ exists and satisfies the ACOE.[2] One representation of the relative value function h^* is obtained through the vanishing discount approach,

$$h^*(y) = \liminf_{\gamma\to 0}(h_\gamma^*(y) - h_\gamma^*(0)), \qquad y \in \mathsf{Y}.$$

As in the discounted-cost setting we set $h^*(y) := \infty$ on Y^c, and define

$$h^{*1}(y) = \mathsf{E}[h^*(y - \mu + L(1))], \qquad y \in \mathsf{Y}. \tag{9.97}$$

We have the following analog of Proposition 9.8.1:

Proposition 9.8.6. *The following hold under the assumptions of Theorem 7.5.2:*

(i) *The functions h^* and h^{*1} are each convex, and the difference $h^* - \overline{c}$ is a monotone function on Y.*

(ii) *The dynamic programming equation holds,*

$$\min_{\iota\geq 0} h^{*1}(y + \iota) = h^*(y) - \overline{c}(y) + \overline{\eta}^*, \qquad y \in \mathsf{Y}. \tag{9.98}$$

(iii) *The h^*-myopic policy is the stationary policy with feedback law,*

$$\phi^*(y) = \arg\min_\iota\{h^{*1}(y + \iota) : \iota \geq 0\}.$$

As a corollary we obtain an analog of Proposition 9.8.2. The switching curve s_* in the average-cost case is defined as in (9.89) by

$$s_*(y_1) := \inf\left\{y_2 : \frac{\partial}{\partial y_2}h^{*1}(y) > 0\right\}. \tag{9.99}$$

Proposition 9.8.7. *Suppose that $y \in \mathsf{Y}$ is given, and that $y^0 := (y_1, s_*(y_1))^{\mathsf{T}}$ satisfies*

$$\frac{\partial}{\partial y_2}h_\gamma^{*1}(y^0) = 0.$$

Then, from this initial condition, $\phi_2^*(y) = (s_*(y_1) - y_2)_+$.

The main result of this section shows that the switching curve under the average-cost criterion converges to an affine policy exponentially fast in y_1. The proof of Proposition 9.8.8 is contained in Lemmas 9.8.12 and 9.8.13 below.

Proposition 9.8.8. *Under the assumptions of Theorem 7.5.2 in Case II**,*

$$\left|s_*(y_1) - s_*^\infty(y_1)\right| \leq \mathcal{E}(y_1),$$

where s_*^∞ is defined in (7.59) via $s_*^\infty(y_1) = m_* y_1 - \overline{y}_2^*(0)$, and the error term is of the form $\mathcal{E}(y_1) = O(\exp(-d_0 y_1))$ for some $d_0 > 0$.

---

[2] See [360] for an extension of the SPP construction to general state space Markov chains.

We begin with a uniform bound on $\tau_\bullet^+$. Lemma 9.8.9 is based on positive recurrence of the height process, as illustrated in Fig. 5.11.

Lemma 9.8.9. *Suppose that $\mathbf{Y}$ is controlled using any stationary policy ϕ defined so that $\iota(t) = \phi(Y(t))$ is zero for $Y(t)$ in a region R satisfying $\mathsf{Y}_{\bar{c}^\infty}^c \cap \{y : y_2 \geq m_* y_1\} \subset R$. In Case II* we then have, for some $\varepsilon_0 > 0$,*

$$\mathsf{E}\big[\mathbf{1}\{\tau_\bullet^+ < \tau_2\}(e^{\varepsilon_0\|Y(\tau_\bullet^+)\|} + e^{\varepsilon_0\tau_\bullet^+})\big] = \mathcal{E}(y).$$

Proof. Let $\delta = \mu - \lambda \in \mathbb{R}_+^2$, and fix a vector $a \in \mathbb{R}^2$ satisfying $a^\mathsf{T}\delta > 0$, of the form

$$a = \bar{c}^+ - \bar{c}^- - \varepsilon_a \mathbf{1}^1,$$

with $\varepsilon_a > 0$. This is possible in Case II*, which requires $(\bar{c}^+ - \bar{c}^-)^\mathsf{T}\delta > 0$. We also have by construction, for some $\varepsilon_a' > 0$,

$$a^\mathsf{T} y \geq \varepsilon_a'\|y\|, \quad y \in \mathsf{Y}_{\bar{c}^\infty}^c \cap \{y : y_2 \geq m_* y_1\}. \tag{9.100}$$

The proof of the lemma is based on a Lyapunov drift analysis using

$$V(y) = e^{\varepsilon a^\mathsf{T} y}, \quad y \in \mathsf{Y}, \tag{9.101}$$

where $\varepsilon > 0$ is a fixed constant. For positive constants d_0, d_1 we have

$$V(y) = O\big(\exp(-d_0 y_1 + d_1|y_2 - m_* y_1|)\big), \quad y \in \mathsf{Y}. \tag{9.102}$$

That is, $V(y) \leq \mathcal{E}(y)$, where $\mathcal{E}$ is defined in (9.88).

Let $\widetilde{L}(t) = L(t) - \lambda$. Whenever $\iota(t) = \mathbf{0}$ we have

$$PV(y) := \mathsf{E}[V(Y(t+1)) \mid Y(t) = y] = V(y)e^{-\varepsilon a^\mathsf{T}\delta}\mathsf{E}[\exp(\varepsilon a^\mathsf{T}\widetilde{L}(t+1))].$$

The mean value theorem gives, for some $\theta \in (0,1)$,

$$e^x = 1 + x + \tfrac{1}{2}x^2 e^{\theta x} \leq 1 + x + \tfrac{1}{2}x^2 e^{|x|}, \quad x \in \mathbb{R}.$$

Hence the expectation is bounded as follows,

$$\mathsf{E}[\exp(\varepsilon a^\mathsf{T}\widetilde{L}(t+1))] \leq 1 + \tfrac{1}{2}b_0(\varepsilon),$$

where $b_0(\varepsilon) = \mathsf{E}[(a^\mathsf{T}\widetilde{L}(t+1))^2 \exp(\varepsilon|a^\mathsf{T}\widetilde{L}(t+1)|)]$. Fixing $\varepsilon_1 > 0$ such that $b_1 := b_0(\varepsilon_1)$ is finite, we obtain for all $\varepsilon \in (0, \varepsilon_1]$

$$\mathsf{E}[\exp(\varepsilon a^\mathsf{T}\widetilde{L}(t+1))] \leq 1 + \tfrac{1}{2}\varepsilon^2 b_1 \leq e^{\frac{1}{2}\varepsilon^2 b_1}.$$

Fix $\varepsilon \in (0, \min(\varepsilon_1, a^\mathsf{T}\mu/b_1))$ so that $\varepsilon_v := \varepsilon a^\mathsf{T}\delta - \tfrac{1}{2}\varepsilon^2 b_1 > 0$. With this choice of ε in the definition of V, we also have, whenever $\iota(0) = \phi(y) = \mathbf{0}$,

$$PV(y) \leq e^{-\varepsilon_v}V(y). \tag{9.103}$$

Consider the stochastic process $M(t) = e^{t\varepsilon_v}V(Y(t))$, $t \geq 0$. If $\iota(t) = \mathbf{0}$ we have

$$\mathsf{E}[M(t+1) \mid Y(0), \ldots, Y(t)] = \mathsf{E}[V(Y(t+1)) \mid Y(t)]e^{\varepsilon_v(t+1)} = e^{\varepsilon_v(t+1)}PV(Y(t)),$$

so that by (9.103),

$$\mathsf{E}[M(t+1) \mid Y(0), \dots, Y(t)] \leq M(t) \qquad \text{if } \tau_\bullet^+ \wedge \tau_2 > t. \qquad (9.104)$$

Equivalently, the stopped process $\{M(t \wedge \tau_\bullet^+ \wedge \tau_2) : t \geq 0\}$ is a super martingale. We have for any t

$$\mathsf{E}[M(t \wedge \tau_\bullet^+ \wedge \tau_2)] \leq M(0) = V(y),$$

and by Fatou's Lemma,

$$\mathsf{E}[M(\tau_\bullet^+ \wedge \tau_2)] = \mathsf{E}[\liminf_{t \to \infty} M(t \wedge \tau_\bullet^+ \wedge \tau_2)] \leq \liminf_{t \to \infty} \mathsf{E}[M(t \wedge \tau_\bullet^+ \wedge \tau_2)] \leq M(0) = V(y).$$

We obviously have $\mathsf{E}[M(\tau_\bullet^+ \wedge \tau_2)] \geq \mathsf{E}[M(\tau_\bullet^+)\mathbf{1}\{\tau_\bullet^+ < \tau_2\}]$, and hence by the definition of M and V,

$$\mathsf{E}\big[\exp\big(\varepsilon_v \tau_\bullet^+ + \varepsilon a^{\mathsf{T}} Y(\tau_\bullet^+)\big)\mathbf{1}\{\tau_\bullet^+ < \tau_2\}\big] \leq V(y).$$

This completes the proof since $V(y) = \mathcal{E}(y)$ by (9.102), and $a^{\mathsf{T}} Y(\tau_\bullet^+) \geq \varepsilon_a' \|Y(\tau_\bullet^+)\|$ by (9.100). □

The following two results are extensions of Proposition 9.7.4 based on the similarity between the height process recursions (7.57) and the simple inventory model (7.34). The proof of each of these results is very similar to the proof of Proposition 9.7.4. However, note that we can enlarge the class of stopping times since the results here concern the optimal policy.

Proposition 9.8.10. *Suppose that τ is a stopping time satisfying the following conditions: It has finite mean, satisfies $0 \leq \tau < \tau_\bullet$ a.s., and also $Y^*(t) + \varepsilon \mathbf{1}^2 \in \mathsf{Y}$ a.s. for $0 \leq t \leq \tau$ and $|\varepsilon| \leq 1$. Then for each initial condition the first and second derivatives satisfy*

$$\frac{\partial}{\partial y_2} P_* h^*(y) = \mathsf{E}_y\left[\sum_{t=1}^{\tau} \chi_{\bar{c}_H}(H^*(t)) + \frac{\partial}{\partial y_2} P_* h^*(Y^*(\tau))\right] \qquad (9.105a)$$

$$\frac{\partial^2}{\partial y_2^2} P_* h^*(y) = |c| \mathsf{E}_y\left[\sum_{t=0}^{\tau \wedge \sigma_2 - 1} p_{L_H}(H^*(t)) + \frac{\partial^2}{\partial y_2^2} P_* h^*(Y^*(\tau \wedge \sigma_2))\right] \qquad (9.105b)$$

where p_{L_H} is the density for L_H appearing in (7.34), $|c| = c^- + c^+$, P_* denotes the transition kernel under the h^*-myopic policy, $\sigma_2 = \inf\{t \geq 0 : \iota_2^*(t) > 0\}$, and the sensitivity function for $\mathbf{H}^\infty$ is defined in (9.80).

Proof. Note that although τ can depend upon $Y(0)$, in the proof we fix one $y \in \mathsf{Y}$ and fix the random variable $\tau = \tau(y)$ when considering other initial conditions.

The random variable $\tau + 1$ is also a stopping time, and hence from the dynamic programming equations,

$$h^*(y) = \inf \mathsf{E}\left[\sum_{t=0}^{\tau} (\bar{c}(Y(t)) - \bar{\eta}^*) + h^*(Y(\tau+1)) \mid Y(0) = y\right],$$

where the infimum is over all policies. The dynamic programming equation gives $P_*h^* = h^* - (\bar{c} - \bar{\eta}^*)$ and the Markov property gives $\mathsf{E}[h^*(Y(\tau+1))] = \mathsf{E}[P_*h^*(Y(\tau))]$. Consequently,

$$P_*h^*(y) = \inf \mathsf{E}\left[\sum_{t=1}^{\tau}(\bar{c}(Y(t)) - \bar{\eta}^*) + P_*h^*(Y(\tau)) \mid Y(0) = y\right]. \qquad (9.106)$$

Based on this representation, replacing τ by $\tau \wedge \sigma_2$ we obtain, exactly as in the proof of Proposition 9.7.4,

$$\frac{\partial}{\partial y_2}P_*h^*(y) = \mathsf{E}\left[\sum_{t=1}^{\tau \wedge \sigma_2}\chi_{\bar{c}_H}(H^*(t)) + \frac{\partial}{\partial y_2}P_*h^*(Y^*(\tau \wedge \sigma_2)) \mid Y(0) = y\right],$$
$$(9.107)$$

and also as in Proposition 9.7.4 we obtain (9.105b).

To complete the proof of (9.105a), on applying (9.107) it remains to show that

$$\mathsf{E}\left[\mathbf{1}\{\tau > \sigma_2\}\left(\sum_{t=\sigma_2+1}^{\tau}\chi_{\bar{c}_H}(H^*(t)) + \frac{\partial}{\partial y_2}P_*h^*(Y^*(\tau))\right) \mid Y(0) = y\right] = 0.$$

By the Markov property, this amounts to establishing the identity

$$\mathsf{E}\left[\sum_{t=1}^{\tau}\chi_{\bar{c}_H}(H^*(t)) + \frac{\partial}{\partial y_2}P_*h^*(Y^*(\tau)) \mid Y(0) = y\right] = 0,$$

whenever $\phi_2(y) > 0$ and τ satisfies the assumptions of the proposition.

Define $Y^\varepsilon(0) := y$ and $Y^\varepsilon(t) := Y^*(t) + \varepsilon 1^2$ for $t \geq 1$. We restrict to $\varepsilon \in [-\varepsilon_0, \varepsilon_0]$, with $\varepsilon_0 = \min(1, s_*(y_1) - y_2) > 0$, to ensure that $\mathbf{Y}^\varepsilon$ is feasible. The identity (9.106) implies the similar identity

$$P_*h^*(y) = \inf \mathsf{E}\left[\sum_{t=1}^{\tau}\bar{c}(Y^\varepsilon(t)) + P_*h^*(Y^\varepsilon(\tau)) \mid Y(0) = y\right],$$

where here the infimum is over all $\varepsilon \in [-\varepsilon_0, \varepsilon_0]$. Since $\varepsilon = 0$ is an interior minimizer we obtain (9.105a) as the first-order necessary condition for optimality. $\square$

The proof of the analogous result for the unconstrained model is identical:

Proposition 9.8.11. *Suppose that* $\mathbf{Y}^\infty$ *is controlled using the optimal (affine) policy, and that* τ *is a stopping time with finite mean. Then, for each initial condition satisfying* $H^\infty(0) \geq -\bar{y}_2^{\infty*} + \mu_H$,

$$\frac{\partial}{\partial y_2}P_{\infty*}h^{\infty*}(y) = \mathsf{E}_y\left[\sum_{t=1}^{\tau}\chi_{\bar{c}_H}(H^\infty(t)) + \frac{\partial}{\partial y_2}h^{\infty*}(Y^\infty(\tau))\right]$$

$$\frac{\partial^2}{\partial y_2^2}P_{\infty*}h^{\infty*}(y) = \mathsf{E}_y\left[|c|\sum_{t=0}^{\tau \wedge \sigma_2^\infty-1}p_{L_H}(H^\infty(t)) + \frac{\partial^2}{\partial y_2^2}h^{\infty*}(Y^\infty(\tau \wedge \sigma_2^\infty))\right],$$

where $\sigma_2^\infty = \inf\{t \geq 0 : \iota_2^{\infty*}(t) > 0\}$.

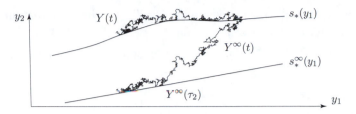

Figure 9.6. Sample paths of the two workload processes.

Lemma 9.8.12. $s_*(y_1) \leq s_*^\infty(y_1) + \mathcal{E}(y_1)$.

Proof. Suppose that $y_1^0 > 0$ is given with $s_*(y_1^0) > s_*^\infty(y_1^0)$, and let $Y^*(0) = Y^{\infty*}(0) = y^0 := (y_1^0, s_*(y_1^0))^\mathsf{T}$. We have $\iota^*(0) = \iota^{\infty*}(0) = 0$, so that

$$P_* h^*(y) = h^{*1}(y), \qquad P_{\infty*} h^{\infty*}(y) = h^{\infty*1}(y).$$

On defining τ_2^∞ in analogy with τ_2 as $\tau_2^\infty = \inf\{t \geq 1 : \iota_2(t) > 0\}$ we obtain the bounds

$$Y_2^*(t) \geq Y_2^{\infty*}(t), \quad Y_1^*(t) = Y_1^{\infty*}(t), \qquad 0 \leq t \leq \tau_2^\infty \wedge \tau_\bullet. \tag{9.109}$$

Typical sample paths of the two processes are shown in Fig. 9.6.

Proposition 9.8.11 implies the representation

$$\frac{\partial}{\partial y_2} h^{\infty*1}(y^0) = \mathsf{E}\left[\sum_{t=1}^{\tau_2^\infty} \chi_{\bar{c}_H}(H^{\infty*}(t))\right].$$

The inequality (9.109) implies that $\chi_{\bar{c}_H}(H^{\infty*}(t)) \leq \chi_{\bar{c}_H}(H^*(t))$ for $0 \leq t \leq \tau_2^\infty \wedge \tau_\bullet$, so that the previous identity combined with Lemma 9.8.9 gives

$$\frac{\partial}{\partial y_2} h^{\infty*1}(y^0) \leq \mathsf{E}\left[\sum_{t=1}^{\tau_2^\infty \wedge \tau_\bullet} \chi_{\bar{c}_H}(H^*(t))\right] + \mathcal{E}(y^0)$$

$$= \frac{\partial}{\partial y_2} h^{*1}(y^0) - \mathsf{E}\left[\frac{\partial}{\partial y_2} h^{*1}(Y^*(\tau_2^\infty \wedge \tau_\bullet))\right] + \mathcal{E}(y^0).$$

It follows that $\frac{\partial}{\partial y_2} h^{\infty*1}(y^0) \leq \mathcal{E}(y^0)$ since $\frac{\partial}{\partial y_2} h^{*1}(y^0) = 0$. Applying the second derivative formula in Proposition 9.8.11 we can conclude that, for some constant $\varepsilon_0 > 0$,

$$\mathcal{E}(y^0) \geq \frac{\partial}{\partial y_2} h^{\infty*1}(y^0) \geq \varepsilon_0 \min(|y_2^0 - s_*^\infty(y_1^0)|, 1), \qquad y_2^0 > s_*^\infty(y_1^0),$$

which completes the proof. $\square$

A lower bound is obtained using similar arguments:

Lemma 9.8.13. $s_*(y_1) \geq s_*^\infty(y_1) - \mathcal{E}(y_1)$.

Proof. Suppose that $y_1^0 > 0$ is given with $s_*(y_1^0) < s_*^\infty(y_1^0)$ so that on setting $y^0 :=$ $(y_1^0, s_*^\infty(y_1^0))^\mathsf{T}$ we have $\frac{\partial}{\partial y_2} h^{*1}(y^0) > 0$ and $\frac{\partial}{\partial y_2} h^{\infty*1}(y^0) = 0$.

Letting $Y^*(0) = Y^{\infty*}(0) = y^0$ we have, exactly as in the proof of the previous lemma,

$$\frac{\partial}{\partial y_2} h^{*1}(y^0) \leq \mathsf{E}\left[\sum_{t=1}^{\tau_2 \wedge \tau_\bullet} \chi_{\bar{c}_H}(H^{\infty*}(t))\right] + \mathcal{E}(y^0)$$

$$\leq \frac{\partial}{\partial y_2} h^{\infty*1}(y^0) + \mathcal{E}(y^0) = \mathcal{E}(y^0).$$

The second derivative formula in Proposition 9.8.10 implies that for some constant $\varepsilon_0 > 0$,

$$\frac{\partial}{\partial y_2} h^{*1}(y^0) \geq \varepsilon_0 \min(|y_2^0 - s_*(y_1^0)|, 1), \qquad y_2^0 > \max(s_*^\infty(y_1^0) - 1, s_*(y_1^0)).$$

This combined with the previous bound completes the proof. $\qquad\square$

9.9 Notes

There is a rich literature on optimization of queueing networks based on Markov decision theory, beginning with the work of Bellman et al. in [39, 37, 38]. Detailed treatments of MDP theory can be found in the monographs [402, 417, 68, 258, 259, 45, 178], and the excellent survey [16].

Much of the material in this chapter is adapted from the work of V. Borkar. The near-monotone assumption for average-cost optimal control was introduced in [65] and refined in [67]. The survey [70] is the most accessible and elegant introduction to the convex analytic approach to MDPs on a countable state space. Theorem 9.2.1 is based on results from this paper, which originates from the author's work in [65, 67, 69]. Recent generalizations to continuous time and general state spaces are contained in [60].

The literature on average-cost optimal control is filled with counter-examples showing that strong assumptions are necessary to move beyond finite state-space MDPs. The counter-examples 1 and 2 of [418, p. 142] imply that some form of irreducibility such as the reachability condition used in Theorem 9.0.4 is in some sense necessary to guarantee the existence of a solution to the ACOE. That is, if irreducibility is relaxed completely, then some other strong conditions must be imposed. The Appendix of [418] provides an example illustrating the value of the coerciveness assumption to ensure that an optimal policy exists. See also [68, p. 87], [430, Chapter 7], and the examples in [139, 67, 16, 402].

The LP approach to dynamic programming was formally introduced by Manne [349]. Some essential ideas are contained in the earlier work of Dantzig [135]. Other early contributions are contained in [142, 270, 13]. For more recent literature on this topic and multiobjective optimization see Altman's monograph [5], Hernández-Lerma and Lasserre [258, 261, 260], and Borkar's survey [70]. Applications to scheduling and "restless bandits" are contained in a series of papers by Bertsimas and Niño-Mora (see [53] and its references).

An early contribution to constrained optimization is Dubins [149], and the monograph of Dubins and Savage with the intriguing title *How to gamble if you must* [150].

A value iteration algorithm for the multiobjective MDP problem was introduced relatively recently in [104]. To avoid randomization, the state process is augmented as $\{X(t), \Psi(t) : t \geq 0\}$, where $\Psi(t)$ denotes the (relative) cost to go for a secondary optimization problem.

The total-cost or shortest-path approach to average-cost optimization is described in Chapter 34 of Whittle's 2-volume monograph [500, 501]. A complete treatment of the finite state-space case appears in [48] (see also Bertsekas' monograph [45]). This approach is applied in

[44] to obtain a refined version of the value iteration algorithm for average-cost optimal control. The convergence proof in [44] is based on the L_∞^V-norm, following Tseng [471].

The most common technique to construct a solution to the average-cost optimality equation is the *vanishing discount approach*. Define for each $\gamma > 0$ the difference $h^\gamma(x) := h_\gamma^*(x) - h_\gamma^*(x^*)$ where h_γ^* is defined in (9.34) and x^* is some fixed state. Under suitable conditions h^γ converges as $\gamma \downarrow 0$ to a solution to the ACOE. In [431] and other papers it is assumed that there is a finite-valued function b on the state space, and a constant $b_0 > 0$ such that

$$-b_0 \leq h^\gamma(x) \leq b(x), \tag{9.110}$$

for all states x, and all γ sufficiently close to zero. Assuming bounds similar to (9.110) together with some additional conditions it is known that policy iteration and value iteration converge to yield optimal policies [258, 432]; the paper [487] uses this approach in its analysis of networks. It is shown in [408] that many common and natural assumptions imply (9.110).

The monograph [268], based on Hordijk's dissertation and Dynkin and Yushkevich [162], develops MDP theory based on connections with Lyapunov theory (or potential theory) in a spirit similar to this chapter. For example, one sufficient condition for (9.110) is the simultaneous Lyapunov condition of Hordijk [268]: For a function $V : \mathsf{X} \to \mathbb{R}_+$, $b < \infty$, and a finite set S, the following bound is assumed to hold for each $x \in \mathsf{X}$ and *every stationary policy*,

$$\sup_\phi \mathsf{E}^\phi[V(X(t+1)) \mid X(t) = x] \leq V(x) - c(x) + b\mathbf{1}_S(x). \tag{9.111}$$

Federgruen, Hordijk, and Tijms introduce a closely related assumption in [173]. It is assumed that under any policy the chain satisfies an x^*-irreducibility condition, and that the value function $h_i(x)$ defined in (9.29) is uniformly bounded over all policies, for any state x. The latter assumption is the key *conclusion* of the uniform Lyapunov condition (9.111).

Similar bounds are assumed in the analysis of recursive algorithms such as value iteration. It is assumed in [90] that (9.111) holds with $S = \{x^*\}$ a singleton. Under this assumption the authors prove that value iteration converges to a solution of the average-cost optimality equations.

The uniform bound (9.111) is not always satisfied in queueing models; under the restriction that all policies are nonidling, a sufficient condition is boundedness of the stability LP described in Section 8.6. Cavazos-Cadena in [89] obtained a significant relaxation of these assumptions. In this paper convergence of value iteration is established under the assumption that the controlled model is x^*-irreducible under any stationary policy, and that the cost satisfies the near-monotone condition of Borkar [65, 66] (see also Section A.3.3).

Harrison introduced the skip-free assumption in [228] to obtain relaxed conditions for the existence of an ϵ-optimal stationary policy.

The VIA is credited to White [494]. Similar successive approximation methods were introduced by Bellman and Ford to solve deterministic optimal routing problems [38, 188]. The treatment of the algorithm presented here (in particular, the emphasis on initialization) is based on [105], following the analysis of policy iteration contained in [356] (see also [358, 357]).

The policy iteration algorithm was first proposed by Howard in [273]. Puterman and Brumelle showed in [403, 404] that the algorithm is equivalent to the Newton–Raphson method.

In simple network models exact optimal policies can be constructed [505], while if the network model is complex, then research typically focuses on characterizations of the structure of optimal policies [39, 460, 222, 208]. Sometimes it is possible to construct exact optimal policies for complex networks by softening the definition of optimality, as in for example [55].

The use of the value iteration in Section 9.6.3 to deduce structural properties of an optimal policy can be found in many papers over the past 30 years [414, 222, 487]. For a survey see [460]. The bounds on the value functions obtained in Proposition 9.6.4 are related to convexity, and a concept known as *multimodularity* [222, 223, 7, 378].

There is a large literature on existence and uniqueness of optimal policies for the CBM model. The most general results to date are contained in [22], which also contains an extensive bibliography.

The approximations obtained in Section 9.8 are based on Chen et al. [103]. A height process is again the focus of analysis, which in this paper takes the form of a one-dimensional reflected Brownian motion.

Exercises

9.1   Solve the ACOE for the single-server queue with the cost function $c(x, u) = x + ru$, $r \geq 0$. That is, find the function h^* and constant η^* satisfying

$$\eta^* + h^*(x) = \min_{u \in \mathsf{U}(x)} [c(x, u) + P_u h^*(x)], \qquad x \in \mathbb{Z}_+.$$

9.2   Consider the homogeneous Klimov model described in Section 5.5.1. The partial workload process $Y^{[m]}$ evolves as a nonidling queue for each $m \geq 1$. Consequently, we can solve for each $m \geq 1$ the ACOE,

$$\min_{u \in \mathsf{U}_\diamond(x)} P_u h_m^*(x) = h_m^*(x) - c_m(x) + \eta_m^*, \qquad x \in \mathbb{Z}_+^\ell,$$

where $c_m(x) = y^{[m]} = x_1 + \cdots + x_m$, and η_m^* is obtained through the Pollaczek–Khintchine formula (3.10). Compute the solution to the ACOE for general linear cost, based on the functions $\{h_m^*\}$.

9.3   Consider a single-server queueing system with finite waiting room $\overline{x}$: The system will reject an arrival if it has $\overline{x}$ or more jobs waiting. There are three separate job classes with different priorities labeled High (H), Medium (M), and Low (L). Assume that the arrival process of each job class is Poisson with rate 1 job per second, and that the service time of each class is exponentially distributed with mean $\mu^{-1} = 0.3$. Obtain a discrete-time Klimov model using uniformization. What is an optimal policy, given a linear cost function that is consistent with these priorities? Under the optimal policy, what is the steady-state distribution of $Q(t)$ and $Q_H(t)$ (where $Q(t)$ denotes the number of customers in queue at time t, and $Q_H(t)$ denotes the number of high-priority customers)? *Suggestion*: Take a look at the previous problem, and at Exercise 4.10. This problem is continued in Exercise 11.2.

9.4   Consider a single-server queue with controlled arrivals: Service times are independent, and exponentially distributed, with mean μ^{-1}. Arrivals come from two Poisson sources: One is uncontrolled, with rate $\alpha_\circ$, and the other is controlled with rate $\alpha_\bullet \zeta$, where $0 \leq \zeta \leq 1$. Moreover, the total waiting room $\overline{x}$ is finite, as in the previous problem. Any arriving customer is rejected if there are $\overline{x}$ customers in queue. Assume that $\alpha_\circ < \mu$ and $\alpha_\circ + \alpha_\bullet > \mu$.

(a) Construct a discrete-time MDP model.

(b) Suppose that there is a linear holding cost c_0 for each customer, and a cost c_r for rejecting any arrival. Write down the average-cost dynamic programming equations, and express the optimal control in terms of the relative value function h^*.

(c) Can you guess the structure of an optimal policy when $\overline{x}$ is large?

(d) Compute h^* using MATLAB. Use value iteration, initialized with $V_0 = 0$. Plot the final value function after 500 steps, and plot the policy obtained using the V_n-myopic policy for $1 \leq n \leq 500$.

(e) Compute a fluid value function J, based on a policy for the fluid model that you feel is consistent with a good policy for the stochastic model. Use J to initialize the value iteration algorithm, and compare the policy plots with (iv).

9.5   Compute using value iteration the optimal policy for the CRW communications model constructed in Exercise 6.7. To apply the algorithm you will have to truncate the state space to obtain a finite set $X_\diamond$ (e.g., for a load of $\rho_\bullet = 0.9$ take $|Q(t)| \leq 5,000$). Initialize your algorithm using $V_0 \equiv 0$, and using the function h defined in Exercise 6.7. Compare your results as in the previous exercise, and plot the final policy on the state space $X_\diamond$.

9.6   Consider the value function for the SPP to a *set* S:

$$h_{\ddagger}^*(x) := \inf \mathsf{E}_x \left[\sum_{t=0}^{\tau_S - 1} \big(c(X(t)) - \eta \big) \right], \qquad x \in X_\diamond,$$

where c is coercive and η is a constant. The function $h_{\ddagger}^*$ solves an ACOE for a coercive function $c_{\ddagger}$. Obtain an expression for the function $c_{\ddagger}$ and its average cost under the optimal policy. *Hint*: Read Section 9.4.3.

9.7   Corollary 9.5.3 requires $V_0 \equiv 0$ to establish that $\limsup_{n \to \infty} n^{-1} V_n^*(x) \leq \eta_x^*$. Prove that this result continues to hold under the assumption that $V_0 \in L_\infty^{V_p}$ with $p \geq 2$, where $V_p = 1 + \|x\|^p$.

9.8   *Policy iteration and Newton's method.* In this exercise you will discover the relation between policy iteration and Newton's method for solving nonlinear equations.

The policy iteration algorithm can be extended to compute the total cost given in (9.59):

Definition 9.9.1 (Policy iteration for total cost). Given an initial stationary policy ϕ_0, the sequence of stationary policies and value functions $\{\phi_{n+1}^*, h_n : n \geq 0\}$ is defined recursively as follows: At stage n based on the policy ϕ_n^*, the function h_n is constructed to solve the equation

$$P_{\phi_n^*} h_n = h_n - c.$$

The next policy is then defined by

$$\phi_{n+1}^*(x) \in \underset{u \in U_\diamond(x)}{\arg\min} \, P_u h_n(x), \qquad x \in X_\diamond. \qquad \blacksquare$$

Suppose that the state space is finite, $X_\diamond = \{1, \ldots, d\}$, so that each h_n can be viewed as a vector in $\mathbb{R}^d$. Let $\mathcal{F} \colon \mathbb{R}^d \to \mathbb{R}^d$ be the function of h defined by

$$[\mathcal{F}(h)]_x = -h(x) + c(x) + \min_{u \in U_\diamond(x)} P_u h(x), \qquad x = 1, \ldots, d.$$

Then V_∞^* solves $\mathcal{F}(V_\infty^*) = 0$.

Newton's method can be used to recursively compute a sequence of estimates of the zero of the function $\mathcal{F}$. Given h_n, the next estimate h_{n+1} solves the linear equation

$$\mathcal{F}(h_n) + \nabla\mathcal{F}(h_n) \cdot (h_{n+1} - h_n) = 0.$$

(a) Show that $\mathcal{F}$ is *concave* as a function of h, in the sense that $\mathcal{F}^*(\theta h^1 + (1-\theta)h^0) \geq \theta\mathcal{F}^*(h^1) + (1-\theta)\mathcal{F}^*(h^0)$ pointwise for each $\{h^0, h^1\}$ and $\theta \in [0,1]$. Hence, for any h there exists at least one $d \times d$ matrix satisfying

$$\mathcal{F}(h+g) \leq \mathcal{F}(h) + M_h g, \qquad g \in \mathbb{R}^d.$$

If $\mathcal{F}$ is differentiable, then M_h is the Jacobian of $\mathcal{F}$ at h.

(b) Suppose that ϕ is a policy satisfying $\mathcal{F}^*(h) = h + c + P_\phi h$. Show that $M_h = P_\phi - I$ satisfies the bound in (a).

(c) Explain why the policy iteration algorithm is an application of Newton's method.

9.9   *Stability of discounted-cost optimal policies*. In this exercise you will consider the CRW scheduling model with c a linear cost function.

(a) Find $\gamma \gg 1$, a cost function c, and a network example for which the discounted-cost optimal policy is transient. The tandem queue or KSRS models are possibilities.

(b) Show that for $\gamma > 0$ sufficiently small, the discounted-cost optimal policy satisfies (V2) with V a constant multiple of h_γ^*. *Hint*: Recall that $h_\gamma^*(x) = \mathsf{E}[c(Q(T;x)]$ under the optimal policy, where T is a geometric random variable with mean γ^{-1}. Can you show that $\mathsf{E}[c(Q(T;x)] \leq c(x) - \varepsilon\gamma^{-1}$ for some $\varepsilon > 0$ and all large x?

(c) Mimic the proof of Proposition 9.6.4 to show that h_γ^* is Lipschitz continuous. Conclude that the function $V_\theta = e^{\theta h_\gamma^*}$ solves (V4) for sufficiently small $\theta > 0$ provided A satisfies the exponential moment bound (8.31).

10

ODE Methods

We remarked in the Introduction that the fluid model can be interpreted as the mean flow for the stochastic network model as suggested by (1.7). Alternatively, the fluid model is motivated by considering the scaled process $q^\kappa(t; x)$ defined in Section 3.2.1 for the single-server queue. In this chapter we investigate the scaled process for the general CRW network model. Among the goals are to expand our set of tools for verifying stability or instability of a given policy. Using similar approaches we also obtain structural properties for the average-cost and discounted-cost value function, as well as structure for the optimal policy.

The collection of techniques introduced in this chapter can be regarded as an extension of the ODE method used in the analysis of dynamical systems and recursive algorithms. To illustrate the general idea, consider a discrete-time dynamical system described by the equations

$$X(t+1) = X(t) + f(X(t), \Phi(t+1)), \qquad t \geq 0, \tag{10.1}$$

where Φ is an ergodic Markov chain on the integers with invariant measure π, and $f \colon \mathbb{R}^\ell \times \mathbb{Z}_+ \to \mathbb{R}^\ell$ is continuous. The ODE method is based on the associated ordinary differential equation

$$\tfrac{d}{dt} x(t) = \overline{f}(x(t)) \tag{10.2}$$

where $\overline{f}$ is the averaged function, $\overline{f}(x) = \sum f(x, n) \pi(n)$, $x \in \mathbb{R}^\ell$. The most common conclusion is that the stochastic model (10.1) is stable in a statistical sense provided the ODE (10.2) is stable. Stability for the deterministic system means that state trajectories converge to some fixed point from each initial condition. This of course requires assumptions on the nonlinear models (10.1), (10.2). The most common assumption is a global Lipschitz condition on each of the functions f and $\overline{f}$.

In this chapter the CRW network model (8.4) plays the role of the dynamical system (10.1), and the associated fluid model is of the general form given in (10.2). The process $\Phi := (A, B)$ is i.i.d. under the assumptions imposed in this book.[1]

---

[1] The CRW model with Φ a suitably stable Markov chain can be analyzed using the methods described in this chapter – see Exercise 10.3 for an example.

To relate the stochastic model and its fluid counterpart we begin with the suggestive representation

$$Q(t) = Q(0) - BZ(t) + \alpha t + N(t), \qquad t \geq 0, \tag{10.3}$$

where Z is the cumulative allocation process

$$Z(t) = \sum_{k=1}^{t} U(k-1), \qquad t \geq 1, \ Z(0) = 0, \tag{10.4}$$

and N is constructed by rearranging terms in the CRW model via

$$N(t) = \sum_{k=1}^{t} [(B(k) - B)U(k-1) + (A(k) - \alpha)], \qquad t \geq 1, \ N(0) = 0.$$

The representation (10.3) was introduced in Section 1.2.1 in the discussion on network modeling.

Throughout this chapter we restrict to the CRW model (8.4) under assumptions (a)–(d) of Theorem 8.0.4. We also relax buffer constraints to facilitate scaling, so that Q evolves on $\mathsf{X}_\diamond = \mathbb{Z}_+^\ell$. As in Section 3.2.1, we extend the definition of (Q, Z) to the time axis $\mathbb{R}_+$ by linear interpolation to form a continuous and piecewise linear function of t. We introduce a scaling parameter denoted $\kappa > 0$, and for a given $x \in \mathsf{X} := \mathbb{R}_+^\ell$ we let $\{x(\kappa) : \kappa > 0\}$ denote a (deterministic) sequence of vectors in $\mathbb{R}_+^\ell$ satisfying $\kappa x(\kappa) \in \mathsf{X}_\diamond$ for each κ, and $x(\kappa) \to x$ as $\kappa \to \infty$. We frequently set $x(\kappa) = x^\kappa$ with

$$x^\kappa := \frac{1}{\kappa} \lfloor \kappa x \rfloor, \qquad x \in \mathbb{R}_+^\ell, \ \kappa > 0, \tag{10.5}$$

where $\lfloor \kappa x \rfloor \in \mathbb{Z}_+^\ell$ denotes the component-wise integer part of the vector κx. The two scaled processes are defined by

$$q^\kappa(t; x(\kappa)) := \frac{1}{\kappa} Q(\kappa t; \kappa x(\kappa)), \quad z^\kappa(t; x(\kappa)) := \frac{1}{\kappa} Z(\kappa t; \kappa x(\kappa)), \qquad t \geq 0. \tag{10.6}$$

Based on a particular version of the Strong Law of Large Numbers (see Theorem 1.3.9), we find that the scaled "disturbance process" vanishes with κ,

$$\lim_{t \to \infty} \kappa^{-1} N(\kappa t) = 0, \qquad \text{a.s.} \tag{10.7}$$

Hence, it is not surprising that the scaled processes in (10.6) are typically convergent to a pair of limiting processes (q, z) satisfying the fluid model equations

$$q(t) = x + Bz(t) + \alpha t, \qquad t \geq 0, \ x \in \mathsf{X}. \tag{10.8}$$

The limiting process (q, z) depends of course upon the particular policy that determines (Q, U).

Convergence of the scaled processes is suggested in several examples considered in earlier chapters. Proposition 3.2.3 establishes convergence to a scalar version of (10.8) for the single-server queue, and this convergence is illustrated in Fig. 1.7. The scaled process q^κ was also used to motivate approximate models for the G/G/1 queue (see

(2.7)). The plot on the left in Fig. 2.13 illustrates convergence in the KSRS model under a priority policy, and convergence is suggested by the plots shown in Fig. 2.16 for two policies in the simple routing model.

The *fluid limit model*, denoted $\mathcal{L}$, is the focus of Section 10.3. For a given policy, this is defined as the set of all possible limits of the scaled processes. In the construction of $\mathcal{L}$ and subsequent analysis it is helpful to isolate a single sample of the probability space, denoted ω, and emphasize the dependence on chance by appending this variable. For example, we let $Q(\,\cdot\,;x,\omega)$ denote the queue length trajectory starting from x with a given sample $\omega \in \Omega$ (see Section 1.3.2). Then, for fixed ω the pair of functions (q, z) belongs to $\mathcal{L}(\omega)$ if for some sequence $\kappa_i \to \infty$ and some convergent sequence $x(\kappa_i)$,

$$q^{\kappa_i}(t; x(\kappa_i), \omega) \to q(t; x, \omega), \quad z^{\kappa_i}(t; x(\kappa_i), \omega) \to z(t; x, \omega), \quad i \to \infty, \qquad (10.9)$$

for each $t \in \mathbb{R}_+$, where $x = \lim_{i \to \infty} x(\kappa_i)$. The complete definition of $\mathcal{L}$ requires other technicalities that are postponed to Definition 10.3.1.

Definition 10.0.1 (Stability of the fluid limit model). The fluid limit model $\mathcal{L}$ is said to be *stable* if there exists $\Omega_0 \subset \Omega$ satisfying $\mathsf{P}\{\Omega_0\} = 1$, and $T_0 > 0$ such that $q(t) = 0$ whenever $t \geq T_0$, $\omega \in \Omega_0$, $q \in \mathcal{L}(\omega)$, and $|q(0)| = 1$. ∎

A function $h\colon \mathbb{R}_+^\ell \to \mathbb{R}_+$ is *equivalent to a quadratic* if $h \in L_\infty^{V_2}$ and $V_2 \in L_\infty^{1+h}$, where V_2 is defined in (8.27). Equivalently, for some $\gamma \geq 1$,

$$-\gamma + \gamma^{-1}|x|^2 \leq h(x) \leq \gamma + \gamma|x|^2, \qquad x \in \mathbb{R}_+^\ell. \qquad (10.10)$$

Recall that trajectory tracking in the L_2 sense (8.47) was a condition for stability in Theorem 8.3.2 for a Discrete Review policy. Theorem 10.0.2 can be regarded as a substantial generalization of this result.

Theorem 10.0.2 (Fluid-scale stability implies stochastic stability). *Suppose that assumptions (a)–(d) of Theorem 8.0.4 hold, and that the allocation sequence U is defined as a stationary policy. If the fluid limit model is stable, then the controlled network satisfies (V3) with $f(x) = 1 + |x|$, $x \in \mathsf{X}$, and a Lyapunov function $V\colon \mathsf{X} \to \mathbb{R}_+$ that is equivalent to a quadratic.*

In typical applications Theorem 10.0.2 provides a far simpler route to establishing stability over direct Lyapunov techniques or linear programming approaches. The construction of a solution to (V3) is based on the fluid limit model:

Proof of Theorem 10.0.2. The solution to (V3) is given explicitly in Proposition 10.4.1 by

$$V_c(x) = \mathsf{E}\left[\sum_{i=0}^{|x|T} c(Q(i; x))\right], \qquad x \in \mathsf{X}_\diamond, \qquad (10.11)$$

where $c\colon \mathbb{R}_+^\ell \to \mathbb{R}_+$ is a linear cost function, and $T \geq 1$ is a sufficiently large fixed integer.

The growth bounds follow from Lemma 10.4.2, which implies that

$$\frac{1}{\kappa^2} V_c(\kappa x(\kappa)) \approx \mathsf{E}\left[\int_0^{|x|T} c(q^\kappa(t; x(\kappa)))\, dt\right], \qquad x \in \mathbb{R}_+^\ell, \ \kappa > 0. \qquad (10.12)$$

$\square$

The idea behind the proof of Theorem 10.0.2 is that the Lyapunov function V_c defined in (10.11) approximates a fluid value function of the form introduced in Part II of the book:

$$J(x) = \int_0^\infty c(q(t; x))\, dt, \qquad x \in \mathsf{X}. \qquad (10.13)$$

This approximation is made precise in Lemma 10.4.2. In its refinement Proposition 10.4.3 it is assumed that $\mathcal{L}_x$ is a singleton for each x. In this case we have

$$\lim_{\kappa\to\infty} \frac{1}{\kappa^2} V_c(\kappa x(\kappa)) = \lim_{N\to\infty} \lim_{\kappa\to\infty} \int_0^N c(q^\kappa(t; x(\kappa)))\, dt = \int_0^\infty c(q(t; x))\, dt, \quad (10.14)$$

with q the unique element in $\mathcal{L}_x$, $x \in \mathsf{X}$.

The value function is always a Lyapunov function for the fluid model in the sense that $\frac{d}{dt} J(q(t)) = -c(q(t)) \leq 0$. The proof follows from the simple formula

$$J(q(t)) = \int_t^\infty c(q(r; x))\, dr = J(q(0)) - \int_0^t c(q(r; x))\, dr.$$

A similar argument leads to the proof that V_c is a Lyapunov function for the CRW model.

A partial converse is established in Section 10.4.2 based on the following form of instability. The set $\mathcal{L}_0$ denotes the fluid limits satisfying $q(0) = 0$.

Definition 10.0.3 (Weak instability of the fluid limit model). The fluid limit model is said to be *weakly unstable* if there exists $\Omega_0 \subset \Omega$ satisfying $\mathsf{P}\{\Omega_0\} = 1$, and for each $\omega \in \Omega_0$ there exists $T_0 = T_0(\omega)$ such that $q(T_0) \neq 0$ for $q \in \mathcal{L}_0(\omega)$. ∎

Theorem 10.0.4 (Weak instability implies transience). *If the fluid limit model is weakly unstable, then the queueing network is unstable in the following sense:*

$$\liminf_{t\to\infty} \frac{1}{t}|Q(t; 0)| > 0 \qquad \text{a.s.}$$

In particular, $\lim_{t\to\infty} |Q(t; 0)| = \infty$.

On pursuing these ideas further we obtain structural results for the relative value function in average-cost optimal control. We will see that Theorem 10.0.5 provides intuition regarding the structure of optimal policies. For example, it implies the *existence* of switching curves resembling dynamic safety stocks in optimal policies. The limit (10.15) is known as *fluid-scale asymptotic optimality* (FSAO).

Theorem 10.0.5 (Fluid-scale asymptotic optimality of optimal policies). *Suppose that h^* solves the ACOE (9.4a) with c a linear cost function, and that h^* has quadratic*

growth, in the sense that $h^* \in L_\infty^{V_2}$. *Then, for each* $x \in \mathsf{X}$ *and any sequence* $\{x(\kappa)\}$
convergent to x,

$$J^*(x) = \lim_{\kappa \to \infty} \frac{1}{\kappa^2} h^*(\kappa x(\kappa)) = \lim_{T \to \infty} \left(\lim_{\kappa \to \infty} \mathsf{E} \left[\int_0^T c(q^\kappa(t; x(\kappa))) \, dt \right] \right), \quad (10.15)$$

where the expectation on the right is with respect to the h^*-*myopic policy.*

Theorems 10.0.2 and 10.0.4 demand that we answer the question, *How can one determine if the limiting process is stable?* One can develop Lyapunov theory or related techniques for this purpose, as shown in Propositions 10.4.6 and 10.4.4 that follow. However, the value of these results is greatest when we take a proactive stance: How can we *design* a policy so that the fluid limit model is stable?

It is natural to take the steps stressed repeatedly in Parts I and II of this book. First spend some time to understand the deterministic fluid model. We have a rich collection of techniques for constructing policies for complex networks based on this idealized model. In particular, the GTO policy and its variants can be computed by solving a finite-dimensional linear program. Once a suitable (and stable) policy is constructed it must be translated for application in a stochastic model or the physical system of interest.

Discrete-review policies are best suited for this translation step. In Section 10.5 we improve upon Theorem 8.3.2 to show how the fluid approximation (8.47) can be attained using safety stocks. This amounts to proving that the scaled processes $\{q^\kappa, z^\kappa\}$ converge to the predetermined solution (q, z). The conclusions in Section 10.5 are also an important foundation for the proof of Theorem 10.0.5.

Section 10.7 contains some extensions of these results to the CBM model. The value functions are defined for the fluid and CBM workload models, respectively, by

$$J(w) := \int_0^\infty \bar{c}(w(t; w)) \, dt, \quad (10.16)$$

$$h(w) := \int_0^\infty \left(\mathsf{E}[\bar{c}(W(t; w))] - \eta \right) dt, \quad w \in \mathsf{W}, \quad (10.17)$$

where $\bar{c}$ is a piecewise linear cost function on W, and η is its steady-state mean.

Theorem 10.0.6. *Suppose that assumptions (a)–(c) of Theorem 5.3.21 hold for the deterministic workload model* w *on the domain* $\mathsf{R} = \mathsf{W}$. *Let* W *denote the* R-*minimal solution for the CBM model. The value functions* h *and* J *differ by a function with linear growth,*

$$\sup_{w \in \mathsf{R}} \frac{|h(w) - J(w)|}{1 + \|w\|} < \infty.$$

Moreoever, the fluid limit holds, analogous to (10.15):

$$J(w) = \lim_{r \to \infty} \kappa^{-2} h(\kappa w), \quad w \in \mathsf{W}. \quad (10.18)$$

We begin with several examples to illustrate how the fluid limit model is constructed, and how it can be used to evaluate a given policy.

10.1 Examples

In the first example we illustrate how to obtain insight without much effort.

Example 10.1.1 (Simple re-entrant line: fluid limits under a priority policy). Consider the simple re-entrant line under the LBFS policy. Approximations of the form given in (4.57) can be used to characterize the fluid limit model $\mathcal{L}$. However, it is easiest to establish a few general properties as follows.

First, if $x_3 > 0$ then we have $z_3^\kappa(t; x(\kappa)) = t$ for each $\kappa > 0$ and all t before the first time that $Q_3(t; \kappa x(\kappa)) = 0$. Hence for any $(\boldsymbol{q}, \boldsymbol{z}) \in \mathcal{L}$ we have $\frac{d}{dt}z_3(t) = 1$ when $q_3(t) > 0$. Similar reasoning implies the nonidleness conditions

$$\tfrac{d^+}{dt}z_1(t) + \tfrac{d^+}{dt}z_3(t) = 1 \quad \text{when } q_1(t) + q_3(t) > 0, \quad \text{and} \quad \tfrac{d^+}{dt}z_2(t) = 1 \quad \text{when } q_2(t) > 0.$$

This is enough to conclude that $\mathcal{L}$ coincides with the possible solutions to the fluid model equations under the LBFS policy.

The analysis in Section 2.8 shows that the LBFS policy is stabilizing for the fluid model whenever $\rho_\bullet < 1$. Hence the assumptions of Theorem 10.0.2 are satisfied, and we conclude that this policy is regular for the CRW model. ∎

We now illustrate the application of Theorem 10.0.4 in the simple routing model.

Example 10.1.2 (Instability in the simple routing model). As discussed in Section 2.10, and illustrated in the simulation on the left in Fig. 6.16, the priority policy can be destabilizing.

For the policy (6.63) the behavior of buffer 2 starting with $Q_2(0) = 0$ is described as follows. For a geometrically distributed time with parameter α_r this buffer remains empty. The mean of this "period of starvation" is α_r^{-1}. At the time of the first arrival, a customer is routed to buffer 2, so that this buffer can begin work. During a geometrically distributed time with parameter μ_2 the buffer is busy, and $Q_2(t)$ remains equal to 1 since any arriving customers are routed to buffer 1. The mean of this busy period is μ_2^{-1}. This cycle repeats when the customer at buffer 2 completes service. Consequently, we have for any $x_1 > 0$, with $x = (x_1, 0)^{\mathsf{T}}$,

$$\zeta_2 := \lim_{\kappa \to \infty} \frac{z_2^\kappa(t; x)}{t} = \frac{\{\text{mean busy period}\}}{\{\text{mean time } Q_2 = 0\} + \{\text{mean busy period}\}} = \frac{\mu_2^{-1}}{\alpha_r^{-1} + \mu_2^{-1}},$$

giving $\zeta_2 = \alpha_r/(\mu_2 + \alpha_r)$.

If $x_1 > 0$ then we obtain using similar (and simpler) arguments that the fluid limit satisfies $\zeta_1 = \zeta_2 = 1$ for $0 \le t < T_2 := x_2/\mu_2$, and $q_2(t) = 0$ for $t > T_2$. Consequently, we obtain $\zeta_2 = \alpha_r/\mu_2/(1 + \alpha_r/\mu_2)$ for $T_2 \le t < T_1$ where $T_1 < \infty$ is the time at which the first buffer empties. We conclude that the first buffer never empties if $\mu_1 \le \alpha_r$ and $x \ne \boldsymbol{0}$.

For the particular numerical values used in Fig. 6.16 we have $\alpha_r/\mu_2 = 9/5$. While $q_1(t) > 0$ and $q_2(t) = 0$ it follows that $\zeta_1 = 1$, $\zeta_2 = \alpha_r/(\mu_2 + \alpha_r) = 9/14$, and hence

$$\tfrac{d^+}{dt}|q(t)| = \alpha_r - (\mu_1\zeta_1 + \mu_2\zeta_2) = \frac{9}{19} - \frac{5}{19}\left(1 + \frac{9}{14}\right) > 0.$$

The assumptions of Theorem 10.0.4 are satisfied, and hence the stochastic model is transient. ∎

In general there may be many fluid limits for a given $x \in \mathbb{R}_+^\ell$.

Example 10.1.3 (KSRS model: bifurcations under a priority policy). We have already observed in Section 2.9 the emergence of a virtual station in the fluid model for the KSRS network, characterized by the inequality constraint (2.29) on the pair (ζ_2, ζ_4). We revisit here the CRW model under the same policy.

Consider the CRW model in which the service rates satisfy (2.27), controlled using the priority policy in which buffers 2 and 4 receive strict priority at their respective stations. A fluid scaling of the controlled process Q leads to the fluid model already analyzed in Section 2.9. Consider the family of initial conditions $\{x(\kappa)\}$ satisfying $\kappa^{-1} x(\kappa) \to x = \mathbf{1}^1$ as $\kappa \to \infty$. While buffer 1 receives service, buffer 2 will grow due to the assumption $\mu_1 > \mu_2$ imposed in (2.27). Moreover, while buffer 2 grows we must have $Q_4(t) = 0$ and $U_3(t) = 0$ due to the priority assumption.

It can be shown using these arguments that $\{q^\kappa, z^\kappa\}$ converge to a solution (q, z) of the fluid model equations. The solution satisfies $q_1(0) = \mathbf{1}^1$ because of our choice of initial condition, and the behavior for $t > 0$ coincides with the fluid model controlled using this priority policy. Applying (2.29) we conclude that the fluid limit (q, z) is stable if and only if the virtual load condition $\rho_v < 1$ holds, with ρ_v defined in (2.30). Theorem 10.0.2 then implies that under the virtual station load condition, along with the distributional assumptions imposed in this theorem, this priority policy is stabilizing.

If $\rho_v > 1$ then the controlled model is transient, by Theorem 10.0.4. This conclusion is consistent with the simulation shown in Fig. 2.13.

We cannot determine stability or instability of the CRW model using these methods when $\rho_v = 1$.

Under this policy the set of fluid limits $\mathcal{L}_x(\omega)$ contains more than one pair of processes (q, z) for certain initial conditions. Suppose that the service rates satisfy (2.27), and moreover the following symmetry conditions hold:

$$\mu_1 = \mu_3 > \mu_2 = \mu_4.$$

Let $x = (1, 0, 1, 0)^\mathsf{T}$, and define the two sequences

$$x^1(\kappa) = \kappa^{-1}(\kappa, \lfloor \log(\kappa) \rfloor, \kappa, 0)^\mathsf{T}, \quad x^2(\kappa) = \kappa^{-1}(\kappa, 0, \kappa, \lfloor \log(\kappa) \rfloor)^\mathsf{T}, \qquad \kappa \in \mathbb{Z}_+.$$

We have $x^i(\kappa) \to x$ as $\kappa \to \infty$ for $i = 1, 2$.

Consider the behavior of the network from the initial condition $Q^1(0) = \kappa x^1(\kappa)$. Fix one sample path ω such that the law of large numbers holds for the service and arrival processes. Since $Q_2^1(0) = \lfloor \log(\kappa) \rfloor$ and buffer 2 receives priority at Station 2, it follows that $U_2(t) = 1$ for some time period $[0, T_2]$. During this time period buffer 4 remains empty, so that $U_1(t) = 1$. Since $\mu_1 > \mu_2$ it is very likely that buffer 2 will grow in length until buffer 1 empties. Based on these observations, it can be shown that

$$\lim_{\kappa \to \infty} q^\kappa(t; x^1(\kappa)) = (1 - (\mu_1 - \alpha_1)t, (\mu_1 - \mu_2)t, 1 + \alpha_3 t, 0)^\mathsf{T}, \qquad 0 \le t \le (\mu_1 - \alpha_1)^{-1}.$$

We can apply parallel arguments to establish the limit using $\{x^2(\kappa)\}$ as follows:

$$\lim_{\kappa \to \infty} q^\kappa(t; x^2(\kappa)) = (1 + \alpha_1 t, 0, 1 - (\mu_3 - \alpha_3)t, (\mu_3 - \mu_4)t)^\mathsf{T}, \qquad 0 \le t \le (\mu_3 - \alpha_3)^{-1}.$$

Each of these limiting processes belongs to $\mathcal{L}_x(\omega)$ for the given ω, and this value of x.   ∎

Theorem 10.0.5 provides motivation for the introduction of unbounded switching curves to guard against starvation of resources.

Example 10.1.4 (Tandem queues: optimal switching curve). Consider the tandem queues in the setting of Example 4.5.2: We restrict to Case 2 in which $\mu_1 > \mu_2$ (i.e., $\rho_1 < \rho_2$), with cost function $c(x) = x_1 + 3x_2$. The infinite-horizon optimal solution for the fluid model is pathwise optimal. It satisfies $\zeta_2(t) = 1$ whenever $|q(t)| > 0$, and $\zeta_1(t) = 0$ when $q_2(t) > 0$.

Theorem 10.0.5 tells us that an optimal policy for the CRW model has a fluid limit model that coincides with this pathwise optimal solution. In particular, taking $x = (0, x_2)^\mathsf{T}$ with $x_2 > 0$ we obtain

$$\lim_{\kappa \to \infty} z^\kappa(t; x^\kappa) = (\mu_2/\mu_1, 1)^\mathsf{T} t, \qquad 0 < t < \frac{x_2}{\mu_2 - \alpha_1}. \tag{10.19}$$

Suppose that the optimal policy is defined by a switching curve $s_* \colon \mathbb{R}_+ \to \mathbb{R}_+$ such that $\phi_1^*(x) = 1$ whenever $x_2 \le s_*(x_1)$, as is the case in the special case illustrated in Fig. 4.7.[2] Then the function s_* must be unbounded: A bounded switching curve will result in excessive starvation at Station 2, so that (10.19) is violated.   ∎

In applications to communications it is possible that the environment is observed, and that $U(t)$ is chosen as a function of environmental variables such as the service process. How does the theory change in this case?

Example 10.1.5 (Klimov model: service-aware scheduling). Consider the Klimov model with two customer classes in which the two service processes $\{M_1(t), M_2(t)\}$ are observed. A stochastic model is given by

$$Q_i(t + 1) = Q_i(t) - M_i(t)U_i(t) + A_i(t + 1), \qquad t \ge 0, \ i = 1, 2,$$

where $\{M_i(t) : t \ge 0\}$ are mutually independent, i.i.d. sequences.

For example, this model describes a wireless communication tower in which two classes of packets are waiting to be sent to one of two destinations. The channel conditions are ideal for broadcast to the first destination if and only if $M_1(t) = 1$.

The scaled processes $\{q^\kappa, z^\kappa\}$ are defined in (10.6), and we again expect that under general conditions on the policy, the scaled queue length process will converge as $\kappa \to \infty$ to a solution to the ODE

$$\tfrac{d^+}{dt} q = -M\zeta + \alpha, \tag{10.20}$$

with $M = \mathrm{diag}(\mu_1, \mu_2)$. However, when $U(t)$ is allowed to depend upon $M(t)$ we find that the set of possible ζ is strictly larger than that in the standard Klimov model in which $\mathsf{U} = \{\zeta : \zeta \ge 0, \ \zeta_1 + \zeta_2 \le 1\}$.

---

[2] The existence of a switching curve follows from a result of Hajek [222].

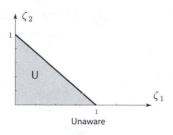
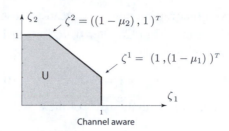

Figure 10.1. The set of possible allocation rates for the standard Klimov model is shown on the left. For any positive values of $\{\mu_i\}$ this is a strict subset of the set of achievable rates in the service-aware model shown on the right.

Let U denote the set of all possible values in the limiting ODE (10.20): $\zeta \in \mathsf{U}$ if there exists some allocation sequence U and some $t > 0$ such that

$$\lim_{\kappa \to \infty} \frac{1}{\kappa} \sum_{0 \le i \le \kappa t - 1} M(\kappa t + 1) U(\kappa t; \kappa x(\kappa)) = t M \zeta, \qquad \text{a.s.} \tag{10.21}$$

Below are policies giving five extreme points in the set U. In these definitions we are ignoring the state-space constraint $Q(t) \ge 0$ since we are only attempting to identify long-term rates as $\kappa \to \infty$.

Policy 1: $U_1(t) = M_1(t)$, $U_2(t) = M_2(t)(1 - M_1(t))$. That is, buffer 1 receives strict priority. This gives $M\zeta^1 = (\mu_1, (1 - \mu_1)\mu_2)^{\mathsf{T}}$ (using independence of M_1 and M_2).

Policy 2: $U_1(t) = M_1(t)(1 - M_2(t))$, $U_2(t) = M_2(t)$, which by symmetry gives $M\zeta^2 = ((1 - \mu_2)\mu_1, \mu_2)^{\mathsf{T}}$.

Policy 3: $U_1(t) = 1$, $U_2(t) = 0$, gives $\zeta^3 = (1, 0)^{\mathsf{T}}$.

Policy 4: $U_1(t) = 0$, $U_2(t) = 1$, gives $\zeta^4 = (0, 1)^{\mathsf{T}}$.

Policy 5: $U(t) = \zeta^5 = (0, 0)^{\mathsf{T}}$.

Any convex combination of these five rates is also achievable as the limit (10.21) using a randomized policy. Moreover, it is not hard to show that the convex hull shown on the right in Fig. 10.1 is precisely the set of possible allocation rates U defined as possible limits in (10.21).

What is the load of this system? To answer this question we first express the set U shown on the right in Fig. 10.1 as the polyhedron

$$\mathsf{U} = \{\zeta \in \mathbb{R}_+^2 : C\zeta \le 1\},$$

where the new constituency matrix is defined by

$$C = \begin{pmatrix} 1 & 0 \\ 0 & 1 \\ k_0\mu_1 & k_0\mu_2 \end{pmatrix}$$

with $k_0 = [1 - (1 - \mu_1)(1 - \mu_2)]^{-1}$.

Given this representation we can represent workload vectors using the general approach described in Chapter 6, following Definition 6.1.2. There are three workload vectors in this model,

$$\xi^1 = (\mu_1^{-1}, 0)^\mathsf{T}, \quad \xi^2 = (0, \mu_2^{-1})^\mathsf{T}, \quad \xi^3 = k_0(1,1)^\mathsf{T},$$

which coincide with the rows of the 3×2 matrix

$$\Xi = CM^{-1}.$$

The vector load is given by $\rho_i := \langle \xi^i, \alpha \rangle$, $i = 1, 2, 3$, and the system load is the maximum,

$$\rho_\bullet = \max \rho_i.$$

This fluid model can be stabilized if and only if $\rho_\bullet < 1$.

For example, in the balanced case $\alpha_1 = \alpha_2$, $\mu_1 = \mu_2$ we have

$$\rho_\bullet = \frac{2\alpha_1}{k_0} = \frac{2\alpha_1}{\mu_1}\left(\frac{1}{2 - \mu_1}\right).$$

The load for the standard Klimov model, given by $2\alpha_1/\mu_1$ in the balanced case, is strictly larger than $\rho_\bullet$ when $\alpha_1 > 0$ and $\mu_1 < 1$.   ∎

To prove the main results surveyed at the start of this chapter we first review some particular mathematical background.

10.2 Mathematical preliminaries

10.2.1 Continuity and convergence

In analyzing the scaled processes $\{q^\kappa\}$ we will apply results concerning equicontinuous functions. In the following definitions we let $\mathcal{K}$ denote an index set that may be uncountably infinite. When speaking of convergence we set $\mathcal{K}$ equal to $\mathbb{Z}_+$ or $\mathbb{R}_+$.

Definition 10.2.1 (Uniform convergence). The following definitions apply to a family of functions $\{f_\kappa\}$, each mapping $\mathbb{R}_+$ to $\mathbb{R}^\ell$:

(i) The family of functions converges *uniformly on compacta* (u.o.c.) to a function $f(\cdot)$ if for each $T > 0$,

$$\lim_{\kappa \to \infty} \sup_{0 \le t \le T} \|f_\kappa(t) - f(t)\| = 0.$$

(ii) The family of functions is *precompact in the uniform topology* if for any subsequence $\{\kappa_m\}$, there exist a further subsequence $\{\kappa_{m_n}\}$ and a function f_∞ such that $f_{\kappa_{m_n}} \to f_\infty$ (u.o.c.) as $n \to \infty$.

(iii) The family is *equicontinuous on compacta* if for each $T > 0$, $\varepsilon > 0$, there exists $\delta > 0$ such that

$$\|f_\kappa(t_1) - f_\kappa(t_2)\| \le \varepsilon, \qquad \text{for all } t_1, t_2 \in [0, T] \text{ s.t. } |t_1 - t_2| \le \delta, \text{ and all } \kappa \in \mathcal{K}.$$

∎

For example, a family of differentiable functions whose derivatives are uniformly bounded on any bounded interval is equicontinuous on compacta. The family of functions $\{z^\kappa\}$ is equicontinuous on compacta since it is uniformly Lipschitz continuous: For some constant $b_L < \infty$,

$$\|z^\kappa(t_1) - z^\kappa(t_2)\| \le b_L|t_1 - t_2|, \qquad t_1, t_2 \in \mathbb{R}_+.$$

The following result will be used to demonstrate that the fluid limit model $\mathcal{L}$ is nonempty.

Theorem 10.2.2 (Arzelà–Ascoli Theorem). *The following are equivalent for a family of functions $\{f_\kappa(\cdot) : \kappa \in \mathcal{K}\}$ on a bounded interval $[0, T]$, where $\mathcal{K}$ is any index set.*

(i) $\{f_\kappa\}$ *is precompact in the uniform topology.*
(ii) $\{f_\kappa\}$ *is equicontinuous, and also uniformly bounded: There exists $b(T) < \infty$ such that*

$$\|f_\kappa(t)\| \le b(T), \qquad \text{for all } t \in [0, T] \text{ and all } \kappa.$$

In a typical application of the Arzelà–Ascoli Theorem we are given a family of functions $\{f_\kappa\}$ that is known to converge pointwise. If the family is equicontinuous on compacta, then necessarily the convergence is u.o.c. The next result provides an alternative route to establishing uniform convergence.

Proposition 10.2.3. *Suppose that a family of functions $\{f_\kappa(\cdot) : \kappa \ge 0\}$ from $\mathbb{R}_+$ to $\mathbb{R}_+$ satisfies the following:*

(a) $\{f_\kappa\}$ *is uniformly bounded on compacta,*
(b) $f_\kappa(t)$ *is a nondecreasing function of t for each κ, and*
(c) *there is a continuous function $f_\infty(\cdot)$ such that*

$$\lim_{\kappa \to \infty} f_\kappa(t) = f_\infty(t) \qquad \text{for each } t \ge 0.$$

Then $f_\kappa \to f_\infty$ u.o.c.

Proof. The proof is by contradiction. If the convergence is not uniform, then there exist $T > 0$, $\varepsilon > 0$, a sequence $\{t_i\} \subset [0, T]$, and an increasing sequence $\{\kappa_i\} \subset \mathbb{Z}_+$ such that for each i,

$$|f_{\kappa_i}(t_i) - f_\infty(t_i)| \ge \varepsilon.$$

Without loss of generality we can assume that t_i is convergent to some $t_\infty \in [0, T]$, and that the sign of $f_{\kappa_i}(t_i) - f_\infty(t_i)$ is either positive for all i, or negative for all i. We henceforth assume that the former condition holds, the case where $f_{\kappa_i}(t_i) - f_\infty(t_i) \le -\varepsilon$ for all i is treated similarly.

We then have, since $f_{\kappa_i}(\cdot)$ is an increasing function of time,

$$f_{\kappa_i}(t) \ge f_\infty(t_i) + \varepsilon, \qquad t \in [t_i, T],$$

or equivalently,

$$f_{\kappa_i}(t) \ge [f_\infty(t_i) + \varepsilon]\mathbf{1}_{[t_i, T]}(t), \qquad t \in \mathbb{R}_+.$$

Letting $i \to \infty$ we obtain

$$f_\infty(t) = \lim_{i \to \infty} f_{\kappa_i}(t) \geq [f_\infty(t_\infty) + \varepsilon]\mathbf{1}_{(t_\infty, T]}(t), \qquad t \in \mathbb{R}_+.$$

This contradicts continuity of f_∞, and completes the proof.   □

10.2.2 Laws of large numbers

Limit theorems for stochastic processes often involve uniform convergence. Consider a real-valued i.i.d. process $\{\mathcal{E}(t) : t \geq 0\}$ with zero mean, define for any integer $T \geq 1$,

$$M_\mathcal{E}(T) = \sum_{t=1}^{T} \mathcal{E}(t), \qquad T \geq 1, \ M_\mathcal{E}(0) = 0, \tag{10.22}$$

and extend the definition to arbitrary $T \in \mathbb{R}_+$ by linear interpolation. We have the following uniform Strong Law of Large Numbers.

Theorem 10.2.4. *Suppose that the i.i.d. process $\mathcal{E} = \{\mathcal{E}(t) : t \geq 0\}$ has zero mean. Then, there exists $\Omega_\mathcal{E} \subset \Omega$ satisfying $\mathsf{P}\{\Omega_\mathcal{E}\} = 1$ such that for each $\omega \in \Omega_\mathcal{E}, T > 0$,*

$$\lim_{T \to \infty} \sup_{0 \leq t \leq T} t^{-1}|M_\mathcal{E}(t)| = 0.$$

Proof. We write $\mathcal{E}(t) = \mathcal{E}_+(t) - \mathcal{E}_-(t)$ where the two terms are nonnegative with finite mean. The process $M_\mathcal{E}$ is similarly decomposed as $M_\mathcal{E} = M_{\mathcal{E}_+} - M_{\mathcal{E}_-}$, with

$$M_{\mathcal{E}_+}(T) = \sum_{t=1}^{T} \mathcal{E}_+(t), \qquad M_{\mathcal{E}_-}(T) = \sum_{t=1}^{T} \mathcal{E}_-(t).$$

By the usual LLN for i.i.d. sequences (a special case of Theorem A.5.8) we have with probability 1

$$\lim_{\kappa \to \infty} |M_{\mathcal{E}_+}^\kappa(t) - e_+ t| = \lim_{\kappa \to \infty} |M_{\mathcal{E}_-}^\kappa(t) - e_- t| = 0, \tag{10.23}$$

where e_+, e_- denote the means of $\mathcal{E}_+(t), \mathcal{E}_-(t)$, and

$$M_{\mathcal{E}_+}^\kappa(t) = \frac{1}{\kappa} M_{\mathcal{E}_+}(\kappa t), \quad M_{\mathcal{E}_-}^\kappa(t) = \frac{1}{\kappa} M_{\mathcal{E}_-}(\kappa t), \qquad t \geq 0, \kappa > 0.$$

Applying Proposition 10.2.3 we conclude that the convergence in (10.23) is u.o.c.   □

The process $M_\mathcal{E}$ defined in (10.22) is a martingale, as defined in Section 1.3.5. We next establish a uniform law of large numbers based on Azuma's Inequality Theorem 1.3.9 (ii).

Theorem 10.2.5. *Suppose that $\{M_\kappa : \kappa \in \mathbb{Z}_+\}$ is a sequence of martingales, each satisfying the conditions of Azuma's Inequality: For some $b_M < \infty$ independent of κ,*

$$|M_\kappa(t) - M_\kappa(t-1)| \leq b_M, \qquad t \geq 1.$$

Then there exists $\Omega_M \subset \Omega$ satisfying $\mathsf{P}\{\Omega_M\} = 1$, and for each $\omega \in \Omega_M, T > 0$,

$$\lim_{\kappa \to \infty} \sup_{0 \leq t \leq T} |M^\kappa(t)| = 0,$$

where the scaled martingale is defined by

$$M^\kappa(t) = \frac{1}{\kappa} M_\kappa(\kappa t), \qquad t \geq 0, \ \kappa \in \mathbb{Z}_+.$$

Proof. We apply the Borel–Cantelli Lemma to the bound obtained in Azuma's Inequality to establish pointwise convergence. Next, we argue that $\{M^\kappa : \kappa > 0\}$ is equicontinuous and uniformly bounded on compacta. Hence uniform convergence follows from the Arzelà–Ascoli Theorem. $\qquad\square$

10.2.3 Differential equations

The scaled process is a perturbation of a solution to the fluid model equations (10.8). When z can be represented in state feedback form $\zeta(t) = \frac{d^+}{dt} z(t) = \phi^0(q(t))$ we obtain the ODE

$$\frac{d^+}{dt} q(t) = B\phi^0(q(t)) + \alpha, \qquad t \geq 0. \tag{10.24}$$

This ODE has a unique solution for each initial condition if ϕ^0 is Lipschitz continuous on $\mathbb{R}_+^\ell$ [59]. Unfortunately, continuity rarely holds in network models. This is why we typically represent the processes using the integral form (10.8).

The *Bellman–Gronwall Lemma* is commonly used to establish existence of solutions to ODEs, or to obtain bounds on the solution when the ODE is perturbed. A proof can be found in [455].

Lemma 10.2.6 (Bellman–Gronwall Lemma). *Suppose that the nonnegative functions of time $\{x(t), \beta(t), b(t) : t \geq 0\}$ satisfy*

$$x(t) \leq b(t) + \int_0^t \beta(s)x(s)\, ds, \qquad 0 \leq t \leq T.$$

Then

$$x(t) \leq \max\{b(s) : s \leq t\} \exp\left(\int_0^t \beta(s)\, ds \right), \qquad 0 \leq t \leq T.$$

10.3 Fluid limit model

We now survey properties of the fluid limit model. We first present a formal definition.

Definition 10.3.1 (Fluid limit model). For each $x \in \mathsf{X}$ and $\omega \in \Omega$ we let $\mathcal{L}_x(\omega)$ denote the set of all possible fluid limits,

$$\mathcal{L}_x(\omega) = \left\{ \begin{array}{l} \text{u.o.c. subsequential limits of } \{q^\kappa(t; x(\kappa), \omega), \ z^\kappa(t; x(\kappa), \omega)\} \\ \text{such that } x(\kappa) \to x \text{ as } \kappa \to \infty \end{array} \right\}.$$

The *fluid limit model* is the union $\mathcal{L} := \cup_{x \in \mathsf{X}} \mathcal{L}_x$. $\qquad\blacksquare$

In the following results we show that Definition 10.3.1 is consistent with the informal definition of $\mathcal{L}$ given at the start of this chapter. In particular, Proposition 10.3.2 implies that the pointwise convergence in (10.9) implies u.o.c. convergence.

10.3.1 Properties

The scaled processes q^κ in (10.6) can be expressed

$$q^\kappa(t; x(\kappa)) = x(\kappa) + Bz^\kappa(t; x(\kappa)) + \alpha t + N^\kappa(t; x(\kappa)), \qquad (10.25)$$

where $N^\kappa(t; x(\kappa)) := \kappa^{-1} N(\kappa t)$. We see in Proposition 10.3.2 that N^κ is negligible for large κ.

Throughout the chapter we take the set Ω_0 in Definition 10.0.1 to be a subset of the set Ω_0 identified in Proposition 10.3.2 (i).

Proposition 10.3.2. *For any policy:*

(i) *There exists* $\Omega_0 \subset \Omega$ *satisfying* $\mathsf{P}\{\Omega_0\} = 1$ *such that for each* $\omega \in \Omega_0$, $x \in \mathsf{X}$, $T > 0$, *and any family of vectors* $x(\kappa) \to x$,

$$\lim_{\kappa \to \infty} \sup_{0 \le t \le T} \|N^\kappa(t; x(\kappa))\| = 0.$$

(ii) *For some constant* b_N *(independent of the policy), any* $x \in \mathsf{X}$, $t > 0$,

$$\mathsf{E}[\|N^\kappa(t; x(\kappa))\|^2] \le \kappa^{-1} b_N t.$$

Proof. We decompose the disturbance N as the sum of two martingales,

$$N(t) = N_A(t) + N_B(t), \qquad t \in \mathbb{Z}_+,$$

$$\text{where } N_A(t) := \sum_{k=1}^{t} (A(k) - \alpha), \quad N_B(t) := \sum_{k=1}^{t} (B(k) - B)U(k - 1). \qquad (10.26)$$

The scaled disturbance is decomposed similarly, $N^\kappa(t; x(\kappa)) = N_A^\kappa(t; x(\kappa)) + N_B^\kappa(t; x(\kappa))$ for $t \in \mathbb{R}$.

Uniform convergence of the first term is simplified since it is independent of U: With probability 1, $N_A^\kappa(\,\cdot\,; x(\kappa)) \to 0$ u.o.c. by Theorem 10.2.4. Uniform convergence of $N_B^\kappa(\,\cdot\,; x(\kappa))$ follows from Theorem 10.2.5 since B and U are bounded sequences. This proves (i).

Part (ii) follows from the triangle inequality,

$$\mathsf{E}[\|N^\kappa(t)\|^2]^{\frac{1}{2}} \le (t\sigma_A^2)^{\frac{1}{2}} + (t\bar{\sigma}_B^2)^{\frac{1}{2}},$$

with σ_A^2 the mean of $\|A(1) - \alpha\|^2$, and $\bar{\sigma}_B^2 = \max\{\mathsf{E}[\|(B(1) - B)u\|^2] : u \in \mathsf{U}\}$. □

Proposition 10.3.3 is a simple corollary.

Proposition 10.3.3. *Suppose that for a given sample* $\omega \in \Omega_0$ *and a subsequence* $\{\kappa_i\}$ *the convergence (10.9) holds pointwise in* t. *Then the convergence is also u.o.c., and hence* $(q, z) \in \mathcal{L}_x(\omega)$.

Proof. Proposition 10.3.2 along with the representation (10.25) implies that the set of limits of $\{q^\kappa\}$ coincides with the limits of the scaled process with disturbance removed,

$$q^{0\kappa}(t; x(\kappa)) := x(\kappa) + Bz^\kappa(t; x(\kappa)) + \alpha t, \qquad t \ge 0. \qquad (10.27)$$

That is, $\|q^\kappa(t; x(\kappa)) - q^{0\kappa}(t; x(\kappa))\| = \|N^\kappa(t; x(\kappa))\|$, so that (10.9) holds u.o.c. $\boldsymbol{q}^\kappa$ if and only if the same holds true for $\boldsymbol{q}^{0\kappa}$.

The Arzelà–Ascoli Theorem implies that pointwise and u.o.c. convergence are equivalent for $\boldsymbol{q}^{0\kappa}$. □

Proposition 10.3.2 implies that with probability 1, any fluid limit is a solution to the fluid model equations. Proposition 10.3.4 summarizes this conclusion and other properties of $\mathcal{L}$.

Proposition 10.3.4. *The following hold for the fluid limit model under any policy:*

(i) $q(t) = x + Bz(t) + \alpha t$ *for* $t \ge 0$, $x \in X$, $(\boldsymbol{q}, \boldsymbol{z}) \in \mathcal{L}_x(\omega)$, $\omega \in \Omega_0$.

(ii) *The set* $\{\mathcal{L}_x(\omega) : |x| \le m\}$ *is compact in the uniform topology for any* $m > 0$ *and* $\omega \in \Omega_0$.

(iii) *The fluid limit model is closed under scaling: For each* $r > 0$,

$$(\boldsymbol{q}^{(r)}, \boldsymbol{z}^{(r)}) \in \mathcal{L}_{r^{-1}x} \text{ whenever } (\boldsymbol{q}, \boldsymbol{z}) \in \mathcal{L}_x,$$

where

$$q^{(r)}(t) = r^{-1}q(rt), \quad z^{(r)}(t) = r^{-1}z(rt), \qquad t \ge 0. \tag{10.28}$$

Proof. Part (i) follows directly from Proposition 10.3.2 applied to the representation (10.25). Part (ii) follows from the Arzelà–Ascoli Theorem since each limit is a solution to the fluid model equations (10.8).

We now prove (iii). If $(\boldsymbol{q}, \boldsymbol{z}) \in \mathcal{L}_x(\omega)$, then (10.9) holds for some $x(\kappa) \to x$ and a subsequence $\{\kappa_i\}$. We have for any $r > 0$

$$q^{\kappa_i}(rt; x(\kappa_i), \omega) = \frac{1}{\kappa_i} Q(\kappa_i rt; \kappa_i x(\kappa_i), \omega)$$

$$= r\frac{1}{\kappa_i'} Q(\kappa_i' t; \kappa_i' x'(\kappa_i'), \omega), \quad i \ge 1,$$

where $\kappa_i' = r\kappa_i$ and $x'(\kappa_i') = r^{-1}x(\kappa_i)$. Consequently,

$$\lim_{i \to \infty} \frac{1}{\kappa_i'} Q(\kappa_i' t; \kappa_i' x'(\kappa_i'), \omega) = \lim_{i \to \infty} r^{-1}q^{\kappa_i}(rt; x(\kappa_i), \omega) = r^{-1}q(rt) = q^{(r)}(t).$$

Moreover, since $x'(\kappa_i') \to r^{-1}x$ as $i \to \infty$, the limit $\boldsymbol{q}^{(r)}$ belongs to $\in \mathcal{L}_{r^{-1}x}$ as claimed. □

The proof of the following semigroup property is similar; its proof is left as an exercise.

Proposition 10.3.5. *Suppose that* $(\boldsymbol{q}, \boldsymbol{z}) \in \mathcal{L}_x$ *for some* $x \in X$ *and* $\omega \in \Omega_0$. *Then for each* $t_1 > 0$ *we have* $(\boldsymbol{q}^{[t_1]}, \boldsymbol{z}^{[t_1]}) \in \mathcal{L}_{x^1}$, *where* $x^1 = q(t_1)$ *and*

$$q^{[t_1]}(t) := q(t + t_1), \quad z^{[t_1]}(t) := z(t + t_1), \qquad t \ge 0.$$

10.3.2 Consequences and characterizations of stability

There are many consequences of stability. From the scaling property given in Proposition 10.3.4 we see that if trajectories $q \in \mathcal{L}$ vanish for $t \geq T$ and initial conditions satisfying $\|x\| \leq 1$, then they vanish for each initial condition.

Proposition 10.3.6. *If the fluid limit model is stable, then there exists $T_0 > 0$ such that for any $x \in \mathsf{X}$,*

$$q(t; x) = 0, \qquad q \in \mathcal{L}_x,\ t \geq T_0|x|, \qquad \omega \in \Omega_0.$$

Proof. We take the value T_0 given in Definition 10.0.1 and apply the scaling formula in Proposition 10.3.4 (iii) to obtain

$$r^{-1}q(rt; rx) = 0, \qquad q \in \mathcal{L}_x,\ |x| = 1,\ t \geq T_0,\ r > 0. \qquad \square$$

If the fluid limit model is stable, then it is "uniformly stable." Recall the definition of $\{x^\kappa\}$ given in (10.5).

Proposition 10.3.7. *Suppose that the fluid model is stable. Then it is uniformly stable in the following two senses:*

(i) Almost surely: *For $T \geq T_0$ and $\omega \in \Omega_0$,*

$$\lim_{\kappa \to \infty} \sup_{|x|=1} |q^\kappa(T; x^\kappa, \omega)| = \lim_{\kappa \to \infty} \sup_{|x|=1} |x + z^\kappa(T; x^\kappa, \omega) + \alpha T| = 0. \quad (10.29)$$

(ii) In the L_2 sense: *For $T \geq T_0$,*

$$\lim_{\kappa \to \infty} \sup_{|x|=1} \mathsf{E}[|q^\kappa(T; x^\kappa)|^2] = 0. \quad (10.30)$$

Proof. We prove (i) by contradiction. If this limit does not hold, then there exist $T \geq T_0, \omega \in \Omega_0, \varepsilon > 0$, sequences $\{x(i)\} \subset \{x \in \mathsf{X} : |x| = 1\}$ and $\{\kappa_i\} \subset \mathbb{R}_+$ such that $\kappa_i \to \infty$ as $i \to \infty$, and for each i,

$$|q^{\kappa_i}(T; x^{\kappa_i}(i), \omega)| \geq \varepsilon.$$

Without loss of generality we can assume that $x(i)$ is convergent to some $x \in \mathsf{X}$ satisfying $|x| = 1$. Moreover, taking a further subsequence if necessary, we can assume that $\{q^{\kappa_i}, z^{\kappa_i} : i \geq 0\}$ is convergent u.o.c., and necessarily the limit (q, z) lies in $\mathcal{L}_x(\omega)$. We have $|q(T)| \geq \varepsilon$ since this bound holds for each i, which is the desired contradiction. This proves (i).

Applying the triangle inequality,

$$\mathsf{E}[|q^\kappa(T; x^\kappa)|^2]^{\frac{1}{2}} \leq \mathsf{E}[|x^\kappa + Bz^\kappa(T; x^\kappa) + \alpha T|^2]^{\frac{1}{2}} + \mathsf{E}[|N^\kappa(T; x^\kappa)|^2]^{\frac{1}{2}},$$

Proposition 10.3.2 gives

$$\limsup_{\kappa \to \infty} \sup_{|x|=1} \mathsf{E}[|q^\kappa(T; x^\kappa)|^2]^{\frac{1}{2}} \leq \limsup_{\kappa \to \infty} \sup_{|x|=1} \mathsf{E}[|x^\kappa + Bz^\kappa(T; x^\kappa) + \alpha T|^2]^{\frac{1}{2}}.$$

We also have for any κ

$$\sup_{|x|=1} \mathsf{E}[|q^\kappa(T; x^\kappa)|^2] \le \mathsf{E}\left[\sup_{|x|=1} |x^\kappa + Bz^\kappa(T; x^\kappa) + \alpha T|^2\right].$$

Hence to prove (ii) it is sufficient to prove that

$$\limsup_{\kappa \to \infty} \mathsf{E}\left[\sup_{|x|=1} |x^\kappa + Bz^\kappa(T; x^\kappa) + \alpha T|^2\right] = 0. \tag{10.31}$$

This follows from (i) and the Dominated Convergence Theorem.   □

The following apparently weaker criterion is useful in establishing stability of the fluid limit model.

Proposition 10.3.8. *Suppose that there exists $\Omega_0 \subset \Omega$ satisfying $\mathsf{P}\{\Omega_0\} = 1$, and $T_0 > 0$ such that for each $\omega \in \Omega_0$ and $q \in \mathcal{L}(\omega)$ satisfying $|q(0)| \le 1$,*

$$\min_{0 \le t \le T_0} |q(t)| = 0.$$

Then the fluid limit model $\mathcal{L}$ is stable.

Proof. Applying Proposition 10.3.4 we have for each $r > 0$, each $x \in \mathsf{X}$ satisfying $|x| \le r$, and $q \in \mathcal{L}_{r^{-1}x}$,

$$\min_{0 \le t \le T_0} |q^{(r)}(t)| = 0.$$

Equivalently,

$$\min_{0 \le t \le rT_0} |q(t)| = 0.$$

Applying the semigroup property Proposition 10.3.5, it follows that for each $t_1 > 0$,

$$\min_{0 \le t \le |q(t_1)|T_0} |q(t_1 + t)| = 0.$$

Since we are working with the ℓ_1-norm we have $|q(t_1 + t)| \le |q(t_1)| + |\alpha|t$ for $t > 0$. These bounds imply that once the origin is reached, the fluid limit must stay there.   □

10.4 Fluid-scale stability

We now take a closer look at the scaled processes defined in (10.6) to provide a proof of the main results summarized at the start of this chapter.

10.4.1 ODE criterion for regularity

Proposition 10.4.1 establishes that the function V_c is a solution to (V3), provided the fluid limit model is stable, and the time T used in its definition is sufficiently large.

Proposition 10.4.1. *Suppose that the fluid limit model is stable for a given stationary policy. Then there exists $T_0 < \infty$ such that the controlled network satisfies (V3) with the Lyapunov function V_c given in (10.11) with $f = 1 + \frac{1}{2}c$, provided $T > T_0$. Moreover, the function V_c is equivalent to a quadratic.*

The proof is based on the approximation (10.12) which asserts that the function V_c can be interpreted as an approximate fluid value function. This is made precise in the following:

Lemma 10.4.2. *For any policy and any $T > 0$ we have*

$$\frac{1}{\kappa^2} \sum_{t=0}^{\kappa T^\kappa} c(Q(t; \kappa x^\kappa)) = \int_0^{T^\kappa} c(q^\kappa(s; x^\kappa)) \, ds + \mathcal{E}^\kappa, \qquad x \in \mathsf{X}, \qquad (10.32)$$

where $T^\kappa = \kappa^{-1} \lfloor \kappa T \rfloor$, and the random variables $\{\mathcal{E}^\kappa\}$ satisfy, for each $r > 0$,

$$\lim_{\kappa \to \infty} \sup_{|x| \le r} \mathsf{E}_x[\|\mathcal{E}^\kappa\|^2] = 0.$$

Proof. Based our convention that Q is piecewise linear as a function of $t \in \mathbb{R}_+$, we have for each initial condition $Q(0)$

$$\int_0^{\kappa T^\kappa} c(Q(t)) \, dt = \sum_{t=1}^{\kappa T^\kappa - 1} \tfrac{1}{2}[c(Q(t+1)) + c(Q(t))]$$

$$= \sum_{t=0}^{\kappa T^\kappa} c(Q(t)) - \tfrac{1}{2}[c(Q(N)) + c(Q(0))]. \qquad (10.33)$$

Setting $Q(0) = \kappa x^\kappa$ we also have

$$\frac{1}{\kappa^2} \int_0^{\kappa T^\kappa} c(Q(t; \kappa x^\kappa)) \, dt = \frac{1}{\kappa} \int_0^{T^\kappa} c(Q(\kappa s; \kappa x^\kappa)) \, ds$$

$$= \int_0^{T^\kappa} c(q^\kappa(s; x^\kappa)) \, ds.$$

The first identity follows from the change of variables $\kappa s = t$, and the second is the definition of the scaled process. Finally, on setting

$$\mathcal{E}^\kappa := \frac{1}{\kappa^2}[c(Q(\kappa T^\kappa)) + c(Q(0))] = \frac{1}{\kappa}[c(q^\kappa(T^\kappa; x^\kappa)) + c(x^\kappa)]$$

the result follows from Proposition 10.3.2.   □

Proposition 10.4.3 provides a refinement of Lemma 10.4.2 under the assumption that q^κ is convergent a.s. for each $x \in \mathbb{R}_+^\ell$.

Proposition 10.4.3. *Suppose that $\mathcal{L}_x$ is a singleton for each $x \in \mathsf{X}$, so that with probability 1 $q^\kappa(t; x) \to q(t; x)$, a deterministic solution to the fluid model equations, for each x and t. Then:*

(i) *For each $x \in \mathsf{X}$ satisfying $|x| = 1$,*

$$\lim_{\kappa \to \infty} \frac{1}{\kappa^2} V_c(\kappa x^\kappa) = \int_0^T c(q(t; x)) \, dt.$$

(ii) *Suppose that the fluid limit model is stable, let $T_0 > 0$ denote the time used in Definition 10.0.1, and assume that the integer T used in the definition (10.11) satisfies $T \geq T_0$. Then the limit (10.14) holds for each $x \in X$,*

$$\lim_{\kappa \to \infty} \frac{1}{\kappa^2} V_c(\kappa x(\kappa)) = \lim_{N \to \infty} \lim_{\kappa \to \infty} \int_0^N c(q^\kappa(t; x(\kappa))) \, dt = \int_0^\infty c(q(t; x)) \, dt.$$

Proof. On taking expectations in (10.32) we obtain

$$\kappa^{-2} V_c(\kappa x^\kappa) = \mathsf{E}\left[\int_0^{T^\kappa} c(q^\kappa(s; x^\kappa)) \, ds + \mathcal{E}^\kappa\right],$$

and hence by Lemma 10.4.2,

$$\lim_{\kappa \to \infty} \frac{1}{\kappa^2} V_c(\kappa x^\kappa) = \lim_{\kappa \to \infty} \mathsf{E}\left[\int_0^{T^\kappa} c(q^\kappa(s; x^\kappa)) \, ds\right] = \int_0^T c(q(s; x)) \, ds. \qquad \square$$

Proof of Proposition 10.4.1. We have seen in Proposition 10.3.7 that stability of the fluid limit model implies L_2 convergence: For $T \geq T_0$,

$$\lim_{\kappa \to \infty} \sup_{|x|=1} \mathsf{E}[|q^\kappa(T; x^\kappa)|^2] = 0. \tag{10.34}$$

This result is used in the main step in the proof.

Before proceeding it is helpful to review the Markov property: Suppose that $\mathcal{C} = \mathcal{C}(Q(0), Q(1), \dots)$ is any random variable with finite mean. We always have

$$\mathsf{E}_{Q(n)}[\mathcal{C}] = \mathsf{E}[\vartheta^n \mathcal{C} \mid Q(0), \dots, Q(n)] = \mathsf{E}[\vartheta^n \mathcal{C} \mid Q(n)], \tag{10.35}$$

where $\mathcal{F}_n := \sigma\{Q(0), \dots, Q(n)\}$, $n \geq 0$, and $\vartheta^n \mathcal{C}$ denotes the random variable

$$\vartheta^n \mathcal{C} = \mathcal{C}(Q(n), Q(n+1), \dots).$$

We apply the Markov property with $n = 1$ and $\mathcal{C} = \sum_{i=0}^{|x|T} c(Q(i))$ to obtain

$$V_c(Q(1)) = \mathsf{E}_{Q(1)}[\mathcal{C}] = \mathsf{E}[\vartheta^1 \mathcal{C} \mid \mathcal{F}_1].$$

Applying the transition matrix to V_c gives $PV_c(x) = \mathsf{E}_x[V_c(Q(1))]$, so that by the definition of $\mathcal{C}$,

$$PV_c(x) = \mathsf{E}\left[\sum_{i=1}^{|Q(1)|T} c(Q(i))\right], \qquad Q(0) = x \in X_\diamond.$$

The sum within the expectation on the righthand side is almost the same as used in the definition of V_c. However, instead of summing from 0 to $T|Q(0)|$, we are summing from 1 to $T|Q(1)|$. Consequently, we have the expression

$$PV_c(x) = V_c(x) - c(x) + b(x), \qquad x \in X_\diamond, \tag{10.36}$$

where $b(x) := \mathsf{E}_x\left[\sum_{i=|x|T+1}^{|y|T} c(Q(i))\right]$ with $y = Q(1)$; the sum within the expectation is interpreted as negative when $|x| \geq |y|$.

To complete the proof we now obtain bounds on this sum. First note that under the assumptions of the theorem we have, for some constant $b_0 < \infty$,

$$c(Q(t+1)) \leq c(Q(t)) + (|A(t+1)| + 1)b_0, \qquad t \geq 1,$$

which gives, for any $i \in (|x|T, |y|T]$,

$$c(Q(i)) \leq c(Q(|x|T)) + \left[\sum_{t=|x|T+1}^{|y|T} b_0(|A(t)| + 1) \right]_+ .$$

This combined with the inequality $|y| \leq |x| + |A(1)|$ gives

$$\sum_{t=|x|T+1}^{|y|T} c(Q(t)) \leq \sum_{t=|x|T+1}^{(|x|+|A(1)|)T} c(Q(t))$$

$$\leq (|A(1)|)T) \left(c(Q(|x|T)) + \sum_{j=|x|T+1}^{(|x|+|A(1)|)T} b_0(|A(j)| + 1) \right).$$

$$(10.37)$$

Note that $j > 1$ in the summation whenever $|x|T \geq 1$ as we assume here. Hence the expectation of the righthand side of (10.37) can be simplified by independence of $A(1)$ and $A(j)$, which gives

$$\mathsf{E} \left[\sum_{\lfloor x|T+1 \rfloor}^{|y|T} c(Q(i)) \right] \leq T\mathsf{E}[c(Q(|x|T))|A(1)|] + (|\alpha|T)b_0(|\alpha| + 1).$$

Now, $c(Q(|x|T))$ and $|A(1)|$ are dependent random variables. Applying the Cauchy Schwartz inequality gives

$$\mathsf{E}[c(Q(|x|T))|A(1)|]^2 \leq \mathsf{E}[c(Q(|x|T))^2]\mathsf{E}[|A(1)|^2].$$

We now consider initial conditions of the form κx^κ for nonzero $x \in \mathbb{R}^\ell_+$. Under the assumptions of the theorem we conclude from (10.37) and (10.34) that

$$\lim_{\kappa \to \infty} \frac{1}{\kappa} b(\kappa x^\kappa) = \lim_{\kappa \to \infty} \frac{1}{\kappa} \sup_{|x|=1} \mathsf{E} \left[\sum_{i=T|\kappa x|+1}^{|y|T} c(Q(i; \kappa x^\kappa)) \right]$$

$$\leq T \lim_{\kappa \to \infty} \sup_{|x|=1} \sqrt{\mathsf{E}[c(q^\kappa(T; x^\kappa))^2]\mathsf{E}[|A(1)|^2]} = 0.$$

That is, $b(x) = o(|x|)$. This combined with (10.36) implies that condition (V3) is indeed satisfied: For some $n_0 \geq 1$, $b_0 < \infty$, we have

$$\mathcal{D}V_c(x) \leq \begin{cases} -(1 + \frac{1}{2}c(x)), & |x| \geq n_0 \\ b_0 & |x| \leq n_0. \end{cases}$$

$\square$

10.4.2 Converse theorems

There are many converse theorems available. We first establish Theorem 10.0.4, which asserts that the stochastic model is unstable provided the fluid limit model is "sufficiently transient."

Proof of Theorem 10.0.4. If the fluid limit is weakly unstable, then for each $\omega \in \Omega_0$, there exists a time $T = T(\omega) > 0$ satisfying $q(T) \neq 0$. We claim that we have for some $\varepsilon = \varepsilon(\omega) > 0$

$$|q(T)| \geq \varepsilon, \qquad \text{for all } q \in \mathcal{L}_0(\omega).$$

This follows from Theorem 10.3.4, which asserts that $\mathcal{L}_0(\omega)$ is compact in the uniform topology.

With this random variable ε we have

$$\liminf_{\kappa \to \infty} |q^\kappa(T; 0)| \geq \varepsilon(\omega), \qquad \omega \in \Omega_0. \tag{10.38}$$

This bound is obtained exactly as in the proof of Proposition 10.3.7 (i): If (10.38) fails, then there exist $\omega \in \Omega_0$, $\varepsilon_1 < \varepsilon$, and a sequence $\{\kappa_i\}$ such that $|q^{\kappa_i}(T; 0)| \leq \varepsilon_1$ for each i. Taking a convergent subsequence we can assume that $\{q^{\kappa_i}\}$ is convergent, and the limit necessarily satisfies $|q^\kappa(T; 0)| \leq \varepsilon_1$, which is a contradiction.

This is enough to complete the proof since we have under (10.38)

$$\varepsilon \leq \liminf_{\kappa \to \infty} \frac{1}{\kappa} Q(\kappa T; 0) = T \liminf_{t \to \infty} \frac{1}{t} Q(t; 0)$$

where in the second equation we used the change of variables $t = \kappa T$. $\square$

A true converse to Theorem 10.0.2 would assert that the fluid limit model is stable whenever the stochastic model is stable. This implication is true provided one chooses an appropriate definition of stability for the stochastic model. In the following results we show that the existence of a suitable Lyapunov function for the stochastic model implies that the fluid limit model is stable.

The simplest result of this kind is based on Foster's criterion. A Lipschitz solution implies stability of the fluid model.

Proposition 10.4.4. *Suppose that the following conditions hold:*

(a) *Assumptions (a)–(c) of Theorem 10.0.2 hold.*
(b) *There exists a solution to Foster's criterion (V2) with the function V Lipschitz continuous on $\mathbb{R}_+^\ell$.*

Then the fluid limit model is stable.

The proof is based on the following extension of Theorem 10.2.5.

Lemma 10.4.5. *Suppose that for each $\kappa \geq 1$ the real-valued sequence $\{\mathcal{E}_\kappa(t) : t \geq 1\}$ is a martingale-difference sequence with respect to some filtration $\{\mathcal{F}_t : t \geq 0\}$. Suppose moreover that there is an i.i.d. sequence G on $\mathbb{R}_+$ with finite mean that* *is also adapted to $\{\mathcal{F}_t : t \geq 0\}$, and such that the following uniform bounds hold:*

$$|\mathcal{E}_\kappa(t)| \leq G(t) \qquad t \geq 1, \ \kappa \geq 1.$$

Then there exists $\Omega_M \subset \Omega$ satisfying $\mathsf{P}\{\Omega_M\} = 1$, and for each $\omega \in \Omega_M, T > 0$,

$$\lim_{\kappa \to \infty} \sup_{0 \leq t \leq T} |M^\kappa(t)| = 0,$$

where the scaled martingale is defined by

$$M^\kappa(T) = \frac{1}{\kappa} \sum_{t=1}^{\kappa t} \mathcal{E}_\kappa(t), \qquad t \geq 0, \ \kappa \in \mathbb{Z}_+.$$

Proof. Define for a given $m < \infty$

$$\mathcal{E}'_\kappa(t) := \mathcal{E}_\kappa(t)\mathbf{1}\{G(t) \leq m\} - \mathsf{E}[\mathcal{E}_\kappa(t)\mathbf{1}\{G(t) \leq m\} \mid \mathcal{F}_{t-1}], \qquad t \geq 1.$$

Under the conditions of the lemma this sequence is uniformly bounded by $2m$, and hence Azuma's Inequality gives

$$0 = \lim_{\kappa \to \infty} \sup_{0 \leq t \leq T} \left| \frac{1}{\kappa} \sum_{t=1}^{\kappa t} \mathcal{E}'_\kappa(t) \right|. \tag{10.39}$$

The martingale-difference property gives

$$\mathsf{E}[\mathcal{E}_\kappa(t)\mathbf{1}\{G(t) \leq m\} \mid \mathcal{F}_{t-1}] = \mathsf{E}[\mathcal{E}_\kappa(t)\mathbf{1}\{G(t) > m\} \mid \mathcal{F}_{t-1}]$$

so that

$$\mathcal{E}_\kappa(t) = \mathcal{E}'_\kappa(t) + \mathcal{E}_\kappa(t)\mathbf{1}\{G(t) > m\} + \mathsf{E}[\mathcal{E}_\kappa(t)\mathbf{1}\{G(t) > m\} \mid \mathcal{F}_{t-1}].$$

This gives the uniform bound

$$\begin{aligned} |\mathcal{E}_\kappa(t) - \mathcal{E}'_\kappa(t)| &\leq G(t)\mathbf{1}\{G(t) > m\} + \mathsf{E}[G(t)\mathbf{1}\{G(t) > m\} \mid \mathcal{F}_{t-1}] \\ &= G(t)\mathbf{1}\{G(t) > m\} + \mathsf{E}[G(t)\mathbf{1}\{G(t) > m\}]. \end{aligned}$$

The limit (10.39) combined with the LLN for G then gives

$$\limsup_{\kappa \to \infty} \sup_{0 \leq t \leq T} \left| \frac{1}{\kappa} \sum_{t=1}^{\kappa t} \mathcal{E}_\kappa(t) \right| \leq 2\mathsf{E}[G(1)\mathbf{1}\{G(1) > m\}].$$

This completes the proof since $m \geq 1$ is arbitrary and $\mathsf{E}[G(1)] < \infty$ by assumption.

$\square$

Proof of Proposition 10.4.4. Let $\mathcal{F}_t = \sigma\{Q(k) : k \leq t\}, t \geq 0$, and define for $t \geq 0$, $N \geq 1$,

$$\mathcal{E}_V(t+1) := V(Q(t+1)) - \mathsf{E}[V(Q(t+1)) \mid \mathcal{F}_t],$$

$$\text{and} \quad \mathcal{M}_V(N) = \sum_{t=1}^{N} \mathcal{E}_V(t). \tag{10.40}$$

The Markov property implies that $\mathcal{E}_V(t+1) = V(Q(t+1)) - PV(Q(t))$, so that we can write Foster's criterion as

$$V(Q(t+1)) \le V(Q(t)) - 1 + b\mathbf{1}_S(Q(t)) + \mathcal{E}_V(t+1), \qquad t \ge 0.$$

On summing over t we obtain for $T \ge 0$

$$V(Q(\kappa T^\kappa; \kappa x^\kappa)) \le V(\kappa x^\kappa) - \kappa T^\kappa + b\sum_{t=0}^{\kappa T^\kappa - 1} \mathbf{1}_S(Q(t)) + \mathcal{M}_V(\kappa T^\kappa; \kappa x^\kappa), \quad (10.41)$$

where $T^\kappa = \kappa^{-1}\lfloor \kappa T \rfloor$, and $\mathcal{M}_V(\kappa T^\kappa; \kappa x^\kappa)$ is defined in (10.40) with $N = \kappa T^\kappa$ and initial condition $Q(0) = \kappa x^\kappa$.

We have the bound

$$|\mathcal{E}_V(t+1)| \le |V(Q(t+1)) - V(Q(t))| + \mathsf{E}[|V(Q(t+1)) - V(Q(t))| \mid \mathcal{F}_t],$$

so that under the Lipschitz assumption on V we have $|\mathcal{E}_V(t+1)| \le b_0(1 + |A(t+1)|$ for some $b_0 < \infty$. Lemma 10.4.5 implies the limit

$$\limsup_{\kappa \to \infty} \kappa^{-1}|\mathcal{M}_V(\kappa T^\kappa; \kappa x^\kappa)| = 0. \qquad (10.42)$$

On dividing each side of (10.41) by κ we obtain for $T \ge 0$

$$V_\kappa(q(T^\kappa; x^\kappa)) \le V_\kappa(x^\kappa) - T^\kappa + b\kappa^{-1}\sum_{t=0}^{\kappa T^\kappa} \mathbf{1}\{q^\kappa(\kappa^{-1}t; x^\kappa) \in \kappa^{-1}S\} + \kappa^{-1}\mathcal{M}_V(\kappa T^\kappa; \kappa x),$$

where $V_\kappa(x) = \kappa^{-1}V(\kappa x)$ and $\kappa^{-1}S = \{x : \kappa x \in S\}$. For each $\varepsilon > 0$ and all sufficiently large κ this gives the bound

$$0 \le V_\kappa(q(T^\kappa; x^\kappa)) \le V_\kappa(x^\kappa) - T^\kappa + b\int_0^{T^\kappa} \mathbf{1}\{\|q^\kappa(t; x^\kappa)\| \le \varepsilon\} + \kappa^{-1}\mathcal{M}_V(\kappa T^\kappa; \kappa x).$$

Letting $\kappa \to \infty$ and applying (10.42) then gives for each $q \in \mathcal{L}_x$

$$0 \le V_\infty(x) - T + b\int_0^T \mathbf{1}\{\|q(t; x)\| \le \varepsilon\},$$

with $V_\infty(x) = \limsup_{\kappa \to \infty} V_\kappa(x)$.

Since $\varepsilon > 0$ is arbitrary we conclude that the conditions of Proposition 10.3.8 hold with $T_0 = \sup\{V_\infty(x) : \|x\| = 1\}$. $\qquad \square$

We can in fact obtain sharper bounds on the fluid limit model even without the Lipschitz assumption based on a solution h to the Poisson inequality (8.12).

Letting $h_\kappa(x) = \kappa^{-2}h(\kappa x)$, $\kappa > 0$, we denote

$$h_\infty(x) := \limsup_{\kappa \to \infty} h_\kappa(x) = \limsup_{\kappa \to \infty} \frac{1}{\kappa^2}h(\kappa x), \qquad x \in \mathbb{R}_+^\ell, \qquad (10.43)$$

which is finite valued provided $h \in L_\infty^{V_2}$. The function h_∞ scales quadratically as follows: For each $r > 0$,

$$h_\infty(rx) = \limsup_{\kappa \to \infty} \frac{1}{\kappa^2} h(\kappa r x) = r^2 \limsup_{\kappa \to \infty} \frac{1}{(\kappa r)^2} h(\kappa r x) = r^2 h_\infty(x).$$

So $h_\infty \in L_\infty^{V_2}$ if and only if it is bounded on bounded subsets of $\mathbb{R}_+^\ell$. Proposition 10.4.6 imposes additional assumptions on the model to simplify discussion.

Proposition 10.4.6. *Consider the network model (8.4) satisfying the following:*

(a) *Assumptions (a)–(d) of Theorem 8.0.4 hold.*
(b) *There exists a function $h \colon X \to \mathbb{R}_+^\ell$ satisfying Poisson's inequality (8.12), and (10.43) holds with $h_\infty \in L_\infty^{V_2}$.*
(c) *$\mathcal{L}_x$ is a singleton for each $x \in X$.*

Then for some $T_0 < \infty$ and each $x \in X$,

$$\lim_{\kappa \to \infty} q^\kappa(T; x^\kappa) \;=\; q(T; x) = 0, \qquad T \geq T_0 \quad \text{a.s.} \tag{10.44a}$$

$$\int_0^\infty c(q(t; x))\, dt \;\leq\; h_\infty(x), \qquad \{q\} = \mathcal{L}_x. \tag{10.44b}$$

Proof. We apply the first bound in Proposition 8.1.5: Replacing r by κT^κ and x by κx^κ in (8.21), and then dividing both sides of this bound by κ^2 gives

$$\mathsf{E}\left[h_\kappa(q^\kappa(T; x^\kappa)) + \int_0^T c(q^\kappa(t; x^\kappa))\, dt \right] \leq h_\kappa(x^\kappa) + \frac{1}{\kappa} \overline{\eta} T,$$

and hence

$$\limsup_{\kappa \to \infty} \mathsf{E}\left[h_\kappa(q^\kappa(T; x^\kappa)) + \int_0^T c(q^\kappa(t; x^\kappa))\, dt \right] \leq h_\infty(x). \tag{10.45}$$

Since $q^\kappa(\,\cdot\,; x^\kappa) \to q(\,\cdot\,; x)$ u.o.c. as $\kappa \to \infty$, and since $h \geq 0$, we obtain

$$\int_0^T c(q(t; x))\, dt = \lim_{\kappa \to \infty} \mathsf{E}\left[\int_0^T c(q^\kappa(t; x^\kappa))\, dt \right]$$

$$\leq \limsup_{\kappa \to \infty} \mathsf{E}\left[h_\kappa(q^\kappa(T; x^\kappa)) + \int_0^T c(q^\kappa(t; x^\kappa))\, dt \right] \leq h_\infty(x).$$

This establishes the bound in (10.44b).

We now apply the semigroup property expressed in Proposition 10.3.5: For each $x \in X$, $q \in \mathcal{L}_x$, and $t_1 > 0$, let $x^1 = q(t_1; x)$ and consider

$$q^{[t_1]}(t) = q(t + t_1), \quad z^{[t_1]}(t) = z(t + t_1), \qquad t \geq 0.$$

Then $q^{[t_1]}$ is the unique element in $\mathcal{L}_{x^1}$. That is, $q(t + t_1; x) = q^{[t_1]}(t) = q(t; x^1)$. Consequently,

$$J(q(t_1; x)) := \int_0^\infty c(q(t; x^1))\, dt = \int_0^\infty c(q(t + t_1; x))\, dt = \int_{t_1}^\infty c(q(t; x))\, dt,$$

so that

$$\tfrac{d^+}{dt} J(q(t;x)) = -c(q(t;x)), \qquad t \geq 0, \ x \in \mathsf{X}. \tag{10.46}$$

Applying part (i) we have $0 \leq J \leq h_\infty$.

The function h_∞ has quadratic growth, which implies that for some $\varepsilon > 0$,

$$J(x) \leq h_\infty(x) \leq [\varepsilon^{-1}c(x)]^2, \qquad x \in \mathsf{X}.$$

Returning to (10.46) we obtain with $G = \sqrt{J}$, whenever $q(t;x) \neq \mathbf{0}$,

$$\begin{aligned}
\tfrac{d}{dt} G(q(t;x)) &= \tfrac{1}{2} \frac{1}{\sqrt{J(q(t;x))}} \tfrac{d}{dt} J(q(t;x)) \\
&= -\tfrac{1}{2} \frac{1}{\sqrt{J(q(t;x))}} c(q(t;x)) \leq -\tfrac{1}{2}\varepsilon.
\end{aligned}$$

Integrating, we obtain

$$0 \leq G(q(T;x)) \leq G(x) - \tfrac{1}{2}\varepsilon T, \quad \text{if } q(t;x) \neq \mathbf{0} \text{ for } 0 \leq t \leq T,$$

which shows that $q(T;x) = \mathbf{0}$ for $T \geq 2\varepsilon^{-1}G(x)$. This proves the result with

$$T_0 = 2\varepsilon^{-1} \max\{G(x) : |x| = 1\} \leq 2\varepsilon^{-1} \max\{\sqrt{h_\infty(x)} : |x| = 1\}. \qquad \square$$

10.5 Safety stocks and trajectory tracking

Safety stocks were introduced in Section 4.6 as a technique to avoid idleness in the CRW scheduling model. In this section we illustrate the application of safety stocks to facilitate tracking a given fluid trajectory. Dynamic safety stocks of the form introduced in Section 4.6.2 are most easily analyzed using the methods of this chapter. We focus on this technique here, and for simplicity we restrict to the scheduling model.

We first present a simple result based on a highly idealized setting: We are given a policy for the fluid model in feedback form $\zeta(t) = \phi^0(q(t))$, where $\phi^0(x) \in \mathsf{U}(x)$ for each $x \in \mathbb{R}_+^\ell$. The policy is assumed to be radially constant and Lipschitz continuous: For some constant $b_L < \infty$,

$$\phi^0(rx) = \phi^0(x), \qquad r > 0, \tag{10.47a}$$

$$\|\phi^0(x) - \phi^0(y)\| \leq b_L \|x - y\|, \qquad \|x\|^2 + \|y\|^2 \geq 1. \tag{10.47b}$$

A randomized policy for Q is defined by

$$\phi_i(x) = \phi_i^0(x)\mathbf{1}\{x_i \geq 1\}. \tag{10.48}$$

Let q^0 denote the solution to the ODE $\tfrac{d}{dt} q^0(t;x) = \phi^0(q^0(t;x))$ with initial condition x.

It is assumed that each buffer increases on average whenever it is small compared to the total customer population. This is captured by an unbounded switching curve $s \colon \mathbb{R}_+ \to \mathbb{R}_+$ that defines a dynamic safety stock. For example, we might use the logarithmic switching curve introduced in Section 4.6.2,

$$s_\theta(r) := \theta \log(1 + r\theta^{-1}), \qquad r \geq 0. \tag{10.49}$$

Proposition 10.5.1. *Suppose that (10.47a), (10.47b) are satisfied for the scheduling model introduced in Section 4.1.1, and that the following bound holds for some $\varepsilon > 0$, each i, and each $t \geq 0$,*

$$\frac{d^+}{dt} q^0(t) \geq \varepsilon \quad \text{when } q_i^0(t) < s(|q(t)|), \tag{10.50}$$

where $s \colon \mathbb{R}_+ \to \mathbb{R}_+$ is increasing and unbounded. Then for each initial condition x, each sequence $\{x(\kappa)\}$ convergent to x, and each $T > 0$,

$$\lim_{\kappa \to \infty} \sup_{0 \leq t \leq T} \|q^\kappa(t; x(\kappa)) - q^0(t; x(\kappa))\| = 0 \qquad a.s.$$

The key step in the proof is to control the amount of time that $\phi_i(Q(t)) \neq \phi_i^0(Q(t))$. Given nonnegative coefficients $\{d_i : i = 1, \ldots, \ell\}$ we let Y denote the weighted sum,

$$Y(t) = \sum d_i Q_i(t), \qquad t \geq 0.$$

For example, we might set $d_i = \mathbf{1}\{i \in \mathcal{I}_s\}$ to represent the buffers at Station s in a scheduling model.

Lemma 10.5.2. *Suppose that for a given policy there exist an unbounded, increasing function $s \colon \mathbb{R}_+ \to \mathbb{R}_+$, an $\varepsilon > 0$, and $t_1 \geq 1$ satisfying for each initial condition, and each $t \geq 0$,*

$$\mathsf{E}[Y(t + t_1) - Y(t) \mid \mathcal{F}_t] \geq \varepsilon \quad \text{when } Y(t) \leq s(|Q(t)|). \tag{10.51}$$

Then, for each nonzero $x \in \mathbb{R}_+^\ell$, and sequence $x(\kappa) \to x$, we have

$$\lim_{\kappa \to \infty} \frac{1}{\kappa} \sum_{t=0}^{\kappa T - 1} \mathbf{1}\{Y(t; \kappa x(\kappa)) = 0\} = 0 \qquad a.s., \text{ for all } T < T^*(x).$$

Proof. We construct a bounded function $V_0 \colon \mathbb{R}_+^\ell \to \mathbb{R}_+$ satisfying, for some ε_0, $\theta > 0$,

$$\mathcal{D}V_0(x) \leq -\varepsilon_0 \mathbf{1}\{y = 0\} + \varepsilon_0^{-1} e^{-\theta s(|x|)}, \qquad x \in \mathsf{X}_\diamond, \ y = \sum d_i x_i. \tag{10.52}$$

This has the equivalent form

$$V_0(Q(t+1)) \leq V_0(Q(t)) - \varepsilon_0 \mathbf{1}\{Y(t) = 0\} + \varepsilon_0^{-1} e^{-\theta s(|Q(t)|)} + \mathcal{E}_{V_0}(t+1),$$

with $\mathcal{E}_{V_0}(t+1) := V_0(Q(t+1)) - \mathsf{E}[V_0(Q(t+1)) \mid \mathcal{F}_t]$. Summing each side over t gives

$$V_0(Q(\kappa T; \kappa x(\kappa))) \leq V_0(\kappa x(\kappa))$$

$$\leq -\varepsilon_0 \sum_{t=0}^{\kappa T - 1} \mathbf{1}\{Y(t) = 0\} + \varepsilon_0^{-1} \sum_{t=0}^{\kappa T - 1} e^{-\theta s(|Q(t; \kappa x(\kappa))|)} + M^\kappa(T)$$

where $M^\kappa(T) = \kappa^{-1} \sum_1^{\kappa T} \mathcal{E}_{V_0}(t)$. Applying Theorem 10.2.5 we can conclude that

$$\limsup_{\kappa \to \infty} \frac{1}{\kappa} \sum_{t=0}^{\kappa T - 1} \mathbf{1}\{Y(t) = 0\} \leq \varepsilon_0^{-2} \limsup_{\kappa \to \infty} \frac{1}{\kappa} \sum_{t=0}^{\kappa T - 1} e^{-\theta s(|Q(t; \kappa x(\kappa))|)} = 0.$$

We now establish (10.52). First suppose $t_1 = 1$ and define $V_0(x) = e^{-\theta y}$, $x \in \mathbb{R}_+^\ell$, with $\theta > 0$ a fixed constant. We have by the mean value theorem, for any $\Delta \in \mathbb{R}$,

$$e^{-\theta\Delta} \leq 1 - \theta\Delta + \tfrac{1}{2}\theta\Delta^2 e^{\theta\Delta_-}$$

where $\Delta_- = \max(0, -\Delta)$. Writing $Y(t+1) = Y(t) + \Delta_Y$ we obtain

$$V_0(Q(t+1)) \leq V_0(Q(t))\left(1 - \theta\Delta_Y + \tfrac{1}{2}\theta^2\Delta_Y^2 e^{\theta(\Delta_Y)_-}\right).$$

We thereby obtain for some $b_0 < \infty$, and all $\theta \in (0,1]$,

$$\mathsf{E}[V_0(Q(t+1)) - V_0(Q(t)) \mid \mathcal{F}_t] \leq \left(-\theta\mathsf{E}[Y(t+1) - Y(t) \mid \mathcal{F}_t] + \tfrac{1}{2}\theta^2 b_0\right)V_0(Q(t)).$$

If $Y(t) \leq s(|Q(t)|)$ then the conditional expectation on the righthand side is no less than ε. If this condition is violated, then it is bounded below by $-\overline{\Delta}_-$, for some constant $\overline{\Delta}_-$. Consequently,

$$\begin{aligned}
\mathcal{D}V_0\left(Q(t)\right) &:= \mathsf{E}[V_0(Q(t+1)) - V_0(Q(t)) \mid \mathcal{F}_t] \\
&\leq \left(-\theta\varepsilon + \tfrac{1}{2}\theta^2 b_0\right)V_0(Q(t))\mathbf{1}\{Y(t) \leq s(|Q(t)|)\} \\
&\quad + \left(\theta\overline{\Delta}_- + \tfrac{1}{2}\theta^2 b_0\right)V_0(Q(t))\mathbf{1}\{Y(t) > s(|Q(t)|)\}.
\end{aligned}$$

Choosing $\theta = \varepsilon/b_0$ we obtain $-\theta\varepsilon + \tfrac{1}{2}\theta^2 b_0 = -\tfrac{1}{2}\theta\varepsilon = -\tfrac{1}{2}\varepsilon^2/b_0$. We can then apply the bounds

$$\begin{aligned}
V_0(Q(t))\mathbf{1}\{Y(t) \leq s(|Q(t)|)\} &\geq \mathbf{1}\{Y(t) = 0\}, \\
V_0(Q(t))\mathbf{1}\{Y(t) > s(|Q(t)|)\} &\leq e^{-\theta s(|Q(t)|)},
\end{aligned}$$

to obtain (10.52) for some $\varepsilon_0 > 0$ suitably small.

The proof for $t_1 > 1$ is similar since we still have, for some $\theta, \varepsilon_0 > 0$,

$$\mathsf{E}[\exp(-\theta Y(t+t_1)) - \exp(-\theta Y(t)) \mid \mathcal{F}_t] \leq -\varepsilon_0\mathbf{1}\{Y(t) = 0\} + \varepsilon_0^{-1}e^{-\theta s(|Q(t)|)}.$$

We define $V_{00}(x) = e^{-\theta y}$ and

$$V_0(x) = \sum_{t=0}^{t_1-1} P^t V_{00}\left(x\right) = \sum_{t=0}^{t_1-1} \mathsf{E}_x[\exp(-\theta Y(t))], \qquad x \in \mathsf{X}_\diamond.$$

By assumption we have

$$P^{t_1}V_{00}\left(x\right) \leq V_{00}(x) - \varepsilon_0\mathbf{1}\{y = 0\} + \varepsilon_0^{-1}e^{-\theta s(|x|)},$$

so that V_0 satisfies the desired bound,

$$\begin{aligned}
PV_0\left(x\right) &= \sum_{t=0}^{t_1-1} P^{t+1}V_{00}\left(x\right) \\
&= V_0(x) - V_{00}(x) + P^{t_1}V_{00}\left(x\right) \\
&\leq V_0(x) - \varepsilon_0\mathbf{1}\{y = 0\} + \varepsilon_0^{-1}e^{-\theta s(|x|)}. \qquad \square
\end{aligned}$$

Proof of Proposition 10.5.1. Under the conditions of the proposition, the bound (10.51) holds with $Y(t) = Q_i(t)$ and *any* $i = 1, \dots, \ell$: For some $\varepsilon > 0$, perhaps smaller than the constant used in (10.50), an integer $t_1 \geq 1$, and a constant $\varrho \in (0, 1)$,

$$\mathsf{E}[Q_i(t + t_1) - Q_i(t) \mid \mathcal{F}_t] \geq \varepsilon \quad \text{when } Q_i(t) \leq \varrho s(|Q(t)|). \tag{10.53}$$

The introduction of ϱ is required since we do not know how $q_i^0(t)$ behaves for $q_i^0(t) \geq s(|q^0(t)|)$.

To compare the scaled processes we first note that the cumulative allocation vector z^0 can be expressed as the integral

$$z^0(t) = \int_0^t \phi^0(q^0(s)) \, ds, \qquad t \geq 0.$$

The scaled cumulative allocation process z^κ for the CRW model can be expressed in a similar form. We first write

$$z^\kappa(t; x(\kappa)) = \frac{1}{\kappa} \sum_{0 \leq j \leq \kappa t - 1} \phi(Q(j; \kappa x(\kappa))) + N_\phi^\kappa(t),$$

$$= \frac{1}{\kappa} \sum_{0 \leq j \leq \kappa t - 1} \phi^0(Q(j; \kappa x(\kappa))) + N_\phi^\kappa(t) - D_z^\kappa(t),$$

where

$$D_z^\kappa(t) := \frac{1}{\kappa} \sum_{0 \leq j \leq \kappa t - 1} \sum_i \phi_i^0(Q(j; \kappa x(\kappa))) \mathbf{1}\{Q_i(j; \kappa x(\kappa)) = 0\} \mathbf{1}^i$$

and $\boldsymbol{N}_\phi^\kappa$ is a scaled martingale,

$$N_\phi^\kappa(t) := \frac{1}{\kappa} \sum_{0 \leq j \leq \kappa t - 1} \big(U(j) - \phi(Q(j; \kappa x(\kappa)))\big).$$

We thus arrive at a representation similar to (10.25),

$$q^\kappa(T; x(\kappa)) = x(\kappa) + \int_0^T [B\phi^0(q^\kappa(t)) + \alpha] \, dt + N^\kappa(T)$$

$$q^0(T; x(\kappa)) = x(\kappa) + \int_0^T [B\phi^0(q^0(t)) + \alpha] \, dt, \qquad T \geq 0,$$

where $\{N^\kappa(t) := B[N_\phi^\kappa(t) - D_z^\kappa(t)] : t \geq 1\}$ converges u.o.c. to zero by Proposition 10.3.2 combined with Lemma 10.5.2.

To apply the Bellman–Gronwall Lemma we define the process x by $x(T) = \|q^\kappa(T; x(\kappa)) - q^0(T; x(\kappa))\|$, $T \geq 0$. Using the Lipschitz bound,

$$x(T) \leq b_L \|B\| \int_0^T x(t) \, dt + \|N^\kappa(T)\|, \qquad T \geq 0,$$

the Bellman–Gronwall Lemma implies the final bound,

$$\lim_{\kappa\to\infty} \sup_{0\le t\le T} \|q^\kappa(t; x(\kappa)) - q^0(t; x(\kappa))\|$$

$$= \lim_{\kappa\to\infty} \sup_{0\le t\le T} x(t)$$

$$\le e^{b_L\|B\|T} \sup_{0\le t\le T} \|N^\kappa(t)\| = 0. \qquad \square$$

The continuity assumption on ϕ^0 is *extremely restrictive*. For example, in the GTO policy and infinite-horizon optimal policies the feedback law ϕ^0 is discontinuous since it is piecewise constant. Without the continuity assumption the Bellman–Gronwall Lemma is no longer available, but other concepts used in the proof of Proposition 10.5.1 will prove valuable.

Consider the following tracking problem: We are given two vectors $x^0, x^1 \in \mathsf{X}$ and we seek a policy for the CRW model for which the fluid limit starting from x^0 reaches x^1 in minimal time. If some components of x^0 or x^1 are zero, then we must find a way to keep $q^\kappa(t)$ away from the boundary of the state space, but not too far. This is achieved by satisfying (10.53) using an unbounded function $s\colon \mathbb{R}_+ \to \mathbb{R}_+$ with sublinear growth.

We define two elements of V,

$$v^* := \frac{1}{T^*(x^0, x^1)}(x^1 - x^0), \quad v^+ := \frac{1}{T^*(0, 1)}\mathbf{1}.$$

That is, if $\frac{d}{dt}q(t; x^0) \equiv v^*$, then x^1 is reached in minimal time from the given initial condition, and if $\frac{d}{dt}q(t; 0) \equiv v^+$, then the vector of ones is reached in minimal time starting from the origin. The velocity vector for the CRW model is chosen as a convex combination of the two,

$$v(x) := \beta(x)v^+ + (1 - \beta(x))v^*, \qquad x \in \mathsf{X}, \tag{10.54}$$

where $\beta(x)$ is constructed so that it is near unity whenever x is near the boundary,

$$\beta(x) = \max_{1\le i\le \ell} \frac{(s(|x|) - x_i)_+}{s(|x|)}, \qquad x \in \mathsf{X}, \, x \ne 0. \tag{10.55}$$

By construction we can find u^*, u^+ contained in $\mathsf{U}(x^0)$ and U, respectively, such that the feedback law $\phi^0(x) = \beta(x)u^+ + (1 - \beta(x))u^*$ satisfies $B\phi^0(x) + \alpha = v(x)$ for each x. We then choose the randomized policy ϕ based on this ϕ^0 and (10.48).

Proposition 10.5.3 establishes convergence:

Proposition 10.5.3. *Suppose that the randomized stationary policy ϕ is constructed so that (10.54) holds with β given in (10.55). The function $s\colon \mathbb{R}_+ \to \mathbb{R}_+$ is assumed to be unbounded and concave, with $s'(r) \to 0$ as $r \to \infty$. Then, for any sequence $x^0(\kappa) \to x^0$, we have $q^\kappa(t; x^0(\kappa)) \to x^0 + v^*t$ for each $0 \le t \le T^*(x^0, x^1)$. The convergence is both a.s., and in L_2.*

The proof is based on an extension of Lemma 10.5.2.

Lemma 10.5.4. *Suppose that the assumptions of Lemma 10.5.2 hold, where the function s satisfies the assumptions of Proposition 10.5.3. Then, for each nonzero $x \in \mathbb{R}_+^\ell$, sequence $x(\kappa) \to x$, and $\varrho \in (0, 1)$, we have*

$$\lim_{\kappa \to \infty} \frac{1}{\kappa} \sum_{t=0}^{\kappa T - 1} \mathbf{1}\{Y(t; \kappa x(\kappa)) \leq \varrho s(|Q(t)|)\} = 0 \qquad \text{a.s., for all } T < T^*(x).$$

Proof. The proof is very similar to the proof of Lemma 10.5.2. We present the details only for $t_1 = 1$ in (10.51) since the extension to larger t_1 follows exactly as before.

The idea is to sharpen the bound in (10.52): We construct a function $V_0 \colon \mathbb{R}_+^\ell \to \mathbb{R}_+$ satisfying, for some $\varepsilon_0, \theta > 0$,

$$\mathcal{D}V_0(x) \leq -\varepsilon_0 \mathbf{1}\{y \leq \varrho s(|x|)\} + \varepsilon_0^{-1} e^{-\theta(y - \varrho s(|x|))} \mathbf{1}\{y \geq s(|x|)\},$$
$$x \in \mathsf{X}_\circ, \ y = \sum d_i x_i. \tag{10.56}$$

The result will then follow from the Comparison Theorem A.4.3 as in the proof of Lemma 10.5.2.

We take $V_0(x) = e^{-\theta(y - \varrho s(|x|))}$, $x \in \mathbb{R}_+^\ell$, with $\theta > 0$ a fixed constant. We have the following bound by concavity of s: For $Q(t) = x$ and $\Delta := Q(t + 1) - x$,

$$s(|Q(t + 1)|) \leq s(|x|) + s'(|x|)|\Delta|.$$

Writing $\varepsilon(x) = \varrho s'(|x|)$, $\Delta_Y = Y(t + 1) - Y(t)$, and $y = Y(t)$ we obtain

$$V_0(Q(t + 1)) = e^{-\theta(Y(t+1) - \varrho s(|Q(t+1)|))}$$
$$\leq e^{-\theta(y + \Delta_Y)} e^{\theta(\varrho s(|x|) + \varepsilon(x)|\Delta|)}$$
$$= V_0(x) e^{-\theta \Delta_Y + \theta \varepsilon(x)|\Delta|}.$$

Taking expectations and applying the mean value theorem as before gives, for some constant $b_0 > 0$ and all $\theta \in (0, 1)$, the following bound on the mean-drift $\mathcal{D}V_0$,

$$\mathsf{E}[V_0(Q(t + 1)) - V_0(Q(t)) \mid Q(t) = x] \leq V_0(x)\left(-1 + \mathsf{E}[e^{-\theta \Delta_Y + \theta \varepsilon(x)|\Delta|}]\right)$$
$$\leq V_0(x)\left(-1 + e^{-\theta \delta_Y(x) + b_0(\theta \varepsilon(x) + \theta^2)}]\right),$$

where $\delta_Y(x) = \mathsf{E}[Y(t + 1) - Y(t) \mid Q(t) = x]$. Under the assumptions on δ_Y we obtain (10.56) for sufficiently small $\theta > 0$.   □

Proof of Proposition 10.5.3. Following the proof of Proposition 10.5.1 we write

$$z^\kappa(t; x(\kappa)) = \frac{1}{\kappa} \sum_{0 \leq j \leq \kappa t - 1} \phi(Q(j; \kappa x(\kappa))) + N_\phi^\kappa(t),$$

where N_ϕ^κ is defined as before. Lemma 10.5.4 implies the limit

$$\lim_{\kappa \to \infty} \frac{1}{\kappa} \sum_{0 \leq j \leq \kappa t - 1} \beta(Q(j; \kappa x(\kappa))) = 0.$$

Since $\phi(x) = \beta(x)u^+ + (1 - \beta(x))u^*$ for each x, convergence of $\mathbf{z}^\kappa$ follows

$$\lim_{\kappa \to \infty} z^\kappa(t; x(\kappa)) = u^* t.$$

The L_2 convergence follows from Proposition 10.3.2. $\qquad\qquad\square$

10.6 Fluid-scale asymptotic optimality

We have seen that the fluid limit model can be used to capture stability of a policy. Here we describe how fluid limits can be used to investigate finer properties. Our main goal is to assemble the ingredients of the proof of Theorem 10.0.5.

We begin with the simpler discounted case.

10.6.1 Discounted cost

The value function h_γ^* defined in (9.34) can be expressed

$$h_\gamma^*(x) = \gamma^{-1} \inf \mathsf{E}[c(Q(T; x))],$$

where T is a geometrically distributed random variable with mean γ^{-1}, independent of Q, and the infimum is over all policies.

To formulate an analog of Theorem 10.0.5 for the discounted-cost optimality criterion we first recall some results obtained for the single-server queue in Section 3.4.5. In this simple model it was shown that h_γ^* can be expressed as a perturbation of the discounted fluid value function. The following result provides an (admittedly weak) generalization.

Proposition 10.6.1. *For each $\gamma > 0$ and $x \in \mathbb{R}_+^\ell$,*

$$\lim_{\kappa \to \infty} \left[h_\gamma^*(\kappa x^\kappa) - J_\gamma^*(\kappa x^\kappa) \right] \geq 0. \tag{10.57}$$

If $x_i > 0$ for each i, then this lower bound is achieved,

$$\lim_{\kappa \to \infty} \left[h_\gamma^*(\kappa x^\kappa) - J_\gamma^*(\kappa x^\kappa) \right] = 0.$$

To prove this result we consider first the impact of scaling on the value function for the fluid model. From the expression for J_γ^* derived in Section 3.4.5 we obtain for the single-server queue

$$\lim_{\kappa \to \infty} \kappa \left[\frac{1}{\kappa} J_\gamma^*(\kappa x^\kappa) - \gamma^{-1} c(x) \right] = -\gamma^{-2}(\mu - \alpha), \qquad x \in \mathbb{R}_+.$$

Proposition 10.6.2 (iii) extends this result to the general ℓ-dimensional fluid model.

Proposition 10.6.2. *The discounted-cost value function for the fluid model satisfies, for each $\gamma > 0$ and $x \in \mathbb{R}_+^\ell$,*

(i) $J_\gamma^*(\kappa x) = \kappa^2 J_{\kappa\gamma}^*(x)$ *for $\kappa > 0$,*

(ii) $\displaystyle \lim_{\kappa \to \infty} \frac{1}{\kappa} J_\gamma^*(\kappa x) = \gamma^{-1} c(x),$

(iii) $\displaystyle\lim_{\kappa\to\infty}\kappa\left[\frac{1}{\kappa}J_\gamma^*(\kappa x)-\gamma^{-1}c(x)\right]=\gamma^{-2}\min_{\zeta\in U(x)}c^{\mathsf{T}}[B\zeta+\alpha].$

Proof. When $\mathsf{X}=\mathbb{R}_+^\ell$ as we assume here, then $\{q(t):t\ge 0\}$ is a solution starting from x if and only if $\{q^{(\kappa)}(t):t\ge 0\}$ is a solution starting from $\kappa^{-1}x$, where $(q^{(\kappa)},z^{(\kappa)})$ are defined in (10.28) for $(q,z)\in\mathcal{L}$.

If q^* achieves the optimal discounted cost from the initial condition $x\in\mathbb{R}_+^\ell$, then

$$J_\gamma^*(x)=\int_0^\infty e^{-\gamma t}c(q^*(t))\,dt=\kappa^2\int_0^\infty e^{-\gamma\kappa s}c(q^{*(\kappa)}(s))\,ds\ge\kappa^2 J_{\kappa\gamma}^*(\kappa^{-1}x)$$

where the second equality follows from the change of variables $s=\kappa t$, and the final inequality is the definition of $J_{\kappa\gamma}^*$. We can use identical reasoning to obtain a bound in the reverse direction. If q is given, with $q^{(\kappa)}$ achieving $J_{\kappa\gamma}^*(\kappa^{-1}x)$, then

$$J_\gamma^*(x)\le\int_0^\infty e^{-\gamma t}c(q(t))\,dt=\kappa^2\int_0^\infty e^{-\gamma\kappa s}c(q^{(\kappa)}(s))\,ds=\kappa^2 J_{\kappa\gamma}^*(\kappa^{-1}x).$$

Combining these two bounds proves (i).

Part (ii) follows from (i) since we can write

$$\kappa\gamma J_{\kappa\gamma}^*(x)=\int_0^\infty \kappa\gamma e^{-\kappa\gamma t}c(q^*(t))\,dt.$$

Even though in general q^* depends upon κ, the right-hand side converges to $c(x)$ as $\kappa\to\infty$.

To see (iii), first note that the right-hand side of the claimed limit can be expressed $c^{\mathsf{T}}[B\zeta^c(x)+\alpha]$, where $\zeta^c(x)$ denotes the value of $\frac{d^+}{dt}z(0)$ when $q(0)=x$ under any c-myopic policy. Let (q^c,z^c) denote a solution to the fluid model equations under any myopic policy, and let J_γ^c denote the value function. We assume that q is piecewise linear, and that ζ^c is radially constant,

$$\zeta^c(rx)=\zeta^c(x),\qquad r>0,\ x\in\mathbb{R}_+^\ell,\ x\ne 0.$$

We have by definition

$$J_\gamma^c(x)\ge J_\gamma^*(x),\qquad x\in\mathbb{R}_+^\ell,$$

but we also have $c(q^*(t;x))\ge c(q^c(t;x))$ for $t\le T_1$, where

$$T_1(x)=\min(\tfrac{d^+}{dt}z^c(t;x)\ne\zeta^c(x)).$$

We have $T_1(\kappa x)=\kappa T_1(x)$ for $x\ne 0$, from which we conclude that

$$\lim_{\kappa\to\infty}\left[J_\gamma^*(\kappa x)-\gamma^{-1}\kappa c(x)\right]=\lim_{\kappa\to\infty}\left[J_\gamma^c(\kappa x)-\gamma^{-1}\kappa c(x)\right]$$

$$=\lim_{\kappa\to\infty}\int_0^{\kappa T_1(x)}e^{-\gamma t}(c^{\mathsf{T}}[B\zeta^c(x)+\alpha]t\,dt.$$

The right-hand side coincides with the value given on the right in (iii), and the left-hand side coincides with the left-hand side of (iii). $\qquad\square$

Proof of Proposition 10.6.1. We again let $\zeta^c(x)$ denote the value of $\frac{d^+}{dt}z(0)$ when $q(0) = x$ under any c-myopic policy. The value $c(x) + c^\intercal[B\zeta^c(x) + \alpha]t$ is a lower bound on $c(q(t; x))$ under any policy, for any x, t, and hence we also have for the CRW model,

$$\mathsf{E}[c(Q(t; x))] \geq c(x) + c^\intercal[B\zeta^c(x) + \alpha]t.$$

See Proposition 4.4.1 for a similar bound. The inequality (10.57) then follows from Proposition 10.6.2 (iii).

To obtain the equality when $x > 0$ we consider the randomized policy $\phi_u(x)\zeta_u^c(x)\mathbf{1}\{x_u \geq 1\}, u \in \{1, \ldots, \ell\}$, for which the value function h_γ satisfies, whenever $x_i > 0$ for each i,

$$\lim_{\kappa \to \infty} \kappa\left[\frac{1}{\kappa}h_\gamma(\kappa x) - \gamma^{-1}c(x)\right] = \gamma^{-2}c^\intercal[B\zeta^c(x) + \alpha]. \qquad \square$$

In Example 4.5.2 we saw that optimal policies are sometimes approximated by a static switching curve in the tandem queues. We now explain this approximation based on a fluid limit model.

Example 10.6.3 (Tandem queues: asymptotic optimality using safety stocks). We consider here a hedging-point policy,

$$\phi_1(x) = \mathbf{1}\{x_1 \geq 1, x_2 < \overline{x}_2\}, \qquad x \in \mathbb{Z}_+^2, \tag{10.58}$$

where the threshold $\overline{x}_2 \geq 1$ is the safety stock level for the second resource. In Example 4.9.2 we saw that the myopic policy based on (4.98) is approximated by a static hedging-point policy of this form, and numerical results presented in Example 4.5.2 show that an optimal policy is also approximated by the policy (10.58) when $c_2 > c_1$. Here we apply a fluid-limit analysis to obtain an approximation of the discounted-cost optimal policy of the form (4.98) for large x_1.

Under the policy (10.58) the resulting process of buffer levels at Station 2 has a simple description that facilitates analysis. Up until the first time that buffer 1 empties, buffer 2 evolves as the M/M/1 queue with finite waiting room $\overline{x}_2$ (see Exercise 7 in Chapter 3). Its steady-state distribution is geometric on $\{0, \ldots, \overline{x}_2\}$ with parameter $\varrho := \mu_1/\mu_2$, and hence for any nonzero initial condition x satisfying $x_2 = 0$,

$$\bar{Q}_2 := \lim_{t \to \infty}\left(\lim_{\kappa \to \infty} \mathsf{E}[Q_2(t; \kappa x(\kappa))]\right) = \frac{\sum_{n=0}^{\overline{x}_2} n\varrho^n}{\sum_{n=0}^{\overline{x}_2} \varrho^n}.$$

We can also estimate the probability that $Q_2(t)$ is nonzero: We have $\{U_2(t) = 1\} = \{Q_2(t) \geq 1\}$ and hence

$$\overline{\zeta}_2 := \lim_{t \to \infty}\left(\lim_{\kappa \to \infty} \mathsf{P}\{U_2(t; \kappa x(\kappa)) = 1\}\right) = \frac{\varrho^{\overline{x}_2+1} - \varrho}{\varrho^{\overline{x}_2+1} - 1}.$$

Moreover, since $Q_2(t+1) - Q_2(t) = S_1(t+1)U_1(t) - S_1(t+1)U_2(t)$ for each t, we can compute the limiting value of $\overline{\zeta}_1 := \mathsf{P}\{U_1(t) = 1\}$,

$$0 = \lim_{t \to \infty}\left(\lim_{\kappa \to \infty} \mathsf{E}[Q_2(t+1; \kappa x^\kappa) - Q_2(t; \kappa x^\kappa)]\right) = \mu_1\overline{\zeta}_1 - \mu_2\overline{\zeta}_2.$$

In particular, from this initial condition the scaled process converges to the fluid limit,

$$\lim_{\kappa\to\infty}\frac{1}{\kappa}Q_1(\kappa t;\kappa x(\kappa))=x_1-(\mu_2\bar\zeta_2-\alpha_1)t,\quad \lim_{\kappa\to\infty}\frac{1}{\kappa}Q_2(\kappa t;\kappa x)=0,\quad (10.59)$$

$t\le x_1/(\mu_2-\alpha_1)$. These approximations can be applied to approximate the discounted-cost value function h_γ^*.

Let T denote a geometrically distributed random variable, independent of Q, with mean γ^{-1}. From the foregoing we obtain the following approximations for the CRW and the fluid model: For all $x\in\mathsf{X}_\diamond$ satisfying $\frac{x_1}{\mu_2-\alpha_1}\gg \mathsf{E}[T]=\gamma^{-1}$ and $x_2=0$,

$$
\begin{aligned}
\mathsf{E}[Q_1(T;x)] &\approx x_1-(\mu_2\bar\zeta_2-\alpha_1)\gamma^{-1}, & \mathsf{E}[Q_2(T;x)] &\approx \bar Q_2 \\
\mathsf{E}[q_1^*(T;x)] &\approx x_1-(\mu_2-\alpha_1)\gamma^{-1}, & \mathsf{E}[q_2^*(T;x)] &= 0.
\end{aligned}
\qquad (10.60)
$$

We only obtain an approximation for $q_1^*(T;x)$ since there is some small probability that $T>x_1/(\mu_2-\alpha_1)$. Based on (10.60) we now construct a static policy that approximately minimizes the discounted cost for large initial conditions.

The value functions for the CRW and fluid models can be expressed

$$h_\gamma(x)=\gamma^{-1}\mathsf{E}[c(Q(T;x))],\qquad J_\gamma^*(x)=\gamma^{-1}\mathsf{E}[c(q^*(T;x))],$$

where T is geometric in the first instance, and exponential in the second. This gives for initial conditions satisfying (10.60)

$$
\begin{aligned}
\gamma h_\gamma(x) &\approx c_1[x_1+(\alpha_1-\bar\zeta_1\mu_1)\gamma^{-1}]+c_2\bar Q_2 \\
\gamma J_\gamma^*(x) &\approx c_1[x_1+(\alpha_1-\mu_1)\gamma^{-1}].
\end{aligned}
\qquad (10.61)
$$

To obtain the optimal threshold it remains to minimize over all $\bar x_2$

$$-c_1\bar\zeta_1\mu_1\gamma^{-1}+c_2\bar Q_2. \qquad (10.62)$$

It is simplest to work with a continuous approximation. Writing $\beta=\log(\varrho)$ gives

$$\bar Q_2\approx\frac{\int_0^{\bar x_2}te^{\beta t}\,dt}{\int_0^{\bar x_2}e^{\beta t}\,dt}=-\beta^{-1}+\bar x_2\frac{e^{\beta\bar x_2}}{e^{\beta\bar x_2}-1},\quad\text{and}\quad \bar\zeta_2\approx\frac{e^{\beta\bar x_2}-e^\beta}{e^{\beta\bar x_2}-1}.$$

Substituting these approximations into (10.62), the minimizer $\bar x_2^*$ is obtained by differentiating the resulting expression with respect to $\bar x_2$ and setting the derivative to zero. This results in the fixed point equation

$$e^{\beta\bar x_2^*}-\beta\bar x_2^*=1+\frac{c_1}{c_2}\gamma^{-1}\mu_1\beta(e^\beta-1).$$

Consider two cases separately, with cost parameters $c_1=1$, $c_2=3$ fixed:

CASE 1: $\mu_1<\mu_2$ When $\mu_1<\mu_2$ then $\beta<0$. If $\gamma>0$ is small then $\bar x_2^*$ will be large, and we obtain the approximation

$$\bar x_2^*=|\beta|^{-1}\left(1+\frac{c_1}{c_2}\gamma^{-1}\mu_1\beta(1-e^\beta)\right).$$

With $\mu_1/\mu_2 = e^\beta = 10/11$ this gives $\overline{x}_2^* \approx 5$ when $\gamma = 0.01$, and $\overline{x}_2^* \approx 21$ when $\gamma = 0.001$.

The discounted optimal policy is shown in Fig. 4.9. We see that these threshold policies nearly coincide with the actual optimal policy in each case.

CASE 2: $\mu_1 \geq \mu_2$ In this case $\beta > 0$, and hence an alternate approximation is required. Assuming again that γ is small, we again conclude that $\overline{x}_2^*$ will be large, which results in the approximation

$$\overline{x}_2^* \approx \beta^{-1} \log\left(1 + \frac{c_1}{c_2}\gamma^{-1}\mu_1\beta(e^\beta - 1)\right).$$

For the special case with parameters $\mu_1/\mu_2 = e^\beta = 11/10$ we obtain $\overline{x}_2^* \approx 5$ when $\gamma = 0.01$ and $\overline{x}_2^* \approx 13$ when $\gamma = 0.001$. The discounted optimal policies shown in Fig. 4.10 again show very close agreement with these asymptotic formulae.

Note that in Case 2 the threshold $\overline{x}_2^*$ scales logarithmically with γ^{-1}, while in Case 1 it scales linearly. This is consistent with the characteristics of the infinite-horizon optimal control problem for the fluid model. In Case 2 this is described by the switching curve $x_2 = m_x^* x_1$ defined in (4.47) with $m_x^* \equiv 0$, while in Case 1 the constant m_x^* is strictly positive. ∎

10.6.2 Infinite horizon

The proof of Proposition 10.6.4 is identical to the proof of Proposition 10.6.2 (i) using optimality of $\{q^{(r)}(t) : t \geq 0\}$ whenever q^* is optimal, where $q^{(r)}$ is defined in (10.28).

Proposition 10.6.4. *The infinite-horizon value function for the fluid model satisfies*

$$J^*(\kappa x) = \kappa^2 J^*(x) \qquad \text{for each } x \in \mathbb{R}_+^\ell, \kappa > 0.$$

Recall that Theorem 10.0.5 claims that $\kappa^{-2}h^*(\kappa x^\kappa) \approx J^*(x)$, which on applying Proposition 10.6.4 can be written

$$\lim_{|x| \to \infty} \frac{h^*(x)}{J^*(x)} = 1.$$

As a first step to proving this result, we consider the more general Poisson's equation (8.2) under a stationary policy.

Proposition 10.6.5. *For any stationary policy, if there is a solution to Poisson's equation $h\colon \mathsf{X}_\diamond \to \mathbb{R}_+$, then for each $T > 0$ and $x \in \mathsf{X}$,*

(i) $\displaystyle\lim_{\kappa \to \infty} \left| \mathsf{E}\left[\frac{1}{\kappa^2}\left(h(\kappa x^\kappa) - h(\kappa q(T^\kappa; x^\kappa))\right) - \int_0^{T^\kappa} c(q^\kappa(s; x^\kappa))\, ds\right] \right| = 0,$

(ii) $\displaystyle\liminf_{\kappa \to \infty} \frac{1}{\kappa^2} h(\kappa x^\kappa) \geq J^*(x).$

Proof. The proof is based on iterating the equation $Ph - h - [c - \eta]$ to obtain for any N

$$P^N h(x) - h(x) = -\sum_{t=0}^{N-1} [P^t c(x) - \eta] = -\mathsf{E}\left[\sum_{t=0}^{N-1} [c(Q(t; x)) - \eta]\right]. \qquad (10.63)$$

Combining Lemma 10.4.2 with (10.63) we obtain (i). Part (ii) follows from (i) using positivity of h and the bound

$$\liminf_{\kappa \to \infty} \mathsf{E}\left[\int_0^{T^\kappa} c(q^\kappa(t; x^\kappa))\, dt\right] \geq \mathsf{E}\left[\liminf_{\kappa \to \infty} \int_0^{T^\kappa} c(q^\kappa(t; x^\kappa))\, dt\right]$$

$$\geq \inf_q \int_0^T c(q(t; x))\, dt$$

where the final bound follows from Proposition 10.3.2, and the infimum is over all fluid trajectories starting from x. The right-hand side coincides with $J^*(x)$ for all $T > 0$ sufficiently large. □

To obtain a complementary bound on $\kappa^{-2} h^*(\kappa x)$ we first appeal to the principle of optimality:

Lemma 10.6.6. *If h^* solves the ACOE, then for any policy and any T,*

$$\limsup_{\kappa \to \infty} \frac{1}{\kappa^2} h^*(\kappa x^\kappa) \leq \limsup_{\kappa \to \infty} \mathsf{E}\left[\frac{1}{\kappa^2} h^*(Q(\kappa T^\kappa; \kappa x^\kappa)) + \int_0^{T^\kappa} c(q^\kappa(s; x^\kappa))\, ds\right].$$

Proof. If h^* solves the ACOE, then (10.63) admits an extension to any policy as an inequality:

$$\mathsf{E}[h^*(Q(N; x))] \geq h^*(x) - \mathsf{E}\left[\sum_{t=0}^{N-1} [c(Q(t; x)) - \eta^*]\right]. \qquad (10.64)$$

Substituting $N = \kappa T^\kappa$ and $x = \kappa x^\kappa$ we obtain the result from Lemma 10.4.2. □

Next we apply Proposition 10.5.1 to conclude that the limit J^* is "almost" achievable using *some* policy.

Lemma 10.6.7. *Under the assumptions of Theorem 10.0.5, for any $\varepsilon > 0$ there exists a policy ϕ such that for all $T > 0$ sufficiently large,*

$$\limsup_{\kappa \to \infty} \int_0^T c(q^\kappa(t; x(\kappa)))\, dt \leq J^*(x) + \varepsilon,$$

$$\limsup_{\kappa \to \infty} \mathsf{E}\left[\int_0^T c(q^\kappa(t; x(\kappa)))\, dt\right] \leq J^*(x) + \varepsilon,$$

$$\limsup_{\kappa \to \infty} \mathsf{E}[\|q^\kappa(T; x(\kappa))\|^2] \leq \varepsilon, \qquad |x| \leq 1.$$

Proof. Fix $\varepsilon_0 \in (0, \varepsilon)$, and choose a piecewise linear trajectory $(q^\varepsilon, z^\varepsilon)$ satisfying $\|q^\varepsilon(t) - q^*(t)\| \le \varepsilon_0$ for all $t \ge 0$. Hence there are times $0 = t_0 < t_1 < \cdots < t_n$ and vectors $\{v^i\} \subset \mathsf{V}$ satisfying

$$\tfrac{d^+}{dt} q^\varepsilon(t) = v^i, \quad t \in [t_i, t_{i+1}), \ 1 \le i \le n - 1.$$

We assume moreover that $q^\varepsilon(t) = \mathbf{0}$ for $t \ge t_n$.

Given $Q(0) = \kappa x(\kappa)$ we set $T_0^\kappa = 0$, $T_i^\kappa = \kappa^{-1} \lfloor \kappa t_i \rfloor$, $1 \le i \le n$, and choose the policy on $[T_i^\kappa, T_{i+1}^\kappa)$ so that $q^\kappa(t; x(\kappa)) \to q^\varepsilon(t)$ as $\kappa \to \infty$ for each $t \le t_n$. It is assumed that on this interval the policy is the randomized stationary policy defined using (10.54), (10.55). On the infinite interval $[T_n^\kappa, \infty)$ the policy reverts to any stationary policy for which the fluid model is stable.

The desired convergence follows from Proposition 10.5.3 since $\varepsilon_0 > 0$ is arbitrary.

□

Proof of Theorem 10.0.5. Following Proposition 10.6.5, to complete the proof of Theorem 10.0.5 it is sufficient to show that the solution to Poisson's equation under the optimal policy satisfies

$$\limsup_{\kappa \to \infty} \frac{1}{\kappa^2} h^*(\kappa x^\kappa) \le J^*(x), \quad x \in \mathbb{R}_+^\ell. \tag{10.65}$$

This follows by combining Lemmas 10.6.6 and 10.6.7, together with the assumption that $h^* \in L_\infty^{V_2}$.

□

We conclude this chapter with some extensions to the CBM model.

10.7 Brownian workload model

Under certain conditions on the model we saw in Section 8.7 that the R-minimal solution on the domain R defined in (8.97) is ergodic. It is in fact V-uniformly ergodic, with V an exponential of the scaled fluid value function. If the assumptions of Theorem 8.7.4 are relaxed we do not know whether such a strong form of ergodicity holds, but we can obtain similar results based on the fluid-scaling approach developed in this chapter.

It will be useful to introduce a scaling parameter $\kappa \ge 0$ to investigate the impact of variability,

$$W(t) = w - \delta t + I(t) + \sqrt{\kappa} N(t), \quad W(0) = w \in \mathsf{W}, \tag{10.66}$$

and modify the region (5.31) as follows,

$$\mathsf{R}(\kappa) = \{ w \in \mathbb{R}_+^n : \langle n^i, w \rangle \ge -\kappa \beta_i, \ 1 \le i \le \ell_R \}, \tag{10.67}$$

where the constant vector $\beta \in \mathbb{R}_+^{\ell_R}$ and vectors $\{n^i\} \subset \mathbb{R}^n$ are as in (5.31).

10.7.1 Value functions

We begin with some structural properties of the value functions associated with the fluid and CBM models. We consider two processes on the domain $R(\kappa)$: the minimal process W satisfying (10.66), and also the deterministic minimal solution $\widehat{w}$ on $R(\kappa)$ defined in (5.27). When we wish to emphasize the dependence on κ and the initial condition $w \in W$ we denote the workload processes by $W(t; w, \kappa)$, $w(t; w, \kappa)$, respectively.

Let η_κ denote the steady-state mean of $\bar{c}(W(t; w, \kappa))$. When $\kappa = 1$ we drop the subscript so that $\eta = \eta_1$. The value functions considered in this section are defined for the fluid and CBM workload models, respectively, by

$$J(w; \kappa) := \int_0^\infty \bar{c}(w(t; w, \kappa))\, dt, \qquad w \in W. \tag{10.68}$$

$$h(w; \kappa) := \int_0^\infty \Big(\mathsf{E}[\bar{c}(W(t; w, \kappa))] - \eta_\kappa \Big)\, dt, \tag{10.69}$$

where $w \in W$ and $\kappa \geq 0$. We again suppress dependence on κ when $\kappa = 1$. We have also removed the "hats" on J, h, etc., to streamline the notation.

General conditions under which the steady-state mean η is well defined are presented in Theorem 10.7.1. The proof is postponed to the end of Section 10.7.2.

Theorem 10.7.1. *Suppose that the following extension of assumption (a) of Theorem 5.3.21 holds: For each $\kappa \geq 0$, the set $R(\kappa)$ has nonempty interior, satisfies $R(\kappa) \subset \mathbb{R}_+^n$, and the pointwise projection $[\,\cdot\,]_R \colon \mathbb{R}^n \to R(\kappa)$ exists.*

The following then hold for the R-minimal process:

(i) *The scaling property holds,*

$$h(w; \kappa) = \kappa^2 h(\kappa^{-1} w; 1), \qquad w \in R(\kappa),\ \kappa > 0. \tag{10.70}$$

(ii) *For some constant $b > 0$, and all $w \in R(\kappa)$,*

$$-b\kappa^2 \leq h(w; \kappa) \leq b(\kappa^2 + \|w\|^2).$$

(iii) *The steady-state mean η and the relative value function h are finite valued, and h solves Poisson's equation,*

$$\mathcal{A}h = -\bar{c} + \eta_\kappa. \tag{10.71}$$

(iv) *If $h' \in L_\infty^{V_2}$ is another solution to Poisson's equation (10.71), then there is a constant b' such that $h'(w; \kappa) = h(w; \kappa) + b'$ on $R(\kappa)$.*

To prove the theorem we begin with some elementary scaling properties. For the fluid model we obtain from the definitions,

Proposition 10.7.2. *The following hold for the fluid model:*

(i) $w(t; w, \kappa) = \kappa w(\kappa^{-1} t; \kappa^{-1} w, 1)$ *for each $t \geq 0$, $\kappa > 0$, and $w \in R(\kappa)$.*

(ii) *If $\rho_\bullet < 1$ we have $J(w; \kappa) < \infty$ for each $w \in W$, $\kappa > 0$, and the following scaling property holds:*

$$J(w; \kappa) = \kappa^2 J(\kappa^{-1} w; 1), \qquad w \in R(\kappa),\ \kappa > 0. \tag{10.72}$$

With slightly more effort we obtain analogous scaling results for the relative value function. The following result follows from the scaling formula for Brownian motion, $\kappa N(\kappa^{-1}t) \overset{\text{dist}}{=} \sqrt{\kappa}N(t)$.

Proposition 10.7.3. $W(t; w, \kappa) \overset{\text{dist}}{=} \kappa W(\kappa^{-1}t; \kappa^{-1}w, 1)$ *for each* $t \geq 0$, $\kappa > 0$, *and* $w \in \mathsf{R}(\kappa)$, *where the equality is in distribution.*

Note that the identity (10.70) implies that $r^{-2}h(rw; 1) = h(w; r^{-1}) \to h(w; 0)$. Similarly, (10.72) implies that $r^{-2}J(rw; 1) = J(w; r^{-1}) \to J(w; 0)$, as $r \to \infty$, and it follows that $r^{-2}|\mathcal{E}(rw)| \to 0$ as $r \to \infty$ since $h(w; 0) = J(w; 0)$. Theorem 10.0.6 is a substantial strengthening of this asymptotic bound.

10.7.2 Regeneration and ergodicity

A stationary version of the workload process can be constructed using the shift-coupling technique introduced in our analysis of the single-server queue. Suppose that the Brownian motion $\boldsymbol{N}$ is defined on the two-sided interval $\mathbb{R}$, with $N(0) = 0$, and construct a process $\boldsymbol{W}^s$ on R, *initialized at time* $-s$. For a given initial condition $w \in \mathsf{R}$, this is defined on the interval $[-s, \infty)$ with initial condition $W^s(-s; w) = w$, and disturbance process $N^s(t) := N(t) - N(-s)$, $t \geq -s$. Suppose that all of the processes are initialized at $w = \mathbf{0}$. If $s' > s$ then $W^{-s}(s'; \mathbf{0}) \geq \mathbf{0} = W^{-s}(s; \mathbf{0})$, and by minimality it then follows that $W^t(s'; \mathbf{0}) \geq \mathbf{0} = W^t(s; \mathbf{0})$ for all $t \geq -s$. That is, for any fixed t, $\{W^s(t; \mathbf{0}) : s \geq 0\}$ is nondecreasing (component-wise) in s for $s \geq -t$. It then follows that the limit exists with probability 1,

$$W^\infty(t) := \lim_{s \to \infty} W^s(t; \mathbf{0}), \qquad -\infty < t < \infty. \tag{10.73}$$

Proposition 10.7.3 combined with Proposition 5.4.7 implies a weak form of stability:

Proposition 10.7.4. *Suppose that assumption (a) of Theorem 5.3.21 holds. Then the minimal process on* R *with* $\kappa = 1$ *satisfies, for each* $p \geq 1$, $w \in \mathsf{W}$,

$$\lim_{r \to \infty} \mathsf{E}[\|r^{-1}W(rt; rw)\|^p] = 0, \quad t \geq T^*(w). \tag{10.74}$$

Proof. From Proposition 5.4.7 we obtain the bound, for each $t \geq 0$, $\kappa > 0$, $w \in \mathsf{W}$,

$$\|W(t; w, \kappa) - w(t; w, \kappa)\| \leq k_R\sqrt{\kappa}\|\boldsymbol{N}\|_{[0,t]}, \tag{10.75}$$

where the norm $\| \cdot \|_{[0,t]}$ is defined in (5.69). If $t \geq T^*(w)$ then $w(t; w, \kappa) = \mathbf{0}$ for each $\kappa > 0$. Combining the bound (10.75) with Proposition 10.7.3 we conclude that, for any $p \geq 1$, $t \geq T^*(w)$,

$$\kappa^p \mathsf{E}[\|W(\kappa^{-1}t; \kappa^{-1}w)\|^p] \leq (k_R\sqrt{\kappa})^p \mathsf{E}[\|\boldsymbol{N}\|_{[0,t]}^p].$$

An application of [413, Corollary 37.12] shows that $\mathsf{E}[\|\boldsymbol{N}\|_{[0,t]}^p]$ is finite for each $t \geq 0$, $p \geq 1$, giving (10.74). $\qquad\square$

For a given function $f \colon \mathsf{R} \to [1, \infty)$, and for a pair of probability distributions μ, ν on $\mathcal{B}(\mathsf{R})$, we define

$$\|\mu - \nu\|_f := \sup_{|g| \leq f} |\mu(g) - \nu(g)|.$$

Define $V_p(w) = \|w\|^p + 1$ for $w \in \mathsf{W}$, and let $\{P^t : t \geq 0\}$ denote the Markov transition group for $\boldsymbol{W}$ with $\kappa = 1$.

Theorem 10.7.5. *Suppose that assumption (a) of Theorem 5.3.21 holds. Then the minimal process on* R *with* $\kappa = 1$ *satisfies the following:*

(i) *The limiting process* $\boldsymbol{W}^\infty$ *exists, and its marginal distribution* π *on* R *is the unique invariant measure for* $\boldsymbol{W}$.

(ii) *The invariant measure has finite moments, and* $\lim_{t \to \infty} t^p \|P^t(w, \cdot) - \pi(\cdot)\|_{V_p} = 0$ *for each* $p \geq 1$, $w \in \mathsf{W}$.

(iii) *There is a compact set* $C_0 \subset \mathsf{W}$ *s.t. for each integer* $p \geq 0$, *there is a finite constant* k_p *satisfying*

$$\mathsf{E}\left[\int_0^{\tau_{C_0}} V_p(W(t; w)) \, dt \right] \leq k_p V_{p+1}(w), \quad w \in \mathsf{W}, \tag{10.76}$$

where τ_{C_0} *denotes the first entrance time to* C_0.

Proof. The proof is outside of the scope of this book, but it is not difficult to describe the main ideas, which are very similar to the proof of Proposition 10.4.1. See the Notes section for references and background.

Define the function V_c in analogy with (10.11) via

$$V_c(w) := \mathsf{E}\left[\int_0^{|w|T} V_p(W(t; w)) \, dt \right] \qquad w \in \mathsf{W}. \tag{10.77}$$

This function is bounded by a fixed multiple of $V_{p+1}(w)$. Moreover, on setting

$$\mathcal{C} := \int_0^{|w|T} V_p(W(t; w)) \, dt$$

we have by the Markov property (10.35), for any $r > 0$,

$$V_c(W(r; w)) = \mathsf{E}\left[\vartheta^r \mathcal{C} \mid \mathcal{F}_r \right] = \mathsf{E}\left[\int_r^{|W(r; w)|T} V_p(W(t; w)) \, dt \mid \mathcal{F}_r \right].$$

Taking expectations of each side then gives

$$\begin{aligned}
P^r V_c(w) &:= \mathsf{E}[V_c(W(r; w))] \\
&= \mathsf{E}\left[\int_r^{|W(r; w)|T} V_p(W(t; w)) \, dt \right] \\
&= V_c(w) - \mathsf{E}\left[\int_0^r V_p(W(t; w)) \, dt \right] + \mathsf{E}\left[\int_{|w|T}^{|W(r; w)|T} V_p(W(t; w)) \, dt \right].
\end{aligned}$$

Choose $T > 0$ sufficiently large so that $\kappa^{-1}W(\kappa t; \kappa w) \to 0$ for $t \geq T$ and $|w| \leq 1$. We can then conclude that for any w satisfying $|w| = 1$

$$\lim_{\kappa \to \infty} \kappa^{-p} \mathsf{E}\left[\int_{\kappa T}^{|W(r;\kappa w)|T} V_p(W(t; \kappa w))\, dt \right] = 0.$$

This implies a bound of the form

$$P^r V_c(w) \leq V_c(w) - \tfrac{1}{2}\mathsf{E}\left[\int_0^r V_p(W(t; w))\, dt \right] + b\mathbf{1}_S(w),$$

where $b < \infty$ and S is a bounded subset of W. This is precisely the sort of Lyapunov drift condition used to establish regularity in the proof of Theorem 10.0.2 (with $p = 1$).

□

Proof of Theorem 10.7.1. The scaling property (10.70) in (i) follows directly from Proposition 10.7.3 and formula (10.69). Without loss of generality we restrict to $\kappa = 1$ in the remainder of the proof.

To prove (ii) first note that Theorem 10.7.5 (iii) implies that W is V_p-regular where $V_p(w) = \|w\|^p + 1$, where the definition in continuous time is identical to the discrete time definition given in Definition A.4.4 (see [367]). We can obtain the following uniform bounds: For each p there exists $k_p < \infty$ such that for any set S with positive π-measure, there is a constant $k_S < \infty$ satisfying for all w,

$$\mathsf{E}\left[\int_0^{\tau_S} V_p(W(s; w))\, ds \right] \leq k_p V_{p+1}(w) + k_S. \tag{10.78}$$

Specializing to $p = 1$, it then follows from [214, Theorem 3.2] that a solution g to Poisson's equation exists with quadratic growth satisfying for each $T \geq 0$

$$\mathsf{E}[g(W(T; w))] = g(w) - \int_0^T \left(\mathsf{E}[\bar{c}(W(t; w))] - \eta \right) dt.$$

Letting $T \to \infty$ and applying Theorem 10.7.5 (ii) then gives $\pi(g) = g(w) - h(w)$, with h defined in (10.69), so that $|h|$ is also bounded by a quadratic function of w. This bound on h allows application of the Markov property to obtain the following representation for each $T > 0$:

$$h(W(T; w)) = \int_T^\infty \left(\mathsf{E}[\bar{c}(W(t; w)) \mid \mathcal{F}_T] - \eta \right) dt.$$

It follows that the stochastic process M_h defined below is a martingale for each initial condition $w \in$ W,

$$M_h(t) := h(W(t; w)) - h(w) + \int_0^t \left[\bar{c}(W(s; w)) - \eta \right] ds, \quad t \geq 0. \tag{10.79}$$

This proves (iii).

We now use the martingale property to prove (ii). Let $S \subset$ R be any compact set with nonempty interior. The optional sampling theorem implies that $\{M_h(t \wedge \tau_S) : t \geq 0\}$ is a martingale, and it can be shown using Theorem 10.7.5 and (10.78) that it is uniformly integrable. Consequently, we obtain the expression

$$E[M_h(\tau_S)] = E\left[h(W(\tau_S; w)) - h(w) + \int_0^{\tau_S} \overline{c}(W(s; w)) - \eta \, ds\right] = 0. \quad (10.80)$$

On setting $S = \{w : \overline{c}(w) \leq \eta\}$ we obtain the lower bound, $h(w) \geq \inf_{w' \in S} h(w')$, completing the proof of (ii).

To prove uniqueness we note first that under the conditions of (iv) the process $M_{h'}$ is again a martingale (as defined in (10.79) using the function h') since $h' \in L_\infty^{V_2}$ solves Poisson's equation. Taking expectations gives, by the martingale property,

$$0 = M_{h'}(0) = E[M_{h'}(t)] = E\left[h'(W(t; w)) - h'(w) + \int_0^t [\overline{c}(W(s; w)) - \eta] \, ds\right].$$

Letting $t \to \infty$ and applying Theorem 10.7.5 (ii) gives $h(w) = h'(w) + b'$ with $b' = \pi(h - h')$. $\square$

Proof of Theorem 10.0.6. We have noted that M_h is a martingale, and M_J is a martingale by Theorem 8.7.2 (iii). Consequently, $M_\mathcal{E} := M_h - M_J$ is a martingale, and can be expressed for $t \geq 0$ by

$$M_\mathcal{E}(t) = \mathcal{E}(W(t; w)) - \mathcal{E}(W(0; w)) + \int_0^t [b_{\mathrm{CBM}}(W(s; w)) - \eta] \, ds. \quad (10.81)$$

Let $C_0 \subset \mathsf{W}$ denote the compact set found in Theorem 10.7.5 (iii). The martingale property for M_h implies the representation (10.80), and combining (10.76) with Itô's formula (8.100) we obtain the analogous expression for M_J. On subtracting, we obtain for each initial condition

$$\mathcal{E}(w) = E\left[\mathcal{E}(W(\tau_{C_0}; w)) + \int_0^{\tau_{C_0}} (b_{\mathrm{CBM}}(W(s; w)) - \eta) \, ds\right]. \quad (10.82)$$

To complete the proof, observe that the function b_{CBM} is bounded and $E_w[\tau_{C_0}]$ has linear growth by Theorem 10.7.5 (iii) with $p = 0$. $\square$

10.8 Notes

For more on the ODE method see the monographs of Kushner and Yin [328], Benveniste, Métivier, and Priouret [41], or Chen [99]. See also the Notes following Chapter 11. In [71] the fluid model approach developed for networks is extended to address stability of stochastic approximation algorithms found in reinforcement learning applications, and Fort et al. [189] show that the same ideas can be used to address stability of MCMC algorithms.

The discussion surrounding Example 10.1.5 on "service-aware scheduling" is motivated by recent research at the intersection of information theory and operations research. A very partial slice of this literature is included in [57, 437, 75, 484].

There are now numerous papers that derive stability conditions for specific network models based on the fluid model. References [130, 51, 98, 132, 129, 321] are just a selected sample. Chen and Zhang [98] develop a general approach to stability verification for priority policies.

Trajectory tracking is the focus of the thesis of Maglaras [346, 345]. Refinements are contained in [33, 359, 133, 56, 50]. Policies for a stochastic network are devised to track a solution to the fluid model, and convergence is established as the initial number of jobs in the system tends to infinity. These results make use of safety stocks in a DR-setting similar to the development in Section 10.5.

The fact that stability of the fluid limit model implies stability of the stochastic network was established in a limited setting by Malyšev and Men′šikov in [347]. This result was applied to the KSRS model by Rybko and Stolyar in [420]. The method was extended to a very broad class of multiclass network models by Dai [124]. A key step in the proof of these results is a multistep version of Foster's criterion introduced in [347] for countable state space models, and generalized in [371, 368].

Converse theorems have appeared in [132, 125, 355] that show that under somewhat strong conditions, instability of the fluid model implies transience of the stochastic network. A perfect converse to Theorem 10.0.2 cannot exist: Bramson has constructed an example of a stable stochastic network whose fluid model is unstable in a sense slightly weaker than Definition 10.0.3 [82]. This result and counterexamples in [195, 127] show that some additional conditions are necessary to obtain a converse.

Proposition 10.3.8 is a special case of a result of Stolyar [461]. Lemmas 10.5.2 and 10.5.4 are related to bounds obtained in [170].

The main result of [124] established positive recurrence only. Moments and rates of convergence to stationarity of the Markovian network model were obtained in [129, 317, 359]. It is shown in [317] that L_2-stability of the fluid limit model is equivalent to a form of regularity for the network. In particular, the conclusions of Theorem 10.7.5 also hold for the CRW model provided the fluid limit model is stable:

Theorem 10.8.1. *Suppose that conditions of Theorem 10.0.2 hold, and that the given stationary policy is weakly nonidling. Suppose moreover that A is a bounded sequence. Then:*

(i) *The policy is regular, with invariant measure π.*
(ii) *The invariant measure has finite moments, and for each $p \geq 1$, $x \in \mathsf{X}$,*

$$\lim_{t \to \infty} t^p \| P^t(x, \cdot) - \pi(\cdot) \|_{V_p} = 0,$$

where $V_p(x) = \|x\|^p + 1$, $x \in \mathbb{R}_+^\ell$.
(iii) *For each integer $p \geq 0$, there is a finite constant k_p satisfying*

$$\mathsf{E}\left[\sum_{t=0}^{\tau_0} f_p(Q(t)) \right] \leq k_p f_{p+1}(x), \quad x \in \mathsf{X}, \tag{10.83}$$

where τ_0 denotes the first entrance time to the empty state $\mathbf{0}$.

Theorem 10.8.1 is taken from Gamarnik and Meyn [196], and it is essentially contained in [129].

The polynomial rate of convergence in (ii) is based on results of Tuominen and Tweedie [476] (now extended and simplified in work of Douc et al. [145]). It is natural to conjecture that the controlled model is geometrically ergodic under the assumptions of Theorem 10.8.1, so that the convergence in (ii) can be strengthened to geometric convergence. Surprisingly, this is *false*. The paper [196] contains an example of a network and stationary policy satisfying the conditions of Theorem 10.8.1, yet the controlled process is not geometrically ergodic.

Dupuis and Williams [160] and Ata et al. [23] establish positive recurrence of the CBM model under assumptions somewhat weaker than Theorem 10.7.5.

Theorem 10.0.5 relating optimality of the CRW and fluid models is based on the two 1997 papers [357, 357]. Refinements of this result have appeared in a series of papers [105, 358, 359, 103, 361, 362, 363]. Much of Section 10.6.1 on fluid-scale asymptotic optimality under the discounted-cost criterion is new.

The logarithmic switching curve (10.49) was introduced in [357] based on numerical experiments obtained in a single example, and heuristic arguments of the form made precise in

Section 10.5. A policy based on (10.49) is analyzed in depth in [192] for a pair of queues in tandem. It is found that the policy is stabilizing, and fluid scale asymptotically optimal. Moreover, a bound is obtained on the rate of convergence in the limit on the right-hand side in (10.15). In this example and several others it is shown in [362] that a similar policy is asymptotically average-cost optimal, with logarithmic regret, in the sense of (6.64). These results were generalized in [365].

Again, there are close parallels with heavy-traffic theory of stochastic networks. Harrison in [234] argues that a safety stock $\bar{x}_i$ at buffer i should satisfy a lower bound of the form

$$\bar{x}_i = K \log\left(\frac{1}{1-\rho_\bullet}\right), \qquad i = 1,\ldots,\ell. \tag{10.84}$$

A safety-stock-based policy of this form is introduced for the processor sharing model in [36]. Using large deviations bounds, among other techniques, it is shown that a scaled version of the process converges in distribution to a reflected Brownian motion. Based on this result, it is argued that the policy is approximately optimal for $\rho_\bullet \approx 1$. Ata and Kumar consider discrete review policies for general network models in [21]. They establish asymptotic optimality assuming $2 + \epsilon$ moments on the interarrival times and processing times based on a similar policy. The main assumption in this paper and [36] is the *complete resource pooling* condition, meaning that there is a single bottleneck in heavy traffic (see Definition 6.2.1).

The development of Section 10.7 for the CBM model is taken from [363]. The uniqueness result Theorem 10.7.1 (iv) is given in [356, Theorem A3] for a version of the CRW model. Although stated in discrete time, Section 6 of [356] contains a roadmap that explains how to translate to continuous time.

Exercises

10.1   Verify using Theorem 10.0.2 that the priority policy, with priority to buffers 1 and 3, is stabilizing for the KSRS model.

10.2   Show that all nonidling policies are in fact stabilizing for the simple re-entrant line, under the assumptions imposed in Example 8.6.4.

10.3   Obtain the fluid model for the *phase-type* queue described in Exercise 18.11.

10.4   The following example concerning a model for breakdown and repair is taken from [129], following [278, 386]. A single machine is operational for a random period of time, and then breaks down and requires service. The cycle of repair and production repeats, where operation time has mean μ_+^{-1}, and the period for repair has mean μ_-^{-1}. This can be modeled using a priority queue as illustrated below:

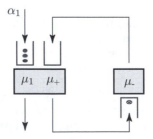

Shown on the left is the station in operation with service rate μ_1. When the single virtual customer at the station on the right completes service, it moves to Station 1 and receives priority. This service period is interpreted as repair.

What is the distribution of the repair time and operation time? How would you generalize to other distributions? Describe the fluid limit model, and discuss implications for stability.

10.5 The following exercise is inspired by the thesis of M. Mitzenmacher, *The power of two choices in randomized load balancing* [373] (see also [374, 27]). Consider the routing model consisting of N buffers and a single arrival stream. For each arriving customer the router chooses two of the N buffers at random (with a uniform distribution) and sends the customer to the buffer with the smaller contents. Suppose that the arrival process A is i.i.d. with mean α and finite variance. The service processes are independent of each other and A, with identical service rate μ.

   (a) Construct a CRW model.
   (b) Conjecture the form of the fluid limit model under this policy.
   (c) Determine the range of μ for which the fluid limit model in (b) is stable.

10.6 Verify the bound on the "time to starvation" given in (4.69) for the KSRS model. One approach is to consider the (conditional) immediate workload at Station 2 given by $V(x) = \mu_2^{-1}x_2 + \mu_3^{-1}x_3$, and obtain a lower bound on the drift,

$$\mathsf{E}[V(Q(t+1)) - V(Q(t)) \mid \mathcal{F}_t] \geq 1, \qquad \text{if } \tau_{02} \geq t.$$

10.7 Consider the CRW model of the tandem queues satisfying (4.60), and controlled using the hedging-point policy (10.58). Assume that $\mu_1 > \frac{1}{3}$, and that the threshold $\overline{x}_2 > 1$ is fixed. Show using Theorem 10.0.4 that there exists $\mu_2^\bullet > (1 - \mu_1)/2$ such that $\rho_\bullet = \rho_2 < 1$ for $\mu_2 \in [\mu_2^\bullet, (1 - \mu_1)/2]$, yet the process Q is transient.

10.8 Suppose that the following bound holds under a stationary policy for some constant b_0 and each $x \in \mathsf{X}_\diamond$,

$$\mathsf{E}_x[\tau_0] \leq b_0|x|, \quad \text{and} \quad \mathsf{E}\left[\sum_{t=0}^{\tau_0-1} |Q(t;x)|\right] \leq b_0 V_2(x).$$

Show that for a constant b_1,

$$\mathsf{E}_x[\tau_0^2] \leq b_1 V_2(x), \qquad x \in \mathsf{X}_\diamond.$$

Hint: By the Markov property, $\mathsf{E}[\tau_0 - t \mid \mathcal{F}_t] = \mathsf{E}_y[\tau_0]$ when $Q(t) = y$ and $t < \tau_0$.

11

Simulation and Learning

The chief motivation for performance evaluation is to compare candidate policies. For example, many of the policies described in Chapters 4 and 10 depend upon static or dynamic safety-stock parameters, and we would like to know how to choose the best parameter values in order to optimize performance.

We have seen in Chapter 8 that linear programming techniques can provide bounds on performance for the CRW model. This approach can be successfully applied in network models with many buffers and stations. However, linear programming techniques are not flexible with respect to the operating policy. For example, in order to distinguish similar policies with different safety-stock levels, constraints must be introduced in the LP specific to each safety-stock parameter. It is not clear how to introduce such constraints in the approaches that have been developed to date.

While not a topic of this book, there are classes of networks for which the invariant measure π is known. The crucial property required is *reversibility*, from which it follows that π has a product form, $\pi(x) = \pi_1(x_1) \cdots \pi_\ell(x_\ell)$ for $x \in \mathsf{X}$ [502, 290].[1] Outside of this very special class of models the computation of π is essentially impossible in large networks. We are thus led to simulation techniques to evaluate performance.

The simulation techniques surveyed in this chapter all involve a Markov chain $\boldsymbol{X}$ on a state space X. Since we plan to develop simulation techniques that can be applied in a wide class of policies, including networks controlled using a discrete-review policy of the form introduced in Section 8.3, we do not assume that $\boldsymbol{X} = \boldsymbol{Q}$ except in special cases.

Throughout the chapter it is assumed that $\boldsymbol{X}$ is x^*-*irreducible*. Recall that this means that the state $x^* \in \mathsf{X}$ is reachable with positive probability from each initial condition.

Monte Carlo and stochastic approximation

Much of this chapter concerns steady-state simulation. In particular, for a given cost function $c \colon \mathsf{X} \to \mathbb{R}_+$ the *Monte Carlo estimates* of the steady-state mean $\eta := \pi(c)$ are defined by

$$\eta(n) := \frac{1}{n} \sum_{t=0}^{n-1} c(X(t)), \quad n \geq 1. \tag{11.1}$$

---

[1] Kelly's monograph [290] is now available online.

452

We also consider briefly in Section 11.1 estimation of transient performance metrics, such as the discounted cost h_γ.

If the chain is c-regular, it then follows from the Strong Law of Large Numbers (LLN) in Theorem A.2.3 that the steady-state mean $\eta := \pi(c)$ is finite, and $\eta(n) \to \eta$ as $n \to \infty$ with probability 1 from each initial condition. Generally, an estimator is called *strongly consistent* or *asymptotically unbiased* when the estimates converge to the true value with probability 1.

Monte Carlo estimation is a special case of the *stochastic approximation* (SA) algorithm of Robbins and Monro. Suppose that $G \colon \mathsf{X} \times \mathbb{R}^d \to \mathbb{R}^d$ is a given function, and we wish to compute a solution $\theta^* \in \mathbb{R}^d$ to the equation

$$\mathsf{E}_\pi[G(X(t), \theta)] = 0. \tag{11.2}$$

In applications G is typically expressed as a gradient $G(x, \theta) = \nabla_\theta g(x, \theta)$, where $g \colon \mathsf{X} \times \mathbb{R}^d \to \mathbb{R}_+$ is some measure of performance, and θ a parameter. In this case, a vector $\theta \in \mathbb{R}^d$ satisfying (11.2) is a candidate minima or maxima of $\mathsf{E}_\pi[g(X(t), \theta)]$.

The SA algorithm is described by the recursion

$$\theta(n+1) = \theta(n) + a_n G(X(n), \theta(n)), \qquad n \geq 0, \ \theta(0) \in \mathbb{R}^d, \tag{11.3}$$

where $\{a_n\}$ is called the *gain sequence*. Usually it is assumed nonnegative, and subject to these two conditions,

$$\sum_n a_n = \infty \quad \text{and} \quad \sum_n a_n^2 < \infty. \tag{11.4}$$

A common choice is $a_n = 1/(n+1)$. The Monte Carlo estimates (11.1) are a special case of the SA recursion (11.3) using this gain sequence, and with

$$\theta(n) = \eta(n), \qquad G(X(n), \theta(n)) = c(X(n)) - \theta(n), \quad n \geq 0.$$

We then ask, *how long must we wait to obtain an accurate estimate of η?* In simulation and in the "learning algorithms" introduced in this chapter we obtain a sequence of estimates $\{\eta(n)\}$ such as (11.1), and we must decide when to stop the algorithm. A natural stopping rule is the *probably almost correct* (PAC) criterion: Given positive constants ε, δ, this requires the existence of a fixed, deterministic value n_0 such that

$$\mathsf{P}\{|\eta(n) - \eta| \geq \varepsilon\} \leq \delta, \qquad n \geq n_0. \tag{11.5}$$

For example, with $\delta = 0.1$, the interval $[\eta(n) - \varepsilon, \eta(n) + \varepsilon]$ is a 90% confidence interval for η when $n \geq n_0$. Bounds on n_0 are described in Section 11.1 for an i.i.d. sequence X, and Markov models are considered in Section 11.2.

The time to obtain an accurate estimate is roughly proportional to the *asymptotic variance*, which under appropriate conditions can be expressed

$$\sigma_{\text{CLT}}^2 = \lim_{n \to \infty} \mathsf{E}_\pi\left[\left(\sqrt{n}(\eta(n) - \eta)\right)^2\right]. \tag{11.6}$$

Under these conditions the asymptotic variance is also expressed in terms of an autocorrelation function, and in terms of a solution to Poisson's equation for X. Some theory is surveyed in Sections 11.2.1 and A.5.

Asymptotic bounds on the estimation error can be obtained using the Central Limit Theorem (CLT), which provides an approximation of the form

$$\eta(n) \approx \eta + \frac{\sigma_{\text{CLT}}}{\sqrt{n}} X^\infty, \qquad n \approx \infty, \tag{11.7}$$

where X^∞ is a standard Gaussian random variable. This approximation is in the sense of weak convergence: see discussion in Sections 1.3 and A.5.4. In particular, we obtain the following limit for each $\varepsilon > 0$:

$$\lim_{n \to \infty} \mathsf{P}\{|\eta(n) - \eta| \ge \varepsilon n^{-\frac{1}{2}}\} = \mathsf{P}\{|X^\infty| \ge \sigma_{\text{CLT}}^{-1}\varepsilon\}. \tag{11.8}$$

Based on (11.8) we obtain the *heuristic approximation*,

$$\mathsf{P}\{|\eta(n) - \eta| \ge \varepsilon\} \approx \mathsf{P}\{|X^\infty| \ge \sigma_{\text{CLT}}^{-1}\varepsilon\sqrt{n}\}.$$

This is a heuristic since weak convergence only implies convergence of expectations of the form $\mathsf{E}[g(\sqrt{n}(\eta(n) - \eta))]$ when g is a *fixed* function, independent of n. The probability on the left-hand side cannot be expressed in this form for a fixed function $g\colon \mathbb{R} \to \mathbb{R}$. However, taking this approximation for granted temporarily, we arrive at the exponential approximation,

$$\mathsf{P}\{|\eta(n) - \eta| \ge \varepsilon\} \approx 2 \int_{\sigma_{\text{CLT}}^{-1}\varepsilon\sqrt{n}}^{\infty} \frac{1}{\sqrt{2\pi}} e^{-\frac{1}{2}x^2} \, dx,$$

which would then imply

$$\frac{1}{n} \log\big(\mathsf{P}\{|\eta(n) - \eta| \ge \varepsilon\}\big) \approx -\frac{1}{2} \frac{\varepsilon^2}{\sigma_{\text{CLT}}^2}, \qquad n \approx \infty. \tag{11.9}$$

Unfortunately, this particular heuristic bound *fails* for almost any Markov chain. It is true that the log-error probability on the left-hand side of (11.9) is typically convergent, but the limiting value is typically *not* a quadratic function of ε.

We say that the sequence of estimates $\{\eta(n) : n \ge 1\}$ satisfies a *large deviations principle* (LDP) with rate function $I\colon \mathbb{R} \to [0, \infty]$ if the following limits hold:

$$\lim_{n \to \infty} \frac{1}{n} \log\Big(\mathsf{P}\{\eta(n) \ge r\}\Big) = -I(r), \qquad r > \eta, \tag{11.10a}$$

$$\lim_{n \to \infty} \frac{1}{n} \log\Big(\mathsf{P}\{\eta(n) \le r\}\Big) = -I(r), \qquad r < \eta. \tag{11.10b}$$

If the LDP holds, then the two bounds (11.10a), (11.10b) taken together imply

$$\lim_{n \to \infty} \frac{1}{n} \log\Big(\mathsf{P}\{|\eta(n) - \eta| \ge \varepsilon\}\Big) = -\min\big(I(\eta + \varepsilon), I(\eta - \varepsilon)\big), \qquad \varepsilon > 0.$$

Moreover, under the typical conditions guaranteeing the existence of LDP limits, the rate function has the following approximation consistent with (11.9),

$$I(\eta + \varepsilon) = \frac{1}{2} \frac{\varepsilon^2}{\sigma_{\text{CLT}}^2} + O(\varepsilon^3), \qquad \varepsilon \approx 0, \tag{11.11}$$

where σ_{CLT}^2 is the asymptotic variance associated with $\{\eta(n)\}$.

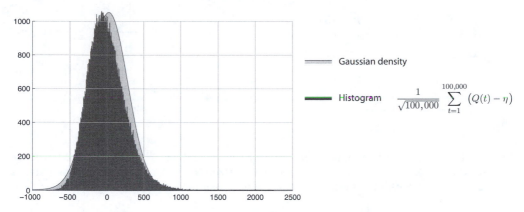

Figure 11.1. Monte Carlo estimates of $\eta := \pi(c)$ with $c(x) \equiv x$ for the M/M/1 queue initialized with $Q(0) \sim \pi$, with load $\rho = 0.9$, and time horizon $T = 10^5$ time steps.

Simulation in networks

Simulation can be difficult in complex networks, especially when there is substantial variability or load. In fact, significant obstacles in simulation are evident in the simplest stochastic model considered in this book, the M/M/1 queue.

Shown in Fig. 11.1 are simulation results for the M/M/1 queue with $\rho = 9/10$. The histogram shows results from 20,000 independent trials. In each trial the Monte Carlo estimate (11.1) was obtained with $\boldsymbol{X} = \boldsymbol{Q}$, $c(x) \equiv x$, and time horizon $T = 10^5$ time steps. The queue was initialized with $Q(0) \sim \pi$ so that $\boldsymbol{Q}$ was stationary. The vertical axis indicates the number of instances that the estimates fall within 1,000 equally spaced bins. The plot suggests that the CLT holds, but the variance seen in this experiment is extremely large. Moreover, the histogram is skewed: The right tail is heavier than anticipated by the CLT, and the left tail is lighter.

Consider now the case of exponential cost, of the form $c(x) = e^{\beta x}$, $x = 0, 1, \ldots$, where $\beta > 0$. Assume that $e^\beta < \rho^{-1}$ so that the steady-state mean for the queue is finite,

$$\eta := \pi(c) = \sum \pi(x) e^{\beta x} = (1 - \rho) \sum (\rho e^\beta)^x = \frac{1 - \rho}{1 - \rho e^\beta}.$$

Shown in Fig. 11.2 are the Monte Carlo estimates (11.1) with $\boldsymbol{X} = \boldsymbol{Q}$, using the specific values $\beta = 0.1$ and $\rho = 9/10$, so that $\eta \approx 18.7$. The queue was initialized at zero, $Q(0) = 0$. The runlength in this simulation extended to $T = 5 \times 10^6$, yet the estimates are significantly larger than the steady-state mean over much of the run.

This chapter contains two explanations for the behavior seen in these experiments. Firstly, the asymptotic variance grows extremely rapidly with $\rho_\bullet$ in most network models, of the order of $\sigma_{\text{CLT}}^2 = O((1 - \rho_\bullet)^{-4})$ when c is a norm. Secondly, on examining large-deviation asymptotics in networks we discover that simulation can be tricky even when the load is not high. Under very general conditions, regardless of load, the two bounds (11.10a), (11.10b) hold when using the estimator (11.1) to compute $\eta = \pi(c)$,

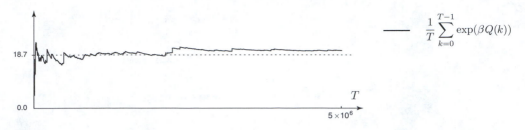

$$\frac{1}{T}\sum_{k=0}^{T-1}\exp(\beta Q(k))$$

Figure 11.2. Monte Carlo estimates of $\eta := \pi(c)$ with $c(x) = e^{0.1x}$ for the M/M/1 queue initialized at zero, with load $\rho = 0.9$, and time horizon $T = 5$ million time steps. After a transient period, the estimates are consistently larger than the steady-state mean of $\eta = (1 - \rho e^{\beta})^{-1}(1 - \rho)$.

but $I(r) \equiv 0$ for $r \geq \eta$. Consequently, simulation leads to overestimates of the steady-state mean, so that the sample-path behavior seen in Fig. 11.2 is typical.

These ideas are illustrated through a detailed analysis of the single-server queue in Section 11.3.

Theorem 11.0.1. *Consider the single-server queue (3.1) with $\rho < 1$ and $0 < \mathsf{E}[A(1)^5] < \infty$. Then:*

(i) *The CLT holds with large asymptotic variance,*

$$\sigma_{\mathrm{CLT}}^2 = \tfrac{1}{2}\frac{[\mathsf{Var}(A(1) - S(1))]^3}{(1 - \rho)^4} + O((1 - \rho)^{-3}).$$

(ii) *The lower LDP (11.10b) holds with $I(r) > 0$ for $r < \eta$; the upper LDP (11.10a) also holds, but $I(r) = 0$ for $r > \eta$.*

Proof. This theorem is necessarily informal since (i) holds for a family of models with increasing load that is introduced in Section 11.3.1.

The LDP (11.10b) follows from Theorem 11.2.4, and (11.10a) with $I(r) = 0$ is established in Proposition 11.3.4. $\qquad\square$

In Section 11.4 we show how to construct improved estimators for the general Markov chain based on the *control variate method*. Simulation can be speeded dramatically through the construction of a control variate based on an approximation to Poisson's equation.

When specialized to network models we find that the fluid value function provides an effective approximation to Poisson's equation leading to significant variance reductions, as well as nontrivial upper and lower LDP bounds.

Suppose that a function $\psi \colon \mathsf{X} \to \mathbb{R}$ is given with quadratic growth. In the notation of Chapter 8, we assume that $\psi \in L_{\infty}^{V_2}$. We then denote $\Delta_{\psi} := \mathcal{D}\psi$, and the *smoothed estimator* is defined by the sample-path averages,

$$\eta_{\psi}^s(n) = \eta(n) + \frac{1}{n}\sum_{t=0}^{n-1}\Delta_{\psi}(X(t)), \qquad n \geq 1. \tag{11.12}$$

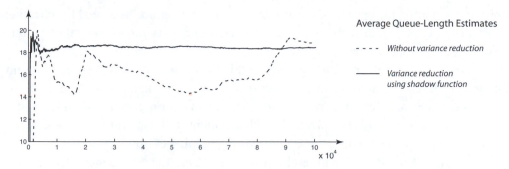

Figure 11.3. Results for a simulation run of length 100,000 steps in the KSRS model initialized at $Q(0) = 0$. The dashed line represents the running average cost using the standard estimator $\eta(n)$ defined in (11.1). The solid line represents the running average cost for the estimator (11.12).

The function Δ_ψ is called a *shadow function* since it is meant to eclipse the function c to be simulated. Proposition 8.2.5 implies that $\pi(\Delta_\psi) = \pi(\mathcal{D}\psi) = 0$ when $X = Q$, so that by Theorem A.5.4 this is an asymptotically unbiased estimator for initial conditions within the support of π.

Fig. 11.3 shows a comparison of the standard estimator (11.1) and the smoothed estimator for the KSRS model, with shadow function based on a fluid value function. Details can be found in Example 11.4.8. The introduction of the zero-mean term Δ_ψ in the smoothed estimator (11.12) results in a 100-fold reduction in variance over the standard estimator in this experiment. We see in the figure that the fluctuations of the standard estimator are tremendous while those of the smoothed estimator are almost nonexistent.

Estimating a value function

The control variate method is just one of many applications of value functions introduced in this book. Recall that an approximate solution to the ACOE can be used to obtain an h-MaxWeight policy that is approximately optimal, and we have seen in Example 6.7.2 how an approximation can be constructed based on a workload relaxation.

This brings us to the final topic in this chapter and in this book: the application of Monte Carlo methods to approximate a value function.

To illustrate the idea, consider again the smoothed estimator to simulate a network. It is likely that we have several candidate approximations to Poisson's equation. Suppose that ℓ_h functions $\{\psi_1, \ldots, \psi_{\ell_h}\}$ are given, and let $\psi \colon X \to \mathbb{R}^{\ell_h}$ denote the corresponding vector-valued function. For each $\theta \in \mathbb{R}^{\ell_h}$ we denote

$$h^\theta = \theta^{\mathrm{T}}\psi, \quad \text{and} \quad \Delta_{h^\theta} := Ph^\theta - h^\theta. \tag{11.13}$$

The goal then is to compute the *best* estimator in this class, in the sense that the resulting simulator using the shadow function Δ_{h^θ} has minimal asymptotic variance.

This problem can be solved using a variant of temporal difference (TD) learning. These algorithms, based on SA, can be used to obtain approximate solutions to

Poisson's equation for optimization or performance evaluation. Section 11.5 contains a survey of TD learning techniques, and related algorithms for various applications involving approximation of value functions.

Section 11.5 is a valuable introduction to this topic even for those readers with no interest in simulation. Sutton and Barto write in their monograph [464], "If one had to identify one idea as central and novel to reinforcement learning, it would undoubtedly be *temporal difference (TD) learning*." Without doubt, this is a valuable algorithmic and conceptual tool that will find use in a growing number of applications.

We begin in Section 11.1 with a closer look at simulating a sequence of i.i.d. random variables where bounds on the integer n_0 used in the PAC criterion (11.5) can be obtained using a variety of methods.

11.1 Deciding when to stop

Here we describe methods for evaluating the estimates obtained when simulating an i.i.d. process denoted $\{\mathcal{C}(n) : n \geq 0\}$. For example, suppose that Q is controlled using some policy, and we wish to estimate the finite-horizon value function,

$$h(x) = \mathsf{E}_x \left[\sum_{t=0}^{T-1} c(Q(t)) \right], \qquad x \in \mathsf{X},$$

with $T > 1$ some fixed integer. We can then simulate an independent sequence of realizations of the network $\{Q^n : n \geq 0\}$, all initialized with $Q^n(0) = x$, and compute

$$\mathcal{C}(n) := \sum_{t=0}^{T-1} c(Q^n(t)), \qquad n \geq 0.$$

Applying the Strong Law of Large Numbers for i.i.d. sequences gives

$$\eta(n) := \frac{1}{n} \sum_{t=0}^{n-1} \mathcal{C}(t) \to h(x), \qquad n \to \infty, \text{ a.s.}$$

This section is devoted to evaluation of these estimates.

11.1.1 Central Limit Theorem

Provided $\{\mathcal{C}(n)\}$ has a second moment, the Central Limit Theorem (11.8) holds for $\{\eta(n)\}$, and the asymptotic variance coincides with the ordinary variance,

$$\sigma_{\text{CLT}}^2 = \mathsf{Var}(\mathcal{C}(0)).$$

For example, choosing $\varepsilon = 2\sigma_{\text{CLT}}$ in (11.8) gives the approximation

$$\mathsf{P}\{|\eta(n) - \eta| \geq 2\sigma_{\text{CLT}} n^{-\frac{1}{2}}\} \approx \mathsf{P}\{|X^\infty| \geq 2\} \approx 0.05.$$

In other words, the interval $[\eta(n) - 2\sigma_{\text{CLT}} n^{-\frac{1}{2}}, \eta(n) + 2\sigma_{\text{CLT}} n^{-\frac{1}{2}}]$ is approximately equal to a 95% confidence interval for η.

However, since the variance is rarely known it too must be estimated, typically via Monte Carlo,

$$\sigma_{\text{CLT}}^2(n) := \frac{1}{n}\sum_{t=0}^{n-1}[\mathcal{C}(t)]^2 - \left[\frac{1}{n}\sum_{t=0}^{n-1}\mathcal{C}(t)\right]^2.$$

The Strong Law of Large Numbers for i.i.d. sequences then gives

$$\lim_{n\to\infty}\sigma_{\text{CLT}}^2(n) = \mathsf{E}[\mathcal{C}^2(0)] - \mathsf{E}[\mathcal{C}(0)]^2 = \mathsf{Var}(\mathcal{C}(0)).$$

Based on this result we obtain the following extension of the CLT. Proposition 11.1.1 confirms that the interval $[\eta(n) - 2\sigma_{\text{CLT}}(n)n^{-\frac{1}{2}}, \eta(n) + 2\sigma_{\text{CLT}}(n)n^{-\frac{1}{2}}]$ is approximately equal to a 95% confidence interval for η when n is large.

Proposition 11.1.1. *Suppose that $\{\mathcal{C}(n) : n \geq 0\}$ is i.i.d. with a finite second moment. Then the two-dimensional process $\{(\sqrt{n}(\eta(n) - \eta), \sigma_{\text{CLT}}^2(n)) : n \geq 1\}$ converges in distribution to $(\sigma_{\text{CLT}}X^\infty, \sigma_{\text{CLT}}^2)$, where X^∞ has a unit-mean Gaussian distribution, and σ_{CLT}^2 is the common variance of $\{\mathcal{C}(n) : n \geq 0\}$.*

11.1.2 Large deviations

The most common approach to obtaining PAC bounds of the form (11.5) is *Chernoff's bound*: For any $\beta > 0$, $r > \eta$,

$$\mathsf{P}\{\eta(n) \geq r\} = \mathsf{P}\{\exp(n\beta\eta(n)) \geq \exp(rn\beta)\} \leq \frac{\mathsf{E}[\exp(n\beta\eta(n))]}{\exp(rn\beta)}. \qquad (11.14)$$

Chernoff's bound is precisely Markov's inequality applied to the random variable $\exp(n\beta\eta(n))$. Under our assumption that the sequence is i.i.d. we obtain from (11.14)

$$\mathsf{P}\{\eta(n) \geq r\} \leq \exp\big(-n(r\beta - \Lambda(\beta))\big), \qquad n \geq 1,\ \beta > 0, \qquad (11.15)$$

where Λ denotes the log-moment-generating function,

$$\Lambda(\beta) := \log \mathsf{E}[\exp(\beta\mathcal{C}(0))], \qquad \beta \in \mathbb{R}. \qquad (11.16)$$

The log-moment-generating function (MGF) was introduced in our treatment of transient events in the single-server queue in Chapter 3.

To obtain useful bounds based on (11.15) we assume that $\Lambda(\beta)$ is finite valued in a neighborhood of the origin. Note that $\Lambda(0) = 0$, and Jensen's inequality gives the lower bound

$$\Lambda(\beta) \geq \log\big(\exp \mathsf{E}[\beta\mathcal{C}(0)]\big) = \beta\eta, \qquad \beta \in \mathbb{R}.$$

This implies that $\Lambda'(0) = \eta$, and we can also obtain formulae for higher order moments.

Proposition 11.1.2. *Suppose that $\Lambda(\beta)$ is finite valued in a neighborhood of the origin. Then*

$$\Lambda'(0) = \eta, \quad \text{and} \quad \Lambda''(0) = \sigma_{\text{CLT}}^2.$$

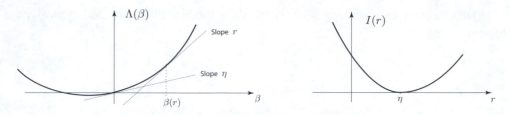

Figure 11.4. Log-MGF and rate function.

Returning to (11.15), the best bound is obtained on maximizing the exponent over all $\beta > 0$. Consider

$$I(r) = \sup_{\beta \in \mathbb{R}} \left(r\beta - \Lambda(\beta) \right), \qquad r \in \mathbb{R}. \tag{11.17}$$

The function $\Lambda \colon \mathbb{R} \to \mathbb{R} \cup \{+\infty\}$ is convex, and the rate function I is precisely its convex dual. When the supremum in (11.17) is attained at some $\beta(r) \in \mathbb{R}$ we then have

$$\frac{d}{d\beta}\left(r\beta - \Lambda(\beta) \right)\Big|_{\beta=\beta(r)} = 0$$

or $r = \Lambda'(\beta(r))$. Convexity of Λ implies that $\beta(r) \geq 0$ whenever $r > \eta$. The relationship between Λ and I is illustrated in Fig. 11.4.

Applying Chernoff's bound (11.15) we obtain the following bounds and asymptotics. Part (iii) of Proposition 11.1.3 follows from Proposition 11.1.2.

Proposition 11.1.3. *Suppose that $\Lambda(\beta)$ is finite valued in a neighborhood of the origin, and that the variance $\mathsf{Var}(\mathcal{C}(0))$ is nonzero and finite. Then*

(i) *For each $n \geq 1$,*

$$\mathsf{P}\{\eta(n) \geq r\} \leq \exp(-nI(r)), \qquad r > \eta$$
$$\mathsf{P}\{\eta(n) \leq r\} \leq \exp(-nI(r)), \qquad r < \eta.$$

(ii) *The LDP (11.10a), (11.10b) also holds for $\{\eta(n) : n \geq 1\}$.*

(iii) *The approximation (11.11) holds.*

By extending Proposition 11.1.2, or through some other technique, we can frequently obtain upper bounds on Λ, and then lower bounds on the rate function I. For example, Proposition 3.6.1 and Theorem 10.2.5 establish bounds on Λ for bounded independent sequences, and martingales with bounded increments.

Suppose that the following holds for some $\beta_0 > 0$ and $\overline{\sigma}_{\text{CLT}} \geq \sigma_{\text{CLT}}$:

$$\Lambda(\beta) \leq \overline{\Lambda}(\beta) := \eta\beta + \tfrac{1}{2}\overline{\sigma}_{\text{CLT}}^2\beta^2, \qquad |\beta| \leq \beta_0. \tag{11.18}$$

Proposition 11.1.2 implies that for *any* $\overline{\sigma}_{\text{CLT}} > \sigma_{\text{CLT}}$ there exists $\beta_0 > 0$ such that (11.18) holds. The following function of r provides a lower bound on the rate function defined in (11.17):

$$\underline{I}(r) = \max_{|\beta| \leq \beta_0} \left(r\beta - \overline{\Lambda}(\beta) \right), \qquad r \in \mathbb{R}. \tag{11.19}$$

Proposition 11.1.4. *We have $I(r) \geq \underline{I}(r)$ whenever $|r - \eta| \leq \overline{\sigma}_{\mathrm{CLT}}^2 \beta_0$, where*

$$\underline{I}(r) := \frac{1}{2} \frac{(r - \eta)^2}{\overline{\sigma}_{\mathrm{CLT}}^2}. \tag{11.20}$$

Proof. When the max in (11.19) is attained at $\beta \in (-\beta_0, \beta_0)$ we must have $r = \overline{\Lambda}'(\beta(r)) = \eta + \overline{\sigma}_{\mathrm{CLT}}^2 \beta$, which gives $I(r) \geq \underline{I}(r)$. $\square$

We are now in a position to compare the approximate bounds obtained from the CLT and the precise bound obtained from the LDP. Suppose that r is fixed and we wish to choose $n \geq 1$ so that (11.5) holds with $\delta = 0.05$:

$$\mathsf{P}\{\eta(n) \in [\eta - \varepsilon, \eta + \varepsilon]^c\} \leq 0.05. \tag{11.21}$$

That is, we seek values of n so that $[\eta(n) - \varepsilon, \eta(n) + \varepsilon]$ is a 95% confidence interval.

Recall that the CLT implies that $[\eta(n) - 2\sigma_{\mathrm{CLT}} n^{-\frac{1}{2}}, \eta(n) + 2\sigma_{\mathrm{CLT}} n^{-\frac{1}{2}}]$ is approximately equal to a 95% confidence interval for η when n is large. Hence if σ_{CLT}^2 is known we can choose $2\sigma_{\mathrm{CLT}} n^{-\frac{1}{2}} \leq \varepsilon$, or

$$n \geq 4 \frac{\sigma_{\mathrm{CLT}}^2}{\varepsilon^2}. \tag{11.22}$$

If the variance is not known, then we can modify (11.22) by introducing an appropriate bound or estimate for σ_{CLT}^2.

We now consider the LDP: Applying (11.20) together with Proposition 11.1.3 we obtain for sufficiently small $\varepsilon > 0$

$$\mathsf{P}\{\eta(n) \in [\eta - \varepsilon, \eta + \varepsilon]^c\} \leq \Big(\exp\big(-nI(\eta + \varepsilon)\big) + \exp\big(-nI(\eta - \varepsilon)\big) \Big)$$

$$\leq 2\exp\Big(-n\frac{1}{2}\frac{\varepsilon^2}{\overline{\sigma}_{\mathrm{CLT}}^2}\Big).$$

To ensure that the right-hand side is no greater than 0.05 we require

$$n \geq -\log\big(0.05/2\big)\Big(\frac{1}{2}\frac{\varepsilon^2}{\overline{\sigma}_{\mathrm{CLT}}^2}\Big)^{-1} \approx 7.4\frac{\overline{\sigma}_{\mathrm{CLT}}^2}{\varepsilon^2}. \tag{11.23}$$

This gives a slightly larger value of n than predicted in (11.22), but for n satisfying (11.23) we are *assured* that the desired bound (11.21) holds.

11.2 Asymptotic theory for Markov models

Recall that in the previous section all of the results apply only to i.i.d. models. In turning to Markov models we take $\mathcal{C}(t) = c(X(t))$ where X is a Markov chain, and $c \colon \mathsf{X} \to \mathbb{R}$ a function with finite steady-state mean $\eta = \pi(c)$.

11.2.1 Central Limit Theorem

When $\{\mathcal{C}(n) : n \geq 0\}$ is not i.i.d. then the asymptotic variance does not coincide with the variance of an individual $c(X(t))$. Denote the auto-correlation function by

$$r(j) = \mathsf{E}_\pi\Big[\tilde{c}(X(i))\tilde{c}(X(i+j))\Big], \qquad i, j \geq 0,$$

with $\tilde{c} := c - \eta$. Consideration of fixed n in (11.6) gives

$$\mathsf{E}_\pi\left[\left(\sqrt{n}(\eta(n) - \eta)\right)^2\right] = \frac{1}{n}\sum_{i=0}^{n-1}\sum_{j=0}^{n-1} r(j - i),$$

and this then leads to the following representation:

$$\sigma_{\text{CLT}}^2 = \sum_{i=-\infty}^{\infty} r(i). \tag{11.24}$$

However, the existence of a CLT and the validity of (11.24) require conditions on X somewhat stronger than (V3). For network models we state a general condition based on a Lyapunov function. Recall that $V_m(x) = 1 + |x|^m$, $x \in \mathbb{R}^\ell$, for $m \in \mathbb{Z}_+$.

Definition 11.2.1 (CLT Drift Condition for networks). X is a Markov chain on $\mathsf{X} \subset \mathbb{R}_+^\ell$, and $\mathcal{D}$ denotes its generator (8.1). There exists a function $V : \mathbb{R}_+^\ell \to \mathbb{R}_+$ satisfying $V \in L_\infty^{V_5}$ and for some $b > 0$,

$$\mathcal{D}V \leq -V_4 + b. \tag{11.25}$$

∎

Under the CLT Drift Condition together with 0-irreducibility the queue-length process Q possesses a unique stationary distribution π with finite fourth moment, $\mathsf{E}_\pi[\|Q(0)\|^4] < b$. The CLT Drift Condition is satisfied for the M/M/1 queue as long as $\rho := \alpha/\mu < 1$: We can take $V(x) = b_0 x^5$, with b_0 a sufficiently large constant.

Recall that the irreducibility condition in Theorem 11.2.2 is satisfied for the CRW scheduling model (8.24) under the assumptions of Theorem 10.0.2.

Theorem 11.2.2. *Suppose that X is a 0-irreducible Markov chain on $\mathsf{X} \subset \mathbb{Z}_+^\ell$, and that the CLT Drift Condition holds. Then, for any function $c \in L_\infty^{V_1}$:*

(i) *$\eta(n) \to \eta$ almost surely, and the CLT (11.8) holds.*
(ii) *There exists a solution to Poisson's equation (8.2) satisfying $h \in L_\infty^{V_2}$.*
(iii) *The asymptotic variance can be expressed as the limit (11.6), the infinite sum (11.24), or*

$$\sigma_{\text{CLT}}^2 = \pi(h^2 - (Ph)^2), \tag{11.26}$$

with h being given in (ii).

Proof. The CLT is given as Theorem 17.2.2 in [368]. Section 17.4.3 of [368] contains various representations of the asymptotic variance, including (11.26).

A bound on the solution to Poisson's equation is obtained as follows: Let $V_c = V^{2/5} \in L_\infty^{V_2}$. An application of Jensen's inequality gives, under (11.25),

$$PV_c = PV^{2/5} \leq (PV)^{2/5} \leq (V - V_4 + b)^{2/5}.$$

By concavity of the function $x^{2/5}$ we obtain

$$(V - V_4 + b)^{2/5} \leq V^{2/5} + (2/5)V^{-3/5}[-V_4 + b],$$

which implies the bound

$$\mathcal{D}V_c \leq -\frac{2}{5}\frac{V_4}{V^{3/5}} + b.$$

We have $V^{3/5} \in L_\infty^{V_3}$, which gives $\mathcal{D}V_c \leq -\varepsilon V_1 + b$ for some $\varepsilon > 0$. It follows from Proposition 8.1.6 that a solution to Poisson's equation exists with $h \in L_\infty^{V_c}$.

The representation (11.24) follows from (11.26) by considering the specific solution to Poisson's equation defined as the infinite sum

$$h(x) = \sum_{t=0}^{\infty} \mathsf{E}_x[\tilde{c}(X(t))], \qquad x \in \mathsf{X}.$$

By Poisson's equation we have

$$h^2 - (Ph)^2 = h^2 - (h - \tilde{c})^2 = 2\tilde{c}h - \tilde{c}^2.$$

Consequently, $\sigma_{\mathrm{CLT}}^2 \leq \pi(|\tilde{c}h|) < \infty$ by (11.26) and finiteness of $\pi(V_4)$.

Moreover, by stationarity and the assumed form for h we have

$$\pi(2\tilde{c}h - \tilde{c}^2) = -\mathsf{E}[\tilde{c}(X(0))\tilde{c}(X(0))] + 2\sum_{t=0}^{\infty}\mathsf{E}[\tilde{c}(X(0))\tilde{c}(X(t))]$$

$$= -\mathsf{E}[\tilde{c}(X(0))\tilde{c}(X(0))] + \sum_{t=0}^{\infty}\mathsf{E}[\tilde{c}(X(0))\tilde{c}(X(t))]$$

$$+ \sum_{t=0}^{\infty}\mathsf{E}[\tilde{c}(X(0))\tilde{c}(X(-t))].$$

It follows that (11.26) implies (11.24). $\qquad\square$

11.2.2 Large deviations

There is no space here to provide a detailed presentation of sample-path limit theory for Markov chains. Here we provide some basic theory on LDPs and some key ideas underlying the foundations of this theory.

Introducing the log-MGF,

$$\Lambda_{n,x}(\beta) := \frac{1}{n}\log\mathsf{E}_x[\exp(n\beta\eta(n))],$$

we arrive at the sequence of bounds, exactly as in the i.i.d. case,

$$\frac{1}{n}\log\mathsf{P}_x\{\eta(n) \geq r\} \leq \Lambda_{n,x}(\beta) - r\beta, \qquad \beta > 0, n \geq 1.$$

Recall that this is a parameterized bound since it depends upon the particular constant $\beta > 0$. On infimizing over all such β we obtain the best possible bound. Define for each n, x and each $r > \eta$,

$$I_{n,x}(r) = \sup_{\beta > 0}\left(r\beta - \Lambda_{n,x}(\beta)\right)$$

so that

$$\frac{1}{n} \log \mathsf{P}_x\{\eta(n) \geq r\} \leq -I_{n,x}(r), \qquad n \geq 1.$$

Under certain conditions on the function c and the Markov chain X we can establish convergence,

$$\Lambda(\beta) := \lim_{n\to\infty} \Lambda_{n,x}(\beta), \qquad (11.27)$$

where the limit is independent of x. Under further conditions the LDP limits (11.10b), (11.10b) hold with rate function given by (11.17).

We are left with several questions. Firstly, what is the form of Λ? When is it finite? Exactly as in the i.i.d. case, a lower bound can be obtained using Jensen's inequality:

Proposition 11.2.3. *Suppose that X is c-regular and that the limit (11.27) holds. Then $\Lambda(\beta) \geq \beta\eta$.*

Proof. Since the exponential is convex we have by Jensen's inequality

$$\Lambda_{n,x}(\beta) \geq \frac{1}{n}\left(\mathsf{E}_x[n\beta\eta(n)]\right), \qquad n \geq 1.$$

The ergodic theorem in Theorem A.5.4 then implies that the right-hand side converges to $\beta\eta$. $\qquad\square$

We can obtain a representation for Λ in analogy with the following representation of the steady-state mean for an x^*-irreducible Markov chain: Recall that $\eta = \pi(c)$ is the solution to the equation

$$\mathsf{E}_{x^*}\left[\sum_{t=0}^{\tau_{x^*}-1} \left(c(X(t)) - \eta\right)\right] = 0.$$

A similar representation holds for the log-MGF under general conditions. For a given $\beta \in \mathbb{R}$, we seek a solution $\Lambda \in \mathbb{R}$ to the identity

$$\log \mathsf{E}_{x^*}\left[\exp\left(\sum_{t=0}^{\tau_{x^*}-1} \left(\beta c(X(t)) - \Lambda\right)\right)\right] = 0. \qquad (11.28)$$

When c is coercive or near-monotone (see the introduction in Chapter 8), then typically there is a unique real number $\Lambda(\beta)$ solving (11.28) *for negative β*. Theorem 11.2.4 is useful for obtaining bounds on the lower error probability (11.10b). It is also a component of our analysis of shadow functions in Section 11.4.

Theorem 11.2.4. *Suppose that (V3) holds with V everywhere finite, and that $c\colon \mathsf{X} \to \mathbb{R}_+$ satisfies $c \in L_\infty^f$. Suppose moreover that the set $S_c(r)$ is finite for some $r > \eta$. Then:*

(i) *The limit (11.27) holds for each initial condition $x \in \mathsf{X}$, and nonpositive β.*

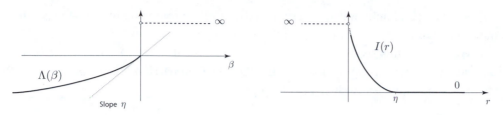

Figure 11.5. Log-moment-generating function and rate function for the CRW queue with $c(x) \equiv x$. The moment-generating function is infinite for positive β, and the rate function is zero for $r > \eta$.

(ii) *There exists $\eta_- < \eta$ such that the supremum in (11.17) is achieved by some $\beta(r) < 0$ whenever $r \in (\eta_-, \eta)$, and*

$$I(r) = r\beta(r) - \Lambda(\beta(r)) > 0, \qquad \eta_- < r < \eta.$$

(iii) *The LDP (11.10b) holds for each initial condition $x \in \mathsf{X}$ and each $r \in (\eta_-, \eta)$, with rate function given in (ii).*

Proof. For fixed $\beta < 0$, the right-hand side of (11.28) is an analytic function of Λ that is strictly negative when $\Lambda = 0$, and diverges to $+\infty$ as $\Lambda \downarrow -\infty$. It follows that (11.28) admits a unique solution $\Lambda(\beta) < 0$.

The limit (11.27) and the LDP are established in [364]. $\qquad\square$

To obtain a useful limit for $\beta > 0$ requires further assumptions. The following result is established in [310]. Geometric ergodicity is defined in Sections 8.2.3 and A.5.3.

Proposition 11.2.5. *Suppose that $\mathbf{X}$ is geometrically ergodic, and that c is a bounded function. Then, there exists $\bar{\beta} > 0$ such that for $\beta \in [-\bar{\beta}, \bar{\beta}]$ Eq. (11.28) admits a unique solution $\Lambda(\beta)$, and the limit (11.27) holds for each initial condition $x \in \mathsf{X}$. Moreover, there exists $\bar{\varepsilon} > 0$ such that the pair of limits (11.10b), (11.10b) hold for each $\varepsilon \in (0, \bar{\varepsilon}]$.*

For the M/M/1 queue we have seen that $\mathbf{Q}$ is geometrically ergodic when $\rho < 1$, so that the LDP does hold for bounded functions. For a coercive function such as $c(x) = x$ or $c(x) = e^{\beta x}$ we obtain from Proposition 11.2.4 the lower LDP (11.10b). However, the upper LDP (11.10a) does not hold for either of these functions, regardless of load: As illustrated in Fig. 11.5, the MGF is infinite for positive β, and $I(r) = 0$ for $r > \eta$.

These conclusions are explained in fuller detail and further developed in the next section where we consider the CLT and LDP for the single-server queue.

11.3 The single-server queue

Here we obtain bounds on the asymptotic variance and large deviation results for the single-server queue. These results are largely negative: The asymptotic variance is very large when $\rho \approx 1$, and the upper LDP (11.10a) holds with $I(r) = 0$ for all $r \geq \eta$ when

c is increasing and unbounded. These negative results provide motivation to search for improved estimators in Section 11.4.

The proofs in this section are unfortunately technical in appearance. They serve to illustrate more refined applications of the MGF, the Comparison Theorem, and the large deviations techniques introduced in Chapter 3.

11.3.1 Asymptotic variance

We consider in Proposition 11.3.1 a family of models parameterized by a variable $\theta \in [0, \theta_0]$ with $0 < \theta_0 < 1$. It is assumed that $\theta = (\mu - \alpha)$, so that $\rho < 1$ for $\theta > 0$, and we define

$$m_n(\theta) = \mathsf{E}[(S(1) - A(1))^n]^{1/n}$$
$$m_{An}(\theta) = \mathsf{E}[A(1)^n]^{1/n}, \qquad n = 1, 2, 3, 4. \tag{11.29}$$

Recall that the asymptotic variance given in Theorem 11.2.2 can be expressed $\sigma_{\text{CLT}}^2 = \pi(h^2 - (Ph)^2) = 2\pi(\tilde{c}h) - \pi(\tilde{c}^2)$. It then follows from the formula for h in Theorem 3.0.1 that the asymptotic variance admits the following approximation for $\theta \approx 0$:

$$\sigma_{\text{CLT}}^2 = 2\pi(J^*\tilde{c}) + O\big((\mu - \alpha)^{-3}\big)$$
$$= \theta^{-1}\mathsf{E}_\pi[Q(0)^3 - \eta Q(0)^2] + O\big(\theta^{-3}\big), \tag{11.30}$$

where J^* denotes the fluid value function, and the "big-oh" notation is used to represent the sum of terms whose growth rate is bounded by a constant times the term in parentheses as $\theta \downarrow 0$. Equation (11.30) is the basis of the asymptotic formulae that follow.

Proposition 11.3.1 gives an asymptotic expression for the asymptotic variance σ_{CLT}^2 in the single-server queue with $c(x) \equiv x$. It is finite when $A(1)$ has a finite fourth moment, but the magnitude is of order $(1 - \rho)^{-4}$ for $\rho \approx 1$ (equivalently, $\theta \approx 0$.) It also grows dramatically with variability: of order $[\mathsf{Var}(A(1) - S(1))]^3$.

The proof is based on Proposition 11.3.3 that follows, which provides the following approximations for the first three moments for $\theta \approx 0$,[2]

$$\eta := \mathsf{E}_\pi[Q(0)] = \tfrac{1}{2}\frac{m_2^2}{\mu - \alpha} + O(1)$$
$$\mathsf{E}_\pi[Q(0)^2] = 2\eta^2 + O\big(\theta^{-1}\big) \tag{11.31}$$
$$\mathsf{E}_\pi[Q(0)^3] = 6\eta^3 + O\big(\theta^{-2}\big).$$

Proposition 11.3.1. *Consider the family of CRW models of the form (3.1), parameterized by* $\theta \in [0, \theta_0]$. *Suppose that* $\mu(\theta) - \alpha(\theta) = \theta$, $m_2(\theta) > 0$, $m_4(\theta) < \infty$ *for*

---

[2] For readers comfortable with the RBM model, observe that the second and third moments of this continuous-time model are exactly as given in (11.31), without the correction terms.

each θ, and that m_n is continuous as a function of θ on $[0, \theta_0]$ for $n = 1, \dots, 4$. Then, $\sigma^2_{\text{CLT}}(\theta) < \infty$ for each $\theta \in (0, \theta_0]$, and we have

$$\lim_{\theta \downarrow 0} \left(\theta^4 \sigma^2_{\text{CLT}}(\theta) \right) = \tfrac{1}{2} [\mathsf{Var}(A(1) - S(1))]^3$$

where $\mathsf{Var}(A(1) - S(1)) = m_2^2(0)$.

Proof. Combining (11.30) and (11.31) gives

$$\sigma^2_{\text{CLT}} = \frac{1}{\theta} [(6\eta^3) - \eta(2\eta^2)] + O(\theta^{-3}).$$

Hence we can compute the limit

$$\lim_{\theta \downarrow 0} \theta^4 \sigma^2_{\text{CLT}}(\theta) = \lim_{\theta \downarrow 0} \theta^4 \frac{1}{\theta} \left(4\eta^3 \right) = \lim_{\theta \downarrow 0} \theta^3 \left(4 \left(\frac{1}{2} \frac{m_2^2(\theta)}{\theta} \right)^3 \right) = \tfrac{1}{2} (m_2^2(0))^3. \qquad \square$$

To establish (11.31) we first consider the M/M/1 queue in which the moments can be computed based on the MGF,

$$m(\beta) := \mathsf{E}_\pi[e^{\beta Q(0)}] = \sum_{i=0}^{\infty} \pi(i) e^{\beta i}, \qquad \beta \in \mathbb{R}.$$

We have for each $n \geq 1$

$$\mathsf{E}_\pi[Q(0)^n] = \left(\frac{d^n}{d\beta^n} m(\beta) \right) \Big|_{\beta = 0}.$$

The invariant measure is given in (3.43) by $\pi(x) = (1 - \rho)\rho^x$, $x \geq 0$, which implies that

$$m(\beta) = \frac{1 - \rho}{1 - \rho e^\beta}, \qquad \beta \in (-\infty, |\log(\rho)|). \tag{11.32}$$

Proposition 11.3.2. *All moments of the M/M/1 queue are finite when $\rho < 1$, and the first three can be expressed*

$$\eta := \mathsf{E}_\pi[Q(0)] = \frac{\rho}{1 - \rho}$$

$$\mathsf{E}_\pi[Q(0)^2] = 2\eta^2 + \eta \tag{11.33}$$

$$\mathsf{E}_\pi[Q(0)^3] = 6\eta^3 + 5\eta^2 + \eta.$$

Proof. The first moment was given previously in Eq. (3.45).

Due to the simple form (11.32), the derivatives of m are easily computed. For example,

$$\frac{d}{d\beta} m(\beta) = \frac{1 - \rho}{(1 - \rho e^\beta)^2} (\rho e^\beta)$$

$$\frac{d^2}{d\beta^2} m(\beta) = 2 \frac{1 - \rho}{(1 - \rho e^\beta)^3} (\rho e^\beta)^2 + \frac{1 - \rho}{(1 - \rho e^\beta)^2} (\rho e^\beta)$$

and the third derivative is given by a similar, though more complex expression. Evaluating these three derivatives at $\beta = 0$ gives the first three moments. $\square$

We now generalize Proposition 11.3.2 to the CRW model via the Comparison Theorem.

Proposition 11.3.3. *Under the assumptions of Proposition 11.3.1, the first three steady-state moments are finite, and can be computed recursively as follows:*

$$\eta := \mathsf{E}_\pi[Q(0)] = \tfrac{1}{2}\frac{\sigma^2}{\mu - \alpha},$$
$$\eta_2^2 := \mathsf{E}_\pi[Q(0)^2] = (3(\mu - \alpha))^{-1}\left(3m_2^2\eta + \rho m_3^3 + (1 - \rho)m_{A3}^3\right)$$
$$\eta_3^3 := \mathsf{E}_\pi[Q(0)^3] = (4(\mu - \alpha))^{-1}\left(6m_2^2\eta_2^2 + 6m_3^3\eta + \rho m_4^4 + (1 - \rho)m_{A4}^4\right)$$

(11.34)

where σ^2 is defined in (3.9).

Proof. The first moment is given in Theorem 3.0.1. Recall that the proof was based upon an application of the refined Comparison Theorem given in Proposition 8.2.5.

Computation of the second and third moments is based on similar arguments.

To compute the second moment we first consider for any $t \geq 0$, $x \in \mathbb{Z}_+$ the conditional expectation

$$\mathsf{E}[Q(t + 1)^3 \mid Q(t) = x,\ U(t) = u]$$
$$= \mathsf{E}[x^3 + 3(-S(t + 1)u + A(t + 1))x^2$$
$$+ 3(-S(t + 1)u + A(t + 1))^2 x + (-S(t + 1)u + A(t + 1))^3]$$
$$= x^3 - 3(\mu - \alpha)x^2 + 3m_2^2 x + m_3^3 u + m_{A3}^3(1 - u).$$

We have $\mathsf{P}_\pi\{Q(0) = 0\} = \mathsf{P}_\pi\{U(0) = 0\} = 1 - \rho$ by Proposition 3.4.4, and hence Proposition 8.2.5 gives

$$0 = -3(\mu - \alpha)\eta_2^2 + 3m_2^2\eta + \rho m_3^3 + (1 - \rho)m_{A3}^3.$$

We conclude that second moment is finite, and expressed as claimed in (11.34).

The third moment is again computed through computation of the conditional expectation of one higher polynomial moment:

$$\mathsf{E}[Q(t + 1)^4 \mid Q(t) = x,\ U(t) = u]$$
$$= \mathsf{E}[x^4 + 4(-S(t + 1)u + A(t + 1))x^3 + 6(-S(t + 1)u + A(t + 1))^2 x^2$$
$$+ 4(-S(t + 1)u + A(t + 1))^3 x + (-S(t + 1)u + A(t + 1))^4]$$
$$= x^4 + 4(\mu - \alpha)x^3 + 6m_2^2\eta_2^2 + 6m_3^3\eta + + m_4^4 u + m_{A4}^4(1 - u).$$

Hence an expression for η_3^3 again follows from Proposition 8.2.5. $\square$

11.3.2 Large deviations

Suppose that Q is a stationary version of the M/M/1 queue with $\rho = \alpha/\mu < 1$. For a given $\delta > 0$, and time $T_0 > 0$, consider the event $\{Q(T_0) \geq \delta T\}$ where $T \geq 1$ is a large constant. The "typical behavior" of Q given this event is illustrated in Fig. 3.2.

This observation generalizes in an obvious way to the single-server queue (3.1) under the assumptions of Theorem 11.0.1, and this leads to the following conclusions:

Proposition 11.3.4. *Consider the single-server queue (3.1) with $\rho < 1$, $0 < \mathsf{E}[A(1)^2] < \infty$, and hence the queue has finite steady-state mean η.*

(i) *The upper error-probability decays subexponentially: For each $r > \eta$,*

$$\lim_{T\to\infty} \frac{1}{T} \log \mathsf{P}\{\eta(T) \geq r\} = 0. \tag{11.35}$$

(ii) *The following lower bound holds for the M/M/1 queue: For each $\beta \in (0, (\mu - \alpha)/4)$,*

$$\liminf_{T\to\infty} \frac{1}{T} \log \mathsf{P}\{\eta(T) \geq \beta T\} \geq \log(\rho)\sqrt{(\mu - \alpha)\beta}. \tag{11.36}$$

Proof. We consider first the M/M/1 queue. For any δ satisfying

$$0 < 2\frac{\delta}{\mu - \alpha} < 1,$$

the area under the "tent" shown in Fig. 3.2 is

$$\text{Area} = \frac{\delta^2}{\mu - \alpha} T^2.$$

On the event $\mathcal{A} := \{Q(t) \text{ lies above the tent for } 0 \leq t \leq T\}$ we have $T\eta(T) \geq \text{Area}$. The probability of the event $\mathcal{A}$ is of order $\rho^{\delta T}$ by Proposition 3.5.2:

$$\lim_{T\to\infty} \frac{1}{T} \log \mathsf{P}\{\mathcal{A}\} = \delta \log(\rho),$$

from which we obtain the asymptotic bound,

$$\lim_{T\to\infty} \frac{1}{T} \log \mathsf{P}\{\eta(T) \geq \frac{\delta^2}{\mu - \alpha} T\} \geq \delta \log(\rho).$$

Letting $\beta = \delta^2/(\mu - \alpha)$ we obtain (ii).

To see (11.35) we apply (11.36) to obtain the following bound for any pair of constants $r > \eta$, $\beta \in (0, (\mu - \alpha)/4)$:

$$\liminf_{T\to\infty} \frac{1}{T} \log \mathsf{P}\{\eta(T) \geq r\} \geq \liminf_{T\to\infty} \frac{1}{T} \log \mathsf{P}\{\eta(T) \geq \beta T\} \geq \log(\rho)\sqrt{(\mu - \alpha)\beta}.$$

Letting $\beta \downarrow 0$ completes the proof of (i) for the M/M/1 queue.

The proof of (i) for the general model is similar, based on Proposition 3.5.4: Only the geometry of the "tent" changes, so that we again obtain, for some $I_0 > 0$ and all $\beta > 0$ sufficiently small,

$$\liminf_{T \to \infty} \frac{1}{T} \log \mathsf{P}\{\eta(T) \geq \beta T\} \geq -I_0 \sqrt{\beta}. \tag{11.37}$$

As in the M/M/1 queue, this implies (11.35). □

We now turn to generalized estimators that can be designed to provide upper *and* lower LDP asymptotics.

11.4 Control variates and shadow functions

We have noted that the effectiveness of the estimator (11.1) depends in part on the magnitude of the asymptotic variance appearing in (11.7). Under the assumptions of Theorem 11.2.2 the asymptotic variance is finite, and can be expressed in terms of Poisson's equation – see (11.26). Unfortunately, as we have seen in Section 11.3.1, the asymptotic variance typically grows very quickly with network load. These results and the numerical results shown in Fig. 11.3 provide ample motivation to find methods for constructing reduced-variance estimators for network models. In this section we explain these numerical results, and introduce several techniques for constructing effective shadow functions.

Suppose that the shadow function in the smoothed estimator (11.12) is chosen so that $\Delta_\psi = -c' + \eta'$, where $c' \approx c$ and η' is a constant satisfying $\eta' = \pi(c') \approx \eta$. Then, the smoothed estimator can be expressed

$$\eta_\psi^s(n) = \eta + \frac{1}{n} \sum_{t=0}^{n-1} \left([c(X(t)) - c'(X(t))] - [\eta - \eta'] \right), \qquad n \geq 1.$$

If these approximations are sufficiently tight, then the estimates $\{\eta_\psi^s(n)\}$ will have low variance. If in fact $\Delta_\psi = -c + \eta$, this means that ψ is a solution to Poisson's equation, and in this case $\eta_\psi^s(n) = \eta$ for all n so the smoothed estimator has *zero* asymptotic variance. In general, we can expect useful variance reductions using the estimator $\eta_\psi^s(n)$ provided ψ is a sufficiently tight approximation to the solution to Poisson's equation.

11.4.1 Optimizing shadow functions

When h^θ is defined in (11.13) based on the ℓ_h functions $\{\psi_1, \ldots, \psi_{\ell_h}\}$, we let $\sigma_{\text{CLT}}^2(\theta)$ denote the asymptotic variance of $c + \Delta_{h^\theta}$ given in (11.26).

Consider first the simpler setting in which we are given two random variables $\mathcal{C}$ and $\mathcal{G}$, taking values in $\mathbb{R}$ and $\mathbb{R}^{\ell_h}$, respectively, with $\mathsf{E}[\mathcal{C}] = \eta$ and $\mathsf{E}[\mathcal{G}] = \mathbf{0} \in \mathbb{R}^{\ell_h}$. We have

$$\mathsf{Var}(\mathcal{C} - \theta^{\mathsf{T}}\mathcal{G}) = \mathsf{Var}(\mathcal{C}) - 2\theta^{\mathsf{T}}\mathsf{Cov}(\mathcal{C}, \mathcal{G}) + \theta^{\mathsf{T}}\mathsf{Cov}(\mathcal{G}, \mathcal{G})\theta,$$

and hence the gradient is given by

$$\nabla \mathsf{Var}(\mathcal{C} - \theta^{\mathsf{T}}\mathcal{G}) = 2\mathsf{Cov}(\mathcal{C}, \mathcal{G}) - 2\mathsf{Cov}(\mathcal{G}, \mathcal{G})\theta. \tag{11.38}$$

Setting this equal to zero we obtain the optimizing parameter

$$\theta^* = [\mathsf{Cov}(\mathcal{G},\mathcal{G})]^{-1}\mathsf{Cov}(\mathcal{C},\mathcal{G}). \qquad (11.39)$$

One could apply this result for the smoothed estimator using h^θ with

$$\mathcal{C} = m^{-1}\sum_{t=0}^{m-1} c(X(t)) \quad \text{and} \quad \mathcal{G} = m^{-1}\sum_{t=0}^{m-1} \Delta_{h^\theta}(X(t)). \qquad (11.40)$$

The *batch-means* method described in Section 11.4.2 is based on (11.38) using the random variables defined in (11.40).

We now construct the parameter optimizing the asymptotic variance of the smoothed estimator. The smoothed estimator is standard Monte Carlo estimation for the function $c_\theta = c + \Delta_{h^\theta}$, where Δ_{h^θ} is defined in (11.13). Under the assumptions imposed below we have $\pi(\Delta_{h^\theta}) = 0$, and hence $\pi(c_\theta) = \pi(c) = \eta$. Moreover, if h solves Poisson's equation for c, then $h - h^\theta$ solves Poisson's equation for c_θ:

$$\mathcal{D}(h - h^\theta) = -c_\theta + \eta.$$

Applying the representation (11.26) then gives

$$\sigma_{\text{CLT}}^2(\theta) = \pi\big((h - h^\theta)^2 - (Ph - Ph^\theta)^2\big). \qquad (11.41)$$

We now introduce a new inner product on the space of square-integrable functions so that we can express the optimizing value of θ in a familiar form. For any two functions g_1, g_2 write

$$\langle g_1, g_2 \rangle_{\text{CLT}} = \pi(g_1 g_2 - (Pg_1)(Pg_2)).$$

The representation (11.41) becomes

$$\sigma_{\text{CLT}}^2(\theta) = \langle h - h^\theta, h - h^\theta \rangle_{\text{CLT}}. \qquad (11.42)$$

On computing the gradient of $\sigma_{\text{CLT}}^2(\theta)$ with respect to θ and setting this equal to zero we obtain the optimal value θ^*, exactly as in the derivation of (11.39):

Proposition 11.4.1. *Suppose that* X *satisfies the conditions of Theorem 11.2.2, so that in particular (V3) holds in the form (11.25), and* $c \in L_\infty^{V_1}$. *Suppose that* $\{\psi_i : 1 \le i \le \ell_h\} \subset L_\infty^{V_2}$ *are given functions, and that the asymptotic covariance* Σ_ψ *is invertible, where*

$$\Sigma_\psi := \langle \psi, \psi \rangle_{\text{CLT}}. \qquad (11.43)$$

That is, Σ_ψ *is the* $\ell_h \times \ell_h$ *matrix whose* i, jth *entry is given by* $\langle \psi_i, \psi_j \rangle_{\text{CLT}}$. *Then, for any function* $c \in L_\infty^{V_1}$ *the vector*

$$\theta^* := \Sigma_\psi^{-1} \langle \psi, h \rangle_{\text{CLT}} \qquad (11.44)$$

is the unique value $\theta^* \in \mathbb{R}^{\ell_h}$ *minimizing the asymptotic variance given in (11.42), where* $h \in L_\infty^{V_2}$ *solves Poisson's equation for* c.

In the most common application of Proposition 11.4.1 we have a single function $\psi \colon \mathsf{X} \to \mathbb{R}$ and rewrite (11.12) as follows:

$$\eta_\psi^s(n) = \eta(n) + \frac{\theta}{n} \sum_{t=0}^{n-1} \Delta_\psi(X(t)), \qquad n \geq 1. \tag{11.45}$$

The optimal value θ^* given in the proposition is the scalar

$$\theta^* = \frac{\pi(\psi h - (P\psi)(Ph))}{\pi(\psi^2 - (P\psi)^2)}. \tag{11.46}$$

11.4.2 Batch-means method

Since the solution to Poisson's equation is not known prior to simulation, in practice it is necessary to estimate the optimizing parameter θ^* given in (11.44). A procedure based on Monte Carlo known as the *batch-means method* can be used to obtain an approximation. For simplicity, here we consider the scalar case (11.45). And, instead of estimating the parameter (11.46) minimizing asymptotic variance, we estimate the parameter θ^* that minimizes the m-step variance (introduced in (11.39)). Techniques for estimating the asymptotic variance based on TD learning are developed in Section 11.5.5.

Suppose that N batches of m observations are recorded,

$$\{X(t) : (i-1)m \leq t \leq im-1\}, \qquad 0 \leq m \leq N-1,$$

let η_c^i, η_ψ^i denote the sample means for the ith batch,

$$\eta_c^i = \frac{1}{m} \sum_{t=(i-1)m}^{im-1} c(X(t)) \quad \text{and} \quad \eta_\psi^i = \frac{1}{m} \sum_{t=(i-1)m}^{im-1} \psi(X(t)),$$

and let $\eta_c^{N,m}$, $\eta_\psi^{N,m}$ denote the overall sample means. Define

$$\Sigma_{\psi\psi}(n) = \frac{1}{N} \sum_{i=0}^{N-1} (\eta_\psi^i - \eta_\psi^{N,m})^2, \qquad \Sigma_{c\psi}(n) = \frac{1}{N} \sum_{i=0}^{N-1} (\eta_c^i - \eta_c^{N,m})(\eta_\psi^i - \eta_\psi^{N,m}).$$

We then take $\theta = \Sigma_{c\psi}(n)/\Sigma_{\psi\psi}(n)$ in the smoothed estimator (11.45). A similar approach can be applied when ψ is vector valued.

11.4.3 Shadow functions for the single-server queue

The single-server queue is a useful starting point to illustrate the construction of a shadow function for simulation.

We consider a generalization of the usual model, described as the reflected random walk,

$$Q(t+1) = [Q(t) + D(t+1)]_+, \qquad t \geq 0, \tag{11.47}$$

with $[x]_+ = \max(x, 0)$ for $x \in \mathbb{R}$, and the scalar sequence D is of the form $D(t) = -S(t) + A(t)$, $t \geq 1$, with (A, S) i.i.d. on $\mathbb{R}_+^2$. It is assumed that $\mathsf{E}[D(t)^2] < \infty$ and $\mathsf{E}[D(t)] < 0$. It follows that a unique invariant measure π exists, and that the steady-state mean $\eta = \pi(c)$ is finite, where $c(x) \equiv x$. This follows from Theorem A.2.3, following the proof of Theorem 3.0.1.

We do not assume that $\mathsf{E}[A(t)^4] < \infty$, so the asymptotic variance may be infinite (see Section 11.3). Moreover, Q may not be geometrically ergodic since we do not assume that the distribution of $A(t)$ has a geometric tail.

However, the standard Monte Carlo estimator of η satisfies the assumptions of Theorem 11.2.4, so the LDP asymptotic (11.10b) holds. To obtain an estimator satisfying a two-sided LDP we introduce a shadow function. Let $\psi = J = \frac{1}{2}(\mu - \alpha)^{-1}x^2$, the fluid value function, so that

$$\Delta_J = PJ - J = -x + R(x) \tag{11.48}$$

where the function R is bounded on $\mathbb{R}_+$. Note that we do not know if $\pi(J) < \infty$ since we have not assumed that A possesses a fourth moment. Nevertheless, $\pi(\Delta_J) = 0$ by Proposition 8.2.5 under the assumption that $m_{A2}^2 < \infty$ so that $\pi(c) < \infty$.

In the following result we consider a pair of estimators of the form (11.12) with the two functions $\psi^- = \theta^- J$, $\psi^+ = \theta^+ J$, where $\theta^+ < \theta^-$ are constants. Proposition 11.4.2 follows from Theorem 11.2.4.

Proposition 11.4.2. *Consider the single-server queue (11.47) with $\mathsf{E}[A(1)^2] < \infty$ and $\rho < 1$. Let $c(x) = x$ for $x \in \mathbb{Z}_+$, fix two parameters $\theta_+ < 1$ and $\theta_- > 1$, and define the pair of estimators $\{\eta^-(n), \eta^+(n)\}$ using the shadow functions $\Delta_{\psi^-}, \Delta_{\psi^+}$, respectively. Then there exist a pair of convex functions $\{I_-, I_+\}$ on $\mathbb{R}$ and constants $\eta_- < \eta < \eta_+$ such that for each initial condition $x \in \mathsf{X}$,*

$$\lim_{n \to \infty} \frac{1}{n} \log \mathsf{P}_x\{\eta^+(n) \leq r\} = -I_+(r) < 0, \quad r \in (\eta_-, \eta)$$

$$\lim_{n \to \infty} \frac{1}{n} \log \mathsf{P}_x\{\eta^-(n) \geq r\} = -I_-(r) < 0, \quad r \in (\eta, \eta_+).$$

An illustration of the smoothed estimator is provided in Fig. 11.6. In this simulation the sequence D was of the form $D(t) = A(t) - S(t)$, where A and S were mutually independent, i.i.d. sequences. Given nonnegative parameters μ, α, κ, the marginal distributions were of the following form:

$$\mathsf{P}\{S(t) = (1 + \kappa)\mu\} = 1 - \mathsf{P}\{S(t) = 0\} = (1 + \kappa)^{-1}$$
$$\mathsf{P}\{A(t) = (1 + \kappa)\alpha\} = 1 - \mathsf{P}\{A(t) = 0\} = (1 + \kappa)^{-1}.$$

This is a worst-case model, as formalized in Proposition 3.6.1.

The variance of $D(t)$ is $\sigma_D^2 = \sigma_A^2 + \sigma_S^2 = (\mu^2 + \alpha^2)\kappa$. The simulation results shown in Fig. 11.6 used $\mu = 4$, $\alpha = 3$, and $\kappa = 2$, so that $\mu - \alpha = 1$ and $\sigma_D^2 = 25$.

The control-variate parameter values $\theta_- = 1.05$ and $\theta_+ = 1$ were used in the construction of $\{\eta^-(n), \eta^+(n)\}$. Note that this value of θ_+ violates the strict inequality $\theta_+ > 1$ required in Proposition 11.4.2.

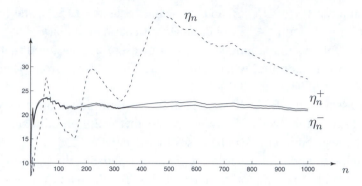

Figure 11.6. Monte Carlo estimates of $\eta := \pi(c)$ with $c(x) = x$ for $x \in \mathbb{R}_+$. The stochastic process Q is the single-server queue (11.47) with $(\mu - \alpha) = -\mathsf{E}[D(0)] = 1$, and $\sigma_D^2 = 25$. The uncontrolled estimator exhibits large fluctuations around its steady-state mean. The upper and lower smoothed estimators show less variability, and the bound $\eta^-(n) < \eta^+(n)$ is maintained throughout the run.

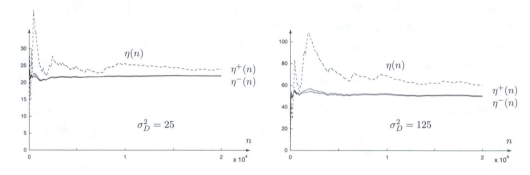

Figure 11.7. The plot on the left shows the same simulation as shown in Fig. 11.6, with the time horizon increased to $T = 20{,}000$. The plot on the right shows the two smoothed estimators along with the uncontrolled estimator when the variance is increased to $\sigma_D^2 = 125$. In each case the estimates obtained from the standard Monte Carlo estimator are significantly larger than those obtained using the smoothed estimator, and the bound $\eta^-(n) < \eta^+(n)$ again holds for all large n.

The plot on the left in Fig. 11.7 illustrates the simulation shown previously in Fig. 11.6, with the time horizon increased to $T = 20{,}000$. The plot on the right shows the controlled and uncontrolled estimators with $\kappa = 5$, and hence $\sigma_D^2 = 125$. The bounds $\eta^-(n) < \eta^+(n) < \eta(n)$ hold for all large n even though all three estimators are asymptotically unbiased.

Sections 11.4.4–11.4.6 provide several techniques for constructing shadow functions in network models.

11.4.4 Quadratic estimator

We now turn to network models. We have seen that the solution to Poisson's equation is equivalent to a quadratic under general conditions – see for example Proposition 8.1.6.

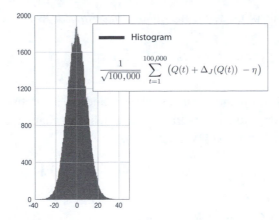

Figure 11.8. Smoothed estimator for the M/M/1 queue with $c(x) \equiv x$ and $h = J$.

This is always true for the single-server queue. So, it is reasonable to approximate the solution to Poisson's equation by a quadratic function.

We first revisit the M/M/1 queue to see what can be generalized to networks.

Example 11.4.3 (Shadow functions and the M/M/1 queue). In this model the identity (11.48) holds with $R(x) = \frac{1}{2}(\mu - \alpha)^{-1}(\mu\phi(x) + \alpha)$, where $\phi(x) = \mathbf{1}(x \geq 1)$. This is bounded, and hence Proposition 11.2.5 implies that the smoothed estimator (11.12) using $\psi = J$ satisfies an LDP.

Shown in Fig. 11.8 are simulation results using this estimator under the same conditions as in the experiment illustrated in Fig. 11.1, again using 20,000 independent trials with $Q(0) \sim \pi$, load $\rho = 0.9$, and time horizon $T = 10^5$ time steps. The CLT variance is reduced by more than two orders of magnitude, and the histogram shown here is more symmetric than that seen previously in the uncontrolled estimator.

To obtain an even better estimator, note that the smoothed estimator is simply averaging $R(Q(t))$, which amounts to estimating the mean of $\phi(Q(t)) = U(t)$. Suppose that we avoid estimation, and replace the average of $U(t)$ by its known steady-state mean $\mathsf{E}_\pi[U(0)] = \alpha/\mu$. The asymptotic variance of the resulting estimator is *zero* since we have removed all randomness.

This ideal estimator can be expressed as the smoothed estimator $\eta(n) + \Delta_h(Q(n))$ in which h is the solution to Poisson's equation, $h(x) = \frac{1}{2}(\mu - \alpha)^{-1}(x^2 + x)$. ∎

The solution to Poisson's equation used in Example 11.4.3 is a version of the quadratic function h_D defined in (8.80), with $D = \frac{1}{2}(\mu - \alpha)^{-1}$. To apply the results of Section 8.6 we restrict to a scheduling model. It is assumed that $\rho_\bullet < 1$, and that the policy ϕ is regular and nonidling.

Theorem 8.5.2 states that h_D is a solution to Poisson's equation for the function defined in (8.75) and recalled here,

$$c_D(x) = -(B\phi(x) + \alpha)^\mathsf{T} Dx, \quad x \in \mathsf{X}.$$

So, given an $\ell \times \ell$ matrix D, define the *quadratic estimator* as the sample-path average,

$$\eta_{h_D}^s(n) := \eta(n) - \frac{\theta}{n} \sum_{t=0}^{n-1} (c_D(Q(t)) - \eta_D), \quad n \geq 1. \tag{11.49}$$

This is precisely (11.45) with $\psi = h_D$, where η_D and h_D are defined in Theorem 8.5.2.

Just as in the construction of the Drift LP we consider a change of variables involving products of the form $U_i(t)Q_j(t)$. We define

$$G_{ij}(t) = U_i(t)Q_j(t) = \phi_i(Q(t))Q_j(t), \quad i, j \in \{1, \ldots, \ell\}, \ t \geq 0,$$

and work with the process $\boldsymbol{G} = \{G(t) : t \geq 0\}$. This is a Markov chain since it is an invertible function of $\boldsymbol{Q}$ when ϕ is nonidling.

Let G denote an arbitrary element of $\mathbb{R}^{\ell^2}$, and consider for a given x the particular element $G(x)$ with entries

$$G_{ij}(x) = \phi_i(x)x_j. \tag{11.50}$$

Recall that $c(x)$ can be expressed as the inner product $c(x) = \langle c^\bullet, G(x) \rangle$, with $c^\bullet$ the ℓ^2-dimensional vector given in (8.93). Similarly, we can express $c_D(x)$ as $c_D(x) = \langle c^D, G(x) \rangle$ for some vector c^D. From the identity $G(t) = G(Q(t))$, the quadratic estimator (11.49) can be expressed

$$\eta_{h_D}^s(n) = \theta\eta_D + \frac{1}{n} \sum_{t=0}^{n-1} \langle c^\bullet - \theta c^D, G(t) \rangle, \quad n \geq 1. \tag{11.51}$$

Based on (11.51), the asymptotic variance of the quadratic estimator (11.49) is given by

$$(c^\bullet - \theta c^D)^\mathsf{T} \Sigma (c^\bullet - \theta c^D), \tag{11.52}$$

where Σ is the $\ell^2 \times \ell^2$ asymptotic covariance matrix for $\boldsymbol{G}$.

To minimize the asymptotic variance we solve a quadratic program,

$$\min_D \|c^\bullet - c^D\|_\Sigma^2 = \min_D (c^\bullet - c^D)^\mathsf{T} \Sigma (c^\bullet - c^D).$$

It is not necessary to assume copositivity of D: The minimum is over all $\ell \times \ell$ matrices. Note that we have eliminated θ since $\theta c^D = c^{\theta D}$. It is not difficult to compute the solution D^* since the vector $c^D \in \mathbb{R}^{\ell^2}$ depends linearly on D. However, since the covariance matrix Σ is unknown prior to the simulation, in practice a suboptimal choice may be necessary. We can instead choose D to minimize $\|c^\bullet - c^D\|$ in some other norm: This problem can be solved using linear programming for the L_1 or L_∞ norm, or using least-squares methods if $\|x\| = \sqrt{x^\mathsf{T} x}$.

Exactly as in the Drift LP, for some policies we may know that $G_{ij}(t) = 0$ for some i and j. In this case, we may modify the norm used in the minimization to ignore the (i, j)th coefficient of $c^\bullet - c^D$.

	Standard		Quadratic			Fluid		
ρ_2	Mean	Var	Mean	Var	Red.	Mean	Var	Red.
0.2	0.48	2.1E−4	0.48	1.7E−6	*120*	0.47	6.1E−5	3.5
0.4	1.26	1.4E−3	1.26	2.7E−5	*52*	1.3	4.0E−4	3.5
0.6	2.8	1.1E−2	2.8	5.1E−4	*22*	2.8	3.5E−3	3.1
0.8	6.9	0.19	6.9	2.6E−2	7.1	6.9	4.3E−2	4.4
0.9	14	2.0	14	0.64	3.1	14	0.17	12
0.95	25	13	25	7.1	1.9	26	0.23	*56*
0.99	70	99	70	95	1.0	110	0.98	*100*

Table 11.1. *Simulation results for the simple re-entrant line. The column marked 'Red.' lists the variance reduction obtained in each case*

In summary, to estimate η:

  (i) Choose D to minimize $\|c^\bullet - c^D\|$ for some suitable norm.

  (ii) Simulate the Markov chain G up until time n. (This amounts to simulating Q since G is a deterministic function of Q.)

  (iii) Compute $\theta(n)$, an estimate of θ^*.

  (iv) Compute the estimator $\eta^s_{h_D}(n)$ defined in (11.49) with $\theta = \theta(n)$.

Example 11.4.4 (Quadratic estimator for the simple re-entrant line). Simulation results for the simple re-entrant line operating under the FBFS (first buffer, first-served) priority policy are given in Table 11.1. In this experiment the Euclidean norm $\|c^\bullet - c^D\|$ was minimized to determine D – the additional information that $G_{31}(t) \equiv 0$ under the chosen policy was ignored.

In these experiments the homogeneous CRW model (4.10) was considered with generator given in (4.62). The parameters used in these simulations were consistent with (2.26) in the sense that Station 2 was the bottleneck, with $\rho_2/\rho_1 = 10/9$. The arrival rate α_1 was chosen to give the values of ρ_2 specified in the table. The simulations were run for 100,000 time steps using 20 batches so that the batch means method surveyed in Section 11.4.2 could be used to choose θ. The experiment was repeated 200 times to obtain an estimate of the error in the estimators. The column labeled "Red." gives the ratio of the observed variances.

The largest variance reduction occurred in light traffic, while smaller variance reductions were observed in moderately loaded systems. These variance reductions are certainly useful, and come at very little additional computational cost, since the only real additional computational cost in computing the estimator $\eta^s_{h_D}(n)$ as opposed to the standard estimator $\eta(n)$ is the solution of the optimization problem to choose D. But this problem is solved once only before the simulation begins, and takes a small amount of time to solve relative to the computational effort devoted to the simulation. ∎

We are more interested in the performance of these simulation estimators in moderately to heavily loaded systems. The results in Table 11.1 suggest that the quadratic estimator is less effective in heavy traffic. This is probably because a pure quadratic is not able to capture the true form of the relative value function.

In several results and examples in this book we have seen that the relative value function is approximated by the fluid value function. In most cases the fluid value function is piecewise quadratic rather than purely quadratic.

11.4.5 Fluid estimator

Under the conditions of Theorem 11.4.5, a solution h to Poisson's equation is approximated by the fluid value function J. This result then strongly motivates the use of a fluid value function as an approximation for the solution to Poisson's equation.

Theorem 11.4.5. *Suppose that for a given nonidling policy ϕ the fluid model is stable and the fluid limit $\mathcal{L}_x$ is a singleton for each x. Then a solution h to Poisson's equation exists, and*

$$\limsup_{\kappa \to \infty} \left[\sup_{|x|=1} \left| \frac{h(\kappa x(\kappa))}{J(\kappa x)} - 1 \right| \right] = 0,$$

where $\{x(\kappa)\}$ is defined in (10.5), and J is the fluid value function,

$$J(x) = \int_0^\infty c(q(t; x)) \, dt, \qquad q(0; x) = x.$$

Proof. The result and its proof are similar to Proposition 10.4.3.

Under the assumptions of the theorem there is a $T > 0$ such that

$$J(x) = \int_0^T c(q(t)) \, dt, \qquad x \in \mathbb{R}_+^\ell, \ |x| = 1, \ \mathcal{L}_x = \{q\}.$$

We shall fix such a T throughout the proof.

From Proposition 10.4.1, a Lyapunov function exists that is equivalent to a quadratic in the sense of (10.10). It follows from Theorem A.2.3 that $\pi(c) < \infty$ for $c \in L_\infty^{V_1}$, and that a solution to Poisson's equation h exists which is bounded from above by a quadratic. Proposition 8.1.6 implies that h is uniformly bounded from below.

We can then take the solution to Poisson's equation and iterate as follows: $P^n h = h - \sum_{i=0}^{n-1} P^i \tilde{c}$, where $\tilde{c}(x) = c(x) - \eta$, to give

$$\frac{\mathsf{E}_x[h(Q(\kappa T; \kappa x))]}{\kappa^2} = \frac{1}{\kappa^2} h(x) - \frac{1}{\kappa} \mathsf{E}_x \left[\sum_{i=0}^{[\kappa T]-1} \frac{1}{\kappa} c(Q(i; \kappa x(\kappa))) \right] - \frac{T\eta}{\kappa}.$$

The result then follows from Lemma 10.4.2. $\square$

The *fluid estimator* is defined as (11.45) with $\psi = J$:

$$\eta_J^s(n) = \eta(n) + \frac{\theta}{n} \sum_{t=0}^{n-1} \Delta_J(Q(t)), \qquad n \geq 1. \tag{11.53}$$

The parameter θ is again a constant that may be chosen to attempt to minimize the variance of the fluid estimator.

To implement the fluid estimator we need to be able to compute Δ_J. This may be moderately time-consuming (computationally speaking) relative to the time taken to simply simulate the process $\mathbf{Q}$. Be that as it may, the time taken to compute Δ_J is relatively insensitive to the congestion in the system.

Example 11.4.6 (Fluid estimator for the simple re-entrant line). Shown on the right in Table 11.1 is a summary of results obtained using the fluid estimator. The best value of θ using the batch-means method was found to be close to unity in each of the simulations, particularly at high loads where it was found to be within $\pm 5\%$ of unity.

Observe that for low traffic intensities, the fluid estimator yields reasonable variance reductions over the standard estimator. However, because it is more expensive to compute than the standard estimator, these results are not particularly encouraging. But as the system becomes more and more congested, the fluid estimator yields *significant* variance reductions over the standard estimator, meaning that the extra computational effort per iteration is certainly worthwhile. ∎

In summary, in the preceding example the quadratic estimator produces useful variance reductions in light to moderate traffic at very little additional computational cost. It is less effective in simulations of heavily loaded networks, but could potentially provide useful variance reductions in this regime if a better choice of D can be employed. The fluid estimator provides modest variance reduction in light to moderate traffic, but appears to be very effective in heavy traffic.

In complex networks the additional computational overhead in computing the fluid estimator can be substantial. In such cases we turn to a workload relaxation to construct an effective shadow function that is easily computed.

11.4.6 Shadow functions and workload

We first consider a special case for which we can obtain exponentially decaying error bounds in the smoothed estimator. Consider the two-dimensional CRW workload model,

$$Y(t+1) = Y(t) - S(t+1)\mathbf{1} + S(t+1)\iota(t) + L(t+1), \qquad Y(0) \in \mathbb{Z}_+^2, \quad (11.54)$$

where $\mathbf{S}$ is a diagonal matrix sequence of the form $\{S(t) = \mathrm{diag}\,(S_1(t), S_2(t)) : t \geq 1\}$ taking values in $\{0,1\}^2$, and $\mathbf{L}$ is a two-dimensional sequence taking values in $\mathbb{Z}_+^2$. The mean values $\mu_i = \mathsf{E}[S_i(1)]$, $\lambda_i = \mathsf{E}[L_i(1)]$ satisfy $\mu_i > \lambda_i$ for $i = 1, 2$.

We restrict to the work-conserving policy, so that (11.54) becomes

$$Y(t+1) = [Y(t) - S(t+1)\mathbf{1} + L(t+1)]_+.$$

The Markov chain $\mathbf{Y}$ is regular since the quadratic $V(x) = \frac{1}{2}\|x\|^2$ solves (V3).

The fluid model under the nonidling policy is expressed

$$y(t) = [y - (\mu - \lambda)t]_+, \qquad t \geq 0, \ y \in \mathbb{R}_+^2.$$

If $\bar{c}\colon \mathbb{R}_+^2 \to \mathbb{R}_+$ is piecewise linear,

$$\bar{c}(y) = \max_{1 \leq i \leq \ell_{\bar{c}}} \langle \bar{c}^i, y \rangle,$$

and if $\bar{c}(\mu - \lambda) = \langle \bar{c}^i, \mu - \lambda \rangle$ for a *unique* $i \in \{1, \ldots, \ell_{\bar{c}}\}$, then the assumptions of Proposition 5.3.21 hold so that the fluid value function is C^1, with

$$\widehat{J}(y) = \int_0^\infty \bar{c}(y(t)) \, dt, \qquad y(0) = y.$$

A smoothed estimator for the steady-state mean $\widehat{\eta}$ of $\bar{c}(Y(t))$ is naturally obtained using $\psi = \widehat{J}$, giving $\Delta_{\widehat{J}}(y) = \mathsf{E}[\widehat{J}(Y(t+1)) - \widehat{J}(Y(t)) \mid Y(t) = y]$, $y \in \mathbb{Z}_+^2$, and

$$\widehat{\eta}_{\widehat{J}}^s(n) = \frac{1}{n} \sum_{t=0}^{n-1} [\bar{c}(Y(t)) + \Delta_{\widehat{J}}(Y(t))], \qquad n \geq 1. \tag{11.55}$$

Proposition 11.4.7. *Consider the two-dimensional CRW workload model (11.54) under the nonidling policy. It is assumed that $\mu_i > \lambda_i$, $i = 1, 2$, and for some $\beta_0 > 0$,*

$$\mathsf{E}[e^{\beta_0 |L(1)|}] < \infty.$$

Suppose that $\bar{c}\colon \mathbb{R}_+^2 \to \mathbb{R}_+$ is piecewise linear, and purely linear in a neighborhood of the vector $\mu - \lambda \in \mathbb{R}_+^2$. Then the smoothed estimator (11.55) satisfies the LDP (11.10a), (11.10b) with nonzero rate function: There exists $\varepsilon > 0$ such that for any $r_+ \in (\widehat{\eta}, \widehat{\eta} + \varepsilon)$, $r_- \in (\widehat{\eta} - \varepsilon, \widehat{\eta})$,

$$\lim_{n \to \infty} \frac{1}{n} \log \left(\mathsf{P}\{\widehat{\eta}_{\widehat{J}}^s(n) \geq r_+\} \right) = -I_{\widehat{J}}(r_+) < 0$$

$$\lim_{n \to \infty} \frac{1}{n} \log \left(\mathsf{P}\{\widehat{\eta}_{\widehat{J}}^s(n) \leq r_-\} \right) = -I_{\widehat{J}}(r_-) < 0.$$

Proof. There are two steps to the proof: First, we show that the function $\bar{c} + \Delta_{\widehat{J}}$ is *uniformly bounded* over all of $\mathbb{R}_+^2$. Next we establish geometric ergodicity: We establish Condition (V3) with $V = \widehat{J}$, and then Condition (V4) with $V_\beta = \exp(\beta\sqrt{\widehat{J}})$ for some $\beta > 0$. Proposition 11.2.5 then implies the desired result.

The smoothed estimator is ordinary Monte Carlo applied to the function

$$\bar{c}(y) + \Delta_{\widehat{J}}(y) = \mathsf{E}[\bar{c}(Y(t)) + \widehat{J}(Y(t+1)) - \widehat{J}(Y(t)) \mid Y(t) = y]. \tag{11.56}$$

To show that this is bounded over $\mathbb{R}_+^2$ we first apply Proposition 5.3.21 to conclude that $\widehat{J}$ is C^1. Moreover, the dynamic programming equation (5.50) holds in the form

$$\mathcal{D}_0 \widehat{J} = \langle \nabla \widehat{J}, -(\mu - \lambda) \rangle = -\bar{c}. \tag{11.57}$$

Consequently we can apply the mean value theorem to obtain

$$\widehat{J}(Y(t+1)) - \widehat{J}(Y(t)) = \langle \nabla \widehat{J}(\bar{Y}), Y(t+1) - Y(t) \rangle$$
$$= \langle \nabla \widehat{J}(Y(t)), Y(t+1) - Y(t) \rangle + \langle \widehat{J}(\bar{Y}) - \nabla \widehat{J}(Y(t)), Y(t+1) - Y(t) \rangle$$

where $\bar{Y}$ lies on the line connecting $Y(t+1)$ and $Y(t)$. Since $\nabla \widehat{J}$ is Lipschitz continuous, to establish that (11.56) is bounded it is sufficient to prove that

$$\mathsf{E}[\bar{c}(Y(t)) + \langle \nabla \widehat{J}(Y(t)), Y(t+1) - Y(t) \rangle \mid Y(t) = y] = 0. \tag{11.58}$$

The proof of this is based on the dynamic programming equation (11.57). We consider three cases.

(i) Suppose that $y_i \geq 1$ for each i. In this case $\mathsf{E}[Y(t+1) - Y(t) \mid Y(t) = y] = -(\mu - \alpha)$, and (11.58) is immediate from the dynamic programming equation.
(ii) Suppose $y = 0$. Then $\nabla \widehat{J}(y)$ and $\bar{c}(y)$ are each zero, so that (11.58) again holds.
(iii) Suppose that $y_i = 0$ for just one i. The boundary conditions in Proposition 5.3.21 (iii) give in this case $\frac{\partial}{\partial y_i} \widehat{J}(y) = 0$. This is illustrated for the KSRS workload model in Fig. 5.4. Consequently, letting i' denote the other index, the dynamic programming equation (11.57) gives

$$\mathsf{E}[\langle \nabla \widehat{J}(Y(t)), Y(t+1) - Y(t) \rangle \mid Y(t) = y]$$
$$= \frac{\partial}{\partial y_{i'}} \widehat{J}(y)(-\mu_{i'} + \lambda_{i'}) = \mathcal{D}_0 \widehat{J}(y) = -\bar{c}(y).$$

Hence, (11.58) holds for all $y \in \mathbb{Z}_+^2$ as claimed, and it follows that (11.56) is bounded over $\mathbb{Z}_+^2$.

To complete the proof we now show that (V4) holds with $V_\beta(x) = \exp(\beta\|x\|)$ for some $\beta > 0$, where $\| \cdot \|$ denotes the standard Euclidean norm. The function $V_0(x) = k_0\|x\|$ is a Lipschitz solution to Foster's criterion for some constant $k_0 > 0$, and thus the drift condition (V4) for V_β follows from Proposition A.5.7 (see also Theorem 8.2.6 for a specialization to the scheduling model). $\qquad \square$

The value of this proposition is largely motivational since our goal is to obtain estimates of statistics for the original network, rather than its relaxation. It is not known whether Proposition 11.4.7 admits an extension to more general CRW models.

However, the result suggests a general approach to the construction of shadow functions for Q, provided the policy is based on a workload relaxation. Recall that in translating a policy from a relaxation to the original network we always attempt to make the queue-length process follow the effective state, so that $\mathcal{P}^*(Q(t)) \approx Q(t)$ for all large t. Under the assumptions of Theorem 6.6.15 we have $J(Q(t)) \approx \widehat{J}(\Xi Q(t))$. Hence a natural choice of ψ in (11.12) is

$$\psi(x) := \widehat{J}(\Xi x), \qquad x \in \mathsf{X}. \tag{11.59}$$

In the next example this technique is illustrated using the KSRS model.

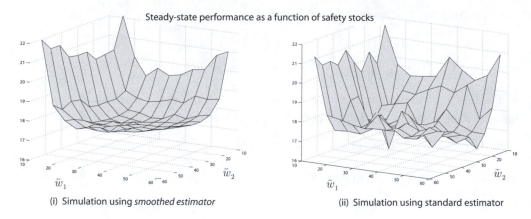

(i) Simulation using *smoothed estimator*   (ii) Simulation using standard estimator

Figure 11.9. Estimates of the steady-state customer population in the KSRS model as a function of 100 different safety-stock levels. Two simulation experiments are shown, where in each case the simulation runlength consisted of 200,000 steps. The left-hand side shows the results obtained using the smoothed estimator; the right-hand side shows results with the standard estimator.

Example 11.4.8 (Variance reduction for the KSRS model). Consider the KSRS model introduced in Section 2.9 in the setting of Example 5.3.16.

We take the cost function c equal to the ℓ_1 norm. We consider only Case I in which the effective cost is monotone. Hence the optimal policy for the two-dimensional relaxation is work-conserving. For Q we choose the policy (4.70). This is a translation of the work-conserving policy for the two-dimensional relaxation, based on a pair of safety-stock levels $\{\overline{w}_s : s = 1, 2\}$. Our goal is to estimate the expected steady-state number of jobs in the system as a function of $\overline{w}$ to determine the most effective safety-stock values.

Fig. 5.4 shows level sets of the value function $\widehat{J}$. It satisfies all of the properties used in the proof of Proposition 11.4.7: It is C^1, and its partial derivatives vanish on the boundary of $\mathbb{R}^2_+$. Consequently, on simulating a workload model Y based on the smoothed estimator we are guaranteed exponential error bounds in (11.10a), (11.10b).

However, our interest is in evaluating the policy (4.70). Hence, instead of simulating the workload model we simulate Q using the smoothed estimator, with shadow function given in (11.59). We take the CRW model with statistics defined in (4.68).

Observe that two translation steps are performed here: First we have constructed a class of candidate policies for Q based on the optimal policy for the relaxation. Second, we have translated the shadow function for the relaxation using (11.59).

Fig. 11.3 shows a single sample path of the smoothed estimator with safety stocks fixed at $\overline{w}_1 = \overline{w}_2 = 50$. The variance reduction obtained in this example is approximately two orders of magnitude.[3]

Shown in Fig. 11.9 are estimates of the steady-state customer population for a range of safety-stock parameters. The estimates were obtained as in Fig. 11.3, with

---

[3] See [256] for further discussion.

simulations run for a family of policies indexed by the safety-stock level $\overline{w} \in \mathbb{R}_+^2$. Fig. 11.9 (i) shows results obtained using the smoothed estimator (11.12), and (ii) contains results obtained from the standard estimator (11.1). The variance reduction observed is again outstanding.   ∎

11.5 Estimating a value function

Value functions have appeared in a surprising range of contexts in this book.

(i) The usual home for value functions is within the field of optimization. In the setting of this book, this means MDPs. Chapter 9 provides many examples, following the introduction for the single-server queue presented in Chapter 3.

(ii) The stability theory for Markov chains and networks in this book is centered around Condition (V3). This is closely related to Poisson's inequality, which is itself a generalization of the average-cost value function.

(iii) Theorem 8.4.1 contains general conditions ensuring that the h-MaxWeight policy is stabilizing. The essence of the proof is that the function h is an approximation to Poisson's equation under the assumptions of the theorem.

(iv) We have just seen how approximate solutions to Poisson's equation can be used to dramatically accelerate simulation.

With the exception of (i), each of these techniques is easily applied in a wide range of settings. The basic reason for this success is that in each of these three cases we are *approximating* a value function. In the case of (iii) the function h in the h-MaxWeight policy is only a crude approximation to the average-cost dynamic programming equation; the simplicity of this policy is a consequence of this modest goal. In contrast, the "curse of dimensionality" arises in optimization when we seek an exact solution.

In this final section of the book we consider methods to construct approximations via simulation. Our goal is to "learn" the value function based on experiments on the network. Of course, learning brings its own curses. This is summarized in the following remark from [350]:

> A large state space presents two major issues. The most obvious one is the storage problem, as it becomes impractical to store the value function (or optimal action) explicitly for each state. The other is the generalization problem, assuming that limited experience does not provide sufficient data for each and every state.

The first issue is resolved by restricting to a parameterized family of approximate value functions. The learning problem is then reduced to finding the best approximation within this class.

If we are lucky, or have some insight on the structure of the value function, then a parameterization can also resolve the second issue. For example, if it is known that the value function is convex, then the family of approximations can be constructed to share this property. This imposes some continuity so that if a great deal is learned about the value function evaluated at a particular state x_0, then this information is useful for learning the value function at nearby points.

In the case of networks, there are natural parameterizations to consider based on results from previous chapters.

(i) The fluid value function J^* is the natural starting point to approximate the solution to the average-cost value function. In the case of the single-server queue, Theorem 3.0.1 can be applied to conclude that the following parameterized family includes the actual value function,

$$h^\theta(x) = \theta_1 J^*(x) + \theta_2 x, \qquad x \in \mathbb{R}_+, \ \theta \in \mathbb{R}^2,$$

where $\theta_1 = 1$ when $h^\theta = h^*$ solves Poisson's equation. The discussion in Section 3.4.5 suggests similar approaches to approximate the discounted-cost value function. Example 11.4.8 illustrates how this approximation technique extends to networks.

(ii) The family of all quadratic functions can be regarded as a parameterized family. Linear programming methods were proposed in Section 8.6 to construct a solution to (V3).

(iii) The perturbed value function introduced in Section 4.9 is another example of a parameterized family of functions that can potentially approximate a value function. For example, given the family of functions $\{h(x) = h_0(\tilde{x})\}$ where h_0 ranges over some class, and the perturbation $\tilde{x}$ defined in (4.93) depends on the parameter $\theta > 0$, what is the best value of θ and h_0?

Each of the parameterizations in (i)–(iii) can be used to obtain an approximate value function for control or simulation. How then can we find the best approximation?

The evaluation criterion will depend on the context. In the case of simulation we might choose the approximation so that the resulting asymptotic variance is minimal. For control, the ultimate goal is to optimize performance over the class of policies considered. The algorithms described here can be used to approximate the value function for application in approximate value iteration or policy iteration. In this case, the metric to evaluate the approximation should reflect our goal to optimize performance.

In the remainder of this section we return to the general Markov setting in which X denotes a Markov chain without control on a state space X with transition matrix P, and unique invariant measure π. It is not necessary to assume that X is countable, but we do assume there is a fixed state $x^* \in X$ satisfying $\pi(x^*) > 0$. A function $c\colon X \to \mathbb{R}$ is given, and our goal is to estimate a value function such as the solution to Poisson's equation, or the discounted-cost value function.

The basic approach to compute the best approximation is stochastic approximation or one of its variants.

11.5.1 Stochastic approximation

There are many different kinds of value functions that we might attempt to approximate. Regardless of its particular form, we assume that a parameterized family of approximations $\{h^\theta : \theta \in \mathbb{R}^{\ell_h}\}$ is given. In the case of a linear parameterization we suppose that we are given ℓ_h functions on X, denoted $\{\psi_i : 1 \leq i \leq \ell_h\}$, and define

$$h^\theta(x) := \theta^\mathsf{T}\psi(x) = \sum_{i=1}^{\ell_h} \theta_i \psi_i(x), \qquad x \in \mathsf{X}. \tag{11.60}$$

We then seek the best approximation in the given class based on a particular metric to describe the distance between h^θ and the value function h of interest.

Throughout most of this section we consider the L_2 *error*,

$$\mathcal{E}(\theta) = \|h - h^\theta\|_\pi^2 := \mathsf{E}_\pi[|h(X(0)) - h^\theta(X(0))|^2], \tag{11.61}$$

or a related *weighted* L_2-norm. On writing (11.61) as the sum

$$\|h - h^\theta\|_\pi^2 = \sum |h(x) - h^\theta(x)|^2 \, \pi(x)$$

we see that this notion of distance penalizes the difference $|h(x) - h^\theta(x)|$ more strongly for states with larger steady-state probability $\pi(x)$. These states are visited more frequently by the chain, and are hence "more important."

This is valid motivation, but the most important reason for considering (11.61) is the fact that we can so easily construct an algorithm to minimize the error over θ.

It is assumed that h^θ is a smooth function of θ, so that the following gradient exists for each x:

$$\psi^\theta(x) := \nabla_\theta h^\theta(x). \tag{11.62}$$

The gradient is independent of θ when the parameterization is linear so that h^θ is expressed as the sum (11.60). We can formulate necessary conditions for optimality by differentiating the error with respect to θ, and setting this equal to zero. In the case of the L_2 error (11.61) the derivative with respect to θ has the probabilistic representation

$$\nabla_\theta\|h^\theta - h\|_\pi^2 = 2\mathsf{E}_\pi[(h^\theta(X) - h(X))\psi^\theta(X)], \tag{11.63}$$

provided one can justify exchanging the derivative and expectation. When the parameterization is linear and both h and ψ have finite second moments then (11.63) is justified, and on setting the derivative equal to zero we obtain the optimal parameter

$$\theta^* = M_\psi^{-1} b_\psi, \qquad \text{where } M_\psi = \mathsf{E}[\psi(X)\psi(X)^\mathsf{T}], \quad b_\psi = \mathsf{E}[h(X)\psi(X)], \tag{11.64}$$

provided the matrix M_ψ is invertible.

The usual steepest descent algorithm to compute a minimum of $\|h^\theta - h\|_\pi^2$ is given by the ODE

$$\frac{d}{d\theta}\theta(t) = -\frac{a}{2}\nabla_\theta\|h^\theta - h_\gamma\|_\pi^2$$
$$= -a\mathsf{E}_\pi[(h^{\theta(t)}(X(k)) - h(X(k)))\psi^{\theta(t)}(X(k))], \qquad t \geq 0, \tag{11.65}$$

where $a > 0$ is a fixed gain. One stochastic approximation algorithm is constructed as the approximation of (11.65)

$$\theta(k+1) - \theta(k) = -a_k[(h^{\theta(k)}(X(k)) - h(X(k)))\psi^{\theta(k)}(X(k))], \quad k \geq 0, \tag{11.66}$$

where $\{a_k\}$ is a positive gain sequence satisfying (11.4).

The recursion (11.66) may very well approximate the ODE (11.65), and both may converge to the optimal value θ^*. Unfortunately there is a fundamental problem with either approach: The value function h appearing on the right-hand side is not available *since this is the function we wish to approximate*! To obtain a practical algorithm we must find an alternative representation for the derivative (11.63). This step is the essential contribution of temporal difference learning.

11.5.2 Least-Squares Temporal Difference learning for discounted cost

So far we have suppressed the precise definition of the value function. To construct a practical algorithm we are forced to be more explicit. The discounted-cost value function h_γ defined in (9.34) is the simplest starting point.

To simplify the discussion further we begin with a linear parameterization. This is the most common situation in practice, and in this case we can construct an efficient algorithm without much trouble. The optimal parameter θ^* has the representation (11.64) in this case, and our goal is to obtain a sequence of matrices $\{M(n)\}$ and vectors $\{b(n)\}$ such that $n^{-1}M(n) \to M_\psi$ and $n^{-1}b(n) \to b_\psi$ as $n \to \infty$, so that $\theta(n) = M(n)^{-1}b(n)$ is an asymptotically consistent estimator of θ^*.

To estimate the matrix M_ψ we can directly apply Monte Carlo. Define the matrix sequence

$$M(n) = M(0) + \sum_{t=0}^{n-1} \psi(X(t))\psi^\mathsf{T}(X(t)), \qquad n \geq 1, \qquad (11.67)$$

where the matrix $M(0) > 0$ is introduced to ensure that $M(n)$ is invertible for each n. Provided the LLN holds for $\boldsymbol{X}$ we will have $n^{-1}M(n) \to M_\psi$ with probability 1.

We assume that the LLN does hold. In fact, since all of the expectations here are in steady state, we can and will assume that $\boldsymbol{X}$ is a stationary process, defined on the two-sided time interval. We also let denote by X a generic random variable distributed according to π.

What about b_ψ? The defintion $b_\psi = \mathsf{E}[h(X(0))\psi(X(0))]$ involves the value function $h = h_\gamma$ we wish to estimate, so we must find an alternate representation. For this we rely on the definition of the value function,

$$h_\gamma(x) = \sum_{t=0}^{\infty} (1+\gamma)^{-t-1}\mathsf{E}[c(X(t)) \mid X(0) = x].$$

On multiplying each side by $\psi(x)$ we obtain

$$h_\gamma(x)\psi(x) = \sum_{t=0}^{\infty} (1+\gamma)^{-t-1}\mathsf{E}[c(X(t))\psi(X(0)) \mid X(0) = x].$$

The vector we are attempting to estimate is precisely

$$b_\psi = \mathsf{E}[h_\gamma(X)\psi(X)] = \sum_{x\in\mathsf{X}} h_\gamma(x)\psi(x)\,\pi(x).$$

Multiplying both sides of the previous equation by $\pi(x)$ and summing over x shows that b_ψ is also expressed as a sum,

$$
\begin{aligned}
b_\psi &= \sum_{t=0}^{\infty} (1+\gamma)^{-t-1} \sum_{x \in \mathsf{X}} \mathsf{E}[c(X(t))\psi(X(0)) \mid X(0) = x]\,\pi(x) \\
&= \sum_{t=0}^{\infty} (1+\gamma)^{-t-1} \mathsf{E}[c(X(t))\psi(X(0))].
\end{aligned}
\tag{11.68}
$$

Each of the expectations in the sum on the right involves functions that are known to us – the cost function c and the basis vector ψ.

The only remaining difficulty is that these expectations involve the process in the future, which complicates the application of Monte Carlo. The *key step* to obtain a practical algorithm is to apply stationarity of $\boldsymbol{X}$, which implies that for any integer i,

$$
\mathsf{E}[c(X(t))\psi(X(0))] = \mathsf{E}[c(X(t+i))\psi(X(i))].
\tag{11.69}
$$

In particular, we can set $i = -t$, so that (11.68) becomes

$$
b_\psi = \sum_{t=0}^{\infty} (1+\gamma)^{-t-1} \mathsf{E}[c(X(0))\psi(X(-t))].
\tag{11.70}
$$

One final transformation is needed. Assume that we have absolute integrability

$$
\sum_{t=0}^{\infty} (1+\gamma)^{-t-1} \mathsf{E}[c(X(0))\|\psi(X(-t))\|] < \infty,
$$

so that Fubini's Theorem can be applied to give

$$
b_\psi = \mathsf{E}\left[c(X(0)) \left(\sum_{t=0}^{\infty} (1+\gamma)^{-t-1} \psi(X(-t)) \right) \right].
$$

Then, if we define for any k the random variable

$$
\varphi(k) = \sum_{t=0}^{\infty} (1+\gamma)^{-t-1} \psi(X(k-t)),
$$

the final expression can be expressed $b_\psi = \mathsf{E}[c(X(0))\varphi(0)]$. In this form we can apply Monte Carlo: We have $n^{-1}b(n) \to b_\psi$ as $n \to \infty$, with

$$
b(n) = b(0) + \sum_{t=0}^{n-1} c(X(t))\varphi(t),
\tag{11.71}
$$

and $b(0) \in \mathbb{R}^{\ell_h}$ an arbitrary initialization. The resulting algorithm is generalized slightly in the following formal definition.

The Matrix Inversion Lemma [219] is used to obtain a recursive algorithm that avoids repeated inversion of $M(n)$. This identity is proven on multiplying each side of (11.72) by $G^{-1} + HK^{\mathsf{T}}$.

Lemma 11.5.1 (Matrix Inversion Lemma). *Suppose that* G, H, *and* K *are respectively* $m \times m$, $m \times n$, *and* $n \times m$ *matrices. If* G *and the sum* $(I + K^\mathsf{T} G H)$ *are invertible, then*

$$(G^{-1} + H K^\mathsf{T})^{-1} = G - GH(I + K^\mathsf{T} GH)^{-1} K^\mathsf{T} G. \tag{11.72}$$

In applying the Matrix Inversion we take $G(n) = M^{-1}(n)$ and $H = K = \psi(X(n))$ to obtain the formula for $G(n+1) = M^{-1}(n+1)$ given in (11.73b).

Definition 11.5.2 (LSTD learning for discounted cost). For given initial conditions $G(0) > 0$, $b(0) \in \mathbb{R}^{\ell_h}$, $\varphi(0) \in \mathbb{R}^{\ell_h}$, the Least-Squares TD (LSTD) algorithm is defined by the sequence of parameter estimates

$$\theta(n) = G(n)b(n), \tag{11.73a}$$

together with the three recursive equations

$$G(n+1) \;=\; G(n) - \frac{G(n)\psi(X(n))\psi(X(n))^\mathsf{T} G(n)}{1 + \psi(X(n))^\mathsf{T} G(n)\psi(X(n))} \tag{11.73b}$$

$$\varphi(n+1) \;=\; (1+\gamma)^{-1}[\varphi(t) + \psi(X(n+1))] \tag{11.73c}$$

$$b(n+1) \;=\; b(n) + \varphi(n)c(X(n)). \tag{11.73d}$$

The vector $\varphi(k)$ is called an *eligibility vector*.   ∎

To establish convergence of LSTD it is necessary to extend the Strong Law of Large beyond static functions of $X(t)$. This can be accomplished by extending the definition of $\boldsymbol{X}$. Define the bivariate process

$$X'(t) = (X(t), \varphi(t)), \qquad t \geq 0. \tag{11.74}$$

This is a general state-space Markov chain, but one that has attractive stability properties provided ψ is π-integrable.

Rather than developing properties of the more complex stochastic process (11.74), in the proof of Theorem 11.5.3 we extend the LLN from $\boldsymbol{X}$ to $\boldsymbol{X}'$ through brute-force calculation in a very special case.

Theorem 11.5.3 (Convergence of LSTD for discounted cost). *Suppose that* $\boldsymbol{X}$ *is an ergodic, finite state-space Markov chain, and that* $M_\psi > 0$. *Then with probability 1, from each initial condition,* $\lim_{n\to\infty} G(n) = M_\psi^{-1}$ *and* $\lim_{n\to\infty} n^{-1}b(n) = b_\psi$. *Hence the LSTD algorithm is convergent:*

$$\theta(n) = G(n)b(n) \to \theta^* \quad \text{as } n \to \infty.$$

Proof. From the Matrix Inversion Lemma we have

$$nG(n) = n\left(M(0) + \sum_{t=0}^{n-1} \psi(X(t))\psi^\mathsf{T}(X(t)) \right)^{-1},$$

so that convergence of $nG(n)$ follows by the LLN for $\boldsymbol{X}$.

We now consider the sequence $\{b(n)\}$. Based on (11.73c) we obtain

$$\varphi(t) = (1+\gamma)^{-t-1}\varphi(0) + \sum_{k=0}^{t}(1+\gamma)^{-k-1}\psi(X(t-k)),$$

so that ignoring transient terms involving the initial conditions,

$$\lim_{n\to\infty}\frac{1}{n}b(n) = \lim_{n\to\infty}\frac{1}{n}\sum_{t=0}^{n-1}c(X(t))\left(\sum_{k=0}^{t}(1+\gamma)^{-k-1}\psi(X(t-k))\right).$$

Following a change in the order of summation, the right-hand side becomes

$$\frac{1}{n}\sum_{k=0}^{n-1}(1+\gamma)^{-k-1}\left(\sum_{t=k}^{n-1}c(X(t))\psi(X(t-k))\right)$$

$$= \sum_{k=0}^{n-1}(1+\gamma)^{-k-1}\frac{n-k}{n}\left(\frac{1}{n-k}\sum_{t=0}^{n-k-1}c(X(t+k))\psi(X(t))\right).$$

For any fixed k the LLN gives

$$\lim_{n\to\infty}\frac{1}{n-k}\sum_{t=0}^{n-k-1}c(X(t+k))\psi(X(t)) = \mathsf{E}[c(X(k))\psi(X(0))].$$

With a bit more book-keeping we obtain from this the desired conclusion,

$$\lim_{n\to\infty}\frac{1}{n}b(n) = \sum_{k=0}^{\infty}(1+\gamma)^{-k-1}\mathsf{E}[c(X(k))\psi(X(0))] = b_\psi. \qquad \square$$

11.5.3 Adjoint equations and TD learning

We now consider nonlinear parameterizations and derive the standard TD algorithm for value function approximation. This is based on transformations similar to those used in the case of a linear parameterization. In particular, a version of the invariance equation (11.69) is again critical.

A more compact, and perhaps more elegant construction of the algorithm is obtained by casting the approximation problem in a vector space setting. The vector space is the *Hilbert space* denoted $L_2(\pi)$, defined as the set of real-valued functions on X whose second moment under π is finite. We define an inner product on this Hilbert space that is consistent with the L_2 error (11.61),

$$\langle f, g\rangle_\pi = \pi(fg) = \sum f(x)g(x)\,\pi(x),$$

so that for any two functions $f, g \in L_2(\pi)$, the L_2 norm of their difference is expressed

$$\|f - g\|_\pi^2 := \langle f - g, f - g\rangle_\pi = \mathsf{E}[(f(X) - g(X))^2].$$

If the state space is finite, with N states, then the functions f and g can be viewed as N-dimensional column vectors, and π an N-dimensional row vector. The inner product is an ordinary inner product in $\mathbb{R}^N$,

$$\langle f, g \rangle_\pi = f^\mathsf{T} \Pi g,$$

where $\Pi = \mathrm{diag}\,(\pi)$.

Next we express the discounted-cost value function as a matrix-vector product. For each t, the t-step transition matrix P^t is the t-fold product of P with itself. For a given discount rate $\gamma > 0$, the resolvent matrix is defined as the infinite sum

$$R_\gamma = \sum_{t=0}^{\infty} (1+\gamma)^{-t-1} P^t.$$

This is actually a power-series expansion for the matrix inverse

$$R_\gamma = [\gamma I - \mathcal{D}]^{-1} = [(1+\gamma)I - P]^{-1}$$

where the generator is defined in (8.1). With the cost function c interpreted as a column vector we can express the value function as the product $h_\gamma = R_\gamma c$.

In this new notation, the derivative (11.63) can be expressed as the inner product

$$\nabla_\theta \|h^\theta - h_\gamma\|_\pi^2 = 2\langle (h^\theta - R_\gamma c), \psi^\theta \rangle_\pi. \tag{11.75}$$

We now interpret the transformations performed to construct the LSTD algorithm as the application of an adjoint operation in $L_2(\pi)$.

The adjoint of a real $N \times N$ matrix is simply its transpose. In the vector space setting of this section, the adjoint of the resolvent $R_\gamma^\dagger$ is characterized by the set of equations

$$\langle R_\gamma f, g \rangle_\pi = \langle f, R_\gamma^\dagger g \rangle_\pi, \qquad f, g \in L_2(\pi). \tag{11.76}$$

A sample-path representation for $\langle R_\gamma f, g \rangle_\pi$ leads to a useful representation for the adjoint. We begin with the definition

$$\langle R_\gamma f, g \rangle_\pi = \mathsf{E}\big[(R_\gamma f\,(X))(g(X))\big] = \mathsf{E}\left[\left(\sum_{t=0}^{\infty}(1+\gamma)^{-t-1}P^t f\,(X(0))\right) g(X(0))\right].$$

We have by the smoothing property of the conditional expectation

$$\mathsf{E}[P^t f\,(X(0))g(X(0))] = \mathsf{E}\big[\mathsf{E}[f(X(t)) \mid X(0)]g(X(0))\big] = \mathsf{E}[f(X(t))g(X(0))]$$

and then applying (11.69) we obtain

$$\langle R_\gamma f, g \rangle_\pi = \sum_{t=0}^{\infty}(1+\gamma)^{-t-1}\mathsf{E}[f(X(0))g(X(-t))].$$

The right-hand side can be expressed $\langle f, R_\gamma^\dagger g \rangle_\pi$, where

$$R_\gamma^\dagger g\,(x) = \sum_{t=0}^{\infty}(1+\gamma)^{-t-1}\mathsf{E}[g(X(-t)) \mid X(0) = x], \qquad x \in \mathsf{X}. \tag{11.77}$$

That is, the adjoint $R_\gamma^\dagger$ is the ordinary resolvent for the time-reversed process $\{X(-t) : t \in \mathbb{Z}_+\}$.

We now obtain an alternate expression for the derivative (11.63) based on the equivalent form (11.75). The resolvent is invertible, with $R_\gamma^{-1} = (1+\gamma)I - P$. Hence the difference $h^\theta - h_\gamma$ can be expressed

$$h^\theta - h_\gamma = h^\theta - R_\gamma c = R_\gamma [R_\gamma^{-1} h^\theta - c] = R_\gamma [(1+\gamma)h^\theta - Ph^\theta - c]. \quad (11.78)$$

Based on this expression, the representation (11.75), and the adjoint equation (11.76), we obtain

$$\tfrac{1}{2}\nabla_\theta \|h^\theta - h_\gamma\|_\pi^2 = \big\langle (1+\gamma)h^\theta - Ph^\theta - c, R_\gamma^\dagger \psi^\theta \big\rangle_\pi, \quad (11.79)$$

or written as an expectation,

$$\tfrac{1}{2}\nabla_\theta \|h^\theta - h_\gamma\|_\pi^2 = \mathsf{E}[d^\theta(t)\varphi^\theta(t)], \quad (11.80)$$

where $d^\theta(t) := (1+\gamma)h^\theta(X(t)) - h^\theta(X(t+1)) - c(X(t))$, and $\varphi^\theta(t) := R_\gamma^\dagger \psi(X(t))$.

We now have sufficient motivation to construct the TD learning algorithm based on the ODE (11.65).

Definition 11.5.4 (TD learning for discounted cost). The TD algorithm constructs recursively a sequence of estimates $\{\theta(n)\}$ based on the following:

(i) *Temporal differences* defined by

$$d(k) := -[(1+\gamma)h^{\theta(k)}(X(k)) - h^{\theta(k)}(X(k+1)) - c(X(k))], \qquad k \geq 1. \quad (11.81)$$

(ii) *Eligibility vectors*: the sequence of ℓ_h-dimensional vectors,

$$\varphi(k) = \sum_{t=0}^{k} (1+\gamma)^{-t-1} \psi^{\theta(k-t)}(X(k-t)), \qquad k \geq 1, \quad (11.82)$$

expressed recursively via

$$\varphi(k+1) = (1+\gamma)^{-1}[\varphi(k) + \psi^{\theta(k+1)}(X(k+1))], \qquad k \geq 0, \ \varphi(0) = 0.$$

The estimates of θ^* are defined by

$$\theta(n+1) - \theta(n) = a_n d(n)\varphi(n+1), \qquad n \geq 0, \quad (11.83)$$

where the nonnegative gain sequence $\{a_n\}$ satisfies (11.4). ∎

Based on (11.80), for large k the following approximation is suggested:

$$\mathsf{E}[d(k)\varphi(k+1)] \approx -\tfrac{1}{2}\nabla_\theta \|h^\theta - h_\gamma\|_\pi^2, \qquad \theta(i) \equiv \theta.$$

The TD algorithm is the stochastic approximation algorithm associated with the ODE (11.65), based on this approximation.

11.5.4 Average cost

Much of Part III of the book has concentrated on average-cost control and average-cost performance evaluation, and most of the approximation results have focused on Poisson's equation rather than the discounted-cost analog. We now generalize the results of Sections 11.5.2 and 11.5.3 to this setting.

Recall that a solution to Poisson's equation, also called the relative value function, can be expressed as a SPP value function with respect to the relative cost $\tilde{c} := c - \eta$. For a fixed state $x^* \in \mathsf{X}$, we take the particular form

$$h_{\tilde{c}}(x) = \mathsf{E}_x \left[\sum_{t=0}^{\sigma_{x^*}} \tilde{c}(X(t)) \right], \qquad x \in \mathsf{X}, \tag{11.84}$$

where σ_{x^*} denotes the first hitting time to x^*.

To extend the Hilbert space framework to this setting we express the value function as an infinite-horizon sum,

$$h_{\tilde{c}}(x) = \sum_{t=0}^{\infty} \mathsf{E}_x \left[\mathbf{1}\{\sigma_{x^*} \geq t\} \tilde{c}(X(t)) \right].$$

Next we express each term in the sum in "matrix-vector product" notation. Letting $\mathbf{1}_{\{x^*\}^c} P$ denote the matrix whose "row" corresponding to x^* has been set to zero, we have the probabilistic interpretation, for any function g,

$$\mathbf{1}_{\{x^*\}^c} P g\,(x) = \mathsf{E}_x \left[\mathbf{1}\{\sigma_{x^*} \geq 1\} g(X(1)) \right].$$

With $(\mathbf{1}_{\{x^*\}^c} P)^t$ the t-fold matrix product we similarly have

$$(\mathbf{1}_{\{x^*\}^c} P)^t g\,(x) = \mathsf{E}_x \left[\mathbf{1}\{\sigma_{x^*} \geq t\} g(X(t)) \right].$$

Hence the relative value function can be expressed as $h_{\tilde{c}} = R_0 \tilde{c}$, where the *potential matrix* is defined for arbitrary functions g via

$$R_0 g\,(x) := \sum_{t=0}^{\infty} (\mathbf{1}_{\{x^*\}^c} P)^t g\,(x), \qquad x \in \mathsf{X}.$$

Just as in our consideration of the resolvent, the potential matrix can be expressed as a matrix inverse

$$R_0 = \sum_{t=0}^{\infty} (\mathbf{1}_{\{x^*\}^c} P)^t = [I - \mathbf{1}_{\{x^*\}^c} P]^{-1}. \tag{11.85}$$

A similar representation for the relative value function is given in (A1.31).

We can follow the derivation of $R_\gamma^\dagger$ in (11.77) to express the adjoint as the potential matrix for the time-reversed process,

$$R_0^\dagger g\,(x) = \mathsf{E} \left[\sum_{\tilde{\sigma}_{x^*}^{[0]} \leq t \leq 0} g(X(t)) \mid X(0) = x \right], \qquad x \in \mathsf{X},\, g \in L_2(\pi), \tag{11.86}$$

where for any k

$$\tilde{\sigma}_{x^*}^{[k]} = \max\{t \leq k : X(t) = x^*\}.$$

Consider the error (11.61) and its gradient (11.63). In the Hilbert space notation introduced in Section 11.5.3, we obtain a representation similar to (11.75),

$$\nabla_\theta \mathcal{E}(\theta) = 2\langle (h^\theta - h_{\tilde{c}}), \psi^\theta \rangle_\pi.$$

Applying the representation (11.85) we obtain, exactly as in (11.78),

$$h^\theta - h_{\tilde{c}} = R_0 [R_0^{-1} h^\theta - \tilde{c}] = R_0 [h^\theta - \mathbf{1}_{\{x^*\}^c} P h^\theta - \tilde{c}].$$

Hence, the first-order condition for optimality of θ becomes

$$0 = \tfrac{1}{2} \nabla_\theta \mathcal{E}(\theta) = \langle [h^\theta - \mathbf{1}_{\{x^*\}^c} P h^\theta - \tilde{c}], R_0^\dagger \psi^\theta \rangle_\pi.$$

Following the derivation of TD learning for the discounted-cost value function we arrive at a TD learning algorithm to estimate $h_{\tilde{c}}$.

Definition 11.5.5 (TD learning for Poisson's equation). The TD algorithm constructs recursively a sequence of estimates $\{\theta(k)\}$ based on the following:

(i) *Temporal differences,*

$$d(k) := -\left[h^{\theta(k)}(X(k)) - \mathbf{1}_{\{x^*\}^c}(X(k)) h^{\theta(k)}(X(k+1)) - \big(c(X(k)) - \eta(k)\big) \right].$$

(ii) *Eligibility vectors,* the sequence of ℓ_h-dimensional vectors,

$$\varphi(k) = \sum_{t=\tilde{\sigma}_{x^*}^{[k]}}^{k} \psi^{\theta(k-t)}(X(k-t)), \qquad k \geq 1, \tag{11.87}$$

or written recursively,

$$\varphi(k+1) = \mathbf{1}\{X(k) \neq x^*\} \varphi(k) + \psi^{\theta(k+1)}(X(k+1)), \qquad k \geq 0.$$

(iii) Estimates $\{\eta(k)\}$ of η are obtained using Monte Carlo (11.1), or any other consistent method.

Estimates $\{\theta(n)\}$ of the optimal parameter are then obtained using the TD recursion (11.83). ∎

When the parameterization is linear we can again use the Monte Carlo estimates (11.73a), where the definition of $\{b(n)\}$ is redefined by

$$b(n) = \sum_{t=0}^{n-1} \big(c(X(t)) - \eta(n)\big) \varphi(t), \qquad n \geq 1,$$

with $\{\varphi(t)\}$ generated using the recursion (11.87).

Unfortunately, neither the TD or LSTD algorithms are effective in queueing models due to the very large variance of the estimates. This will be seen in Example 11.5.7, but first we introduce a method to reduce the variance. For this we modify the norm through the introduction of a weighting function $\Omega \colon \mathsf{X} \to [1, \infty]$. Define for two functions f, g the new inner product

$$\langle f, g \rangle_{\pi, \Omega} = \sum \big(f(x) g(x) / \Omega(x) \big) \pi(x),$$

with associated weighted norm

$$\|f - g\|_{\pi,\Omega}^2 := \langle f - g, f - g \rangle_{\pi,\Omega} = \mathsf{E}[(f(X) - g(X))^2/\Omega(X)]. \qquad (11.88)$$

We use this norm to define the error between h^θ and $h_{\tilde{c}}$ for a given parameter θ:

$$\mathcal{E}(\theta) = \|h^\theta - h_{\tilde{c}}\|_{\pi,\Omega}^2. \qquad (11.89)$$

As in the foregoing, we obtain a representation for the derivative of $\mathcal{E}(\theta)$; setting this equal to zero gives the first-order necessary condition for optimality of θ. Under general conditions (to justify taking the derivative inside the inner product) the derivative can be expressed $\nabla_\theta \mathcal{E}(\theta) = 2\langle h^\theta - h_{\tilde{c}}, \psi^\theta \rangle_{\pi,\Omega}$, where ψ^θ is the gradient of h^θ. For a linear parameterization this is independent of θ, giving

$$\tfrac{1}{2}\nabla_\theta \mathcal{E}(\theta) = \langle h^\theta, \psi \rangle_{\pi,\Omega} - \langle h_{\tilde{c}}, \psi \rangle_{\pi,\Omega}.$$

On denoting

$$M_\psi = \mathsf{E}[\psi(X)\psi(X)^{\mathsf{T}}\Omega^{-1}(X)], \qquad (11.90)$$

we have $\langle h^\theta, \psi \rangle_{\pi,\Omega} = M_\psi \theta$. From this expression and the definition of the adjoint we have

$$\tfrac{1}{2}\nabla_\theta \mathcal{E}(\theta) = M_\psi \theta - \langle \tilde{c}, R_0^\dagger \psi \rangle_{\pi,\Omega}$$

and setting this equal to zero gives the unique optimizer, provided the matrix M_ψ is invertible.

To obtain an algorithm we must first interpret these inner products. Based on (11.90), the matrix M_ψ can be estimated using an obvious modification of (11.67).

The inner product $\langle \tilde{c}, R_0^\dagger \psi \rangle_{\pi,\Omega}$ plays the role of the vector b_ψ that was introduced in the construction of LSTD for discounted cost. The adjoint $R_0^\dagger$ is again given by (11.86), so that the LSTD algorithm for average cost is obtained as a minor modification of Definition 11.5.2.

Definition 11.5.6 (LSTD learning for average cost with state weighting). For given initial conditions $G(0) > 0$, $b(0) \in \mathbb{R}^{\ell_h}$, $\varphi(0) \in \mathbb{R}^{\ell_h}$, the sequence of parameter estimates is defined by

$$\theta(n) = G(n)b(n), \qquad (11.91a)$$

where the sequences G and b are defined by the recursive equations

$$G(n+1) \;=\; G(n) - \frac{G(n)\psi(X(n))\psi(X(n))^{\mathsf{T}}G(n)}{\Omega(X(n)) + \psi(X(n))^{\mathsf{T}}G(n)\psi(X(n))} \qquad (11.91b)$$

$$b(n+1) \;=\; b(n) + \varphi(n)\big(c(X(n)) - \eta(n)\big) \qquad (11.91c)$$

$$\varphi(n+1) \;=\; \mathbf{1}\{X(n) \neq x^*\}\varphi(n) + \psi(X(n+1))/\Omega(X(n+1)). \qquad (11.91d)$$

∎

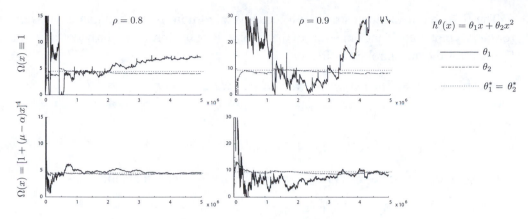

Figure 11.10. LSTD estimates for the relative value function in the M/M/1 queue based on Definition 11.5.6.

Example 11.5.7 (LSTD for the M/M/1 queue). The solution to Poisson's equation for the M/M/1 queue is the quadratic given in Proposition 3.4.2. For arbitrary $\theta \in \mathbb{R}^2$ define

$$h^\theta(x) = \theta_1 x + \theta_2 x^2, \qquad x \in \mathbb{R}_+. \tag{11.92}$$

Then, with $\theta_1^* = \theta_2^* = \frac{1}{2}(\mu - \alpha)^{-1}$ this is the solution given in Proposition 3.4.2.

To estimate θ^* we first apply the LSTD algorithm with $\Omega \equiv 1$. Observe that the recursion (11.91b) is designed to estimate the inverse of M_ψ. This involves estimating indirectly the mean of the *fourth* moment of the queue-length process since

$$M_\psi = M_\psi = \mathsf{E}[\psi(Q)\psi(Q)^\mathsf{T}] = \mathsf{E}\left[\begin{pmatrix} Q & Q^3 \\ Q^3 & Q^4 \end{pmatrix}\right].$$

The asymptotic variance of the standard estimator of $Q(t)^4$ is of order $(1 - \rho)^{-10}$ (!) Hence we can expect high variances when using the LSTD algorithm.

Shown in Fig. 11.10 are results from several experiments using the LSTD algorithm. In one set of experiments the weighting function was set to unity, and in the other the polynomial

$$\Omega(x) = (1 + (\mu - \alpha)x)^p, \qquad x \geq 0.$$

Several values of p were tried in experiments; the best value of $p = 4$ was chosen in the figure.

Consequences of the high variance are evident in Fig. 11.10. For loads of $\rho = 0.8$ or higher the estimates show high variability even after 5 million iterations. The introduction of the weighting function *significantly* reduces variability. ∎

11.5.5 Optimizing shadow functions

The last performance criterion to be considered is the asymptotic variance. For simplicity we restrict to linear parameterizations with $h^\theta = \theta^\mathsf{T}\psi$.

The first step toward constructing a recursive algorithm is to obtain an expression for the asymptotic variance in terms of the adjoint. Proposition 11.5.8 provides a useful formula for the uncontrolled estimator.

Proposition 11.5.8. *The asymptotic variance in (11.26) can be expressed*

$$\sigma_{\text{CLT}}^2 = \pi(2\tilde{c}h_{\tilde{c}}^\dagger - \tilde{c}^2), \tag{11.93}$$

where $h_{\tilde{c}}^\dagger := R_0^\dagger\tilde{c}$, and the adjoint is defined in (11.86).

Proof. The representation (11.26) combined with Poisson's equation gives

$$\sigma_{\text{CLT}}^2 = \pi(h_{\tilde{c}}^2 - (Ph_{\tilde{c}})^2) = \pi(h_{\tilde{c}}^2 - (h_{\tilde{c}} - \tilde{c})^2) = \pi(2h_{\tilde{c}}\tilde{c} - \tilde{c}^2).$$

From the definition of the adjoint and $h_{\tilde{c}}$ we have $\pi(h_{\tilde{c}}\tilde{c}) = \langle R_0\tilde{c}, \tilde{c}\rangle_\pi = \langle \tilde{c}, R_0^\dagger\tilde{c}\rangle_\pi$, and (11.93) then follows. $\square$

The value of Proposition 11.5.8 is that the steady-state mean $\pi(\tilde{c}h_{\tilde{c}}^\dagger)$ is easily estimated using standard Monte Carlo since $h_{\tilde{c}}^\dagger\tilde{c}$ can be expressed in terms of the history of the process. Indeed, define

$$\varphi_c(k+1) = \mathbf{1}\{X(k) \neq x^*\}\varphi_c(k) + \big(c(X(k+1)) - \eta(k+1)\big), \qquad k \geq 0,$$

where $\{\eta(k)\}$ are consistent estimates of η. Then, under general conditions,

$$\pi(\tilde{c}h_{\tilde{c}}) = \lim_{k\to\infty} \pi\big(\tilde{c}(X(k))\varphi_c(k)\big) = \lim_{n\to\infty} \frac{1}{n}\sum_{k=0}^{n-1}\big(c(X(k)) - \eta(k)\big)\varphi_c(k) \qquad \text{a.s.}$$

We now turn to shadow functions. Proposition 11.4.1 tells us that the optimal parameter θ^* is the solution to

$$\Sigma_\psi\theta = \langle\psi, h_{\tilde{c}}\rangle_{\text{CLT}}, \tag{11.94}$$

where Σ_ψ is defined in (11.43). Applying the adjoint technique once more we obtain the following expression for the θ^*:

Proposition 11.5.9. *Under the assumptions of Proposition 11.4.1 the optimal parameter (11.44) can be expressed*

$$\theta^* = \Sigma_\psi^{-1}b_\psi,$$

where

$$b_\psi = \langle\tilde{c}, R_0^\dagger(\psi - \psi^1) + \psi^1\rangle_\pi, \tag{11.95}$$

$R_0^\dagger$ *is defined in (11.86), and $\psi^1 := P\psi$.*

Proof. In view of (11.94), it is enough to establish that the vector b_ψ defined in (11.95) coincides with the vector $\langle\psi, h_{\tilde{c}}\rangle_{\text{CLT}}$. Poisson's equation for $h_{\tilde{c}}$ gives

$$\langle\psi, h_{\tilde{c}}\rangle_{\text{CLT}} = \langle\psi, h_{\tilde{c}}\rangle_\pi - \langle P\psi, Ph_{\tilde{c}}\rangle_\pi = \langle\psi, h_{\tilde{c}}\rangle_\pi - \langle\psi^1, h_{\tilde{c}} - \tilde{c}\rangle_\pi,$$

or $\langle \psi, h_{\tilde{c}} \rangle_{\text{CLT}} = \langle \psi - \psi^1, R_0 \tilde{c} \rangle_\pi + \langle \psi^1, \tilde{c} \rangle_\pi$. From the defining property of the adjoint,

$$\langle \psi - \psi^1, R_0 \tilde{c} \rangle_\pi = \langle R_0^\dagger (\psi - \psi^1), \tilde{c} \rangle_\pi,$$

we obtain the desired conclusion that $b_\psi = \langle \psi, h_{\tilde{c}} \rangle_{\text{CLT}}$.   □

To estimate the optimizer θ^* we can separately estimate Σ_ψ and $\eta := \pi(c)$ (required to construct $\tilde{c}$). It is also necessary to estimate expectations involving the two functions $R_0^\dagger \psi$ and $R_0^\dagger \psi^1$.

The random variables $\{\varphi(t)\}$ defined in (11.87) will be used for estimating expectations involving $R_0^\dagger \psi$, and we introduce a second sequence of eligibility vectors for $R_0^\dagger \psi^1$:

$$\varphi^1(k) = \sum_{t=\tilde{\sigma}_{x^*}^{[k]}}^{k} \psi^1(X(k-t)), \qquad k \geq 1. \tag{11.96}$$

We can then estimate the vector b_ψ defined in (11.95) as $n^{-1}b(n)$ with

$$b(n) = \sum_{t=0}^{n-1} \Big[\big(\varphi(t) - \varphi^1(t) + \psi^1(X(t)) \big) \big(c(X(t)) - \eta(n) \big) \Big]. \tag{11.97}$$

The matrix Σ_ψ can be estimated using standard Monte Carlo $n^{-1}\Sigma(n)$, with

$$\Sigma(n) = \sum_{t=0}^{n-1} \Big[\psi(X(t))\psi^{\mathsf{T}}(X(t)) - \psi^1(X(t))\psi^{1\mathsf{T}}(X(t)) \Big]. \tag{11.98}$$

Hence we obtain estimates of θ^* using

$$\theta(n) = \Sigma(n)^{-1}b(n), \qquad n \geq 1.$$

Once again we obtain a recursive algorithm based on the Matrix Inversion Lemma 11.5.1. Define the two $\ell_h \times 2$ matrices $\Psi(n) = [\psi(X(n)) \mid \psi^1(X(n))]$, $\Psi^-(n) = [\psi(X(n)) \mid -\psi^1(X(n))]$ so that

$$\Psi(n)\Psi^-(n)^{\mathsf{T}} = \psi(X(n))\psi(X(n))^{\mathsf{T}} - \psi^1(X(n))\psi^1(X(n))^{\mathsf{T}}.$$

Then, with $G(n) = \Sigma(n)^{-1}$, the inverse $G(n+1) = [G(n)^{-1} + \Psi(n)\Psi^-(n)^{\mathsf{T}}]^{-1}$ is expressed as (11.99) using the Matrix Inversion Lemma.

Definition 11.5.10 (LSTD learning for shadow functions). For given initial conditions $G(0) > 0$, $b(0) \in \mathbb{R}^{\ell_h}$, $\varphi(0), \varphi^1(0) \in \mathbb{R}^{\ell_h}$, the LSTD algorithm is defined by the sequence of parameter estimates

$$\begin{aligned}
\theta(n+1) &= G(n+1)b(n+1) \\[4pt]
G(n+1) &= G(n) - G(n)\Psi(n)[I + \Psi^-(n)^{\mathsf{T}}G(n)\Psi(n)]^{-1}\Psi^-(n)^{\mathsf{T}}G(n) \\[4pt]
b(n+1) &= b(n) + \big(\varphi(n) - \varphi^1(n) + \psi^1(X(n)) \big) \big(c(X(n)) - \eta(n+1) \big) \\[4pt]
\varphi(n+1) &= \mathbf{1}\{X(n) \neq x^*\}\varphi(n) + \psi(X(n+1)) \\[4pt]
\varphi^1(n+1) &= \mathbf{1}\{X(n) \neq x^*\}\varphi^1(n) + \psi^1(X(n+1)).
\end{aligned}$$
                                        ∎

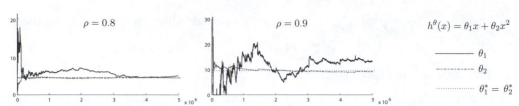

Figure 11.11. LSTD estimates for optimizing shadow functions in the M/M/1 queue using Definition 11.5.10.

Example 11.5.11 (LSTD for the M/M/1 queue). Fig. 11.11 shows results from the LSTD algorthim based on the basis functions $\psi_1(x) \equiv x$, $\psi_2(x) \equiv x^2$. The two experiments shown in the figure are typical results for $\rho = 0.8$ and 0.9.

Note that the sequence $\{\theta(n)\}$ is convergent in this setting, but the variance is again high. This can be seen in the figure, and it can be shown analytically that it is even more variable than the estimates of the mean $\eta = \rho/(1 - \rho)$. Nevertheless, convergence of $\{\theta_2(n)\}$ to a value reasonably close to θ_2^* occurs within 500,000 iterations in each experiment.

Convergence of $\{\theta_1(n)\}$ is *much slower*. This is to be expected since the asymptotic variance of the smoothed estimator is less sensitive to this coefficient.   ∎

11.6 Notes

This chapter spans several disciplines, and the topics surveyed here cover only a small fraction of two fields, simulation and machine learning. *Modern simulation and modeling* by Rubinstein and Melamed [419] provides a broad and accessible introduction to simulation. Machine learning is a rapidly evolving discipline. Some of the most elegant recent work has emerged from Tsitsiklis and his progeny. The monographs [464] and [49] contain introductions to this field, and some specific papers in the area of TD learning are discussed below.

The stochastic approximation algorithm (11.3) was introduced in Robbins and Monro's celebrated 1951 paper [409]. One year later, Kiefer and Wolfowitz introduced a variation intended to solve optimization problems in which the objective function is not smooth, or the derivative is not easily computed [297]. It is remarkable that this highly applicable methodology was developed almost 50 years before computers became fast enough to make these algorithms widely implementable. Reference [384] is a valuable classic text; see [328], [41], or [99] for modern treatments.

A highly influential paper on simulation of Markov chains with an emphasis consistent with this chapter is [269]. In particular, an emphasis of the paper is estimation of confidence bounds, as well as the asymptotic variance (11.6). Heidelberger's thesis on related topics [250, 249] contains a kernel of the control variate techniques based on the shadow functions described in Section 11.4.

The fantastically large runlengths required for accurate simulation in queueing models were first recognized by Whitt [496] and Asmussen [20]. For more on related themes see the work of Glynn, Iglehart, and Whitt [275, 210, 212, 496, 216], and especially the survey [213].

Section 11.2.1 on the CLT for Markov models is based on [214, 368]. Much of Section 11.1 is adapted from the survey [252]. In particular, Proposition 11.1.1 is taken from [252], following [61, Exercise 29.4] (1986 edition).

One interpretation of Proposition 11.3.1 is that the moments and asymptotic variance of the CRW model converge to those of an associated RBM as $\rho \uparrow \infty$, which follows from Kingman's original 1961 result [301]. Related results for generalized Jackson networks are contained in [197, 85].

There are several excellent treatments of large deviations and overflow probabilities for queues. See [217, 153, 446], and in particular the book *Big Queues* [198].

The failure of the LDP for the single-server queue as summarized in Proposition 11.3.4 (as well as Theorem 11.2.4) is taken from [364]. The proof presented here based on Fig. 3.2 is inspired by the papers of Guillemin and Pinchon [221], and Borovkov, Boxma, and Palmowski [74] where the *most likely area* under a tent is computed in an asymptotic setting in which the area tends to infinity. This result can be used to establish a nonzero limit in (11.37), rather than a lower bound.

A valuable reference on control variates and other variance reduction techniques is the *Handbook of Simulation* by Henderson and Nelson [257]. See also Nelson [382], Rubinstein and Melamed [419], and the survey by Glynn and Szechtman [209]. Much of Section 11.4 is based on joint work with Henderson, and his thesis [251]. The quadratic estimator was introduced in [254], which was inspired by Henderson's thesis and prior work with Glynn on control variates for the GI/G/1 queue [251, 253]. Section 11.4.5 on the *fluid estimator* is adapted from [255, 256]. Theory supporting these algorithms is contained in [364, 363, 312].

The histograms shown in Figs. 11.1 and 11.8 were provided as a gift from Prof. Ken Duffy.

Veatch in [482] explores bounded perturbations of the fluid value function in approximate dynamic programming. Related techniques are considered in current research to refine the fluid estimator.

Proposition 11.4.1 is a variant on established control variate techniques (see [331]). The selection of an optimal control variate requires knowledge of a covariance matrix Σ, such as given in (11.43). While it is true that this can be estimated using Monte Carlo or a version of TD learning, there is the danger of increased variance associated with the additional estimation of Σ [331]. The "loss factor" is discussed in [330, 382] for terminating simulations, and in [340] for steady-state simulation. It is argued that estimation should be performed "off-line" based on a large amount of data. Once an estimate of Σ is obtained, this is held fixed for application in subsequent simulations.

The reader is referred to the monographs [464, 49] for general background on TD learning. More detailed treatments can be found in [474, 479, 381, 308, 46] and the references that follow.

The Least-Squares Temporal Difference learning algorithm (LSTD) was introduced for the discounted-cost value function in [78]. The regeneration approach to average-cost TD learning in Section 11.5.4 is based on [308]. Methods to *construct* a basis $\{\psi_i\}$ based on observations of a Markov model are described in [474, 350].

The TD learning algorithm is typically presented in a modified form. For example, in the discounted-cost case the eligibility vectors defined in (11.82) are modified through the introduction of a "forgetting factor" $\lambda \in [0, 1]$,

$$\varphi(k+1) = (1+\gamma)^{-1}[\lambda \varphi(k) + \psi(X(k+1))], \qquad \varphi(0) = 0.$$

The resulting algorithm (11.83) is called TD(λ), where the definition of the temporal differences remain unchanged. Under general conditions, the algorithm remains convergent to some $\theta(\infty) \in \mathbb{R}^{\ell_h}$, but it is no longer consistent. That is, in general $\theta(\infty) \neq \theta^*$, although bounds on the error $\|\theta(\infty) - \theta^*\|_\pi$ as a function of λ can be constructed [474, 288].

The introduction of λ is difficult to justify, given the fact that variance reduction can be achieved through state weighting, as in (11.88), and this approach does not introduce bias.

Optimization of control variates in i.i.d. models is part of the original formulation of this technique. In the Markov setting this step is substantially more difficult and solutions are more recent. The first algorithms for this purpose were introduced in [465, 299]. The LSTD algorithm for shadow function optimization contained in Section 11.5.5 is new, and this is the first such algorithm that is asymptotically consistent.

Exercises

11.1   Simulate the M/M/n queueing system described in Exercise 13.9. Find the mean time spent by a job in the queue when $\alpha = 0.9$ for the following four values: $\mu = n^{-1}$, with $n = 1, 2, 3, 4$. Estimate a 95% confidence interval associated with your results using the batch-means method.

11.2 Consider again the single-server queueing system introduced in Exercise 3 consisting of three job classes. Estimate the mean and variance of the time spent by each job class in the queue, together with the server utilization. Estimate the asymptotic covariance of the vector process (Q_H, Q_M, Q_L). Estimate the fraction of jobs of each job class that are rejected.

11.3 *Ordinal estimation.* We have seen that simulation to estimate the steady-state cost can be difficult in a stochastic network model. Even in the single-server queue the variance of the standard estimator grows rapidly with load, and the LDP fails as shown in Theorem 11.0.1.

Comparison of policies can be relatively easy. In this exercise you will explore this principle using the one-dimensional inventory model,

$$Y(t+1) = Y(t) - S(t+1) + S(t+1)\iota(t) + L(t+1), \qquad t \geq 0, \ Y(0) \in \mathbb{Z},$$

where (S, L) is an i.i.d. process, S is Bernoulli, and the marginal of L is supported on a bounded interval in $\mathbb{Z}_+$ with $\mathsf{P}\{L(t) > S(t)\} > 0$, and $\mathsf{E}[L(t)] = \lambda < \mu = \mathsf{E}[S(t)]$.

Suppose that $\{\phi^i : i = 1, 2\}$ are two hedging point policies defined by two integer-valued thresholds $\{\overline{y}_i : i = 1, 2\}$,

$$\phi^i(y) = \max(0, \overline{y}_i + 1 - y).$$

The resulting workload processes are driven by the same sequence (S, L) and hence satisfy

$$Y^1(t+1) = \max(Y^1(t) - S(t+1), \overline{y}^1) + L(t+1)$$
$$Y^2(t+1) = \max(Y^2(t) - S(t+1), \overline{y}^2) + L(t+1).$$

Let $\{\eta^i(n) : n \geq 1\}$ denote the Monte Carlo estimates of the steady-state mean $\eta^i = \lim_{t\to\infty} \mathsf{E}[c(Y^i(t))]$, where c is some Lipschitz continuous cost function.

(a) Explain why the LDP fails, in the sense that for each initial condition,

$$\lim_{n\to\infty} \frac{1}{n} \log\left(\mathsf{P}\{\eta^i(n) \geq r\}\right) = 0, \qquad r > \eta^i, \ i = 1, 2.$$

(b) Show that there is a fixed constant b_0 independent of t such that $|Y^1(t) - Y^2(t)| \leq b_0$ for all $t \geq 1$ whenever $Y^1(0) = Y^2(0)$.

(c) Show that if $\eta^1 < \eta^2$ then the probability of error decays geometrically fast:

$$\limsup_{n\to\infty} \frac{1}{n} \log\left(\mathsf{P}\{\eta^1(n) \geq \eta^2(n)\}\right) < 0.$$

Suggested approach: First, explain why $\Phi(t) := (Y^1(t), Y^2(t))$ defines a Markov chain. Next, show that Φ is geometrically ergodic. The result then follows from Proposition 11.2.5.

11.4   Simulate the model introduced in Exercise 9.4 under several threshold policies. In each of your simulations use *identical* arrival and service processes, and plot the average cost as a function of the threshold parameter. Repeat the experiment several times, with a time window of $T = 10,000$. Do the plots give consistent predictions of the best policy?

11.5   Consider again the queue with phase-type arrivals and service introduced in Exercise 4.10. Choose a model, and simulate to obtain an approximation for the asymptotic variance σ^2 of $A(t) - S(t)$. For example, you can use the batch-means method (note that σ^2 is *not* the ordinary variance of $A(t) - S(t)$.) Next, simulate the queue to obtain an estimate of its mean. Does the formula (3.10) provide a reasonable approximation for the mean you obtained via simulation? Explain why it should for high load.

11.6   Simulate the model introduced in Exercise 6.1 using a policy based on a translation of the optimal policy for a one-dimensional relaxation. Try a control variate of your choosing, and experiment with variations on your policy.

11.7   Simulate the policy obtained for the ALOHA communications model in Exercise 6.1. You should try a control variate of your choosing, and experiment with variations on your policy.

11.8   Estimate the moment-generating function $\mathsf{E}_\pi[e^{\beta Q(t)}]$ for a range of $\beta \in [-1, 1]$ for the M/M/1 queue using,

(a) Standard Monte Carlo.

(b) The smoothed estimator based on the fluid value function,

$$\psi_\beta(x) = \int_0^\infty \left(e^{\beta q(t;x)} - 1 \right) dt.$$

Note that this function is C^1 as a function of x.

(c) Verify that $e^{\beta x} + \Delta_{\psi_\beta}(x)$ is not bounded over $x \geq 0$, when ψ_β is given in (ii). Find a simple modification of ψ_β to make this function bounded, and use the resulting shadow function to obtain a third set of estimates.

Compare your results using a plot of the form shown in Section 11.4.3.

11.9   In this exercise you will simulate the infinite-server queue introduced in Exercise 3.10. Estimate the mean $\mathsf{E}_\pi[Q(t)]$, as well as the moment generating function $\mathsf{E}_\pi[e^{\beta Q(t)}]$ for $\beta \in \{\pm 0.25, \pm 0.5, \pm 0.75, \pm 1\}$ using *identical* arrival and service processes. Obtain these results using standard Monte Carlo, and using the smoothed estimator based on nine different fluid value functions. See if you can construct a better shadow function, as in the previous problem.

11.10   Construct a control variate for the simple inventory model based on a fluid value function. Try each of the functions introduced in Exercise 7.6, and check to see whether $c + \Delta_\psi$ is a bounded function on $\mathbb{R}$. Perform a simulation with and without the control variate and comment on the variance reduction observed.

11.11 Consider Proposition 11.3.4 (i): If $(\mu - \alpha)/4 < \beta < (\mu - \alpha)/2$ then the "tent" is truncated at time T. How would the bound (11.36) change in this case? What happens if $\beta > (\mu - \alpha)/2$?

11.12 Suppose that $g \colon \mathbb{R}_+ \to \mathbb{R}_+$ is increasing and unbounded. Following the ideas in Proposition 11.3.4 (iii), show that for each $\varepsilon > 0$,

$$\liminf_{n \to \infty} \frac{1}{n} \log \mathsf{P}\{\eta(n) \geq \eta + \varepsilon\} = 0, \qquad (\text{E}11.1)$$

where $\eta(n) = n^{-1} \sum_{t=0}^{n-1} g(Q(t))$.

11.13 Consider a CRW model for the tandem queues with $\rho_\bullet < 1$ and $c(x) = x_1 + x_2$. Is the function $c - \Delta_J$ bounded when J is the fluid value function? How about $c - \Delta_h$, with $h(x) = J(\tilde{x})$ and $\tilde{x}$ defined in (4.93)?

11.14 *Control variates for a queue in heavy traffic.* If $\rho \approx 1$ then the scaled queue-length process $Q(nt; \sqrt{n}x)/\sqrt{n}$ for the CRW queue can be approximated by a reflected Brownian motion starting from x, where $n = (1 - \rho)^{-1}$. The solution to Poisson's equation for the RBM is precisely the fluid value function $J(x) = \frac{1}{2}x^2/(\mu - \alpha)$. For the CRW model an expression of the form (11.48) holds, and hence the smoothed estimator has the form

$$\eta_J^s(n) = \frac{1}{n} \sum_{t=0}^{n-1} R(Q(t)), \qquad n \geq 1.$$

In this exercise you will estimate the asymptotic variance for $\{\eta_J^s(n)\}$. Each part below concerns the CRW queue (3.1) satisfying $\mathsf{E}[A(k)^2] < \infty$.

  (a) Verify (11.48) with $R(x) = b\mathbf{1}_0(x)$, for some constant b.

  (b) Verify that $h(x) = b_R x$ solves Poisson's equation $Ph = h - R + \pi(R)$ for some constant $b_R < 0$, where $\pi(R) = (1 - \rho)b$.

  (c) Compute the asymptotic variance for the smoothed estimator using Theorem 11.2.2.

11.15 *Nonlinear control variates.* In this exercise you will consider nonlinear-control variates for variance reduction in simulation. Suppose that c and f are two functions on X and that the mean $\pi(f)$ is known. Let $\eta(n)$ and $\eta_f(n)$, $n \geq 1$, denote the Monte Carlo estimates of $\eta = \pi(c)$ and $\eta_f = \pi(f)$, respectively. It is assumed that $G \colon \mathbb{R}^2 \to \mathbb{R}$ is a smooth function satisfying $G(r, \eta_f) = r$ for all real r, and the smoothed estimator is defined by

$$\eta_\psi^s(n) = G(\eta(n), \eta_f(n)), \qquad n \geq 1. \qquad (\text{E}11.2)$$

  (a) First, let us consider an abstraction: Suppose that $\mathcal{W}$ is a two-dimensional Gaussian random variable with covariance Σ. Suppose that the first and second derivatives of G have polynomial growth. Compute the limit

$$\sigma_s^2 = \lim_{n \to \infty} n \mathsf{Var}\big(G(\eta^2 + n^{-\frac{1}{2}}\mathcal{W})\big).$$

where $\eta^2 = (\eta, \eta_f)^\mathsf{T}$. *Hint*: Apply the mean value theorem,

$$G(\eta^2 + \frac{1}{\sqrt{n}}w) = G(\eta^2) + \frac{1}{\sqrt{n}}w^{\mathsf{T}}\nabla G(\eta^2) + \frac{1}{n}w^{\mathsf{T}}\nabla G(\eta^2 + \theta n^{-\frac{1}{2}}w)w,$$

$w \in \mathbb{R}^2$, where $\theta \in (0, 1)$.

(b) Find an expression for the asymptotic variance of the smoothed estimator (E11.2) based on the asymptotic variance of the bivariate process $\{c(X(t)), f(X(t)) : t \geq 0\}$. Explain what assumptions on c, f are needed when X is i.i.d. Any thoughts on the Markov case?

(c) Consider the special case: X is Markov chain on $\mathbb{Z}_+$ with known mean $m_1 = \mathsf{E}[X]$, and we wish to estimate $m_2 = \mathsf{E}[X^2]$. Construct a function G satisfying the required conditions, such as $G(x, y) = y((m_1 + 1)/(1 + x))^2$. Work out an expression for the asymptotic variance based on the asymptotic variance of the bivariate process $\{X(t), X^2(t) : t \geq 0\}$.

11.16 The queue-length process Q is not Markov for the model with phase-type arrival and service. However, this does not stop us from applying the TD learning algorithm. Based on the model you have considered in Exercise 11.5 run one of the algorithms introduced in Example 11.5.7. Compare the function obtained to the actual solution to Poisson's equation for a CRW queue with identical first- and second-order statistics, based on (3.11).

11.17 Consider the "Bellman error" for the discounted-cost value function defined by

$$\mathsf{E}[(Ph^{\theta}(X) - (1 + \gamma)h^{\theta}(X) + c(X))^2].$$

Show that the parameter θ^* minimizing the error satisfies

$$\mathsf{E}[(Ph^{\theta}(X) - (1 + \gamma)h^{\theta}(X) + c(X))(P\psi^{\theta}(X) - (1 + \gamma)\psi^{\theta}(X))] = 0.$$

Based on this formula, obtain an expression for θ^* analogous to (11.64) when the parameterization is linear. How would you use simulation to estimate θ^*?

11.18 The L_1 Bellman error is defined by

$$\mathsf{E}[|Ph^{\theta}(X) - (1 + \gamma)h^{\theta}(X) + c(X)|].$$

Find an expression for the minimizing θ for a linear parameterization. How would you use simulation to estimate θ^*?

Appendix

Markov Models

This appendix describes stability theory and ergodic theory for Markov chains on a countable state space that provides foundations for the development in Part III of this book. It is distilled from Meyn and Tweedie [368], which contains an extensive bibliography (the monograph [368] is now available online).

The term "chain" refers to the assumption that the time parameter is discrete. The Markov chains that we consider evolve on a countable state space, denoted X, with transition law defined as follows:

$$P(x,y) := \mathsf{P}\{X(t+1) = y \mid X(t) = x\} \qquad x,y \in \mathsf{X},\ t = 0,1,\dots.$$

The presentation is designed to allow generalization to more complex general state space chains as well as reflected Brownian motion models.

Since the publication of [368] there has been a great deal of progress on the theory of geometrically ergodic Markov chains, especially in the context of Large Deviations theory. See [310, 311, 309] and [364] for some recent results. The website [444] also contains online surveys on Markov and Brownian models.

A.1 Every process is (almost) Markov

Why do we focus so much attention on Markov chain models? An easy response is to cite the powerful analytical techniques available, such as the operator-theoretic techniques surveyed in this appendix. A more practical reply is that most processes can be approximated by a Markov chain.

Consider the following example: Z is a stationary stochastic process on the non-negative integers. A Markov chain can be constructed that has the same steady-state behavior, and similar short-term statistics. Specifically, define the probability measure on $\mathbb{Z}_+ \times \mathbb{Z}_+$ via

$$\Pi(z_0, z_1) = \mathsf{P}\{Z(t) = z_0, Z(t+1) = z_1\}, \qquad z_0, z_1 \in \mathbb{Z}_+.$$

Note that Π captures the steady-state behavior by construction. By considering the distribution of the pair $(Z(t), Z(t+1))$ we also capture some of the dynamics of Z.

505

The first and second marginals of Π agree, and are denoted π,

$$\pi(z_0) = \mathsf{P}\{Z(t) = z_0\} = \sum_{z_1 \in \mathbb{Z}_+} \Pi(z_0, z_1), \qquad z_0, \in \mathbb{Z}_+.$$

The transition matrix for the approximating process is defined as the ratio

$$P(z_0, z_1) = \frac{\Pi(z_0, z_1)}{\pi(z_0)}, \qquad z_0, z_1 \in \mathsf{X},$$

with $\mathsf{X} = \{z \in \mathbb{Z}_+ : \pi(z) > 0\}$.

The following simple result is established in [108], but the origins are undoubtedly ancient. It is a component of the model reduction techniques pioneered by Mori and Zwanzig in the area of statistical mechanics [375, 509].

Proposition A.1.1. *The transition matrix P describes these aspects of the stationary process $\mathbf{Z}$:*

(i) *One-step dynamics:* $P(z_0, z_1) = \mathsf{P}\{Z(t+1) = z_1 \mid Z(t) = z_0\}$, $z_0, z_1 \in \mathsf{X}$.
(ii) *Steady-state: The probability π is invariant for P,*

$$\pi(z_1) = \sum_{z_0 \in \mathsf{X}} \pi(z_0) P(z_0, z_1), \qquad z_1, \in \mathsf{X}.$$

Proof. Part (i) is simply Bayes' rule

$$\mathsf{P}\{Z(t+1) = z_1 \mid Z(t) = z_0\} = \frac{\mathsf{P}\{Z(t+1) = z_1, Z(t) = z_0\}}{\mathsf{P}\{Z(t) = z_0\}} = \frac{\Pi(z_0, z_1)}{\pi(z_0)}.$$

The definition of P gives $\pi(z_0) P(z_0, z_1) = \Pi(z_0, z_1)$, and stationarity of $\mathbf{Z}$ implies that $\sum_{z_0} \pi(z_0) P(z_0, z_1) = \sum_{z_0} \Pi(z_0, z_1) = \pi(z_1)$, which is (ii). $\square$

Proposition A.1.1 is just one approach to approximation. If $\mathbf{Z}$ is not stationary, an alternative is to redefine Π as the limit

$$\Pi(z_0, z_1) = \lim_{N \to \infty} \frac{1}{N} \sum_{t=0}^{N-1} \mathsf{P}\{Z(t) = z_0, Z(t+1) = z_1\},$$

assuming that this exists for each $z_0, z_1 \in \mathbb{Z}_+$. Similar ideas are used in Section 9.2.2 to prove that an optimal policy for a controlled Markov chain can be taken stationary without loss of generality.

Another common technique is to add some history to $\mathbf{Z}$ via

$$X(t) := [Z(t), Z(t-1), \dots, Z(t-n_0)],$$

where $n_0 \in [1, \infty]$ is fixed. If $n_0 = \infty$ then we are including the entire history, and in this case $\mathbf{X}$ is Markov: For any possible value x_1 of $X(t+1)$,

$$\mathsf{P}\{X(t+1) = x_1 \mid X(t), X(t-1), \dots\} = \mathsf{P}\{X(t+1) = x_1 \mid X(t)\}.$$

A.2 Generators and value functions

The main focus of the Appendix is performance evaluation, where performance is defined in terms of a cost function $c\colon \mathsf{X} \to \mathbb{R}_+$. For a Markov model there are several performance criteria that are well motivated and are also conveniently analyzed using tools from the general theory of Markov chains:

Discounted cost For a given discount parameter $\gamma > 0$, recall that the discounted-cost value function is defined as the sum

$$h_\gamma(x) := \sum_{t=0}^{\infty}(1+\gamma)^{-t-1}\mathsf{E}_x[c(X(t))], \qquad X(0) = x \in \mathsf{X}. \tag{A1.1}$$

Recall from (1.18) that the expectations in (A1.1) can be expressed in terms of the t-step transition matrix via

$$\mathsf{E}[c(X(t)) \mid X(0) = x] = P^t c\,(x), \qquad x \in \mathsf{X},\ t \geq 0.$$

Consequently, denoting the *resolvent* by

$$R_\gamma = \sum_{t=0}^{\infty}(1+\gamma)^{-t-1}P^t, \tag{A1.2}$$

the value function (A1.1) can be expressed as the "matrix-vector product,"

$$h_\gamma(x) = R_\gamma c\,(x) := \sum_{y\in\mathsf{X}} R_\gamma(x,y)c(y), \qquad x \in \mathsf{X}.$$

Based on this representation, it is not difficult to verify the following dynamic programming equation. The discounted-cost value function solves

$$\mathcal{D}h_\gamma = -c + \gamma h_\gamma, \tag{A1.3}$$

where the *generator* $\mathcal{D}$ is defined as the difference operator

$$\mathcal{D} = P - I. \tag{A1.4}$$

The dynamic programming equation (A1.3) is a first step in the development of dynamic programming for controlled Markov chains contained in Chapter 9.

Average cost The average cost is the limit supremum of the Cesaro averages,

$$\eta_x := \limsup_{r\to\infty}\frac{1}{r}\sum_{t=0}^{r-1}\mathsf{E}_x\big[c(X(t))\big], \qquad X(0) = x \in \mathsf{X}.$$

A probability measure is called invariant if it satisfies the invariance equation

$$\sum_{y\in\mathsf{X}}\pi(x)\mathcal{D}(x,y) = 0, \qquad x \in \mathsf{X}. \tag{A1.5}$$

Under mild stability and irreducibility assumptions we find that the average cost coincides with the spatial average $\pi(c) = \sum_{x'}\pi(x')c(x')$ for each initial condition. Under these conditions the limit supremum in the definition of the average cost becomes a limit, and it is also the limit of the normalized discounted cost for vanishing discount rate,

$$\eta_x = \pi(c) = \lim_{r \to \infty} \frac{1}{r} \sum_{t=0}^{r-1} \mathsf{E}_x \big[c(X(t)) \big] = \lim_{\gamma \downarrow 0} \gamma h_\gamma(x). \qquad (A1.6)$$

In a queueing network model the following x^*-*irreducibility* assumption frequently holds with $x^* \in \mathsf{X}$ taken to represent a network free of customers.

Definition A.2.1 (Irreducibility). The Markov chain X is called

(i) x^*-*Irreducible* if $x^* \in \mathsf{X}$ satisfies for one (and hence any) $\gamma > 0$,

$$R_\gamma(x, x^*) > 0 \qquad \text{for each } x \in \mathsf{X}.$$

(ii) The chain is simply called *irreducible* if it is x^*-irreducible for each $x^* \in \mathsf{X}$.

(iii) An x^*-irreducible chain is called *aperiodic* if there exists $n_0 < \infty$ such that $P^n(x^*, x^*) > 0$ for all $n \geq n_0$. ∎

When the chain is x^*-irreducibile, we find that the most convenient sample-path representations of η are expressed with respect to the *first return time* τ_{x^*} to the fixed state $x^* \in \mathsf{X}$. From Proposition A.3.1 we find that η is independent of x within the support of π, and has the form

$$\eta = \pi(c) = \left(\mathsf{E}_{x^*} \big[\tau_{x^*} \big] \right)^{-1} \mathsf{E}_{x^*} \left[\sum_{t=0}^{\tau_{x^*}-1} c(X(t)) \right]. \qquad (A1.7)$$

Considering the function $c(x) = \mathbf{1}\{x \neq x^*\}$ gives

Theorem A.2.2 (Kac's Theorem). *If X is x^*-irreducible then it is positive recurrent if and only if $\mathsf{E}_{x^*}[\tau_{x^*}] < \infty$. If positive recurrence holds, then letting π denote the invariant measure for X, we have*

$$\pi(x^*) = (\mathsf{E}_{x^*}[\tau_{x^*}])^{-1}. \qquad (A1.8)$$

Total cost and Poisson's equation For a given function $c\colon \mathsf{X} \to \mathbb{R}$ with steady-state mean η, denote the centered function by $\tilde{c} = c - \eta$. Poisson's equation can be expressed

$$\mathcal{D}h = -\tilde{c}. \qquad (A1.9)$$

The function c is called the *forcing function*, and a solution $h\colon \mathsf{X} \to \mathbb{R}$ is known as a *relative value function*. Poisson's equation can be regarded as a dynamic programming equation; note the similarity between (A1.9) and (A1.3).

Under the x^*-irreducibility assumption we have various representations of the relative value function. One formula is similar to the definition (A1.6):

$$h(x) = \lim_{\gamma \downarrow 0} \big(h_\gamma(x) - h_\gamma(x^*) \big), \qquad x \in \mathsf{X}. \qquad (A1.10)$$

Alternatively, we have a sample-path representation similar to (A1.7),

$$h(x) = \mathsf{E}_x \left[\sum_{t=0}^{\tau_{x^*}-1} \left(c(X(t)) - \eta \right) \right], \qquad x \in \mathsf{X}. \tag{A1.11}$$

This appendix contains a self-contained treatment of Lyapunov criteria for stability of Markov chains to validate formulae such as (A1.11). A central result known as the *Comparison Theorem* is used to obtain bounds on η or any of the value functions described above.

These stability criteria are all couched in terms of the generator for X. The most basic criterion is known as condition (V3): for a function $V : \mathsf{X} \to \mathbb{R}_+$, a function $f : \mathsf{X} \to [1, \infty)$, a constant $b < \infty$, and a finite set $S \subset \mathsf{X}$,

$$\mathcal{D}V(x) \leq -f + b\mathbf{1}_S(x), \qquad x \in \mathsf{X}, \tag{V3}$$

or equivalently,

$$\mathsf{E}[V(X(t+1)) - V(X(t)) \mid X(t) = x] \leq \begin{cases} -f(x) & x \in S^c \\ -f(x) + b & x \in S. \end{cases} \tag{A1.12}$$

Under this *Lyapunov drift condition* we obtain various ergodic theorems in Section A.5. The main results are summarized as follows:

Theorem A.2.3. *Suppose that X is x^*-irreducible and aperiodic, and that there exist $V : \mathsf{X} \to (0, \infty)$, $f : \mathsf{X} \to [1, \infty)$, a finite set $S \subset \mathsf{X}$, and $b < \infty$ such that condition (V3) holds. Suppose moreover that the cost function $c : \mathsf{X} \to \mathbb{R}_+$ satisfies $\|c\|_f := \sup_{x \in \mathsf{X}} c(x)/f(x) \leq 1$.*

Then there exists a unique invariant measure π satisfying $\eta = \pi(c) \leq b$, and the following hold:

(i) *Strong Law of Large Numbers: For each initial condition,* $\dfrac{1}{n} \sum_{t=0}^{n-1} c(X(t)) \to \eta$ *a.s. as $n \to \infty$.*

(ii) *Mean Ergodic Theorem: For each initial condition, $\mathsf{E}_x[c(X(t))] \to \eta$ as $t \to \infty$.*

(iii) *Discounted-cost value function h_γ: Satisfies the uniform upper bound,*

$$h_\gamma(x) \leq V(x) + b\gamma^{-1}, \qquad x \in \mathsf{X}.$$

(iv) *Poisson's equation h: Satisfies, for some $b_1 < \infty$,*

$$|h(x) - h(y)| \leq V(x) + V(y) + b_1, \qquad x, y \in \mathsf{X}.$$

Proof. The Law of Large Numbers is given in Theorem A.5.8, and the mean ergodic theorem is established in Theorem A.5.4 based on coupling X with a stationary version of the chain.

The bound $\eta \leq b$ along with the bounds on h and h_γ are given in Theorem A.4.6.

$\square$

These results are refined elsewhere in the book in the construction and analysis of algorithms to bound or approximate performance in network models.

A.3 Equilibrium equations

In this section we consider in greater detail representations for π and h, and begin to discuss existence and uniqueness of solutions to equilibrium equations.

A.3.1 Representations

Solving either Eq. (A1.5) or (A1.9) amounts to a form of inversion, but there are two difficulties. One is that the matrices to be inverted may not be finite dimensional. The other is that these matrices are *never invertable*! For example, to solve Poisson's equation (A1.9) it appears that we must invert $\mathcal{D}$. However, the function f which is identically equal to 1 satisfies $\mathcal{D}f \equiv 0$. This means that the null space of $\mathcal{D}$ is nontrivial, which rules out invertibility.

On iterating the formula $Ph = h - \tilde{c}$ we obtain the sequence of identities

$$P^2 h = h - \tilde{c} - P\tilde{c} \implies P^3 h = h - \tilde{c} - P\tilde{c} - P^2\tilde{c} \implies \cdots .$$

Consequently, one might expect a solution to take the form

$$h = \sum_{i=0}^{\infty} P^i \tilde{c}. \tag{A1.13}$$

When the sum converges absolutely, then this function does satisfy Poisson's equation (A1.9).

A representation which is more generally valid is defined by a random sum. Define the first entrance time and first return time to a state $x^* \in \mathsf{X}$ by, respectively,

$$\sigma_{x^*} = \min(t \geq 0 : X(t) = x^*), \qquad \tau_{x^*} = \min(t \geq 1 : X(t) = x^*). \tag{A1.14}$$

Proposition A.3.1 (i) is contained in [368, Theorem 10.0.1], and (ii) is explained in Section 17.4 of [368].

Proposition A.3.1. *Let $x^* \in \mathsf{X}$ be a given state satisfying $\mathsf{E}_{x^*}[\tau_{x^*}] < \infty$. Then:*

(i) *The probability distribution defined below is invariant:*

$$\pi(x) := \left(\mathsf{E}_{x^*}[\tau_{x^*}]\right)^{-1} \mathsf{E}_{x^*}\left[\sum_{t=0}^{\tau_{x^*}-1} \mathbf{1}(X(t) = x)\right], \qquad x \in \mathsf{X}. \tag{A1.15}$$

(ii) *With π defined in (i), suppose that $c\colon \mathsf{X} \to \mathbb{R}$ is a function satisfying $\pi(|c|) < \infty$. Then the function defined below is finite valued on $\mathsf{X}_\pi :=$ the support of π,*

$$h(x) = \mathsf{E}_x\left[\sum_{t=0}^{\tau_{x^*}-1} \tilde{c}(X(t))\right] = \mathsf{E}_x\left[\sum_{t=0}^{\sigma_{x^*}} \tilde{c}(X(t))\right] - \tilde{c}(x^*), \qquad x \in \mathsf{X}. \tag{A1.16}$$

Moreover, h solves Poisson's equation on X_π.

The formulae for π and h given in Proposition A.3.1 are perhaps the most commonly known representations. In this section we develop operator-theoretic representations that are truly based on matrix inversion. These representations help to simplify the stability theory that follows, and they also extend most naturally to general state-space Markov chains, and processes in continuous time.

Operator-theoretic representations are formulated in terms of the resolvent *resolvent matrix* defined in (A1.2). In the special case $\gamma = 1$ we omit the subscript and write

$$R(x, y) = \sum_{t=0}^{\infty} 2^{-t-1} P^t(x, y), \qquad x, y \in \mathsf{X}. \tag{A1.17}$$

In this special case, the resolvent satisfies $R(x, \mathsf{X}) := \sum_y R(x, y) = 1$, and hence it can be interpreted as a transition matrix. In fact, it is precisely the transition matrix for a sampled process. Suppose that $\{t_k\}$ is an i.i.d. process with geometric distribution satisfying $\mathsf{P}\{t_k = n\} = 2^{-n-1}$ for $n \geq 0$, $k \geq 1$. Let $\{T_k : k \geq 0\}$ denote the sequence of partial sums

$$T_0 = 0, \text{ and } T_{k+1} = T_k + t_{k+1} \text{ for } k \geq 0.$$

Then, the sampled process,

$$Y(k) = X(T_k), \qquad k \geq 0, \tag{A1.18}$$

is a Markov chain with transition matrix R.

Solutions to the invariance equations for Y and X are closely related:

Proposition A.3.2. *For any Markov chain X on X with transition matrix P:*

(i) *The resolvent equation holds,*

$$\mathcal{D}R = R\mathcal{D} = \mathcal{D}_R, \qquad \text{where } \mathcal{D}_R = R - I. \tag{A1.19}$$

(ii) *A probability distribution π on X is P-invariant if and only if it is R-invariant.*

(iii) *Suppose that an invariant measure π exists, and that $g \colon \mathsf{X} \to \mathbb{R}$ is given with $\pi(|g|) < \infty$. Then, a function $h \colon \mathsf{X} \to \mathbb{R}$ solves Poisson's equation $\mathcal{D}h = -\tilde{g}$ with $\tilde{g} := g - \pi(g)$, if and only if*

$$\mathcal{D}_R h = -R\tilde{g}. \tag{A1.20}$$

Proof. From the definition of R we have

$$PR = \sum_{t=0}^{\infty} 2^{-(t+1)} P^{t+1} = \sum_{t=1}^{\infty} 2^{-t} P^t = 2R - I.$$

Hence $\mathcal{D}R = PR - R = R - I$, proving (i).

To see (ii) we premultiply the resolvent equation (A1.19) by π,

$$\pi \mathcal{D}R = \pi \mathcal{D}_R.$$

Obviously then, $\pi \mathcal{D} = 0$ if and only if $\pi \mathcal{D}_R = 0$, proving (ii). The proof of (iii) is similar. $\qquad \square$

The operator-thoretic representations of π and h are obtained under the following *minorization condition*: Suppose that $s\colon \mathsf{X} \to \mathbb{R}_+$ is a given function, and ν is a probability on X such that

$$R(x, y) \geq s(x)\nu(y), \qquad x, y \in \mathsf{X}. \tag{A1.21}$$

For example, if ν denotes the probability on X which is concentrated at a singleton $x^* \in \mathsf{X}$, and s denotes the function on X given by $s(x) := R(x, x^*)$, $x \in \mathsf{X}$, then we do have the desired lower bound,

$$R(x, y) \geq R(x, y)\mathbf{1}_{x^*}(y) = s(x)\nu(y), \qquad x, y \in \mathsf{X}.$$

The inequality (A1.21) is a matrix inequality that can be written compactly as

$$R \geq s \otimes \nu \tag{A1.22}$$

where R is viewed as a matrix, and the right-hand side is the outer product of the column vector s, and the row vector ν. From the resolvent equation and (A1.22) we can now give a roadmap for solving the invariance equation (A1.5). Suppose that we already have an invariant measure π, so that

$$\pi R = \pi.$$

Then, on subtracting $s \otimes \nu$ we obtain

$$\pi(R - s \otimes \nu) = \pi R - \pi[s \otimes \nu] = \pi - \delta\nu,$$

where $\delta = \pi(s)$. Rearranging gives

$$\pi[I - (R - s \otimes \nu)] = \delta\nu. \tag{A1.23}$$

We can now attempt an inversion. The point is that the operator $\mathcal{D}_R := I - R$ is not invertible, but by subtracting the outer product $s \otimes \nu$ there is some hope in constructing an inverse. Define the *potential matrix* as

$$G = \sum_{n=0}^{\infty} (R - s \otimes \nu)^n. \tag{A1.24}$$

Under certain conditions we do have $G = [I - (R - s \otimes \nu)]^{-1}$, and hence from (A1.23) we obtain the representation of π,

$$\pi = \delta[\nu G]. \tag{A1.25}$$

We can also attempt the "forward direction" to construct π: Given a pair s, ν satisfying the lower bound (A1.22), we *define* $\mu := \nu G$. We must then answer two questions: (i) when is μ invariant? (ii) when is $\mu(\mathsf{X}) < \infty$? If both are affirmative, then we do have an invariant measure, given by

$$\pi(x) = \frac{\mu(x)}{\mu(\mathsf{X})}, \qquad x \in \mathsf{X}.$$

We will show that μ always exists as a finite-valued measure on X, and that it is always *subinvariant*,

$$\mu(y) \geq \sum_{x \in \mathsf{X}} \mu(x) R(x,y), \qquad y \in \mathsf{X}.$$

Invariance and finiteness both require some form of *stability* for the process.

The following result shows that the formula (A1.25) coincides with the representation given in (A1.15) for the sampled chain $\mathbf{Y}$.

Proposition A.3.3. *Suppose that* $\nu = \delta_{x^*}$, *the point mass at some state* $x^* \in \mathsf{X}$, *and suppose that* $s(x) := R(x, x^*)$ *for* $x \in \mathsf{X}$. *Then we have for each bounded function* $g \colon \mathsf{X} \to \mathbb{R}$,

$$(R - s \otimes \nu)^n g\,(x) = \mathsf{E}_x[g(Y(n))\mathbf{1}\{\tau_{x^*}^Y > n\}], \qquad x \in \mathsf{X}, \; n \geq 1, \qquad \text{(A1.26)}$$

where $\tau_{x^*}^Y$ *denotes the first return time to* x^* *for the chain* $\mathbf{Y}$ *defined in (A1.18). Consequently,*

$$Gg\,(x) := \sum_{n=0}^{\infty} (R - s \otimes \nu)^n g\,(x) = \mathsf{E}_x \left[\sum_{t=0}^{\tau_{x^*}^Y - 1} g(Y(t)) \right].$$

Proof. We have $(R - s \otimes \nu)(x,y) = R(x,y) - R(x,x^*)\mathbf{1}_{y=x^*} = R(x,y)\mathbf{1}_{y \neq x^*}$. Or, in probabilistic notation,

$$(R - s \otimes \nu)(x,y) = \mathsf{P}_x\{Y(1) = y, \tau_{x^*}^Y > 1\}, \qquad x,y \in \mathsf{X}.$$

This establishes the formula (A1.26) for $n = 1$. The result then extends to arbitrary $n \geq 1$ by induction. If (A1.26) is true for any given n, then

$$(R - s \otimes \nu)^{n+1}(x,g)$$
$$= \sum_{y \in \mathsf{X}} \left[(R - s \otimes \nu)(x,y)\right]\left[(R - s \otimes \nu)^n(y,g)\right]$$
$$= \sum_{y \in \mathsf{X}} \mathsf{P}_x\{Y(1) = y, \tau_{x^*}^Y > 1\}\mathsf{E}_y[g(Y(n))\mathbf{1}\{\tau_{x^*}^Y > n\}]$$
$$= \mathsf{E}_x\left[\mathbf{1}\{\tau_{x^*}^Y > 1\}\mathsf{E}[g(Y(n+1))\mathbf{1}\{Y(t) \neq x^*, \, t = 2, \ldots, n+1\} \mid Y(1)]\right]$$
$$= \mathsf{E}_x\left[g(Y(n+1))\mathbf{1}\{\tau_{x^*}^Y > n+1\}\right]$$

where the second equation follows from the induction hypothesis, and in the third equation the Markov property was applied in the form (1.19) for $\mathbf{Y}$. The final equation follows from the smoothing property of the conditional expectation. $\qquad\square$

A.3.2 Communication

The following result shows that one can assume without loss of generality that the chain is irreducible by restricting to an *absorbing* subset of X. The set $\mathsf{X}_{x^*} \subset \mathsf{X}$ defined in Proposition A.3.4 is known as a *communicating class*.

Proposition A.3.4. *For each $x^* \in X$ the set defined by*

$$X_{x^*} = \{y : R(x^*, y) > 0\} \tag{A1.27}$$

is absorbing: $P(x, X_{x^*}) = 1$ for each $x \in X_{x^*}$. Consequently, if X is x^*-irreducible then the process may be restricted to X_{x^*}, and the restricted process is irreducible.

Proof. We have $DR = R - I$, which implies that $R = \frac{1}{2}(RP + I)$. Consequently, for any $x_0, x_1 \in X$ we obtain the lower bound,

$$R(x^*, x_1) \geq \frac{1}{2} \sum_{y \in X} R(x^*, y) P(y, x_1) \geq \frac{1}{2} R(x^*, x_0) P(x_0, x_1).$$

Consequently, if $x_0 \in X_{x^*}$ and $P(x_0, x_1) > 0$, then $x_1 \in X_{x^*}$. This shows that X_{x^*} is always absorbing. $\qquad\square$

The resolvent equation in Proposition A.3.2 (i) can be generalized to any one of the resolvent matrices $\{R_\gamma\}$:

Proposition A.3.5. *Consider the family of resolvent matrices (A1.2). We have the two resolvent equations:*

(i) $[\gamma I - \mathcal{D}] R_\gamma = R_\gamma [\gamma I - \mathcal{D}] = I, \gamma > 0.$
(ii) *For distinct $\gamma_1, \gamma_2 \in (1, \infty)$,*

$$R_{\gamma_2} = R_{\gamma_1} + (\gamma_1 - \gamma_2) R_{\gamma_1} R_{\gamma_2} = R_{\gamma_1} + (\gamma_1 - \gamma_2) R_{\gamma_2} R_{\gamma_1}. \tag{A1.28}$$

Proof. For any $\gamma > 0$ we can express the resolvent as a matrix inverse,

$$R_\gamma = \sum_{t=0}^{\infty} (1 + \gamma)^{-t-1} P^t = [\gamma I - \mathcal{D}]^{-1}, \qquad x \in X, \tag{A1.29}$$

and from (A1.29) we deduce (i). To see (ii) write

$$[\gamma_1 I - \mathcal{D}] - [\gamma_2 I - \mathcal{D}] = (\gamma_1 - \gamma_2) I.$$

Multiplying on the left by $[\gamma_1 I - \mathcal{D}]^{-1}$ and on the right by $[\gamma_2 I - \mathcal{D}]^{-1}$ gives

$$[\gamma_2 I - \mathcal{D}]^{-1} - [\gamma_1 I - \mathcal{D}]^{-1} = (\gamma_1 - \gamma_2)[\gamma_1 I - \mathcal{D}]^{-1}[\gamma_2 I - \mathcal{D}]^{-1},$$

which is the first equality in (A1.28). The proof of the second equality is identical. $\quad\square$

When the chain is x^*-irreducible then one can solve the minorization condition with s positive everywhere:

Lemma A.3.6. *Suppose that X is x^*-irreducible. Then there exist $s \colon X \to [0, 1]$ and a probability distribution ν on X satisfying*

$$s(x) > 0 \text{ for all } x \in X \text{ and } \nu(y) > 0 \text{ for all } y \in X_{x^*}.$$

Proof. Choose $\gamma_1 = 1$, $\gamma_2 \in (0, 1)$, and define $s_0(x) = \mathbf{1}_{x^*}(x)$, $\nu_0(y) = R_{\gamma_2}(x^*, y)$, $x, y \in X$, so that $R_{\gamma_2} \geq s_0 \otimes \nu_0$. From (A1.28),

$$R_{\gamma_2} = R_1 + (1 - \gamma_2) R_1 R_{\gamma_2} \geq (1 - \gamma_2) R_1 [s_0 \otimes \nu_0].$$

Setting $s = (1 - \gamma_2)R_1 s_0$ and $\nu = \nu_0$ gives $R = R_1 \geq s \otimes \nu$. The function s is positive everywhere due to the x^*-irreducibility assumption, and ν is positive on X_{x^*} since $R_{\gamma_2}(x^*, y) > 0$ if and only if $R(x^*, y) > 0$. $\qquad\square$

The following is the key step in establishing subinvariance, and criteria for invariance. Note that Lemma A.3.7 (i) only requires the minorization condition (A1.22).

Lemma A.3.7. *Suppose that the function $s \colon \mathsf{X} \to [0, 1)$ and the probability distribution ν on X satisfy (A1.22). Then:*

(i) $Gs(x) \leq 1$ *for every $x \in \mathsf{X}$.*
(ii) $(R - s \otimes \nu)G = G(R - s \otimes \nu) = G - I$.
(iii) *If $\boldsymbol{X}$ is x^*-irreducible and $s(x^*) > 0$, then $\sup_{x \in \mathsf{X}} G(x, y) < \infty$ for each $y \in \mathsf{X}$.*

Proof. For $N \geq 0$, define $g_N \colon \mathsf{X} \to \mathbb{R}_+$ by

$$g_N = \sum_{n=0}^{N}(R - s \otimes \nu)^n s.$$

We show by induction that $g_N(x) \leq 1$ for every $x \in \mathsf{X}$ and $N \geq 0$. This will establish (i) since $g_N \uparrow Gs$, as $N \uparrow \infty$.

For each x we have $g_0(x) = s(x) = s(x)\nu(\mathsf{X}) \leq R(x, \mathsf{X}) = 1$, which verifies the induction hypothesis when $N = 0$. If the induction hypothesis is true for a given $N \geq 0$, then

$$
\begin{aligned}
g_{N+1}(x) &= (R - s \otimes \nu)g_N(x) + s(x) \\
&\leq (R - s \otimes \nu)\mathbf{1}(x) + s(x) \\
&= [R(x, \mathsf{X}) - s(x)\nu(\mathsf{X})] + s(x) = 1,
\end{aligned}
$$

where in the last equation we have used the assumption that $\nu(\mathsf{X}) = 1$.

Part (ii) then follows from the definition of G.

To prove (iii) we first apply (ii), giving $GR = G - I + Gs \otimes \nu$. Consequently, from (i),

$$GRs = Gs - s + \nu(s)Gs \leq 2 \qquad \text{on } \mathsf{X}. \tag{A1.30}$$

Under the conditions of the lemma we have $Rs(y) > 0$ for every $y \in \mathsf{X}$, and this completes the proof of (iii), with the explicit bound

$$G(x, y) \leq 2(Rs(y))^{-1} \text{ for all } x, y \in \mathsf{X}. \qquad\square$$

It is now easy to establish subinvarance:

Proposition A.3.8. *For an x^*-irreducible Markov chain, and any small pair (s, ν), the measure $\mu = \nu G$ is always subinvariant. Writing $p_{(s,\nu)} = \nu Gs$, we have*

(i) $p_{(s,\nu)} \leq 1$;
(ii) *μ is invariant if and only if $p_{(s,\nu)} = 1$;*
(iii) *μ is finite if and only if $\nu G(\mathsf{X}) < \infty$.*

Proof. Result (i) follows from Lemma A.3.7 and the assumption that ν is a probability distribution on X. The final result (iii) is just a restatement of the definition of μ. For (ii), write

$$
\begin{aligned}
\mu R &= \sum_{n=0}^{\infty} \nu (R - s \otimes \nu)^n R \\
&= \sum_{n=0}^{\infty} \nu (R - s \otimes \nu)^{n+1} + \sum_{n=0}^{\infty} \nu (R - s \otimes \nu)^n s \otimes \nu \\
&= \mu - \nu + p_{(s,\nu)} \nu \leq \mu.
\end{aligned}
$$

It turns out that the case $p_{(s,\nu)} = 1$ is equivalent to a form of recurrence.   □

Definition A.3.9 (Recurrence). An x^*-irreducible Markov chain X is called:

(i) *Harris recurrent*, if the return time (A1.14) is finite almost surely from each initial condition,

$$
\mathsf{P}_x\{\tau_{x^*} < \infty\} = 1, \qquad x \in \mathsf{X}.
$$

(ii) *Positive Harris recurrent*, if it is Harris recurrent, and an invariant measure π exists.   ∎

For a proof of the following result the reader is referred to [388]. A key step in the proof is the application of Proposition A.3.3.

Proposition A.3.10. *Under the conditions of Proposition A.3.8:*

(i) *$p_{(s,\nu)} = 1$ if and only if $\mathsf{P}_{x^*}\{\tau_{x^*} < \infty\} = 1$. If either of these conditions hold, then $Gs(x) = \mathsf{P}_x\{\tau_{x^*} < \infty\} = 1$ for each $x \in \mathsf{X}_{x^*}$.*
(ii) *$\mu(\mathsf{X}) < \infty$ if and only if $\mathsf{E}_{x^*}[\tau_{x^*}] < \infty$.*

To solve Poisson's equation (A1.9) we again apply Proposition A.3.2. First note that the solution h is not unique since we can always add a constant to obtain a new solution to (A1.9). This gives us some flexibility: *Assume that $\nu(h) = 0$, so that $(R - s \otimes \nu)h = Rh$.* This combined with the formula $Rh = h - Rf + \eta$ given in (A1.20) leads to a familiar looking identity,

$$
[I - (R - s \otimes \nu)]h = R\tilde{c}.
$$

Provided the inversion can be justified, this leads to the representation

$$
h = [I - (R - s \otimes \nu)]^{-1} R\tilde{c} = GR\tilde{c}. \tag{A1.31}
$$

Based on this we define the *fundamental matrix*

$$
Z := GR(I - \mathbb{1} \otimes \pi), \tag{A1.32}
$$

so that the function in (A1.31) can be expressed $h = Zc$.

Proposition A.3.11. *Suppose that $\mu(\mathsf{X}) < \infty$. If $c \colon \mathsf{X} \to \mathbb{R}$ is any function satisfying $\mu(|c|) < \infty$, then the function $h = Zc$ is finite valued on the support of ν and solves Poisson's equation.*

Proof. We have $\mu(|\tilde{c}|) = \nu(GR|\tilde{c}|)$, which shows that $\nu(GR|\tilde{c}|) < \infty$. It follows that h is finite valued a.e. $[\nu]$. Note also from the representation of μ that

$$\nu(h) = \nu(GR\tilde{c}) = \mu(R\tilde{c}) = \mu(\tilde{c}) = 0.$$

To see that h solves Poisson's equation we write

$$Rh = (R - s \otimes \nu)h = (R - s \otimes \nu)GR\tilde{c} = GR\tilde{c} - R\tilde{c},$$

where the last equation follows from Lemma A.3.7 (ii). We conclude that h solves the version of Poisson's equation (A1.20) for the resolvent with forcing function Rc, and Proposition A.3.2 then implies that h is a solution for P with forcing function c. $\square$

A.3.3 Near-monotone functions

A function $c \colon \mathsf{X} \to \mathbb{R}$ is called *near monotone* if the sublevel set $S_c(r) := \{x : c(x) \leq r\}$ is finite for each $r < \sup_{x \in \mathsf{X}} c(x)$. In applications the function c is typically a cost function, and hence the near-monotone assumption is the natural condition that large states have relatively high cost.

The function $c = \mathbf{1}_{\{x^*\}^c}$ is near monotone since $S_c(r)$ consists of the singleton $\{x^*\}$ for $r \in [0,1)$, and it is empty for $r < 0$. A solution to Poisson's equation with this forcing function can be constructed based on the sample-path formula (A1.16),

$$h(x) = \mathsf{E}_x \left[\sum_{t=0}^{\tau_{x^*}-1} \mathbf{1}_{\{x^*\}^c}(X(t)) - \pi(\{x^*\}^c) \right]$$
(A1.33)

$$= (1 - \pi(\{x^*\}^c)\mathsf{E}_x[\tau_{x^*}] - \mathbf{1}_{x^*}(x) = \pi(x^*)\mathsf{E}_x[\sigma_{x^*}].$$

The last equality follows from the formula $\pi(x^*)\mathsf{E}_{x^*}[\tau_{x^*}] = 1$ (see (A1.15)) and the definition $\sigma_{x^*} = 0$ when $X(0) = x^*$.

The fact that h is bounded from below is a special case of the following general result.

Proposition A.3.12. *Suppose that c is near monotone with $\eta = \pi(c) < \infty$. Then:*

(i) *The relative value function h given in (A1.31) is uniformly bounded from below, finite valued on X_{x^*}, and solves Poisson's equation on the possibly larger set $\mathsf{X}_h = \{x \in \mathsf{X} : h(x) < \infty\}$.*

(ii) *Suppose there exists a nonnegative-valued function satisfying $g(x) < \infty$ for some $x \in \mathsf{X}_{x^*}$, and the Poisson inequality*

$$\mathcal{D}g(x) \leq -c(x) + \eta, \qquad x \in \mathsf{X}.$$
(A1.34)

Then $g(x) = h(x) + \nu(g)$ for $x \in \mathsf{X}_{x^*}$, where h is given in (A1.31). Consequently, g solves Poisson's equation on X_{x^*}.

Proof. Note that if $\eta = \sup_{x \in \mathsf{X}} c(x)$ then $c(x) \equiv \eta$ on X_{x^*}, so we may take $h \equiv 1$ to solve Poisson's equation.

We henceforth assume that $\eta < \sup_{x \in \mathsf{X}} c(x)$, and define $S = \{x \in \mathsf{X} : c(x) \leq \eta\}$. This set is finite since c is near monotone. We have the obvious bound $\tilde{c}(x) \geq -\eta \mathbf{1}_S(x)$ for $x \in \mathsf{X}$, and hence

$$h(x) \geq -\eta GR\mathbf{1}_S(x), \qquad x \in \mathsf{X}.$$

Lemma A.3.7 and (A1.30) imply that $GR\mathbf{1}_S$ is a bounded function on X. This completes the proof that h is bounded from below, and Proposition A.3.11 establishes Poisson's equation.

To prove (ii) we maintain the notation used in Proposition A.3.11. On applying Lemma A.3.6 we can assume without loss of generality that the pair (s, ν) used in the definition of G is nonzero on X_{x^*}. Note first of all that by the resolvent equation,

$$Rg - g = RDg \leq -R\tilde{c}.$$

We thus have the bound

$$(R - s \otimes \nu)g \leq g - R\tilde{c} - \nu(g)s,$$

and hence for each $n \geq 1$,

$$0 \leq (R - s \otimes \nu)^n g \leq g - \sum_{i=0}^{n-1}(R - s \otimes \nu)^i R\tilde{c} - \nu(g)\sum_{i=0}^{n-1}(R - s \otimes \nu)^i s.$$

On letting $n \uparrow \infty$ this gives

$$g \geq GR\tilde{c} + \nu(g)Gs = h + \nu(g)h_0,$$

where $h_0 := Gs$. The function h_0 is identically 1 on X_{x^*} by Proposition A.3.10, which implies that $g - \nu(g) \geq h$ on X_{x^*}. Moreover, using the fact that $\nu(h) = 0$,

$$\nu(g - \nu(g) - h) = \nu(g - \nu(g)) - \nu(h) = 0.$$

Hence $g - \nu(g) - h = 0$ a.e. $[\nu]$, and this implies that $g - \nu(g) - h = 0$ on X_{x^*} as claimed. $\qquad \square$

Bounds on the potential matrix G are obtained in the following section to obtain criteria for the existence of an invariant measure as well as explicit bounds on the relative value function.

A.4 Criteria for stability

To compute the invariant measure π it is necessary to compute the mean random sum (A1.15), or invert a matrix, such as through an infinite sum as in (A1.24). To verify the *existence* of an invariant measure is typically far easier.

In this section we describe Foster's criterion to test for the existence of an invariant measure, and several variations on this approach which are collectively called the *Foster–Lyapunov criteria* for stability. Each of these stability conditions can be interpreted as a relaxation of the Poisson *inequality* (A1.34).

A.4.1 Foster's criterion

Foster's criterion is the simplest of the "Foster–Lyapunov" drift conditions for stability. It requires that for a nonnegative-valued function V on X, a finite set $S \subset$ X, and $b < \infty$,

$$\mathcal{D}V(x) \leq -1 + b\mathbf{1}_S(x), \qquad x \in \mathsf{X}. \tag{V2}$$

This is precisely condition (V3) (introduced at the start of this chapter) using $f \equiv 1$. The construction of the *Lyapunov function* V is illustrated using the M/M/1 queue in Section 3.3.

The existence of a solution to (V2) is equivalent to positive recurrence. This is summarized in the following.

Theorem A.4.1 (Foster's criterion). *The following are equivalent for an x^*-irreducible Markov chain:*

(i) *An invariant measure π exists.*
(ii) *There is a finite set $S \subset$ X such that $\mathsf{E}_x[\tau_S] < \infty$ for $x \in S$.*
(iii) *There exists $V : \mathsf{X} \to (0, \infty]$, finite at some $x_0 \in \mathsf{X}$, a finite set $S \subset \mathsf{X}$, and $b < \infty$ such that Foster's criterion (V2) holds.*

If (iii) holds then there exists $b_{x^*} < \infty$ such that

$$\mathsf{E}_x[\tau_{x^*}] \leq V(x) + b_{x^*}, \qquad x \in \mathsf{X}.$$

Proof. We just prove the implication (iii) $\implies$ (i). The remaining implications may be found in [368, Chapter 11].

Take any pair (s, ν) positive on X_{x^*} and satisfying $R \geq s \otimes \nu$. On applying Proposition A.3.8 it is enough to show that $\mu(\mathsf{X}) < \infty$ with $\mu = \nu G$.

Letting $f \equiv 1$ we have under (V2) $\mathcal{D}V \leq -f + b\mathbf{1}_S$, and on applying R to both sides of this inequality we obtain using the resolvent equation (A1.19), $(R - I)V = R\mathcal{D}V \leq -Rf + bR\mathbf{1}_S$, or on rearranging terms,

$$RV \leq V - Rf + bR\mathbf{1}_S. \tag{A1.35}$$

From (A1.35) we have $(R - s \otimes \nu)V \leq V - Rf + g$, where $g := bR\mathbf{1}_S$. On iterating this inequality we obtain

$$\begin{aligned}
(R - s \otimes \nu)^2 V &\leq (R - s \otimes \nu)(V - Rf + g) \\
&\leq V - Rf + g \\
&\quad -(R - s \otimes \nu)Rf \\
&\quad +(R - s \otimes \nu)g.
\end{aligned}$$

By induction we obtain for each $n \geq 1$

$$0 \leq (R - s \otimes \nu)^n V \leq V - \sum_{i=0}^{n-1}(R - s \otimes \nu)^i Rf + \sum_{i=0}^{n-1}(R - s \otimes \nu)^i g.$$

Rearranging terms then gives

$$\sum_{i=0}^{n-1}(R - s\otimes\nu)^i Rf \le V + \sum_{i=0}^{n-1}(R - s\otimes\nu)^i g,$$

and thus from the definition (A1.24) we obtain the bound

$$GRf \le V + Gg. \qquad (A1.36)$$

To obtain a bound on the final term in (A1.36) recall that $g := bR\mathbf{1}_S$. From its definition we have

$$GR = G[R - s\otimes\nu] + G[s\otimes\nu] = G - I + (Gs)\otimes\nu,$$

which shows that

$$Gg = bGR\mathbf{1}_S \le b[G\mathbf{1}_S + \nu(S)Gs].$$

This is uniformly bounded over X by Lemma A.3.7. Since $f \equiv 1$ the bound (A1.36) implies that $GRf(x) = G(x,\mathsf{X}) \le V(x) + b_1$, $x \in \mathsf{X}$, with b_1 an upper bound on Gg.

Integrating both sides of the bound (A1.36) with respect to ν gives

$$\mu(\mathsf{X}) = \sum_{x\in\mathsf{X}}\nu(x)G(x,\mathsf{X}) \le \nu(V) + \nu(g).$$

The minorization and the drift inequality (A1.35) give

$$s\nu(V) = (s\otimes\nu)(V) \le RV \le V - 1 + g,$$

which establishes finiteness of $\nu(V)$, and the bound

$$\nu(V) \le \inf_{x\in\mathsf{X}}\frac{V(x) - 1 + g(x)}{s(x)}. \qquad \square$$

The following result illustrates the geometric considerations that may be required in the construction of a Lyapunov function, based on the relationship between the gradient $\nabla V(x)$, and the *drift vector field* $\Delta : \mathsf{X} \to \mathbb{R}^\ell$ defined by

$$\Delta(x) := \mathsf{E}[X(t+1) - X(t) \mid X(t) = x], \qquad x \in \mathsf{X}. \qquad (A1.37)$$

This geometry is illustrated in Fig. A1.1 based on the following proposition.

Proposition A.4.2. *Consider a Markov chain on $\mathsf{X} \subset \mathbb{Z}_+^\ell$, and a C^1 function $V : \mathbb{R}^\ell \to \mathbb{R}_+$ satisfying the following conditions:*

(a) *The chain is skip free in the mean, in the sense that*

$$b_{\mathsf{X}} := \sup_{x\in\mathsf{X}}\mathsf{E}[\|X(t+1) - X(t)\| \mid X(t) = x] < \infty.$$

(b) *There exists $\varepsilon_0 > 0$, $b_0 < \infty$, such that*

$$\langle\Delta(y), \nabla V(x)\rangle \le -(1+\varepsilon_0) + b_0(1+\|x\|)^{-1}\|x - y\|, \qquad x, y \in \mathsf{X}. \quad (A1.38)$$

Then the function V solves Foster's criterion (V2).

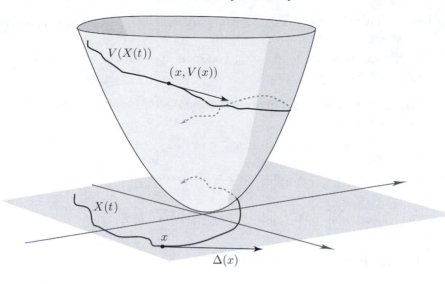

Figure A1.1. $V(X(t))$ is decreasing outside of the set S.

Proof. This is an application of the mean value theorem which asserts that there exists a state $\bar{X} \in \mathbb{R}^\ell$ on the line segment connecting $X(t)$ and $X(t+1)$, with

$$V(X(t+1)) = V(X(t)) + \langle \nabla V(\bar{X}), (X(t+1) - X(t)) \rangle,$$

from which the following bound follows:

$$V(X(t+1)) \le V(X(t)) - (1 + \varepsilon_0) + b_0(1 + \|X(t)\|)^{-1}\|X(t+1) - X(t)\|$$

Under the skip-free assumption this shows that

$$\begin{aligned}
\mathcal{D}V(x) &= \mathsf{E}[V(X(t+1)) - V(X(t)) \mid X(t) = x] \\
&\le -(1 + \varepsilon_0) + b_0(1 + \|x\|)^{-1}b_X, \qquad \|x\| \ge n_0.
\end{aligned}$$

Hence, Foster's criterion is satisfied with the finite set $S = \{x \in \mathsf{X} : (1 + \|x\|)^{-1}b_X \ge \varepsilon_0\}$. $\qquad\square$

A.4.2 Criteria for finite moments

We now turn to the issue of performance bounds based on the discounted cost defined in (A1.2) or the average cost $\eta = \pi(c)$ for a cost function $c \colon \mathsf{X} \to \mathbb{R}_+$. We also introduce martingale methods to obtain performance bounds. We let $\{\mathcal{F}_t : t \ge 0\}$ denote the filtration, or history generated by the chain

$$\mathcal{F}_t := \sigma\{X(0), \ldots, X(t)\}, \qquad t \ge 0.$$

Recall that a random variable τ taking values in $\mathbb{Z}_+$ is called a *stopping time* if for each $t \ge 0$,

$$\{\tau = t\} \in \mathcal{F}_t.$$

That is, by observing the process X on the time interval $[0, t]$ it is possible to determine whether or not $\tau = t$.

The Comparison Theorem is the most common approach to obtaining bounds on expectations involving stopping times.

Theorem A.4.3 (Comparison Theorem). *Suppose that the nonnegative functions V, f, g satisfy the bound*

$$\mathcal{D}V \leq -f + g, \qquad x \in \mathsf{X}. \tag{A1.39}$$

Then for each $x \in \mathsf{X}$ and any stopping time τ we have

$$\mathsf{E}_x\left[\sum_{t=0}^{\tau-1} f(X(t))\right] \leq V(x) + \mathsf{E}_x\left[\sum_{t=0}^{\tau-1} g(X(t))\right].$$

Proof. Define $M(0) = V(X(0))$, and for $n \geq 1$,

$$M(n) = V(X(n)) + \sum_{t=0}^{n-1} (f(X(t)) - g(X(t))).$$

The assumed inequality can be expressed

$$\mathsf{E}[V(X(t+1)) \mid \mathcal{F}_t] \leq V(X(t)) - f(X(t)) + g(X(t)), \qquad t \geq 0,$$

which shows that the stochastic process M is a *super-martingale*,

$$\mathsf{E}[M(n+1) \mid \mathcal{F}_n] \leq M(n), \qquad n \geq 0.$$

Define for $N \geq 1$,

$$\tau^N = \min\{t \leq \tau : t + V(X(t)) + f(X(t)) + g(X(t)) \geq N\}.$$

This is also a stopping time. The process M is uniformly bounded below by $-N^2$ on the time interval $(0, \ldots, \tau^N - 1)$, and it then follows from the super-martingale property that

$$\mathsf{E}[M(\tau^N)] \leq \mathsf{E}[M(0)] = V(x), \qquad N \geq 1.$$

From the definition of M we thus obtain the desired conclusion with τ replaced by τ^N: For each initial condition $X(0) = x$,

$$\mathsf{E}_x\left[\sum_{t=0}^{\tau^N-1} f(X(t))\right] \leq V(x) + \mathsf{E}_x\left[\sum_{t=0}^{\tau^N-1} g(X(t))\right].$$

The result then follows from the Monotone Convergence Theorem since we have $\tau^N \uparrow \tau$ as $N \to \infty$. □

In view of the Comparison Theorem, to bound $\pi(c)$ we search for a solution to (V3) or (A1.39) with $|c| \leq f$. The existence of a solution to either of these drift inequalities is closely related to the following stability condition.

Definition A.4.4 (**Regularity**). Suppose that X is an x^*-irreducible Markov chain, and that $c\colon \mathsf{X} \to \mathbb{R}_+$ is a given function. The chain is called *c-regular* if the following *cost over a y-cycle* is finite for each initial condition $x \in \mathsf{X}$, and each $y \in \mathsf{X}_{x^*}$:

$$\mathsf{E}_x \left[\sum_{t=0}^{\tau_y - 1} c(X(t)) \right] < \infty. \qquad \blacksquare$$

Proposition A.4.5. *Suppose that the function* $c\colon \mathsf{X} \to \mathbb{R}$ *satisfies* $c(x) \geq 1$ *outside of some finite set. Then:*

(i) *If* X *is c-regular then it is positive Harris recurrent and* $\pi(c) < \infty$.

(ii) *Conversely, if* $\pi(c) < \infty$ *then the chain restricted to the support of* π *is c-regular.*

Proof. The result follows from [368, Theorem 14.0.1]. To prove (i) observe that X is Harris recurrent since $\mathsf{P}_x\{\tau_{x^*} < \infty\} = 1$ for all $x \in \mathsf{X}$ when the chain is *c*-regular. We have positivity and $\pi(c) < \infty$ based on the representation (A1.15). $\qquad \square$

Criteria for *c*-regularity will be established through operator manipulations similar to those used in the proof of Theorem A.4.1 based on the following refinement of Foster's criterion: For a nonnegative-valued function V on X, a finite set $S \subset \mathsf{X}$, $b < \infty$, and a function $f\colon \mathsf{X} \to [1, \infty)$,

$$\mathcal{D}V(x) \leq -f(x) + b\mathbf{1}_S(x), \qquad x \in \mathsf{X}. \qquad (\mathbf{V3})$$

The function f is interpreted as a bounding function. In Theorem A.4.6 we consider $\pi(c)$ for functions c bounded by f in the sense that

$$\|c\|_f := \sup_{x \in \mathsf{X}} \frac{|c(x)|}{f(x)} < \infty. \qquad (\text{A1.40})$$

Theorem A.4.6. *Suppose that* X *is* x^*-*irreducible, and that there exists* $V\colon \mathsf{X} \to (0, \infty)$, $f\colon \mathsf{X} \to [1, \infty)$, *a finite set* $S \subset \mathsf{X}$, *and* $b < \infty$ *such that (V3) holds. Then for any function* $c\colon \mathsf{X} \to \mathbb{R}_+$ *satisfying* $\|c\|_f \leq 1$:

(i) *The average cost satisfies the uniform bound*

$$\eta_x = \pi(c) \leq b < \infty, \qquad x \in \mathsf{X}.$$

(ii) *The discounted-cost value function satisfies the following uniform bound, for any given discount parameter* $\gamma > 0$,

$$h_\gamma(x) \leq V(x) + b\gamma^{-1}, \qquad x \in \mathsf{X}.$$

(iii) *There exists a solution to Poisson's equation satisfying, for some* $b_1 < \infty$,

$$h(x) \leq V(x) + b_1, \qquad x \in \mathsf{X}.$$

Proof. Observe that (ii) and the definition (A1.6) imply (i).

To prove (ii) we apply the resolvent equation,

$$PR_\gamma = R_\gamma P = (1 + \gamma)R_\gamma - I. \qquad (\text{A1.41})$$

Equation (A1.41) is a restatement of Eq. (A1.29). Consequently, under (V3),

$$(1 + \gamma) R_\gamma V - V = R_\gamma P V \leq R_\gamma [V - f + b\mathbf{1}_S].$$

Rearranging terms gives $R_\gamma f + \gamma R_\gamma V \leq V + b R_\gamma \mathbf{1}_S$. This establishes (ii) since $R_\gamma \mathbf{1}_S (x) \leq R_\gamma (x, \mathsf{X}) \leq \gamma^{-1}$ for $x \in \mathsf{X}$.

We now prove (iii). Recall that the measure $\mu = \nu G$ is finite and invariant since we may apply Theorem A.4.1 when the chain is x^*-irreducible. We shall prove that the function $h = GR\tilde{c}$ given in (A1.31) satisfies the desired upper bound.

The proof of the implication (iii) $\Longrightarrow$ (i) in Theorem A.4.1 was based upon the bound (A1.36),

$$GRf \leq V + Gg,$$

where $g := bR\mathbf{1}_S$. Although it was assumed there that $f \equiv 1$, the same steps lead to this bound for general $f \geq 1$ under (V3). Consequently, since $0 \leq c \leq f$,

$$GR\tilde{c} \leq GRf \leq V + Gg.$$

Part (iii) follows from this bound and Lemma A.3.7 with $b_1 := \sup Gg(x) < \infty$. □

Proposition A.4.2 can be extended to provide the following criterion for finite moments in a skip-free Markov chain:

Proposition A.4.7. *Consider a Markov chain on $\mathsf{X} \subset \mathbb{R}^\ell$, and a C^1 function $V : \mathbb{R}^\ell \to \mathbb{R}_+$ satisfying the following conditions:*

(i) *The chain is skip free in mean square:*

$$b_{X2} := \sup_{x \in \mathsf{X}} \mathsf{E}[\|X(t+1) - X(t)\|^2 \mid X(t) = x] < \infty.$$

(ii) *There exists $b_0 < \infty$ such that*

$$\langle \Delta(y), \nabla V(x) \rangle \leq -\|x\| + b_0 \|x - y\|^2, \qquad x, y \in \mathsf{X}. \tag{A1.42}$$

Then the function V solves (V3) with $f(x) = 1 + \frac{1}{2} \|x\|$.

A.4.3 State-dependent drift

In this section we consider consequences of state-dependent drift conditions of the form

$$\sum_{y \in \mathsf{X}} P^{n(x)}(x, y) V(y) \leq g[V(x), n(x)], \qquad x \in S^c, \tag{A1.43}$$

where $n(x)$ is a function from X to $\mathbb{Z}_+$, g is a function depending on which type of stability we seek to establish, and S is a finite set.

The function $n(x)$ here provides the state dependence of the drift conditions, since from any x we must wait $n(x)$ steps for the drift to be negative.

In order to develop results in this framework we work with a sampled chain $\widehat{X}$. Using $n(x)$ we define the new transition law $\{\widehat{P}(x, A)\}$ by

$$\widehat{P}(x, A) = P^{n(x)}(x, A), \qquad x \in \mathsf{X}, \ A \subset \mathsf{X}, \qquad (A1.44)$$

and let $\widehat{X}$ denote a Markov chain with this transition law. This Markov chain can be constructed explicitly as follows. The time $n(x)$ is a (trivial) stopping time. Let $\{n_k\}$ denote its iterates: That is, along any sample path, $n_0 = 0$, $n_1 = n(x)$, and

$$n_{k+1} = n_k + n(X(n_k)).$$

Then it follows from the strong Markov property that

$$\widehat{X}(k) = X(n_k), \qquad k \geq 0 \qquad (A1.45)$$

is a Markov chain with transition law $\widehat{P}$.

Let $\widehat{\mathcal{F}}_k = \mathcal{F}_{n_k}$ be the σ-field generated by the events "before n_k": that is,

$$\widehat{\mathcal{F}}_k := \{A : A \cap \{n_k \leq n\} \in \mathcal{F}_n, n \geq 0\}.$$

We let $\hat{\tau}_S$ denote the first return time to S for the chain $\widehat{X}$. The time n_k and the event $\{\hat{\tau}_S \geq k\}$ are $\widehat{\mathcal{F}}_{k-1}$-measurable for any $S \subset \mathsf{X}$.

The integer $n_{\hat{\tau}_S}$ is a particular time at which the original chain visits the set S. Minimality implies the bound

$$n_{\hat{\tau}_S} \geq \tau_S. \qquad (A1.46)$$

By adding the lengths of the sampling times n_k along a sample path for the sampled chain, the time $n_{\hat{\tau}_S}$ can be expressed as the sum

$$n_{\hat{\tau}_S} = \sum_{k=0}^{\hat{\tau}_S - 1} n(\widehat{X}(k)). \qquad (A1.47)$$

These relations enable us to first apply the drift condition (A1.43) to bound the index at which $\widehat{X}$ reaches S, and thereby bound the hitting time for the original chain.

We prove here a state-dependent criterion for positive recurrence. Generalizations are described in the Notes section in Chapter 10, and Theorem 10.0.2 contains strengthened conclusions for the CRW network model.

Theorem A.4.8. *Suppose that X is an x^*-irreducible chain on X, and let $n(x)$ be a function from X to $\mathbb{Z}_+$. The chain is positive Harris recurrent if there exist some finite set S, a function $V : \mathsf{X} \to \mathbb{R}_+$, and a finite constant b satisfying*

$$\sum_{y \in \mathsf{X}} P^{n(x)}(x, y)V(y) \leq V(x) - n(x) + b\mathbf{1}_S(x), \qquad x \in \mathsf{X}, \qquad (A1.48)$$

in which case for all x

$$\mathsf{E}_x[\tau_S] \leq V(x) + b. \qquad (A1.49)$$

Proof. The state-dependent drift criterion for positive recurrence is a direct consequence of the f-regularity results of Theorem A.4.3, which tell us that without any irreducibility or other conditions on X, if f is a nonnegative function and

$$\sum_{y \in \mathsf{X}} P(x, y) V(y) \leq V(x) - f(x) + b\mathbf{1}_S(x), \qquad x \in \mathsf{X} \tag{A1.50}$$

for some set S, then for each $x \in \mathsf{X}$

$$\mathsf{E}_x \left[\sum_{t=0}^{\tau_S - 1} f(X(t)) \right] \leq V(x) + b. \tag{A1.51}$$

We now apply this result to the chain $\widehat{X}$ defined in (A1.45). From (A1.48) we can use (A1.51) for $\widehat{X}$, with $f(x)$ taken as $n(x)$, to deduce that

$$\mathsf{E}_x \left[\sum_{k=0}^{\hat{\tau}_S - 1} n(\widehat{X}(k)) \right] \leq V(x) + b. \tag{A1.52}$$

Thus from (A1.46,A1.47) we obtain the bound (A1.49). Theorem A.4.1 implies that X is positive Harris.   □

A.5 Ergodic theorems and coupling

The existence of a Lyapunov function satisfying (V3) leads to the ergodic theorems (1.23), and refinements of this drift inequality lead to stronger results. These results are based on the coupling method described next.

A.5.1 Coupling

Coupling is a way of comparing the behavior of the process of interest X with another process Y which is already understood. For example, if Y is taken as the stationary version of the process, with $Y(0) \sim \pi$, we then have the trivial mean ergodic theorem,

$$\lim_{t \to \infty} \mathsf{E}[c(Y(t))] = \mathsf{E}[c(Y(t_0))], \qquad t_0 \geq 0.$$

This leads to a corresponding ergodic theorem for X provided the two processes couple in a suitably strong sense.

To precisely define coupling we define a bivariate process,

$$\Psi(t) = \begin{pmatrix} X(t) \\ Y(t) \end{pmatrix}, \qquad t \geq 0,$$

where X and Y are two copies of the chain with transition probability P, and different initial conditions. It is assumed throughout that X is x^*-irreducible, and we define the *coupling time* for Ψ as the first time both chains reach x^* simultaneously,

$$T = \min(t \geq 1 : X(t) = Y(t) = x^*) = \min\left(t : \Psi(t) = \begin{pmatrix} x^* \\ x^* \end{pmatrix}\right).$$

To give a full statistical description of $\mathbf{\Psi}$ we need to explain how $\mathbf{X}$ and $\mathbf{Y}$ are related. We assume a form of conditional independence for $k \leq T$:

$$\mathsf{P}\{\Psi(t+1) = (x_1, y_1)^{\mathsf{T}} \mid \Psi(0), \ldots, \Psi(t); \Psi(t) = (x_0, y_0)^{\mathsf{T}}, T > t\}$$
$$= P(x_0, x_1)P(y_0, y_1). \tag{A1.53}$$

It is assumed that the chains coellesce at time T, so that $X(t) = Y(t)$ for $t \geq T$.

The process $\mathbf{\Psi}$ is not itself Markov since given $\Psi(t) = (x, x)^{\mathsf{T}}$ with $x \neq x^*$ it is impossible to know if $T \leq t$. However, by appending the indicator function of this event we obtain a Markov chain denoted

$$\Psi^*(t) = (\Psi(t), \mathbf{1}\{T \leq t\}),$$

with state space $\mathsf{X}^* = \mathsf{X} \times \mathsf{X} \times \{0, 1\}$. The subset $\mathsf{X} \times \mathsf{X} \times \{1\}$ is absorbing for this chain.

The following two propositions allow us to infer properties of $\mathbf{\Psi}^*$ based on properties of $\mathbf{X}$. The proof of Proposition A.5.1 is immediate from the definitions.

Proposition A.5.1. *Suppose that $\mathbf{X}$ satisfies (V3) with f coercive. Then (V3) holds for the bivariate chain $\mathbf{\Psi}^*$ in the form*

$$\mathsf{E}[V_*(\Psi(t+1)) \mid \Psi(t) = (x, y)^{\mathsf{T}}] \leq V_*(x, y) - f_*(x, y) + b_*,$$

with $V_*(x, y) = V(x) + V(y)$, $f_*(x, y) = f(x) + f(y)$, and $b_* = 2b$. Consequently, there exists $b_0 < \infty$ such that

$$\mathsf{E}\left[\sum_{t=0}^{T-1} \left(f(X(t)) + f(Y(t))\right)\right] \leq 2[V(x) + V(y)] + b_0, \qquad x, y \in \mathsf{X}.$$

A necessary condition for the mean ergodic theorem for arbitrary initial conditions is aperiodicity. Similarly, aperiodicity is both necessary and sufficient for x^{**}-irreducibility of $\mathbf{\Psi}^*$ with $x^{**} := (x^*, x^*, 1)^{\mathsf{T}} \in \mathsf{X}^*$:

Proposition A.5.2. *Suppose that $\mathbf{X}$ is x^*-irreducible and aperiodic. Then the bivariate chain is x^{**}-irreducible and aperiodic.*

Proof. Fix any $x, y \in \mathsf{X}$, and define

$$n_0 = \min\{n \geq 0 : P^n(x, x^*)P^n(y, x^*) > 0\}.$$

The minimum is finite since $\mathbf{X}$ is x^*-irreducible and aperiodic. We have $\mathsf{P}\{T \leq n\} = 0$ for $n < n_0$ and by the construction of $\mathbf{\Psi}$,

$$\mathsf{P}\{T = n_0\} = \mathsf{P}\{\Psi(n_0) = (x^*, x^*)^{\mathsf{T}} \mid T \geq n_0\} = P^{n_0}(x, x^*)P^{n_0}(y, x^*) > 0.$$

This establishes x^{**}-irreducibility.

For $n \geq n_0$ we have

$$\mathsf{P}\{\Psi^*(n) = x^{**}\} \geq \mathsf{P}\{T = n_0, \Psi^*(n) = x^{**}\} = P^{n_0}(x, x^*)P^{n_0}(y, x^*)P^{n-n_0}(x^*, x^*).$$

The right-hand side is positive for all $n \geq 0$ sufficiently large since $\mathbf{X}$ is aperiodic. $\square$

A.5.2 Mean ergodic theorem

A mean ergodic theorem is obtained based upon the following *coupling inequality*:

Proposition A.5.3. *For any given* $g \colon \mathsf{X} \to \mathbb{R}$ *we have*

$$\left| \mathsf{E}[g(X(t))] - \mathsf{E}[g(Y(t))] \right| \le \mathsf{E}[(|g(X(t))| + |g(Y(t))|)\mathbf{1}(T > t)].$$

If $Y(0) \sim \pi$ *so that* $\boldsymbol{Y}$ *is stationary, we thus obtain*

$$|\mathsf{E}[g(X(t))] - \pi(g)| \le \mathsf{E}[(|g(X(t))| + |g(Y(t))|)\mathbf{1}(T > t)].$$

Proof. The difference $g(X(t)) - g(Y(t))$ is *zero* for $t \ge T$.

The *f-total variation norm* of a signed measure μ on X is defined by

$$\|\mu\|_f = \sup\{|\mu(g)| : \|g\|_f \le 1\}.$$

When $f \equiv 1$ then this is exactly twice the *total-variation norm*: For any two probability measures π, μ,

$$\|\mu - \pi\|_{tv} := \sup_{A \subset \mathsf{X}} |\mu(A) - \pi(A)|. \qquad \square$$

Theorem A.5.4. *Suppose that* $\boldsymbol{X}$ *is aperiodic, and that the assumptions of Theorem A.4.6 hold. Then:*

(i) $\|P^t(x, \,\cdot\,) - \pi\|_f \to 0$ *as* $t \to \infty$ *for each* $x \in \mathsf{X}$.

(ii) *There exists* $b_0 < \infty$ *such that for each* $x, y \in \mathsf{X}$,

$$\sum_{t=0}^{\infty} \|P^t(x, \,\cdot\,) - P^t(y, \,\cdot\,)\|_f \le 2[V(x) + V(y)] + b_0.$$

(iii) *If in addition* $\pi(V) < \infty$, *then there exists* $b_1 < \infty$ *such that*

$$\sum_{t=0}^{\infty} \|P^t(x, \,\cdot\,) - \pi\|_f \le 2V(x) + b_1.$$

The coupling inequality is only useful if we can obtain a bound on the expectation $\mathsf{E}[|g(X(t))|\mathbf{1}(T > t)]$. The following result shows that this vanishes when $\boldsymbol{X}$ and $\boldsymbol{Y}$ are each stationary.

Lemma A.5.5. *Suppose that* $\boldsymbol{X}$ *is aperiodic, and that the assumptions of Theorem A.4.6 hold. Assume moreover that* $X(0)$ *and* $Y(0)$ *each have distribution* π, *and that* $\pi(|g|) < \infty$. *Then:*

$$\lim_{t \to \infty} \mathsf{E}[(|g(X(t))| + |g(Y(t))|)\mathbf{1}(T > t)] = 0.$$

Proof. Suppose that $\boldsymbol{X}, \boldsymbol{Y}$ are defined on the two-sided time interval with marginal distribution π. It is assumed that these processes are independent on $\{0, -1, -2, \dots\}$. By stationarity we can write

$$\mathsf{E}_\pi[|g(X(t))|\mathbf{1}(T > t)] = \mathsf{E}_\pi[|g(X(t))|\mathbf{1}\{\Psi(i) \ne (x^*, x^*)^{\mathsf{T}},\ i = 0, \dots, t\}]$$

$$= \mathsf{E}_\pi[|g(X(0))|\mathbf{1}\{\Psi(i) \ne (x^*, x^*)^{\mathsf{T}},\ i = 0, -1, \dots, -t\}].$$

The expression within the expectation on the right-hand side vanishes as $t \to \infty$ with probability 1 by $(x^*, x^*)^{\mathsf{T}}$-irreducibility of the stationary process $\{\Psi(-t) : t \in \mathbb{Z}_+\}$. The Dominated Convergence Theorem then implies that

$$\lim_{t\to\infty} \mathsf{E}[|g(X(t))|\mathbf{1}(T > t)] = \mathsf{E}_\pi[|g(X(0))|\mathbf{1}\{\Psi(i) \neq (x^*, x^*)^{\mathsf{T}}, \; i = 0, -1, \ldots, -t\}] = 0.$$

Repeating the same steps with X replaced by Y we obtain the analogous limit by symmetry. □

Proof of Theorem A.5.4. We first prove (ii). From the coupling inequality we have, with $X(0) = x$, $X^\circ(0) = y$,

$$|P^t g(x) - P^t g(y)| = |\mathsf{E}[g(X(t))] - \mathsf{E}[g(Y(t))]|$$

$$\leq \mathsf{E}\big[(|g(X(t))| + |g(Y(t))|)\mathbf{1}(T > t)\big]$$

$$\leq \|g\|_f \mathsf{E}\big[(f(X(t)) + f(Y(t)))\mathbf{1}(T > t)\big].$$

Taking the supremum over all g satisfying $\|g\|_f \leq 1$ then gives

$$\|P^t(x, \cdot) - P^t(y, \cdot)\|_f \leq \mathsf{E}\big[(f(X(t)) + f(Y(t)))\mathbf{1}(T > t)\big], \qquad (A1.54)$$

so that on summing over t,

$$\sum_{t=0}^{\infty} \|P^t(x, \cdot) - P^t(y, \cdot)\|_f \leq \sum_{t=0}^{\infty} \mathsf{E}\big[(f(X(t)) + f(Y(t)))\mathbf{1}(T > t)\big]$$

$$= \mathsf{E}\left[\sum_{t=0}^{T-1}(f(X(t)) + f(Y(t)))\right].$$

Applying Proposition A.5.1 completes the proof of (ii).

To see (iii) observe that

$$\sum_{y\in\mathsf{X}} \pi(y)|P^t g(x) - P^t g(y)| \geq \left|\sum_{y\in\mathsf{X}} \pi(y)[P^t g(x) - P^t g(y)]\right| = |P^t g(x) - \pi(g)|.$$

Hence by (ii) we obtain (iii) with $b_1 = b_0 + 2\pi(V)$.

Finally we prove (i). Note that we only need establish the mean ergodic theorem in (i) for a single initial condition $x_0 \in \mathsf{X}$. To see this, first note that we have the triangle inequality

$$\|P^t(x, \cdot) - \pi(\cdot)\|_f \leq \|P^t(x, \cdot) - P^t(x_0, \cdot)\|_f + \|P^t(x_0, \cdot) - \pi(\cdot)\|_f, \qquad x, x_0 \in \mathsf{X}.$$

From this bound and part (ii) we obtain

$$\limsup_{t\to\infty} \|P^t(x, \cdot) - \pi(\cdot)\|_f \leq \limsup_{t\to\infty} \|P^t(x_0, \cdot) - \pi(\cdot)\|_f.$$

Exactly as in (A1.54) we have, with $X(0) = x_0$ and $Y(0) \sim \pi$,

$$\|P^t(x_0, \cdot) - \pi(\cdot)\|_f \leq \mathsf{E}\big[(f(X(t)) + f(Y(t)))\mathbf{1}(T > t)\big]. \qquad (A1.55)$$

We are left to show that the right-hand side converges to zero for some x_0. Applying Lemma A.5.5 we obtain

$$\lim_{t\to\infty} \sum_{x,y} \pi(x)\pi(y)\mathsf{E}\big[[f(X(t)) + f(Y(t))]\mathbf{1}(T > t) \mid X(0) = x,\ Y(0) = y\big] = 0.$$

It follows that the right-hand side of (A1.55) vanishes as $t \to \infty$ when $X(0) = x_0$ and $Y(0) \sim \pi$. $\square$

A.5.3 Geometric ergodicity

Theorem A.5.4 provides a mean ergodic theorem based on the coupling time T. If we can control the tails of the coupling time T, then we obtain a rate of convergence of $P^t(x, \cdot)$ to π.

The chain is called *geometrically recurrent* if $\mathsf{E}_{x^*}[\exp(\varepsilon\tau_{x^*})] < \infty$ for some $\varepsilon > 0$. For such chains it is shown in Theorem A.5.6 that for a.e. $[\pi]$ initial condition $x \in \mathsf{X}$, the total variation norm vanishes geometrically fast.

Theorem A.5.6. *The following are equivalent for an aperiodic, x^*-irreducible Markov chain:*

(i) *The chain is geometrically recurrent.*
(ii) *There exists $V : \mathsf{X} \to [1,\infty]$ with $V(x_0) < \infty$ for some $x_0 \in \mathsf{X}$, $\varepsilon > 0$, $b < \infty$, and a finite set $S \subset \mathsf{X}$ such that*

$$\mathcal{D}V(x) \le -\varepsilon V(x) + b\mathbf{1}_S(x), \qquad x \in \mathsf{X}. \tag{V4}$$

(iii) *For some $r > 1$,*

$$\sum_{n=0}^{\infty} \|P^n(x^*,\cdot) - \pi(\cdot)\|_1 r^n < \infty.$$

If any of the above conditions hold, then with V given in (ii), we can find $r_0 > 1$ and $b < \infty$ such that the stronger mean ergodic theorem holds: For each $x \in \mathsf{X}, t \in \mathbb{Z}_+$,

$$\|P^t(x,\cdot) - \pi(\cdot)\|_V := \sup_{|g|\le V}\big|\mathsf{E}_x[g(X(t)) - \pi(t)]\big| \le br_0^{-t}V(x). \tag{A1.56}$$

In applications Theorem A.5.6 is typically applied by constructing a solution to the drift inequality (V4) to deduce the ergodic theorem in (A1.56). The following result shows that (V4) is not that much stronger than Foster's criterion.

Proposition A.5.7. *Suppose that the Markov chain $\mathbf{X}$ satisfies the following three conditions:*

(i) *There exist $V : \mathsf{X} \to (0,\infty)$, a finite set $S \subset \mathsf{X}$, and $b < \infty$ such that Foster's criterion (V2) holds.*
(ii) *The function V is uniformly Lipschitz,*

$$l_V := \sup\{|V(x) - V(y)| : x, y \in \mathsf{X},\ \|x - y\| \le 1\} < \infty.$$

(iii) *For some $\beta_0 > 0$, $b_1 < \infty$,*

$$b_1 := \sup_{x \in X} \mathsf{E}_x[e^{\beta_0 \|X(1) - X(0)\|}] < \infty.$$

Then there exists $\varepsilon > 0$ such that the controlled process is V_ε-uniformly ergodic with $V_\varepsilon = \exp(\varepsilon V)$.

Proof. Let $\widetilde{\Delta}_V = V(X(1)) - V(X(0))$, so that $\mathsf{E}_x[\widetilde{\Delta}_V] \le -1 + b_1 \mathbf{1}_S(x)$ under (V2). Using a second-order Taylor expansion we obtain for each x and $\varepsilon > 0$

$$[V_\varepsilon(x)]^{-1} P V_\varepsilon(x) = \mathsf{E}_x[\exp(\varepsilon \widetilde{\Delta}_V)]$$

$$= \mathsf{E}_x[1 + \varepsilon \widetilde{\Delta}_V + \tfrac{1}{2}\varepsilon^2 \widetilde{\Delta}_V^2 \exp(\varepsilon \vartheta_x \widetilde{\Delta}_V)] \tag{A1.57}$$

$$\le 1 + \varepsilon(-1 + b_1 \mathbf{1}_S(x)) + \tfrac{1}{2}\varepsilon^2 \mathsf{E}_x[\widetilde{\Delta}_V^2 \exp(\varepsilon \vartheta_x \widetilde{\Delta}_V)]$$

where $\vartheta_x \in [0, 1]$. Applying the assumed Lipschitz bound and the bound $\tfrac{1}{2}z^2 \le e^z$ for $z \ge 0$ we obtain, for any $a > 0$,

$$\tfrac{1}{2}\widetilde{\Delta}_V^2 \exp(\varepsilon \vartheta_x \widetilde{\Delta}_V) \le a^{-2}\exp((a + \varepsilon)|\widetilde{\Delta}_V|)$$
$$\le a^{-2}\exp((a + \varepsilon)l_V \|X(1) - X(0)\|).$$

Setting $a = \varepsilon^{1/3}$ and restricting $\varepsilon > 0$ so that $(a + \varepsilon)l_V \le \beta_0$, the bound (A1.57) and (iii) then give

$$[V_\varepsilon(x)]^{-1} P V_\varepsilon(x) \le (1 - \varepsilon) + \varepsilon b \mathbf{1}_S(x) + \varepsilon^{4/3} b_1.$$

This proves the theorem, since we have $1 - \varepsilon + \varepsilon^{4/3} b_1 < 1$ for sufficiently small $\varepsilon > 0$, and thus (V4) holds for V_ε. $\square$

A.5.4 Sample paths and limit theorems

We conclude this section with a look at the sample path behavior of partial sums,

$$S_g(n) := \sum_{t=0}^{n-1} g(X(t)). \tag{A1.58}$$

We focus on two limit theorems under (V3):

LLN The *Strong Law of Large Numbers* holds for a function g if for each initial condition,

$$\lim_{n \to \infty} \frac{1}{n} S_g(n) = \pi(g) \qquad \text{a.s.} \tag{A1.59}$$

CLT The *Central Limit Theorem* holds for g if there exists a constant $0 < \sigma_g^2 < \infty$ such that for each initial condition $x \in X$,

$$\lim_{n \to \infty} \mathsf{P}_x\left\{(n\sigma_g^2)^{-1/2} S_{\tilde{g}}(n) \le t\right\} = \int_{-\infty}^{t} \frac{1}{\sqrt{2\pi}} e^{-x^2/2}\, dx$$

where $\tilde{g} = g - \pi(g)$. That is, as $n \to \infty$,

$$(n\sigma_g^2)^{-1/2} S_{\tilde{g}}(n) \xrightarrow{\text{w}} N(0, 1).$$

The LLN is a simple consequence of the coupling techniques already used to prove the mean ergodic theorem when the chain is aperiodic and satisfies (V3). A slightly different form of coupling can be used when the chain is periodic. There is only room for a survey of theory surrounding the CLT, which is most elegantly approached using martingale methods. A relatively complete treatement may be found in [368], and the more recent survey [282].

The following versions of the LLN and CLT are based on Theorem 17.0.1 of [368].

Theorem A.5.8. *Suppose that X is positive Harris recurrent and that the function g satisfies $\pi(|g|) < \infty$. Then the LLN holds for this function.*

If moreover (V4) holds with $g^2 \in L_\infty^V$, then:

(i) *Letting $\tilde{g}$ denote the centered function $\tilde{g} = g - \int g \, d\pi$, the constant*

$$\sigma_g^2 := \mathsf{E}_\pi[\tilde{g}^2(X(0))] + 2 \sum_{t=1}^{\infty} \mathsf{E}_\pi[\tilde{g}(X(0))\tilde{g}(X(t))] \qquad (A1.60)$$

is well defined, nonnegative, and finite, and

$$\lim_{n\to\infty} \frac{1}{n} \mathsf{E}_\pi\left[\left(S_{\tilde{g}}(n)\right)^2\right] = \sigma_g^2. \qquad (A1.61)$$

(ii) *If $\sigma_g^2 = 0$ then for each initial condition,*

$$\lim_{n\to\infty} \frac{1}{\sqrt{n}} S_{\tilde{g}}(n) = 0 \qquad \text{a.s.}$$

(iii) *If $\sigma_g^2 > 0$ then the CLT holds for the function g.*

The proof of the theorem in [368] is based on consideration of the martingale

$$M_g(t) := \hat{g}(X(t)) - \hat{g}(X(0)) + \sum_{i=0}^{t-1} \tilde{g}(X(i)), \qquad t \geq 1,$$

with $M_g(0) := 0$. This is a martingale since Poisson's equation $P\hat{g} = \hat{g} - \tilde{g}$ gives

$$\mathsf{E}[\hat{g}(X(t)) \mid X(0), \dots, X(t-1)] = \hat{g}(X(t-1)) - \tilde{g}(X(t-1)),$$

so that

$$\mathsf{E}[M_g(t) \mid X(0), \dots, X(t-1)] = M_g(t-1).$$

The proof of the CLT is based on the representation $S_{\tilde{g}}(t) = M_g(t) + \hat{g}(X(t)) - \hat{g}(X(0))$, combined with limit theory for martingales, and the bounds on solutions to Poisson's equation given in Theorem A.4.6.

An alternate representation for the asymptotic variance can be obtained through the alternate representation for the martingale as the partial sums of a martingale difference sequence,

$$M_g(t) = \sum_{i=1}^{t} \widetilde{\Delta}_g(i), \qquad t \geq 1,$$

with $\{\widetilde{\Delta}_g(t) := \hat{g}(X(t)) - \hat{g}(X(t-1)) + \tilde{g}(X(t-1))\}$. Based on the martingale difference property,

$$\mathsf{E}[\widetilde{\Delta}_g(t) \mid \mathcal{F}_{t-1}] = 0, \qquad t \geq 1,$$

it follows that these random variables are uncorrelated, so that the variance of M_g can be expressed as the sum

$$\mathsf{E}[(M_g(t))^2] = \sum_{i=1}^{t} \mathsf{E}[(\widetilde{\Delta}_g(i))^2], \qquad t \geq 1.$$

In this way it can be shown that the asymptotic variance is expressed as the steady-state variance of $\widetilde{\Delta}_g(i)$. For a proof of (A1.62) (under conditions much weaker than assumed in Proposition A.5.9) see [368, Theorem 17.5.3].

Proposition A.5.9. *Under the assumptions of Theorem A.5.8 the asymptotic variance can be expressed*

$$\sigma_g^2 = \mathsf{E}_\pi[(\widetilde{\Delta}_g(0))^2] = \pi(\hat{g}^2 - (P\hat{g})^2) = \pi(2g\hat{g} - g^2). \tag{A1.62}$$

A.6 Converse theorems

The aim of Section A.5 was to explore the application of (V3) and the coupling method. We now explain why (V3) is *necessary* and sufficient for these ergodic theorems to hold.

Converse theorems abound in the stability theory of Markov chains. Theorem A.6.1 contains one such result: If $\pi(f) < \infty$ then there is a solution to (V3), defined as a certain "value function." For an x^*-irreducible chain the solution takes the form

$$PV_f = V_f - f + b_f \mathbf{1}_{x^*}, \tag{A1.63}$$

where the Lyapunov function V_f defined in (A1.64) is interpreted as the "cost to reach the state x^*." The identity (A1.63) is a dynamic programming equation for the *shortest path problem* described in Section 9.4.1.

Theorem A.6.1. *Suppose that $\mathbf{X}$ is an x^*-irreducible, positive recurrent Markov chain on X and that $\pi(f) < \infty$, where $f : \mathsf{X} \to [1, \infty]$ is given. Then, with*

$$V_f(x) := \mathsf{E}_x\left[\sum_{t=0}^{\sigma_{x^*}} f(X(t))\right], \qquad x \in \mathsf{X}, \tag{A1.64}$$

the following conclusions hold:

(i) *The set* $\mathsf{X}_f = \{x : V_f(x) < \infty\}$ *is nonempty and absorbing:*

$$P(x, \mathsf{X}_f) = 1 \quad \text{for all } x \in \mathsf{X}_f.$$

(ii) *The identity (A1.63) holds with* $b_f := \mathsf{E}_{x^*}\left[\sum_{t=1}^{\tau_{x^*}} f(X(t))\right] < \infty.$

(iii) *For* $x \in \mathsf{X}_f,$

$$\lim_{t \to \infty} \frac{1}{t} \mathsf{E}_x[V_f(X(t))] = \lim_{t \to \infty} \mathsf{E}_x[V_f(X(t))\mathbf{1}\{\tau_{x^*} > t\}] = 0.$$

Proof. Applying the Markov property, we obtain for each $x \in \mathsf{X}$

$$
\begin{aligned}
PV_f(x) &= \mathsf{E}_x\left[\mathsf{E}_{X(1)}\left[\sum_{t=0}^{\sigma_{x^*}} f(X(t))\right]\right] \\
&= \mathsf{E}_x\left[\mathsf{E}\left[\sum_{t=1}^{\tau_{x^*}} f(X(t)) \mid X(0), X(1)\right]\right] \\
&= \mathsf{E}_x\left[\sum_{t=1}^{\tau_{x^*}} f(X(t))\right] = \mathsf{E}_x\left[\sum_{t=0}^{\tau_{x^*}} f(X(t))\right] - f(x), \qquad x \in \mathsf{X}.
\end{aligned}
$$

On noting that $\sigma_{x^*} = \tau_{x^*}$ for $x \neq x^*$, the identity above implies the desired identity in (ii).

Based on (ii) it follows that X_f is absorbing. It is nonempty since it contains x^*, which proves (i).

To prove the first limit in (iii) we iterate the idenitity in (ii) to obtain

$$\mathsf{E}_x[V_f(X(t))] = P^t V_f(x) = V_f(x) + \sum_{k=0}^{t-1}[-P^k f(x) + b_f P^k(x, x^*)], \quad t \geq 1.$$

Dividing by t and letting $t \to \infty$ we obtain, whenever $V_f(x) < \infty,$

$$\lim_{t \to \infty} \frac{1}{t} \mathsf{E}_x[V_f(X(t))] = \lim_{t \to \infty} \frac{1}{t} \sum_{k=0}^{t-1}[-P^k f(x) + b_f P^k(x, x^*)].$$

Applying (i) and (ii) we conclude that the chain can be restricted to X_f, and the restricted process satisfies (V3). Consequently, the conclusions of the Mean Ergodic Theorem A.5.4 hold for initial conditions $x \in \mathsf{X}_f$, which gives

$$\lim_{t \to \infty} \frac{1}{t} \mathsf{E}_x[V_f(X(t))] = -\pi(f) + b_f \pi(x^*),$$

and the right-hand side is zero for by (ii).

By the definition of V_f and the Markov property we have for each $m \geq 1$

$$
\begin{aligned}
V_f(X(m)) &= \mathsf{E}_{X(m)} \left[\sum_{t=0}^{\sigma_{x^*}} f(X(t)) \right] \\
&= \mathsf{E} \left[\sum_{t=m}^{\tau_{x^*}} f(X(t)) \mid \mathcal{F}_m \right], \quad \text{on } \{\tau_{x^*} \geq m\}.
\end{aligned}
\tag{A1.65}
$$

Moreover, the event $\{\tau_{x^*} \geq m\}$ is $\mathcal{F}_m$ measurable. That is, one can determine if $X(t) = x^*$ for some $t \in \{1, \ldots, m\}$ based on $\mathcal{F}_m := \sigma\{X(t) : t \leq m\}$. Consequently, by the smoothing property of the conditional expectation,

$$
\begin{aligned}
\mathsf{E}_x[V_f(X(m))\mathbf{1}\{\tau_{x^*} \geq m\}] &= \mathsf{E} \left[\mathbf{1}\{\tau_{x^*} \geq m\}\mathsf{E} \left[\sum_{t=m}^{\tau_{x^*}} f(X(t)) \mid \mathcal{F}_m \right] \right] \\
&= \mathsf{E} \left[\mathbf{1}\{\tau_{x^*} \geq m\} \sum_{t=m}^{\tau_{x^*}} f(X(t)) \right] \leq \mathsf{E} \left[\sum_{t=m}^{\tau_{x^*}} f(X(t)) \right].
\end{aligned}
$$

If $V_f(x) < \infty$, then the right-hand side vanishes as $m \to \infty$ by the Dominated Convergence Theorem. This proves the second limit in (iii). □

Proposition A.6.2. *Suppose that the assumptions of Theorem A.6.1 hold: X is an x^*-irreducible, positive recurrent Markov chain on X with $\pi(f) < \infty$. Suppose that there exist $g \in L_\infty^f$ and $h \in L_\infty^{V_f}$ satisfying*

$$
Ph = h - g.
$$

Then $\pi(g) = 0$, so that h is a solution to Poisson's equation with forcing function g. Moreover, for $x \in \mathsf{X}_f$,

$$
h(x) - h(x^*) = \mathsf{E}_x \left[\sum_{t=0}^{\tau_{x^*}-1} g(X(t)) \right].
\tag{A1.66}
$$

Proof. Let $M_h(t) = h(X(t)) - h(X(0)) + \sum_{k=0}^{t-1} g(X(k))$, $t \geq 1$, $M_h(0) = 0$. Then M_h is a zero-mean martingale,

$$
\mathsf{E}[M_h(t)] = 0, \quad \text{and} \quad \mathsf{E}[M_h(t+1) \mid \mathcal{F}_t] = M_h(t), \quad t \geq 0.
$$

It follows that the stopped process is a martingale,

$$
\mathsf{E}[M_h(\tau_{x^*} \wedge (r+1)) \mid \mathcal{F}_r] = M_h(\tau_{x^*} \wedge r), \quad r \geq 0.
$$

Consequently, for any r,

$$
0 = \mathsf{E}_x[M_h(\tau_{x^*} \wedge r)] = \mathsf{E}_x \left[h(X(\tau_{x^*} \wedge r)) - h(X(0)) + \sum_{t=0}^{\tau_{x^*} \wedge r - 1} g(X(t)) \right].
$$

On rearranging terms and subtracting $h(x^*)$ from both sides,

$$h(x) - h(x^*) = \mathsf{E}_x \left[[h(X(r)) - h(x^*)]\mathbf{1}\{\tau_{x^*} > r\} + \sum_{t=0}^{\tau_{x^*} \wedge r - 1} g(X(t)) \right], \quad \text{(A1.67)}$$

where we have used the fact that $h(X(\tau_{x^*} \wedge t)) = h(x^*)$ on $\{\tau_{x^*} \leq t\}$.

Applying Theorem A.6.1 (iii) and the assumption that $h \in L_\infty^{V_f}$ gives

$$\limsup_{r \to \infty} \left| \mathsf{E}_x \left[(h(X(r)) - h(x^*))\mathbf{1}\{\tau_{x^*} > r\} \right] \right|$$

$$\leq (\|h\|_{V_f} + |h(x^*)|) \limsup_{r \to \infty} \mathsf{E}_x [V_f(X(r))\mathbf{1}\{\tau_{x^*} > r\}] = 0.$$

Hence by (A1.67), for any $x \in \mathsf{X}_f$,

$$h(x) - h(x^*) = \lim_{r \to \infty} \mathsf{E}_x \left[\sum_{t=0}^{\tau_{x^*} \wedge r - 1} g(X(t)) \right].$$

Exchanging the limit and expectation completes the proof. This exchange is justified by the Dominated Convergence Theorem whenever $V_f(x) < \infty$ since $g \in L_\infty^f$. $\qquad \square$

Bibliography

[1] D. Adelman. A price-directed approach to stochastic inventory/routing. *Oper. Res.*, 52(4): 499–514, 2004.

[2] R. Agrawal, R. L. Cruz, C. Okino, and R. Rajan. Performance bounds for flow control protocols. *IEEE/ACM Trans. Netw.*, 7:310–323, 1999.

[3] R. Ahlswede. Multi-way communication channels. In *Second International Symposium on Information Theory (Tsahkadsor, 1971)*, pages 23–52. Akadémiai Kiadó, Budapest, 1973.

[4] M. S. Akturk and F. Erhun. An overview of design and operational issues of KANBAN systems. *Int. J. Prod. Res.*, 37(17):3859–3881, 1999.

[5] E. Altman. *Constrained Markov Decision Processes*. Stochastic Modeling. Chapman & Hall/CRC, Boca Raton, FL, 1999.

[6] E. Altman. Applications of Markov decision processes in communication networks: a survey. In E. Feinberg and A. Shwartz, editors, *Markov Decision Processes: Models, Methods, Directions, and Open Problems*, pages 489–536. Kluwer, Holland, 2001.

[7] E. Altman, B. Gaujal, and A. Hordijk. Multimodularity, convexity, and optimization properties. *Math. Oper. Res.*, 25(2):324–347, 2000.

[8] E. Altman, B. Gaujal, and A. Hordijk. *Discrete-Event Control of Stochastic Networks: Multimodularity and Regularity*. Lecture Notes in Mathematics. Springer-Verlag, New York, 2004.

[9] E. Altman, T. Jiménez, and G. Koole. On the comparison of queueing systems with their fluid limits. *Probab. Eng. Inf. Sci.*, 15(2):165–178, 2001.

[10] V. Anantharam. The stability region of the finite-user, slotted ALOHA system. *IEEE Trans. Inf. Theory*, 37:535–540, 1991.

[11] E. J. Anderson. *A Continuous Model for Job-Shop Scheduling*. PhD thesis, University of Cambridge, Cambridge, UK, 1978.

[12] E. J. Anderson. A new continuous model for job-shop scheduling. *Int. J. Syst. Sci.*, 12:1469–1475, 1981.

[13] E. J. Anderson and P. Nash. *Linear Programming in Infinite-Dimensional Spaces*. Wiley-Interscience Series in Discrete Mathematics and Optimization. John Wiley Chichester, UK, 1987. Theory and applications, A Wiley-Interscience Publication.

[14] L.-E. Andersson, G. Z. Chang, and T. Elfving. Criteria for copositive matrices using simplices and barycentric coordinates. In *Proceedings of the Workshop "Nonnegative Matrices, Applications and Generalizations" and the Eighth Haifa Matrix Theory Conference (Haifa, 1993)*, volume 220, pages 9–30, 1995.

[15] APX homepage. http://www.apx.nl/home.html, 2004. Formally Amsterdam Power Exchange.

[16] A. Arapostathis, V. S. Borkar, E. Fernandez-Gaucherand, M. K. Ghosh, and S. I. Marcus. Discrete-time controlled Markov processes with average cost criterion: a survey. *SIAM J. Control Optim.*, 31:282–344, 1993.

[17] K. J. Arrow, T. Harris, and J. Marschak. Optimal inventory policy. *Econometrica*, 19:250–272, 1951.

[18] K. J. Arrow, S. Karlin, and H. E. Scarf. *Studies in the Mathematical Theory of Inventory and Production*. Stanford Mathematical Studies in the Social Sciences, I. Stanford University Press, Stanford, CA, 1958.

[19] S. Asmussen. *Applied Probability and Queues*. John Wiley, New York, 1987.

[20] S. Asmussen. Queueing simulation in heavy traffic. *Math. Oper. Res.*, 17:84–111, 1992.

[21] B. Ata and S. Kumar. Heavy traffic analysis of open processing networks with complete resource pooling: asymptotic optimality of discrete review policies complete resource pooling: asymptotic optimality of discrete review policies. *Ann. Appl. Probab.*, 15(1A):331–391, 2005.

[22] R. Atar and A. Budhiraja. Singular control with state constraints on unbounded domain. *Ann. Probab.*, 34(5):1864–1909, 2006.

[23] R. Atar, A. Budhiraja, and P. Dupuis. On positive recurrence of constrained diffusion processes. *Ann. Probab.*, 29(2):979–1000, 2001.

[24] D. Atkins and H. Chen. Performance evaluation of scheduling control of queueing networks: fluid model heuristics. *Queueing Syst. Theory Appl.*, 21(3–4):391–413, 1995.

[25] F. Avram, D. Bertsimas, and M. Ricard. Fluid models of sequencing problems in open queueing networks: an optimal control approach. Technical Report, Massachussetts Institute of Technology, MA, 1995.

[26] F. Avram, D. Bertsimas, and M. Ricard. An optimal control approach to optimization of multi-class queueing networks. In F. Kelly and R. Williams, editors, *Volume 71 of IMA Volumes in Mathematics and its Applications*, New York, 1995. Springer-Verlag. Proceedings of Workshop on Queueing Networks of the Mathematical Institute, Minneapolis, 1994.

[27] Y. Azar, A. Z. Broder, A. R. Karlin, and E. Upfal. Balanced allocations. *SIAM J. Comput.*, 29(1):180–200 (electronic), 1999.

[28] K. Azuma. Weighted sums of certain dependent random variables. *Tôhoku Math. J.*, 19:357–367, 1967.

[29] F. Baccelli and P. Brémaud. *Elements of Queueing Theory: Palm Martingale Calculus and Stochastic Recurrences*. Springer, Berlin, 2003.

[30] F. Baccelli and S. Foss. Ergodicity of Jackson-type queueing networks. *Queueing Syst. Theory Appl.*, 17:5–72, 1994.

[31] T. Başar and G. J. Olsder. *Dynamic Noncooperative Game Theory*, 2nd edition. Academic Press, London, 1995.

[32] F. Baskett, K. M. Chandy, R. R. Muntz, and F. G. Palacios. Open, closed, and mixed networks of queues with different classes of customers. *J. ACM*, 22:248–260, 1975.

[33] N. Bäuerle. Asymptotic optimality of tracking policies in stochastic networks. *Ann. Appl. Probab.*, 10(4):1065–1083, 2000.

[34] P. H. Baxendale. Renewal theory and computable convergence rates for geometrically ergodic Markov chains. *Adv. Appl. Probab.*, 15(1B):700–738, 2005.

[35] S. L. Bell and R. J. Williams. Dynamic scheduling of a system with two parallel servers: Asymptotic optimality of a continuous review threshold policy in heavy traffic. In *Proceedings of the 38th Conference on Decision and Control*, pages 1743–1748, Phoenix, Arizona, 1999.

[36] S. L. Bell and R. J. Williams. Dynamic scheduling of a system with two parallel servers in heavy traffic with complete resource pooling: asymptotic optimality of a continuous review threshold policy. *Ann. Appl. Probab.*, 11:608–649, 2001.

[37] R. Bellman. *Dynamic Programming*. Princeton University Press, Princeton, NJ, 1957.

[38] R. Bellman. On a routing problem. *Q. Appl. Math.*, 16:87–90, 1958.

[39] R. Bellman, I. Glicksberg, and O. Gross. On the optimal inventory equation. *Manage. Sci.*, 2:83–104, 1955.

[40] G. Bennett. Probability inequalities for the sum of independent random variables. *J. Am. Stat. Assoc.*, 57:33–45, 1962.

[41] A. Benveniste, M. Métivier, and P. Priouret. *Adaptive Algorithms and Stochastic Approximations. Applications of Mathematics (New York)*, volume 22. Springer-Verlag, Berlin, 1990. Translated from the French by Stephen S. Wilson.

[42] D. S. Bernstein, R. Givan, N. Immerman, and S. Zilberstein. The complexity of decentralized control of Markov decision processes. *Math. Oper. Res.*, 27(4):819–840, 2002.

[43] D. Bertsekas and R. Gallager. *Data Networks*. Prentice Hall, Englewood Cliffs, NJ, 1992.

[44] D. P. Bertsekas. A new value iteration method for the average cost dynamic programming problem. *SIAM J. Control Optim.*, 36(2):742–759 (electronic), 1998.

[45] D. P. Bertsekas. *Dynamic Programming and Optimal Control*, 3rd edition. Athena Scientific, Cambridge, MA, 2007.

[46] D. P. Bertsekas, V. Borkar, and A. Nedic. Improved temporal difference methods with linear function approximation. In J. Si, A. Barto, W. Powell, and D. Wunsch, editors, *Handbook of Learning and Approximate Dynamic Programming*, pages 690–705. Wiley-IEEE Press, Piscataway, NJ, 2004.

[47] D. P. Bertsekas and J. N. Tsitsiklis. *Parallel and Distributed Computation: Numerical Methods*. Prentice Hall, Upper Saddle River, NJ, USA, 1989.

[48] D. P. Bertsekas and J. N. Tsitsiklis. An analysis of stochastic shortest path problems. *Math. Oper. Res.*, 16(3):580–595, 1991.

[49] D. P. Bertsekas and J. N. Tsitsiklis. *Neuro-Dynamic Programming*. Athena Scientific, Cambridge, MA, 1996.

[50] D. Bertsimas, D. Gamarnik, and J. Sethuraman. From fluid relaxations to practical algorithms for high-multiplicity job-shop scheduling: the holding cost objective. *Oper. Res.*, 51(5):798–813, 2003.

[51] D. Bertsimas, D. Gamarnik, and J. N. Tsitsiklis. Stability conditions for multiclass fluid queueing networks. *IEEE Trans. Autom. Control*, 41(11):1618–1631, 1996.

[52] D. Bertsimas, D. Gamarnik, and J. N. Tsitsiklis. Correction: "Stability conditions for multiclass fluid queueing networks" *IEEE Trans. Autom. Control*, 41(11):1618–1631, 1996; MR1419686 (97f:90028)]. *IEEE Trans. Autom. Control*, 42(1):128, 1997.

[53] D. Bertsimas and J. Niño-Mora. Restless bandits, linear programming relaxations, and a primal-dual index heuristic. *Oper. Res.*, 48(1):80–90, 2000.

[54] D. Bertsimas, I. Paschalidis, and J. N. Tsitsiklis. Optimization of multiclass queueing networks: polyhedral and nonlinear characterizations of achievable performance. *Ann. Appl. Probab.*, 4:43–75, 1994.

[55] D. Bertsimas and I. Ch. Paschalidis. Probabilistic service level guarantees in make-to-stock manufacturing systems. *Oper. Res.*, 49(1):119–133, 2001.

[56] D. Bertsimas and J. Sethuraman. From fluid relaxations to practical algorithms for job shop scheduling: the makespan objective. *Math. Program.*, 92(1A):61–102, 2002.

[57] P. Bhagwat, P. P. Bhattacharya, A. Krishna, and S. K. Tripathi. Enhancing throughput over wireless LANs using channel state dependent packet scheduling. *INFOCOM* 3:1133–1140, 1996.

[58] S. Bhardwaj, R. J. Williams, and A. S. Acampora. On the performance of a two-user mimo downlink system in heavy traffic. *IEEE Trans. Inf. Theory*, 53(5), 2007.

[59] N. P. Bhatia and G. P. Szegö. *Stability Theory of Dynamical Systems*. Springer-Verlag, New York, 1970.

[60] A. G. Bhatt and V. S. Borkar. Existence of optimal Markov solutions for ergodic control of Markov processes. *Sankhyā*, 67(1):1–18, 2005.

[61] P. Billingsley. *Probability and Measure*. John Wiley, New York, 1995.

[62] F. Black and M. Scholes. The pricing of options and corporate liabilities. *J. Pol. Econ.*, 81(3):637–654, 1973.

[63] G. Bolch, S. Greiner, H. de Meer, and K. S. Trivedi. *Queueing Networks and Markov Chains: Modeling and Performance Evaluation with Computer Science Applications* 2nd edition. Wiley-Interscience, New York, 2006.

[64] A. M. Bonvik, C. Couch, and S. B. Gershwin. A comparison of production-line control mechanisms. *Int. J. Prod. Res.*, 35(3):789–804, 1997.

[65] V. S. Borkar. Controlled Markov chains and stochastic networks. *SIAM J. Control Optim.*, 21(4):652–666, 1983.

[66] V. S. Borkar. On minimum cost per unit time control of Markov chains. *SIAM J. Control Optim.*, 22(6):965–978, 1984.

[67] V. S. Borkar. Control of Markov chains with long-run average cost criterion: the dynamic programming equations. *SIAM J. Control Optim.*, 27(3):642–657, 1989.

[68] V. S. Borkar. *Topics in Controlled Markov Chains*. Pitman Research Notes in Mathematics Series 240, Longman Scientific & Technical, UK, 1991.

[69] V. S. Borkar. Ergodic control of Markov chains with constraints the general case. *SIAM J. Control Optim.*, 32:176–186, 1994.

[70] V. S. Borkar. Convex analytic methods in Markov decision processes. In *Handbook of Markov Decision Processes. International Series in Operations Research and Management Science*, volume 40, pages 347–375. Kluwer Academic Publisher, Boston, MA, 2002.

[71] V. S. Borkar and S. P. Meyn. The O.D.E. method for convergence of stochastic approximation and reinforcement learning. *SIAM J. Control Optim.*, 38(2):447–469, 2000. Also presented at the *IEEE CDC*, December 1998.

[72] A. N. Borodin and P. Salminen. *Handbook of Brownian Motion – Facts and Formulae*. Probability and its Applications 1st edition. Birkhäuser Verlag, Basel, 1996. Second edition published 2002.

[73] A. A. Borovkov. Limit theorems for queueing networks. *Theory Probab. Appl.*, 31:413–427, 1986.

[74] A. A. Borovkov, O. J. Boxma, and Z. Palmowski. On the integral of the workload process of the single server queue. *J. Appl. Probab.*, 40(1):200–225, 2003.

[75] S. C. Borst and P. A. Whiting. Dynamic rate control algorithms for HDR throughput optimization. *IEEE INFOCOM*, 2:976–985, 2001.

[76] D. D. Botvich and A. A. Zamyatin. Ergodicity of conservative communication networks. Technical Report, Institut National de Recherche en Informatique et en Automatique, Rocquencourt, 1993.

[77] J. R. Bradley and P. W. Glynn. Managing capacity and inventory jointly in manufacturing systems. *Manage. Sci.*, 48(2):273–288, 2002.

[78] S. J. Bradtke and A. G. Barto. Linear least-squares algorithms for temporal difference learning. *Mach. Learn.*, 22(1–3):33–57, 1996.

[79] M. Bramson. Correction: Instability of FIFO queueing networks. *Ann. Appl. Probab.*, 4(3):952, 1994.

[80] M. Bramson. Instability of FIFO queueing networks. *Ann. Appl. Probab.*, 4(2):414–431, 1994.

[81] M. Bramson. State space collapse with application to heavy traffic limits for multiclass queueing networks. *Queueing Syst. Theory Appl.*, 30:89–148, 1998.

[82] M. Bramson. A stable queueing network with unstable fluid model. *Ann. Appl. Probab.*, 9(3):818–853, 1999.

[83] M. Bramson and R. J. Williams. On dynamic scheduling of stochastic networks in heavy traffic and some new results for the workload process. In *Proceedings of the 39th Conference on Decision and Control*, Sydney, Australia, page 516, 2000.

[84] M. Bramson and R. J. Williams. Two workload properties for Brownian networks. *Queueing Syst. Theory Appl.*, 45(3):191–221, 2003.

[85] A. Budhiraja and C. Lee. Stationary distribution convergence for generalized Jackson networks in heavy traffic. Submitted for publication, 2007.

[86] C. Buyukkoc, P. Varaiya, and J. Walrand. The c-μ rule revisited. *Adv. Appl. Probab.*, 17(1):237–238, 1985.

[87] California's Electricity, UIC Nuclear Issues Briefing Paper No. 61. Uranium Information Centre. http://www.uic.com.au/nip61.htm, May 2004.

[88] C. G. Cassandras and S. Lafortune. *Introduction to discrete event systems*. The Kluwer International Series on Discrete Event Dynamic Systems, 11. Kluwer Academic Publishers, Boston, MA, 1999.

[89] R. Cavazos-Cadena. Value iteration in a class of communicating Markov decision chains with the average cost criterion. *SIAM J. Control Optim.*, 34(6):1848–1873, 1996.

[90] R. Cavazos-Cadena and E. Ferndndez-Gaucherand. Value iteration in a class of controlled Markov chains with average criterion: unbounded costs case. In *Proceedings of the 34th Conference on Decision and Control*, New Orleans, LA, page TP05 3:40, 1995.

[91] CBS. More Enron tapes, more gloating. http:// www.cbsnews.com/stories/ 2004/06/08/eveningnews/~main621856.shtml, June 8, 2004.

[92] C.-S. Chang, X. Chao, M. Pinedo, and R. Weber. On the optimality of LEPT and c-μ rules for machines in parallel. *J. Appl. Probab.*, 29(3):667–681, 1992.

[93] C.-S. Chang, J. A. Thomas, and S.-H. Kiang. On the stability of open networks: a unified approach by stochastic dominance. *Queueing Syst. Theory Appl.*, 15:239–260, 1994.

[94] H. Chen and A. Mandelbaum. Discrete flow networks: bottleneck analysis and fluid approximations. *Math. Oper. Res.*, 16(2):408–446, 1991.

[95] H. Chen and D. D. Yao. Dynamic scheduling of a multiclass fluid network. *Oper. Res.*, 41(6):1104–1115, November–December 1993.

[96] H. Chen and D. D. Yao. *Fundamentals of Queueing Networks: Performance, Asymptotics, and Optimization*. Stochastic Modelling and Applied Probability. Springer-Verlag, New York, 2001.

[97] H. Chen and H. Zhang. Stability of multiclass queueing networks under FIFO service discipline. *Math. Oper. Res.*, 22(3):691–725, 1997.

[98] H. Chen and H. Zhang. Stability of multiclass queueing networks under priority service disciplines. *Oper. Res.*, 48(1):26–37, 2000.

[99] H.-F. Chen. *Stochastic Approximation and Its Applications*. Nonconvex Optimization and its Applications, volume 64. Kluwer Academic Publishers, Dordrecht, 2002.

[100] M. Chen. *Modelling and Control of Complex Stochastic Networks, with Applications to Manufacturing Systems and Electric Power Transmission Networks*. PhD thesis, University of Illinois at Urbana Champaign, University of Illinois, Urbana, IL, 2005.

[101] M. Chen, I.-K. Cho, and S. P. Meyn. Reliability by design in a distributed power transmission network. *Automatica*, 42:1267–1281, August 2006. Invited.

[102] M. Chen, R. Dubrawski, and S. P. Meyn. Management of demand-driven production systems. *IEEE Trans. Autom. Control*, 49(2):686–698, May 2004.

[103] M. Chen, C. Pandit, and S. P. Meyn. In search of sensitivity in network optimization. *Queueing Syst. Theory Appl.*, 44(4):313–363, 2003.

[104] R. C. Chen and G. L. Blankenship. Dynamic programming equations for discounted constrained stochastic control. *IEEE Trans. Autom. Control*, 49(5):699–709, 2004.

[105] R.-R. Chen and S. P. Meyn. Value iteration and optimization of multiclass queueing networks. *Queueing Syst. Theory Appl.*, 32(1–3):65–97, 1999.

[106] R. Chitashvili and M. Mania. Generalized Itô formula and derivation of Bellman's equation. In *Stochastic Processes and Related Topics (Siegmundsberg, 1994)*. Stochastics Monographs, volume 10, pages 1–21. Gordon and Breach, Yverdon, 1996.

[107] I.-K. Cho and S. P. Meyn. The dynamics of the ancillary service prices in power networks. In *Proceedings of the 42nd IEEE Conference on Decision and Control*, volume 3, Maui, H1, pages 2094– 2099, December 9–12, 2003.

[108] A. J. Chorin. Conditional expectations and renormalization. *J. Multiscale Model. Simul.*, 1(1):105–118, 2003.

[109] K. L. Chung. *A Course in Probability Theory*, 2nd edition. Academic Press, New York, 1974.

[110] A. J. Clark and H. E. Scarf. Optimal policies for a multi-echelon inventory problem. *Manage. Sci.*, 6:465–490, 1960.

[111] E. G. Coffman Jr., D. S. Johnson, P. W. Shor, and R. R. Weber. Markov chains, computer proofs, and average-case analysis of best fit bin packing. In *STOC '93: Proceedings of the Twenty-Fifth Annual ACM Symposium on Theory of Computing*, pages 412–421, New York, 1993. ACM Press.

[112] E. G. Coffman Jr. and I. Mitrani. A characterization of waiting times realizable by single server queues. *Oper. Res.*, 28:810–821, 1980.

[113] E. G. Coffman Jr. and M. I. Reiman. Diffusion approximations for storage processes in computer systems. In *SIGMETRICS '83: Proceedings of the 1983 ACM SIGMETRICS Conference on Measurement and Modeling of Computer Systems*, pages 93–117, New York, 1983. ACM Press.

[114] J. W. Cohen. *The Single Server Queue*, 2nd edition. North-Holland, Amsterdam, 1982.

[115] T. H. Cormen, C. E. Leiserson, and R. L. Rivest. *Introduction to Algorithms*. MIT Press, MA, 1990.

[116] C. Courcoubetis and R. Weber. Stability of on-line bin packing with random arrivals and long-run average constraints. *Probab. Eng. Inf. Sci.*, 4:447–460, 1990.

[117] T. M. Cover. Comments on broadcast channels. *IEEE Trans. Inf. Theory*, 44(6):2524–2530, 1998. Information theory: 1948–1998.

[118] T. M. Cover and J. A. Thomas. *Elements of information theory*. John Wiley, New York, 1991. A Wiley-Interscience Publication.

[119] D. R. Cox and W. L. Smith. *Queues*. Methuen's Monographs on Statistical Subjects. Methuen, London, 1961.

[120] J. Cox, S. Ross, and M. Rubinstein. Option pricing: a simplified approach. *J. Financ. Econ.*, 7:229–263, 1979.

[121] R. L. Cruz. A calculus for network delay. I. Network elements in isolation. *IEEE Trans. Inf. Theory*, 37(1):114–131, 1991.

[122] J. Csirik, D. S. Johnson, C. Kenyon, J. B. Orlin, P. W. Shor, and R. R. Weber. On the sum-of-squares algorithm for bin packing. *J. Assoc. Comput. Mach.*, 53(1):1–65, 2006.

[123] M. Dacre, K. Glazebrook, and J. Niño-Mora. The achievable region approach to the optimal control of stochastic systems. *J. R. Stat. Soc. Ser. B*, 61(4):747–791, 1999.

[124] J. G. Dai. On positive Harris recurrence of multiclass queueing networks: a unified approach via fluid limit models. *Ann. Appl. Probab.*, 5(1):49–77, 1995.

[125] J. G. Dai. A fluid-limit model criterion for instability of multiclass queueing networks. *Ann. Appl. Probab.*, 6:751–757, 1996.

[126] J. G. Dai and J. M. Harrison. Reflected Brownian motion in an orthant: numerical methods for steady-state analysis. *Ann. Appl. Probab.*, 2:65–86, 1992.

[127] J. G. Dai, J. J. Hasenbein, and J. H. Vande Vate. Stability and instability of a two-station queueing network. *Ann. Appl. Probab.*, 14(1):326–377, 2004.

[128] J. G. Dai and W. Lin. Maximum pressure policies in stochastic processing networks. *Oper. Res.*, 53(2):197–218, 2005.

[129] J. G. Dai and S. P. Meyn. Stability and convergence of moments for multiclass queueing networks via fluid limit models. *IEEE Trans. Aut. Control*, 40:1889–1904, November 1995.

[130] J. G. Dai and J. H. Vande Vate. The stability of two-station multi-type fluid networks. *Oper. Res.*, 48:721–744, 2000.

[131] J. G. Dai and Y. Wang. Nonexistence of Brownian models of certain multiclass queueing networks. *Queueing Syst. Theory Appl.*, 13:41–46, May 1993.

[132] J. G. Dai and G. Weiss. Stability and instability of fluid models for reentrant lines. *Math. Oper. Res.*, 21(1):115–134, 1996.

[133] J. G. Dai and G. Weiss. A fluid heuristic for minimizing makespan in job shops. *Oper. Res.*, 50(4):692–707, 2002.

[134] Y. Dallery and S. B. Gershwin. Manufacturing flow line systems: a review of models and analytical results. *Queueing Syst. Theory Appl.*, 12(1–2):3–94, 1992.

[135] G. B. Dantzig. Linear programming under uncertainty. *Manage. Sci.*, 7(3):197–206, April–July 1995.

[136] G. B. Danzig, J. Folkman, and N. Shapiro. On the continuity of the minimum set of a continuous function. *J. Math. Anal. Appl.*, 17:519–548, 1967.

[137] D. P. de Farias and B. Van Roy. The linear programming approach to approximate dynamic programming. *Oper. Res.*, 51(6):850–865, 2003.

[138] Y. De Serres. Simultaneous optimization of flow control and scheduling in a single server queue with two job classes. *Oper. Res. Lett.*, 10(2):103–112, 1991.

[139] R. Dekker. Counterexamples for compact action Markov decision chains with average reward criteria. *Comm. Stat. Stoch. Models*, 3(3):357–368, 1987.

[140] A. Dembo and O. Zeitouni. *Large Deviations Techniques And Applications*, 2nd edition. Springer-Verlag, New York, 1998.

[141] E. Denardo. Contraction mappings underlying the theory of dynamic programming. *SIAM Rev.*, 9:165–177, 1967.

[142] E. V. Denardo and B. L. Fox. Multichain Markov renewal programs. *SIAM J. Appl. Math.*, 16:468–487, 1968.

[143] J. L. Doob. *Stochastic Processes*. John Wiley, New York, 1953.

[144] B. T. Doshi. Optimal control of the service rate in an M/G/1 queueing system. *Adv. Appl. Probab.*, 10:682–701, 1978.

[145] R. Douc, G. Fort, E. Moulines, and P. Soulier. Practical drift conditions for subgeometric rates of convergence. *Ann. Appl. Probab.*, 14(3):1353–1377, 2004.

[146] D. Down and S. P. Meyn. Stability of acyclic multiclass queueing networks. *IEEE Trans. Autom. Control*, 40:916–919, 1995.

[147] D. Down and S. P. Meyn. Piecewise linear test functions for stability and instability of queueing networks. *Queueing Syst. Theory Appl.*, 27(3–4):205–226, 1997.

[148] D. Down, S. P. Meyn, and R. L. Tweedie. Exponential and uniform ergodicity of Markov processes. *Ann. Probab.*, 23(4):1671–1691, 1995.

[149] L. E. Dubins. On extreme points of convex sets. *J. Math. Anal. Appl.*, 5:237–244, 1962.

[150] L. E. Dubins and Leonard J. Savage. *How to Gamble If You Must: Inequalities for Stochastic Processes*. McGraw-Hill, New York, 1965.

[151] R. Dubrawski. *Myopic and Far-Sighted Strategies for Control of Demand-Driven Networks*. Master's thesis, Department of Electrical Engineering, University of Illinois at Urbana Champaign, Urbana, Illinois, 2000.

[152] I. Duenyas. A simple release policy for networks of queues with controllable inputs. *Oper. Res.*, 42(6):1162–1171, 1994.

[153] N. G. Duffield and Neil O'Connell. Large deviations and overflow probabilities for the general single-server queue, with applications. *Math. Proc. Camb. Phil. Soc.*, 118(2):363–374, 1995.

[154] W. T. M. Dunsmuir, S. P. Meyn, and G. Roberts. Obituary: Richard Lewis Tweedie. *J. Appl. Probab.*, 39(2):441–454, 2002.

[155] P. Dupuis and R. S. Ellis. *A Weak Convergence Approach to the Theory of Large Deviations*. Wiley Series in Probability and Statistics: Probability and Statistics. John Wiley, New York, 1997. A Wiley-Interscience Publication.

[156] P. Dupuis and I. Hitoshi. On Lipschitz continuity of the solution mapping to the Skorokhod problem, with applications. *Stochastics*, 35(1):31–62, 1991.

[157] P. Dupuis and I. Hitoshi. SDEs with oblique reflection on nonsmooth domains. *Ann. Probab.*, 21(1):554–580, 1993.

[158] P. Dupuis and K. Ramanan. Convex duality and the Skorokhod problem. I, II. *Probab. Theory Relat. Fields*, 115(2):153–195, 197–236, 1999.

[159] P. Dupuis and K. Ramanan. An explicit formula for the solution of certain optimal control problems on domains with corners. *Theory Probab. Math. Stat.* 63:33–49, 2001; *Teor. Jmovirn. Mat. Stat.* 63:32–48, 2000.

[160] P. Dupuis and R. J. Williams. Lyapunov functions for semimartingale reflecting Brownian motions. *Ann. Appl. Probab.*, 22(2):680–702, 1994.

[161] A. Dvoretzky, J. Kiefer, and J. Wolfowitz. The inventory problem. I. Case of known distributions of demand. *Econometrica*, 20:187–222, 1952.

[162] E. B. Dynkin and A. A. Yushkevich. *Controlled Markov Processes*. Grundlehren der Mathematischen Wissenschaften [Fundamental Principles of Mathematical Sciences], volume 235. Springer-Verlag, Berlin, 1979. Translated from the Russian original by J. M. Danskin and C. Holland.

[163] D. Eng, J. Humphrey, and S. P. Meyn. Fluid network models: linear programs for control and performance bounds. In J. Cruz, J. Gertler, and M. Peshkin, editors, *Proceedings of the 13th IFAC World Congress*, volume B, pages 19–24, San Francisco, California, 1996.

[164] A. Ephrimedes, P. Varaiya, and J. Walrand. A simple dynamic routing problem. *IEEE Trans. Autom. Control*, 25:690–693, 1980.

[165] A. Eryilmaz and R. Srikant. Joint congestion control, routing, and MAC for stability and fairness in wireless networks. *IEEE J. Sel. Areas Commun.*, 24(8):1514–1524, 2006.

[166] A. Eryilmaz, R. Srikant, and J. R. Perkins. Stable scheduling policies for fading wireless channels. *IEEE/ACM Trans. Netw.*, 13(2):411–424, 2005.

[167] S. N. Ethier and T. G. Kurtz. *Markov Processes: Characterization and Convergence*. John Wiley, New York, 1986.

[168] P. Fairley. The unruly power grid. *IEEE Spectr.*, 41(8):22–27, August 2004.

[169] G. Fayolle. On random walks arising in queueing systems: ergodicity and transience via quadratic forms as Lyapounov functions – part I. *Queueing Syst.*, 5:167–183, 1989.

[170] G. Fayolle, V. A. Malyshev, and M. V. Men'shikov. *Topics in the Constructive Theory of Countable Markov Chains*. Cambridge University Press, Cambridge, 1995.

[171] G. Fayolle, V. A. Malyshev, M. V. Menshikov, and A. F. Sidorenko. Lyapunov functions for jackson networks. *Math. Oper. Res.*, 18(4):916–927, 1993.

[172] A. Federgruen and H. Groenevelt. Characterization and control of achievable performance in general queueing systems. *Oper. Res.*, 36:733–741, 1988.

[173] A. Federgruen, A. Hordijk, and H. C. Tijms. Denumerable state semi-Markov decision processes with unbounded costs, average cost criterion unbounded costs, average cost criterion. *Stoch. Proc. Appl.*, 9(2):223–235, 1979.

[174] A. Federgruen, P. J. Schweitzer, and H. C. Tijms. Denumerable undiscounted semi-Markov decision processes with unbounded rewards. *Math. Oper. Res.*, 8(2):298–313, 1983.

[175] A. Federgruen and P. Zipkin. Computing optimal (s, S) policies in inventory models with continuous demands. *Adv. Appl. Probab.*, 17(2):424–442, 1985.

[176] A. Federgruen and P. Zipkin. An inventory model with limited production capacity and uncertain demands. I. The average-cost criterion. *Math. Oper. Res.*, 11(2):193–207, 1986.

[177] A. Federgruen and P. Zipkin. An inventory model with limited production capacity and uncertain demands. II. The discounted-cost criterion. *Math. Oper. Res.*, 11(2):208–215, 1986.

[178] E. A. Feinberg and A. Shwartz, editors. *Handbook of Markov Decision Processes: Methods and Applications*. International Series in Operations Research & Management Science, 40. Kluwer Academic Publishers, Boston, MA, 2002.

[179] Final report on the August 14, 2003 blackout in the United States and Canada: causes and recommendations. https://reports.energy.gov/, April 2004.

[180] L. K. Fleischer. Faster algorithms for the quickest transshipment problem. *SIAM J. Control Optim.*, 12(1):18–35 (electronic), 2001.

[181] L. K. Fleischer. Universally maximum flow with piecewise-constant capacities. *Networks*, 38(3):115–125, 2001.

[182] S. Floyd and V. Jacobson. Random early detection gateways for congestion avoidance. *IEEE/ACM Trans. Netw.*, 1(4):397–413, 1993.

[183] R. D. Foley and D. R. McDonald. Join the shortest queue: stability and exact asymptotics. *Ann. Appl. Probab.*, 11(3):569–607, 2001.

[184] R. D. Foley and D. R. McDonald. Large deviations of a modified Jackson network: Stability and rough asymptotics. *Ann. Appl. Probab.*, 15(1B):519–541, 2005.

[185] H. Föllmer and P. Protter. On Itô's formula for multidimensional Brownian motion. *Probab. Theory Relat. Fields*, 116(1):1–20, 2000.

[186] N.-T. Fong and X. Y. Zhou. Hierarchical feedback controls in two-machine flow shops under uncertainty. In *Proceedings of the 35th Conference on Decision and Control*, pages 1743–1748, Kobe, Japan, 1996.

[187] H. Ford. *My Life and Work*. Kessinger Publishing, Montana, 1922. Reprint January 2003.

[188] L. R. Ford Jr. Network flow theory. Paper P-923, The RAND Corporation, Santa Moncia, CA, August 1956.

[189] G. Fort, S. Meyn, E. Moulines, and P. Priouret. ODE methods for Markov chain stability with applications to MCMC. In *Proceedings of VALUETOOLS' 06*, Pisa Italy, 2006.

[190] G. Foschini and J. Salz. A basic dynamic routing problem and diffusion. *IEEE Trans. Commun.*, 26:320–327, 1978.

[191] F. G. Foster. On the stochastic matrices associated with certain queuing processes. *Ann. Math. Stat.*, 24:355–360, 1953.

[192] A. Gajrat, A. Hordijk, and A. Ridder. Large deviations analysis of the fluid approximation for a controllable tandem queue. *Ann. Appl. Probab.*, 13:1423–1448, 2003.

[193] R. G. Gallager. An inequality on the capacity region of multiaccess multipath channels. In U. Maurer, R. E. Blahut, D. J. Costello Jr. and T. Mittelholzer, editors, *Communications and Cryptography: Two Sides of One Tapestry*. The International Series in Engineering and Computer Science, pages 129–139. Kluwer, Norwell, MA, 1994.

[194] D. Gamarnik. On deciding stability of constrained homogeneous random walks and queueing systems. *Math. Oper. Res.*, 27(2):272–293, 2002.

[195] D. Gamarnik and J. J. Hasenbein. Instability in stochastic and fluid queueing networks. *Ann. Appl. Probab.*, 15(3):1652–1690, 2005.

[196] D. Gamarnik and S. P. Meyn. On exponential ergodicity in multiclass queueing networks. Asymptotic Analysis of Stochastic Systems, invited session at the INFORMS Annual Meeting, November 13–16, 2005.

[197] D. Gamarnik and A. Zeevi. Validity of heavy traffic steady-state approximations in generalized jackson networks. *Adv. Appl. Probab.*, 16(1):56–90, 2006.

[198] A. Ganesh, N. O'Connell, and D. Wischik. *Big Queues*. Lecture Notes in Mathematics, volume 1838. Springer-Verlag, Berlin, 2004.

[199] N. Gans, G. Koole, and A. Mandelbaum. Commissioned paper: telephone call centers: tutorial, review, and research prospects. *Manuf. Serv. Oper. Manage.*, 5(2):79–141, 2003.

[200] E. Gelenbe and I. Mitrani. *Analysis and Synthesis of Computer Systems*. Academic Press, New York-London, 1980.

[201] L. Georgiadis, W. Szpankowski, and L. Tassiulas. A scheduling policy with maximal stability region for ring networks with spatial reuse. *Queueing Syst. Theory Appl.*, 19(1–2):131–148, 1995.

[202] S. B. Gershwin. A hierarchical framework for discrete event scheduling in manufacturing systems. In *Proceedings of IIASA Conference on Discrete Event System: Models and Applications (Sopron. Hungary, 1987)*, pages 197–216, 1988.

[203] S. B. Gershwin. *Manufacturing Systems Engineering*. Prentice Hall, Englewood Cliffs, NJ, 1993.

[204] S. B. Gershwin. Production and subcontracting strategies for manufacturers with limited capacity and volatile demand. *IIE Trans. Des. Manuf.*, 32(2):891–906, 2000. Special Issue on decentralized control of manufacturing systems.

[205] R. J. Gibbens and F. P. Kelly. Dynamic routing in fully connected networks. *IMA J. Math. Cont. Inf.*, 7:77–111, 1990.

[206] R. J. Gibbens and F. P. Kelly. Measurement-based connection admission control, 1997. In Fifteenth International Teletraffic Congress, Amsterdam, June 1997.

[207] J. C. Gittins. Bandit processes and dynamic allocation indices. *J. R. Stat. Soc. Ser. B*, 41:148–177, 1979.

[208] P. Glasserman and D. D. Yao. *Monotone Structure in Discrete-Event Systems*. Wiley Series in Probability and Mathematical Statistics: Applied Probability and Statistics. John Wiley, New York, 1994. A Wiley-Interscience Publication.

[209] P. Glynn and R. Szechtman. Some new perspectives on the method of control variates. In K. T. Fang, F. J. Hickernell, and H. Niederreiter, editors, *Monte Carlo and Quasi-Monte Carlo Methods 2000: Proceedings of a Conference held at Hong Kong Baptist University, Hong Kong SAR, China*, pages 27–49, Berlin, 2002. Springer-Verlag.

[210] P. W. Glynn. Some asymptotic formulas for markov chains with applications to simulation. *J. Stat. Comput. Simul.*, 19:97–112, 1984.

[211] P. W. Glynn. Diffusion approximations. In *Handbooks in Operations Research and Management Science, 2: Stochastic Models*, pages 145–198. North-Holland, Amsterdam, 1990.

[212] P. W. Glynn and D. L. Iglehart. A joint Central Limit Theorem for the sample mean and the regenerative variance estimator. *Ann. Oper. Res.*, 8:41–55, 1987.

[213] P. W. Glynn and D. L. Iglehart. Simulation methods for queues: an overview. *Queueing Syst. Theory Appl.*, 3(3):221–255, 1988.

[214] P. W. Glynn and S. P. Meyn. A Liapounov bound for solutions of the Poisson equation. *Ann. Probab.*, 24(2):916–931, 1996.

[215] P. W. Glynn and D. Ormoneit. Hoeffding's inequality for uniformly ergodic Markov chains. *Stat. Probab. Lett.*, 56:143–146, 2002.

[216] P. W. Glynn and W. Whitt. The asymptotic efficiency of simulation estimators. *Oper. Res.*, 40(3):505–520, 1992.

[217] P. W. Glynn and W. Whitt. Logarithmic asymptotics for steady-state tail probabilities in a single-server queue. *Stud. Appl. Probab.*, 17:107–128, 1993.

[218] G. C. Goodwin, M. M. Seron, and J. A. De Doná. *Constrained Control and Estimation: An Optimisation Approach*. Communications and Control Engineering. Springer-Verlag, London, 2005.

[219] G. C. Goodwin and K. S. Sin. *Adaptive Filtering Prediction and Control*. Prentice Hall, Englewood Cliffs, NJ, 1984.

[220] T. C. Green and S. Stidham. Sample-path conservation laws, with applications to scheduling queues and fluid systems. *Queueing Syst. Theory Appl.*, 36:175–199, 2000.

[221] F. Guillemin and D. Pinchon. On the area swept under the occupation process of an $M/M/1$ queue in a busy period. *Queueing Syst. Theory Appl.*, 29(2–4):383–398, 1998.

[222] B. Hajek. Optimal control of two interacting service stations. *IEEE Trans. Autom. Control*, AC-29:491–499, 1984.

[223] B. Hajek. Extremal splittings of point processes. *Math. Oper. Res.*, 10(4):543–556, 1985.

[224] B. Hajek and R. G. Ogier. Optimal dynamic routing in communication networks with continuous traffic. *Networks*, 14(3):457–487, 1984.

[225] P. Hall and C. C. Heyde. *Martingale Limit Theory and Its Application*. Probability and Mathematical Statistics. Academic Press. [Harcourt Brace Jovanovich Publishers], New York, 1980.

[226] S. V. Hanly and D. N. C. Tse. Multiaccess fading channels. II. Delay-limited capacities. *IEEE Trans. Inf. Theory*, 44(7):2816–2831, 1998.

[227] R. Hariharan, M. S. Moustafa, and S. Stidham Jr. Scheduling in a multi-class series of queues with deterministic service times. *Queueing Syst. Theory Appl.*, 24:83–89, 1997.

[228] J. M. Harrison. Discrete dynamic programming with unbounded rewards. *Ann. Math. Statist.*, 43:636–644, 1972.

[229] J. M. Harrison. A limit theorem for priority queues in heavy traffic. *J. Appl. Probab.*, 10:907–912, 1973.

[230] J. M. Harrison. Dynamic scheduling of a multiclass queue: discount optimality. *Oper. Res.*, 23(2):370–382, March–April 1975.

[231] J. M. Harrison. *Brownian Motion and Stochastic Flow Systems*. John Wiley, New York, 1985.

[232] J. M. Harrison. Brownian models of queueing networks with heterogeneous customer populations. In *Stochastic Differential Systems, Stochastic Control Theory and Applications (Minneapolis, Minn., 1986)*, pages 147–186. Springer, New York, 1988.

[233] J. M. Harrison. *Brownian Motion and Stochastic Flow Systems*. Robert E. Krieger, Malabar, FL, 1990. Reprint of the 1985 original.

[234] J. M. Harrison. In F. P. Kelly, S. Zachary, and I. Ziedins, editors, *The BIGSTEP approach to flow management in stochastic processing Networks*, pages 57–89. In *Stochastic Networks Theory and Applications*. Clarendon Press, Oxford, 1996.

[235] J. M. Harrison. Heavy traffic analysis of a system with parallel servers: asymptotic optimality of discrete-review policies. *Ann. Appl. Probab.*, 8(3):822–848, 1998.

[236] J. M. Harrison. Brownian models of open processing networks: canonical representations of workload. *Ann. Appl. Probab.*, 10:75–103, 2000.

[237] J. M. Harrison. Stochastic networks and activity analysis. In Y. Suhov, editor, *Analytic Methods in Applied Probability. In Memory of Fridrih Karpelevich*. American Mathematical Society, Providence, RI, 2002.

[238] J. M. Harrison. Correction: "Brownian models of open processing networks: canonical representation of workload" [*Ann. Appl. Probab.* 10(1): 75–103, 2000; MR1765204 (2001g:60230)]. *Ann. Appl. Probab.*, 13(1):390–393, 2003.

[239] J. M. Harrison and M. J. López. Heavy traffic resource pooling in parallel-server systems. *Queueing Syst. Theory Appl.*, 33:339–368, 1999.

[240] J. M. Harrison and M. I. Reiman. On the distribution of multidimensional reflected Brownian motion. *SIAM J. Appl. Math.*, 41(2):345–361, 1981.

[241] J. M. Harrison and J. A. Van Mieghem. Dynamic control of Brownian networks: state space collapse and equivalent workload formulations. *Ann. Appl. Probab.*, 7(3):747–771, 1997.

[242] J. M. Harrison and L. M. Wein. Scheduling networks of queues: heavy traffic analysis of a simple open network. *Queueing Syst. Theory Appl.*, 5(4):265–279, 1989.

[243] J. M. Harrison and L. M. Wein. Scheduling networks of queues: heavy traffic analysis of a two-station closed network. *Oper. Res.*, 38(6):1052–1064, 1990.

[244] J. M. Harrison and R. J. Williams. Brownian models of open queueing networks with homogeneous customer populations. *Stochastics*, 22(2):77–115, 1987.

[245] J. M. Harrison and R. J. Williams. Multidimensional reflected Brownian motions having exponential stationary distributions. *Ann. Probab.*, 15(1):115–137, 1987.

[246] J. M. Harrison and R. J. Williams. On the quasireversibility of a multiclass Brownian service station. *Ann. Probab.*, 18(3):1249–1268, 1990.

[247] J. M. Harrison and R. J. Williams. Brownian models of feedforward queueing networks: quasireversibility and product form solutions. *Ann. Appl. Probab.*, 2(2):263–293, 1992.

[248] M. Hauskrecht and B. Kveton. Linear program approximations to factored continuous-state Markov decision processes. In *Advances in Neural Information Processing Systems*, volume 16. MIT Press, Cambridge, MA, pp. 895–902, 2004.

[249] P. Heidelberger. Variance reduction techniques for simulating markov chains. In *WSC '77: Proceedings of the 9th Conference on Winter Simulation*, pages 160–164. Winter Simulation Conference, Gaitherburg, MD, 1977.

[250] P. Heidelberger. *Variance Reduction Techniques for the Simulation of Markov Processes*. PhD thesis, Stanford University, Palo Alto, CA, 1977.

[251] S. G. Henderson. *Variance Reduction Via an Approximating Markov Process*. PhD thesis, Stanford University, Stanford, CA, 1997.

[252] S. G. Henderson. Mathematics for simulation. In S. G. Henderson and B. L. Nelson, editors, *Handbook of Simulation*. Handbooks in Operations Research and Management Science. Elsevier, 2006.

[253] S. G. Henderson and P. W. Glynn. Approximating martingales for variance reduction in Markov process simulation. *Math. Oper. Res.*, 27(2):253–271, 2002.

[254] S. G. Henderson and S. P. Meyn. Efficient simulation of multiclass queueing networks. In S. Andradottir, K. Healy, D. H. Withers, and B. L. Nelson, editors, *Proceedings of the 1997 Winter Simulation Conference*, pages 216–223, Piscataway, NJ, 1997. IEEE.

[255] S. G. Henderson and S. P. Meyn. Variance reduction for simulation in multiclass queueing networks. *IIE Trans. Oper. Eng.*, to appear.

[256] S. G. Henderson, S. P. Meyn, and V. B. Tadić. Performance evaluation and policy selection in multiclass networks. *Discrete Event Dyn. Syst. Theory Appl.*, 13(1–2):149–189, 2003. Special issue on learning, optimization and decision making. Invited.

[257] S. G. Henderson and B. L. Nelson, editors. *Handbook of Simulation*. Handbooks in Operations Research and Management Science, XII. Elsevier, Cambridge, MA, 2005.

[258] O. Hernández-Lerma and J. B. Lasserre. *Discrete-Time Markov Control Processes. Basic Optimality Criteria*. Applications of Mathematics (New York), volume 30. Springer-Verlag, New York, 1996.

[259] O. Hernández-Lerma and J. B. Lasserre. *Further Topics on Discrete-Time Markov Control Processes*. Stochastic Modeling and Applied Probability. Springer-Verlag, Berlin, Germany, 1999.

[260] O. Hernández-Lerma and J. B. Lasserre. The linear programming approach. In *Handbook of Markov Decision Processes. International Series in Operations Research and Management Science*, volume 40. Kluwer Academic Publisher, Boston, MA, pp. 377–407, 2002.

[261] O. Hernández-Lerma and J. B. Lasserre. *Markov Chains and Invariant Probabilities*. Progress in Mathematics, volume 211. Birkhäuser Verlag, Basel, 2003.

[262] D. P. Heyman and M. J. Sobel. *Stochastic Models in Operations Research, 1: Stochastic Processes and Operating Characteristics*. Dover Publications, Mineola, NY, 2004. Reprint of the 1982 original.

[263] D. P. Heyman and M. J. Sobel. *Stochastic Models in Operations Research, 2: Stochastic Optimization*. Dover Publications, Mineola, NY, 2004. Reprint of the 1984 original.

[264] D. P. Heyman and M. J. Sobel, editors. In *Handbooks in Operations Research and Management Science, 2: Stochastic Models*. North-Holland, Amsterdam, 1990.

[265] F. S. Hillier and G. J. Lieberman. *Introduction to Operations Research*, 7th edition. McGraw-Hill, New York, 2002.

[266] W. Hoeffding. Probability inequalities for sums of bounded random variables. *J. Am. Stat. Assoc.*, 58:13–30, 1963.

[267] W. J. Hopp and M. L. Spearman. *Factory Physics: Foundations of Manufacturing Management*. Irwin/McGraw-Hill, New York, 2001.

[268] A. Hordijk. *Dynamic Programming and Markov Potential Theory, Second Edition*. Mathematical Centre Tracts 51, Mathematisch Centrum, Amsterdam, 1977.

[269] A. Hordijk, D. L. Iglehart, and R. Schassberger. Discrete time methods for simulating continuous time Markov chains. *Adv. Appl. Probab.*, 8:772–788, 1979.

[270] A. Hordijk and L. C. M. Kallenberg. Linear programming and Markov decision processes. In *Second Symposium on Operations Research, Teil 1, Aachen 1977 (Rheinisch-Westfälische Tech. Hochsch. Aachen, Aachen, 1977), Teil 1*, pages 400–406, Hain, 1978.

[271] A. Hordijk and N. Popov. Large deviations bounds for face-homogeneous random walks in the quarter-plane. *Probab. Eng. Inf. Sci.*, 17(3), 2003.

[272] A. Hordijk and F. M. Spieksma. On ergodicity and recurrence properties of a Markov chain with an application. *Adv. Appl. Probab.*, 24:343–376, 1992.

[273] R. A. Howard. *Dynamic Programming and Markov Processes*. John Wiley/MIT Press, New York, 1960.

[274] C. Huitema. *Routing in the Internet*. Prentice Hall, Englewood Cliffs, NJ, 1999.

[275] D. L. Iglehart and G. S. Shedler. *Regenerative Simulation of Response Times in Networks of Queues*. Lecture Notes in Control and Information Sciences, volume 26. Springer-Verlag, New York, 1980.

[276] D. L. Iglehart and W. Whitt. Multiple channel queues in heavy traffic. I. *Adv. Appl. Probab.*, 2:150–177, 1970.

[277] D. L. Iglehart and W. Whitt. Multiple channel queues in heavy traffic. II. Sequences, networks, and batches. *Adv. Appl. Probab.*, 2:355–369, 1970.

[278] M. J. Ingenoso. *Stability Analysis for Certain Queuing Systems and Multi-Access Communication Channels*. PhD thesis, University of Wisconsin, Madison, WI, 2004.

[279] ITRS public home page. Provides a roadmap of the needs and challenges facing the semiconductor industry. http://public.itrs.net/.

[280] J. R. Jackson. Jobshop-like queueing systems. *Manage. Sci.*, 10:131–142, 1963.

[281] P. A. Jacobson and E. D. Lazowska. Analyzing queueing networks with simultaneous resource possession. *Commun. ACM*, 25(2):142–151, 1982.

[282] G. L. Jones. On the Markov chain central limit theorem. *Probab. Surv.*, 1:299–320 (electronic), 2004.

[283] I. Karatzas and S. Shreve. *Brownian Motion and Stochastic Calculus*, 2nd edition. Springer-Verlag, New York, 1988.

[284] M. J. Karol, M. G. Hluchyj, and S. P. Morgan. Input versus output queueing on a space-division packet switch. *IEEE Trans. Commun.*, 35:1347–1356, 1987.

[285] N. V. Kartashov. Criteria for uniform ergodicity and strong stability of Markov chains with a common phase space. *Theory Probab. Appl.*, 30:71–89, 1985.

[286] N. V. Kartashov. Inequalities in theorems of ergodicity and stability for Markov chains with a common phase space. *Theory Probab. Appl.*, 30:247–259, 1985.

[287] H. Kaspi and A. Mandelbaum. Regenerative closed queueing networks. *Stoch. Stoch. Rep.*, 39(4):239–258, 1992.

[288] M. Kearns and S. Singh. Bias-variance error bounds for temporal difference updates. In *Proceedings of the 13th Annual Conference on Computational Learning Theory*, Stanford University, pages 142–147, 2000.

[289] J. B. Keller and H. P. McKean, editors. *Stochastic Differential Equations*. American Mathematical Society, Providence, RI, 1973.

[290] F. P. Kelly. *Reversibility and Stochastic Networks*. Wiley Series in Probability and Mathematical Statistics. John Wiley, Chichester, UK, 1979.

[291] F. P. Kelly. Dynamic routing in stochastic networks. In F. Kelly and R. Williams, editors, *IMA Volumes in Mathematics and Its Applications*, volume 71, pages 169–186, Springer-Verlag, New York, 1995.

[292] F. P. Kelly and C. N. Laws. Dynamic routing in open queueing networks: Brownian models, cut constraints and resource pooling. *Queueing Syst. Theory Appl.*, 13:47–86, 1993.

[293] D. G. Kendall. Some problems in the theory of queues. *J. R. Stat. Soc. Ser. B*, 13:151–185, 1951.

[294] D. G. Kendall. Stochastic processes occurring in the theory of queues and their analysis by means of the imbedded Markov chain. *Ann. Math. Stat.*, 24:338–354, 1953.

[295] I. G. Kevrekidis, C. W. Gear, and G. Hummer. Equation-free: the computer-assisted analysis of complex, multiscale systems. *Ann. Inst. Chem. Eng. J.*, 50:1346–1354, 2004.

[296] R. Z. Khas'minskii. *Stochastic Stability of Differential Equations*. Sijthoff & Noordhoff, Netherlands, 1980.

[297] J. Kiefer and J. Wolfowitz. Stochastic estimation of the maximum of a regression function. *Ann. Math. Stat.*, 23:462–466, 1952.

[298] E. Kim and M. P. Van Oyen. Beyong the c-μ rule: dynamic scheduling of a two-class loss queue. *Math. Methods Oper. Res.*, 48(1):17–36, 1998.

[299] S. Kim and S. G. Henderson. Adaptive control variates. In R. Ingalls, M. Rossetti, J. Smith, and B. Peters, editors, *Proceedings of the 2004 Winter Simulation Conference*, pages 621–629, Piscataway, NJ, 2004. IEEE.

[300] T. Kimura. A bibliography of research on heavy traffic limit theorems for queues. *Econ. J. Hokkaido Univ.*, 22:167–179, 1993.

[301] J. F. C. Kingman. The single server queue in heavy traffic. *Proc. Camb. Phil. Soc.*, 57:902–904, 1961.

[302] J. F. C. Kingman. On queues in heavy traffic. *J. R. Stat. Soc. Ser. B*, 24:383–392, 1962.

[303] L. Kleinrock. *Queueing Systems Vol. 1: Theory*. John Wiley, New York, 1975.

[304] L. Kleinrock. *Queueing Systems Vol. 2: Computer Applications*. John Wiley, New York, 1976.

[305] G. P. Klimov. Time-sharing queueing systems. I. *Teor. Veroyatn. Primen.*, 19:558–576, 1974.

[306] G. P. Klimov. *Procesy obsługi masowej (Polish) [Queueing Theory]*. Wydawnictwa Naukowo-Techniczne (WNT), Warsaw, 1979. Translated from the Russian by E. Fidelis and T. Rolski.

[307] P. V. Kokotovic, J. O'Reilly, and J. K. Khalil. *Singular Perturbation Methods in Control: Analysis and Design*. Academic Press, Orlando, FL, 1986.

[308] V. R. Konda and J. N. Tsitsiklis. On actor-critic algorithms. *SIAM J. Control Optim.*, 42(4):1143–1166 (electronic), 2003.

[309] I. Kontoyiannis, L. A. Lastras-Montaño, and S. P. Meyn. Relative entropy and exponential deviation bounds for general Markov chains. In *Proceedings of the IEEE International Symposium on Information Theory*, Adelaide, Australia, 4–9 September, 2005.

[310] I. Kontoyiannis and S. P. Meyn. Spectral theory and limit theorems for geometrically ergodic Markov processes. *Ann. Appl. Probab.*, 13:304–362, 2003. Presented at the IN-FORMS Applied Probability Conference, NYC, July 2001.

[311] I. Kontoyiannis and S. P. Meyn. Large deviations asymptotics and the spectral theory of multiplicatively regular Markov processes. *Electron. J. Probab.*, 10(3):61–123 (electronic), 2005.

[312] I. Kontoyiannis and S. P. Meyn. Computable exponential bounds for screened estimation and simulation. To appear *Ann. Appl. Prob.*, 2008.

[313] L. Kruk, J. Lehoczky, K. Ramanan, and S. Shreve. An explicit formula for the skorohod map on [0,a]. Submitted for publication, March 2006.

[314] N. V. Krylov. *Controlled Diffusion Processes*. Applications of Mathematics, volume 14. Springer-Verlag, New York, 1980. Translated from the Russian by A. B. Aries.

[315] N. V. Krylov. On a proof of Itô's formula. *Trudy Mat. Inst. Steklov.*, 202:170–174, 1993.

[316] P. R. Kumar and S. P. Meyn. Stability of queueing networks and scheduling policies. *IEEE Trans. Autom. Control*, 40(2):251–260, February 1995.

[317] P. R. Kumar and S. P. Meyn. Duality and linear programs for stability and performance analysis queueing networks and scheduling policies. *IEEE Trans. Autom. Control*, 41(1):4–17, 1996.

[318] P. R. Kumar and T. I. Seidman. Dynamic instabilities and stabilization methods in distributed real-time scheduling of manufacturing systems. *IEEE Trans. Autom. Control*, AC-35(3):289–298, March 1990.

[319] S. Kumar and P. R. Kumar. Performance bounds for queueing networks and scheduling policies. *IEEE Trans. Autom. Control*, AC-39:1600–1611, August 1994.

[320] S. Kumar and P. R. Kumar. Closed queueing networks in heavy traffic: fluid limits and efficiency. In P. Glasserman, K. Sigman, and D. Yao, editors, *Stochastic Networks: Stability and Rare Events*. Lecture Notes in Statistics, volume 117, pages 41–64. Springer-Verlag, New York, 1996.

[321] S. Kumar and P. R. Kumar. Fluctuation smoothing policies are stable for stochastic re-entrant lines. *Discrete Event Dyn. Syst. Theory Appl.*, 6(4):361–370, October 1996.

[322] H. J. Kushner. Numerical methods for stochastic control problems in continuous time. *SIAM J. Control Optim.*, 28(5):999–1048, 1990.

[323] H. J. Kushner. *Heavy Traffic Analysis of Controlled Queueing and Communication Networks*. Stochastic Modelling and Applied Probability. Springer-Verlag, New York, 2001.

[324] H. J. Kushner and Y. N. Chen. Optimal control of assignment of jobs to processors under heavy traffic. *Stochastics*, 68(3–4):177–228, 2000.

[325] H. J. Kushner and P. G. Dupuis. *Numerical Methods for Stochastic Control Problems in Continuous Time*. Springer-Verlag, London, 1992.

[326] H. J. Kushner and L. F. Martins. Numerical methods for stochastic singular control problems. *SIAM J. Control Optim.*, 29(6):1443–1475, 1991.

[327] H. J. Kushner and K. M. Ramchandran. Optimal and approximately optimal control policies for queues in heavy traffic. *SIAM J. Control Optim.*, 27:1293–1318, 1989.

[328] H. J. Kushner and G. G. Yin. *Stochastic Approximation Algorithms and Applications*. Applications of Mathematics (New York), volume 35. Springer-Verlag, New York, 1997.

[329] H. Kwakernaak and R. Sivan. *Linear Optimal Control Systems*. Wiley-Interscience, New York, 1972.

[330] S. S. Lavenberg and P. D. Welch. A perspective on the use of control variables to increase the efficiency of Monte Carlo simulations. *Manage. Sci.*, 27:322–335, 1981.

[331] A. M. Law and W. D. Kelton. *Simulation Modeling and Analysis*, 3rd edition. McGraw-Hill, New York, 2000.

[332] C. N. Laws and G. M. Louth. Dynamic scheduling of a four-station queueing network. *Probab. Eng. Inf. Sci.*, 4:131–156, 1990.

[333] N. Laws. *Dynamic Routing in Queueing Networks*. PhD thesis, Cambridge University, Cambridge, UK, 1990.

[334] D. Leith and V. Subramanian. Draining time based scheduling algorithm. *IEEE Conf. Dec. and Control*. Submitted for publication, 2007.

[335] H. Liao. *Multiple-Access Channels*. PhD thesis, Unversity of Hawaii, Manoa, Hawaii, 1972.

[336] G. Liberopoulos and Y. Dallery. A unified framework for pull control mechanisms in multi-stage manufacturing systems. *Ann. Oper. Res.*, 93(1):325–355, 2000.

[337] W. Lin and P. R. Kumar. Optimal control of a queueing system with two heterogeneous servers. *IEEE Trans. Autom. Control*, AC-29:696–703, August 1984.

[338] S. Lippman. Applying a new device in the optimization of exponential queueing systems. *Oper. Res.*, 23:687–710, 1975.

[339] Steven A. Lippman. On dynamic programming with unbounded rewards. *Manage. Sci.*, 21(11):1225–1233, 1974/75.

[340] W. W. Loh. *On the Method of Control Variates*. PhD thesis, Department of Operations Research, Stanford University, Stanford, CA, 1994.

[341] S. H. Lu and P. R. Kumar. Distributed scheduling based on due dates and buffer priorities. *IEEE Trans. Autom. Control*, 36(12):1406–1416, December 1991.

[342] D. G. Luenberger. *Linear and Nonlinear Programming*, 2nd edition. Kluwer Academic Publishers, Norwell, MA, 2003.

[343] R. B. Lund, S. P. Meyn, and R. L. Tweedie. Computable exponential convergence rates for stochastically ordered Markov processes. *Ann. Appl. Probab.*, 6(1):218–237, 1996.

[344] X. Luo and D. Bertsimas. A new algorithm for state-constrained separated continuous linear programs. *SIAM J. Control Optim.*, 37:177–210, 1998.

[345] C. Maglaras. Dynamic scheduling in multiclass queueing networks: stability under discrete-review policies. *Queueing Syst. Theory Appl.*, 31:171–206, 1999.

[346] C. Maglaras. Discrete-review policies for scheduling stochastic networks: trajectory tracking and fluid-scale asymptotic optimality. *Ann. Appl. Probab.*, 10(3):897–929, 2000.

[347] V. A. Malyšev and M. V. Men'šikov. Ergodicity, continuity and analyticity of countable Markov chains. *Trudy Moskov. Mat. Obshch.*, 39:3–48, 235, 1979; *Trans. Mosc. Math. Soc.*, 1:1–48, 1981.

[348] A. Mandelbaum and A. L. Stolyar. Scheduling flexible servers with convex delay costs: heavy-traffic optimality of the generalized $c\mu$-rule. *Oper. Res.*, 52(6):836–855, 2004.

[349] A. S. Manne. Linear programming and sequential decisions. *Manage. Sci.*, 6(3):259–267, 1960.

[350] S. Mannor, I. Menache, and N. Shimkin. Basis function adaptation in temporal difference reinforcement learning. *Ann. Oper. Res.*, 134(2):215–238, 2005.

[351] D. Q. Mayne, J. B. Rawlings, C. V. Rao, and P. O. M. Scokaert. Constrained model predictive control: stability and optimality. *Automatica*, 36(6):789–814, June 2000.

[352] R. D. McBride. Progress made in solving the multicommodity flow problem. *SIAM J. Control Optim.*, 8(4):947–955, 1998.

[353] H. P. McKean. *Stochastic Integrals*. AMS Chelsea Publishing, Providence, RI, 2005. Reprint of the 1969 edition, with errata.

[354] M. Medard, S. P. Meyn, J. Huang, and A. J. Goldsmith. Capacity of time-slotted aloha systems. *Trans. Wirel. Commun.*, 3(2):486–499, 2004/03. *In Proceedings of ISIT*, page 407, 2000.

[355] S. P. Meyn. Transience of multiclass queueing networks via fluid limit models. *Ann. Appl. Probab.*, 5:946–957, 1995.

[356] S. P. Meyn. The policy iteration algorithm for average reward Markov decision processes with general state space. *IEEE Trans. Autom. Control*, 42(12):1663–1680, 1997.

[357] S. P. Meyn. Stability and optimization of queueing networks and their fluid models. In *Mathematics of Stochastic Manufacturing Systems (Williamsburg, VA, 1996)*, pages 175–199. American Mathematical Society, Providence, RI, 1997.

[358] S. P. Meyn. Algorithms for optimization and stabilization of controlled Markov chains. *Sādhanā*, 24(4–5):339–367, 1999. Special invited issue: *Chance as necessity*.

[359] S. P. Meyn. Sequencing and routing in multiclass queueing networks. Part I: Feedback regulation. *SIAM J. Control Optim.*, 40(3):741–776, 2001.

[360] S. P. Meyn. Stability, performance evaluation, and optimization. In E. Feinberg and A. Shwartz, editors, *Markov Decision Processes: Models, Methods, Directions, and Open Problems*, pages 43–82. Kluwer, Holland, 2001.

[361] S. P. Meyn. Sequencing and routing in multiclass queueing networks. Part II: Workload relaxations. *SIAM J. Control Optim.*, 42(1):178–217, 2003.

[362] S. P. Meyn. Dynamic safety-stocks for asymptotic optimality in stochastic networks. *Queueing Syst. Theory Appl.*, 50:255–297, 2005.

[363] S. P. Meyn. Workload models for stochastic networks: value functions and performance evaluation. *IEEE Trans. Autom. Control*, 50(8):1106–1122, August 2005.

[364] S. P. Meyn. Large deviation asymptotics and control variates for simulating large functions. *Ann. Appl. Probab.*, 16(1):310–339, 2006.

[365] S. P. Meyn. Stability and asymptotic optimality of generalized MaxWeight policies. Submitted for publication, 2006.

[366] S. P. Meyn and D. G. Down. Stability of generalized Jackson networks. *Ann. Appl. Probab.*, 4:124–148, 1994.

[367] S. P. Meyn and R. L. Tweedie. Generalized resolvents and Harris recurrence of Markov processes. *Contemp. Math.*, 149:227–250, 1993.

[368] S. P. Meyn and R. L. Tweedie. *Markov Chains and Stochastic Stability*, 2nd edition. Springer-Verlag, London, 1993. `http://black.csl.uiuc.edu/~meyn/pages/book.html`.

[369] S. P. Meyn and R. L. Tweedie. Stability of Markovian processes III: Foster-Lyapunov criteria for continuous time processes. *Adv. Appl. Probab.*, 25:518–548, 1993.

[370] S. P. Meyn and R. L. Tweedie. Computable bounds for convergence rates of Markov chains. *Ann. Appl. Probab.*, 4:981–1011, 1994.

[371] S. P. Meyn and R. L. Tweedie. State-dependent criteria for convergence of Markov chains. *Ann. Appl. Probab.*, 4:149–168, 1994.

[372] B. Mitchell. Optimal service-rate selection in an $M/G/\hat{1}$ queue. *SIAM J. Appl. Math.*, 24(1):19–35, 1973.

[373] M. Mitzenmacher. *The Power of Two Choices in Randomized Load Balancing*. PhD thesis, University of California, Berkeley, CA, 1996.

[374] M. Mitzenmacher. The power of two choices in randomized load balancing. *IEEE Trans. Parallel Distrib. Syst.*, 12(10):1094–1104, 2001.

[375] H. Mori. Transport, collective motion, and brownian motion. *Prog. Theor. Phys.*, 33:423–455, 1965.

[376] J. R. Morrison and P. R. Kumar. New linear program performance bounds for queueing networks. *J. Optim. Theory Appl.*, 100(3):575–597, 1999.

[377] A. Muharremoglu and J. N. Tsitsiklis. A single-unit decomposition approach to multi-echelon inventory systems. Under revision for Operations Research. Preprint available at `http://web.mit.edu/jnt/www/publ.html`, 2003.

[378] K. Murota. Note on multimodularity and L-convexity. *Math. Oper. Res.*, 30(3):658–661, 2005.

[379] K. G. Murty and S. N. Kabadi. Some np-complete problems in quadratic and nonlinear programming. *Math. Program.*, 39(2):117–129, 1987.

[380] P. Nash. *Optimal Allocation of Resources between Research Projects*. PhD thesis, Cambridge University, Cambridge, England, 1973.

[381] A. Nedic and D. P. Bertsekas. Least squares policy evaluation algorithms with linear function approximation. *Discrete Event Dyn. Syst. Theory Appl.*, 13(1–2):79–110, 2003.

[382] B. L. Nelson. Control-variate remedies. *Oper. Res.*, 38(4):974–992, 1990.

[383] M. F. Neuts. *Matrix-Geometric Solutions in Stochastic Models: An Algorithmic Approach*. Dover Publications, New York, 1994. Corrected reprint of the 1981 original.

[384] M. B. Nevel'son and R. Z. Has'minskiĭ. *Stochastic Approximation and Recursive Estimation*. American Mathematical Society, Providence, RI, 1973. Translated from the Russian by the Israel Program for Scientific Translations, Translations of Mathematical Monographs, volume 47.

[385] G. F. Newell. *Applications of Queueing Theory*, 2nd edition. Monographs on Statistics and Applied Probability. Chapman & Hall, London, 1982.

[386] V. Nguyen. Fluid and diffusion approximations of a two-station mixed queueing network. *Math. Oper. Res.*, 20(2):321–354, 1995.

[387] J. Niño-Mora. Restless bandit marginal productivity indices, diminishing returns, and optimal control of make-to-order/make-to-stock $M/G/1$ queues. *Math. Oper. Res.*, 31(1): 50–84, 2006.

[388] E. Nummelin. *General Irreducible Markov Chains and Nonnegative Operators*. Cambridge University Press, Cambridge, 1984.

[389] J. Ou and L. M. Wein. Performance bounds for scheduling queueing networks. *Ann. Appl. Probab.*, 2:460–480, 1992.

[390] C. H. Papadimitriou and J. N. Tsitsiklis. Intractable problems in control theory. *SIAM J. Control Optim.*, 24(4):639–654, 1986.

[391] C. H. Papadimitriou and J. N. Tsitsiklis. The complexity of Markov decision processes. *Math. Oper. Res.*, 12(3):441–450, 1987.

[392] C. H. Papadimitriou and J. N. Tsitsiklis. The complexity of optimal queueing network control. *Math. Oper. Res.*, 24(2):293–305, 1999.

[393] I. Ch. Paschalidis D. Bertsimas and J. N. Tsitsiklis. Scheduling of multiclass queueing networks: bounds on achievable performance. In *Workshop on Hierarchical Control for Real–Time Scheduling of Manufacturing Systems*, Lincoln, New Hampshire, October 16–18, 1992.

[394] A. Peña-Perez and P. Zipkin. Dynamic scheduling rules for a multiproduct make-to-stock. *Oper. Res.*, 45:919–930, 1997.

[395] J. Perkins. *Control of Push and Pull Manufacturing Systems*. PhD thesis, University of Illinois, Urbana, IL, September 1993. Technical report no. UILU-ENG-93-2237 (DC-155).

[396] J. R. Perkins and P. R. Kumar. Stable distributed real-time scheduling of flexible manufacturing/assembly/disassembly systems. *IEEE Trans. Autom. Control*, AC-34:139–148, 1989.

[397] J. R. Perkins and P. R. Kumar. Optimal control of pull manufacturing systems. *IEEE Trans. Autom. Control*, AC-40:2040–2051, 1995.

[398] J. R. Perkins and R. Srikant. Failure-prone production systems with uncertain demand. *IEEE Trans. Autom. Control*, 46:441–449, 2001.

[399] M. C. Pullan. An algorithm for a class of continuous linear programs. *SIAM J. Control Optim.*, 31:1558–1577, 1993.

[400] M. C. Pullan. Existence and duality theory for separated continuous linear programs. *Math. Model. Syst.*, 3(3):219–245, 1995.

[401] M. C. Pullan. Forms of optimal solutions for separated continuous linear programs. *SIAM J. Control Optim.*, 33:1952–1977, 1995.

[402] M. L. Puterman. *Markov Decision Processes*. John Wiley, New York, 1994.

[403] M. L. Puterman and S. L. Brumelle. The analytic theory of policy iteration. In *Dynamic Programming and Its Applications (Proc. Conf., Univ. British Columbia, Vancouver, BC, 1977)*, pages 91–113. Academic Press, New York, 1978.

[404] M. L. Puterman and S. L. Brumelle. On the convergence of policy iteration in stationary dynamic programming. *Math. Oper. Res.*, 4(1):60–69, 1979.

[405] K. Ramanan. Reflected diffusions defined via the extended Skorokhod map. *Electron. J. Probab.*, 11:934–992 (electronic), 2006.

[406] R. R. Rao and A. Ephremides. On the stability of interacting queues in a multiple-access system. *IEEE Trans. Inf. Theory*, 34(5, part 1):918–930, 1988.

[407] M. I. Reiman. Open queueing networks in heavy traffic. *Math. Oper. Res.*, 9:441–458, 1984.

[408] R. K. Ritt and L. I. Sennott. Optimal stationary policies in general state space Markov decision chains with finite action sets. *Math. Oper. Res.*, 17(4):901–909, 1992.

[409] H. Robbins and S. Monro. A stochastic approximation method. *Ann. Math. Stat.*, 22:400–407, 1951.

[410] J. Robinson. Fabtime – cycle time management for wafer fabs. http://www.FabTime.com.

[411] J. K. Robinson, J. W. Fowler, and E. Neacy. Capacity loss factors in semiconductor manufacturing (working paper). http://www.fabtime.com/bibliogr.shtml, 1996.

[412] S. Robinson. The price of anarchy. *SIAM Newsl.*, 37(5):1–4, 2004.

[413] L. C. G. Rogers and D. Williams. *Diffusions, Markov Processes, and Martingales*, volume 2. Cambridge Mathematical Library. Cambridge University Press, Cambridge, 2000. Ito calculus, Reprint of the second (1994) edition.

[414] Z. Rosberg, P. P. Varaiya, and J. C. Walrand. Optimal control of service in tandem queues. *IEEE Trans. Autom. Control*, 27:600–610, 1982.

[415] J. S. Rosenthal. Correction: "Minorization conditions and convergence rates for Markov chain Monte Carlo". *J. Am. Stat. Assoc.*, 90(431):1136, 1995.

[416] J. S. Rosenthal. Minorization conditions and convergence rates for Markov chain Monte Carlo. *J. Am. Stat. Assoc.*, 90(430):558–566, 1995.

[417] S. M. Ross. *Introduction to Stochastic Dynamic Programming*. Academic Press, New York, 1984.

[418] S. M. Ross. *Applied Probability Models with Optimization Applications*. Dover Publications, New York, 1992. Reprint of the 1970 original.

[419] R. Y. Rubinstein and B. Melamed. *Modern Simulation and Modeling*. John Wiley, Chichester, UK, 1998.

[420] A. N. Rybko and A. L. Stolyar. On the ergodicity of random processes that describe the functioning of open queueing networks. *Probl. Pereda. Inf.*, 28(3):3–26, 1992.

[421] C. H. Sauer and E. A. MacNair. Simultaneous resource possession in queueing models of computers. *SIGMETRICS Perform. Eval. Rev.*, 7(1–2):41–52, 1978.

[422] H. E. Scarf. The optimality of (S, s) policies in the dynamic inventory problem. In *Mathematical Methods in the Social Sciences, 1959*, pages 196–202. Stanford University Press, Stanford, CA, 1960.

[423] H. E. Scarf. A survey of analytic techniques in inventory theory. In *Multistage Inventory Models and Techniques*, pages 185–225. Stanford University Press, Stanford, CA, 1963.

[424] M. Schreckenberg and S. D. Sharma, editors. *Pedestrian and Evacuation Dynamics*. Springer, Berlin, 2002.

[425] P. J. Schweitzer. Aggregation methods for large Markov chains. In *Mathematical Computer Performance and Reliability (Pisa, 1983)*, pages 275–286. North-Holland, Amsterdam, 1984.

[426] E. Schwerer. *A Linear Programming Approach to the Steady-state Analysis of Markov Processes*. PhD thesis, Stanford University, Stanford, CA, 1997.

[427] E. Schwerer. A linear programming approach to the steady-state analysis of reflected Brownian motion. *Stoch. Models*, 17(3):341–368, 2001.

[428] T. I. Seidman. First come first serve can be unstable. *IEEE Trans. Autom. Control*, 39(10):2166–2170, October 1994.

[429] T. I. Seidman and L. E. Holloway. Stability of pull production control methods for systems with significant setups. *IEEE Trans. Autom. Control*, 47(10):1637–1647, 2002.

[430] L. I. Sennott. *Stochastic Dynamic Programming and the Control of Queueing Systems*. Wiley Series in Probability and Statistics: Applied Probability and Statistics. John Wiley, New York, 1999. A Wiley-Interscience Publication.

[431] L. I. Sennott. Average cost optimal stationary policies in infinite state Markov decision processes with unbounded cost. *Oper. Res.*, 37:626–633, 1989.

[432] L. I. Sennott. The convergence of value iteration in average cost Markov decision chains. *Oper. Res. Lett.*, 19:11–16, 1996.

[433] S. P. Sethi and G. L. Thompson. *Optimal Control Theory: Applications to Management Science and Economics*. Kluwer Academic Publishers, Boston, 2000.

[434] S. P. Sethi and Q. Zhang. *Hierarchical Decision Making in Stochastic Manufacturing Systems*. Birkhauser Verlag, Basel, Switzerland, 1994.

[435] D. Shah and D. Wischik. Optimal scheduling algorithm for input-queued switch. *IEEE INFOCOM*, 2006. 25th IEEE Intl. Conf. on Computer Comm., pp. 1–11.

[436] M. Shaked and J. G. Shanthikumar. *Stochastic Orders and Their Applications*. Probability and Mathematical Statistics. Academic Press, Boston, MA, 1994.

[437] S. Shakkottai and R. Srikant. Scheduling real-time traffic with deadlines over a wireless channel. In *Proceedings of the 2nd ACM International Workshop on Wireless Mobile Multimedia*, pages 35–42, Seattle, WA, 1999.

[438] S. Shakkottai, R. Srikant, and A. L. Stolyar. Pathwise optimality of the exponential scheduling rule for wireless channels. *Adv. Appl. Probab.*, 36(4):1021–1045, 2004.

[439] S. Shamai and A. Wyner. Information theoretic considerations for symmetric, cellular, multiple-access fading channels – part I . *IEEE Trans. Inf. Theory*, 43(6):1877–1894, 1997.

[440] J. G. Shanthikumar and U. Sumita. Convex ordering of sojourn times in single-server queues: extremal properties of fifo and lifo service disciplines. *J. Appl. Probab.*, 24:737–748, 1987.

[441] J. G. Shanthikumar and D. D. Yao. Stochastic monotonicity in general queueing networks. *J. Appl. Probab.*, 26:413–417, 1989.

[442] J. G. Shanthikumar and D. D. Yao. Multiclass queueing systems: polymatroid structure and optimal scheduling control. *Oper. Res.*, 40:S293–S299, 1992.

[443] X. Shen, H. Chen, J. G. Dai, and W. Dai. The finite element method for computing the stationary distribution of an SRBM in a hypercube with applications to finite buffer queueing networks. *Queueing Syst. Theory Appl.*, 42(1):33–62, 2002.

[444] S. Shreve. Lectures on stochastic calculus and finance. `http://www-2.cs.cmu. edu/~chal/shreve.html`, 2003.

[445] S. E. Shreve. *Stochastic Calculus for Finance II: Continuous-Time Models*. Springer Finance. Springer-Verlag, New York, 2004.

[446] A. Shwartz and A. Weiss. *Large Deviations for Performance Analysis: Queues, Communications, and Computing*. Stochastic Modeling Series. Chapman & Hall, London, 1995. With an appendix by Robert J. Vanderbei.

[447] K. Sigman. Queues as harris-recurrent Markov chains. *Queueing Syst.*, 3:179–198, 1988.

[448] K. Sigman. The stability of open queueing networks. *Stoch. Proc. Appl.*, 35:11–25, 1990.

[449] K. Sigman. A note on a sample-path rate conservation law and its relationship with $h = \lambda g$. *Adv. Appl. Probab.*, 23:662–665, 1991.

[450] R. Simon, R. Alonso-Zldivar, and T. Reiterman. Enron memos prompt calls for a wider investigation electricity: regulators order all energy trading companies to preserve documents on tactics. *Los Angeles Times*, July 2002.

[451] A. V. Skorokhod. Stochastic equations for diffusions in a bounded region. *Theory Probab. Appl.*, 6:264–274, 1961.

[452] J. M. Smith and D. Towsley. The use of queuing networks in the evaluation of egress from buildings. *Environ. Plann. B Plann. Des.*, 8(2):125–139, 1981.

[453] W. E. Smith. Various optimizers for single-stage production. *Nav. Res. Log. Q.*, 3:59–66, 1956.

[454] H. M. Soner and S. E. Shreve. Regularity of the value function for a two-dimensional singular stochastic control problem. *SIAM J. Control Optim.*, 27(4):876–907, 1989.

[455] E. D. Sontag. *Mathematical Control Theory*: Deterministic Finite-Dimensional Systems. Texts in Applied Mathematics, volume 6, 2nd edition. Springer-Verlag, New York, 1998.

[456] M. L. Spearman, W. J. Hopp, and D. L. Woodruff. A hierarchical control architecture for CONWIP production systems. *J. Manuf. Oper. Manage.*, 3:147–171, 1989.

[457] M. L. Spearman, D. L. Woodruff, and W. J. Hopp. CONWIP: a pull alternative to KANBAN. *Int. J. Prod. Res.*, 28:879–894, 1990.

[458] F. M. Spieksma. *Geometrically Ergodic Markov Chains and the Optimal Control of Queues*. PhD thesis, University of Leiden, Leiden, the Netherlands, 1991.

[459] R. Srikant. *The Mathematics of Internet Congestion Control*. Systems & Control: Foundations & Applications. Birkhäuser Boston, Boston, MA, 2004.

[460] S. Stidham Jr. and R. Weber. A survey of Markov decision models for control of networks of queues. *Queueing Syst. Theory Appl.*, 13(1–3):291–314, 1993.

[461] A. L. Stolyar. On the stability of multiclass queueing networks: a relaxed sufficient condition via limiting fluid processes. *Markov Process. Relat. Fields*, 1(4):491–512, 1995.

[462] A. L. Stolyar. Maxweight scheduling in a generalized switch: state space collapse and workload minimization in heavy traffic. *Adv. Appl. Probab.*, 14(1):1–53, 2004.

[463] S. H. Strogatz. Exploring complex networks. *Nature*, 410:268–276, 2001.

[464] R. S. Sutton and A. G. Barto. *Reinforcement Learning: An Introduction*. MIT Press, MA, 1998. http://www.cs.ualberta.ca/~sutton/book/the-book.html.

[465] V. Tadic and S. P. Meyn. Adaptive Monte-Carlo algorithms using control-variates. In *Proceedings of the American Control Conference*, Denver, CO, June 2003.

[466] H. Takagi. *Queueing Analysis: A Foundation of Performance Evaluation: Finite Systems*, volume 2. North-Holland, Amsterdam, 1993.

[467] L. Tassiulas. Adaptive back-pressure congestion control based on local information. *IEEE Trans. Autom. Control*, 40(2):236–250, 1995.

[468] L. Tassiulas and A. Ephremides. Stability properties of constrained queueing systems and scheduling policies for maximum throughput in multihop radio networks. *IEEE Trans. Autom. Control*, 37(12):1936–1948, 1992.

[469] L. M. Taylor and R. J. Williams. Existence and uniqueness of semimartingale reflecting Brownian motions in an orthant. *Prob. Theory Relat. Fields*, 96(3):283–317, 1993.

[470] D. N. C. Tse and S. V. Hanly. Multiaccess fading channels. I. Polymatroid structure, optimal resource allocation and throughput capacities. *IEEE Trans. Inf. Theory*, 44(7):2796–2815, 1998.

[471] P. Tseng. Solving H-horizon, stationary Markov decision problems in time proportional to $\log(H)$. *Oper. Res. Lett.*, 9(5):287–297, 1990.

[472] J. N. Tsitsiklis. Periodic review inventory systems with continuous demand and discrete order sizes. *Manage. Sci.*, 30(10):1250–1254, 1984.

[473] J. N. Tsitsiklis. A short proof of the Gittins index theorem. *Ann. Appl. Probab.*, 4(1):194–199, 1994.

[474] J. N. Tsitsiklis and B. Van Roy. An analysis of temporal-difference learning with function approximation. *IEEE Trans. Autom. Control*, 42(5):674–690, 1997.

[475] P. Tsoucas and J. Walrand. Monotonicity of throughput in non-Markovian networks. *J. Appl. Probab.*, 26:134–141, 1989.

[476] P. Tuominen and R. L. Tweedie. Subgeometric rates of convergence of f-ergodic Markov chains. *Adv. Appl. Probab.*, 26:775–798, 1994.

[477] J. A. Van Mieghem. Dynamic scheduling with convex delay costs: the generalized c-μ rule. *Ann. Appl. Probab.*, 5(3):809–833, 1995.

[478] J. A. E. E. van Nunen and J. Wessels. A note on dynamic programming with unbounded rewards. *Management Sci.*, 24(5):576–580, 1977/78.

[479] B. Van Roy. Neuro-dynamic programming: overview and recent trends. In E. Feinberg and A. Shwartz, editors, *Markov Decision Processes: Models, Methods, Directions, and Open Problems*, pages 43–82. Kluwer, Holland, 2001.

[480] J. S. Vandergraft. A fluid flow model of networks of queues. *Manage. Sci.*, 29:1198–1208, 1983.

[481] M. H. Veatch. Using fluid solutions in dynamic scheduling. In S. B. Gershwin, Y. Dallery, C. T. Papadopoulos, and J. M. Smith, editors, *Analysis and Modeling of Manufacturing Systems*. Operations Research and Management Science, pages 399–426. Kluwer-Academic International, New York, 2002.

[482] M. H. Veatch. Approximate dynamic programming for networks: fluid models and constraint reduction. Submitted for publication, 2004.

[483] A. F. Veinott Jr. Discrete dynamic programming with sensitive discount optimality criteria. *Ann. Math. Stat.*, 40(5):1635–1660, 1969.

[484] P. Viswanath, D. Tse, and R. Laroia. Opportunistic beam-forming using dumb antennas. *IEEE Trans. Inf. Theory*, 48(6):1277–1294, 2002.

[485] J. Walrand and P. Varaiya. *High-Performance Communication Networks*. The Morgan Kaufmann Series in Networking. Morgan Kaufman, San Francisco, CA, 2000.

[486] R. Weber. On the optimal assignment of customers to parallel servers. *J. Appl. Probab.*, 15:406–413, 1978.

[487] R. Weber and S. Stidham. Optimal control of service rates in networks of queues. *Adv. Appl. Probab.*, 19:202–218, 1987.

[488] L. M. Wein. Dynamic scheduling of a multiclass make-to-stock queue. *Oper. Res.*, 40(4):724–735, 1992.

[489] L. M. Wein and P. B. Chevalier. A broader view of the job-shop scheduling problem. *Manage. Sci.*, 38(7):1018–1033, 1992.

[490] L. M. Wein and M. H. Veatch. Scheduling a make-to-stock queue: index policies and hedging points. *Oper. Res.*, 44:634–647, 1996.

[491] G. Weiss. Approximation results in parallel machines stochastic scheduling. *Ann. Oper. Res.*, 26(1–4):195–242, 1990.

[492] G. Weiss. On the optimal draining of re-entrant fluid lines. Technical Report, Georgia Institute of Technology and Technion, Atlanta, Georgia, 1994.

[493] J. Wessels. Markov programming by successive approximations with respect to weighted supremum norms. *J. Math. Anal. Appl.*, 58(2):326–335, 1977.

[494] D. J. White. Dynamic programming, Markov chains, and the method of successive approximations. *J. Math. Anal. Appl.*, 6:373–376, 1963.

[495] W. Whitt. Some useful functions for functional limit theorems. *Math. Oper. Res.*, 5(1):67–85, 1980.

[496] W. Whitt. Simulation run length planning. In *WSC '89: Proceedings of the 21st Conference on Winter Simulation*, pages 106–112, New York, 1989. ACM Press.

[497] W. Whitt. An overview of Brownian and non-Brownian FCLTs for the single-server queue. *Queueing Syst. Theory Appl.*, 36(1–3):39–70, 2000.

[498] W. Whitt. *Stochastic-Process Limits*. Springer Series in Operations Research. Springer-Verlag, New York, 2002.

[499] P. Whittle. Multi-armed bandits and the Gittins index. *J. R. Stat. Soc. Ser. B*, 42:143–149, 1980.

[500] P. Whittle. *Optimization over Time: Dynamic Programming and Stochastic Control*. Wiley Series in Probability and Mathematical Statistics: Applied Probability and Statistics, volume 1. John Wiley, Chichester, 1982.

[501] P. Whittle. *Optimization over Time: Dynamic Programming and Stochastic Control*. Wiley Series in Probability and Mathematical Statistics: Applied Probability and Statistics, volume 2. John Wiley, Chichester, 1983. Dynamic programming and stochastic control.

[502] P. Whittle. *Networks: Optimisation and Evolution*. Cambridge Series in Statistical and Probabilistic Mathematics. Cambridge University Press, Cambridge, 2007.

[503] R. J. Williams. Diffusion approximations for open multiclass queueing networks: sufficient conditions involving state space collapse. *Queueing Syst. Theory Appl.*, 30(1–2):27–88, 1998.

[504] W. Willinger, V. Paxson, R. H. Riedi, and M. S. Taqqu. Long-range dependence and data network traffic. In *Theory and Applications of Long-Range Dependence*, pages 373–407. Birkhäuser Boston, Boston, MA, 2003.

[505] W. Winston. Optimality of the shortest line discipline. *J. Appl. Probab.*, 14(1):181–189, 1977.

[506] P. Yang. *Pathwise Solutions for a Class of Linear Stochastic Systems*. PhD thesis, Department of Operations Research, Stanford University, Stanford, CA, 1988.

[507] G. G. Yin and Q. Zhang. *Discrete-Time Markov Chains: Two-Time-Scale Methods and Applications*: Stochastic Modelling and Applied Probability. Applications of Mathematics, volume 55. Springer-Verlag, New York, 2005.

[508] Y. Zou, I. G. Kevrekidis, and D. Armbruster. Multiscale analysis of re-entrant production lines: an equation-free approach. *Phys. A Stat. Mech. Appl.*, 363:1–13, 2006.

[509] R. Zwanzig. *Nonequilibrium Statistical Mechanics*. Oxford University Press, Oxford, England, 2001.

Index